Eric Lichtfouse • Jan Schwarzbauer
Didier Robert
Editors

# Environmental Chemistry for a Sustainable World

## Volume 2: Remediation of Air and Water Pollution



Springer

*Editors*
Eric Lichtfouse
INRA, UMR Agroécologie
Dijon, France
eric.lichtfouse@dijon.inra.fr
www.researcherid.com/rid/F-4759-2011

Didier Robert
Laboratory of de Chimie Industrielle
rue Victor Demange
Université de Metz
57500 Saint-Avold
France
didier.robert@iut.univmetz.fr

Jan Schwarzbauer
RWTH Aachen
Institute für Geologie, Geochemie u.
Lagerstätten d. Erdöls u. d. Kohle
Lochnerstr. 4-20, 52056 Aachen
Germany
schwarzbauer@lek.rwth-aachen.de

ISBN 978-94-007-9196-1 ISBN 978-94-007-2439-6 (eBook)
DOI 10.1007/978-94-007-2439-6
Springer Dordrecht Heidelberg London New York



Printed on acid-free paper

Springer is part of Springer Science+Business Media (www.springer.com)

Environmental Chemistry for a Sustainable World

# Preface

*I was at a conference where someone said something about the Holocene. I suddenly thought this was wrong. The world has changed too much. So I said: 'No, we are in the Anthropocene.' I just made up the word on the spur of the moment. Everyone was shocked. But it seems to have stuck.*

Nobel Prize-winner Paul Crutzen

*To find rapidly chapters of interest in this book please see list of topics in Table 1 page ix.*

## Fukushima, Chernobyl and Climate Change

### *Nuclear Plants on Earthquake Zones*

Everybody has been recently shocked by the major accident of the nuclear power plant on March 31, 2011 at Fukushima, Japan. Such a failure was both unexpected and expected. Unexpected because most thought that the 1986 nuclear disaster at the power plant of Chernobyl, Ukraine, could never happen again, especially in wealthy, high technology countries such as Japan. Expected because geology tells us that Japan lies on the cusp of the Pacific-Philippine-Eurasian triple plate junction, where the complex interactions of three tectonic plates is unpredictable and loaded with potential activity. As a consequence, Japan experiences regular, high intensity earthquakes, tsunamis and volcanoes since centuries. Similarly to the Chernobyl global 'event', the release of radioactive pollutants from Fukushima nuclear plant in water, air, and soil will most probably severely affect human health, food security and economy worldwide for decades. Therefore, one might just ask why nuclear plants are built on such high-risk areas.

## *Human Errors Are Repeatable*

The global warming event has similar features – though less rapid and catastrophic – as the Chernobyl and Fukushima events on several rationales. First, the global warming is a worldwide event due notably to worldwide $CO_2$ emissions (Lichtfouse 2009a), a fact that is nicely coined by the popular saying "pollutants have no borders". Second, the effects of global warming are now clearly proven by many scientific trends (Feehan et al. 2009; Jones et al. 2009; Lavalle et al. 2009). Third, despite all such scientific evidence humans still use cars and planes that emit $CO_2$; they practice intensive, industrial agriculture that decrease soil carbon, and in turn emits $CO_2$; they cut forest, which in turn emits $CO_2$, and so on. Recent essays reports on global issues of and solution for society (Lichtfouse 2009a, b, 2010). From the scientist view, one might say ironically that "human errors are repeatable".

## *Social Chemistry*

From those global issues several conclusions and advices can be drawn to improve society and the life of further generations. First, overwhelming scientific evidence is not sufficient to convince humans, notably decision-makers. Such a failure is due in particular to the lack of communication between science and society. In other words, scientists should not only publish in scholarly journals and attend high-level scientific meetings, but also communicate with the "real world". Here, a "social impact factor" to measure the impact of science on society – based for instance on web, facebook and blog usage data – would be very appreciated.

A second advice is that classical natural sciences such as chemistry, physics, biology, geology and medicine should integrate social, human and political sciences. In other words, the real world should be involved in the process of scientific discovery to bridge the gap between science and people. The integration of social disciplines is already occurring in agrosciences (Fleming and Vanclay 2010; Karami and Keshavarz 2010; Lichtfouse 2010; Lichtfouse et al. 2009, 2010). Indeed agriculture has always been historically closer to citizens, e.g. farmers, than chemistry and physics. The need for analysis of citizen discourses is nicely shown by the following survey answer: 'What's sustainable? You've got to look at our world as we know it. We're not in a sustainable position at the moment. That's why I say what is sustainable – I don't know' (Fleming and Vanclay 2010).

The concept of discourse was introduced in the 1960s by the French philosopher Michel Foucault. Foucault (1972) maintained that the way language is used has consequences for a whole range of things that go beyond the level of individuals or disciplines, to the very structures of society that shape and limit how people are able to speak, think, and act, and to the social structures that are developed accordingly. Now, environmental chemistry should use techniques of social sciences such as discourse analysis to discover novel findings that will be both innovative and accepted by citizens.

## The Success of Environmental Chemistry

### *Association of Chemistry and the Environment*

We founded the Association of Chemistry and the Environment in 2000 with a group of environmental scientists (www.europeanace.com). A "chemistry flower" logo was designed to symbolise positive benefits of chemistry for Nature (Fig. 1). The association was launched by the organisation of the two first European Meeting on Environmental Chemistry by Eric Lichtfouse, Brigitte Elbisser and co-workers in 2000 in Nancy, France, and in 2001 in Dijon, France. Meetings were immediately a success, with more than 300 attendants, due to several factors such as hard work from the organising team, willingness to create a new science community in a highly conservative science system, and gathering in a friendly location scientists from various isolated disciplines such as soil science, toxicology and chemistry. The first presidents of the association were Eric Lichtfouse (2000–2004) and Jan Schwarzbauer (2005–2009). Noteworthy, well established associations, in particular the Division of Environmental Chemistry and the Division of Geochemistry of the American Chemical Society (ACS), supported us, in particular by organising joint symposia sponsored by the petroleum research fund. We therefore thank very much the American Chemical Society.

**Fig. 1** The "chemistry flower" was designed by Eric Lichtfouse and Guillaume Decaux to symbolise positive benefits of chemistry for Nature. Guillaume Decaux is a professional drawer living in Strasbourg, France (http://www.alcide.fr/)

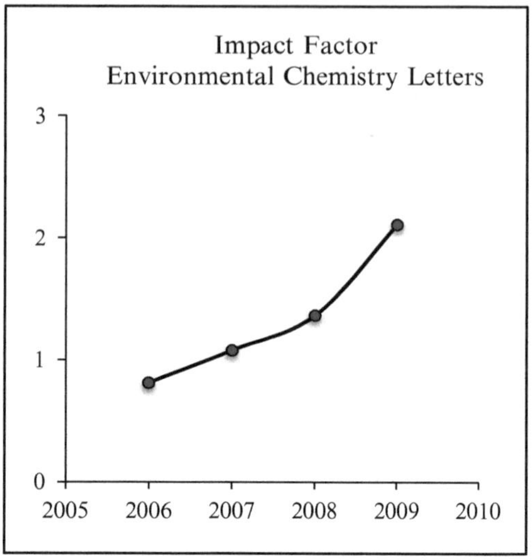


**Fig. 2** The impact factor of the journal Environmental Chemistry Letters (*ECL*) increased from 0.814 in 2006 to 2.109 in 2009. Article pdf downloads reached 36,549 in 2009 (100 per day)

## *Environmental Chemistry Letters*

We founded the journal Environmental Chemistry Letters in 2003 to fill the science gap between chemistry and environment (www.springer.com/10311). Despite a tough selection by the Thomson Reuters agency to enter the Science Citation Index, we got our first impact factor of 0.814 for 2006, only 3 years later . The impact factor increased steadily to reach 2.109 in 2009 (Fig. 2) (www.thomsonreuters.com). The increase of the impact factor can be explained by a higher quality of articles, as a result of higher rejection that reached 73% in 2009. Given the rapid increase of submitted articles, more that 70% of articles are now declined at pre-screening stage. Journal articles are highly viewed as proven by the number of pdf downloads that reached 36,549 in 2009 (100 per day). This finding is both unexpected and expected. Unexpected because most articles are not in open access. Expected because Springer has about 30 millions scientists who access articles. Here, contrarily to the common thinking, articles published in restricted access by a major publisher are probably much more visible than open access articles published by a minor publisher.

## *Environmental Chemistry*

We published the book Environmental Chemistry in 2005 (Lichtfouse et al. 2005b). The book includes 69 chapters sorted in 7 sections: Analytical Chemistry, Toxic Metals, Organic Pollutants, Polycyclic Aromatic Hydrocarbons, Pesticides, Green Chemistry, and Ecotoxicology. The book is a success with over 35,000 chapters downloads from 2007 to 2010. Book chapters are still highly downloaded with 639

**Table 1** Chapters of environmental chemistry for a sustainable world (Lichtfouse et al. 2011a, b)

| 1st author | Keywords | | | DOI |
|---|---|---|---|---|
| Tan | Nanotubes | Energy | Water | 10.1007/978-94-007-2442-6_1 |
| Landy | Cyclodextrin | Sequestration | Water | 10.1007/978-94-007-2442-6_2 |
| Chauhan | Nanotubes | Energy | Sensors | 10.1007/978-94-007-2442-6_3 |
| Liu | Photocatalysis | Remediation | Water | 10.1007/978-94-007-2442-6_4 |
| Visinescu | Polysaccharides | Metal oxides | Green chemistry | 10.1007/978-94-007-2442-6_5 |
| Ricking | DDT isomers | Remediation | Pesticides | 10.1007/978-94-007-2442-6_6 |
| Heim | Geochronolgy | Pollution history | Sediments | 10.1007/978-94-007-2442-6_7 |
| Anupama | Endocrine disruptors | Toxicity | Pesticides | 10.1007/978-94-007-2442-6_8 |
| Stankovic | Heavy metal | Toxicity | Seafood | 10.1007/978-94-007-2442-6_9 |
| Sarkar | DDT, PCBs | Toxicity | Seafood | 10.1007/978-94-007-2442-6_10 |
| Ribeiro | Pharmaceuticals | Remediation | Water | 10.1007/978-94-007-2439-6_1 |
| Sanjurjo | Buildings | $CO_2$, $SO_2$, $NO_x$ | Air | 10.1007/978-94-007-2439-6_2 |
| Rene | Bioreactors | Remediation | VOCs | 10.1007/978-94-007-2439-6_3 |
| Etxebarria | Extraction | Pollutants | Water | 10.1007/978-94-007-2439-6_4 |
| Sharma | $ClO_2$, R-Cl | Oxidation | Water | 10.1007/978-94-007-2439-6_5 |
| Mondal | Dyes | Biodegradation | Water | 10.1007/978-94-007-2439-6_6 |
| Shukla | Heavy metals | Lichens | Air | 10.1007/978-94-007-2439-6_7 |
| Brillas | Pharmaceuticals | Remediation | Water | 10.1007/978-94-007-2439-6_8 |
| López | Biogas | Cleaning | $CO_2$, $H_2S$, R-Cl | 10.1007/978-94-007-2439-6_9 |
| Mudhoo | Heavy metals | Toxicity | Biosorption | 10.1007/978-94-007-2439-6_10 |
| Gasparatos | Heavy metals | Sequestration | Nodules, soil | 10.1007/978-94-007-2439-6_11 |
| Dabrowska | Arsenic | Remediation | Biosorption | 10.1007/978-94-007-2439-6_12 |
| Mahmood | Photocatalysis | Microbes | Disinfection | 10.1007/978-94-007-2439-6_13 |

Keywords and DOI for fast access. *DDT* 2,2,-bis(4-chlorophenyl)-1,1,1-trichlorethane, *PCBs* polychorinated biphenyls, *VOCs* volatile organic compounds downloads in January 2011 (20 per day). The highest recent downloads can be freely viewed at Springer Realtime (www.realtime.springer.com). Here the most popular topics are heavy metals, bioremediation and green chemistry.

## *Environmental Chemistry for a Sustainable World*

This new book series presents 23 chapters published into 2 volumes: Nanotechnology and Health Risk (Lichtfouse et al. 2011a), and Remediation of Air and Water Pollution (Lichtfouse et al. 2011b). Table 1 allows fast access to chapters by topics. All chapters have been reviewed and the rejection rate was 15%. The Nanotechnology section highlights carbon nanotubes for energy and detection; cyclodextrins for pollutant trapping; magnetic nanophotocatalysts for pollutant degradation; and polysaccharides for metal oxide green synthesis. The Health Risk section describes new findings on the old DDT pesticide; geochronology of river pollutants; toxic effects of endocrine disruptors; and heavy metals in seafood. The Air and Water Pollution section presents the selective degradation of chiral pharmaceuticals; the alteration of housing walls by $CO_2$, $SO_2$ and NOx; cleaning industrial waste gas and dyes wastewater; methods to extract and detect pollutants; and harmful chlorinated pollutants. The Remediation section highlights the electrochemical degradation of pharmaceuticals; methods to treat biogas $CO_2$, $CH_4$, $H_2S$ and $NH_4$; heavy metal sequestration on biomass and soil nodules; As phytoremediation; and photocatalytic inactivation of water microbial pathogens.

Eric Lichtfouse,
Jan Schwarzbauer,
Didier Robert

## References

Feehan J, Harley M, Minnen J (2009) Climate change in Europe. 1. Impact on terrestrial ecosystems and biodiversity. A review. Agron Sustain Dev 29:409–421. doi:10.1051/agro:2008066

Fleming A, Vanclay F (2010) Farmer responses to climate change and sustainable agriculture. A review. Agron Sustain Dev 30:11–19. doi:10.1051/agro/2009028

Foucault M (1972) The archaeology of knowledge: and the discourse on language. Tavistock Publications Ltd, London, 249 p. ISBN 0-415-28753-7

Jones A, Stolbovoy V, Rusco E, Gentile AR, Gardi C, Marechal B, Montanarella L (2009) Climate change in Europe. 2. Impact on soil. A review. Agron Sustain Dev 29:423–432. doi:10.1051/agro:2008067

Karami E, Keshavarz M (2010) Sociology of sustainable agriculture. In: Lichtfouse E (ed) Sociology, organic farming, climate change and soil science. Sustain Agric Rev 3:19–40. doi:10.1007/978-90-481-3333-8

Lavalle C, Micale F, Durrant Houston T, Camia A, Hiederer R, Lazar C, Conte C, Amatulli G, Genovese G (2009) Climate change in Europe. 3. Impact on agriculture and forestry. A review. Agron Sustain Dev 29 (2009):433–446. doi:10.1051/agro/2008068

Lichtfouse E (2009a) Climate change, society issues and sustainable agriculture. In: Lichtfouse E (ed) Climate change, intercropping, pest control and beneficial microorganisms. Sustain Agric Rev 2:1–7. doi:10.1007/978-90-481-2716-0_1

Lichtfouse E (2009b) Sustainable agriculture as a central science to solve global society issues. In: Lichtfouse E (ed) Organic farming, pest control and remediation of soil pollutants. Sustain Agric Rev 1:1–3. doi:10.1007/978-1-4020-9654-91

Lichtfouse E (2010) Society issues, painkiller solutions, dependence and sustainable agriculture. In: Lichtfouse E (ed) Sociology, organic farming, climate change and soil science. Sustain Agric Rev 3:1–17. doi:10.1007/978-90-481-3333-8_1

Lichtfouse E, Schwarzbauer J, Robert D (2005a) Preface. In: Lichtfouse E, Schwarzbauer J, Robert D (eds) Environmental chemistry, vol 1, p V. Springer, Berlin. doi:10.1007/b137751

Lichtfouse E, Schwarzbauer J, Robert D (2005b) (eds) Environmental chemistry, vol 1, p V. Springer, Berlin. ISBN 978-3-540-22860-8. doi:10.1007/b137751

Lichtfouse E, Navarrete M, Debaeke P, Souchère V, Alberola C, Ménassieu J (2009) Agronomy for sustainable agriculture. A review. Agron Sustain Dev 29:1–6. doi:10.1051/agro:2008054

Lichtfouse E, Hamelin M, Navarrete M, Debaeke P, Henri A (2010) Emerging agroscience. Agron Sustain Dev 29:1–10. doi:10.1051/agro/200955

Lichtfouse E, Schwarzbauer J, Robert D (2011a) Environmental chemistry for a sustainable world. Volume 1: nanotechnology and health risk. Springer, Dordrecht, 458 p. doi: 10.1007/978-94-007-2442-6

Lichtfouse E, Schwarzbauer J, Robert D (2011b) Environmental chemistry for a sustainable world. Volume 2: remediation of air and water pollution. Springer, Dordrecht, 562 p. doi: 10.1007/978-94-007-2439-6

# Other publications by the Editors

## Books

Environmental Chemistry
ISBN: 978-3-540-22860-8

Organic contaminants in riverine and ground water systems. Aspects of the anthropogenic contribution
ISBN: 978-3-540-31169-0

Sustainable Agriculture
ISBN: Vol. 1: 978-90-481-2665-1
ISBN: Vol. 2: 978-94-007-0393-3

Rédiger pour être publié! Conseils pratiques pour les scientifiques
ISBN: 978-2-287-99395-4

## Journals

Environmental Chemistry Letters
ISSN: 1610–3653
http://www.springer.com/10311

Agronomy for Sustainable Development
ISSN: 1774–0746
http://www.springer.com/13593

## Book series

Sustainable Agriculture Reviews
ISSN: 2210–4410
http://www.springer.com/series/8380

# Contents

# Contributors

**Carlos Alves** Centro de Investigação Geológica, Ordenamento e Valorização de Recursos, Campus de Gualtar, Universidade do Minho, 4710-057, Braga, Portugal, casaix@dct.uminho.pt

**Anil Kumar Anal** Center of Excellence in Nanotechnology, School of Engineering and Technology, Asian Institute of Technology, P.O. Box 4, Klong Luang, Pathumthani, 12120, Thailand

**Sunandan Baruah** Center of Excellence in Nanotechnology, School of Engineering and Technology, Asian Institute of Technology, P.O. Box 4, Klong Luang, Pathumthani, 12120, Thailand

**A. Bhattacharya** Department of Marine Science, University of Calcutta, 35 Ballygunge Circular Road, Calcutta 700 019, India

**Prosun Bhattacharya** KTH-International Groundwater Arsenic Research Group, Department of Land and Water Resources Engineering, Royal Institute of Technology, SE-100 44, Stockholm, Sweden

**Enric Brillas** Laboratori d'Electroquímica dels Materials i del Medi Ambient, Departament de Química Física, Facultat de Química, Universitat de Barcelona, Martí i Franquès 1-11 Barcelona 08028, Spain, brillas@ub.edu

**Paula M.L. Castro** Centro de Biotecnologia e Química Fina, Escola Superior de Biotecnologia, Universidade Católica Portuguesa, Porto, Portugal

**Brajesh Chauhan** Environmental Engineering Section, Department of Civil Engineering, Faculty of Engineering and Technology, Aligarh Muslim University, Aligarh, UP, 202 002, India

**Beata B. Dabrowska** Department of Botany, Stockholm University, SE-106 91, Stockholm, Sweden

**Joydeep Dutta** Center of Excellence in Nanotechnology, School of Engineering & Technology, Asian Institute of Technology, P.O. Box 4, Klong Luang, Pathumthani, 12120, Thailand, joy@ait.asia

**N. Etxebarria** Department of Analytical Chemistry, University of the Basque Country, PK 644 48080, Bilbao, Spain

**L.A. Fernández** Department of Analytical Chemistry, University of the Basque Country, PK 644 48080, Bilbao, Spain

**Vinod Kumar Garg** Department of Environmental Science and Engineering, Guru Jambheshwar University of Science and Technology, Hisar, Haryana, 125001, India

**Dionisios Gasparatos** Laboratory of Soils and Agricultural Chemistry, Agricultural University of Athens, Iera Odos 75 Athens 11855, Greece, gasparatos@aua.gr

**K.R. Gunaratna** Department of Environmental Microbiology, School of Biotechnology, AlbaNova, Royal Institute of Technology, SE-100 44, Stockholm, Sweden

**Christian Kennes** Chemical Engineering Laboratory, Faculty of Sciences, University of La Coruña, Rua da Fraga, 1 E-15008, La Coruña, Spain, Kennes@udc.es

**Eric Lichtfouse** INRA, UMR Agroécologie, BP 86510 21065, Dijon Cedex, France, Eric.Lichtfouse@dijon.inra.fr

**M. Estefanía López** Chemical Engineering Laboratory, Faculty of Sciences, University of La Coruña, Rua da Fraga, 1 E-15008, La Coruña, Spain

**Mohammad Abbas Mahmood** Center of Excellence in Nanotechnology, School of Engineering and Technology, Asian Institute of Technology, Klong Luang, Pathumthani, 12120, Thailand, mohammadabbas33@gmail.com

**Romeela Mohee** Department of Chemical and Environmental Engineering, Faculty of Engineering, University of Mauritius, Reduit, Mauritius

**Pijush Kanti Mondala** Environmental Research laboratory, Department of Applied Chemistry, Faculty of Engineering and Technology, Aligarh Muslim University, Aligarh, UP, 202 002, India, pijushamu@gmail.com

**María Montes** Chemical Engineering Laboratory, Faculty of Sciences, University of La Coruña, Rua da Fraga, 1 E-15008, La Coruña, Spain

**Ackmez Mudhoo** Department of Chemical & Environmental Engineering, Faculty of Engineering, University of Mauritius, Reduit, Mauritius, ackmezchem@yahoo.co.uk

**Arun B. Mukherjee** Department of Biological and Environmental Sciences, Helsinki University, P.O. Box 27 FI-00014, Helsinki, Finland

**P. Navarro** Laboratoire de Chimie Analytique Bio-Inorganique et Environnement, UMR 5254 CNRS-UPPA, 2 Avenue Président P. Angot 64053, Pau, France

**M. Olivares** Department of Analytical Chemistry, University of the Basque Country, PK 644 48080, Bilbao, Spain

**A. Prieto** Department of Analytical Chemistry, Helmholtz Centre for Environmental Research, UFZ Permoserstraße 15 D-04318, Leipzig, Germany

**Eldon R. Rene** Chemical Engineering Laboratory, Faculty of Sciences, University of La Coruña, Rua da Fraga, 1 E-15008, La Coruña, Spain

**Ana R. Ribeiro** Centro de Investigação em Ciências da Saúde (CICS), Instituto Superior de Ciências da Saúde-Norte, Gandra, Portugal

Centro de Biotecnologia e Química Fina, Escola Superior de Biotecnologia, Universidade Católica Portuguesa, Porto, Portugal

**Didier Robert** Equipe 'Photocatalyse et Nanostructures, LMSPC-CNRS-UMR 7515 et ELCLASS, European Laboratory for Catalysis and, Surface Sciences, Antenne de Saint-Avold UPV, Metz Rue Victor Demange 57500, Saint-Avold, France, drobert@univ-metz.fr

**Jorge Sanjurjo-Sánchez** Instituto Universitario de Xeoloxía "Isidro Parga Pondal", Edificio Servizos Centrais de Investigación, Universidade da Coruña, Campus de Elviña 15071, A Coruña (España), Spain, jsanjurjo@udc.es

**Jan Schwarzbauer** Institute of Geology and Geochemistry of Petroleum and Coal, RWTH Aachen University, Aachen, Germany, schwarzbauer@lek.rwth-aachen.de

**Virender K. Sharma** Chemistry Department, Florida Institute of Technology, 150 West University Boulevard, Melbourne, FL, 32901, USA, vsharma@fit.edu

**Anuradha Shukla** Traffic Planning and Environment Division, Central Road Research Institute (CSIR), New Delhi 110020, India

**Vertika Shukla** Department of Environmental Science, Bhimrao Ambedkar Central University, Vidya Vihar Raebareli Road Lucknow, UP, 226 025, India, vertika_shukla@rediffmail.com

**Ignasi Sirés** Laboratori d'Electroquímica dels Materials i del Medi Ambient, Departament de Química Física, Facultat de Química, Universitat de Barcelona, Martí i Franquès 1-11, Barcelona 08028, Spain, isiresa@gmail.com; i.sires@ub.edu

**Mary Sohn** Chemistry Department, Florida Institute of Technology, 150 West University Boulevard, Melbourne, FL, 32901, USA

**Maria Elizabeth Tiritan** Centro de Investigação em Ciências da Saúde (CICS), Instituto Superior de Ciências da Saúde-Norte, Rua Central de Gandra, 13174585-116, Gandra PRD, Portugal

Centro de Química Medicinal da Universidade de Porto (CEQUIMED-UP), Porto, Portugal, elizabeth.tiritan@iscsn.cespu.pt

**D.K. Upreti** Lichenology Laboratory, National Botanical Research Institute, Rana Pratap Marg, Lucknow 226 001, India, upretidk@rediffmail.com

**A. Usobiaga** Department of Analytical Chemistry, University of the Basque Country, PK 644 48080, Bilbao, Spain

**María C. Veiga** Chemical Engineering Laboratory, Faculty of Sciences, University of La Coruña, Rua da Fraga, 1 E-15008, La Coruña, Spain

**Meththika Vithanage** Chemical and Environmental Systems Modeling Research Group, Institute of Fundamental Studies, Hantana Road, Kandy, 20000, Sri Lanka, meththikavithanage@gmail.com; meththika@ifs.ac.lk

**Shaobin Wang** Department of Chemical Engineering, Curtin University of Technology, G.P.O. Box U1987, Perth, WA, 6845, Australia

**Olatz Zuloaga** Department of Analytical Chemistry, University of the Basque Country, PK 644 48080, Bilbao, Spain, olatz.zuloaga@ehu.es

# Part I
# Air and Water Pollution

# Chapter 1
# Environmental Fate of Chiral Pharmaceuticals: Determination, Degradation and Toxicity

**Ana R. Ribeiro, Paula M.L. Castro, and Maria Elizabeth Tiritan**

## Contents



A.R. Ribeiro
Centro de Investigação em Ciências da Saúde (CICS), Instituto Superior de Ciências da Saúde-Norte, Rua Central de Gandra, 1317, 4585-116, Gandra PRD, Portugal

Centro de Biotecnologia e Química Fina (CBQF), Escola Superior de Biotecnologia, Universidade Católica Portuguesa, Porto, Portugal

P.M.L. Castro
Centro de Biotecnologia e Química Fina (CBQF), Escola Superior de Biotecnologia, Universidade Católica Portuguesa, Porto, Portugal

E. Lichtfouse et al. (eds.), *Environmental Chemistry for a Sustainable World: Volume 2: Remediation of Air and Water Pollution*, DOI 10.1007/978-94-007-2439-6_1,

**Abstract** Pollution of the aquatic environment by pharmaceuticals is of major concern. Indeed pharmaceutical pollutants have several undesirable effects for many organisms, such as endocrine disruption and bacterium resistance. They are resistant to several degradation processes, making their removal difficult and slow. Pharmaceuticals reach the environment due to their inefficient removal by waste water treatment plants (WWTP), and by improper disposal of unused medicines. In aquatic environments pharmaceuticals reach concentrations at trace levels of $ngL^{-1}$–$\mu gL^{-1}$ range. Many pharmaceutical pollutants are chiral. They occur in nature as a single enantiomer or as mixtures of the two enantiomers, which have different spatial configuration and can thus be metabolized selectively. In spite of similar physical and chemical properties, enantiomers have different interactions with enzymes, receptors or other chiral molecules, leading to different biological response. Therefore they can affect living organisms in a different manner. The fate and effects of enantiomers of chiral pharmaceuticals in the environment are still largely unknown. Biodegradation and toxicity can be enantioselective, in contrast to abiotic degradation. Thus accurate methods to measure enantiomeric fractions in the environment are crucial to better understand the biodegradation process and to estimate toxicity of chiral pharmaceuticals.

We review (1) general properties of chiral compounds, (2) current knowledge on chiral pharmaceuticals in the environment, (3) chiral analytical methods to determine the enantiomers composition in environmental matrices, (4) degradation and removal processes of chiral pharmaceuticals in the environment and (5) their toxicity to aquatic organisms. The major analytical methods discussed are gas chromatography (GC), high performance liquid chromatography (HPLC), electrochemical sensors and biosensors. These chiral methods are crucial for the correct quantification of the enantiomers regarding that if an enantiomer with more or less toxic effects is preferentially degraded, the assessed exposure based on measurements of achiral methodologies would overestimate or underestimate ecotoxicity. The degradation and biodegradation is discussed using few examples of important therapeutic classes usually detected in the aquatic environment. Few examples of ecotoxicity studies are also given on the occurrence of enantiomers and their fate in the environment which differs with regard to undesirable effects and to biochemical processes.

M.E. Tiritan (✉)
Centro de Investigação em Ciências da Saúde (CICS), Instituto Superior de Ciências da Saúde-Norte, Rua Central de Gandra, 1317, 4585-116, Gandra PRD, Portugal

Centro de Química Medicinal da Universidade de Porto (CEQUIMED-UP), Porto, Portugal

Instituto Superior de Ciências da Saúde-Norte, Rua Central de Gandra, 1317, 4585-116, Gandra PRD, Portugal
e-mail: elizabeth.tiritan@iscsn.cespu.pt; ritalado@hotmail.com

**Keywords** Chiral chromatography • Enantioselective • Biodegradation • Enantiomeric fraction • Enantioseparation • Racemates • Enantiomers • Chiral switching • Chiral analysis • Waste water treatment plant • Beta-blockers • Antidepressants • Non steroid anti-inflammatory drugs • Enantiospecific toxicity • Chiral stationary phases • Fluoxetine • Propranolol • Atenolol • Ibuprofen • Naproxen • Chiral ecotoxicity

## 1.1 Introduction – Basic Concepts of Chirality

Chiral compounds are substances with a similar chemical structure that, in general, confers them the same physical and chemical properties like melting point, solubility and reactivity. However they differ in the deviation of polarized light due to the different spatial configuration originated by planes, axis or centers of asymmetry given two nonsuperposable left-handed and right-handed mirror images compounds, called enantiomers. The planar chirality refers to a planar unit connected to an adjacent part of the structure by a bond which results in restricted torsion avoiding the symmetry plane as in the monosubstituted paracyclophane (Fig. 1.1a). When a set of ligands is held around an axis originates a spatial arrangement which is nonsuperposable on its mirror image known as axial chirality. Such cases include *ortho*-substituted biphenyls and substituted allenes, a group of stereoisomers called atropoisomers which have a restricted rotation about a single bond, being the absolute configuration called *R* or *S* preceded by an a (*aR* or *aS*) to distinguish them by other optical active compounds (Fig. 1.1b, c, d) (Moss 1996; Santos et al. 2007). The center chirality consists in a stereogenic center holding a set of four different substituents in a spatial arrangement which also origins two nonsuperposable left-handed and right-handed mirror image compounds (Fig. 1.2). The most common type of center chirality is a stereogenic carbon (Fig. 1.3), but other atoms such as sulphur or phosphorus can also be a stereogenic center (Fig. 1.4). Enantiomers can be identified by the manner to rotate the polarized light, to the right (clockwise) are called dextrorotatory, (d) or (+), and to the left (counter clockwise) are denominated levorotatory, (l) or (−). Concerning their chemical configuration relative to the spatial orientation of the substituent of the chiral center, enantiomers can be (*R*) from the Latin rectus or (*S*) from the Latin sinister. The equimolar mixture of two enantiomers is a racemate or a racemic mixture and does not rotate the polarized light (Eliel and Wilen 1994). Despite the similar thermodynamic properties in achiral medium, enantiomers normally have a different behaviour when the environment is also chiral. Amino acids, carbohydrates, deoxyribose and ribose are chiral compounds which are the unity of important natural molecules such as proteins, glycoproteins, DNA and RNA, respectively (Müller and Kohler 2004; Hühnerfuss and Shah 2009). As such, in biological processes, enzymes, receptors and other binding-molecules have the capacity to recognize enantiomers in a different way. The model that demonstrate the recognition of an enantiomer by a receptor due to their complementary configuration is illustrated in Fig. 1.5 demonstrating that the

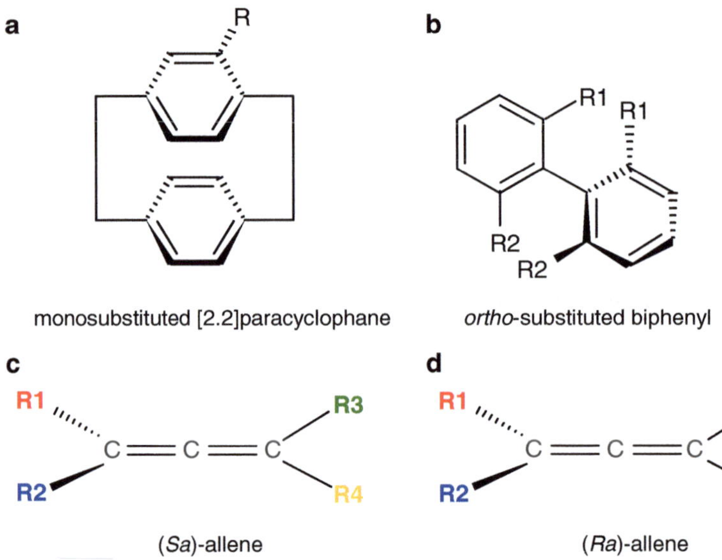


**Fig. 1.1** (**a**) Structure of monosubstituted [2.2] paracyclophane possessing planar chirality; (**b**) Structure of ortho-substituted biphenyl with axial chirality; (**c, d**) Structures of (*aS*) and (*aR*)-allenes showing its axial chirality, supposing that the sequence order is R4>R3>R2>R1, originating the ranking 1 for R4, 2 for R3, 3 for R2 and 4 for R1(Eliel and Wilen 1994)

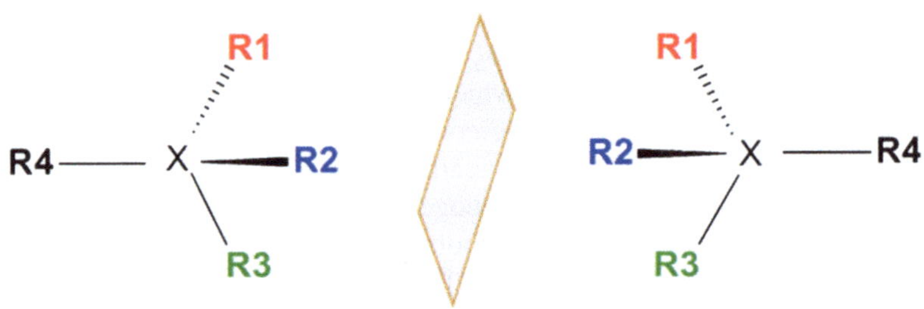


**Fig. 1.2** Stereogenic atom (X) with four different substituents (R1, R2, R3, R4) originating the nonsuperposable structures

other enantiomer does not match the same receptor and thus do not have the same activity in this receptor. However it can be a ligand for other receptor, leading to another activity, side effects, or even toxic effects. When enantiomers of one pharmaceutical compound have the same type of activity but differ in the potency, eutomer is the enantiomer whit higher potency whereas distomer is the one that have such activity reduced, being the ratio of the potencies termed eudismic ratio (Caldwell et al. 1988; Cox 1994).

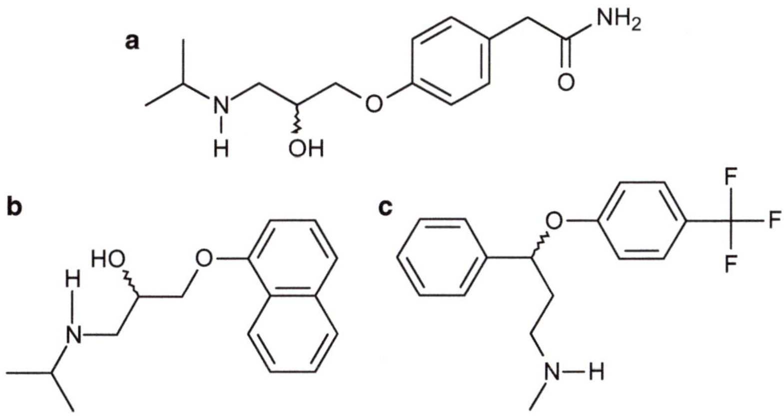


**Fig. 1.3** Structures of three pharmaceuticals with a carbon stereogenic center. (**a**) atenolol; (**b**) propranolol; (**c**) fluoxetine

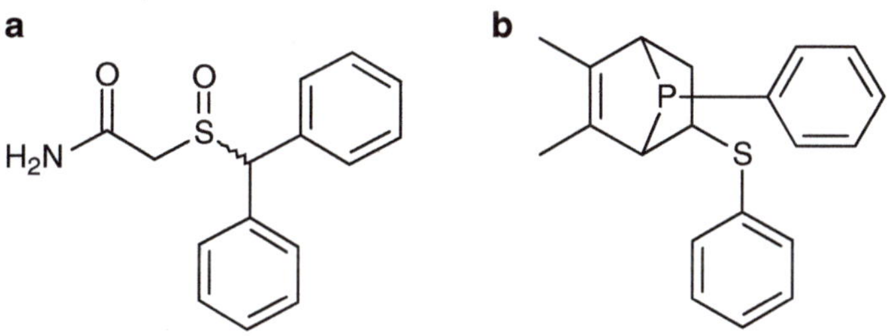


**Fig. 1.4** (**a**) Structure of modafinil with a sulphur stereogenic center; (**b**) Structure of 2,3-dimethyl-7-phenyl-5-(phenylsulfanyl)-7-(*R/S*)-phosphabicyclo[2.2.1]hept-2-ene with a phosphorus stereogenic center

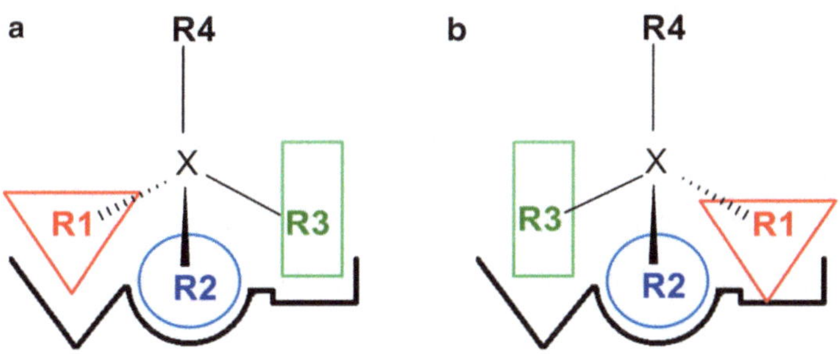


**Fig. 1.5** Schematic process of the complementary recognition of one enantiomer (**a**) rather than the other (**b**)

## 1.2 Chiral Pharmaceuticals

Pharmaceuticals are chemicals used for diagnosis, treatment, alteration or prevention of disease, health condition, or structure/function of the human body, including veterinary drugs. They act by interaction with the binding site of the drug receptor which is a macromolecule such as enzymes, nucleic acids or membrane-bounded proteins, promoting the pharmacological action by the formation of a drug-receptor complex (Stringer 2006). Their action in humans is well described, but their mechanism and cumulative effects are mostly unknown in non-target organisms (Daughton and Ternes 1999).

Chiral pharmaceuticals are becoming of environmental concern because of the different pharmacokinetic, pharmacodynamics, toxicological and ecotoxicological properties occurring between enantiomers which may differently affect aquatic organisms. Concerning to pharmacokinetics, absorption, distribution (e.g. transport by passive or active way, in which the protein binding is important), metabolism and excretion can be different for each enantiomer since they represent biological processes in which binding proteins, membrane proteins, enzymes and other chiral molecules have an important role. Likewise, the pharmacodynamics involves the interaction with receptors like membrane proteins or enzymes which can be different for each enantiomer. Because chiral pharmaceuticals have normally different pharmacokinetic and pharmacodynamics properties, they have different dissociation constant from the binding site and different attachment to it leading to different biological response in quality or quantity (Campo et al. 2009).

In a review concerning the chirality in pharmaceuticals, Lima (1997) related few examples of chiral pharmaceuticals that can be used as racemates, as in the following cases:

(a) The pharmacologic activity of both enantiomers is the same (e.g., prometazine, antihistaminic). This effect is due to the equal pharmacokinetic and pharmacodynamics properties;
(b) One enantiomer is biologically more active than the other (e.g., propranolol, β-blocker). In this case both of enantiomers can reach the receptor, binding to it at the binding site with different dissociation constant, leading to a stronger attachment of one enantiomer;
(c) One of the enantiomers antagonizes the side effects of the other (e.g., indacrinone).

In others cases there are drugs that should be commercialized only as a single enantiomer, mostly due to the association of the enantiomers to different receptors, leading to different responses where (Mannschreck et al. 2007):

(a) One enantiomer is biologically active and the other has no activity (e.g., α-metildopa);
(b) One enantiomer is biologically active and the other is antagonist (e.g., picenadol) (Franz et al. 1990);

(c) The pharmacologic activity of enantiomers is different (e.g., propoxyphene: analgesic and antitussive effects);
(d) One enantiomer has side effects (e.g., L-Dopa).

Table 1.1 shows few chiral pharmaceuticals that can be used as racemates and/or enantiopure formulations, depending on the effects of each enantiomer.

The recent advances in stereoselective synthesis and chiral analysis led to an increase of the single enantiomers as drugs available in the market (Hutt 1998). The re-evaluation of the license of the enantiomeric pure drugs which were produced in a racemic mixture, called as chiral switching process, also collaborated to increase the use of single enantiomer (Cordato et al. 2003; Hutt and Valentová 2003; Caner et al. 2004). However the chiral switching is related to economical, legislation and patent aspects, apart from the evident pharmacologic and toxicological details. Some pharmaceutical companies promote the patenting of new enantiopure pharmaceuticals at the expiring of the patent of the racemic drug (Somogyi et al. 2004). This aspect is highlighted by these authors especially when companies start the marketing approval of the new enantiopure drug when the racemic drug patent is expiring avoiding the release of generics from other companies (Barreiro et al. 1997; Somogyi et al. 2004; Mannschreck et al. 2007). Besides the advantages of enantiopure pharmaceuticals, like lower therapeutic doses, more safety margin, less interindividual variability, less drug interactions and fewer side effects, chiral switching can lead to unexpected toxicity (Baker et al. 2002). As such, the therapeutic advantage of an enantiopure drug must be clinically proven to justify the cost increase of the new treatment instead of the use of racemates. The single enantiomer and the relative racemate should be compared concerning to pharmacokinetics and toxicological studies (Hutt and Valentová 2003; Somogyi et al. 2004; Orlando et al. 2007). Fluoxetine is a noted example of a failed chiral switching. Both enantiomers of fluoxetine are potent inhibitors of the serotonin reuptake pump but the enantiomers of the metabolite norfluoxetine have differences in this activity, being (*R*)-norfluoxetine an inactive metabolite and (*S*)-norfluoxetine the active metabolite (Henry et al. 2005). The affinity of (*R*)-fluoxetine for human 5-$HT_{2A}$ and 5-$HT_{2C}$ receptor subtypes is high, unlike (*S*)-fluoxetine, being (*R*)-fluoxetine an antagonist for these receptors leading to an increase of extracellular catecholamines, such as serotonine, dopamine and norepinephrine. (*S*)-norfluoxetine has a similar binding to the transporter and is recognized as an active metabolite whereas (*R*)-norfluoxetine is an inactive metabolite (Koch et al. 2002). Besides, (*S*)-norfluoxetine is a potent inhibitor of its metabolic enzyme and contributes significantly to the half-life of the racemate, being (*R*)-norfluoxetine less active regarding this enzyme (Stevens and Wrighton 1993). The chiral switching would be the way to develop a new license market of (*R*)-fluoxetine with the same indications as racemic fluoxetine, with a better half-life, a better selectivity for important receptor subtypes in depression mechanism and less drug-drug interactions, being a good improvement for clinical strategies. However, the development of the (*R*)-isomer of fluoxetine was abandoned after an unexpected cardiac side effect, at higher doses, identified at Stage II clinical trials (McConathy and Owens 2003). The company

**Table 1.1** Examples of chiral pharmaceuticals used as racemates or as single enantiomers

| | Qualitative activity | | | | |
|---|---|---|---|---|---|
| Drug | Equal | Different | Mode of action | Description | References |
| Promethazine | X | | Equal pharmacologic potency | Promethazine | Chen et al. (1992) |
| Propranolol | X | | Different pharmacologic potency | Propranolol: (−)-propranolol has a pharmacological activity 100 times superior to (+)-propranolol | Pavlinov et al. (1990) |
| Warfarin | X | | | Warfarin: anticoagulant, with greater anticoagulant potency of (S)-warfarin | Choonara et al. (1986) |
| Methadone | X | | | Methadone: (R)-methadone has higher affinity for the μ-opioid receptor and longer plasma elimination half-life | Huq (2007) |
| Bupivacaine | X | | Both enantiomers have pharmacological activity but one of them has less toxicity and less side effects | Bupivacaine: S (−)-bupivacaine has the same neural blocking characteristics, but has a higher margin of safety | Huang et al. (1998) |
| Fluoxetine | | X | Enantiomers have different pharmacological activity | Fluoxetine: (R)-fluoxetine is an antidepressant and (S)-fluoxetine was tested for migraines prophylaxis | Steiner et al. (1998) |
| Propoxyphene | | X | | Propoxyphene: (+)-propoxyphene is analgesic and (−)-propoxyphene is antitussive | Cooper and Anders (1974) |
| Indacrinone | | X | One enantiomer antagonize the side effects of the other | Indacrinone: (S)-indacrinone is natriuretic and (R)-indacrinone is uricosuric | Jain et al. (1984) |

(continued)

**Table 1.1** (continued)

| Drug | Qualitative activity | | Mode of action | Description | References |
|---|---|---|---|---|---|
| | Equal | Different | | | |
| Ibuprofen | | X | One enantiomer is pharmacologically active and the other doesn't have activity | Ibuprofen: its anti-inflammatory activity is almost due to (S)-ibuprofen | Mayer and Testa (1997) |
| Cetirizine | | X | | Cetirizine: its antiallergic activity is due to (R)-cetirizine | Mannschreck et al. (2007) |
| Dopa | | X | One enantiomer is pharmacologically active and the other has side effects | Dopa: L-dopa is used in Parkinson's disease and D-dopa has side effects like nausea, anorexia, involuntary movements and granulocytopenia | Hutt and Valentová (2003) |
| Thalidomide | | X | One enantiomer is pharmacologically active and the other is toxic | Thalidomide: (R)-thalidomide was used for insomnia and nausea therapy; (S)-thalidomide is teratogenic | Smith (2009) |
| Picenadol | | X | One enantiomer is pharmacologically active and the other is antagonist | Picenadol: (+)-picenadol is an opioid agonist and (-)-picenadol is a weak agonist/ antagonist | Franz et al. (1990) |

This table resumes the benefits and disadvantages depending on the pharmacological or side effect

had hoped that (*R*)-fluoxetine would replace fluoxetine, which was facing imminent expiration of its patent protection.

Thus there are many pharmaceutical drugs that are both commercialized at racemic and enantiopure forms, as described in Table 1.2 (Tucker 2000; Hutt and Valentová 2003; Orlando et al. 2007).

There are also some pharmaceuticals, such as Non Steroid Anti-Inflammatory Drugs (NSAIDs), which suffer chiral inversion, e.g., propionic acid derivatives. These drugs act by inhibition of cyclo-oxygenase (COX) and consequently the synthesis of prostaglandins and thromboxanes. Except for naproxen (Fig. 1.6) they are

**Table 1.2** Examples of new enantiopure pharmaceuticals, which racemates were not withdrawn from the market

| Racemate | Therapeutic class | Enantiopure pharmaceutical | Trade name |
|---|---|---|---|
| Citalopram | Antidepressant | (*S*)-citalopram (escitalopram) | Lexapro, Cipralex |
| Ibuprofen | Anti inflammatory | (*S*)-ibuprofen (dexibuprofen) | Seractil |
| Ketoprofen | Anti inflammatory | (*S*)-ketoprofen (dexketoprofen) | Sympal, Ketesse |
| Bupivacaine | Anaesthetic | (*S*)-bupivacaine (levobupivacaine) | Chirocaine |
| Cetirizine | Antihistaminic | (*R*)-cetirizine (levocetirizine) | Xyzal |
| Omeprazole | Proton pump inhibitor | (*S*)-omeprazole (esomeprazole) | Nexium |
| Ofloxacin | Antimicrobial | (*S*)-ofloxacin (levofloxacin) | Levaquin,Tavanic |
| Salbutamol | β2-adrenergic receptor agonist | (*S*)-salbutamol (levalbuterol) | Xopenex |

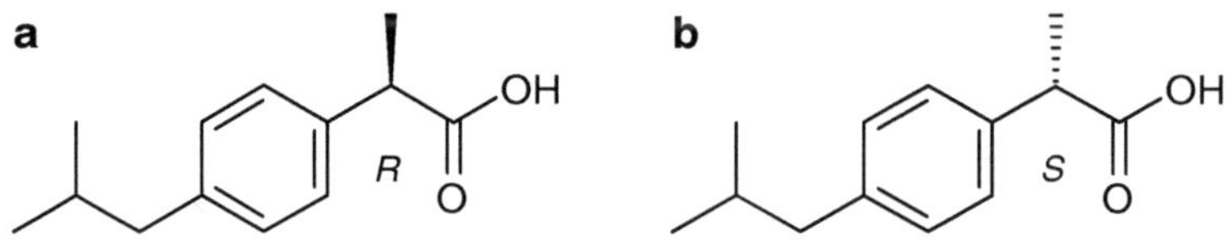


**Fig. 1.6** Structure of (*R*)-naproxen (**a**) and (*S*)–naproxen (**b**)

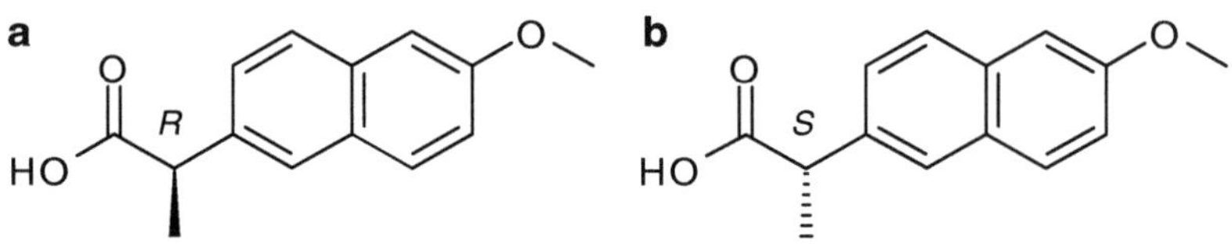


**Fig. 1.7** Structure of (*R*)-ibuprofen (**a**) and (*S*)–ibuprofen (**b**)

commercialized as racemates in most countries. Ibuprofen (Fig. 1.7) is a common example of a NSAIDs used for pain, inflammation, fever and arthritis, which is normally sold as racemate although *S*-(+)-ibuprofen is practically the responsible for the pharmacologic action. Moreover this drug, as other profens, suffers unidirectional conversion to the pharmacologic active *S*-(+)-enantiomer *in vivo* which difficult to obtain the enantiopure (*S*)-ibuprofen with a good percentage of conversion justifying the use of racemates. The disadvantage of the racemic ibuprofen is that (*S*)-ibuprofen has less gastric side effects than the racemic mixture; is more hydrosoluble leading to a more rapid action; the chiral inversion can originate interindividual variability in analgesia; and the accumulation of the intermediate (*R*)-ibuprofen-CoA in the chiral inversion, which can react with triglycerides leading to the accumulation at fatty tissue in the organism (Lima 1997; Tucker 2000; Bonabello et al. 2003; Carvalho et al. 2006; Chávez-Flores and Salvador 2009).

Another therapeutic option is the use of chiral drugs with an enantiomeric ratio different from 1 to improve the benefits of each enantiomer, when applicable. Enantiomeric Fraction (EF) is the proportion of the concentration of one enantiomer to the total concentration, expressing the relative concentration of an enantiomer's compound. Racemate exhibits an EF of 0.5 while an enantiomerically pure compound has a value of 0 or 1. Thus, a mixture of two enantiomers in a proportion different to 1:1 has an EF between 0 and 1 and different to 0.5. As an example, the advantage of the administration of a mixture of *S*0.75:*R*0.25 bupivacaíne rather than the racemate was reported, with the same anaesthetic properties and with less toxicity (Gonçalves et al. 2003).

In this review we will focus on three pharmaceutical classes: Beta-blockers, Antidepressants and Non Steroid Anti-Inflammatory Drugs (NSAIDs), due to their persistence in the environment and respective ecological effects (Ternes 1998; Trenholm et al. 2006; Vanderford and Snyder 2006; Pérez and Barceló 2008; Fernández et al. 2010; Santos et al. 2010).

## 1.3 Analytical Methods for Quantification/Identification of Chiral Pharmaceuticals in the Environment

### *1.3.1 Chromatography*

The basis of analytical methods for analyses of chiral compounds is the formation of diastereoisomers between the enantiomers and other optical pure compound which provides different retention times, leading to the separation of the enantiomers. The diastereoisomers, which are non-mirror image compounds, can be separated by chromatography due to their different physical and chemical properties (Carvalho et al. 2006). In chromatography methods, indirect and direct experimental approaches can be used (Bojarski 2002; Hühnerfuss and Shah 2009).

#### 1.3.1.1 Indirect Methods

Indirect methods require the derivatization of the racemic mixture with an optical pure reagent, leading to the formation of diastereoisomers. Since diastereoisomers have different physical and chemical properties, they can be resolved by an achiral stationary phase. After the separation in an achiral column, enantiomers can be recovered by inversing the derivatization reaction. This approach is possible if the racemate has at least one suitable functional group to allow the reaction. As advantages, indirect methods can be performed in conventional chromatographic columns as RP-$C_{18}$, it is possible to introduce chromophores or fluorophores and also it is easy to predict the elution order. However, the complex and time-consuming derivatization step with different reaction rates for each enantiomer, the poor availability of optical pure chiral reagents, the possibility of racemization and the difficulty to apply to environmental samples turns this method weak for environmental analysis (Bojarski 2002; Carvalho et al. 2006; Hühnerfuss and Shah 2009).

### 1.3.1.2 Direct Methods

Direct methods use a chiral selector as an additive in the mobile phase (Chiral Mobile Phase Additive) or on the stationary phase (Chiral Stationary Phase or CSP). In both cases, the chiral selector has more affinity with one of the enantiomers. Chiral Mobile Phase Additive is a more complex and expensive method because it requires higher amounts of chiral selector to prepare the mobile phase, which is not recovered and thus it is wasted in a large scale. This kind of direct method can also interfere with the detection mode (Lämmerhofer 2010).

Chiral Stationary Phase (CSP) consists in a chiral selector adsorbed or covalently bonded to a solid support, which forms transitory diastereoisomeric complexes with the two enantiomers with different stability (Francotte 2001; Lämmerhofer 2010). The more stable complex will elute with larger retention time. There are many different type of commercial chiral columns: Pirkle type or donor-acceptor CSP, crown ethers CSP, ligand-exchange CSP, polysaccharide derivatives CSP, cyclodextrin CSP, protein CSP, macrocyclic glycopeptides antibiotics CSP and others such as CSPs based on synthetic polymers (Pirkle and Pochapsky 1989; Haginaka 2008; Lämmerhofer 2010). They are performed in open or tubular columns, open tubular capillaries and thin layer plates (Wistuba and Schurig 2000; Wistuba 2010). The major advantage of this method is the ability of the analyte to remain unmodified. Despite the numerous of commercial CSP, an universal column is not available, thus making the trial versus error a challenge. The choice of the CSP is based on the experience and on the available literature, being often empirical by trial and error evaluation. However some authors start this with polysaccharides or macrocyclic antibiotics because of their broad application and the possibility to operate in normal, reverse or polar organic mode (Perrin et al. 2002a, b; Andersson et al. 2003; Cass et al. 2003a; Sousa et al. 2004; Ates et al. 2008; Pirzada et al. 2010). Thus, the method developing time is reduced and the probability of success is enhanced. The more usual separation mode is normal-phase, however this tendency is turning to reversed-phase and polar organic mode, being all applicable to the most used polysaccharides or macrocyclic antibiotics CSPs (Cass and Batigalhia 2003; Cass et al. 2003b; Montanari et al. 2006).

Andersson et al. (2003) performed a screening of polysaccharides or macrocyclic antibiotics CSPs to compounds used in pharmaceutical industry. They reported that when using polysaccharides CSPs in normal-phase, the alcohol used as modifier affects selectivity, elution order and retention time. The additives used for separation of basic compounds in normal-phase mode must be basic to reduce tailing, whereas for separation of acidic compounds the additive must be acid to reduce the excessive retention of these compounds. Neutral compounds are unaffected by these additives so they can be processed with any of them. Thus, besides the type of column and the alcohol chosen, the additive used is of great importance for the success of the resolution of enantiomers. Concerning to polar organic mode, better resolution is achieved with neutral and weak acidic and basic compounds. Using macrocyclic antibiotics CSPs, the number of compounds enantioresolved is lower than using polysaccharides CSPs, although they complement the enantioselectivity and can resolve important classes of pharmaceutical compounds like β-blockers, bronchodilators and some local anaesthetics (Andersson et al. 2003).

Molecularly Imprinted Polymers are a synthetic alternative in which the target compound is used in the polymerization process as a template, to produce a specific receptor to one enantiomer which is known to be more retained. The removal of the template by a solvent or a chemical reaction provides a complementary site in the receptor to the target molecule, providing a Molecularly Imprinted Polymer which can be used as CSP avoiding the choice based on the trial and error method. Commonly these Molecularly Imprinted Polymers are incorporated in conventional HPLC columns. These materials are easy to obtain, highly selective and stable, and the elution order is predictable. The template is the more retained enantiomer being the last to eluate. The drawback of Molecularly Imprinted Polymers is the peak tailing produced due to the heterogeneous binding site. Molecularly Imprinted Polymers can be coupled to HPLC, Capillary Electro-Chromatography, Thin Layer Chromatography and Analytical Sensors (Wistuba and Schurig 2000; Sancho and Minguillón 2009).

### 1.3.2 Gas Chromatography (GC)

Gas Chromatography (GC) can be performed in the indirect and direct mode, having the advantages of high efficiency, sensitivity, speed of separation, and ability to separate analytes from impurities. The indirect mode requires the derivatization of the chiral compound with an optical pure reagent, resulting in diastereoisomers that are then separated on an achiral stationary phase. The direct mode is performed with a CSP. GC has the great advantage of not needing optimization of mobile phase concerning solvents, pH, modifiers and gradients (Eljarrat et al. 2008). However, there are only a few chiral columns that can be used in GC, which limits the application of this technique. The most used columns are cyclodextrin-based and have demonstrated a great applicability in organochlorine pesticides (Wong and Garrison 2000; Eljarrat et al. 2008). Another useful CSP are based on metal complexes such as metal-β-diketonate polymers (Rykowska and Wasiak 2009). One disadvantage of GC is the need of derivatization, in some cases, to increase the volatility, to prevent the peak tailing and thus to improve detection limits by the peak shaping (Zhang et al. 2009). The high temperature used in GC is a drawback when the analytes are not volatile and when the chiral compound can suffer racemization or decomposition (Schurig 2001; Jiang and Schurig 2008). Thus GC has several limitations to environmental analyses of chiral pharmaceuticals (Huggett et al. 2003; Lamas et al. 2004; Jones et al. 2007).

### 1.3.3 High Performance Liquid Chromatography (HPLC)-Liquid Chromatography/Mass Spectrometry (LC/MS)

High Performance Liquid Chromatography (HPLC) has become a technique of routine analysis in replacement of GC because of the numerous combinations of the available columns with the potential mobile phases, and the detection methods that can be coupled, such as Mass Spectrometry (MS) or optical detectors when the analytes are UV transparent (Aboul-Enein and Ali 2004; Görög 2007). The high variety of chiral commercial columns, like Pirkle, crown ethers, ligand exchangers, cyclodextrins, polysaccharides, proteins, macrocyclic glycopeptides antibiotics, and others, usually design for HPLC analysis and the different elution mode makes this technique a powerful tool for chiral analysis (Perrin et al. 2002a, b; Andersson et al. 2003; Zhang et al. 2010). HPLC with different type of detection has been used for the determination of the EF of many chiral pharmaceuticals in a variety of matrix, including environmental samples (Francotte 2009; Kasprzyk-Hordern et al. 2010). Few examples are shown in Table 1.3.

Liquid Chromatography/Mass Spectrometry (LC/MS) and LC–tandem MS (LC/MS/MS) have played a crucial role in environmental analysis (Niessen 2003; Kot-Wasik et al. 2007; Pérez and Barceló 2008) mainly due to their versatility, sensitivity and selectivity (González et al. 2007). Recent trends in environmental Mass Spectrometry methods have been emerging with special focus on hybrid mass spectrometers such as Quadrupole-Time Of Flight (Qq-TOF) (Farré et al. 2008) and Quadrupole-Linear Ion Trap (Qq-LIT) (Díaz-Cruz et al. 2008). However, Triple Quadrupole (QqQ) mass analysers are the most used analytical technology in the environmental analyses (Gros et al. 2006). Few publications have also employed Ion Trap (IT) mass spectrometers for environmental determinations (Feitosa-Felizzola et al. 2007; Madureira et al. 2009; Barreiro et al. 2010). Despite the largely application to environmental analyses the ion suppression of the interest compounds peaks with the components of the matrix may consist in a limitation of the method. So, the MS/MS detectors are a powerful tool that must be used in environmental analysis because of their high selectivity (Pérez and Barceló 2008; Madureira et al. 2009).

EF is an important tool in biodegradation, toxicological studies and wastewater monitoring (Fono and Sedlak 2005; Hashim et al. 2010). It was demonstrated that the determination of EF can be affected by the peak integration method (Asher et al. 2009). They found that the Deconvolution Method, a method which considers two independent Gaussian-based mathematic functions, is more accurate especially when peaks overlap (poor resolution or asymmetry). Instead, the commonly used method, called Valley Drop Method (VDM), showed significant bias in EF determination. Thus, for environmental analysis the peak integration method must be evaluated and Deconvolution Method is preferred when poor resolution or tailing occurs.

#### 1.3.3.1 Capillary Electro-Chromatography (CEC)

Capillary Electro-Chromatography (CEC) is a hybrid separation technique of HPLC and Capillary Electrophoresis (CE) since it is based on the partition of solutes between two phases (HPLC) and on the different mobility of the same solutes towards an electric field (CE) and has the advantages of both CE and HPLC. Likewise HPLC it can be applied also to neutral compounds and has high

**Table 1.3** Analytical methods of separation of several enantiomers in different matrices

| Drug | Enantioselective method | Elution mode | Chiral stationary phase | Mobile phase | Detection limit | Matrix application | Observations | Refs |
|---|---|---|---|---|---|---|---|---|
| Omeprazole | 2D-LC/UV or IT-MS/MS | Reverse-phase | RAM BSA C8 and amylose tris-(3,5-dimethylphenylcarbamate) of coated onto APS-Nucleosil | Phosphate buffer (pH 6.8; 0.1 M NaCl) | 5.0 $\mu gL^{-1}$ (UV-vis); 0.025 $\mu gL^{-1}$ (IT-MS/MS) | Waste and estuarine water samples | It was developed a two-dimensional liquid chromatography method which allows the use of 1 mL of sample | Barreiro et al. (2010) |
| Lansoprazole | 2D-HPLC/UV | Reverse-phase | RAM BSA C8 and amylose tris (3,5-dimethoxyphenylcarbamate) coated onto APS-Nucleosil | (35:65, V/V) ACN:water | 0.025 $\mu g\ mL^{-1}$ | Human plasma | It was performed an automatized method which allows the sample cleanup and quantification of the analyte in the sample | Gomes et al. (2010) |
| 10 β-blockers | HPLC/UV | Normal-phase | Cellucoat | n-heptane/EtOH/DEA | 3.0–297.0 $\mu g\ mL^{-1}$ | Screening study | Good enantioresolution of ten β-blockers in a single column | Ali et al. (2009a) |
| Fluoxetine and norfluoxetine | HPLC/FD | Reverse-phase | Chiralcel OD | (75:25, V/V) potassium hexafluorophosphate 7.5 mM and sodium phosphate 0.25 M solution (pH 3.0)/ACN | 5.0 $ng\ mL^{-1}$ | Human plasma | A sensitive method was developed to further analysis of plasma samples, following the therapy with antidepressants | Silva et al. (2009) |

(continued)

**Table 1.3** (continued)

| Drug | Enantioselective method | Elution mode | Chiral stationary phase | Mobile phase | Detection limit | Matrix application | Observations | Refs |
|---|---|---|---|---|---|---|---|---|
| Venlafaxine (antidepressant) and 11 analogs | HPLC/UV | Normal-phase | Chiralpak AD and IA | 0.5%TFA or DEA in n-hexane/2-propanol or EtOH | n.a | Semi-preparative HPLC for in vitro pharmaco-logical essays | The quality of separation is affected by the pH and alcoholic modifier of mobile phase | Caccamese et al. (2009) |
| Methadone | LC/MS | Reverse-phase | Chiralcel OJ | (65:35, V/V) ACN/0.02% TEA in water | 1 ng $mL^{-1}$ | Human plasma | It was used an isotope-labelled internal standard to avoid matrix effects | Ansermot et al. (2009) |
| Dechloroethylifosfamide metabolites of Ifosfamide | LC/MS/MS | Reverse-phase | Chiralpak-AGP | Gradient of 10 mM ammonium acetate in water (pH 7.00) and 30 mM (pH 4.00) | 1 ng $mL^{-1}$ | Human plasma | To consider the application of the less toxic enantiomer | Aleksa et al. (2009) |
| 62 pharmaceuticals | HPLC/UV | Polar-organic phase | Chiralcel AD-RH, OD-RH, AS-RH and OJ-R | (100:0.1:0,1, V/V/V) ACN or MeOH/DEA/TFA | n.a | Screening study | ACN/DEA/TFA is the first choice of solvent system followed by MeOH/DEA/TFA | Ates et al. (2008) |
| 62 pharmaceuticals | HPLC/UV | Polar-organic phase | Sepapak-2 and Sepapak-3 | (100:0.1:0.1, V/V/V) ACN or MeOH or EtOH/DEA/TFA | n.a. | Screening study | The CSP preference is Sepapak-2>Chiralcel OD-RH>AS-RH>OJ-R | Ates et al. (2008) |
| Ketoprofen | HPLC/UV | Reverse-phase | Diamonsil C18 | (71:29, V/V) MeOH/0.01 M $KH_2PO_4$ (pH 4.5) | 0.03 μg $mL^{-1}$ (LQ) | Rat plasma (in vitro release test) | It was observed a stereoselective release of ketoprofen, being faster for (S)-ketoprofen | Wang et al. (2007) |

| | | | | | | | | |
|---|---|---|---|---|---|---|---|---|
| β-blockers, selective serotonine re-uptake inhibitors and salbutamol | LC/MS/MS | Reverse-phase | Chirobiotic V | (90:10, V/V) MeOH/20 mM $NH_4OAc$ in water, 0.1% formic acid (pH 4) | 0.2–7.5 $ngL^{-1}$ | WWTP influents and effluents | Able to quantify β-blockers at $ngL^{-1}$ levels in aqueous samples with little sample pre-treatment | MacLeod et al. (2007) |
| Bufurarol | HPLC/UV | Polar ionic phase | Chirobiotic V | (100:0.015:0.010, V/V/V) MeOH/acetic acid/TEA | 2 $ngL^{-1}$ | Plasma samples and pharmaceutical formulations | Good recovery which allows therapeutic drug monitoring | Hefnawy et al. (2007) |
| Fluoxetine and norfluoxetine | HPLC/FD | Normal-phase | Extrasil CN/Apex Silica | 75% tetramethyl ammonium chloride in ACN; (40:60, V/V) tetrahydrofurane/isooctane | 0.2–0.25 ng $mL^{-1}$ | Plasma and cerebral cortex | This method allows the assessment of the relationships between the concentration of individual enantiomers. | Unceta et al. (2007) |
| Fluoxetine | HPLC/UV | Normal-phase//Polar-organic phase | Chiralcel AD, OD, OJ, Cyclobond I 2000 DM and Kromasil CHI-TBB | Hexane/isopropanol/DEA or MeOH/0.2% triethylamine acetic acid (Cyclobond) | n.a | Screening study | The better resolution was obtained with Chiralcel AD, OD and Cyclobond. | Zhou et al. (2007) |
| β-blockers | LC/MS/MS | Reverse-phase | Chirobiotic V | (90:10, V/V) MeOH/0.1% TEAA in water, acetic acid (pH 4) | 3–17 $ngL^{-1}$ (effluent); 17–110 $ngL^{-1}$ (influent) | WWTP influents and effluents | Enough resolution to quantify β-blockers at ng/L levels in aqueous samples | Nikolai et al. (2006) |

(continued)

**Table 1.3** (continued)

| Drug | Enantioselective method | Elution mode | Chiral stationary phase | Mobile phase | Detection limit | Matrix application | Observations | Refs |
|---|---|---|---|---|---|---|---|---|
| β-blockers (atenolol, betaxolol, metoprolol, nadolol) | HPLC/UV | Normal-phase | Chiralcel OD | Hexane/EtOH or 2-propanol/DEA and, or acetic acid | n.a. | Biological samples and pharmaceutical formulations | Rapid and efficient enantioseparation which provide the simultaneous determination of the β-blockers | Singh et al. (2006) |
| β-blockers and profens | HPLC/UV | Reverse-phase | Chirobiotic V or V2 | MeOH/0.1–1.0% triethylamine acetate buffer | n.a | Screening study | Chirobiotic V had higher retention and enantioresolution for the majority of analytes | Bosáková et al. (2005) |
| β-blockers and profens | HPLC/UV | Polar-organic phase | Chirobiotic V or V2 | MeOH/0.005-0.1% acetic acid/0.005-0.1% TEA | n.a | Screening study | Polar-organic separation mode showed better enantioseparation | (Bosáková et al. 2005) |
| β-blockers | HPLC/UV | Reverse-phase | Synthesized resin m-[(+)-α-methyl benzyl carboxamide] XAD-4 | (60:40, V/V) ACN/sodium acetate-acetic acid buffer (pH 4.1) | n.a | Screening study | New stationary phase based on a chiral selector attached to polymeric support achieved a better separation with shorter retention times | Agrawal and Patel (2005) |
| Bupivacaine | HPLC/UV | Normal-phase | Kromasil CHI-TBB | (98:2:0.3:0.05, V/V) hexane/2-propanol/acetic acid/TEA | n.a | To develop continuous chromatographic SMB unit | The column has a great enantioresolution for bupivacaine | Da Silva Júnior et al. (2005) |

| | | | | | | | | |
|---|---|---|---|---|---|---|---|---|
| Fluoxetine and norfluoxetine | HPLC/UV | Reverse-phase | Chiralcel OD-R | (65:35, V/V) potassium hexafluorophosphate 100 mM/ACN (pH 3.0) | 10 ng $mL^{-1}$ (LQ) | Human plasma | A simple, accurate and precise method of separation of four enantiomers in less than 25 min | Gatti et al. (2003) |
| 53 compounds used in pharmaceutical industry | HPLC/UV-DAD | Normal-phase | Chiralcel AD, OD, AS and OJ | (15:85:0.1, V/V) EtOH or 2-propanol/isohexane/TFA or DEA (for acidic or basic and neutral compounds, respectively) | n.a | Screening study | The four columns showed enantioselectivity for 85% of the tested compounds. The additive was DEA for basic compounds and TFA for acidic compounds | Andersson et al. (2003) |
| | | Polar-organic phase | | (50:50, V/V) MeOH EtOH (for basic and neutral compounds) | | | | |
| 53 compounds used in pharmaceutical industry | HPLC/UV-DAD | Polar-organic phase | Chirobiotic R, V and T | (100:0.02:0.01, V/V) MeOH/acetic acid/TEA (for basic and acidic compounds) | n.a | Screening study | The three columns showed enantioselectivity for 65% of the tested compounds | Andersson et al. (2003) |
| | | Reverse-phase | | (25:75, V/V) MeOH/TEA (0.1%, pH 6) (for neutral and acidic compounds) | | | | |

(continued)

**Table 1.3** (continued)

| Drug | Enantioselective method | Elution mode | Chiral stationary phase | Mobile phase | Detection limit | Matrix application | Observations | Refs |
|---|---|---|---|---|---|---|---|---|
| 8 β-blockers | HPLC/UV and FD | Reverse-phase | Kromasil C8 | Gradient of 0.03% TFA in water and 0.03% TFA in (50:50, V/V) water/ACN | 0.7– 1.3 nM | corneal permeability studies | The gradient HPLC associated to UV and FD detection allowed the determination of all β-blockers | Ranta et al. (2002) |
| 36 pharmaceuticals | HPLC/ UV-DAD | Normal-phase | Chiralcel AD, OD and OJ | n-hexane/2-propanol or n-hexane/ EtOH | n.a | Screening study | The three columns showed complementary and enantioselectivity for the tested compounds | Perrin et al. (2002b) |
| Fluvastatin | LC/FD | Reverse-phase | Chiralcel OD-R | (40:60, V/V) ACN/ phosphate buffer pH 2.5 | 1 nM (LQ) | Blood plasma | The aim was develop an on-line automated method to determinate fluvastatin and its enantiomers | Toreson and Eriksson (1997) |

Note that these methods are widely used in pharmaceutical, biomedical and environmental applications
*ACN* acetonitrile, *EtOH* ethanol, *MeOH* methanol, *DEA* diethylamine, *TEA* triethylamine, *TFA* trifluoracetic acid, *LQ* limit of quantification , *WWTP* Waste Water Treatment Plants, *n.a.* not applicable

selectivity because of the wide range of mobile and stationary phases available. As CE it has high peak efficiency without the necessity for pump systems to provide high pression (Sancho and Minguillón 2009). So, it makes this an easy technique with a short time of analyses and a low consumption of sample and electrolyte, providing a high potential in the pharmaceutical, biomedical and environmental analysis (Li et al. 2010). Enantiomers can be separated within a wall coated open tubular capillary, being the chiral selectors bound to the capillary wall. In the case of packed capillaries, they are obtained by packing chiral selectors in a gel. Otherwise, packed capillaries can be obtained by polymerization in situ providing monolithic phases that can be classified in two major categories, the silica-based and the organic-polimer-based (polyacrylate, polyacrylamide or polystyrene) (Scriba 2003; Gübitz and Schmid 2004). Alternatively, the chiral selectors are directly added into the separation electrolyte. The chiral selectors used in Capillary Electrophoresis include crown ethers, cyclodextrins, polysaccharides, proteins, and macrocyclic glycopeptides antibiotics, the same selectors used in HPLC as reviewed by Scriba (2003).

#### 1.3.3.2 Other Techniques

There are other chromatographic related techniques that are used in a less extent.

Micellar Electrokinetic Capillary Chromatography (MEKC) is a technique which consists in the addition of a surfactant molecule (above its critical micellar concentration) in Capillary Electrophoresis, leading to the formation of a micellar pseudo-stationary phase in the solution. Thus, the chromatographic principle is present. The surfactant forms micelle occurring partition mechanisms between the micellar pseudo-stationary phase and the mobile phase. Enantioresolution can be achieved using chiral surfactants or using chiral agents that are added to an achiral micellar buffer, such as combination of micelles and cyclodextrins or modified cyclodextrins. To improve the enantioresolution it can be also combined chiral surfactant and cyclodextrins. A recent approach is the use of polymeric surfactants, a group of high-molecular-mass molecules. Despite the advantages of the so called molecular micelles, such as lower concentrations that can be used, more stability and rigidity, the latter allows slower mass transfer of the analytes between the pseudo-stationary phase and the mobile phase, leading to a lower resolution compared to the conventional micelles (Ha et al. 2006).

Super Critical Fluid Chromatography (SFC) is a hybrid technique of GC and LC and combines the better characteristics of each. Because of the equal density, dissolving capacity, the intermediate viscosity and diffusion coefficient of its mobile phase, it has the advantages of providing higher flow rates and faster separation than LC and a lower dispersion. The mobile phase is compared with those used in GC and HPLC due to the ability to carry compounds and to dissolve them, respectively. Open tubular columns can be used with Flame Ionisation Detection (FIA), Electron Capture, nitrogen phosphorus, and sulphur chemiluminescence. Otherwise packed columns are usually coupled to an UV detector (Taylor 2009). This technique is an alternative to HPLC when the enantioresolution is partial by normal phase or is not achieved by reverse phase, providing the lower consumption of solvents and the use of non-toxic and non-explosive solvents. However, there are some limitations like high cost of the equipment, complexity of the hardware and the weak experience in this technique (Maftouh et al. 2005).

Thin Layer Chromatography (TLC) is preferentially used in indirect mode but it is limited because of its low resolution and weak power to detect low concentrations like those found in the environment. TLC can be used as complementary to HPLC since it is less expensive and allows the optimization of the separation parameters in an inferior time and with less costs (Bhushan and Martens 1997), so the application on environmental analysis is very limited.

### *1.3.4 Electrochemical Sensors and Biosensors*

Electrochemical Sensors and Biosensors provide an enantioselective analysis using an electrochemical cell coupled to a chiral receptor (Potentiometric Enantioselective Membrane Electrodes, PEME), an enzyme (Chiral Amperometric Biosensors) or an antibody (Enantioselective Immunosensors), in which the sample flows without separation steps. The principle of PEME is based on the thermodynamics of the reaction between each enantiomer and the chiral selector, with different stability in the complexes formed with each enantiomer, leading to different reaction energies (Izake 2007). These techniques have advantages like direct determination in the matrix, high precision, fastness and the possibility of on-line detection with Flow Injection Analysis and Sequential Injection Analysis systems (Izake 2007). It was developed a method for detection of (*S*)- and (*R*)-captopril using two amperometric biosensors based on L- and D-amino acid oxidase with an on-line SIA system (Stefan et al. 2000a). The same researchers developed a PEME based on maltodextrin for the assay of *(S)*-captopril and an amperometric biosensor for the assay of (*R*)-captopril with an on-line SIA system (Stefan et al. 2000b). Both of them provide more than 30 assays per hour.

## 1.4 Biotic and Abiotic Degradation and Removal Processes of Chiral Pharmaceuticals in the Environment

Removal of pharmaceuticals in surface waters can be due to photo-transformation, biodegradation, hydrolysis and partition to sediment (Liu et al. 2009). Pharmaceuticals are generally highly polar compounds since they are produced in a way which promotes their transport and excretion in the organism. Thus, their removal in Waste Water Treatment Plants (WWTP) is mostly restricted to biodegradation and to abiotic processes as oxidation and sorption. Air-stripping and photo-transformation are also abiotic processes but normally are insignificant in the removal of pharmaceuticals in WWTP. Sipma et al. (2010) reviewed the comparison of Conventional Activated Sludge systems and Membrane Bioreactors, where the removal efficiencies of many pharmaceuticals were higher in Membrane Bioreactors. These compounds include racemates of chiral pharmaceuticals like β-blockers (atenolol, metoprolol, propranolol), antidepressants (fluoxetine, paroxetine) and NSAIDs (ibuprofen, indomethacin, ketoprofen, mefenamic acid, naproxen, propyphenazone).

Air-stripping can be potentiated using Membrane Bioreactors in the case of some pharmaceuticals with high Henry air/water coefficient (Sipma et al. 2010). However, regarding photo-degradation it can occur at a higher extent in WWTP than in Membrane Bioreactors since photo-transformation requires the exposition of the wastewater to sunlight which occurs in a moderate way at the superficial layer of wastewater and in the secondary clarifiers; conversely the Membrane Bioreactors do not have a second clarifier and have a high biomass concentration which difficult the sunlight penetration. Concerning sorption, pharmaceuticals normally have low sorption constants due to their high polarity leading to a slight contribution to the global removal in WWTP, justifying their low efficiency during the primary treatment. In the case of the few pharmaceuticals with a higher sorption constant, Membrane Bioreactors may remove them more efficiently due to the absence of suspended solids in the final effluent (Sipma et al. 2010). So, oxidation and biodegradation are the treatments with more promising results. However, oxidation has some disadvantages like process costs and may lead to the production of metabolites more resistant to biodegradation (Sipma et al. 2010).

### *1.4.1 Biodegradation of Non Steroid Anti-inflammatory Drugs (NSAIDs), Beta-blockers and Antidepressants as Individual Enantiomers*

There are many studies on microbial degradation of chiral pharmaceuticals in the environment as racemates (Trautwein et al. 2008; Benotti and Brownawell 2009; Calisto and Esteves 2009; Mascolo et al. 2010; Santos et al. 2010), however biodegradation studies of individual enantiomers are scarcer (Table 1.4) (Buser et al. 1999; Fono and Sedlak 2005; Fono et al. 2006; Matamoros et al. 2009; Winkler et al. 2001).

The distribution of worldwide approved drugs indicates that the use of single enantiomers has increased during the 1990s reaching about 60% in the 2001, exceeding the achirals. The racemic drugs represented the minority category and many of the top-selling drugs, are marketed as single enantiomers (Caner et al. 2004). According to a recent survey, the distribution of the 15 Food and Drug Administration approved drugs in the period of January–August 2003 was 64% of single enantiomers, 14% racemates and 22% achirals (Fig. 1.8) (Caner et al. 2004). Taking this into account both the biodegradation studies and the method quantification are crucial to be done for single enantiomers.

**Table 1.4** Biodegradation studies of individual enantiomers in the environmental

| Method | Matrix | Biodegradation experiment | Observation | Sampling local | Refs |
|---|---|---|---|---|---|
| GC/MS/MS | Lakes, rivers and North Sea; WWTP influent and effluent | Incubation of fortified lake water; Incubation of WWTP' influent with activated sludge (aerobic conditions) | Degradation of ibuprofen was rapid and mostly biological mediated, with S-form being faster degraded in both experiments | Switzerland and North Sea | Buser et al. (1999) |
| GC/MS | Rivers | Incubation of a biofilm with raw river water | (R)-ibuprofen was degraded faster than the (S)-form of ibuprofen, the active one | Canada | Winkler et al. (2001) |
| GC/MS/MS | WWTP influent and effluent | Microcosms experiments with filtered secondary effluent | EF decreased in microcosms inoculated with activated sludge but remained constant in not inoculated or sterilized treatments | USA (California and New York) | Fono and Sedlak (2005) |
| GC/MS/MS | Trinity River; WWTP effluent | Microcosms experiments with river water to assess phototransformation and degradation in the dark | EF decreased from the effluent to downstream, suggesting the biological mediated degradation | Texas | Fono et al. (2006) |
| HPLC/MS/MS | WWTP | n.a | EF changed from influent to effluent for all the tested drugs, except for metoprolol, salbutamol and sotalol | Canada | MacLeod et al. (2007) |
| GC/MS | WWTP influent and effluent | Microcosms experiments with synthetic and real wastewater | (S)-ibuprofen was degraded faster under aerobic conditions, depending on the oxidation status of WWTP. In anaerobic conditions, EF remains constant. Enantioselective degradation of naproxen is similar under aerobic and anaerobic conditions | Spain and Denmark | Matamoros et al. (2009) |
| HPLC/MS/MS | Aerated lagoon and three terciary WWTP | n.a | EF changed over time for all drugs except to sotalol in one of the WWTP'effluent. EF differed among the three WWTP suggesting variations on biodegradation, except for metoprolol | Canada | MacLeod and Wong (2010) |

Biodegradation needs attention concerning the enantiomeric degradation ratio, since microorganisms can degrade selectively one enantiomer, can alter their enantioselectivity, or can degrade both enantiomers which is more unlikely to happen. Microbial degradation can also promote the enantiomerization (Richardson 2006; Pérez and Barceló 2008)

*EF* enantiomeric fraction, *WWTP* waste water treatment plant, *n.a.* not applicable

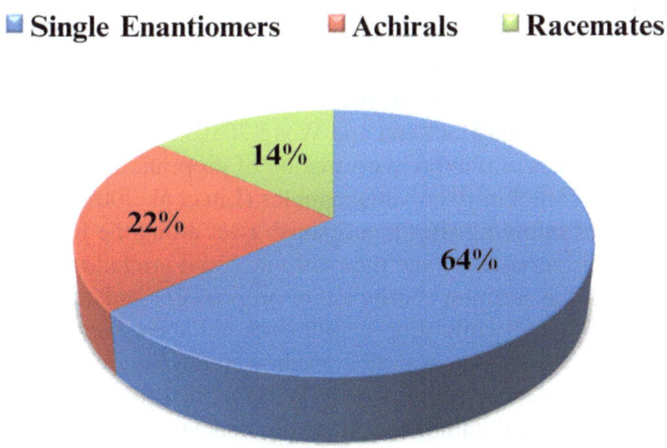


**Fig. 1.8** Distribution of the 15 Food and Drug Administration-approved drugs in the period of January–August 2003. Note that the majority of the chiral drugs approved are single enantiomers (64%)

Propranolol, a β-blocker used for treatment of cardiovascular diseases, with annual sales of the brand Inderal® of about $30 billion (Intercontinental Marketing Services-Health report 2006; Bartholow 2010), is normally detected in surface waters and in WWTP effluents (Ternes 1998; Huggett et al. 2003; Fono and Sedlak 2005). Propranolol is an important tool to distinguish the raw and treated sewage since propranolol has an EF of 0.5 (racemic) in WWTP influent and significantly below 0.5 in WWTP effluent, suggesting that this pharmaceutical is enantiomerically selective degraded through the biological treatment (Fono and Sedlak 2005). In the same study, EF decreased in microcosms inoculated with activated sludge but remain constant in not inoculated or sterilized treatments. The same researchers found in microcosms experiments that the EF of metoprolol, an analogous β-blocker, decreased from the effluent to downstream, suggesting selective biological mediated degradation of enantiomers (Fono et al. 2006).

Atenolol is a β-blocker also used for treatment of cardiovascular diseases with annual sales of the brand Tenormin® of approximately $102 million (Intercontinental Marketing Services-Health report 2006) included in the Top 200 Prescription Drugs of 2009 of United States (Bartholow 2010) and was one of the 11 most frequently detected compounds in United States in drinking water, being classified as an indicator of Endocrine Disrupting Compounds and other organic pollutants contamination and an indicator of the treatment success of a WWTP (Benotti et al. 2009). In the same study, atenolol demonstrated to be persistent and poorly removed by WWTP when traditional chlorine disinfection was applied, however it was easy removed when the treatment occurred with ozone. Vanderford and Snyder (2006) reported the presence of atenolol in the final effluent of WWTP at concentrations above 800 $ngL^{-1}$. The detection of atenolol in finished drinking water has also been reported (Trenholm et al. 2006). In a enantioselective evaluation of effluents of three different WWTP in Canada, the EF of atenolol was different for the different effluents, indicating that the different microbial communities can affect the enantioselectivity of the biodegradation (MacLeod and Wong 2010).

Despite the evidences of phototransformation of propranolol and atenolol and of biodegradation of atenolol in river water samples (Liu et al. 2009), they were found in more than 80% of sampling sites in a Spanish river along the four seasons of the year, remarking their persistence and their chronic use (Fernández et al. 2010). Due to the differences in the sorption coefficient of propranolol and atenolol, atenolol is likely to adsorb to sludge while propranolol does not show significant adsorption (Maurer et al. 2007). Reungoat et al. (2010) showed the improvement of the pharmaceuticals removal such as atenolol in WWTP with further treatments such as coagulation, flocculation, flotation and filtration, ozonation, activated carbon filtration and ozonation for disinfection.

Fluoxetine, an antidepressant of the group of Selective Serotonin Reuptake Inhibitors, is one of the most dispensed drugs in the world (Stanley et al. 2007). It was one of the most prescribed drugs included in the Top 200 of the United States (Bartholow 2010). Fluoxetine was found in surface waters, treated wastewater, raw influent and even in finished drinking water in United States and South Korea at $ngL^{-1}$ levels (Trenholm et al. 2006; Vanderford and Snyder 2006). Fluoxetine and also its active metabolite norfluoxetine have been detected in WWTP effluents and in surface waters (MacLeod et al. 2007). Fluoxetine was reported as a persistent pharmaceutical after chlorine treatment rather than ozone because of the more potent oxidant effect attributed to ozone (Westerhoff et al. 2005). In a dissipation study of five Selective Serotonin Reuptake Inhibitors in aquatic microcosms, fluoxetine was the most persistent (Johnson et al. 2005). In a study using batch incubation of seawater samples, Benotti and Brownawell (2009) included fluoxetine in the more labile group and considered that adsorption to suspended sediment was minimal. Paterson and Metcalfe (2008) studied the uptake and depuration of fluoxetine in freshwater fish species, Japanese medaka (*Oryzias latipes*), and found a half-life three times higher than in mammalian species and a lower ability to biodegrade and eliminate such drug and its metabolite norfluoxetine. So, fluoxetine can indicate a recalcitrant behaviour in biological tissues, suggesting the possible chronic effects (Paterson and Metcalfe 2008).

Ibuprofen, a NSAID used for pain, fever and rheumatism, is the third most popular clinically used drug in the world and one of the 200 drugs most prescribed in the Unites States in 2009 (Bartholow 2010). Some researchers reviewed the occurrence of ibuprofen in numerous sediment and aquatic compartments (Ali et al. 2009b). Buser et al. (1999) detected ibuprofen in river, lakes and influents of WWTP, with a higher concentration of the pharmacologically active (*S*)-enantiomer. In the same study, the degradation behaviour was similar during incubation experiments in WWTP influent incubated with activated sludge and in lake water fortified with racemic ibuprofen, with a faster dissipation of the (*S*)-enantiomer, mostly biological mediated. Hijosa-Valero et al. (2010) reported the decrease of EF of ibuprofen, which is according to the biological-depending degradation (Hijosa-Valsero et al. 2010). Jones et al. (2007) studied a WWTP located in southern England with activated sludge in the secondary biological treatment. Ibuprofen was found in all samples with a removal efficiency of ca 80–90% indicating a biological mediated degradation. Matamoros et al. (2009) found differences in the behaviour of ibuprofen EF's under aerobic and anaerobic conditions. In predominantly aerobic wastewater treatment systems ibuprofen EF's decreased, showing that the (*S*)-form was predominantly degraded, depending on the oxidation status of WWTP. In contrast, the anaerobic conditions led to a similar degradation of both enantiomers. In the same study, the behaviour of naproxen EF's showed a similar pattern in degradation at both aerobic and anaerobic conditions making it a good indicator for removal efficiency in aerobic or anaerobic wastewater treatment systems (Matamoros et al. 2009). Mascolo et al. (2010) reported biodegradation of racemic-naproxen, with degradation rates slower in wastewater samples than in synthetic wastewater, probably due to the presence of other organic compounds.

Marco-Urrea et al. (2010) studied the biodegradation of naproxen by the white-rot fungus *Trametes vesicolor* reporting degradation of 95% from an initial concentration of 55 $\mu gL^{-1}$, a concentration similar to that found in the environment (Marco-Urrea et al. 2010).

Musson et al. (2010) reported chemical degradation as possible mechanism of degradation of racemic-ibuprofen. In the same study, racemic-ibuprofen was more resistant to biological removal than in other studies probably due to the higher concentration tested (Musson et al. 2010).

The occurrence and removal of ibuprofen, naproxen, fluoxetine and its metabolite norfluoxetine were studied in a tertiary sewage treatment plant in Sweden (Zorita et al. 2009), being reported removal rates superior to 90% for these compounds and the higher load of ibuprofen (6.9 $\mu gL^{-1}$) and naproxen (4.9 $\mu gL^{-1}$) due to the great consumption. Fluoxetine and norfluoxetine were below their detection limits after primary treatment, being removed by adsorption to fat and primary sludge. The additional chemical treatment step improved the removal of these pharmaceuticals. Apart from the common mechanical and biological treatments, the tertiary treatment such as ozonation, flocculation, advanced oxidation, osmosis or ultrafiltration, is unusual in Europe because it is expensive (Zorita et al. 2009).

In a WWTP in Spain it was reported removal rates below 20% for propranolol and atenolol, 60% for fluoxetine and naproxen and 95% for ibuprofen (Rosal et al. 2010). After ozonation, all of them were below the quantification limits in a few minutes of contact with low ozone doses (<50 μM, except for naproxen which was <220 μM), providing a superior removal of compounds that are relatively refractory to biological treatment such as propranolol and atenolol.

Thus, there are very few reports concerning enantioselective degradation of pharmaceuticals in the environment. It is important to include this issue mostly in the biodegradation studies due to the enantioselectivity of the biological processes to assess the more recalcitrant enantiomers and to relate with their ecotoxicity.

## 1.5 Toxicity of Emergent Chiral Pharmaceuticals in Aquatic Organisms

Although there are few studies about ecological toxicity of isolated enantiomers of chiral pharmaceuticals, many studies have been developed reporting the toxicity of these pharmaceuticals to aquatic organisms (Khetan and Collins 2007), mostly at concentrations higher than those found in the environment (Fent et al. 2006), which show the need of research on the effects of each enantiomer separately (Table 1.5).

Briefly, ecotoxicity evaluation of propranolol using the marine bacterium *Vibrio fischeri*, the water flea *Daphnia magna*, the green alga *Desmodesmus subspicatus,* the duck weed *Lemna minor*, the freshwater crustacean *Thamnocephalus platyurus* and a fish species *Oryzias latipes* classified it as very toxic (Cleuvers 2005; Escher et al. 2005; Kim et al. 2009). Long-term exposure to *C. dubia* and *H. azteca* showed that propranolol affects sex hormone blood plasma concentrations and reduces fecundity (Huggett et al. 2002). Christen et al. (2010) classified propranolol and fluoxetine as compounds with a high potential of environmental risk.

Enantiospecific toxicity of fluoxetine to *Daphnia magna* and *Pimephales promelas* was demonstrated, being (*S*)-fluoxetine more toxic to *P. promelas* after 7-day exposure concerning to survival, growth, and reproduction endpoints and also to *D. magna* regarding the feeding and growth endpoints (Stanley et al. 2007).

Winter et al. (2008) reported the low chronic toxicity of atenolol to a freshwater fish, the fathead minnow (*Pimephales promelas*), at the concentrations normally found in the environment. They suggested that atenolol is less toxic than propranolol in fish probably due to the different target receptor since atenolol is a β1-selective blocker and propranolol is a non-selective β–blocker, acting in both β1 and β2 receptors (Owen et al. 2007). It was reported that (*R*)-Atenolol is more toxic than the (*S*)-enantiomer, although both of them have low toxicity (De Andrés et al. 2009).

Fluoxetine in the environment has antimicrobial properties, mainly against Gram positive bacteria. Fluoxetine has a synergic activity with some antibiotics, even when there is some resistance, or can increase the activity of them, exerting toxicity by inhibiting cellular efflux pumps (Munoz-Bellido et al. 2000). The accumulation of fluoxetine in three fish species (*Lepomis macrochirus*, *Ictalurus punctatus*, and *Pomoxis nigromaculatus*) was reported in brain, liver and muscle tissues at $ngg^{-1}$ levels, supporting the theorical human exposure to pharmaceuticals in fish tissue (Brooks et al. 2005). The induction of spawn in zebra mussels (*Dreissena polymorpha*) by fluoxetine at low concentrations as 5 μM or even less was also reported (Fong and Molnar 2008). It was demonstrated that, as opposed to chronic effects that can affect aquatic organisms, the acute effects of fluoxetine in three aquatic species were achieved with concentrations above the levels found in the environment shielding the possible occurrence of these concentrations in municipal effluents which do not have dilution from upstream (Brooks et al. 2003). Low concentrations slowed female and male sexual development in *Gambusia affinis* (Henry and Black 2008), and a chronic exposure to *Daphnia* increases fecundity (Flaherty and Dodson 2005) as well as increased production of egg masses accompanied by higher mortality in

**Table 1.5** Ecotoxicological data on racemic pharmaceuticals as racemates and as enantiopure compounds, showing the effects caused in aquatic organisms by the compounds focused in this review

| Drug | Test organism | Test | Effects | Observations | Refs |
|---|---|---|---|---|---|
| Propranolol | *Oncorhynchus mykiss* (rainbow trout) | 10-days $NOEC^{growth}$; 10-days $LOEC^{growth}$ | 1.0 $mgL^{-1}$; 10 $mgL^{-1}$ | The growth was retarded after 10 days of an exposure of 10 $mgL^{-1}$, however after more 30 days of exposure the effect was recovered, suggesting a possible adaptation of this species to propranolol | Owen et al. (2009) |
| Propranolol and ibuprofen | *Thaemnocephalus platyrus* (freshwater crustacean) | 24 h LC50 | 10.31 $mgL^{-1}$; 19.59 $mgL^{-1}$ | 24 h LC50 (24 h median lethal concentration) was calculated based on the immobilization of T. platyrus | Kim et al. (2009) |
| Atenolol | *Thaemnocephalus platyrus* (freshwater crustacean) | 24 h LC50 | No acute toxicity (at the tested levels >100 $mgL^{-1}$) | | |
| Propranolol | *Oryzias latipes* (fish) | 96 h LC50 | 11.40 $mgL^{-1}$ | 96 h LC50 (96 h median lethal concentration) was calculated based on the stereoscopic microscope observation and removal of dead larvae on the absence of heartbeat | |
| Atenolol and ibuprofen | *Oryzias latipes* (fish) | 96 h LC50 | No acute toxicity (at the tested levels >100 $mgL^{-1}$) | | |
| Fluoxetine | *Cyprinodon variegatus* (sheepshead minnow) | 96 h LOEC; 96 h NOEC | 2 $mgL^{-1}$; 1.250 $mgL^{-1}$ (levels above the environmental concentrations) | The effects on neurotransmitter pathways are observed at concentrations one order of magnitude higher the reported in the environment. These effects can lead to behaviour changes in the organisms which can potentiate the inability towards a predator or the failure on food providing | Winder et al. (2009) |
| | | Sublethal effects (serotonergic system) | Effects at 0.300 and 0.030 $mgL^{-1}$ | | |

(continued)

**Table 1.5** (continued)

| Drug | Test organism | Test | Effects | Observations | Refs |
|---|---|---|---|---|---|
| Fluoxetine | Sexually mature female zebrafish | Estardiol ovarian levels | Estradiol levels decreased three fold when exposed to 32 $\mu gL^{-1}$ of fluoxetine (7 days) | These results suggest there was no alteration in the ovulatory pathway, being the decreasing of aromatase responsible for the decreased estradiol levels. This effect together with the reduced follicle-stimulating hormone receptor and luteinizing hormone receptor gene expression reduced may led to the decline in spawned eggs | Lister et al. (2009) |
| | | Egg production | Decline in spawned eggs when exposed to 32 $\mu gL^{-1}$ of fluoxetine or to 50% municipal effluent | | |
| | | Gene expression | Prostaglandins gene expression constant; ovarian aromatase, follicle-stimulating hormone receptor and luteinizing hormone receptor gene expression was reduced | | |
| Fluoxetine | *Physa acuta* (Gastropoda, Pulmonata) | Mortality and reproduction effects | Increased mortality and decreased reproduction effects (250 $\mu gL^{-1}$) | The adsorption of fluoxetine to biomass and sediments must be considered to the effects on reproduction of freshwater molluscs | Sánchez-Argüello et al. (2009) |
| Fluoxetine and atenolol | *Daphnia magna* (crustacean) | Acute immobiliza-tion test – EC50 | (*R*)- and (*S*)-fluoxetine: 8.1 $mgL^{-1}$ and 6.9 $mgL^{-1}$; (*R*)- and (*S*)-atenolol:1.450 $mgL^{-1}$ and 755 $mgL^{-1}$ | Both enantiomers of fluoxetine were toxic to D. magna. Atenolol enantiomers showed stereoselectivity but did not show a high toxicity. (R)-fluoxetine is considered harmful to P. subcapita. (S)-atenolol is more toxic than the (R)-enantiomer. (S)-fluoxetine is more toxic in an order of magnitude. (R)-atenolol is more toxic than the (S)-enantiomer | De Andrés et al. (2009) |
| | *Pseudokirchneriella subcapitata* (microalga) | Growth inhibition test – EC50 | (*R*)-fluoxetine: 34 $mgL^{-1}$; (*R*)- and (*S*)-atenolol: 190 $mgL^{-1}$ and 143 $mgL^{-1}$ | | |
| | *Tetrahymena thermophila* (protozoan) | Growth inhibition test – EC50 | (*R*)- and (*S*)-fluoxetine: 30.5 $mgL^{-1}$ and 3.2 $mgL^{-1}$; (*R*)- and (*S*)-atenolol: 13.7 $mgL^{-1}$ and 55.7 $mgL^{-1}$ | | |

| | | | | | |
|---|---|---|---|---|---|
| Atenolol | *Pimephales promelas* (fish) | Early life stage – NOEC/LOEC | 4 days embryo: $NOEC^{hatching}$ – 10 $mgL^{-1}$; $LOEC^{hatching}$ – >10 $mgL^{-1}$; after 28 days $NOEC^{growth}$ – 3.2 $mgL^{-1}$; $LOEC^{growth}$ – 10 $mgL^{-1}$ | Atenolol has been reported to be low toxic to fish respecting to the chronic toxicity at low concentrations found in the environment | Winter et al. (2008) |
| | | Reproduction effects - NOEC/LOEC | Short-term: $NOEC^{reprodution}$ – 10 $mgL^{-1}$; $LOEC^{reproduction}$ – >10 $mgL^{-1}$; increase in male condition index (CI) $NOEC^{CI}$ – 1.0 $mgL^{-1}$; $LOEC^{CI}$ – 3.2 $mgL^{-1}$ | | |
| Ibuprofen | *Oncorhynchus mykiss* (rainbow trout) cell lines | Cytotoxicity and cytostatic action | Ibuprofen at concentrations higher than 15 $\mu gmL^{-1}$ is citotoxic to fish cells. The cytostatic effect was verified between 15 and 150 $\mu gmL^{-1}$ | The cytotoxicity of ibuprofen occurred at concentrations 10,000 times greater than the higher value reported in the environment, being the bioaccumulation an important factor to be studied. | Schnell et al. (2008) |
| Fluoxetine | *Oryzias latipes* (fish) | 96 h LC50 | 5.5 $mgL^{-1}$ (pH 7); 1.3 $mgL^{-1}$ (pH8); 0.20 $mgL^{-1}$ (pH 9) | At higher pH, there was an increased bioconcentration factor due to the higher fraction of lipophilic no ionized species, leading to more toxicity | Nakamura et al. (2008) |
| Fluoxetine and norfluoxetine | *Spirostomum ambiguum* (protozoan) | Lethality tests – 24 h LC50 | ca 0.55 $mgL^{-1}$ | Fluoxetine and its metabolite norfluoxetine showed a high toxicity. | Nalecz-Jawecki (2007) |
| | *Thamnocephalus platyurus* (crustacean) | | ca 0.76 $mgL^{-1}$ | | |

(continued)

**Table 1.5** (continued)

| Drug | Test organism | Test | Effects | Observations | Refs |
|---|---|---|---|---|---|
| Fluoxetine | *Pimephales promelas* (fish) | Sublethal and behaviour effects – LOEC | (R)- and (S)-fluoxetine: 170 $\mu gL^{-1}$ and 101 $\mu gL^{-1}$ | (S)-fluoxetine was more toxic for both organisms. The primary metabolite (S)-norfluoxetine is more potent than the metabolite (R)-norfluoxetine in mammals. Thus, the aquatic vertebrates can predict better the mammalian behaviour than the invertebrate organisms | Stanley et al. (2007) |
| | | Growth inhibition test – EC10 | (R)- and (S)-fluoxetine revealed a NOEC of 132.9 and 14.1 $\mu gL^{-1}$, respectively | | |
| β-blockers | *Daphnia magna* (crustacean) | Immobilization test - EC50 | Propranolol 7.7 $mgL^{-1}$; atenolol 313 $mgL^{-1}$; metoprolol 438 $mgL^{-1}$ | The log P differences are responsible for the different toxicity. Higher log P, higher toxicity, with propranolol being very toxic, metoprolol being toxic and atenolol seeming to be non-toxic to aquatic organisms. The toxicity is related with the bioconcentration, which depends on log P | Cleuvers (2005) |
| | *Desmodesmus subspicatus* (green algae) | Growth inhibition test – EC50 | Propranolol 0.7 $mgL^{-1}$; atenolol 620 $mgL^{-1}$; metoprolol 7.9 $mgL^{-1}$ | | |
| | *Lemma minor* (duckweed) | Growth inhibition test – EC50 | Propranolol 113 $mgL^{-1}$; atenolol and metoprolol (no effects up to 320 $mgL^{-1}$) | | |
| Ibuprofen and naproxen | *Desmodesmus subspicatus* (green algae) | Growth inhibition test – EC50 | 342.2 $mgL^{-1}$ for ibuprofen; 625.5 $mgL^{-1}$ for naproxen | Ibuprofen and naproxen were not classified as harmful to aquatic organisms since the EC50 values were superior than 100 $mgL^{-1}$. These concentrations are unlikely to cause acute effects in aquatic organisms | Cleuvers (2004) |
| | *Daphnia magna* (crustacean); *Ceriodaphnia dubia* | Immobilization test – EC50 | 101.2 $mgL^{-1}$ for ibuprofen; 166.3 $mgL^{-1}$ for naproxen | | |

| | | | | | |
|---|---|---|---|---|---|
| Fluoxetine | *Pseudokirchneriella promelas* (algae) | Growth inhibition test – EC50 | 24–39 $\mu gL^{-1}$ | The lowest effect value is one order of magnitude higher the highest reported environmental concentration of fluoxetine | Brooks et al. (2003) |
| | *Daphnia magna* (crustacean); Ceriodaphnia dubia | 48 h LC50 | 820 $\mu gL^{-1}$; 234 $\mu gL^{-1}$ respectively | | |
| | Pimephales promelas (fish) | 48 h LC50 | 705 $\mu gL^{-1}$ | | |
| Fluoxetine | *Dreissena polymorpha* (zebra mussels) | Gametes spawning | $5 \times 10^{-7}$–$5 \times 10^{-4}$ M (males); $5 \times 10^{-7}$–$5 \times 10^{-4}$ M (females) | Fluoxetine was an effective spawning inducer, since it induced spawning to 100% of males of zebra mussels that were exposed to $5 \times 10^{-6}$ M | Fong (1998) |

*LOEC* lowest observed effect concentration, *NOEC* no observed effect concentration, *LC50* lethal concentration of 50%, *EC50* effective concentration that causes 50% of maximal response, *EC10* effective concentration that causes 10% of maximal response adults of a mollusc specie (*Physa acuta*) (Sánchez-Argüello et al. 2009). In Japanese medaka it led to a developmental abnormality in offspring, increasing female circulating plasma estradiol levels (Foran et al. 2004). In studies involving zebrafish and sheepshead minnow behavioural changes and disturb of serotonergic pathway that may result in implications in ecological responses were also observed (Lister et al. 2009; Winder et al. 2009). Bioaccumulation in fish (Nakamura et al. 2008) and enantiospecific differences in sublethal responses to a model aquatic vertebrates and invertebrates were also detected (Stanley et al. 2007).

Kim et al. (2009) reported acute toxic effects of ibuprofen in the freshwater crustacean *Thamnocephalus platyurus*, unlike the fish species *Oryzias latipes*. The different behaviour is related with the more sensitive of *T. platyurus* to pharmaceuticals in the environment. Schnell et al. (2008) reported ibuprofen as cytotoxic and cytostatic to rainbow trout (*Oncorhynchus mykiss*) cell lines (liver and hepatocellular carcinoma), affecting the viability and proliferation of fish cells. Ibuprofen was suggested to affect population growth rate, survival and reproduction in *Daphnia magna* (Heckmann et al. 2007). It was reported the effects of ibuprofen on reproduction of Japanese medaka (*Oryzias latipes*), which induced the less frequent spawning and more eggs production when spawn occurred (Flippin et al. 2007).

Isidori et al. (2005) found the phototransformation of naproxen in four products with less polarity, and as such with an EC50 (Effective Concentration that causes 50% of maximal response) lower than naproxen, making these photoderivates more toxic than the parent compound. The same authors reported no mutagenesis and no genotoxicity for naproxen and its photoderivates (Isidori et al. 2005).

The studies involving the occurrence and removal of pharmaceuticals should contemplate bioassays as the reduction of the concentrations does not demonstrate the absence of toxicity (Reungoat et al. 2010). Another important feature that is not considered in almost studies is the food quality of the organisms used in toxicological tests. It was recently demonstrated for *Daphnia magna* that the elemental food quality has interactive effects with fluoxetine which change the survival, growth and reproduction of this species. The decrease of ingestion of phosphorous decreased the effects caused by fluoxetine (Hansen et al. 2008).

## 1.6 Conclusions

Methods for quantification and biodegradation studies of single enantiomers of chiral pharmaceuticals in the environment are scarce. Chiral analytical methods for quantification of chiral pharmaceuticals in the environment in an accurate way are imperative for a better knowledge of the environmental fate and behaviour of individual enantiomers. Studies concerning the degradation performance of racemic mixtures and the individual enantiomers of chiral pharmaceuticals in the environment are clearly needed and require further research. The toxicity of isolated enantiomers on non-target organism is also a field to be investigated. This is well stated in the recent review which calls for the attention to the registration, evaluation, authorisation and restriction of chemicals (REACH) and addresses the continued use of chemical "substances of very high concern" (Union 2006). Depending on the substance in question and its application, producers and importers may be obliged to investigate its effects on human health and in the environment. As a consequence, detailed insight into the fate, transformation and effects of pharmaceuticals in the environment is required.

**Acknowledgments** To FCT for PhD grant attributed to Ana Rita Ribeiro (SFRH/BD/64999/2009) and to QREN-POPH, European Social Fund and MCTES. Authors also wish to acknowledge the support from CESPU (09-GCQF-CICS-09) and FCT (FLUOROPHARMA, PTDC/EBB-EBI/111699/2009).

## References

Aboul-Enein HY, Ali I (2004) Analysis of the chiral pollutants by chromatography. Toxicol Environ Chem 86(1):1–22

Agrawal YK, Patel RN (2005) Chiral chromatographic separation of β-blockers. J Chromatogr B 820(1):23–31

Aleksa K, Nava-Ocampo A, Koren G (2009) Detection and quantification of (R) and (S)-dechloroethylifosfamide metabolites in plasma from children by enantioselective LC/MS/MS. Chirality 21(7):674–680

Ali I, Gaitonde VD, Aboul-Enein HY, Hussain A (2009a) Chiral separation of β-adrenergic blockers on CelluCoat column by HPLC. Talanta 78(2):458–463

Ali I, Singh P, Aboul-Enein HY, Sharma B (2009b) Chiral analysis of ibuprofen residues in water and sediment. Anal Lett 42(12):1747–1760

Andersson ME, Aslan D, Clarke A, Roeraade J, Hagman G (2003) Evaluation of generic chiral liquid chromatography screens for pharmaceutical analysis. J Chromatogr A 1005(1–2):83–101

Ansermot N, Rudaz S, Brawand-Amey M, Fleury-Souverain S, Veuthey J-L, Eap CB (2009) Validation and long-term evaluation of a modified on-line chiral analytical method for therapeutic drug monitoring of (R, S)-methadone in clinical samples. J Chromatogr B 877(23):2301–2307

Asher BJ, D'Agostino LA, Way JD, Wong CS, Harynuk JJ (2009) Comparison of peak integration methods for the determination of enantiomeric fraction in environmental samples. Chemosphere 75(8):1042–1048

Ates H, Mangelings D, Vander Heyden Y (2008) Chiral separations in polar organic solvent chromatography: updating a screening strategy with new chlorine-containing polysaccharide-based selectors. J Chromatogr B 875(1):57–64

Baker GB, Prior TI, Coutts RT (2002) Chirality and drugs used to treat psychiatric disorders. J Psychiatry Neurosci 27(6):401–403

Barreiro EJ, Ferreira VF, Costa PRR (1997) Substâncias enantiomericamente puras (SEP): a questão dos fármacos quirais. Quím Nova 20(6):647–656

Barreiro JC, Vanzolini KL, Madureira TV, Tiritan ME, Cass QB (2010) A column-switching method for quantification of the enantiomers of omeprazole in native matrices of waste and estuarine water samples. Talanta 82(1):384–391

Bartholow M (2010) Top 200 prescription drugs of 2009. Retrieved 27 May 2010

Benotti MJ, Brownawell BJ (2009) Microbial degradation of pharmaceuticals in estuarine and coastal seawater. Environ Pollut 157(3):994–1002

Benotti MJ, Trenholm RA, Vanderford BJ, Holady JC, Stanford BD, Snyder SA (2009) Pharmaceuticals and endocrine disrupting compounds in U.S. drinking water. Environ Sci Technol 43(3):597–603

Bhushan R, Martens J (1997) Direct resolution of enantiomers by impregnated TLC. Biomed Chromatogr 11(5):280–285

Bojarski J (2002) Stereoselective chromatography of cardiovascular drugs: an update. J Biochem Biophys Methods 54(1–3):197–220

Bonabello A, Galmozzi MR, Canaparo R, Isaia GC, Serpe L, Muntoni E, Zara GP (2003) Dexibuprofen (S(+)-isomer ibuprofen) reduces gastric damage and improves analgesic and antiinflammatory effects in rodents. Anesth Analg 97(2):402–408

Bosáková Z, Curínová E, Tesarová E (2005) Comparison of vancomycin-based stationary phases with different chiral selector coverage for enantioselective separation of selected drugs in high-performance liquid chromatography. J Chromatogr A 1088(1–2):94–103

Brooks BW, Turner PK, Stanley JK, Weston JJ, Glidewell EA, Foran CM, Slattery M, La Point TW, Huggett DB (2003) Waterborne and sediment toxicity of fluoxetine to select organisms. Chemosphere 52(1):135–142

Brooks BW, Chambliss CK, Stanley JK, Ramirez A, Banks KE, Johnson RD, Lewis RJ (2005) Determination of select antidepressants in fish from an effluent-dominated stream. Environ Toxicol Chem 24(2):464–469

Buser H-R, Poiger T, Muller MD (1999) Occurrence and environmental behavior of the chiral pharmaceutical drug ibuprofen in surface waters and in wastewater. Environ Sci Technol 33(15):2529–2535

Caccamese S, Bianca S, Carter GT (2009) Direct high-performance liquid chromatographic separation of the enantiomers of venlafaxine and 11 analogs using amylose-derived chiral stationary phases. Chirality 21(6):569–577

Caldwell J, Hutt AJ, Fournel-Gigleux S (1988) The metabolic chiral inversion and dispositional enantioselectivity of the 2-arylpropionic acids and their biological consequences. Biochem Pharmacol 37(1):105–114

Calisto V, Esteves VI (2009) Psychiatric pharmaceuticals in the environment. Chemosphere 77(10):1257–1274

Campo VL, Bernardes LSC, Carvalho I (2009) Stereoselectivity in drug metabolism: molecular mechanisms and analytical methods. Curr Drug Metab 10(2):188–205

Caner H, Groner E, Levy L, Agranat I (2004) Trends in the development of chiral drugs. Drug Discov Today 9(3):105–110

Carvalho PO, Cass QB, Calafatti SA, Contesini FJ, Bizaco R (2006) Review – alternatives for the separation of drug enantiomers: ibuprofen as a model compound. Braz J Chem Eng 23(3):291–300

Cass QB and Batigalhia F (2003) Enantiomeric resolution of a series of chiral sulfoxides by high-performance liquid chromatography on polysaccharide-based columns with multimodal elution. J Chromatogr A 987(1-2): 445-452

Cass QB, Degani ALG, Cassiano NM (2003a) Effects on enantioselectivity by the use of polysaccharide-based columns by multimodal elution. J Liquid Chromatogr Relat Technol 26(13):2083–2101

Cass QB, Watanabe CSF, Rabi JA, Bottari PQ, Costa MR, Nascimento RM, Cruz JED, Ronald RC (2003b) Polysaccharide-based chiral phase under polar organic mode of elution in the determination of the enantiomeric purity of emtricitabine an anti-HIV analogue nucleoside. J Pharm Biomed Anal 33(4):581–587

Chávez-Flores D, Salvador JM (2009) Commercially viable resolution of ibuprofen. Biotechnol J 4(8):1222–1224

Chen C-Y, Shieh W-R, Chen C-S (1992) Clinical implications of using racemic drugs. Clin Res Regul Aff 9(4):247–259

Choonara IA, Haynes BP, Cholerton S, Breckenridge AM, Park BK (1986) Enantiomers of warfarin and vitamin K1 metabolism. Br J Clin Pharmacol 22(6):729–732

Christen V, Hickmann S, Rechenberg B, Fent K (2010) Highly active human pharmaceuticals in aquatic systems: a concept for their identification based on their mode of action. Aquat Toxicol 96(3):167–181

Cleuvers M (2004) Mixture toxicity of the anti-inflammatory drugs diclofenac, ibuprofen, naproxen, and acetylsalicylic acid. Ecotoxicol Environ Saf 59(3):309–315

Cleuvers M (2005) Initial risk assessment for three β-blockers found in the aquatic environment. Chemosphere 59(2):199–205

Cooper MJ, Anders MW (1974) Metabolic and pharmacodynamic interactions of enantiomers of propoxyphene and methorphan. Life Sci 15(9):1665–1672

Cordato DJ, Mather LE, Herkes GK (2003) Stereochemistry in clinical medicine: a neurological perspective. J Clin Neurosci 10(6):649–654

Cox PJ (1994) Drug stereochemistry – analytical methods and pharmacology. Talanta 41(2):343

Da Silva Júnior I, De Veredas V, Carpes MJS, Santana CCP (2005) Chromatographic separation of bupivacaine enantiomers by HPLC: parameters estimation of equilibrium and mass transfer under linear conditions. Adsorption 11(2):123–129

Daughton CG, Ternes TA (1999) Pharmaceuticals and personal care products in the environment: agents of subtle change? Environ Health Perspect Suppl 107(S6):907–938

De Andrés F, Castañeda G, Ríos Á (2009) Use of toxicity assays for enantiomeric discrimination of pharmaceutical substances. Chirality 21(8):751–759

Díaz-Cruz MS, García-Galán MJ, Barceló D (2008) Highly sensitive simultaneous determination of sulfonamide antibiotics and one metabolite in environmental waters by liquid chromatography-quadrupole linear ion trap-mass spectrometry. J Chromatogr A 1193(1–2):50–59

Eliel EL, Wilen SH (1994) Stereochemistry of organic compounds. Wiley, New York

Eljarrat E, Guerra P, Barceló D (2008) Enantiomeric determination of chiral persistent organic pollutants and their metabolites. TrAC Trends Anal Chem 27(10):847–861

Escher BI, Bramaz N, Maurer M, Richter M, Sutter D, Känel Cv, Zschokke M (2005) Screening test battery for pharmaceuticals in urine and wastewater. Environ Toxicol Chem 24(3): 750–758

Farré M, Gros M, Hernández B, Petrovic M, Hancock P, Barceló D (2008) Analysis of biologically active compounds in water by ultra-performance liquid chromatography quadrupole time-of-flight mass spectrometry. Rapid Commun Mass Spectrom 22(1):41–51

Feitosa-Felizzola J, Temime B, Chiron S (2007) Evaluating on-line solid-phase extraction coupled to liquid chromatography-ion trap mass spectrometry for reliable quantification and confirmation of several classes of antibiotics in urban wastewaters. J Chromatogr A 1164(1–2):95–104

Fent K, Weston AA, Caminada D (2006) Ecotoxicology of human pharmaceuticals. Aquat Toxicol 76(2):122–159

Fernández C, González-Doncel M, Pro J, Carbonell G, Tarazona JV (2010) Occurrence of pharmaceutically active compounds in surface waters of the henares-jarama-tajo river system (madrid, spain) and a potential risk characterization. Sci Total Environ 408(3):543–551

Flaherty CM, Dodson SI (2005) Effects of pharmaceuticals on *Daphnia* survival, growth, and reproduction. Chemosphere 61(2):200–207

Flippin JL, Huggett D, Foran CM (2007) Changes in the timing of reproduction following chronic exposure to ibuprofen in Japanese medaka, *Oryzias latipes*. Aquat Toxicol 81(1): 73–78

Fong PP (1998) Zebra Mussel Spawning Is Induced in Low Concentrations of Putative Serotonin Reuptake Inhibitors. The Biol Bull 194: 143–149

Fong PP, Molnar N (2008) Norfluoxetine induces spawning and parturition in estuarine and freshwater bivalves. Bull Environ Contam Toxicol 81(6):535–538

Fono LJ, Sedlak DL (2005) Use of the chiral pharmaceutical propranolol to identify sewage discharges into surface waters. Environ Sci Technol 39(23):9244–9252

Fono LJ, Kolodziej EP, Sedlak DL (2006) Attenuation of wastewater-derived contaminants in an effluent-dominated river. Environ Sci Technol 40(23):7257–7262

Foran CM, Weston J, Slattery M, Brooks BW, Huggett DB (2004) Reproductive assessment of Japanese medaka (*oryzias latipes*) following a four-week fluoxetine (SSRI) exposure. Arch Environ Contam Toxicol 46(4):511–517

Francotte ER (2001) Enantioselective chromatography as a powerful alternative for the preparation of drug enantiomers. J Chromatogr A 906(1–2):379–397

Francotte E (2009) Enantioselective chromatography: from its emergence to its successful implementation in the pharmaceutical environment. CHIMIA Int J Chem 63:867–871

Franz PM, Anliker SL, Callaghan JT, DeSante KA, Dhahir PH, Nelson RL, Rubin A (1990) Disposition in humans of racemic picenadol, an opioid analgesic. Drug Metab Dispos 18(6):968–973

Gatti G, Bonomi I, Marchiselli R, Fattore C, Spina E, Scordo G, Pacifici R, Perucca E (2003) Improved enantioselective assay for the determination of fluoxetine and norfluoxetine enantiomers in human plasma by liquid chromatography. J Chromatogr B 784(2):375–383

Gonçalves RF, Lauretti GR, Mattos ALd (2003) Comparative study between 0.5% Bupivacaine and 0.5% enantiomeric mixture of bupivacaine (S75-R25) in epidural anesthesia. Rev Bras Anestesiol 53(2):169–176

González S, Petrovic M, Barceló D (2007) Advanced liquid chromatography-mass spectrometry (LC-MS) methods applied to wastewater removal and the fate of surfactants in the environment. TrAC Trends Anal Chem 26(2):116–124

Gomes RF, Cassiano NM, Pedrazzoli Jr. J, Cass QB (2010) Two-dimensional chromatography method applied to the enantiomeric determination of lansoprazole in human plasma by direct sample injection. Chirality 22(1):35–41

Görög S (2007) The changing face of pharmaceutical analysis. TrAC Trends Anal Chem 26(1):12–17

Gros M, Petrovic M, Barceló D (2006) Development of a multi-residue analytical methodology based on liquid chromatography-tandem mass spectrometry (LC-MS/MS) for screening and trace level determination of pharmaceuticals in surface and wastewaters. Talanta 70(4): 678–690

Gübitz G, Schmid MG (2004) Recent advances in chiral separation principles in capillary electrophoresis and capillary electrochromatography. Electrophoresis 25(23):3981–3996

Haginaka J (2008) Recent progresses in protein-based chiral stationary phases for enantioseparations in liquid chromatography. J Chromatogr B 875(1):12–19

Ha PTT, Hoogmartens J, Van Schepdael A (2006) Recent advances in pharmaceutical applications of chiral capillary electrophoresis. J Pharm Biomed Anal 41(1):1–11

Hansen LK, Frost PC, Larson JH, Metcalfe CD (2008) Poor elemental food quality reduces the toxicity of fluoxetine on Daphnia magna. Aquat Toxicol 86(1):99–103

Hashim NH, Shafie S, Khan SJ (2010) Enantiomeric fraction as an indicator of pharmaceutical biotransformation during wastewater treatment and in the environment – a review. Environ Technol 31(12):1349–1370

Heckmann L-H, Callaghan A, Hooper HL, Connon R, Hutchinson TH, Maund SJ, Sibly RM (2007) Chronic toxicity of ibuprofen to *daphnia magna*: effects on life history traits and population dynamics. Toxicol Lett 172(3):137–145

Hefnawy MM, Sultan MA, Al-Shehri MM (2007) HPLC separation technique for analysis of bufuralol enantiomers in plasma and pharmaceutical formulations using a vancomycin chiral stationary phase and UV detection. J Chromatogr B 856(1–2):328–336

Henry TB, Black MC (2008) Acute and chronic toxicity of fluoxetine (selective serotonin reuptake inhibitor) in western mosquitofish. Arch Environ Contam Toxicol 54(2):325–330

Henry ME, Schmidt ME, Hennen J, Villafuerte RA, Butman ML, Tran P, Kerner LT, Cohen B, Renshaw PF (2005) A comparison of brain and serum pharmacokinetics of R-fluoxetine and racemic fluoxetine: a 19-F MRS study. Neuropsychopharmacology 30(8):1576–1583

Hijosa-Valsero M, Matamoros V, Martín-Villacorta J, Bécares E, Bayona JM (2010) Assessment of full-scale natural systems for the removal of PPCPs from wastewater in small communities. Water Res 44(5):1429–1439

Huang YF, Pryor ME, Mather LE, Veering BT (1998) Cardiovascular and central nervous system effects of intravenous levobupivacaine and bupivacaine in sheep. Anesth Analg 86(4):797–804

Huggett DB, Brooks BW, Peterson B, Foran CM, Schlenk D (2002) Toxicity of select beta adrenergic receptor-blocking pharmaceuticals (B-blockers) on aquatic organisms. Arch Environ Contam Toxicol 43(2):229–235

Huggett DB, Khan IA, Foran CM, Schlenk D (2003) Determination of beta-adrenergic receptor blocking pharmaceuticals in United States wastewater effluent. Environ Pollut 121(2):199–205

Hühnerfuss H, Shah MR (2009) Enantioselective chromatography-a powerful tool for the discrimination of biotic and abiotic transformation processes of chiral environmental pollutants. J Chromatogr A 1216(3):481–502
Huq F (2007) Molecular modelling analysis of the metabolism of methadone. Asian J Biochem 2(1):19–30
Hutt AJ (1998) Drug chirality and its pharmacological consequences. In: Smith HJ, Williams' H (eds) Introduction to the principles of Drug design and action, 3rd edn. Elsevier, Amsterdam
Hutt AJ, Valentová J (2003) The chiral switch: the development of single enantiomer drugs from racemates. Acta Fac Pharm Univ Comen 50:7–23
IMS Health report (2006) http://www.imshealth.com
Isidori M, Lavorgna M, Nardelli A, Parrella A, Previtera L, Rubino M (2005) Ecotoxicity of naproxen and its phototransformation products. Sci Total Environ 348(1–3):93–101
Izake EL (2007) Chiral discrimination and enantioselective analysis of drugs: an overview. J Pharm Sci 96(7):1659–1676
Jain AK, Michael R, Ryan JR, McMahon FG (1984) Antihypertensive and biochemical effects of indacrinone enantiomers. Pharmacotherapy 4(5):278–283
Jiang Z, Schurig V (2008) Existence of a low isoenantioselective temperature in complexation gas chromatography: profound change of enantioselectivity of a nickel(II) chiral selector either bonded to, or dissolved in, poly(dimethylsiloxane). J Chromatogr A 1186(1–2):262–270
Johnson DJ, Sanderson H, Brain RA, Wilson CJ, Bestari KT, Solomon KR (2005) Exposure assessment and microcosm fate of selected selective serotonin reuptake inhibitors. Regul Toxicol Pharmacol 42(3):313–323
Jones OAH, Voulvoulis N, Lester JN (2007) The occurrence and removal of selected pharmaceutical compounds in a sewage treatment works utilising activated sludge treatment. Environ Pollut 145(3):738–744
Kasprzyk-Hordern B, Kondakal VVR, Baker DR (2010) Enantiomeric analysis of drugs of abuse in wastewater by chiral liquid chromatography coupled with tandem mass spectrometry. J Chromatogr A 1217(27):4575–4586
Khetan SK, Collins TJ (2007) Human pharmaceuticals in the aquatic environment: a challenge to green chemisty. Chem Rev 107(6):2319–2364
Kim J-W, Ishibashi H, Yamauchi R, Ichikawa N, Takao Y, Hirano M, Koga M, Arizono K (2009) Acute toxicity of pharmaceutical and personal care products on freshwater crustacean (*Thamnocephalus platyurus*) and fish (*Oryzias latipes*). J Toxicol Sci 34(2):227–232
Koch S, Perry KW, Nelson DL, Conway RG, Threlkeld PG, Bymaster FP (2002) R-fluoxetine increases extracellular DA, NE, as well as 5-HT in Rat prefrontal cortex and hypothalamus: an in vivo microdialysis and receptor binding study. Neuropsychopharmacology 27(6):949–959
Kot-Wasik A, Debska J, Namiesnik J (2007) Analytical techniques in studies of the environmental fate of pharmaceuticals and personal-care products. TrAC Trends Anal Chem 26(6):557–568
Lamas JP, Salgado-Petinal C, García-Jares C, Llompart M, Cela R, Gómez M (2004) Solid-phase microextraction-gas chromatography-mass spectrometry for the analysis of selective serotonin reuptake inhibitors in environmental water. J Chromatogr A 1046(1–2):241–247
Lämmerhofer M (2010) Chiral recognition by enantioselective liquid chromatography: mechanisms and modern chiral stationary phases. J Chromatogr A 1217(6):814–856
Li L, Zhou S, Jin L, Zhang C, Liu W (2010) Enantiomeric separation of organophosphorus pesticides by high-performance liquid chromatography, gas chromatography and capillary electrophoresis and their applications to environmental fate and toxicity assays. J Chromatogr B 878(17–18):1264–1276
Lima VLE (1997) Drugs and chirality: a brief overview. Quím Nova 20(6):657–663
Lister A, Regan C, Van Zwol J, Van Der Kraak G (2009) Inhibition of egg production in zebrafish by fluoxetine and municipal effluents: a mechanistic evaluation. Aquat Toxicol 95(4):320–329
Liu Q-T, Cumming RI, Sharpe AD (2009) Photo-induced environmental depletion processes of β-blockers in river waters. Photochem Photobiol Sci 8(6):768–777

MacLeod SL, Wong CS (2010) Loadings, trends, comparisons, and fate of achiral and chiral pharmaceuticals in wastewaters from urban tertiary and rural aerated lagoon treatments. Water Res 44(2):533–544

MacLeod SL, Sudhir P, Wong CS (2007) Stereoisomer analysis of wastewater-derived β-blockers, selective serotonin re-uptake inhibitors, and salbutamol by high-performance liquid chromatography-tandem mass spectrometry. J Chromatogr A 1170(1–2):23–33

Madureira TV, Barreiro JC, Rocha MJ, Cass QB, Tiritan ME (2009) Pharmaceutical trace analysis in aqueous environmental matrices by liquid chromatography-ion trap tandem mass spectrometry. J Chromatogr A 1216(42):7033–7042

Maftouh M, Granier-Loyaux C, Chavana E, Marini J, Pradines A, Heyden YV, Picard C (2005) Screening approach for chiral separation of pharmaceuticals: part III. Supercritical fluid chromatography for analysis and purification in drug discovery. J Chromatogr A 1088(1–2):67–81

Mannschreck A, Kiesswetter R, von Angerer E (2007) Unequal activities of enantiomers via biological receptors: examples of chiral drug, pesticide, and fragrance molecules. J Chem Educ 84(12):2012–2018

Montanari MLC, Cass QB, Leitão A, Andricopulo AD and Montanari CA (2006) The role of molecular interaction fields on enantioselective and nonselective separation of chiral sulfoxides. J Chromatogr A 1121(1): 64–75

Marco-Urrea E, Pérez-Trujillo M, Blánquez P, Vicent T, Caminal G (2010) Biodegradation of the analgesic naproxen by *Trametes versicolor* and identification of intermediates using HPLC-DAD-MS and NMR. Bioresour Technol 101(7):2159–2166

Mascolo G, Balest L, Cassano D, Laera G, Lopez A, Pollice A, Salerno C (2010) Biodegradability of pharmaceutical industrial wastewater and formation of recalcitrant organic compounds during aerobic biological treatment. Bioresour Technol 101(8):2585–2591

Matamoros V, Hijosa M, Bayona JM (2009) Assessment of the pharmaceutical active compounds removal in wastewater treatment systems at enantiomeric level. Ibuprofen and naproxen. Chemosphere 75(2):200–205

Maurer M, Escher BI, Richle P, Schaffner C, Alder AC (2007) Elimination of β-blockers in sewage treatment plants. Water Res 41(7):1614–1622

Mayer JM, Testa B (1997) Pharmacodynamics, pharmacokinetics and toxicity of ibuprofen enantiomers. Drugs Future 22:1347–1366

McConathy J, Owens MJ (2003) Stereochemistry in drug action. Prim Care J Clin Psychiatry Companion 5(2):70–73

Moss GP (1996) Basic terminology of stereochemistry (IUPAC recommendations 1996). Pure Appl Chem 68(12):2193–2222

Müller TA, Kohler HPE (2004) Chirality of pollutants–effects on metabolism and fate. Appl Microbiol Biotechnol 64(3):300–316

Munoz-Bellido JL, Munoz-Criado S, Garcìa-Rodrìguez JA (2000) Antimicrobial activity of psychotropic drugs: Selective serotonin reuptake inhibitors. Int J Antimicrob Agents 14(3): 177–180

Musson SE, Campo P, Tolaymat T, Suidan M, Townsend TG (2010) Assessment of the anaerobic degradation of six active pharmaceutical ingredients. Sci Total Environ 408(9):2068–2074

Nakamura Y, Yamamoto H, Sekizawa J, Kondo T, Hirai N, Tatarazako N (2008) The effects of pH on fluoxetine in Japanese medaka (*Oryzias latipes*): acute toxicity in fish larvae and bioaccumulation in juvenile fish. Chemosphere 70(5):865–873

Nalecz-Jawecki G (2007) Evaluation of the *in vitro* biotransformation of fluoxetine with HPLC, mass spectrometry and ecotoxicological tests. Chemosphere 70(1):29–35

Niessen WMA (2003) Progress in liquid chromatography-mass spectrometry instrumentation and its impact on high-throughput screening. J Chromatogr A 1000(1–2):413–436

Nikolai LN, McClure EL, MacLeod SL, Wong CS (2006) Stereoisomer quantification of the -blocker drugs atenolol, metoprolol, and propranolol in wastewaters by chiral high-performance liquid chromatography-tandem mass spectrometry. J Chromatogr A 1131(1–2):103–109

Orlando RM, Cardoso Filho N, Gil ES, Stringhetta JPS (2007) The pharmaceutical importance of chiral drugs. Rev Electrôn Farm IV(1):8–14

Owen SF, Giltrow E, Huggett DB, Hutchinson TH, Saye J, Winter MJ, Sumpter JP (2007) Comparative physiology, pharmacology and toxicology of β-blockers: mammals versus fish. Aquat Toxicol 82(3):145–162

Owen SF, Huggett DB, Hutchinson TH, Hetheridge MJ, Kinter LB, Ericson JF, Sumpter JP (2009) Uptake of propranolol, a cardiovascular pharmaceutical, from water into fish plasma and its effects on growth and organ biometry. Aquat Toxicol 93(4):217–224

Paterson G, Metcalfe CD (2008) Uptake and depuration of the anti-depressant fluoxetine by the Japanese medaka (*Oryzias latipes*). Chemosphere 74(1):125–130

Pavlinov SA, Belolipetskaya VG, Piotrovskii VK, Metelitsa VI, Filatova NP, Bochkareva EV (1990) Enantiomers of propranolol in the blood of patients with cardiovascular diseases during long-term therapy. Pharm Chem J 24(1):17–19

Pérez S, Barceló D (2008) Applications of LC-MS to quantitation and evaluation of the environmental fate of chiral drugs and their metabolites. TrAC Trends Anal Chem 27(10):836–846

Perrin C, Matthijs N, Mangelings D, Granier-Loyaux C, Maftouh M, Massart DL, Vander Heyden Y (2002a) Screening approach for chiral separation of pharmaceuticals: part II. Reversed-phase liquid chromatography. J Chromatogr A 966(1–2):119–134

Perrin C, Vu VA, Matthijs N, Maftouh M, Massart DL, Vander Heyden Y (2002b) Screening approach for chiral separation of pharmaceuticals: part I. Normal-phase liquid chromatography. J Chromatogr A 947(1):69–83

Pirkle WH, Pochapsky TC (1989) Considerations of chiral recognition relevant to the liquid chromatographic separation of enantiomers. Chem Rev 89(2):347–362

Pirzada Z, Personick M, Biba M, Gong X, Zhou L, Schafer W, Roussel C, Welch CJ (2010) Systematic evaluation of new chiral stationary phases for supercritical fluid chromatography using a standard racemate library. J Chromatogr A 1217(7):1134–1138

Ranta V-P, Toropainen E, Talvitie A, Auriola S, Urtti A (2002) Simultaneous determination of eight β-blockers by gradient high-performance liquid chromatography with combined ultraviolet and fluorescence detection in corneal permeability studies in vitro. J Chromatogr B Analyt Technol Biomed Life Sci 772(1):81–87

Reungoat J, Macova M, Escher BI, Carswell S, Mueller JF, Keller J (2010) Removal of micropollutants and reduction of biological activity in a full scale reclamation plant using ozonation and activated carbon filtration. Water Res 44(2):625–637

Richardson SD (2006) Environmental mass spectrometry: emerging contaminants and current issues. Anal Chem 78(12):4021–4045

Rosal R, Rodríguez A, Perdigón-Melón JA, Petre A, García-Calvo E, Gómez MJ, Agüera A, Fernández-Alba AR (2010) Occurrence of emerging pollutants in urban wastewater and their removal through biological treatment followed by ozonation. Water Res 44(2):578–588

Rykowska I, Wasiak W (2009) Recent advances in gas chromatography for solid and liquid stationary phases containing metal ions. J Chromatogr A 1216(10):1713–1722

Sánchez-Argüello P, Fernández C, Tarazona JV (2009) Assessing the effects of fluoxetine on *Physa acuta* (Gastropoda, Pulmonata) and *Chironomus riparius* (Insecta, Diptera) using a two-species water-sediment test. Sci Total Environ 407(6):1937–1946

Sancho R, Minguillón C (2009) The chromatographic separation of enantiomers through nanoscale design. Chem Soc Rev 38(3):797–805

Santos AR, Pinheiro AC, Sodero ACR, Cunha AS, Padilha MC, Sousa PM, Fontes SP (2007) Atropisomerism: the effect of the axial chirality in bioactive compounds. Quím Nova 30(1):125–135

Santos LHMLM, Araújo AN, Fachini A, Pena A, Delerue-Matos C, Montenegro MCBSM (2010) Ecotoxicological aspects related to the presence of pharmaceuticals in the aquatic environment. J Hazard Mater 175(1–3):45–95

Schnell S, Kawano A, Porte C, Lee LEJ, Bols NC (2008) Effects of ibuprofen on the viability and proliferation of rainbow trout liver cell lines and potential problems and interactions in effects assessment. Environ Toxicol 24(2):157–165

Scriba GKE (2003) Pharmaceutical and biomedical applications of chiral capillary electrophoresis and capillary electrochromatography: An update. Electrophoresis 24(15):2409–2421

Schurig V (2001) Separation of enantiomers by gas chromatography. J Chromatogr A 906(1–2):275–299

Silva BJG, Lanças FM, Queiroz MEC (2009) Determination of fluoxetine and norfluoxetine enantiomers in human plasma by polypyrrole-coated capillary in-tube solid-phase microextraction coupled with liquid chromatography-fluorescence detection. J Chromatogr A 1216(49):8590–8597

Singh AK, Kedor-Hackmann ÉRM, Santoro MIRM (2006) Applied chiral phase liquid chromatography in enantiomeric separation of cardiovascular drugs. Braz J Pharm Sci 42(4):553–566

Sipma J, Osuna B, Collado N, Monclús H, Ferrero G, Comas J, Rodriguez-Roda I (2010) Comparison of removal of pharmaceuticals in MBR and activated sludge systems. Desalination 250(2):653–659

Smith SW (2009) Chiral toxicology: it's the same thing only different. Toxicol Sci 110(1):4–30

Somogyi A, Bochner F, Foster D (2004) Inside the isomers: the tale of chiral switches. Aus Prescriber 27(2):47–49

Sousa EP, Tiritan ME, Oliveira RV, Afonso CMM, Cass QB, Pinto MMM (2004) Enantiomeric resolution of kielcorin derivatives by HPLC on polysaccharide stationary phases using multimodal elution. Chirality 16(5):279–285

Stanley JK, Ramirez AJ, Chambliss CK, Brooks BW (2007) Enantiospecific sublethal effects of the antidepressant fluoxetine to a model aquatic vertebrate and invertebrate. Chemosphere 69(1):9–16

Stefan R-I, van Staden JF, Aboul-Enein HY (2000a) Amperometric biosensors/sequential injection analysis system for simultaneous determination of S- and R-captopril. Biosens Bioelectron 15(1–2):1–5

Stefan R-I, van Staden JF, Aboul-Enein HY (2000b) Simultaneous detection of S and R captopril using sequential injection analysis. Talanta 51(5):969–975

Steiner TJ, Ahmed F, Findley LJ, MacGregor EA, Wilkinson M (1998) S-fluoxetine in the prophylaxis of migraine: a phase II double-blind randomized placebo-controlled study. Cephalalgia 18(5):283–286

Stevens JC, Wrighton SA (1993) Interaction of the enantiomers of fluoxetine and norfluoxetine with human liver cytochromes P450. J Pharmacol Exp Ther 266(2):964–971

Stringer JL (2006) Basic concepts in pharmacology: a student's survival guide, 3rd edn. McGraw Hill, New York

Taylor LT (2009) Supercritical fluid chromatography for the 21st century. J Supercrit Fluids 47(3):566–573

Ternes TA (1998) Occurrence of drugs in German sewage treatment plants and rivers. Water Res 32(11):3245–3260

Toreson H, Eriksson BM (1997) Liquid chromatographic determination of fluvastatin and its enantiomers in blood plasma by automated solid-phase extraction. Chromatographia 45:29–34

Trautwein C, Kümmerer K, Metzger JW (2008) Aerobic biodegradability of the calcium channel antagonist verapamil and identification of a microbial dead-end transformation product studied by LC-MS/MS. Chemosphere 72(3):442–450

Trenholm RA, Vanderford BJ, Holady JC, Rexing DJ, Snyder SA (2006) Broad range analysis of endocrine disruptors and pharmaceuticals using gas chromatography and liquid chromatography tandem mass spectrometry. Chemosphere 65(11):1990–1998

Tucker GT (2000) Chiral switches. Lancet 355(9209):1085–1087

Unceta N, Barrondo S, de Azúa IR, Gómez-Caballero A, Goicolea MA, Sallés J, Barrio RJ (2007) Determination of fluoxetine, norfluoxetine and their enantiomers in rat plasma and brain samples by liquid chromatography with fluorescence detection. J Chromatogr B 852(1–2):519–528

Union, E.P.a.t.C.o.t.E (2006). Regulation (EC) No 1907/2006 of the European Parliament and of the Council of 18 December 2006 concerning the Registration, Evaluation, Authorisation and Restriction of Chemicals (REACH), establishing a European Chemicals Agency, amending Directive 1999/45/EC and repealing Council Regulation (EEC) No 793/93 and Commission

Regulation (EC) No 1488/94 as well as Council Directive 76/769/EEC and Commission Directives 91/155/EEC, 93/67/EEC, 93/105/EC and 2000/21/EC Official Journal of the European Union: 396/391-396/849
Vanderford BJ, Snyder SA (2006) Analysis of pharmaceuticals in water by isotope dilution liquid chromatography/tandem mass spectrometry. Environ Sci Technol 40(23):7312–7320
Wang SH, Liang ZH, Zeng S (2007) Monitoring release of ketoprofen enantiomers from biodegradable poly(D, L-lactide-co-glycolide) injectable implants. Int J Pharm 337(1–2):102–108
Westerhoff P, Yoon Y, Snyder S, Wert E (2005) Fate of endocrine-disruptor, pharmaceutical, and personal care product chemicals during simulated drinking water treatment processes. Environ Sci Technol 39(17):6649–6663
Winder VL, Sapozhnikova Y, Pennington PL, Wirth EF (2009) Effects of fluoxetine exposure on serotonin-related activity in the sheepshead minnow (*Cyprinodon variegatus*) using LC/MS/MS detection and quantitation. Comp Biochem Physiol C Toxicol Pharmacol 149(4):559–565
Winkler M, Lawrence JR, Neu TR (2001) Selective degradation of ibuprofen and clofibric acid in two model river biofilm systems. Water Res 35(13):3197–3205
Winter MJ, Lillicrap AD, Caunter JE, Schaffner C, Alder AC, Ramil M, Ternes TA, Giltrow E, Sumpter JP, Hutchinson TH (2008) Defining the chronic impacts of atenolol on embryo-larval development and reproduction in the fathead minnow (*Pimephales promelas*). Aquat Toxicol 86(3):361–369
Wistuba D (2010) Chiral silica-based monoliths in chromatography and capillary electrochromatography. J Chromatogr A 1217(7):941–952
Wistuba D, Schurig V (2000) Enantiomer separation of chiral pharmaceuticals by capillary electrochromatography. J Chromatogr A 875(1–2):255–276
Wong CS, Garrison AW (2000) Enantiomer separation of polychlorinated biphenyl atropisomers and polychlorinated biphenyl retention behavior on modified cyclodextrin capillary gas chromatography columns. J Chromatogr A 866(2):213–220
Zhang H, Spiteller M, Guenther K, Boehmler G, Zuehlke S (2009) Degradation of a chiral nonylphenol isomer in two agricultural soils. Environ Pollut 157(6):1904–1910
Zhang T, Nguyen D, Franco P (2010) Reversed-phase screening strategies for liquid chromatography on polysaccharide-derived chiral stationary phases. J Chromatogr A 1217(7):1048–1055
Zhou J, Yang Y-W, Wei F, Ping-dong W (2007) Comparison of the performance of chiral stationary phase for separation of fluoxetine enantiomers. J Zhejiang Univ Sci B 8(1):56–59
Zorita S, Mårtensson L, Mathiasson L (2009) Occurrence and removal of pharmaceuticals in a municipal sewage treatment system in the south of Sweden. Sci Total Environ 407(8):2760–2770Lobore essi. Re volore et, sum nonsectet wis dolorti scilit, sum in vel ulluptat ecte etuer sequamcommod

# Chapter 2
# Pollutant-Induced Decay of Building Materials

**Jorge Sanjurjo-Sánchez and Carlos Alves**

## Contents



Financial support from Fundação para a Ciência e Tecnologia (Portugal) with FEDER (European Union) and Portuguese funds

J. Sanjurjo-Sánchez (✉)
Instituto Universitario de Xeoloxía "Isidro Parga Pondal". Edificio Servizos Centrais de Investigación, Universidade da Coruña, Campus de Elviña, 15071 A Coruña (España), Spain
e-mail: jsanjurjo@udc.es

C. Alves
Centro de Investigação Geológica, Ordenamento e Valorização de Recursos, Campus de Gualtar, Universidade do Minho, 4710-057 Braga, Portugal
e-mail: casaix@dct.uminho.pt

E. Lichtfouse et al. (eds.), *Environmental Chemistry for a Sustainable World: Volume 2: Remediation of Air and Water Pollution*, DOI 10.1007/978-94-007-2439-6_2,

**Abstract** There are major concerns on the effect of environmental pollution on humankind, other animals, ecosystems, and also on the built environment, where the action of pollutants can affect diverse materials, harming historical elements and modern constructions. Here we review the effects of pollutants on materials used in the built environment, e.g. housing, walls, stones, wood, plastics and coatings. Compositional characteristics that make materials susceptible to pollutants are presented. The role of materials as pollution sources and notions on decay features that affect building materials are shown. Then we review the decay effects of gases, particulate matter and solutions from wet deposition to capillary-rising and including circulation waters such as run-off and seawater. The effect of pollutants on materials is discussed, as well as the role of water in decay processes and the damage caused by neoformation minerals on the surface and pore system of the building materials. The main pollutants discussed are $CO_2$, $CH_4$, CO, hydrocarbons, $SO_2$, $NO_x$, $NH_3$, $NO_2$, $O_3$, $F_2$, HF and atmospheric particles. Afterwards we discuss the use of tracers, e.g. elements, ions and isotopes, for the study of pollutants sources and migration pathways. Main isotopes include $^{2}H$, $^{13}C$, $^{15}N$, $^{18}O$, $^{34}S$, $^{3}H$, $^{14}C$, $^{238}U$, $^{232}Th$ and $^{40}K$. We show the possible use of decay features to study main sources of air pollution, e.g. transport, industrial, domestic, since building materials can fix and interact with pollutants. Proposals regarding the use of some stable and radioactive isotopes for the study of pollutants that have strong potential but have not been tested yet are also presented. At the end of this chapter we discuss the durability assessment of materials, and the use of tracers to assess the origin of damaging compounds.

**Keywords** Building materials • Built environment • Atmospheric pollution • Decay tracers • Decay features • Pollution sources • Pollution pathways • Deposition • Capillary-rise pollution • Geochemical studies • Infiltrations • Air pollutants • Gaseous pollutants • Particulate pollutants • Soluble salts • Tracers • Isotopes

## 2.1 Introduction

There are major concerns in relation to environmental pollution regarding its effects on humankind, other animals, ecosystems and also on the built environment, where the action of pollutants can affect diverse materials, harming historical elements and modern constructions (Amoroso and Fassina 1983; Manning 1988; Arnold and Zehnder 1991; Winkler 1994; Goudie and Viles 1997; Grossi and Brimblecombe 2002; Brimblecombe and Grossi 2010).

The main aim of this chapter is to present an overview of the decay effects of diverse pollutants on the main materials used in the built environment. Studies on the decay of building materials have been abundant in the last 30 years. These studies have been focused on three questions: (a) decay processes of particular materials (field and experimental studies); (b) origin, pathways and mechanisms of agents of this decay; (c) procedures for the restoration and conservation of materials and buildings. Given the diversity of disciplines that include this type of studies, e.g. environmental chemistry, geochemistry, atmospheric sciences, hydrogeology, and architecture, and the scope of the book, this chapter has focused on the first and second point that are critical for the third one (Amoroso and Fassina 1983; Arnold and Zehnder 1991; Saiz-Jimenez 1993; Sabbioni et al. 1998). Atmospheric pollution has deserved most of the attention, but pollutants from the ground or resulting from some materials can have marked effects on the built environment and will be an important part of this chapter. Due to the research interests of the authors, this work is clearly dominated by stony materials but some general considerations, with appropriated references, are made in relation to other materials.

There is a large variety of issues involved on the characterization and assessment of materials decay. Also, a multiplicity and wide dispersion of studies addressing those issues can be found in the literature. Therefore, it is attempted here to prepare a synthesis of the state of the knowledge on this subject focusing on two main aims:

- decay action of diverse pollutants, mainly atmospheric and ground-related, on several building materials considering also the interactions between the pollutants,
- procedures for the characterization of pollution sources based on the identification and quantification of chemical substances, including what has been used and proposals to test new tracers that could potentially help to identify decay processes and sources of polluting agents.

## 2.2 Building Materials: Characteristics and Decay Features

In this section is made an introduction to the main materials found in the built environment highlighting the characteristics that render them susceptible to the effects of pollutants. It is also included a brief presentation of the decay features that affect the materials as consequence of the action of the pollutants.

### *2.2.1 Building Materials*

Chemical and physical properties of materials are important as they determine the result of the interactions with gases, particles and solutions. Thus, solubility, corrosion resistance, weathering resistance, etc. are chemical properties strongly related to density, porosity, hardness, and other physical properties, but also to environmental factors. In this section are presented some of the features that render materials susceptible to the action of pollutants, with special focus on the chemical characteristics, considering diverse types of natural and human-made products.

Stone has been used from the Neolithic to built different types of human-made building structures, e.g. housing, tombs. The use of stone materials has been related to their local availability but also to their properties. Most of them are polymineral rocks, e.g. granite and shale, but there are also rocks that are mostly monomineralogical, e.g. gypsum and limestone. The variable chemical and physical properties of igneous, sedimentary and metamorphic rocks are decisive in the decay processes.

Igneous rocks are considered less reactive to dissolution than other rocks. Sedimentary rocks, i.e. shale, sandstone and limestone are usually considered as more susceptible, as they are commonly soluble. Metamorphic rocks, i.e. gneiss, slate, schist, marble, quartzite and hornfels, result of the transformation of other rock types due to temperature or pressure changes (metamorphism). According to the characteristics of the metamorphic process and the mineralogy of the previous rock, they can be more or less sensitive to decay.

Round and small-size stones used in building structures are united into structural elements with the aid of cementing materials such as clay, lime and others. Such binders or mortars have been used from early human civilizations (Elert et al. 2002) to fill the gaps between stony blocks. Mortars are typically composed of mixture of an aggregate (typically sand) and a binder agent. Depending on the mortar type, binders can be lime, clay, gypsum, cement or others.

Modern processes for the manufacturing of binders usually include burning of the raw materials. Burning cause dehydration of minerals forming minerals, e.g. gypsum, or dissociation, as occurs in air-hardening lime. In the case of lime mortars, crushed limestone is burned in a kiln at 900–1,000°C to form quicklime (calcium oxide). The quicklime is then slaked with water, forming calcium hydroxide, to give lime putty or hydrated lime. Before use, hydrated non-hydraulic lime is usually left in the absence of carbon dioxide, usually under water, to mature. Lime putty can be matured in air for anything from 24 h to many years (increasing maturation time improves the quality of the putty). Hydraulic lime mortars (have the ability to set under water) are produced in a similar manner but the raw limestone contains clay and other impurities (as was already noted by Vitruvius, see http://penelope.uchicago.edu/Thayer/E/Roman/Texts/Vitruvius/home.html, consulted in August/2010).

Ceramic materials are artificial burned stony materials at 900–1,000°C to achieve strength, manufactured from loams or clays. The ancient methods to produce bricks started with the raw clay, preferably in a mix with 25–30% sand to reduce shrinkage. The burning of clay is accompanied by other essential changes in the constituents of the raw materials (Shestoperov 1988). Other building materials, such as tiles, are also ceramic materials. Before developing such procedure, mudbrick were used in ancient times and are still used currently in some parts of the World (being even considered presently a kind of greener material). Mudbricks are unfired bricks made of a mixture of clay, mud, sand, and water mixed with a binding material such as rice husks or straw. They were let to dry in the sun for about 30 days.

Other man-made or natural materials such as glass, metals, paintings, wood bituminous pavements, etc. are common in buildings and can suffer decay due to pollutants. The studies on such materials are not very frequent but some data exist on the decay features and effects of pollutants.

**Fig. 2.1** Differential erosion in limestones associated with the presence of heterogeneities

## *2.2.2 Weathering and Decay Features of Building Materials*

The weathering of rocks (Ollier 1979) is the breakdown and alteration of such materials near the Earth's surface resulting in products that are more in equilibrium with newly imposed physical and chemical conditions. Two main classes of weathering are usually considered: physical, that is the breakdown of material by entirely mechanical processes, and chemical: breakdown by chemical processes.

Weathering is also the effect of atmospheric exposure to man-made structures and materials. Buildings made of any stony material (rocks, brick or concrete) are susceptible to the same weathering agents that any exposed rock surface. Limestone and soluble rocks are considered as very susceptible to air pollution damages due to the reactive effect of gaseous, liquid and solid pollutants. Silicate rich rocks can be damaged but they are considered as less reactive (Amoroso and Fassina 1983). Simão et al. (2006) found greater reactivity in syenite and grabro than in granite (resulting from the different mineralogical characteristics). Several other intrinsic characteristics of stones can have an impact on the effects of pollutants (see Matias and Alves 2002; Alves 2009) namely the characteristics of the pore network and the presence of heterogeneities (see illustration in Fig. 2.1). The characteristics of the built elements (morphology, structure, exposition) and the environment, namely atmospheric conditions and pollution, must also be considered.

The effects of pollutants on the built environment frequently result in macroscopic evidences – sometimes called "pathologies" of materials – that can be referred by different designations that are more or less self-explanatory such as "flakes", "granular disintegration", "biological colonisation", etc. An expanded scheme of classification of the different weathering features for stony materials, that can also be applied to other materials, is found in Fitzner and Heinrichs (2002). These decay features are not limited to historical ("old") elements but can also affect modern constructions.

Decay of building materials could be related to colour and textural changes resulting from absorption of or reactions with exogenous matter as well as the deposition of matter. The simple absorption of water can cause drastic and unpleasant chromatic changes in some materials such as wood materials: see example in Fig. 2.6; water can additionally promote other decay processes such as biological activity. Colour and textural changes result in coatings that might cover and hide extensive areas of the original substrate. These processes can affect porous materials but also materials such as glass (Lefèvre et al. 2005; Lombardo et al. 2010) and paintings (De-Bock et al. 1996; Saunders 2000). Studies with glass samples (that have a low reactivity) have evidence soiling under polluted atmospheres (Lefèvre et al. 2005; Lombardo et al. 2010), including airborne particles (diverse types of particles) and salt crystals (dominated by sulphates). While being most frequent outdoors, coatings can also be observed indoors associated with air pollution (De-Bock et al. 1996; Saunders 2000) or water infiltration (Alves et al. 2009).

One of the most (in)famous types of coatings are so-called black crusts (Fig. 2.2) made of gypsum aggregates and atmospheric particles, a dark type of coating that might affect great extensions of buildings in urban and industrial locals. A chromatic symmetrical type of coating are whitish carbonate coatings (Fig. 2.2) that have received less attention but that might be an important and in some places the major coating type affecting diverse types of modern constructions (Alves 2010).

Decay features of materials also include erosive features that result in material loss such as detachment of planar fragments (Fig. 2.3) or grains (granular disintegration). These erosive processes might imply extensive destruction of the element surface and loss of ornamental value (Fig. 2.4).

Besides the relevance for the evaluation of susceptibility, the characterization of materials is also important to understand its possible role as pollution sources for the surrounding environment (in relation to cementitious materials see, e.g. van-der-Sloot (2000); regarding metallic materials several examples can be found in Townsend (2002)). Pollutants present in materials either as pore content or resulting from dissolution can migrate causing decay on the same or other materials. The presence of important amounts of soluble salts on natural outcrops of rocks has been referred (Arnold and Zehnder 1991; Heiss et al. 1991; Wüst and Schlüchter 2000). Another frequent problem is pollutants resulting from chemical reactions with materials and that can affect the same or other materials (in this last case sometimes referred as a problem of incompatibility of building materials). In relation to natural stone, this type of process is frequently mentioned associated with weathering of calcite (Cooper et al. 1991; Zannini et al. 1991; Weaver 1991; Haneef et al. 1992; Duffy and O'Brien 1996; Smith 1999; Figueiredo et al. 2007) but there also references to reactions with dolomite (Rodríguez-Navarro et al. 1997b), silicates (Smith et al. 2002; Lopez-Arce et al. 2008), iron sulphides (Stocksiefen 1982; Benea 1996; Robert et al. 1996; Honeyborne 1998; Dreesen et al. 2007) or oxides (Bhargav et al. 1999;

**Fig. 2.2** Examples of common coatings on the built environment: (**a**) black crusts; (**b**) white crusts. Both on granitic stones

Storemyr 2004; Trujillano et al. 1996; Alonso et al. 2008) as well as organic components (Lea 1970; see also several references in Hartog and Mckenzie 2004). Another classical example is the migration of products of metals corrosion commonly copper or iron that causes stains on other materials (Winkler 1994).

Cement-based materials (mortars, concrete) could be an important source of pollutants in the initial stages of setting and after setting. Several studies have been

**Fig. 2.3** Example of planar detachments in limestones applied in a modern construction near the sea

**Fig. 2.4** Erosion of stone elements with marked depth and disappearance of decorative elements in granite

done regarding the chemical composition of pore-water in the first stages after mixing of water with cement and a summary of the main characteristics (with several references) can be found in Odler (1998). This author refers the dominance of alkalis (sodium and potassium), calcium, sulphate and hydroxide in the first stages after mixing. After setting, components of the hydrated cement ionic pollutants could be dissolved and released (Lea 1970; Dow and Glasser 2003; Anstice et al. 2005). To the complexity resulting from the initial cement characteristics it is necessary to add considerations on the interacting solutions and the flow regime. In recent constructions, as it has been verified by the present authors, it is common to found carbonate-rich crusts (Alves 2010) as well as alkaline carbonates and sulphates (Figg et al. 1976; Figg et al. 1978; Halle et al. 1978), results that seem to be related to the contribution of recent cements. Lea (1970) related the occurrence of efflorescences to the alkali content on cement (this author also indicated the possible occurrence of efflorescences of magnesium sulphate related to cement). Besides the solid components, the water used in the preparation of some mixed building materials could also be an important source of pollutants that cause decay effects (see example in Netterberg and Bennet 2004).

Whatever the source of pollutants, decay features varies according to the materials used in the built environment. Although other mortars decay by the same process, the relatively simple composition of lime and gypsum mortars (calcite and sand, or gypsum and sand) can be used to assess the damaging progress caused by the air pollution load in a given local setting. The decay of ceramic materials is similar to the decay of other porous materials. The destructive effect of air pollution on ceramic materials has been demonstrated (Cultrone et al. 2000). Decay of ceramic bricks is also associated with the appearance of stresses due to the crystallization of various salts or the action of water pressure as it freezes partly.

Some other materials can be damaged by pollutant compounds. We will present here briefly some of the main features that render those materials susceptible to the action of the pollutants.

The chemical composition of glasses influences the behaviour of these materials in relation to pollutants. According to Doremus (1994), leaching affects preferentially alkali ions in multicomponent silicate glasses with alkali oxides. This author also presents data indicating that glasses with alumina have lower dissolution rates. Lefèvre et al. (2005) reported that the composition of ancient glass (potassium-rich) makes this material less durable than modern glasses (sodium-rich) under pollution conditions. However, Leissner (2003) highlights that medieval glasses can have significative variations in composition as is evidenced by very different behaviour showed by glasses placed in the same local. Kunicki-Goldfinger (2008) reviews some examples of the effects of the composition of historical glass in its stability.

Metallic materials are typically affected by corrosion, since uncombined metals have higher free energies than most chemical compounds of metals (Trethewey and Chamberlain 1995). The corrosion behaviour of the metals depends on the behaviour of the thin metal oxide layer that is formed rapidly when the metal is exposed to air (Trethewey and Chamberlain 1995; Roberge 2000; Tidblad and Kucera 2003).

**Fig. 2.5** Detachments of impermeable wall paint related to salt efflorescences

A distinction can be made between the corrosion processes of one metal elements and what is designated as dissimilar metal corrosion resulting from the coupling of different metals (details can be found in e.g. Trethewey and Chamberlain 1995; Roberge 2000). Galvanic series based on free corrosion potentials show the tendency to corrosion of the metals as well as the corrosion hazard related to the coupling of different metals (corrosion hazard being higher for higher differences of potential). Some metals can be used as sacrificial materials to protect other metals (Trethewey and Chamberlain 1995; Roberge 2000) such as corrosion of steel to protect copper elements. The occurrence of corrosion can be favoured by heterogeneities such as grain interfaces, scratches, crevices or inclusions that promote selective attack or electrolyte concentration (Trethewey and Chamberlain 1995; Roberge 2000).

Durability of painting materials can be related to the characteristics of its components and to its formulation depending on the pollutants considered (Lea 1970; Koleske 1995; Wypych 1995; Weldon 2009). Impermeable paints are especially susceptible to detachments (Fig. 2.5) due to moisture and salt crystallisation in the substrate beneath, processes that promote the disruption and erosion of the paint, both in common wall paints and in artistic paintings (several examples could be found in Arnold and Zehnder 1991).

Susceptibility of polymers (see Delre and Miller 1988; Seymour 1988) depend on aspects such as structures (diffusion and absorption are greater on amorphous materials; the presence of branches, pendant groups or cross links decreases the rate of diffusion), the susceptibility of specific components in the case of copolymers and the presence of additives (that might be more susceptible to chemical attack and

**Fig. 2.6** Stains in wood materials due to water absorption

promote the permeation of pollutants). The presence of heavy metals and adventitious photosensitizers can promote degradation by oxidation processes (Seymour 1988). Specific characteristics of polymers increase susceptibility to diverse oxygen compounds (Delre and Miller 1988; Seymour 1988).

Wood materials can be markedly affected by simple water absorption (Fig. 2.6). Different types of woods show different dimensional changes related to wetting (see, e.g., Douglas and Ransom 2007) that can cause distortions and cracking. Sandberg and Söderström (2006), in a study of wood weathering by outdoor exposition, reported higher total crack length on tangential sections (compared to radial sections) but similar colour changes due to weathering on both types of sections. Simpson and TenWolde (1999) presented a division of wood types according to average heartwood decay resistance based on service records, laboratory tests and general experience. Feist (1992) referred the presence of a silvery gray layer in woods exposed in coastal areas. He observed that, in general, less dense woody materials experience faster weathering and erosion. The substances that are naturally present on woods and the products that are used in its treatment can also affect the performance of wood materials. According to Feist (1992), tannates present in woods can react with iron salt solutions producing chromatic alterations. Additives such as fire retardants have been considered potential sources of soluble salts that might crystallise on the wood surface (see Östman et al. 2001).

Regarding bituminous pavements, different aspects can affect the deleterious effects of pollutants such as soluble salts including maximum salt content of the materials, terrain characteristics and the procedures for the application of the pavement (Obika et al. 1989; Netterberg and Bennet 2004).

## 2.3 Decay Effects of Pollutants

This section would deal with the sources, pathways and decay mechanisms of the main pollutants affecting materials in the built environment. The diverse pollutants will be discussed in specific sections but here are presented some transversal concepts that are relevant for diverse forms of the polluting substances.

In the built environment, water (even in its purest form) can be considered a pollutant as it promotes decay reactions on materials such as paints (Koleske 1995; Wypych 1995), corrosion in metals (Trethewey and Chamberlain 1995; Roberge 2000; Tidblad and Kucera 2003), dissolution, alkali leaching and reaction with silicon-oxygen network in glass (Doremus 1994; Leissner 2003) and hydrolysis of polymers (Delre and Miller 1988). Water can produce visual stains in porous materials (see illustration in Fig. 2.6 above) and volumetric changes that are especially significative in the case of wood (Douglas and Ransom 2007) but that have been referred as causing damage in other materials such as natural stone (see example in Rodriguez-Navarro et al. 1997a). There are also references that indicate that rock strength is lowered by the presence of water in the rock pores (Winkler 1994; Topal and Sözmen 2003).

Air pollutants induce the weakening of the porous structure and the surface layers of stony materials. The susceptibility of natural rocks to decay processes induced by air pollution strongly depend on the pollution, climatic factors and the intrinsic characteristics of the rock forming minerals.

Atmospheric chemical constituents can precipitate by wet occult and dry deposition. In humid climates, where important rainfalls are recorded, wet deposition must be considered as an important source of ions and particles. Although no doubt exists about acid rain damage on most building materials, rain can also wash away accumulated pollutants. Various studies have considered that wet deposition is not important compared to dry deposition (Charola and Ware 2002). Only in rural areas it has been considered as an important factor in the decay of building materials. This asseveration is related to the fact that atmospheric pollution is low in rural areas, and also to the conclusion that the pollutants affecting building materials in rural areas are related to distant emission sources (Torraca 1988). However, it must be considered that wet deposition by rainfall will affect only the exposed surfaces of a building, while dry and occult deposition can affect all surfaces (Furlan and Girardet 1992).

Occult deposition or precipitation refers to precipitation of cloud/fog water. It can be more damaging when compared to dry and wet deposition, providing moisture to activate the chemical attack of dry deposited gases and particles. Fog water has been shown to be an important collector and carrier of pollutants in urban areas as it can concentrate chlorides and sulphates originated from local emission sources (Fassina and Stevan 1992). At seaside areas, seawater sulphate and salts due to sea spray may contribute to the wet and occult deposition on the building materials (Chabas and Jeannette 2001).

Dry deposition is an important mechanism for removing gaseous and particulate pollutants from the atmosphere in the absence of precipitation, above all in arid and semiarid regions (Kumar et al. 2005). It is determined by two factors: the concentration of pollutant gases and particles in the atmosphere, and the interaction between the atmosphere (and the air pollutants) and the building surface (including building features and materials) under dry conditions.

Solid surfaces, such as building materials, attract atmospheric gases and particles to their surface due to molecular or atomic forces. If the surface is wet, attraction and chemical reactivity of the surface is enhanced. Dry deposition is considered the most important deposition process in highly polluted areas (Charola and Ware 2002). Although some studies have related dry deposition with distant industrial gaseous and particulate emissions (Schiavon 1993), others consider that it is related to nearby sources (Hicks 1986; Sanjurjo-Sánchez et al. 2011). Understanding the air-surface dry exchange process is very important to estimate amounts delivered from the air. In fact, the proportion of atmospheric particles and gases in a measured sampling point will depend on the emission characteristics, the distance from sources and meteorological conditions.

Fossil fuel combustion by industrial facilities and vehicular engines is a major source of anthropogenic emissions into the atmosphere, and the effect of such pollution is stronger in urban areas. The main studied gaseous pollutants are sulphur, nitrogen and carbon compounds and other organic pollutants (Judeikis and Stewart 1976; Johansson et al. 1988; Lipfert 1989; Cobourn et al. 1993; Saiz-Jimenez 1993; Ruijgrok et al. 1995).

While some authors (Sabbioni 2003) defend that atmospheric pollution is generally the predominant contributor for decay of building stones, important and very marked decay features can arise from other processes, namely those related to capillary-rising solutions, processes that can be dominant in certain regions and for certain application of materials. Soluble salts are among the pollutants with higher impact in the built environment affecting diverse materials of historical and modern constructions. They can result from dissolution of gases on moisture films of the surfaces, dry and wet deposition and the contribution of solutions (infiltration, capillary rising, circulation). Several extended reviews on the decay effects of soluble salts on materials of the built environment can be found in Doornkamp and Ibrahim, (1990), Arnold and Zehnder, (1991), Goudie and Viles (1997), Rodriguez-Navarro and Doehne (1999), Alves and Sequeira-Braga (2000), Charola (2000), Doehne (2002), and Charola et al. (2007). Diverse types of soluble salts can be related to simple formation of coatings, physical disintegration due to salt crystallisation or hydration and chemical reactions causing visual and erosive deleterious effects (unlike what seems to be suggested by Douglas and Ransom 2007, other soluble salts besides magnesium sulphate can cause disintegration of surfaces). Besides the several examples on porous stony materials (see previous references), soluble salts could affect bituminous binders either by physical disintegration related to salt crystallization (Obika et al. 1989; French 1997; Netterberg and Bennet 2004) or by dissolution of bituminous binders (French 1997).

From the quantitative point of view, damage or dose-response functions based in laboratory experiments have been proposed to determine the influence of diverse factors (namely pollutants load) on the degradation processes of materials (Steiger and Dannecker 1994; Roberge et al. 2002; Sabbioni 2003; Tidblad and Kucera 2003;

Brimblecombe and Grossi 2010; Lombardo et al. 2010). However, it must be considered that factors such as wind action, relative humidity, etc., are usually kept constant in these laboratory simulations.

The question of dose-response is also relevant for assessing the possible consequences of reduction of emission of certain pollutants, such as $CO_2$ and specially $SO_2$. The main questions is whether reduction will result in lesser decay rates, specially since some questions have been raised in relation to answers of systems in relation to reduction of pollutants emissions such as those in relation to the necessary level of $CO_2$ reduction (see Lovelock 2008). One must also consider the problems resulting from what is, for modelling effects, a natural system with diverse interdependences. For example, it has been proposed that the reduction of sulphur oxide could contribute to global warming and, based on this, the stratospheric injection of $SO_2$ has even been considered as a measured against global warming (see discussion in Jones et al. 2010).

In relation to $SO_2$, there are models that suggest that decrease of atmospheric levels would imply a lesser decay action of this pollutant. Butlin et al. (1995) proposed a model were the use of dose-response functions would allow elaborating maps to assess the impact of $SO_2$ reductions. Grossi et al. (2008) studied diverse models whose results suggested that erosion rates would be kept at lower values or decline under low levels of pollution levels. Some results of field studies by Trudgill et al. (1991) registered the existence of sustainable erosion rates after the decrease of $SO_2$, a result that could be attributed to high levels of $NO_x$ or to some sort of "memory effect". Vleugels et al. (1993), in an experiment where stone slabs where switched between places with different pollution levels, concluded that there was no evidence of a memory effect since the decay rates clearly changed according to the new environment. Trudgill et al. 2001 found a decrease of erosion rates in horizontal surfaces coinciding with the decrease of $SO_2$ but the same studies found higher erosion rates in a vertical surface. These authors also found that erosion rates decreased in a period where there was no decrease in $SO_2$ levels. It is clear that the matter needs further empirical research in real structures and in more locations. Questions such as the influence of climatic variations on decay processes must also be taken in consideration (Trudgill et al. 2001; Brimblecombe and Grossi 2008; Grossi et al. 2008).

The climatic factors of the area where the studied buildings are located have an important role on the effect of atmospheric pollutants on the building materials. Thus, weather conditions and turbulence strongly influence the transport and deposition of pollutants on building façades. The urban arrangement plays a decisive role on the distribution of pollutants depending on the geometry of the street of square where a studied building is located. In canyon streets higher concentrations of pollutants from traffic are usually found closer to the ground but pollution particles from distant sources (e.g. industrial sources in the outskirts of large cities) are more concentrated at the upper level (Gilbert et al. 2003; Janhäll et al. 2009; Taseiko et al. 2009). In addition, when the wind direction is perpendicular to the street direction a vortex is generated in a street canyon (Berkowicz et al. 1996; Taseiko et al. 2009).

Climatic conditions will also influence the action of soluble salts, affecting the crystallisation position and the existence of cycles of crystallisation-precipitation. Drier conditions will promote the crystallisation inside the porous media (Lewin 1982; Hammecker 1995) and the erosion of the materials. Variations in temperature and relative humidity conditions affect processes that enhanced the decay action of soluble salts such as dissolution-crystallisation (depending on deliquescence point) and hydration (Arnold and Zehnder 1991). Episodes of wet deposition have been associated with new occurrences of salt crystallisation spots and the spreading of decay (Alves and Sequeira Braga 1994).

Some publications had considered the possible effects of climate change on materials decay. Such a discussion will need to address the future characteristics of the climate in a scale relevant for the built surfaces and the interdependences between the pollutants and climatic conditions. Inkpen (2003) refers the difficulty of predicting the impact of climate on a specific built element and that the response of the built materials might not have a very marked expression nor be immediate. Some general trends have been considered. Viles (2002), admitting projections of an increasing NW-SE climatic gradient for the United Kingdom, predicts increasing salt weathering in the SE. Smith et al. (2004) considered increasing "time-of-wetness" for North Ireland that according to these authors would promote, among other changes, deeper penetration of soluble salts (promoting erosion) and facilitation of dust deposition. Grossi et al. (2008) discussing projections of different models indicate a possible increase in deposition of pollutants resulting from higher damp in winter. The models analysed by Brimblecombe and Grossi (2008) for London indicate a decrease in frost weathering and an increase in salt weathering.

### 2.3.1 *Gaseous Pollutants*

While not being usually considered among the gaseous pollutants, this section would begin with a brief reference to water vapour. Depending on relative humidity and temperature, water vapour can cause the formation of moisture films by condensation, propitiating the action of other pollutants (dissolution of gases and the fixation of particulate matter) and biological decay. For example, the oxidation of $SO_2$ in moisture films on iron with formation of sulphates leads to autocatalytic reactions of corrosion (Roberge 2000). The action of water vapour can be especially relevant in sheltered areas that are not affected also by rising or infiltration solutions.

The main gaseous pollutants resulting from anthropogenic sources are presented in Table 2.1. Among gaseous pollutants, nitrogen and sulphur oxides, carbon oxides and ozone are those of main concern.

The sources and properties of different gaseous pollutants in the atmosphere will be considered in the following sections. They will be mostly considered to assess their effects on stony materials. However, gaseous pollutants also affect other materials such as polymers (Delre and Miller 1988; Seymour 1988; Wypych 1995), metals (Trethewey and Chamberlain 1995; Oesch and Faller 1997; Roberge 2000;

**Table 2.1** Yearly emissions of gaseous compounds (From Fellenberg 2000)

| Emission | Natural (million T) | Anthropogenic (million T) |
|---|---|---|
| $CO_2$ | 600,000 | 22,000 |
| CO | 3,800 | 550 |
| Hydrocarbons | 2,600 | 90 |
| $CH_4$ | 1,600 | 110 |
| $NH_3$ | 1,200 | 7 |
| $NO_2$ | 770 | 53 |
| $SO_2$ | 20 | 150 |

Tidblad and Kucera 2003), glasses (Doremus 1994; Leissner 2003), paints (Haynie and Spence 1984; Koleske 1995) and wood materials (Hon and Feist 1993), but less effort has been made to study their effect on such materials.

#### 2.3.1.1 Sulphur Oxides

Gaseous sulphur is naturally produced by volcanic activity, sea spray and different biogenic processes, mainly in swamps and tidal flats. Anthropogenic emissions are considered as the most important in urban areas. They come from different combustion processes being both coal combustion in coal-fired power plants and refining processes of crude oil the main sources (Gaffney and Marley 2009). Such industrial facilities are usually located in the outskirts of urban settings. However, some authors consider that sources nearby damaged buildings are the main responsible of dry deposition of sulphur oxides on building façades (Torraca 1988). Sulphur gaseous compounds are mainly present in the lower troposphere as sulphur dioxide ($SO_2$), sulphur trioxide ($SO_3$) and hydrogen sulphide ($H_2S$).

Hydrogen sulphide is an occasional atmospheric pollutant. It can be harmful for marbles and limestones, since it can be transformed into sulphuric acid and react with calcium carbonates of the materials to give calcium sulphate. It has been suggested that such attack can be catalysed by a large number of sulphur bacteria (mainly of the genus *Thiobacillus*) that oxidize sulphur providing an acid solution effective in the attack on stone (Amoroso and Fassina 1983).

Sulphur oxides and particularly $SO_2$ are considered as the most important gaseous compound for stone decay processes. Gaseous sulphur dioxide is mainly adsorbed by calcareous building materials and oxidized on their surface (Lipfert 1989). However, a fundamental question is how $SO_2$ interacts with building materials. A first unsolved question is whether sulphites are formed and subsequently oxidized or whether sulphates are the only reaction products. The oxidation of $SO_2$ can occur in the gas phase or in the gas-liquid or gas-solid phase (Amoroso and Fassina 1983; Fassina 1988; Laurenzi-Tabasso and Marabelli 1992). In the gaseous phase, the reaction is a homogeneous process of oxidation of $SO_2$ in air that involves photoexcitation and absorption of solar radiation, and it is strongly influenced by low concentrations of nitrogen oxides and some hydrocarbons in polluted air and smog

(Amoroso and Fassina 1983; Johansson et al. 1988). On wet surfaces the gas-liquid reaction rate depends on the pH of the solution, the presence of salts (as they affect the pH), ammonia, calcite particles and heavy metals with catalytic effect (Beilke and Gravenhorst 1978; Fassina et al. 1979). Such catalytic particles are frequently small droplets (<100 μm diameter) that have higher absorption rates than larger ones (Charola and Ware 2002). Skoulikidis and Papakonstantinou-Ziotis (1981) considered that the sulphation reaction is due to oxidation of the adsorbed $SO_2$ (catalysed by surface particles) on dry surfaces and subsequent sulphation. However, Amoroso and Fassina (1983) proposed that the adsorption of $SO_2$ is followed of neutralization to sulphite and oxidation to sulphate. In fact, the catalytic effect of some particles by nucleation has been observed in experimental studies (Cultrone et al. 2000; Simão et al. 2006).

The amount of sulphur deposition measured in urban areas has been found to be approximately twice the sulphur deposited in rural areas (Charola and Ware 2002). The amount of deposited 'dry sulphur' is increased due to particles such as fly-ash and calcium carbonate contained in dry particle aerosols. This is an important mechanism on non-calcareous building materials. Particle aerosols can also contain calcium sulphate particles (generally 1–2 μm) generated from sulphur deposition on calcareous stones or from wind erosion of exposed gypsum beds (Amoroso and Fassina 1983). Laboratory studies in simulation chambers confirmed that the reactivity of stones to $SO_2$ could increase significantly with increased relative humidity but weakly with increased temperature (Girardet and Furlan 1996). In the case of rock building materials, molasses, limestones, calcareous sandstones and marble are considered as the most reactive, while the siliceous sandstones, granite and gneiss could be classified as non-reactive (Furlan and Girardet 1992).

The decay of Ca-rich stony materials depends on various factors. Regarding wet deposition, the dissolution and sulphation rate of calcite appears to be transport-controlled below pH 4, while between pH 4 and 6 the reaction could be controlled by surface kinetics. Other factors influence the dissolution, such as hydrodynamic flow and pressure of $CO_2$ (Guidobaldi and Mecchi 1985; Charola 1988). Although the low pH of rain is often referred to explain stone decay, particularly in calcareous stones, it has been found that acidity frequently decreases as rainfall continues (Camuffo et al. 1984, 1988). There is a remarkable discrepancy when the effect of acid rain and dry deposition of $SO_2$ on calcareous materials are compared. It has been assessed that the relative contribution of acid rain to the total $Ca^{2+}$ loss of various stones is approximately 74% in rural areas and only 21% in polluted areas (Steiger et al. 1993). However, a study carried out by Irving (1991) concluded that dry deposition of $SO_2$ causes near 5–20% of chemical weathering while wet deposition causes about 10%. Other authors report that dry deposition is the major cause of stone decay in urban areas (Livingstone 1992).

The most obvious damage caused by gaseous sulphur is the formation of gypsum by the reaction with calcium carbonate. A large number of studies on the decay of Ca-rich materials have been published (Amoroso and Fassina 1983; Lipfert 1989; Charola and Ware 2002; Bityukova 2006). Multiple works have studied the decay patterns of calcareous building rocks, particularly limestone and marble (Marchesini 1969;

Camuffo et al. 1982, 1983; Del-Monte and Sabbioni 1984; Charola 1988; Mirwald et al. 1988; Kozlowski et al. 1990; Rossi-Manaresi 1991; Török 2004). Camuffo (1990) indicated three mechanisms of decay. The first one is the run-off associated with 'white areas' where the surface is eroded and covered with reprecipitated spatic calcite crystals. The second one refers to 'black areas' formed on surfaces without run-off but where percolation and windborne droplet deposition occurred. The third one concerns 'grey areas' formed in absence of runoff, percolation and/or droplet deposition. These last are not chemically affected but covered by a layer of dust and particles. However, studies comparing different rock types (from limestone to sandstones) concluded that highly variable mechanisms can be dominant in the rock deterioration depending on the rock chemistry and porosity, the air pollution, climatic and micro-climatic factors, and architectural features (Amoroso and Fassina 1983; Charola 1988; Livingstone 1992; Steiger et al. 1993; Charola and Ware 2002). In fact, gypsum crystals formed due to the effect of sulphur compounds can be found either on the surface of building materials (as efflorescences or gypsum layers) or into their pore system, where gypsum crystallizes causing important erosive damage features such as flakes or scales (see section on decay features).

A similar pattern of decay is observed for lime mortars, present on buildings as joint mortars, plasters and renders. They undergo dissolution and sulphation due to sulphur attack (Sabbioni et al. 1998, 2001). Mortars of different composition are also affected by the 'sulphate attack' as occurs with cementious mortars and Portland cement (Santhanam et al. 2001, 2002). In some cases, ettringite ($(CaO)_6(Al_2O_3)(SO_3)_3{\cdot}32H_2O$) is also found due to the effect of air pollution (Sabbioni et al. 2001).

A very visible effect of sulphate oxides on Ca-rich stony materials is the formation of gypsum-rich black crusts on building façades due to the transformation of the calcareous surface. These black crusts are typically formed in areas of the building façades protected from rainwater washing. They have a strong aesthetical effect, but can be also very damaging to the underlying stony materials. Differences in the thermal expansion of gypsum and calcite crystals have been suggested as an additional damaging mechanism (Camuffo et al. 1982, 1983).

The case of siliceous-rich stony materials (mainly siliceous stones) is different as Ca-rich minerals are not present or abundant in the rock. Gypsum crusts can be commonly found on monuments built with Si-rich stony materials as sandstones or granitic rocks (Nord and Holenyi 1999; Begonha and Sequeira Braga 1996). Some authors have specifically postulated that these crusts are formed by combination of sulphates from air pollution and Ca from feldspars (Schiavon 1993). For others workers, gypsum crusts are produced in general by sulphation and nucleation by clay and airborne particles (Del-Monte and Sabbioni 1984; Nord and Ericsson 1993; Begonha et al. 1996) or by microorganisms involved in their formation (Blázquez et al. 1997). Some authors have attributed them to the oxidation of rock pyrites (Schiavon et al. 1994), although in most cases, the S in the rock is very low or absent. However, although the existence of most of these processes has been proven, they cannot explain the formation of thick and compact crusts, due to the low proportion of Ca in Si-rich rocks and the low crystallization of gypsum by nucleation, even in experiments in polluted atmospheres with high sulphate concentrations

(Cultrone et al. 2000). This limitation has encouraged other authors to propose alternative sources of calcium sulphate, as the sulphation of calcium from mortars (Smith et al. 1994). The mortars (jointing and plaster) are very prone to sulphation, especially in urban environments with high sulphate pollution, because they are a huge source of calcium and salts, easily soluble. Another alternative explanation is the use of gypsum plaster on the granite ashlars of the buildings, whose deterioration and blackening, by deposition of pollution particles, could explain his appearance as crusts (Sanjurjo-Sanchez et al. 2004, 2009; Pavía and Caro 2006).

#### 2.3.1.2 Nitrogen Oxides

Natural sources of nitrogen oxides (NO and $NO_2$, usually referred as $NO_x$) include volcanoes, oceans, biological decay and lightning strikes. Vehicle engines are the main anthropogenic source of nitrogen oxides in the urban environment although, emissions from industrial combustion of coal in coal-fired power plants, boilers, warm air furnaces, gas turbines, open burning, structural fires, explosives and fertilizers must be considered (Carslaw and Beevers 2004; Carslaw et al. 2007).

In urban settings, gasoline and diesel engines emit NO, with smaller amounts of $NO_2$ and other nitrogen species. Studies have estimated that diesel engines produce five times more $NO_x$ (per mass of fuel burned) when compared to gasoline engines. At high relative humidity, NO is very rapidly oxidised to $NO_2$ (Gaffney and Marley 2009). Direct emissions of $NO_2$ ('primary $NO_2$') are generally accepted to account for about 5% of $NO_x$ emissions, by volume, although this proportion is variable (Jenkin 2004b). Studies near major highways show $NO_2$ concentrations positively correlated with traffic density (Roorda-Knape et al. 1999; Gilbert et al. 2003). Although the total emissions of nitrogen oxides in Europe have declined in the last years (European Environment Agency 2007), other factors may counteract the effect of these emission reductions. There is strong evidence of recent increases in the fractional contribution of $NO_2$ to $NO_x$ emissions from vehicle emissions in Europe, to levels around 10% around 2,000 and up to 20% in 2005 in the UK (Pleijel et al. 2009).

$NO_x$ strongly react in the presence of sunlight by way of a series of photochemical reactions (Gaffney and Marley 2009) with ozone ($O_3$) to form 'secondary $NO_2$'. Although a large number of chemical processes potentially oxidize NO to $NO_2$ in the atmosphere (Jenkin and Clemitshaw 2000), the most important source of secondary $NO_2$ is the reaction of NO with $O_3$. Throughout the diurnal cycle, $NO_2$ is slowly converted into $NO_3$ by reaction with $O_3$ (Jenkin and Clemitshaw 2000). The $NO_3$ radical leads to the formation of gaseous $HNO_3$. Nitric acid shows the highest deposition velocity and will result in the eventual formation of highly soluble nitrates (Behlen et al. 1996).

There is evidence that background $O_3$ has been rising in the last two decades (Pleijel et al. 2009), although not necessarily at a constant rate. If background ozone concentrations increase, this can lead to a faster oxidation of NO to $NO_2$, enhancing the $NO_2$ concentration at certain level of $NO_x$ emission (Jenkin et al. 2008). Nitrogen dioxide ($NO_2$) and nitric oxide (NO) play a key role in determining concentrations of $O_3$, nitric acid ($HNO_3$), nitrous acid ($HNO_2$), organic nitrates, nitrate aerosols and other species in the troposphere (Martin et al. 2010). The photolysis of nitrogen dioxide in the presence of volatile organic compounds (VOCs) is considered as the only key initiator of the photochemical formation of ozone and photochemical smog (WHO 2000). Volatile organic compounds are a major group of pollutants (Parmar and Rao 2009) that undergo complex photochemical reactions in the lower troposphere (as occurs with peroxyacetylnitrate, PAN). Other chemical transformations of $NO_x$ lead to the generation of a variety of inorganic and organic oxidised nitrogen compounds (Jenkin and Clemitshaw 2000). On a regional scale a large number of reactions, including volatile organic compounds and CO, are important to determine the $O_3$ concentration, while on a local scale the dynamics of $NO–NO_2–O_3$ is dominated by a limited number of fast reactions (Jenkin et al. 2008). Studies have demonstrated that annual mean $NO_2$ levels show substantial variations from one site to another, even for sites for which the annual mean $NO_x$ level is comparable (Jenkin and Clemitshaw 2000).

The deposition of nitrogen oxides on building materials is considered as low under dry conditions (Charola and Ware 2002). Amoroso and Fassina (1983) referred that nitrogen oxides enhance the sulphate attack to Ca-rich building materials. Nitrogen compounds are very soluble and usually do not form stable compounds with stony building materials (Sabbioni et al. 1998). Studies in atmospheric simulation chambers have been dedicated to the interplay between $SO_2$ and $NO_2$ in urban settings and its role in the decay of some stony building materials, as. Different works (Haneef et al. 1992; Grossi and Esbert 1994) established that for Ca-rich stony materials the salt formation reaction is greater in wet/dry cycles that in dry conditions. They point out the importance of other catalysing agents such as transition metals (i.e. Fe, Mn, Ti). In the same conditions, the presence of $O_3$ in the chambers is less important than the presence of water (Haneef et al. 1992). On the other hand, these authors point out the importance of porosity in the development of a moisture film and thus active reaction points.

Strong synergistic effect on stone degradation has been reported for $NO_2$ and $SO_2$ mixtures in humid air. It has been proposed that $NO_2$ acts as a catalyst for the oxidation of $SO_2$ on the stone surface (Johansson et al. 1988; Johnson et al. 1990; Haneef et al. 1992). In the case of calcareous stony materials, calcium nitrate can be formed by $NO_x$ attack (Allen et al. 2000). However, Allen et al. (2000) concluded that $NO_2$ increases the rate of attack by $SO_2$ to Ca-rich components of buildings but did not strictly play a catalytic role (based on the reactions that will be here briefly summarized). According to these authors, the presence of $NO_2$ accelerates the rate of sulphation of Ca-rich materials by increasing the oxidation of sulphur dioxide to sulphur trioxide. The sulphur trioxide reacts with water vapour producing sulphuric acid that afterwards reacts with the calcareous materials to form calcium sulphate. $NO_2$ is initially consumed but, in the presence of sulphuric acid, the reaction proceeds by the formation of nitrosonium hydrogen sulphate that reacts with $SO_2$ forming NO, which may then be readily oxidised back to $NO_2$. This sequence of reactions explains, according to Allen et al. (2000), the possible confusion in relation to the role of $NO_2$ as catalyst. The greatest increases in surface sulphur concentration are found when both pollutant gases ($NO_2 + SO_2$) were present.

For other stony materials, the effect of $NO_x$ has been also studied (Martínez-Ramírez et al. 1997). The reaction between lime mortars and NO pollutant gas was observed as low in atmospheric simulation chambers, but surface water enhances dry deposition effects mainly in the presence of ozone.

#### 2.3.1.3 Ozone

Ozone is mainly generated in photochemical reactions. Episodes of high $O_3$ concentration are frequent in densely populated areas and they have been related to forest decline and industrial emissions (Virgarzan 2004). Atmospheric $O_3$ in the lower troposphere is produced by photolysis of nitrogen dioxide ($NO_2$) and subsequent recombination between atomic and molecular oxygen in the presence of a third-body molecule. $O_3$ is also produced by natural sources (e.g. trees, thunderstorms). However, the presence of VOCs in the atmosphere will interrupt this process by being more attractive to the nitric oxide that the $O_3$ molecule and therefore will contribute to $O_3$ accumulation (Haagen-Smit and Fox 1954, 1956; Leighton 1961). It has been found that the photochemical $O_3$ production in urban areas rises with the $NO_x$ emissions and is less sensitive to the VOC emissions (Sillman and Samson 1995). Partially oxidised organic species such as aldehydes, ketones and carbon monoxide are produced as intermediate oxidation products, with $O_3$ formed as a by-product. An enormous variety of VOC classes may be emitted from numerous anthropogenic and biogenic sources (e.g. Rudd 1995; Guenther et al. 1995) and, depending on location, either or both categories can make a major contribution to photochemical $O_3$ formation (e.g., Sillman 1999, and references therein) (Jenkin and Clemitshaw 2000).

In areas moderately contaminated, $O_3$ sensitivity to emission of nitrogen oxides depends on the season and on the emission rates (Kleinman 1991). The non-linearity in the $NO_2$ vs. $NO_x$ concentration relationship results mainly from the chemical coupling of $O_3$ and $NO_x$. Another consequence of this is that any reduction in the level of $NO_2$ is accompanied by an increase in the level of $O_3$ (Jenkin 2004a, b; Itano et al. 2007).

Besides its known effects on polymers and paints, it is very difficult to estimate the direct effect of ozone on building materials. It is considered that the effect of the $O_3$ on stony materials would be related to the enhancement of effect of sulphur and nitrogen oxides (Amoroso and Fassina 1983; Martínez-Ramírez et al. 1997; Charola and Ware 2002). In fact, it is considered that the rate of formation of sulphate from $SO_2$ is 1,000 times faster if ozone is present (Penkett et al. 1979). $O_3$ can increase $SO_2$ oxidation in the absence of water, and according to Martínez-Ramírez et al. (1997) the most aggressive environment for lime mortar is $SO_2$ in presence of water and $O_3$.

#### 2.3.1.4 Carbon Oxides

Carbon oxides (monoxide and dioxide) are normal constituents of the atmosphere. The exchange processes between carbon dioxide in the atmosphere, ocean, and biosphere are key steps in the exogenous carbon cycle. $CO_2$ concentrations are highly variably depending on the area and the existence of vegetation, solar radiation, etc. (Amoroso and Fassina 1983). Natural sources of carbon dioxide in the atmosphere include combustion of organic matter, respiration processes in organisms and volcanic outgassing. Anthropogenic sources are related to the burning of fossil fuels for power generation, transport and heating and other industrial processes. Due to such additional sources in urban areas their concentration can be high. Anthropogenic emissions of $CO_2$ have constantly increased from the nineteenth century. The concentration of carbon dioxide in the Earth's atmosphere has been measured in approximately 387 ppm by volume in the year 2009, rising about 1.9 ppm/year in average during the period 2000–2008 (data on carbon Budget available online at http://www.globalcarbonproject.org/carbonbudget/08/hl-full.htm; accessed on August of 2010). This supposes an average concentration of 0.0387% by volume. This is 103 ppmv (36%) above the 1,832 antarctic ice core levels of 284 ppmv (Etheridge et al. 1998). Estimated emissions of traffic exhaust are responsible by 10% of the $CO_2$ global emissions (Gaffney and Marley 2009).

Carbon monoxide (CO) exists in lower concentration than $CO_2$ in air. Natural sources of CO are natural gases, volcanic outgassing, forest fires, bacterial activity and methane oxidation in the atmosphere. Anthropogenic CO is a final combustion product if the amount of $O_2$ is not sufficient for complete oxidation, sometimes due to a bad mixing of the air and fuel. In large urban centres CO concentrations can reach 100 ppm and higher (Fellenberg 2000). It is considered that about 25% of CO in the atmosphere was produced by direct anthropogenic emissions (Amoroso and Fassina 1983).

The effect of the CO on building materials is not easy to assess and no evidence exist on the direct damage of the CO to them. The lack of studies does not mean that no effect exist on the building materials. CO is an indirect source of $CO_2$, as it reacts quickly with OH to give $CO_2$ (Jenkin and Clemitshaw 2000). CO must be considered as an air pollutant, which damages building materials, as increases the concentration of $CO_2$ in rainwater. Despite this fact, the indirect effect of CO emissions on the decay of building materials has not been studied yet.

$CO_2$ in air can be dissolved in rainwater acidifying the pH of the solution. It has been calculated that the increase of $CO_2$ in rainwater is 1% for an increase of 12.5% in $CO_2$ in the air, depending on the water temperature. An increase of water temperature causes a decrease in the amount of carbon dioxide dissolved. Thus, water at temperatures near 0°C dissolves about twice the quantity of $CO_2$ of water at about 25°C (Winkler 1994). The resulting acid solution can cause dissolution of calcium and magnesium carbonates in calcareous rocks (limestone, dolomite, marble) and other stony materials (lime mortars) to form calcium and magnesium bicarbonates respectively. Both bicarbonate salts are much more soluble than their respective carbonates (Amoroso and Fassina 1983).

Amoroso and Fassina (1983) proposed that the dissolution of building materials due to high $CO_2$ concentration in air is more important in an indoor environment with high water vapour content and near saturation conditions. Water can also leach out some metallic ions contained in silicate minerals (micas, feldspars, chlorites) or carry foreign ions (from atmospheric particles). These authors defend that the increase of porosity caused by the recrystallization of water-soluble calcium bicarbonate is harmful to the building materials (including calcareous and non-calcareous stony materials), since it allows acidic solutions containing sulphuric acid and soluble salts to penetrate more deeply inside the stone and to accelerate the decay.

#### 2.3.1.5 Other Gaseous Compounds

Although the effect on the building materials of the main gaseous compounds emitted by different anthropogenic sources has been showed above, other minor gaseous components have been shortly investigated. Among them, gaseous fluorine has been briefly studied. Gaseous fluorine is mainly emitted into the air, including fluorine ($F_2$) and hydrogen fluoride (HF). They result from industrial processes related to the production of aluminium, steel, glass, brick, tile and fertilizers (Amoroso and Fassina 1983). Although their atmospheric concentration is rarely measured, their effect on building materials has been studied but the role of the hydrogen fluoride was not clarified (Luckat 1976). Experimental studies in simulation atmospheric chambers showed that deterioration on siliceous stone surfaces was due to nucleation and growth of new crystalline phases (silicon tetrafluoride and other compounds) and genesis of fissures, microcracks and pores (Cortes and Martin 1982a). Similar damage was observed in calcareous stones, but new crystalline phases were mainly composed by fluorite ($CaF_2$) and a surface layer of chlorite and silica was observed (Cortes and Martin 1982b).

Gaseous chlorine has also been considered as a cause of stone decay in buildings, as can be dissolved in rainwater to cause crystallization of chlorine salts. It can be formed as a result of several industrial processes (plastic production, burning of coal and water treatment plants). However, the formation of chlorine salts can be more easily related to sea-salt aerosols and rising damp (Amoroso and Fassina 1983; Begonha et al. 1996; Chabas and Jeannette 2001; Sabbioni et al. 2001).

### *2.3.2 Particulate Matter*

The term "atmospheric particles" is used to refer to fine solid or liquid particles suspended in air. Depending on the research field and interest (e.g. human health, atmospheric chemistry), atmospheric particles can be classified considering their size. The smallest particles have very short lifetimes in air because their attachment to larger particles. The largest particles are short-lived and remain airborne near to their source due to their high rate of sedimentation. Atmospheric particulate content tends to be modified with time due to coagulation, adsorption of gases, dry and wet removal and chemical reaction within particles in the presence of liquid water (Amoroso and Fassina 1983).

Atmospheric particles may be originated by some natural and anthropogenic sources. The main natural sources of atmospheric particles are volcanic outgassing, forest fires, sea salt (directly emitted), and gas-phase conversion of other atmospheric compounds. Anthropogenic sources are mainly burning of fossil fuels (industrial, transport and domestic burning), diverse industrial processes, mining and agriculture (Collings and Graskow 2000). Natural sources greatly exceed anthropogenic emissions, but anthropogenic particle emissions are frequently concentrated. Atmospheric particles are either emitted directly into the atmosphere or originated from gas-phase conversions within the atmosphere. Due to climatic factors, the cyclic human activity and the diurnal cycles, the concentration of suspended particulate matter in the air oscillates diary, weakly and yearly (Amoroso and Fassina 1983; Fellenberg 2000).

In highly populated urban settings fossil fuel combustion by industrial facilities and vehicular engines can be responsible by high concentration of particles in the air. Few research works on dry deposition of particles have been performed in comparison to gaseous compounds (Judeikis and Stewart 1976; Johansson et al. 1988; Lipfert 1989; Cobourn et al. 1993; Saiz-Jimenez 1993; Ruijgrok et al. 1995). This is because it is very difficult to assess the sources (natural or anthropogenic) of atmospheric particles from their composition.

Regarding the chemical composition of particles, some sulphur-rich, Cl, Mg, K, Na, Ti, Al, Fe, etc. can be attributed to both natural and anthropogenic sources as sea spray in coastal areas or wind-blow of crustal dust. Others (Van-Grieken and Torfs 1996; Pio et al. 1998; Bityukova et al. 2000; Chabas and Lefèvre 2000; Song et al. 2001; Krupa 2002; Singh and Kumar 2009; Maas et al. 2010) are attributed to oil combustion (Ni or V), urban solid waste (carbonaceous particles, Pb and Zn), industrial burning of coal or other industrial emissions (As, Cr, Cd, Pb, V, Zn, Sn, Cu, B and carbonaceous particles) and gasoline and diesel combustion (sulphur- and nitrogen-rich salts, Pb, Cl, Br, Cd, Zn, Al, Si, carbonaceous particles, volatile compounds and fly ash).

According to Gaffney and Marley (2009), the combustion of fuel in transport is a significant source of carbonaceous particles (carbon soot or fly ash by incomplete combustion) and metallic particles (Al, Si, Zn, Cd, Pb). Simão et al. (2006) refers a study of Colebeck published in 1995 that indicates that, in the UK, on-road diesel combustion contribution to airborne elemental carbon level is though to be between 80% and 95%. Motor Vehicle emission tests carried out on different types of vehicles have revealed important concentrations of Pb, Na, Cl, Cu, Zn, Mn, Fe, Ca, P, $SO_2$, Si, Al, carbonaceous particles and organic carbon (Pierson and Brachaczek 1983; Hildemann et al. 1991). Other particle emissions caused by car traffic such as brake dust ($SO_2$, Mg, Fe, Si, carbonaceous particles and organic carbon), tire dust ($SO_2$, Fe, Si, carbonaceous particles and organic carbon) or paved road dust (Si, Al, K, Ca, Ti, Fe, $SO_2$, carbonaceous particles and organic carbon) are not usually considered in pollution studies (Hildemann et al. 1991). Human activities in urban areas also contribute to particle emissions through (Hildemann et al. 1991) oil boilers ($SO_2$, $NH_3$ and carbonaceous particles), wood fires (Cl, carbonaceous particles and organic carbon), synthetic fires (Cl, Al, $SO_2$, carbonaceous particles and organic carbon), natural gas home appliances ($SO_2$, $NO_3$, Cl, Na, Ca, carbonaceous particles and organic carbon), meat cooking (Cl, Na, SO, $NO_3$, Cu and organic carbon) or vegetative detritus (Si, Al, Cl, K, Ca, Fe, Cu, Zn and organic carbon) (Hildemann et al. 1991).

**Table 2.2** Deliquescence points at 20°C (From Amoroso and Fassina 1983)

| Compound | Deliquescence point (relative humidity, %) | Compound | Deliquescence point (relative humidity, %) |
|---|---|---|---|
| $NH_4HSO_4$ | 30–40 | $NaNO_2$ | 66 |
| $NaHSO_4$ | 52 | $NaNO_3$ | 74 |
| $(NH_4)_3H(SO_4)_2$ | 68 | $KNO_2$ | 45 |
| $(NH_4)_2SO_4$ | 80 | $KNO_3$ | 92 |
| $KHSO_4$ | 86 | $Pb(NO_3)_2$ | 88 |
| $Na_2SO_4$ | 85–93 | NaCl | 76 |
| $MgSO_4$ | 91 | $MgCl_2$ | 33 |
| $ZnSO_4{\cdot}7{\cdot}H_2O$ | 87 | $CaCl_2$ | 18 |
| $ZnSO_4$ | 88 | LiCl | 13 |
| $CuSO_4$ | 98 | $BaCl_2$ | 90 |
| $K_2SO_4$ | 97 | $NH_4Cl$ | 77 |
| $CaSO_4$ | 23 | KCl | 85 |
| $CaSO_4{\cdot}H_2O$ | 68 | KI | 84 |
| $HH_4NO_3$ | 62 | KBr | 81 |
| $Zn(NO_3)_2$ | 45 | NaBr | 58 |

The question of which pollution sources are the main contributors to damage caused on building materials is a key question to assess the role of each pollution source in urban areas. An open discussion exists regarding nearby or surrounding emission sources as responsible for the effect of gaseous and particulate pollutants on building materials. However, just in relation to particles, the concentration of particle sizes <10 μm in diameter in air, when emitted by car engines in urban streets, decreases between 60% and 80% of the maximum occurring at 2 m above the ground (Wu et al. 2002).

Atmospheric particles are responsible by damage on metallic, glassy and stony building materials as they can form coatings (soiling) even on smooth surfaces. They can dirt surfaces, causing both aesthetic damage (Grossi et al. 2007) and creating ideal conditions for oxidation processes catalyzed by carbon particles and other metals present in dust. Transformation of sulphur oxides to sulphuric acid has been linked to hygroscopic components of dust with ionic content (heavy metals and salts) containing gaseous sulphur, nitrogen or carbon. Simão et al. (2006), in a study with silicate rocks, concluded that elemental C and Fe-rich particles of diesel particulate matter promote the formation of $H_2SO_4$ by catalytic oxidation of $SO_2$, while organic-C and Pb-rich particles did not have a significant contribution on sulphation of stones (sulphation of stones was much less significative in stones with gasoline particulate matter). The acid solution and the saline and metallic ions can penetrate into the building materials that corrode far faster (Fassina et al. 1979; Fellenberg 2000; Charola and Ware 2002). The effect of saline compounds depends on their abundance and deliquescent point (Table 2.2). The deposition and effects of saline ions have been studied and highlighted for $HCO^-$, $Cl^-$, $H^+$, $Mg^{+2}$, $K^+$ and $Na^+$ (Van-Grieken and Torfs 1996) with different contributions depending on the time of the year and the type of deposition (dry or wet).

Sulphur-rich atmospheric particles have been considered in studies of decay of building materials due to air pollution, as they are strongly damaging and frequent in urban areas (Amoroso and Fassina 1983). It has been also proposed that major gypsum crystallization on the surface of some Ca-rich stone materials can be due to environmental deposition of gypsum (Toniolo et al. 2009) although this mechanism cannot explain the sulphation of calcium carbonate minerals. The most common sulphate particles are $H_2SO_4$, $NH_4HSO_4$, $MgSO_4$, $CaSO_4$, $NaSO_4$, $ZnSO_4$ and $(NH_4)_2SO_4{\cdot}ZnSO_4$. Some of them can also form metal-sulphur complexes if metal ions are present. They are generally associated with particles<3 μm size. These sulphur-rich atmospheric particles cause similar damage than gaseous sulphur to building materials (Amoroso and Fassina 1983). Metallic ions (Fe, Al, Ti, Cr, etc.) can also be found as part of the atmospheric particulate components (Van-Grieken and Torfs 1996), on the surface of building materials or forming black coatings and crusts (Del-Monte and Sabbioni 1984; Maravelaki-Kalaitzaki 2005; Sanjurjo-Sánchez et al. 2011).

In urban environments, different types of coatings formed on façade areas shielded from rainwater of historic buildings have allowed to study indirectly the qualitative air content in the environment of these buildings (Del-Monte and Sabbioni 1984; Johansson et al. 1988; Jones et al. 1996). Dust coatings (soiling), silica coatings and crusts, Fe-rich and heavy metal films, and saline crusts have been attributed to particulate air pollution (Dorn 1998; Vazquez-Calvo et al. 2008; Sanjurjo-Sánchez et al. 2008, 2011) although other scenarios can explain their origin. Dust coatings or soiling result of the accumulation of particles transported by wind on rock surfaces. They are common in both arid and urban or industrialized areas. Particles, salts and carbonates accumulate hygroscopic water that captures $CO_2$ acidifying the coating (Dorn 1998; Charola and Ware 2002). Microorganisms and organic detritus, pollen and fly ash are also found (Sanjurjo-Sánchez et al. 2008). Fe-rich coatings have been related to abiotic or biogenic processes, but also to deposition of Fe and other metallic oxides from the atmosphere (Del-Monte and Sabbioni 1984). Si-rich coatings are frequent on building materials and are essentially formed by amorphous Si, Al and clasts to form an amorphous or detritic matrix (Curtiss et al. 1985; Watchman 1991; Dorn 1998). They have been related to deposition of $SO_2$, although other hypothesis exists, such as precipitation of silica from mortars, chemical weathering or biological weathering of silicate minerals (Dorn 1998; Vidal-Romaní et al. 2003; Magee et al. 1988; Sulovsky et al. 1996).

Carbonaceous atmospheric particles are considered as very active in the formation of black crusts because they contain sulphur compounds, they have a catalytic effect and they can act as nucleation points for growing of gypsum crystals (Del-Monte and Sabbioni 1984; Camuffo 1990; Sabbioni et al. 1996; Derbez and Lefèvre 1996). Carbonaceous particles and organic matter show high specific surface area and thus they set particles on the material surfaces and are strongly adsorbed to minerals. In fact, carbonaceous particles act as nucleating agents for the growing of gypsum crystals (Christwell et al. 1988; Del-Monte and Sabbioni 1984; Saiz-Jimenez 1993; Cultrone et al. 2000). On Ca-rich stony materials they play a catalytic effect in the transformation of calcium carbonate into gypsum. In fact, gypsum-rich black crusts usually contain abundant carbonaceous particles (Benner et al. 1982; Camuffo et al. 1983; Leysen et al. 1989; Ghedini et al. 2000; Sanjurjo-Sánchez et al. 2009). Roberge (2000) refers that the deposition on a metallic surface of particles of less reactive substances could promote corrosion of that metal due to galvanic effects (the author emphasize a special attention to carbonaceous deposits). Carbon compounds on the surface of building materials are due to organic atmospheric particles, biological colonization or surface paints and treatments (Saiz-Jimenez 1993, 1995; Turpin et al. 1991; Turpin and Huntzicker 1995; Ghedini et al. 2000; Ho et al. 2002).

Fly ash particles, generated in the combustion of coal and diesel, are among the main carbonaceous particles relevant to materials decay. The components of the fly ash particles vary considerably depending on the source, but all fly ash include substantial amounts of silicon dioxide (both amorphous and crystalline) and calcium oxide (Pryor and Barthelmie 2000; Gaffney and Marley 2009). Some works have compared fly ash derived from oil and coal (Alliksaar et al. 1998) by using scanning electron microscopy (SEM) and energy dispersive X-ray analysis (EDX). Fly ash particles can be used to assess spatial distribution and temporal trends of atmospheric deposition of pollutants derived from high temperature combustion of fossil fuels. Studies of fly ash particles in sediments, soils and surfaces are relevant because they can be used as indicators of atmospheric deposition of pollutants from fossil fuel combustion, as other pollutants such as $NO_x$ and $SO_2$ do not leave reliable historical records on them (Wik and Renberg 1996; Sanjurjo-Sánchez et al. 2011).

In general, research on organic particulate compounds in air has been carried out since the nineteenth century, but few papers have been devoted to the effect of them on building materials. Particulate organic compounds are molecular particles of very variable origin (Hildemann et al. 1991; Simoneit and Mazuret 1981; Mazuret and Simoneit 1986). Macromolecular organic components in the atmosphere range from submicron (diameter <0.1 μm) to supermicron (diameter ~100 μm) dimensions. In general, they can be classified in aliphatic, aromatic and halogenated hydrocarbons, hydroxy and carbonyl compounds, esters and carboxylic acids (Simoneit and Mazuret 1981; Mazuret and Simoneit 1986; Hildemann et al. 1991).

Aliphatic hydrocarbons and lipids (e.g. methane, butane, pentane) are produced in industrial processes but also combustion of gasoline and diesel or domestic heating systems (Simoneit and Mazuret 1981; Mazuret and Simoneit 1986; Schauer et al. 1999). Aromatic hydrocarbons as mono-, di- and trimethylbenzenes, benzene, toluene and xylenes have been detected in rainwater and attributed to automobile emissions but also to forestation of the outskirts or urban areas. Polynuclear aromatic hydrocarbons (PAHs) have been directly linked to combustion of fossil fuel due to car engines. In fact, PAHs have been detected in soil associated with heavy metals (Pb, Cu, Zn) (Morillo et al. 2008). High molecular weight aromatic species are primarily associated with the particulate phase and low molecular weight aromatics are enriched in rainwater (Lunde et al. 1977; Simoneit and Mazuret 1981; Schauer et al. 1999; Mazuret and Simoneit 1986).

Other compounds such as halogenated hydrocarbons are frequently due to organochloride pesticide residues (e.g. DDT, DDE), polychlorinated biphenyls (PCBs) and solvents (e.g. chloroform, carbontetrachloride) emitted due to industrial processes. They are frequent in rainwater in urban settings. Hydroxy- compounds (phenolic, n-alkanol and sterol constituents) have been related to unknown industrial ambient emission sources whereas sterols and n-alkanol homologues are found in association with biogenic atmospheric emanations (plant wax, microbial lipids). Phenols have been found in urban settings in dry and wet fallout but they also have related to diffuse biogenic sources. In addition, n-alkanols and steroids are found in semirural settings and remote marine areas (Simoneit and Mazuret 1981; Kawamura and Kaplan 1987; Mazuret and Simoneit 1986). Carbonyl compounds have been observed in studies of wet and occult deposition. Aldehydes and ketones are more frequent in urban samples when compared with rural and remote areas due to road traffic emissions. Few studies have identified ester components and only the higher molecular weight esters belonging to the lipid fraction have been reported in urban, rural and remote settings. Carboxylic acids contribute a small and irregular fraction of the total precipitation acidity. Low molecular weight organic acids are frequently detected in urban environments (Simoneit and Mazuret 1981; Mazuret and Simoneit 1986; Schauer et al. 1999).

Airborne organic compounds have been found on the surface of building materials and marked as partly responsible by the damage caused on building materials (Gillberg et al. 1977; Hoke 1978; Saiz-Jimenez 1993; Sabbioni et al. 1998, 2001). They usually are deposited in urban areas and other carbonaceous particles (as carbon soot) can also be found. In fact, inorganic and organic carbon carbonaceous particles are found in dust layers (soiling) and gypsum crusts, and different studies considered their catalytic effect on the sulphation of Ca-rich minerals in building materials (Del-Monte and Sabbioni 1984; Dorn 1998; Saiz-Jimenez 1993; Sabbioni et al. 1998, 2001; Sanjurjo-Sánchez et al. 2009). They also can strongly corrode metallic structures (Saiz-Jimenez 1993).

Organic compounds found on the building materials can be similar in the same area but different when different settings are considered. Differences depend on the concentration of pollutants and the closeness of pollution sources. In addition, identical compounds are present in crusts and dust layers on buildings whatever the building material analyzed (Saiz-Jimenez 1993). All kind of organic compounds reported above can be frequently found on black crusts and damaged building materials (Boyer and Laitinen 1975; Mazuret and Simoneit 1986; Saiz-Jimenez 1993; Sabbioni et al. 1998). When stronger pollution is observed, complex mixtures of branched and cyclic hydrocarbons can be found (Saiz-Jimenez 1993). In addition, biological colonization of building materials can be enhanced due to the deposition of carbonaceous particles and organic compounds. This fact can causes degradation of some organic compounds. The overprint of biological activity can be observed if fatty acids, some hydrocarbons (plant waxes) and organic acids are found on building materials (Saiz-Jimenez 1993).

### 2.3.3 Solutions

Solutions are an important way of transport of pollutants in either ionic or solid form. Several deleterious processes of materials can arise from reactions between the constituents of this materials and ionic pollutants such as:

- corrosion of metals – the presence of electrolytes promotes corrosion reactions (Trethewey and Chamberlain; 1995; Roberge 2000; Tidblad and Kucera 2003). The effects of pH and redox (related to oxygen levels) of the solutions on metal corrosion can be assessed by Eh-pH diagrams, also called Pourbaix diagrams (see examples in e.g. Trethewey and Chamberlain 1995; Roberge 2000);
- alkali-aggregate reactions in concrete – related to alkaline solutions and some types of aggregates, specially some forms of reactive silica and some carbonates (Sims and Brown 1998; Farny and Kerkhoff 2007);
- sulphate attack – when sulphates in solution react with components of the set cement causing several deleterious phenomena. This is a classical subject that is still a subject of research and discussion (for reviews see Eglinton 1998; Santhanam et al. 2001; Bensted 2002; Santhanam et al. 2002; Neville 2004);
- dissolution – the dissolving effects of waters with low pH (acid rain) on carbonates is nowadays material of highschool books. There are also references to the effects of acids on cement phases (Eglinton 1998) and on siliceous constituents of sandstones (Ashurst 1998). Dreesen et al. (2007) referred the contribution of sulphate solutions to dolomite dissolution (due to the strong ion pair sulphate-magnesium). Doremus (1994) refer that silicate glasses are more rapidly dissolved by basic solutions. Some authors have referred that solutions with high pH could affect silicates (Winkler 1994; Bernabé et al. 1995; Pye and Mottershead 1995; García-Talegón et al. 1996; Lopez-Arce et al. 2008). According to French (1997) bituminous binders can be partially dissolved by alkaline solutions. Another factor that might contribute to dissolution is the higher ionic strength of salt solutions that affect the solubility of the minerals (Gillott 1978; Dibb et al. 1983; Magee et al. 1988; Zezza and Macrì 1995; Galan et al. 1996; Skoulikidis et al. 1996; Chabas and Jeannette 2001; Delalieux et al. 2001; Cardell et al. 2008; see also review by Turkington and Paradise 2005);
- sulphatation – rock phases are replaced by sulphates. Most studies concern the substitution of calcite (Camuffo et al. 1983; Verges-Belmin 1994; Camuffo 1995; Goudie and Viles 1997; Charola 2000; Charola and Ware 2002; Charola et al. 2007) but there are also references to pseudomorphic replacement of silicate minerals in granites (Schiavon 1996);
- dedolomitisation, where dolomite is replaced by calcite, a process attributed in Valls-del-Barrio et al. (2000) to circulation of Ca-rich solutions from mortars through dolostone (this seems to favour granular disintegration). A similar process has been considered by Dreesen et al. (2007) that related dedolomitisation to calcium sulphate solutions associated with oxidation of sulphides in the stones.

Solutions promote the migration and decay action of soluble salts. Solutions with soluble salts have also been referred as promoters of the deleterious effects of some rock constituents, namely clay minerals, by favouring (due to hydroscopicity) wetting-drying cycles (Rodriguez-Navarro et al. 1997a) and by increasing the volumetric expansion of clays (Pye and Mottershead 1995). Some authors also defend that rock strength can be affected by the composition of pore solutions (Feucht and Logan 1990; Feng et al. 2001).

Several pathways for the actions of solutions can be considered for the built environment such as atmospheric wet deposition, surface circulating waters (including seawater) and both descending (infiltration) and ascending (capillary-rising solutions) solutions inside the pore space of the materials.

It was referred above that the wet deposition of sulphur gaseous compounds is generally considered less important in the acid decay of building materials than dry deposition. However, in rural areas where low atmospheric pollution exists, wet deposition can be more important (Furlan and Girardet 1992). The combination of rainout and washout is called 'acid rain'. It is considered that the pollutants incorporated in wet deposition are usually produced by distant sources more than local sources (Torraca 1988).

Rainfall can be regarded as the dominant pathway of wet deposition. Compared to dry deposition, the action of rainfall is more complicated in the case of porous building materials (Steiger and Dannecker 1994). Direct attack to the materials can be due to rainfall acidity. When rainfall hits the surface of a porous material it is first absorbed by capillary action. It is considered that the dissolution of sulphur oxides in water droplets comprises several stages: dissolution of gaseous sulphur dioxide in water droplets in clouds or fog followed by oxidation of $SO_2$ to $H_2SO_4$. The sulphuric acid formation increases the acidity of the droplets and stops at pH values below two (Amoroso and Fassina 1983; Steiger and Dannecker 1994). When these acid droplets are deposited on the stone surface they promote the transformation of calcium carbonate into gypsum. The efficiency of 'acid rain' depends on the intensity of precipitation, the vertical distribution of substances in the lower troposphere and the droplet formation. The chemical composition of the droplets depends on various factors: nucleation scavenging, scavenging of gas and particles during the evolution of the dispersed system, chemical reactions in the liquid phase, and microphysical evolution of the system depending on atmospheric conditions (Camuffo 1990, 1991).

Acidity can be responsible to the attack to building materials, although strong rain could imply washing of dry deposits on the building façades with a positive effect. On calcareous stones it has been suggested that water mobilizes any compounds on the stone surface (due to dry deposition) increasing the pH of the solution. Condensation (due to frog and night dew) can be relevant, particularly in springtime when warm and humid air masses reach stony surfaces at temperatures below the dew point (Camuffo et al. 1982, 1983).

Before considering the effects of water that circulates on the porous media (either by infiltration or by capillary rising from the ground) a few comments on possible effects of other free circulating waters are made. Besides being a potentially important

**Fig. 2.7** Stains on limestones related to metal migration on a modern construction

source of pollutants, seawater can have disastrous consequences when in direct contact with materials due to its high salinity (Trethewey and Chamberlain 1995; Dibb et al. 1983; Roberge 2000). Another classical example of the effects on building materials of pollutants in circulating waters is the corrosion of sewers related to sulphuric acid production due to microbial action (for a description of the process see Herman and Maier 2000). Also among the free circulating waters, run-off waters have been considered an important factor in the dissolution of materials. They contribute, consequently, to the transport of pollutants to other portions of the built environment. Water pathways on the built environment could imply that some materials constitute pollution sources for other materials such as the stains related to metals (Fig. 2.7) or the distribution of white coatings relate to carbonate precipitation (such as those showed in Fig. 2.2) whose distribution is affected by hydrodynamical features related to the characteristics of the structures (Alves 2010).

**Fig. 2.8** Effects of water infiltration in historical (**a**) and modern constructions (**b**)

Infiltration of water can lead to solutions with different and more aggressive characteristics due to long-time contact between the rain-derived solutions and the building materials or pollutants that have accumulated on places such as cornices. For example, pigeons that can leave important quantities of excreta on potentially infiltrating areas of buildings. This constitute an important pollution source since pigeon droppings have been related to the formation of salts (Gómez-Heras et al. 2004; Begonha 2009)

Examples of decay features related to infiltrating solutions are presented in Fig. 2.8 including an example from a historical construction and another from an unfinished modern construction (this last one highlight a salt pollution risk). Downwards movement of solutions in building works related to hydraulic properties of natural terrains can also be considered here. For example, Robbins et al. (1997) refer that infiltrations of acid groundwater were associated with the corrosion of railway tunnel linings and Puppala et al. (2010) refer heave of shotcrete tunnel linings due to the presence of sulphates.

Pollutants related to groundwater, interaction between groundwaters and the soils, interactions between soils and atmospheric deposition or all of them can accumulate in the ground below built elements. Composition of groundwaters is determined by natural and anthropogenic sources (Appelo and Postma 2005). It has been found that urbanization affects the composition of groundwaters (Appleyard 1995; Lerner and Barret 1996). Besides the contribution of rising groundwater, soilwater can be an important source of ionic components due to inorganic components in the soil (Doner and Lynn 1977). Water balance and leaching conditions can also be relevant: periods of drying conditions favour concentration of solutions (see reference in Domenico and Schwartz (1990) to a study by Hendry and collaborators published in 1986); solutions can be concentrated on sheltered zones (Hawkins and Pinches 1987; Oyama et al. 1996). Organic activity promotes lower pH due to higher $CO_2$ content (resulting from respiration and organic matter degradation) and to organic acids (Appelo and Postma 2005). Human and animal excreta, due to their high salt content (Stevenson and Cole 1999; Tjandraatmadja and Diaper 2006) can be an important source of pollutants affecting groundwaters and soilwater.

These pollutants at the base of the built elements can cause deleterious effects by heave movements (Hawkins and Pinches 1987; Oyama et al. 1996) or impregnate the materials by upward movements related to capillary rising processes (the sometimes called "wick effect", see Goudie 1986). Decay features associated with capillary-rising solutions will result from water balance between capillary feeding and drying that promotes salt crystallisation (Lewin 1982; Hammecker 1995) as well as from chemical reactions between the rising solutions and the materials (such as those presented above). Capillary-rising solutions can have an important and generalized impact in regions of high salt content (Blackburn and Hutton 1980; Goudie and Viles 1997) and $SO_3$ content (total and water-soluble) has been even proposed as criteria to classify soils in relation to the hazard of sulphate attack (see Eglinton 1998). Decay features related to capillary rising solutions tend to show height-related spatial patterns (Fig. 2.9) such as concentration of detachments features on portions nearer the ground or at an intermediate height position related to the equilibrium between moisture feeding and drying.

The composition of capillary rising solutions might change along height due to reactions with the building materials and to chemical differentiation processes such as precipitation. A classical example is the general model of distribution of soluble salts associated with capillary rising (see, e.g. Arnold and Zehnder 1991) where less soluble salts are precipitated nearer the ground while more soluble salts are found in higher positions. Therefore, salt weathering related to capillary rising depends on the properties of the porous materials (that control the attained height and the moisture migration rate), the chemical characteristics of the solutions (that control the

**Fig. 2.9** Salt efflorescences related to capillary-rising solutions in limestones at a modern construction: note the distribution of salt efflorescences at the top of the first block, typical of water rising from the ground and where the position of the salt efflorescences results from the equilibrium between moisture feeding rate, drying rate and of solubility of the salt present at the efflorescences

supersaturation conditions that lead to salt crystallisation) and the environmental conditions (that affect drying rate). An example of the use of these data on the modelling of salt weathering of different salt minerals can be found in Hammecker (1995). Geochemical differentiation related to solubility could also happen horizontally (see model in Hammecker 1995), a process that would explain the distribution of soluble salts in depth (some examples can be seen in Wendler et al. 1990; Casal-Porto et al. 1991; Alves and Sequeira Braga 2000; Steiger 2003).

## 2.4 Signatures of Pollution Sources

As was seen from the previous sections, several pollution sources can act on a given local, sometimes simultaneously. In the study of the effects of pollution sources on the built environment it is important to identify the pollution sources that actually affect a given element, both to understand the decay processes and to plan the intervention procedures. Mineralogical and geochemical (elements, ions, isotopes) studies have been an important factor for this aim. New tracers can be considered and are proposed here based in its previous applications on environmental studies of other media (air, water, soil) and its relation to the decay processes of materials in the built environment.

## 2.4.1 Mineralogy

Several substances with a crystalline structure (that will be considered as minerals since they are not the result of intended processes) can be formed after emplacement of the building materials, arising from pollution sources or from the interaction between these and the building materials or from both. Occurrences of these mineral phases are frequently macroscopically detectable by visual inspection and are usually easy no collect by low impact procedures (a big issue in the study of materials that are part of public or private property or cultural heritage). Its macroscopic characteristics allow the study of changes related to variations in the climatic characteristics (Arnold and Zehnder 1991) or to infiltration episodes (Alves and Sequeira Braga 1994). They can constitute stable markers of the general chemical features of the decay sources and its characterization is useful for the interpretation of chemical results since the presence of a given mineral phase implies specific relations between elements or ions. The characterization of the mineral content of decay features is furthermore essential to delineate intervention procedures, namely removal operations (whose success is strongly affected by the solubility of the minerals in a given medium).

In relation to soluble salts due to atmospheric and ground pollution, as well as to the contribution of stony materials, an extensive list (as well as discussion of major sources) can be found in Arnold and Zehnder (1991). Metallic elements also give origin to a rich variety of mineral substances (some examples might be found in Graedel 1987; Roberge 2000) that might form in very short time (see Watanabe et al. 2009).

While some more peculiar minerals might be specifically linked with a few pollution sources, it must be noted that decay-related minerals suffer of the epistemological problem of equifinality that generally affects studies in the natural sciences (see, e.g., Turkington and Paradise 2005). For example, the presence of halite (cubic NaCl) is usually related to seawater influence. However, sodium chloride could result from organic contributions and can be found very far away from seacoast areas. It would be appropriated to quote Arnold (1996) that alerting for the problems related to the origin of soluble salts ended its text with "Chlorides near the sea may not be from the sea." The problem of relation between neoformation minerals and pollution sources is not only affected by the existence of multiple potential sources but also by geochemical differentiation processes (such as are described for natural evaporites in, for example, Risacher and Fritz 2009; regarding the built environment see also Steiger 2003) and the interaction with the substrate.

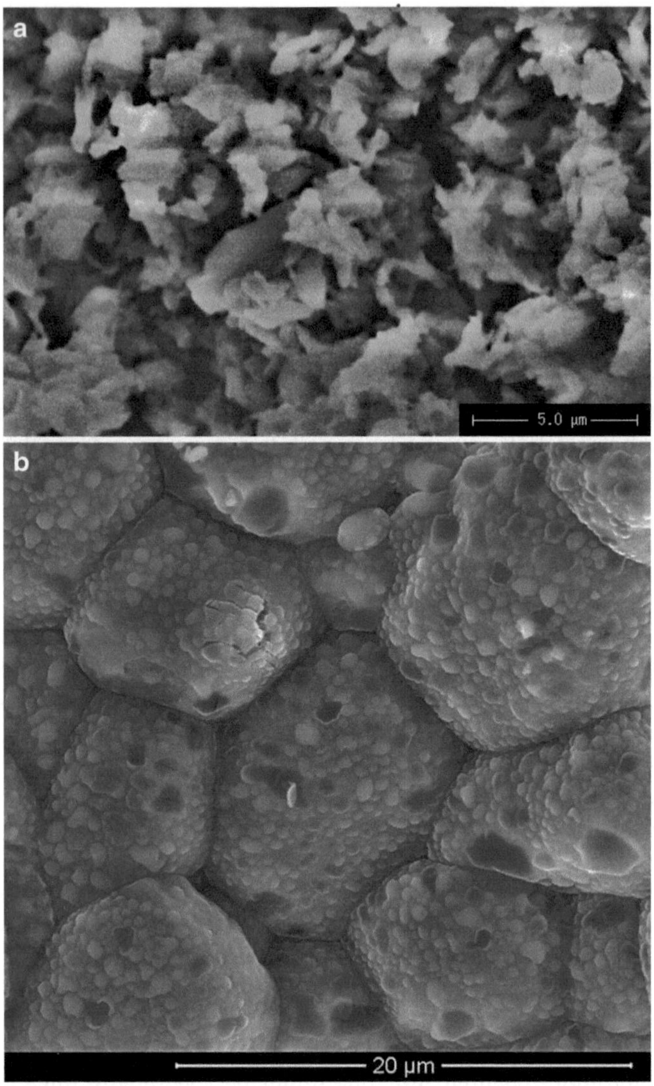


**Fig. 2.10** Scanning electron microscopy observations of soluble salts on efflorescences on modern constructions: (**a**) alkaline sulphates; (**b**) potassium nitrate

In general some associations have been proposed (they are discussed in more detail in Arnold and Zehnder 1991; Arnold 1996), such as potassium nitrate with animal's stables, alkaline sulphates and carbonates with modern building materials and oxalates with organic activity (but in this last case there is also the possibility of being the product of paintings). Examples of scanning electron microscopy observations of soluble salt occurrences on modern constructions are presented in Fig. 2.10 consisting of alkaline sulphates (which, with the alkaline carbonates, constitute the usual minerals in efflorescences found in modern constructions) and potassium nitrate (niter), typical of historical monuments but that can also be found in modern constructions.

Given the problems indicated above, it is useful to map the occurrence of these minerals and its associations (see Arnold and Zehnder 1991) in order to identify patterns of distribution that might help to discuss the possible pollution sources. In Alves and Sequeira Braga (2000), this methodology allowed to separate different salt systems that the authors attributed to human activities, organic sources and building materials as well as to identify the widespread occurrence of gypsum in the stones affected by capillary rising. Antunes et al. (2004) present an example of a more ambitious study where the characterization of efflorescences at the scale of a town suggested to these authors the possible existence of regional trends.

### 2.4.2 *Elements and Ions*

In this section, the use elements and ions will be considered in the assessment of pollution sources (isotopic data will be discussed in the next section "Isotopic Tracers (Stable Isotopes)"). Qualitative studies about the presence of certain elements or ions could give indications about the presence of pollutants (see Sabbioni et al. 2003 on organic ions) and can give relevant information for the typology of decay features (Sanjurjo-Sánchez et al. 2008). The present authors defend that qualitative chemical data regarding presence/absence of substances in decay features, such as those obtained from scanning electron microscopy, can be used for statistical assessment (e.g., contingency tables) of pollution conditions, for example, to compare pollution loads in different façades.

Absolute values of contents of elements and ions have been used to compare decay features and materials (Maravelaki-Kalaitzaki and Biscontin 1999; Valls-del-Barrio et al. 2002; Sabbioni et al. 2003) and to study patterns of distribution of substances in order to discuss the sources of the pollutants and differentiation processes (see, e.g., Wendler et al. 1990; Casal-Porto et al. 1991; Dorn 1998; Maravelaki-Kalaitzaki and Biscontin 1999; Alves and Sequeira Braga 2000; Steiger 2003). Quantification of elements and ions also allows the comparison of contents in different fractions by selective extraction (McAlister et al. 2006).

Enrichment factors are classically used in geochemistry to compare a certain sample with possible sources and evaluate the contribution of these sources. The enrichment factor for a given element or ion compares the concentration of that element or ion in the sample and the concentration of the same element or ion in the possible source (Reimann and Caritat 2000). Such comparison can be normalized by using a reference element or ion that is considered immobile or conservative (Reimann and Caritat 2000). A classical example is the use of chloride or sodium as a conservative ion for the evaluation of seawater and non-seawater fraction in rain composition (Appelo and Postma 2005). However, caution is needed in the use of enrichment factors (see Reimann and Caritat 2000) due to problems arising from the

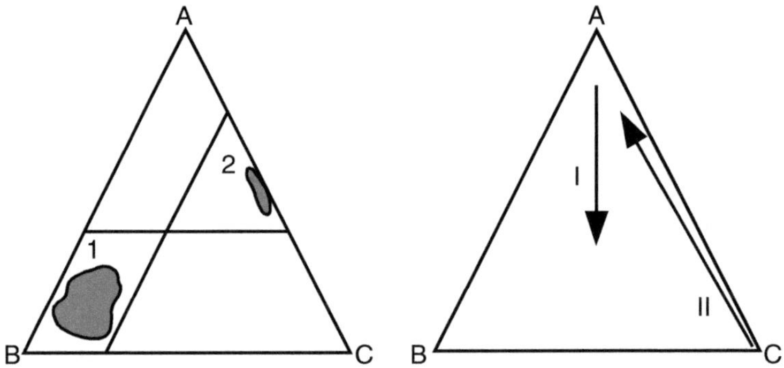


**Fig. 2.11** Schematic representation of the use of ternary diagrams in studies of pollutants effects on building materials: (**a**) comparison of groups of samples (see examples in Derbez and Lefèvre 1996; Begonha and Sequeira Braga 1996; Alves and Sequeira Braga 2000) where "1" represents a set of samples in which constituent "B" is predominant while "2" represents a set of samples with similar proportions of "A" and "C" and "*B*" is the minor component (*lines* could be used for definition of classification fields; see, e.g., Derbez and Lefèvre 1996); (**b**) trends of evolution, where "I" represents depletion in "A" (keeping similar proportions between "B" and "C") and "II" represents enrichment in "A" over "C" while "B" remains a minor constituent (for examples considering evolution with spatial position of samples see Alves and Sequeira Braga 2000; Moreno et al. 2006)

representativeness of the contents considered as reference as well as to the effects of the fractioning processes. In the case of the built environment, there are examples of using enrichment factors to compare chemical characteristics of decay features with possible pollution sources and with the affected substrate in order to distinguish their contributions (Sabbioni 1995; Torfs and Van-Grieken 1997). Enrichment factors have also been used to compare different zones and different depths of decay features (Maravelaki-Kalaitzaki and Biscontin 1999).

Graphical procedures used in the study of both sources and evolution of pollutants include both bidimensional dispersion plots and ternary plots. Dispersion plots allow the comparison of elements or ions considering signatures of pollution sources and mineralogical relations (Begonha and Sequeira Braga 1996; Begonha et al. 1996). They are also useful to study variations in the amounts of elements/ions against a spatial parameter, such as height, in order to study the existence of possible processes of differentiation (Alves and Sequeira Braga 2000; Moreno et al. 2006).

Ternary plots have been applied in the study of pollution of the built environment (see illustration in Fig. 2.11) concerning both atmospheric pollution and salt contamination. Some examples of the use of ternary diagrams in the study of atmospheric pollution include the grouping of atmospheric particles deposited on surfaces (Derbez and Lefèvre 1996) and the study of decay features related to atmospheric pollution (Begonha and Sequeira Braga 1996). In relation to salt pollution, ternary diagrams have been used to discuss pollution sources (Alves and Sequeira Braga 2000; Moreno et al. 2006) as well as to assess evolution with spatial position (horizontal/height) of samples (Alves and Sequeira Braga 2000; Moreno et al. 2006).

### 2.4.3 Isotopic Tracers (Stable Isotopes)

The distribution of the different isotopes of a same element between reacting chemical compounds of coexisting phases is not uniform because of the slightly different geochemical and physical behaviour resulting from differences in mass and energy contents. Such differences result in isotope fractionation. Fractionation is more pronounced in light elements (H, C, N, O and S) and is widely used in environmental research, although fractionation of some heavy elements can also be used (e.g., U).

Stable isotopes are not radioactive. In geochemical systems, as those that occur in building materials, chemical equilibrium is necessary but not sufficient for isotopic equilibrium because isotope exchange reactions can proceed without any apparent change in the distribution of the chemical species (Fritz and Fontes 1980). Stable isotope geochemistry is concerned with the relative portioning of stable isotopes among substances. Differences in fractionation are due to equilibrium and kinetic effects. In general, heavier isotopes form more stable bounds and molecules of different masses that react at different rates. In surface or near-surface environments, many processes are unidirectional and thus isotopic disequilibrium exists between reactants and products (e.g. evaporation). Reaction paths and environmental conditions may be identifiable by measuring partitioning of isotopes between different chemical compounds, because they are more of less reproducible under given environmental conditions (Fritz and Fontes 1980).

Stable isotopes in minerals are especially useful for unravelling the geochemical history of a system. It is difficult to determine accurately the absolute abundance of an isotope in every compound. For most geochemical purposes it suffices to know the relative abundances with respect to a standard value. These relative isotope concentrations can be accurately determined through differential isotope ratio measurements using double-collecting mass spectrometers. This relative difference is called δ-value defined as:

$$\delta \mathrm{x} = \left(\left(R_{\mathrm{x}} - R_{\mathrm{std}}\right) - 1\right) \times 10^{3} \tag{2.1}$$

As example, if an isotope fractionation of a sample has a $\delta = 10‰$ that means that is enriched in such isotope by 10‰ or 1% in relation to the standard. The partitioning of stable isotopes in a reaction such as $A \Leftrightarrow B$, can be quantitatively expressed by a fractionation factor on the basis of isotope ratios:

$$\alpha_{A-B} = R_A / R_B \tag{2.2}$$

Where $R$ is the ratio of the less abundant isotope to the more abundant one, such as e.g., $^{34}S/^{32}S$. In the case of building materials, $A$ and $B$ can be a water solution and a growing mineral. Introducing the δ definition we can re-write the equation in terms of δ values:

$$\alpha_{A-B} = (1000 + \delta_A) / (1000 + \delta_B) \tag{2.3}$$

**Table 2.3** Main internationally accepted standards for light stable isotopes

| Element | Standard acronym | Standard name |
|---|---|---|
| H | V-SMOW | Vienna Standard Mean Ocean Water |
| | NBS-1 | National Bureau of Standards |
| | NBS-1A | (National Bureau of Standards) |
| | SLAP | Standard Light Antarctic Precipitation |
| C | NSB-20 | Olenhofen limestone (National Bureau of Standards) (National Bureau of Standards) |
| | NSB-21 | Spectrographic carbon |
| N | NBS-14 | Nitrogen-gas sample (National Bureau of Standards) |
| O | V-SMOW (for waters) | Vienna Standard Mean Ocean Water |
| | VPDB (for carbonates) | Carbonate standard derived from *Belemnitella Americana* from the Pee Dee Formation of South Carolina (USA) |
| S | VCDT | Vienna Canyon Diablo Triolite |

Most factors of equilibrium isotope fractionation are close to unity, with variations in the third decimal place. If the isotopes are randomly distributed in the compounds $A$ and $B$ then under isotopic equilibrium conditions the fractionation factor ($\alpha$) is related to the equilibrium constant ($K$) by:

$$\alpha = K^{1/n} \tag{2.4}$$

where $n$ is the number of atoms exchanged and $K = \alpha$ in a monoatomic reaction. Thus, for simplicity isotope exchange reactions are usually written such that only one atom is exchanged.

There are internationally accepted standards for the different stable isotopes. Those of hydrogen, carbon, nitrogen, oxygen and sulphur are shown in the Table 2.3.

It is possible to follow the processes of the formation, dissolution or recrystallization of carbonates, nitrates and sulphates, using only isotope variations linked to the chemical reactions involving stable isotopes. The variations in natural isotope abundance are linked to the fact that isotopes with the heavier mass react slightly slower than low-mass isotopes according to the square-root of the ratio of the mass of the two isotopes. Such variations also occur during the decay of some building materials. When solid carbonates, sulphates or nitrates precipitates slowly from a solution exactly at its saturation point, there exist an "isotopic equilibrium" between the isotope concentrations in the solution and the solid, which is dependent only on temperature. Thus, although studies on the stable isotopes of light elements in building materials are scarce, they have strong potential as tracers of the damage caused by diverse pollution sources.

#### 2.4.3.1 Hydrogen

There are variations in the natural abundance of stable hydrogen isotope species ($^1H$ and $^2H$, H and D, or hydrogen and deuterium, respectively) due to the result of fractionation caused by phase transitions, chemical reactions and transport processes (surface evaporation, atmospheric condensation, reevaporation and precipitation). Values of $\delta^{18}O$ and $\delta D$ in water vapour and precipitation (HDO, $H_2{}^{18}O$ and $H_2{}^{16}O$) suffer large fractionations as a result of condensation and precipitation and show regular seasonal variations generally correlated with temperature changes (precipitation during winter is depleted in heavy isotope species). Thus, the water isotope fractionation is a sensitive tracer for diagnosing transport and dehydration mechanisms (Webster and Heymsfield 2003; Jouzel et al. 2000). In addition, fractionation depends on the latitude, altitude, distance from the coast and the amount of precipitation (Dansgaard 1964; Gat 1980; Araguas-Araguas et al. 2000). As an example, progressive depletion of precipitation with increasing distance from the coast has been observed (Rozanski et al. 1993). In continental areas the surface air seems close to isotopic equilibrium with local precipitation. In coastal areas, the situation is more variable (Gat 1980). Excess of deuterium in precipitation provides information on the climatic conditions at the oceanic source region, but water evaporated in continental basins can modify the original signal. This could also occur if the D excess were measured in pore water of building materials or neo-forming minerals due to damage caused by atmospheric pollution.

The atmospheric $H_2$ has an unusually high D content, due to photochemical production from methane and hydrocarbons, and minor emissions coming from volcanoes, microbial nitrogen fixation, fossil fuel and biomass burning. Fractionation has provided information on sources and sinks (Rahn et al. 2002; Ehhalt and Rohrer 2009). Isotopic measurements in polluted air of urban areas have reported depleted $\delta D$ values from −180‰ to −340‰ (Gerst and Quay 2001; Rahn et al. 2002; Ehhalt and Rohrer 2009), although measurements in winter suggested cleaner air. An enriched $\delta D$ global average of 130‰ has been proposed by Gerst and Quay (2000). Measurements on fossil fuel showed $\delta D$ values between −196‰ and −270‰ (Ehhalt and Rohrer 2009). Automobile exhaust is believed to dominate anthropogenic $H_2$ emissions. According to Vollmer et al. (2010), the isotopic signatures are controlled by a temperature-dependent isotope equilibration between $H_2$ and $H_2O$. A similar process was suggested for the oxygen isotopes in $CO_2$ and $H_2O$ in car exhaust (Affek and Eiler 2006), and for $H_2$–$H_2O$ in high temperature steam reforming and low-temperature photobiological processes (Rahn et al. 2002).

Variations of D in atmospheric $H_2$ and $H_2O$ occur in the air and solutions but remain unchanged in minerals. This is very important to use H/D as tracers in building materials. Oxygen and hydrogen isotope fractionations occur between the water of crystallization in hydrated salts and the solutions from which they were formed (Savin 1980). For most minerals, the isotopic composition of the water depends almost exclusively on that of the solution from which the salt crystallized. Mineral phases can be formed and remain out of equilibrium with the environment for long periods of time Thus, $\delta^{18}O$ and $\delta D$ values of the crystallization water in salts formed in buildings will allow inferences about the waters from which they have precipitated. The hydration water of gypsum is enriched, relative to the solution from which it crystallizes, by about 4‰ in $^{18}O$ and is depleted in D by 15‰ (Jouzel et al. 2000). Such fractionations do not vary with temperature. The H/D fractionation is very useful for carbonate and sulphate minerals but also for siliceous compounds (including opal-A) that are very frequent in building materials as neo-forming mineral growths due to damage caused by atmospheric pollution (Savin 1980).

Although measurements of H/D signatures in rainwater, water vapour and $H_2$ exist and have been used in geochemical studies, data are not found on building materials. They could be useful as tracers of pollutants and salts that damage building materials as they should provide important information of the sources of solutions carrying pollutants.

#### 2.4.3.2 Carbon

Carbon can be found in the carbon cycle of the Earth as two stable isotopes: $^{12}C$ and $^{13}C$. The isotopic $\delta^{13}C$ value in atmospheric $CO_2$ is estimated in −7‰ or −8‰ (Deines 1980; Brenninkmeijer et al. 1999). The concentration of $CO_2$ in air is negatively correlated with the $^{13}C$ ratio. Based on anthropogenic emissions a decline in the $\delta^{13}C$ of atmospheric $CO_2$ by 2.6% since the last century is expected but analyses of wood samples indicate much smaller change (Deines 1980). In addition, $CH_4$ oxidation may lead to the formation of CO as an intermediate product, hence, the $CH_4$, CO and $H_2$ cycles appear to be closely related.

Results of carbon isotopes in $CO_2$ emissions produced by different anthropogenic sources in urban environments have been reported in the literature. Deines (1980) remarked that $CO_2$ from the combustion of fossil fuel have an isotopic composition between −25‰ and −30‰and $CO_2$ from road traffic ranges from: −27.6‰ to −29.3‰. Other combustible sources show ranges between −24.6‰ and −25.6‰ for coal, −29.1‰ to −29.5‰ for fuel oil and −37.7‰ to −40.5‰ for natural gas. In general, depleted $\delta^{13}C$ values of anthropogenically emitted $CO_2$ vary widely from −40.5‰ to −24.6‰ and urban air values range from −8.6‰ to −24.4‰ (Widory and Javoy 2003). The combination of $CO_2$ concentration and $\delta^{13}C$ measurements of $CO_2$ in air allow the identification of the main sources in urban, rural and marine settings. The $\delta^{13}C$ from pollution varies due to mixture of fossil fuel emissions and clean air $CO_2$. The main pollutant $CO_2$ source are vehicle exhausts, with diesel exhausts showing much lower $CO_2$ concentration than gasoline exhausts (Widory and Javoy 2003).

CO is also a good indicator of the effects of human activity on the atmosphere since fossil fuel combustion is one of its sources, but its isotope ratio is modified during by reaction with OH radicals. The magnitude of the isotopic fractionation depends on pressure, altitude (removal of CO in the upper troposphere is due to the lower pressure) and diffusion. In recent years, several data have been reported on isotopic composition of CO in the exhaust of automobiles. Estimated averages of isotopic $\delta^{13}C$ of CO emitted vary: $-23.8 \pm 0.8$‰ for gasoline engines and $-19.5 \pm 0.7$‰ for diesel engines(Tsunogai et al. 2003). A good linear correlation has been observed for the CO content and the isotopic compositions in road and urban atmosphere, irrespective of their sampling points and time (Stevens et al. 1972; Kato et al. 1999; Tsunogai et al. 2003). Other anthropogenic combustion sources of CO show different

$\delta^{13}C$ values, such as bituminous coal that ranges from −22‰ to −27‰ (Elswick et al. 2007). Assuming steady state, anthropogenic CO from fossil fuel burning will be enriched in $^{13}C$ during its atmospheric lifetime. The $\delta^{13}C$ values in regional studies have showed that local sources contribute CO that is about 5‰ enriched (Brenninkmeijer et al. 1999). Air samples collected from cities showed that $\delta^{13}C$ variation exceeded that for $\delta^{18}O$. This suggests that $\delta^{13}C$ values for CO in each urban area reflected the variation in the predominant source of petroleum (Stevens et al. 1972; Conny 1998).

The isotopic fractionation of gaseous carbon oxides will be recorded in neoforming minerals in buildings, as gases are dissolved and incorporated to the building environment. Chemical interactions at or near the solid-solution interface have major influence on the behaviour of carbonate minerals. Capture of inorganic carbon occurs in carbonates, mainly $CaCO_3$. These processes are affected by pH, temperature and the presence of other ions in solution. Some substances that are frequent in air pollution are potential inhibitors of the precipitation of calcium carbonate, as occurs with sulphur compounds, such as several metallic ions and organic compounds (Morse and Mackenzie 1990). The carbon isotopic composition of carbonates, dissolved carbonic acid species, methane and organic matter exhibit considerable variation. In such compounds, the degree of fractionation also depends on the effect of the diffusion in a solid media. Brenninkmeijer et al. (1999) asseverate that because most CO and $CO_2$ are absorbed in the top few cm of soils or building materials, diffusion effects may be suppressed in some cases (e.g. surface coatings). However, several studies reveal that as carbon crosses the gas-water interface a kinetic isotope fractionation takes place, caused by the different diffusion velocities of the light and heavy $CO_2$ molecules since lighter isotopes diffuse more quickly (Létolle et al. 1988; MacLeod et al. 1991).

In buildings, the ultimate carbon source of surface carbonation in concretes and mortars is atmospheric $CO_2$, which is dissolved into infiltrating water as meteoric rainwater, groundwater, or both. Thus, $^{13}C$ from carbonates can be used as tracers of primary, secondary of further carbonation processes. In building materials, carbonate ions are in equilibrium with $CaCO_3$. Lime mortars are composed of both an aggregate (usually sand) and a $CaCO_3$ binder, formed due to reaction of $Ca(OH)_2$. Neoformation of calcium carbonate can occurs when high air $CO_2$ concentrations exist and $^{13}C$ can be used as a tracer.

Strong fluctuations in the isotopic fractionation from the atmospheric value are evident in mortars. From experimental studies it has been found that $^{13}C$ fractionation during the formation of $CaCO_3$ varies considerably over the first 48 h of exposure. The absorbed $CO_2$ measured after 48 h decreases with increasing weight of CaO (Pachiaudi et al. 1986). However, considerable variations in the absorption of atmospheric $CO_2$ are probably due to uncontrollable atmospheric effects (weather conditions). $CaCO_3$ from lumps in the binder (originated from incomplete combustion of $CaCO_3$ to obtain CaO) must be considered separately, as its $\delta^{13}C$ value corresponds to the limestone used to obtain the lime mortar. Pachiaudi et al. (1986) calculated a value close to −21‰ for mortars that set in air containing $CO_2$ of $\delta^{13}C = -7$‰. In the case of concrete and Portland cement, carbonates can be studied in a similar way to what is done with carbonate deposits in nature. Freshwater travertines produced by the interaction of $CO_2$ with hyper-alkaline waters are geochemically comparable to calcites produced in the concrete environment. The formation of calcites in alkaline freshwater solutions involves a kinetic carbon isotope fractionation between calcite and $CO_2$.

One of the causes of concrete decay is carbonate mineral growth from water flowing across and through concrete structures as dissolved gaseous carbon react with portlandite (Lea 1970; Pye and Schiavon 1989; MacLeod et al. 1990). This decarbonation process is important since it has consequences for the mechanical properties of concrete (e.g. alteration of steel reinforcements). Such processes have been successfully studied by stable C and O isotopes. The pH values of the pore waters of concrete are thought to be in the range 12–14 (Glasser 1986), caused by the amount of dissolved portlandite ($Ca(OH)_2$) which is a major component of the cement paste within concrete (Lea 1970). At high pH-values (>10.3) the dominant specie is the carbonate ion and deposition of calcite is instantaneous. In addition, calcite crusts can be formed due to the reaction of portlandite coatings (of the concrete surface) with carbon dioxide in solution (Lea 1970; MacLeod et al. 1991). The mechanism of crusts or efflorescent carbonate growths has been related to leaching of portlandite due to moisture migration through the concrete. The $\delta^{13}C$ from calcite growths in concrete will indicate the source of carbon. Assuming the $\delta^{13}C$ value of −7‰ in the atmosphere, a range of $\delta^{13}C$ values in the concrete can be calculated (Macleod et al. 1991; Iden and Hagelia 2003). However, measured values fall outside the calculated fractionation ranges. Craig (1954) suggested that further kinetic fractionation could possibly result from secondary diffusive processes. The earlier models were based on the migration of moisture through the structure depositing calcite on the surface (Lea 1970). It has been suggested that pH may have chemically enhanced the fractionation process (Létolle et al. 1988; Macleod et al. 1991) and a certain depletion of $^{13}C$ will occur in the reaction product. Létolle et al. (1992) and Rafai et al. (1991, 1992), proposed that a considerable quantity of atmospheric $CO_2$ is trapped by $Ca(OH)_2$ to form $CaCO_3$ and by calcium silicates to give a complex $CSH–CO_3^{-2}$ combination, during the concrete strengthening. Such minerals show very low $\delta^{13}C$ and $\delta^{18}O$ and a high pH zone is created in contact with water saturated $CO_2$. In a second step, $CO_2$ will be slowly released giving further calcite, while $Ca^{+2}$ is released by the same hydrolysis process and a new calcite is formed with a different isotopic composition. As time goes by, calcite shows smaller $\delta^{13}C$ due to the slow diffusion of $CO_2$ in the water-filled porosity of cement. When a wave of carbonation from the periphery to the depth of the concrete occurs and is followed by decarbonation, very low isotope ratios can be measured at the carbonation front. In the outmost part of concrete structures, the $\delta^{13}C$ will be depleted compared to inner parts. However, interactions between water molecules and dissolved substances will reduce this effect. Kinetic fractionation will also result in $^{13}C$ depletion of the products, but values in the literature are highly variable (Iden and Hagelia 2003).

Dietzel (2000) described two major ways of calcite sinter formation in concretes: by absorption of $CO_2$ in alkaline calcium hydroxide solutions within the concrete, or by deposition from ground water with dissolved $CaCO_3$. The first reaction will result in $\delta^{13}C$ values of about −25‰, and the second in values about −13‰.

Intermediate values indicate overlap of the two processes. However, Iden and Hagelia (2003) suggest that the dominating reaction that causes the growth of the calcite crusts on concrete surfaces open to the atmosphere is the interaction of hyper-basic fluids with $CO_2$. The rate of growth of calcite within the cement paste itself is increased when water percolates into the structure and $CO_2$ enters into the concrete to be fixed as calcite. There are no similar studies of stable carbon isotopes on the carbonation and decarbonation of other materials (such as Ca-rich rocks) but the processes related above must be considered in future studies of $\delta^{13}C$ as a tracer in building materials.

Other studies have used $\delta^{13}C$ to assess stone decay in Heritage buildings. Aberg et al. (1995) studied the decay of carbonaceous sandstones in buildings of Sweden, exposed to outdoor air for several decades to centuries. They found a lowering in the $\delta^{13}C$ values at or close to the surface of the rocks, while the overall $\delta^{13}C$ value in the unweathered sandstones was homogeneous ($\delta^{13}C$ values of +3–+4‰). They suggested that such lowering was caused by an isotopic exchange between carbon in the atmospheric $CO_2$ and carbon in calcite during sulphation processes affecting the rock.

#### 2.4.3.3 Nitrogen

Two stable isotopes of nitrogen ($^{14}N$ and $^{15}N$) exist. Air is used as the $\delta^{15}N$ standard ($\delta^{15}N=0‰$) because of the constant average abundance of $^{15}N$ (0.366%). The $\delta^{15}N$ values of atmospheric gaseous nitrogen usually range between −15‰ and +15‰ (Kendall 1998) depending on the nitrogen compound: $N_2$ commonly ranges between −1‰ and 1‰, $NH_4$ between −3‰ and 10‰, and $NO_3$ between −10‰ and 8‰ (Létolle 1980).

Anthropogenic sources include chemical processing and combustion of fossil fuels in automobiles and power plants, and they show extremely low $\delta^{15}N$ values. Isotope shifts can occur due to selective rainwater washout of N-rich materials (Heaton 1987) and the total range observed at any single location can be as large as 20‰. In general, the $NO_3^-$ in rain appears to have a higher $\delta^{15}N$ value than the co-existing $NH_4$. Equilibrium exchange reaction between gaseous NO or $NO_2$ and dissolved $NO_3^-$ could explain the resulting $^{15}N$ enrichment of the $NO_3^-$. However, other studies have illustrated various complicated relations (Moore 1977; Heaton 1987) and there is considerable inter-storm, site-specific and seasonal variability.

Special attention has been given to nitrogen oxides in polluted air because of their contributions to acid rain. Values of $\delta^{18}O$ and $\delta^{15}N$ of nitrate may allow better separation of different anthropogenic sources of gaseous nitrogen in the atmosphere (Kendall et al. 1995; Kendall 1998; Widory 2007). Measurements of the $\delta^{15}N$ of anthropogenic sources of gaseous $NO_x$ (Heaton 1990) provided values from −11‰ to −2‰ for gasoline vehicles, from −13‰ to −2‰ for diesel vehicles, and from 6‰ to 13‰ for coal-fired power stations.

Regarding atmospheric particles, few studies have used nitrogen isotope though preliminary results showed they could be a potential tracer of their origin (Widory 2007). Nitrogen in atmospheric particulate matter mainly occurs as nitrate and ammonium

(Seinfeld 1986; Seinfeld and Pandis 1997). It is known that aerosol nitrogen species in Europe are mainly anthropogenic, arising from the oxidation of gaseous $NO_x$ produced in high-temperature combustion process due to car engines and industrial activities. General data of $\delta^{15}N$ of atmospheric particles from pollution sources provided by Widory (2007) show values of 4.6‰ for unleaded gasoline, 3.9–5.4‰ for diesel combustion, 3.0–15.4‰ for natural gas, −19.4–2.9‰ from fuel oil, and 5.5–8.0‰ for waste incinerators. For coal burning, values depend on the type of coal: 1.9‰ for bituminous coal and 2.7–5.35‰ for anthracites. Seasonal variations have been observed and explained as resulting from negative fractionation in the $NO_2$/OH reaction in summer compared to winter by the daytime and nighttime removal reactions of $NO_x$ (Freyer et al. 1993).

Although isotopic data of $\delta^{15}N$ fractionation exist in air, water, soils, sediments and rocks for studies of pollution there is a lack of research on fractionation of N stable isotopes in building materials. Values of $\delta^{15}N$ combined with heavy metal concentrations in mosses growing in walls and roofs have been used to monitor traffic exposure (Pearson et al. 2000). Fractionation of $\delta^{15}N$ has been successfully used to study the decay of wood in ancient tombs (Filley et al. 2001). Different physicochemical processes such as diffusion through porous and aqueous media, dissolution of molecular nitrogen in water, evaporation of volatile nitrogen compounds and ion exchange, are the principal inorganic process responsible of isotope fractionating. The diffusion of nitrogen in porous materials must be considered in the of study $\delta^{15}N$ fractionation in building materials. The diffusion of molecular nitrogen in water is also important as it will occur in the porous system of building materials.

#### 2.4.3.4 Oxygen

Although three stable oxygen isotopes exist ($^{16}O$, $^{17}O$ and $^{18}O$) only the lighter and the heavier are commonly used as pollution tracers. Variable $\delta^{18}O$ values in the atmosphere can be found in $H_2O$, $O_2$, $CO_2$, CO, $NO_x$ and $SO_2$ in the atmosphere.

In continental areas the surface air seems to be close to isotopic equilibrium with local precipitation, while in coastal areas variable situations can be observed (as marine air is not at isotopic equilibrium with surface waters). Variation of de $\delta^{18}O$ with altitude (Gat 1980), due to the lowering of temperature with increasing elevation, leads to enhanced condensation and therefore to a progressive depletion in heavy isotopes of precipitation. The stable isotope composition of precipitation samples has been intensely studied in the world. As a rule they are depleted in the heavy isotope and can be correlated with the temperature changes of precipitation during rains. Values of $\delta^{18}O$ and $\delta D$ in precipitation are linearly correlated. However, the $\delta D$ vs. $\delta^{18}O$ relationship for precipitation in any region often differs from the global equation (Gat 1980). Latitudinal variations also exist in precipitation (Araguas-Araguas et al. 2000).

The $\delta^{18}O$ of atmospheric $CO_2$ is influenced by several processes (Friedli et al. 1987) such as: exchange and isotopic equilibration with cloud droplets, exchange with the ocean, with leaves of plants, soil respiration, and anthropogenic emissions.

As important anthropogenic emissions of CO exist in urban environments, oxygen stable isotopes of CO from combustion sources have been more studied than $CO_2$. CO is enriched in $^{18}O$ (Brenninkmeijer et al. 1999). Brenninkmeijer and Röckmann (1997) and Huff and Thiemens (1998) reported large $\delta^{18}O$ variations, while Kato et al. (1999) reported homogeneous values (+22.3‰) in exhaust from gasoline automobiles. Huff and Thiemens (1998) attribute the $\delta^{18}O$ variation to the role of catalytic converters in recent engines. However, Brenninkmeijer and Röckmann (1997) found $\delta^{18}O$ variations (from +6‰ to +19.5‰) in a non-catalyst automobile. In addition, a linear correlation between $\delta^{13}C$ and $\delta^{18}O$ measured in CO is observed for gasoline automobiles but not for the old, non-catalyst automobiles. An average isotopic composition of CO from automobiles was calculated as $+23.7 \pm 1.8$‰ (Tsunogai et al. 2003).

$\delta^{18}O$ in CO has two major causes of variability (i.e. combustion sources and kinetic fractionation). However, $\delta^{18}O$ values of CO from methane and hydrocarbon oxidation are not well known (Brenninkmeijer et al. 1999). Important variations of $\delta^{18}O$ values are observed in non-urban air when compared with other isotope fractionation values (such as $\delta^{13}C$), while air samples collected from cities do not show such variability (Conny 1998; Stevens et al. 1972). Stevens et al. (1972) suggested that seasonal variation in $\delta^{13}C$ and $\delta^{18}O$ values were driven by variation in anthropogenic CO sources. Therefore, there is a dual oxygen/carbon isotopic signature for local CO sources in some cases.

$\delta^{18}O$ values have been studied in other atmospheric gaseous compounds. Although no much data exist, it is known that the $^{18}O$ of nitrous oxide emitted from anthropogenic sources is variable. Nitrous oxide in the flue gas during the staged combustion of coal became depleted in $^{18}O$, although enriched relative to the nitrous oxide in the background air (Ogawa and Yoshida 2005). Oxygen isotope ratios have proved useful for distinguishing $N_2O$ formation pathways (Wahlen and Yoshinari 1985). The fractionation of O isotopes in atmospheric $SO_2$ is also variable. The isotopic $\delta^{18}O$ of rainfall sulphate ranges between +8‰ and +14‰, with a mean of +10.2‰, close to the value of seawater sulphate +9.6‰. Seasonal variations have also been observed, parallel to seasonal $\delta^{18}O$ changes in the rainwater (Pearson and Rightmire 1980). Anthropogenic emissions of particulate and gaseous sulphur compounds have a wide range of isotopic S and O compositions, depending into the source of fuel and temperature of combustion. During the oxidation of sulphide to sulphate fractionation occurs in two steps. When sulphide is produced, molecular $^{16}O$ is consumed in preference to $^{18}O$ with a kinetic fractionation of 8.7‰. Sulphite oxidation (second step) is catalyzed by trace amounts of transition metals (Pearson and Rightmire 1980) and the isotopic fractionation occurred in this step must be considered when O isotopes are used as tracers in building materials.

Equilibrium isotope fractionation of oxygen between minerals and water has been studied by experimental and field measurements. It can be important in salts and soluble minerals (e.g. carbonates and sulphates) but is usually low between water and silicate minerals. Oxygen fractionations occur in hydrated salts and the solutions from which they formed, depending on the crystallographic sites occupied by the water, as occurs with $^{1}H/^{2}H$.

Values of $\delta^{18}O$ reported from common building materials have mainly been measured in stony materials, such as concrete and mineral growths due to concrete decay. They have been useful for the identification of various phenomena occurring during concrete carbonation, as with $\delta^{13}C$. However, the use of oxygen isotopic features to identify the pathways of pollutants through concrete is more complicated, since $CO_2$ readily exchanges isotopes with carbonate ions in water. After a few hours of concrete setting isotopic equilibrium is reached and dissolved $CO_2$ is richer in $^{18}O$ by 41% relative to the water medium at 20–25°C (Rafai et al. 1991, 1992). If dissolved $CO_2$ has not equilibrated with $H_2O$ before it contacts the high pH zone, its $^{18}O$ content is much less than in the equilibrium case, and the $^{18}O$ content of precipitated carbonate will be much less (Rafai et al. 1991, 1992; Létolle et al. 1992). Kinetic fractionation of $^{18}O$ occurs during precipitation, but the fractionation factor is unknown (Rafai et al. 1991, 1992; Létolle et al. 1992). Létolle et al. (1992) and Rafai et al. (1992) proposed very low $\delta^{18}O$ for the complex combinations of calcium silicates and calcite, the last one formed as a result of reaction of portlandite with atmospheric $CO_2$ during the concrete strengthening. As time goes by, calcite shows an important shift in $^{18}O$ due to the slow diffusion of $CO_2$ in the water-filled porosity of cement.

Macleod et al. (1991) measured $\delta^{18}O$ values in calcite growths in damaged concrete. Values ranging from +8‰ to +17‰ suggested re-equilibration after or during initial precipitation. Iden and Hagelia (2003) measured fractionation of oxygen isotopes in different minerals of concrete structures, such as thaumasite, calcite and calcium silicate hydrates. They found high variable $\delta^{18}O$ values specific for the porous reaction zone (available for influxes of diverse origins), extending to atmospheric values. Oxygen fractionation values have also been measured in gypsum crystals that growth in stony materials (e.g. mortars, plasters and rocks). Values of $\delta^{18}O$ in anhydrite and gypsum from marine deposits of a single age generally fall within a fairly narrow range (Savin 1980). Vallet et al. (2006) showed that $\delta^{18}O$ values are low for original sulphur-rich materials (rock pyrite, gypsum plasters and mortars) and that gypsum from mortars showed $\delta^{18}O$ values ranging from +15.8‰ to +20.8‰. However, $\delta^{18}O$ values are lower for gypsum crystals formed as a result of the alteration of materials (sulphated calcite, weathering of pyrite), ranging from +8‰ to +16‰. Thus, the use of O stable isotopes shows optimistic future prospects to assess the sources of pollutants and the pathways of carrier solutions in the building environment.

#### 2.4.3.5 Sulphur

Sulphur has four stable isotopes: $^{32}S$, $^{33}S$, $^{34}S$ and $^{36}S$, with approximate terrestrial abundances of 95.02, 0.75, 4.21 and 0.02%, respectively. The S isotopic fractionations of sulphates and sulphites are controlled by reaction kinetics of the $H_2O$ and $H_2S$ exchange, although other less abundant species can be important. Anthropogenic emissions of gaseous and particulate sulphur compounds come from coal and fuel combustion in power plants and refining processes of crude oil (Gaffney and Marley 2009).

Sulphur isotopes in the atmosphere show important variations depending on the region and time of year. Values of $\delta^{34}S$ in gaseous and particulate $H_2S$, $SO_2$ and organic S range from −20‰ to +30‰ in the air (Krouse 1980). The fractionation of S isotopes should provide a good measure of the extent of $SO_2$ oxidation with distance from the emission source. Diffusion of $SO_2$ results in a small decrease in $^{34}S$. In coal-fired plumes the oxidation exceed 5% for distances up to 50 km, while in oil-fired plumes, conversions of up to 13% have been found over shorter distances (Krouse 1980).

In general, when isotopic fractionation is measured in sulphate minerals (e.g. gypsum, anhydrite, thaumasite), the heavier isotope is enriched in the higher oxidation state. Thus, under equilibrium conditions sulphate minerals are enriched in the heavy isotope relative to disulfide minerals, which in turn are enriched relative to monosulphide minerals (Krouse 1980; Seal et al. 2000). Exchange rates also depend on pH and temperature. As a rule, high temperature isotopic signals are preserved during the cooling process and stable isotopes commonly record environmental conditions such as the temperature and the fluid composition. By contrast, when low temperature and more basic pH exist, disequilibrium conditions remain (Seal et al. 2000).

Attempts to relate the origin of sulphur compounds in building materials to different sources of sulphur oxides have been made by using S stable isotope fractionation (Longinelli and Bartelloni 1978; Pye and Schiavon 1989; Rösch and Schwarz 1993; Torfs et al. 1997; Nord et al. 2001; Klemm and Siedel 2002; Iden and Hagelia 2003; Přikryl et al. 2004; Vilela et al. 2005; Vallet et al. 2006; Hosono et al. 2006; Schleicher and Recio-Hernández 2010). Such studies have provided valuable information about the decay mechanism and pollution sources of sulphur in building materials. Sulfur isotopes have been the most studied stable isotopes in the built environment, a trend that is related to the frequent presence of sulphur in decay features.

Studies on concrete have shown that one of the causes of decay is sulphate and carbonate mineral growth from water flowing across and through concrete structures (Pye and Schiavon 1989; Macleod et al. 1991) as dissolved gaseous carbon and sulphates react with portlandite. In addition, gypsum crusts can be formed due to the reaction of portlandite with sulphur dioxide from solutions. Studies of S isotopes in concrete revealed that thaumasite is due not only to $SO_2$ resulting from to air pollution but also to the possible contribution from oxidation of sulphides without excluding possible organic contributions (Iden and Hagelia 2003).

Several studies of $\delta^{34}S$ have shown this to be an interesting tracer for the study of alteration features such as black crusts in masonry buildings (Longinelli and Bartelloni 1978; Buzek and Srámek 1985; Pye and Schiavon 1989; Torfs et al. 1996, 1997; Siedel 2000; Vallet et al. 2006; Vilela et al. 2005). Other degradation patterns like erosive features have been less investigated. Due to the difficulties in the characterization of the isotopic signature of sources (that might mix in variable proportions in unknown periods of time), Vilela et al. (2005) essayed the comparison of samples of decay features that field studies indicated were related to different sources. These authors found differences in the $\delta^{34}S$ values of sulphates between samples related to atmospheric processes and samples related to capillary rising solutions.

## 2.4.4 Radioactive Isotopes

Some light radioactive isotopes are also interesting as tracers, such as occurs with $^{3}H$ (tritium) and $^{14}C$ (radiocarbon). We show below some additional information about the sources and migration of pollutants in the building environment. Future use of radioactive tracers for the study of the effects of air pollution in building materials will be possible due to the increased knowledge on the behaviour of such isotopes in the atmosphere and in other geochemical systems.

### 2.4.4.1 Tritium

Tritium is a radioactive isotope of H ($^{3}H$) with a half-life of 12.26 years. All three isotopes of H occur naturally in atmospheric $H_2$. Atmospheric $H_2$ contains tritium in very low ratios being often expressed in Tritium Units, TU, where 1 TU corresponds to a T/H ratio of $10^{-18}$ (Ehhalt and Rohrer 2009). Tritium is naturally produced due to the interaction of cosmic radiation with the atmosphere. However, anthropogenic tritium has overshadowed the natural production by orders of magnitude (Happell et al. 2004) due to release resulting from thermonuclear bombs in the 1950s and emissions from the nuclear power industry beginning in the 1970s. Early investigations of tritium produced during atmospheric nuclear testing suggested that D/H ratios of atmospheric $H_2$ in clean air are enriched relative to VSMOW standard by approximately 70–180‰ (Rahn et al. 2002).

In the early years of research on hydrogen isotopes in the atmosphere, tritium was the isotope of major interest, because of its enhanced presence in the atmosphere as a result of anthropogenic emissions (Matsuoka et al. 1994; Araguas-Araguas et al. 2000). Vertical and latitudinal gradients, and variations form coastal to continental regions of tritium have been observed and seemed to vary year-to-year and with location (Mason 1977; Gat 1980; Ehhalt and Rohrer 2009). Concentrations are higher at higher latitudes, with values lower by a factor of 5 or so at low-latitude and tropical areas (Athavale et al. 1967; Gat 1980).

Tritium has been very useful as a water tracer in the hydrologic cycle (Gat 1980).

Tritium levels have also been studied in Municipal Solid Waste landfills. The propensity of tritium to substitute readily the hydrogen in both inorganic and organic molecules in waste landfills suggests that tritium should also be present in the methane and other hydrocarbons found in landfill gas (Coleman, et al. 1993). More recently, attention has been focused on gaseous tritium lighting devices (used in buildings, ships, and aircraft) as likely being responsible of tritium production (Mutch et al. 2007). Thus, water migrations in the built environment should be traced by measuring tritium in solutions, hygroscopic and hydration water of mineral growths and crusts. Tritium geophysics is controlled by the timing, location and intensity of exchange of air masses with other atmospheric components such as ozone. Thus, it should be a useful tracer to understand wet deposition of atmospheric pollutants on building façades, the rising of underground water in buildings and the different effect and penetration of water within building materials of different solubility and porosity. Despite such potential applications, examples of the use of tritium as a tracer in the built environment have not been found.

#### 2.4.4.2 Radiocarbon

The radioactive isotope of carbon $^{14}C$ is continuously being produced in the Earth's atmosphere, by nuclear reaction between the secondary cosmic ray neutrons and nitrogen. Changes in radiocarbon concentrations in the atmosphere can be attributed to natural and anthropogenic causes. Anthropogenic sources of gaseous carbon can have either a global or a local range. Human activity can lead to an increase as well as decrease of $^{14}C$ concentration in the environment. After 1945, nuclear weapons testing caused increases in the concentrations of radiocarbon in the atmosphere, but radiocarbon concentrations have been decreasing since 1963, due to the cessation of atmospheric nuclear weapons testing and the transfer of radiocarbon into the deep, terrestrial biosphere and dilution by fossil fuel $CO_2$ (Mook 1980; Pazdur et al. 2007).

Other anthropogenic-related changes of radiocarbon concentration are caused by the Suess effect. This is due to the increase of mining and burning of fossil fuels in industrial areas since the nineteenth century that caused the emission of $^{14}C$ depleted $CO_2$ to the atmosphere and changes of carbon isotopic composition in both the atmosphere and other carbon reservoirs (Levin and Hesshaimer 2000; Pazdur et al. 2007; Rakowski et al. 2001, 2005). It is possible to determine the local Suess effect in polluted areas (Pawelczyk and Pazdur 2004). It can be assumed that lower values of $\delta^{13}C$ in organic material (e.g. some plants, three rings) are connected with lower values of $^{14}C$ and with the concentration of $CO_2$ in air at industrial areas. Several investigators have tried to establish an atmospheric record of this fraction. The $^{14}C$ depletion in the first half of the twentieth century has been shown to be larger in more polluted areas, due to extra $CO_2$ from the burning of fossil fuels detectable as $^{14}C$ depletion relative to background levels. Such effect is important in investigations of the degree of atmosphere contamination, air mass migration, and spread of the contamination (Levin and Hesshaimer 2000).

Levels of $^{14}C$ in atmospheric CO are much higher than in $CH_4$ and $CO_2$. Thus, CO is considered as a very useful tracer of fossil fuel sources, as low $^{14}C$ can be measured in CO in polluted areas (Brenninkmeijer et al. 1999). The $^{14}CO$ concentration in air is dependent upon both the altitude and latitude due to the cosmogenic production of $^{14}C$. Seasonal variations have been observed due to variations of $OH^-$. Since CO originated from fossil fuel combustion contains virtually no $^{14}C$, $^{14}CO$ is a useful tracer for the anthropogenic CO sources and $^{14}CO$ has been used to estimate the strength of the non-fossil source of CO in urban environments (Conny 1998).

Radiocarbon can be a useful tracer to study the effects of pollution sources on building materials. However, such kind of studies have not yet performed or considered. Information on the effect of $^{14}CO_2$ on organic materials and minerals, usually deposited or formed on building materials can be useful to perform such kind of studies. Levels of $^{14}C$ measured in dissolved $CaCO_3$ has been used as a tracer in groundwater studies. In addition, it can be used to date neomineralizations such as speleothems. Thus, it could be useful as a tracer in building materials containing $CaCO_3$ or other carbonates such as calcareous rocks, lime mortars and concrete (Mook 1980). Measurements of $^{14}C$ in $CaCO_3$ from lime mortars have been performed to date historical buildings (Heinemeier et al. 1997). From such studies some information was obtained on the decarbonation/carbonation process due to the ageing of the lime mortars.

#### 2.4.4.3 Radioactive Isotopes of Heavy Elements

The levels and patterns of atmospheric radioactivity are strongly related to the geographical location. The natural radioactivity observed at a given setting will be related to the area of land surrounding the site, the distance to the sea, the varying speed and directions of the winds, etc. The interaction of such parameters with removal and dilution determine the local radioactive composition of the air (Lockhart 1962). The use of radioactive isotopes as tracers of urban pollution for the study of building materials decay has not been tried, but the present authors propose that they could be useful tracers of these processes. The deposition of such radioactive isotopes on the surface of damaged materials could provide interesting information on the effect of air pollution on decay. If such approach is considered, it must be taken into account that building materials can also contain radioactive isotopes. All building materials contain varying amounts of natural radioactive nuclides: materials derived from rock and soil mainly contain natural $^{40}K$ and radionuclides of the $^{238}U$ and $^{232}Th$ series (Stojanovska et al. 2010; Veiga et al. 2006).

In terms of natural radioactivity, granite rocks used in buildings exhibit an enhanced elemental concentration of uranium (U) and thorium (Th) compared to the very low abundance of these elements in other rock types (Tzortzis et al. 2003). Lower radiation levels are associated with sedimentary rocks such as sand, gravel, gypsum, etc. (Sorantin and Steger 1984; Hizem et al. 2005). However, there are exceptions like some shale and phosphate rocks, which have relatively high content of radionuclides. Some other materials, such as mortars, contain measurable contents of radioactive isotopes of the series of $^{238}U$, $^{232}Th$ and $^{40}K$. In the case of the $^{238}U$ series, the contribution of the radionuclides in the first half of the series between $^{238}U$ and $^{232}Th$ is negligible relative to the second half comprising $^{226}Ra$ to $^{210}Pb$ (Table 2.4).

Although natural sources of radioactivity in air are important, anthropogenic emissions can be decisive in industrialized areas. Nuclear power plants can cause light radioactive emissions and other industrial activities provide greater quantities of radioactive materials to the air. As coal contains low levels (impurities) of uranium ($^{238}U$), thorium ($^{232}Th$) and daughter isotopes (such as radium, radon, polonium, bismuth, and lead) and potassium ($^{40}K$), the combustion of coal causes the release of significant amounts of these isotopes to the air. In fact, the effective radioactive dose emitted from coal plants is near 100 times that from nuclear plants (McBride et al. 1978). Naturally occurring radioactive species released by coal

**Table 2.4** Main radioactive isotopes of uranium and thorium

| Uranium series | | | Thorium series | | |
|---|---|---|---|---|---|
| Isotope | Half-life | Decay | Isotope | Half-life | Decay |
| $^{238}$U | $4.468\times10^{9}$ years | Alpha | $^{232}$Th | $1.405\times10^{10}$ years | Alpha |
| $^{234}$Th | 24.1 days | Beta | $^{228}$Ra | 5.75 years | Beta |
| $^{234}$Pa | 1.17 min | Beta | $^{228}$Ac | 6.25 h | Beta |
| $^{234}$U | $2.48\times10^{5}$ years | Alpha | $^{228}$Th | 1.9116 years | Alpha |
| $^{230}$Th | $7.7\times10^{4}$ years | Alpha | $^{224}$Ra | 3.6319 days | Alpha |
| $^{226}$Ra | 1600 years | Alpha | $^{220}$Rn | 55.6 days | Alpha |
| $^{222}$Rn | 3.82 days | Alpha | $^{216}$Po | 0.145 s | Alpha |
| $^{218}$Po | 3.05 min | Alpha | $^{212}$Pb | 10.64 h | Beta |
| $^{214}$Pb | 26.8 min | Beta | $^{212}$Bi | 60.55 min | Alpha + beta |
| $^{214}$Bi | 19.8 min | Beta | $^{212}$Po | 299 nsec | Alpha |
| $^{214}$Po | 162 μsec | Alpha | $^{208}$Tl | 3.053 min | Beta |
| $^{210}$Pb | 22.3 years | Beta | $^{208}$Pb | Stable | |
| $^{210}$Bi | 5.01 days | Beta | | | |
| $^{210}$Po | 138.4 days | Alpha | | | |
| $^{206}$Pb | Stable | | | | |

combustion are accumulating in the environment along with particles containing Hg, As, Si, Ca, Cl, Al, Fe, Pb, Mg, Ti, B, Cr and other elements that are continually dispersed in millions of tons of coal combustion by-products.

Some radioactive isotopes have been used as pollution tracers in sediments, as radioactive elements precipitate and remain trapped in sediment layers. For example, there is a $^{210}$Pb fraction supported by the $^{226}$Ra content in the sediment grain structure and another fraction unsupported. This unsupported fraction is due to the fact that $^{222}$Rn partially escapes to the atmosphere once is generated by $^{226}$Ra disintegration (Quintana et al. 2006).

## 2.5 Final Considerations

Research carried out in the last decades has revealed important information about decay of building materials, especially stony materials. Such interest is justified due to the importance of the conservation of Heritage Buildings and Monuments, but also to solve important degradation problems in modern buildings. Pollutants can affect the built environment through a diversity of sources, pathways and mechanisms. In the present chapter, the origin and decay mechanisms of pollutants as well as the chemical fingerprints that can be used to trace the pollution sources have been reviewed.

As it has been visible in this review, there are marked interdependences between the different pollutants, sources and pathways. Gaseous pollutants show interactions between them in the development of decay processes and influence the composition of other sources such as dry, occult and wet deposition. Special attention was paid to the progress in the matter of atmospheric diffusion and processes in urban environments carried out in the last years. Such progress has been possible due to intense research but there is some delay in transferring these advances to study the degradation of materials. The main achievements in this subject and their possible employment in the study of the decay of materials have been reviewed.

Particulate matter is an important medium for the transport of pollutants that can contribute to the composition of solutions derived from wet deposition. This transport mechanism also plays a major role in dry deposition and can lead to high concentrations of pollutants in sheltered portions or in exposed portions during rainless periods. In this last case, high concentrations can be built up increasing the pollution effect that result when rain episodes occur.

In studies focused on the decay of building materials water solutions can be considered as pollutants and as important agents of transport of pollutants in either ionic or solid form into the surface and pore system of materials. Condensation promotes chemical reactions especially in sheltered portions contributing to the decay action of the gaseous pollutants and particulate matter. Wet deposition plays a major role in the source of pollutants in exterior zones. Wet deposition can further interact with the built environment trough run-off and infiltration, and contributes to capillary-rising solutions. Run-off and infiltration solutions become more polluted due to contributions coming from building materials and accumulated pollutants (especially in the case of longer contact times). Capillary rising solutions can concentrate atmospheric deposition (due to water balance conditions) and add pollutants from such diverse sources as natural weathering products and organic waste.

Decay processes affecting materials where reviewed in this chapter mostly from a qualitative point of view. This knowledge is indispensible for any attempt to develop a quantitative assessment of the decay process for durability predictions. Some examples of quantitative assessment were mentioned but they are confined to laboratory or field exposure tests. Questions regarding size and time effects have to be considered in the extrapolation to actual built elements. Another relevant factor, and one that was ignored in this mostly chemical review, is the way in which the design of the built environment (both the morphological features and the spatial relations between components) influence the action of the pollutants on the materials. A related subject is the potential consequences of the present reduction of levels of some gaseous pollutants such as $CO_2$ and $SO_2$ that have been addressed by several models but where further empirical studies are required. We propose that the regular study of new built elements constitutes a more promising methodology for this aim when compared to the study of older built elements where diverse agents have left imprinted their effects along decades, centuries or even millennia.

Besides the different sources, the chemical composition of solutions in the built environment is susceptible to be modified by solubility-related differentiation processes. The study of these varied, multi-component and complex systems would benefit of the use of classical geochemical techniques to identify sources of pollution and characterize the diffusion and distribution of pollutants on the built environment. It is proposed that procedures for quantitative assessments of durability would be always very limited given not only the variety, but also the unpredictability of actual conditions. Qualitative modelling approaches (for general considerations on qualitative modelling on the Earth Sciences see Pilkey and Pilkey-Jarvis 2007) that take into account the diverse factors are reviewed here. The diverse aspects of the design of built elements and field assessments of previous performances would be the best way to make predictions on the usability of a given material under a given scope of exposition conditions.

This review also aimed to update the existing knowledge in this field for future application in studies of the deterioration of building materials due to pollution. It is important to identify the pollution sources that actually affect a given building element, both to understand the decay processes and to plan the intervention procedures. Different tracers has been considered in order to attribute the damage on building materials to sources such as dry and wet deposition, groundwater or the same (or other) building materials, and to elucidate if damaging atmospheric pollutants are generated by distant industrial gaseous and particulate emissions or nearby sources. This is important as urban planning could contribute to prevent deterioration problems in buildings derived from different air pollution sources.

Atmospheric particulate materials, minor gaseous compounds and ions in solution can be used as tracers. The analysis of trace elements on some damaged materials has provided valuable information just in some cases. However, in the last years the use of some stable isotopes has been especially successful in geochemistry and atmospheric sciences, and we reported here the main achievements of its use in the matter of circulation of solutions, interaction with minerals and damage on building materials. Stable isotopes of some light elements such as C, O and S have been used to understand the diffusion on pollutants in air, solutions and decay of building materials. Although other stable isotopes of light elements have been studied (e.g. H, N), there is a great scarcity of data concerning building materials but it is proposed that these isotopes have a great potential for the fringerprinting of pollution sources. In this review were also proposed some radioactive isotopes that have been very useful in the study of pollutants cycles in air and water in other fields (e.g. deposition on sediments) and that might have promising applications in the study of the damage of building materials. In particular, tritium, radiocarbon and radioactive isotopes of heavy elements (e.g. U, Th and K) have potential application as tracers of pollution and damage on the building environment.

## References


Aberg G, Löfvendahi R, Stijfhoorn D, Raheim A (1995) Provenance and weathering depth of carbonaceous Gotland sandstone by use of carbon and oxygen isotopes. Atmos Environ 29:781–789. doi:10.1016/1352-2310(94)00324-E

Affek HP, Eiler JM (2006) Abundance of mass 47 $CO_2$ in urban air, car exhaust, and human breath. Geochim Cosmochim Acta 70:1–12. doi:10.1016/j.gca.2005.08.021

Allen GC, El-Turki A, Hallam KR, McLaughlin D, Stacey M (2000) Role of $NO_2$ and $SO_2$ in degradation of limestone. Br Corrosion J 35:35–38. doi:10.1179/000705900101501047

Alliksaar T, Hörstedt P, Renberg I (1998) Characteristic fly-ash particles from oil-shale combustion found in lake sediments. Water Air Soil Pollut 104:149–160. doi:10.1023/A:1004918419356

Alonso FJ, Vázquez P, Esbert RM, Ordaz J (2008) Ornamental granite durability: evaluation of damage caused by salt crystallization. Materiales de Construcción 58:191–201. doi:10.3989/mc.2008.v58.i289-290.78

Alves C (2009) Salt weathering of natural building stones: a review of the influence of rock characteristics. In: Cornejo DN, Haro JL (eds) Building materials: properties, performance and applications. Nova, New York, pp 57–94

Alves C (2010) "White" crusts on recent buildings. Mater Sci Forum 636–637:1300–1305. doi:10.4028/www.scientific.net/MSF.636-637.1300

Alves C, Sequeira Braga MA (1994) Niter and gypsum and their decay effects in a granitic monument of Braga (Portugal). In: Proceedings of the 16th general meeting IMA, Pisa, p 9

Alves CAS, Sequeira Braga MA (2000) Decay effects associated with soluble salts on granite buildings of Braga (NW Portugal). In: Cotter-Howells JD, Campbell LS, Valsami-Jones E, Batchelder M (eds) Environmental mineralogy: microbial interactions, anthropogenic influences, contaminated land and waste management. Mineralogical Society of Great Britain & Ireland, London, pp 181–199

Alves C, Figueiredo C, Maurício A, Figueiredo P, Aires-Barros L (2009) Contribution of exogenous substances to alteration of stone surfaces in metropolitan stations. In: Proceedings IRF'2009, 3 rd international conference on integrity, reliability & failure, Oporto, CD-ROM. Paper Ref: S2006_P0316

Amoroso GG, Fassina V (1983) Stone decay and conservation: atmospheric pollution, cleaning, consolidation and protection. Elsevier, Amsterdam

Anstice DJ, Page CL, Page MM (2005) The pore solution phase of carbonated cement pastes. Cement Concr Res 35:377–383. doi:10.1016/j.cemconres.2004.06.041

Antunes SG, Moreno F, Alves CAS, Pamplona JMV (2004) Erosive decay features of granitic stones monuments and characterisation of salt efflorescences in Évora. In: Aires-Barros L, Zezza F, Dionísio A, Rodrigues M (eds) Proceedings 6th international symposium on the conservation of monuments in the Mediterranean Basin, Lisbon, pp 223–228

Appelo CAJ, Postma D (2005) Geochemistry, groundwater and pollution. A.A Balkema Publishers, Leiden

Appleyard SJ (1995) The impact of urban development on recharge and groundwater quality in coastal aquifer near Perth, Western Australia. Hydrogeol J 3:65–75. doi:10.1007/s100400050072

Araguas-Araguas L, Froehlich K, Rozanski K (2000) Deuterium and oxygen-18 isotope composition of precipitation and atmospheric moisture. Hydrol Process 14:1341–1355. doi:10.1002/1099-1085(20000615)14:8<1341::AID-HYP983>3.0.CO;2-Z

Arnold A (1996) Origin and behaviour of some salts in context of weathering on monuments. In: Zezza F (ed) Origin, mechanisms, and effects of salts on degradation of monuments in marine and continental environments. Protection and conservation of the European cultural heritage research report no. 4. European Commission, Brussels, pp 131–139

Arnold A, Zehnder K (1991) Monitoring wall paintings affected by soluble salts. The conservation of wall paintings, Getty Conservation Institute, pp 103–135. Available online at http://getty.edu/conservation/publications/pdf_publications/wall_paintings.pdf. Accessed Aug 2010

Ashurst J (1998) Cleaning masonry buildings. In: Ashurst J, Dimes FG (eds) Conservation of building and decorative stone, part II. Butterworth-Heinemann, Oxford, pp 125–154

Athavale RN, Lal D, Rama S (1967) The measurement of tritium activity in natural waters. Part II. Characteristics of global fallout of $^{3}H$ and $^{90}Sr$. Proc Indian Acad Sci Sect A 65:73–103

Begonha A (2009) Mineralogical study of the deterioration of granite stones of two Portuguese churches and characterization of the salt solutions in the porous network by the presence of diatoms. Mater Charact 60(7):621–635. doi:10.1016/j.matchar.2008.12.019

Begonha A, Sequeira Braga MA (1996) Black crusts and thin black layers in granitic monuments: their characterization and the role of air pollution. In: Riederer J (ed) Proceedings of the 8th international congress on deterioration of conservation of stone. Möller Druck und Verlag, Berlin, pp 371–375

Begonha A, Sequeira Braga MA, Gomes Da Silva F (1996) Rain water as a source of the soluble salts responsible for stone decay in the granitic monuments of Oporto and Braga-Portugal. In: Riederer J (ed) Proceedings of the 8th international congress on deterioration of conservation of stone. Möller Druck und Verlag, Berlin, pp 481–487

Behlen A, Wittenburg C, Steiger M, Dannecker W (1996) Dry deposition of NO, NO2, HONO, HNO3 and PAN on historical building stones. In: Riederer J (ed) Proceedings of the 8th international congress on deterioration of conservation of stone. Möller Druck und Verlag, Berlin, pp 377–385

Beilke S, Gravenhorst G (1978) Heterogeneous $SO_2$-oxidation in the droplet phase. Atmos Environ 12:231–241. doi:10.1016/0004-6981(78)90203-2

Benea B (1996) Representative stones and weather- ing forms at Histria Fortress, Romania. In: Zezza F (ed) Origin, mechanisms, and effects of salts on degradation of monuments in marine and continental environments. Protection and conservation of the European cultural heritage research report no. 4. European Commission, Brussels, pp 205–216

Benner WH, Brodzinsky R, Novakov T (1982) Oxidation of $SO_2$ in droplets which contain soot particles. Atmos Environ 16:1333–1339. doi:10.1016/0004-6981(82)90054-3

Bensted J (2002) A discussion of the review paper "Sulphate attack research—whither now?" by M. Santhanam, M.D.Cohen, and J. Olek. Cement Concr Res 32:995-1001. doi:10.1016/S0008-8846(02)00731–7

Berkowicz R, Palmgren F, Hertel O, Vignati E (1996) Using measurements of air pollution in street for evaluation of urban air quality – meteorological analysis and model calculations. Sci Total Environ 189–190:259–265. doi:10.1016/0048-9697(96)05217-5

Bernabé E, Bromblet P, Robert M (1995) Role of natron crystallization on granular disintegration of a granitic monument in Brittany. C R Acad Sci Paris Série IIa 320:571–578

Bhargav JS, Mishra RC, Das CR (1999) Environmental deterioration of stone monuments of Bhubaneswar, the temple city of India. Stud Conserv 44(1):1–11

Bityukova L (2006) Air pollution effect on the decay of carbonate building stones in old town of Tallinn. Water Air Soil Pollut 172:239–271. doi:10.1007/s11270-006-9078-1

Bityukova L, Shogenova A, Birke M (2000) Urban geochemistry: a study of elements distribution in the topsoil's of Tallinn (Estonia). Environ Geocheistry Health 22:173–195. doi:10.1023/A:1006754326260

Blackburn G, Hutton JT (1980) Soil conditions and the occurrence of salt damp in buildings of metropolitan Adelaide. Aust Geographer 14:360–365. doi:10.1080/00049188008702784

Blázquez F, García-Vallès M, Krumbein W, Sterflinger K, Vendrell-Saz M (1997) Microstromatolitic deposits on granitic monuments: development and decay. Eur J Mineral 9:889–901

Boyer KW, Laitinen HA (1975) Automobile exhaust particulates. Properties of environment significance. Environ Sci Technol 9:457–469. doi:10.1021/es60103a003

Brenninkmeijer CAM, Röckmann T (1997) Principal factors determining the $^{18}O/^{16}O$ ratio of atmospheric CO as derived from observations in the southern hemispheric troposphere and lowermost stratosphere. J Geophys Res 102:25477–25485. doi:10.1029/97JD02291

Brenninkmeijer CAM, Röckmann T, Braunlich M, Jöckel P, Bergamaschi P (1999) Review of progress in isotope studies of atmospheric carbon monoxide. Chemosphere: Global Change Sci 1:33–52. doi:10.1016/S1465-9972(99)00018-5

Brimblecombe P, Grossi CM (2008) Millennium-long recession of limestone Facades in London. Environ Geol 56:463–471. doi:10.1007/s00254-008-1465-z

Brimblecombe P, Grossi CM (2010) Potential damage to modern building materials from 21st century air pollution. Sci World J 10:116–125

Butlin RN, Yates TJS, Chakrabarti B (1995) Mapping of critical loads and levels for pollution damage to building materials in the United Kingdom. Water Air Soil Pollut 85:2693–2699. doi:10.1007/BF01186241

Buzek F, Srámek J (1985) Sulfur isotopes in the study of stone monument conservation. Stud Conserv 30:171–176

Camuffo D (1990) Acidic precipitation research in Italy. In: Bresser AH, Salomons W (eds) Acidic precipitation 5, international overview and assessment. Springer, New York, pp 229–265
Camuffo D (1991) Physical weathering of monuments. In: Zezza F (ed) Weathering and air pollution, first course, community of Mediterranean universities, university school of monument conservation. Mario Adda Editore, Bari, pp 21–66
Camuffo D (1995) Physical weathering of stones. Sci Total Environ 167:1–14. doi:10.1016/0048-9697(95)04565-I
Camuffo D, Del-Monte M, Sabbioni C, Vittori O (1982) Wetting, deterioration and visual features of stone surfaces in an urban area. Atmos Environ 16:253–2259. doi:10.1016/0004-6981(82)90296-7
Camuffo D, Del-Monte M, Sabbioni C (1983) Origin and growth mechanisms of the sulfated crusts on urban limestone. Water Air Soil Pollut 19:351–359. doi:10.1007/BF00159596
Camuffo D, Del-Monte M, Ongaro A (1984) The pH of the atmospheric precipitation in Venice, related to both the dynamics of precipitation events and the weathering of monuments. Sci Total Environ 40:125–139. doi:10.1016/0048-9697(84)90347-4
Camuffo D, Bernardi A, Zanetti M (1988) Analysis of the real-time measurement of the pH of rainfall at Padova, Italy: seasonal variation and meteorological aspects. Sci Total Environ 71:187–200. doi:10.1016/0048-9697(88)90166-0
Cardell C, Benavente D, Rodríguez-Gordillo J (2008) Weathering of limestone building material by mixed sulfate solutions. Characterization of stone microstructure, reaction products and decay forms. Mater Charact 59:1371–1385. doi:10.1016/j.matchar.2007.12.003
Carslaw DC, Beevers SD (2004) Investigating the potential importance of primary $NO_2$ emissions in a street canyon. Atmos Environ 38:3585–3594. doi:10.1016/j.atmosenv.2004.03.041
Carslaw DC, Beevers SD, Bell MC (2007) Risks of exceeding the hourly EU limit value for nitrogen dioxide resulting from increased road transport emissions of primary nitrogen dioxide. Atmos Environ 41:2073–2082. doi:10.1016/j.atmosenv.2006.10.074
Casal-Porto M, Silva B, Rodrigues JD (1991) Agents and forms of weathering in granitic rocks used in monuments. In: Baer NS, Sabbioni C, Sors AI (eds) Science, technology and European cultural heritage. Butterworth Heinemann, Oxford, pp 439–442
Chabas A, Jeannette D (2001) Weathering of marbles and granites in marine environment: petrophysical properties and special role of atmospheric salts. Environ Geol 40:359–368. doi:10.1007/s002540000157
Chabas A, Lefèvre R (2000) Chemistry and microscopy of atmospheric particulate at Delos (Cyclades, Greece). Atmos Environ 34:225–238. doi:10.1016/S1352-2310(99)00255-1
Charola AE (1988) Chemical-physical factors in stone deterioration. In: Rossvall J, Aleby S (eds) Air pollution and conservation. Safeguarding our architectural heritage. Elsevier, Amsterdam, pp 125–132
Charola AE (2000) Salts in the deterioration of porous materials: an overview. J Am Inst Conserv 39:327–343. Available (Aug 2010) at http://cool.conservation-us.org/coolaic/jaic/articles/jaic39-03-002.html
Charola AE, Ware R (2002) Acid deposition and the deterioration of stone: a brief review of a broad topic. In: Siegesmund GS, Vollbrecht A, Weiss T (eds) Natural stone, weathering phenomena, conservation strategies and case studies, Special publications 205. Geological Society, London, pp 393–406. doi:10.1144/GSL.SP.2002.205.01.28
Charola AE, Pühringer J, Steiger M (2007) Gypsum: a review of its role in the deterioration of building materials. Environ Geol 52:339–352. doi:10.1007/s00254-006-0566-9
Christwell CG, Ogawa I, Tschetter MJ, Markuszewski R (1988) Effect of hydrofluoric or hydrochloric acid pretreatment on the ultrasonic extraction of organic materials from fly ash for chromatographic analysis. Environ Sci Technol 22:1506–1508. doi:10.1021/es00177a019
Cobourn WG, Gauri KL, Tambe S, Suhan L, Saltik E (1993) Laboratory measurements of sulfur dioxide deposition velocity on marble and dolomite stone surfaces. Atmos Environ 27:193–201. doi:10.1016/0957-1272(93)90005-Q
Coleman DD, Liu CL, Hackley KC, Benson, LJ (1993) Identification of landfill methane using carbon and hydrogen isotope analysis. Presented at the sixteenth international Madison waste

conference, Madison, 22–23 Sept 1993, Department of Engineering Professional Development, University of Wisconsin-Madison
Collings N, Graskow BR (2000) Particles from internal combustion engines – what we need to know. Philos Trans R Soc Lond A 358:2611–2623. doi:10.1098/rsta.2000.0672
Conny JM (1998) The isotopic characterization of carbon monoxide in the troposphere. Atmos Environ 32:2669–2683. doi:10.1016/S1352-2310(97)00398-1
Cooper TP, Dowding P, Lewis JO, Mulvin L, O'Brien P, Olley J, O' Daly G (1991) Contribution of calcium from limestone and mortar to the decay of granite walling. In: Baer NS, Sabbioni C, Sors AI (eds) Science, technology and European cultural heritage. Butterworth Heinemann, Oxford, pp 456–461
Cortes V, Martin A (1982a) The effects of atmospheric pollution by fluorides on stone alteration – I: siliceous materials. In: Proceedings of the 3rd international symposium on the deterioration and conservation of stones, Venice, 1979. Libreria Progetto, Padova, pp 1–12
Cortes V, Martin A (1982b) The effects of atmospheric pollution by fluorides on stone alteration – II: calcareous materials. In: Proceedings of the 3rd international symposium on the deterioration and conservation of stones, Venice, 1979. Libreria Progetto, Padova, pp 13–21
Craig H (1954) Carbon 13 and carbon 14 in plants and the relationships between carbon 13 and carbon 14 variations in nature. J Geol 62:115–149. doi:10.1086/626141
Cultrone G, De-la-Torre M, Sebastian JEM, Cazalla O (2000) Behavior of brick samples in aggressive environments. Water Air Soil Pollut 119:191–207. doi:10.1023/A:1005142612180
Curtiss B, Adams JB, Ghiorso MS (1985) Origin, development and chemistry of silica-alumina rock coatings from the semiarid regions of the island of Hawaii. Geochim Cosmochim Acta 49:49–56. doi:10.1016/0016-7037(85)90190-5
Dansgaard W (1964) Stable isotopes in precipitation. Tellus 16:436–468. doi:10.1111/j.2153-3490.1964.tb00181.x
De-Bock LA, Van Grieken RE, Camuffo D, Grime GW (1996) Microanalysis of museum aerosols to elucidate the soiling of paintings: case of the correr museum, Venice, Italy. Environ Sci Technol 30:3341–3350. doi:10.1021/es9602004
Deines P (1980) The isotopic composition of reduced organic carbon. In: Fritz P, Fontes JC (eds) Handbook of environmental geochemistry. Elsevier, Amsterdam, pp 329–405
Delalieux F, Cardell C, Todorov V, Dekov V, VanGrieken RJ (2001) Environmental conditions controlling the chemical weathering of the Madara Horseman monument, NE Bulgaria. J Cultural Herit 2:43–54. doi:10.1016/S1296-2074(01)01105-0
Del-Monte M, Sabbioni C (1984) Gypsum crusts and fly ash particles on carbonatic outcrops. Arch Meteorol Geophy Bioclimatol Ser B 35:105–111. doi:10.1007/BF02269413
Delre LC, Miller RW (1988) Characterization of weather aging and radiation susceptibility. In: Epel JN, Margolis JM, Neuman S, Seymour RB (eds) Engineered materials handbook vol 2, engineering plastics. ASM International, Metals Park, pp 575–580
Derbez M, Lefèvre RA (1996) Le contenu microparticulaire des croûtes gypseuses de la Cathédrale Saint-Gatien de Tours: comparaison avec l'air et la pluie. In: Riederer J (ed) Proceedings 8th international congress on deterioration and conservation of stone. Möller Druck und Verlag, Berlin, pp 359–370
Dibb TE, Hughes DW, Poole ABQ (1983) The identification of critical factors affecting rock durability in marine environments. Q J Eng Geol Hydrogeol 16:149–161. doi:10.1144/GSL.QJEG.1983.016.02.08
Dietzel M (2000) Measurements of stable carbon isotopes in calcite sinters on concrete. ZKG Int 9:544–548
Doehne E (2002) Salt weathering: a selective review. In: Siegesmund GS, Vollbrecht A, Weiss T (eds) Natural stone, weathering phenomena, conservation strategies and case studies, Special publications 205. Geological Society, London, pp 51–64. doi:10.1144/GSL.SP.2002.205.01.05
Domenico PA, Schwartz FW (1990) Physical and chemical hydrogeology. Wiley, New York
Doner HE, Lynn WC (1977) Carbonate, halide, sulfate and sulfide minerals. In: Dixon JB, Weed SB (eds) Minerals in soil environments. Soil Science Society of America, Madison, pp 75–98

Doornkamp JC, Ibrahim HAM (1990) Salt weathering. Prog Phys Geogr 14:335–348. doi:10.1177/030913339001400303

Doremus RH (1994) Glass science. Wiley-Interscience, New York

Dorn RI (1998) Rock coatings. Elsevier Science B.V, Amsterdam

Douglas J, Ransom B (2007) Understanding building failures. Taylor & Francis, London

Dow C, Glasser FP (2003) Calcium carbonate efflorescence on Portland cement and building materials. Cement Concr Res 33:147–154. doi:10.1016/S0008-8846(02)00937-7

Dreesen R, Nielsen P, Lagrou D (2007) The staining of blue stone limestones petrographically unraveled. Mater Charact 58(11–12):1070–1081. doi:10.1016/j.matchar.2007.03.015

Duffy AP, O'Brien PF (1996) A basis for evaluating the durability of new building stone. In: Smith BJ, Warke PA (eds) Processes of urban stone decay. Donhead Publishing Ltd, London, pp 253–260

Eglinton M (1998) Resistance of concrete to destructive agencies. In: Hewlett PC (ed) Lea's Chemistry of cement and concrete. Elsevier, Oxford, pp 299–342

Ehhalt DH, Rohrer F (2009) The tropospheric cycle of $H_2$: a critical review. Tellus Ser B 61:500–535. doi:10.1111/j.1600-0889.2009.00416.x

Elert K, Rodriguez-Navarro C, Sebastian Pardo E, Hansen E, Cazalla O (2002) Lime mortars for the conservation of historic buildings. Stud Conserv 47, p. 62

Elswick ER, Hower JC, Carmo AM, Sun T, Mardon SM (2007) Sulfur and carbon isotope geochemistry of coal and derived coal-combustion by-products: an example from an eastern Kentucky mine and power plant. Appl Geochem 22:2065–2077. doi:10.1016/j.apgeochem. 2007.04.018

Etheridge DM, Steele LP, Francey RJ, Langenfields RL (1998) Atmospheric methane between 1000 A.D. and present: evidence of anthropogenic emissions and climatic variability. J Geophys Res 103:15979–15993. doi:10.1029/98JD00923

European Environment Agency (2007) Air pollution in Europe 1990–2004. EEA Report 2/2007. ISSN 1725–9177

Farny JA, Kerkhoff B (2007) Concrete technology: diagnosis and control of alkali-aggregate reactions in concrete. Portland Cement Association, Skokie, 25 p

Fassina V (1988) Environmental pollution in relation to stone decay. Air pollution and conservation. In: Rossvall J, Aleby S (eds) Safeguarding our architectural heritage. Elsevier, Amsterdam, pp 133–174

Fassina V, Stevan A (1992) Fogwater composition in Venice in relation to stone decay. In: Delgado Rodrigues J, Henriques F, Telmo Jeremias F (eds) Proceedings of the 7th international congress on deterioration and conservation of stone. Laboratório Nacional de Engenharia Civil, Lisbon, pp 365–374

Fassina V, Lazzarini, Biscontin G, Calogero S (1979) Influenza del materiale particellare atmosferico sui processi di degradazione della pietra a Venezia (Influence of atmospheric particulates on the deterioration process of the stones in Venice) In: 3rd international congress on deterioration and conservation of stone (Venice) Università degli Studi di Padova, Padua, pp 43–53

Feist WC (1992) Natural weathering of wood and its control by water-repellent preservatives. Am Paint Contract 69:18–25

Fellenberg G (2000) The chemistry of pollution. Wiley, Chichester, p 174

Feng X-T, Chen S, Li S (2001) Effects of water chemistry on microcracking and compressive strength of granite. Int J Rock Mech Min Sci 38:557–568. doi:10.1016/S1365-1609(01)00016-8

Feucht LJ, Logan JM (1990) Effects of chemically active solutions on shearing behavior of a sandstone. Tectonophysics 175:159–176. doi:10.1016/0040-1951(90)90136-V

Figg J, Moore AE, Gutteridge WA (1976) On the occurrence of the mineral trona ($Na_2CO_3$. $NaHCO_3$*$2H_2O$) in concrete deterioration products. Cement Concr Res 6:691–696. doi:10.1016/0008-8846(76)90033-8

Figg J, Moore AE, Gutteridge WA (1978) A reply to Halle's et al. discussion of the paper "On the occurrence of the mineral trona ($Na_2CO_3$.$NaHCO_3$*$2H_2O$) in concrete deterioration products". Cement Concr Res 8:255. doi:10.1016/0008-8846(78)90015-7

Figueiredo CAM, Aires-Barros L, Basto MJ, Graça RC, Maurício A (2007) The weathering and weatherability of Basílica da Estrela stones, Lisbon, Portugal. In: Přikryl R, Smith BJ (eds) Building stone decay: from diagnosis to conservation. Special Publications, 271. Geological Society, London pp 99–107. doi:0.1144/GSL.SP.2007.271.01.11

Filley RT, Blanchette RA, Simpson E, Fogel ML (2001) Nitrogen cycling by wood decomposing soft-rot fungi in the "King Midas tomb", Gordion, Turkey. PNAS 98:13346–13350. doi:10.1073/pnas.221299598

Fitzner B, Heinrichs K (2002) Damage diagnosis on stone monuments – weathering forms, damage categories and damage indices. In: Prikryl R, Viles HA (eds) Understanding and managing stone decay. The Karolinum Press, Prague, pp 11–56. Available online at http://www.stone.rwth-aachen.de/decay_diagnosis.pdf. Accessed Aug 2010

French WJ (1997) Influence of groundwater chemistry and motion on highway construction materials. In: Hawkins AB (ed) Ground chemistry: implications for construction. Balkema, Rotterdam, pp 199–210

Freyer HD, Kley D, Volz-Thomas A, Kobel K (1993) On the interaction of isotopic exchange processes with photochemical reactions in atmospheric oxides of nitrogen. J Geophys Res 98:14791–14796. doi:10.1029/93JD00874

Friedli H, Siegenthaler U, Rauber D, Oeschger H (1987) Measurements of concentration, $^{13}C/^{12}C$ and $^{18}O/^{16}O$ ratios of tropospheric carbon dioxide over Switzerland. Tellus 39B:80–88. doi:10.1111/j.1600-0889.1987.tb00272.x

Fritz P, Fontes JCh (1980) Introduction. In: Fritz P, Fontes JC (eds) Handbook of environmental geochemistry. Elsevier, Amsterdam, pp 4–17

Furlan V, Girardet F (1992) Pollution atmosphérique et réactivité de pierres. In: Delgado Rodrigues J, Henriques F, Telmo Jeremias F (eds) Proceedings of the 7th international congress on deterioration and conservation of stone. Laboratório Nacional de Engenharia Civil, Lisbon, pp 153–161

Gaffney JS, Marley NA (2009) The impacts of combustion emissions on air quality and climate – From coal to biofuels and beyond. Atmos Environ 43:23–36. doi:10.1016/j.atmosenv.2008.09.016

Galan, E, Guerrero MA, Vázquez MA, Carretero MI, Ortiz P (1996) The Cathedral of Cádiz (Spain) Environmental study and stone damage evaluation. In: Zezza F (ed) Origin, mechanisms, and effects of salts on degradation of monuments in marine and continental environments. Protection and conservation of the European cultural heritage research report no. 4. European Commission, Brussels, pp 361–372

García-Talegón J, Molina E, Vicente MA (1996) Behaviour of granitic materials under a non-polluted semiarid climate (Avila cathedral, Spain). In: Vicente MA, Delgado RJ, Acevedo J (eds) Degradation and conservation of granitic rocks in monument. Protection and conservation of the European cultural heritage research report no. 5. European Commission, Brussels, pp 61–66

Gat JR (1980) The isotopes of hydrogen and oxygen in precipitation. In: Fritz P, Fontes JC (eds) Handbook of environmental geochemistry. Elsevier, Amsterdam, pp 21–47

Gerst S, Quay P (2000) The deuterium content of atmospheric molecular hydrogen: method and initial measurements. J Geophys Res 105:26433–26445. doi:10.1029/2000JD900387

Gerst S, Quay P (2001) Deuterium component of the global molecular hydrogen cycle. J Geophys Res 106:5021–5031. doi:10.1029/2000JD900593

Ghedini N, Gobbi G, Sabbioni C, Zappia G (2000) Determination of elemental and organic carbon on damaged stone monuments. Atmos Environ 34:4383–4391. doi:10.1016/S1352-2310(00)00250-8

Gilbert NL, Woodhouse S, Stieb DM, Brook JR (2003) Ambient nitrogen dioxide and distance from a major highway. Sci Total Environ 312:43–46. doi:10.1016/S0048-9697(03)00228-6

Gillberg G, Gezelius LH, Lööf G, Ringblom H, Rosenquist K (1977) Cleaning of external surfaces of buildings. In: Proceedings of RILEM/ASTM/CIB symposium evaluation of the performance of external vertical surface of buildings, Espoo, pp 144–153

Gillott JE (1978) Effect of deicing agents and sulphate solutions on concrete aggregate. Q J Eng Geol Hydrogeol 11:177–192. doi:10.1144/GSL.QJEG.1978.011.02.06

Girardet F, Furlan V (1996) Réactivité des pierres au SO2 atmosphérique, etude en chamber de simulation et correlation avec les mesures en site réel. In: Riederer J (ed) Proceedings of the 8th international congress on deterioration of conservation of stone, Berlin, 1996, pp 341–347

Glasser FP (1986) Chemical and geochemical basis for the immobilization of nuclear waste materials in cements. Fortschr Der Mineralogie 64:19–35

Gómez-Heras M, Benavente D, Álvarez de Buergo M, Fort R (2004) Soluble salt minerals from pigeon droppings as potential contributors to the decay of stone based cultural heritage. Eur J Mineral 16(3):505–509. doi:10.1127/0935-1221/2004/0016-0505

Goudie AS (1986) Laboratory simulation of 'the wick effect' in salt weathering of rock. Earth Surf Processes Landforms 11:275–285. doi:10.1002/esp. 3290110305

Goudie AS, Viles HA (1997) Salt weathering hazards. Wiley, Chichester, 241 p

Graedel TE (1987) Copper patinas formed in the atmosphere-II. A qualitative assessment of mechanisms. Corrosion Sci 27:721–740. doi:10.1016/0010-938X(87)90053-9

Grossi CM, Esbert RM (1994) Las sales solubles en el deterioro de rocas monumentales. Revisión bibliográfica. Materiales de Construcción 44(235):1994.

Grossi CM, Brimblecombe P (2002) The effect of atmospheric pollution on building materials. J Phys IV France 12:197–210. doi:10.1051/jp4:20020460

Grossi CM, Brimblecombe P, Esbert RM, Alonso FJ (2007) Color changes in architectural limestones from pollution and cleaning. Color Res Appl 32:320–331. doi:10.1002/col.20322

Grossi C, Bonazza A, Brimblecombe P, Harris I, Sabbioni C (2008) Predicting twenty-first century recession of architectural limestone in European cities. Environ Geol 56:455–461. doi:10.1007/s00254-008-1442-6

Guenther A, Hewitt CN, Erickson D, Fall R, Geron C, Graedel T, Harley P, Klinger L, Lerdau M, McKay WA, Pierce T, Scholes B, Steinbrecher R, Tallamraju R, Taylor J, Zimmerman P (1995) A global model of natural volatile organic compound emissions. J Geophys Res 100:8873–8892. doi:10.1029/94JD02950

Guidobaldi F, Mecchi AM (1985) Corrosion of marble by rain. The influence of surface roughness, rain intensity and additional washing. In: Proceedings of the 5th international congress on deterioration and conservation of stone, Lausanne. Presses Polytechniques Romandes, Lausanne, pp 467–474

Haagen-Smit AJ, Fox MM (1954) Photochemical ozone formation with hydrocarbons and automobile exhaust. J Air Pollut Control Assoc 4:105–109

Haagen-Smit AJ, Fox MM (1956) Ozone formation in photochemical oxidation of organic substances. Indian Eng Chem 48:1484

Halle R, Carin V, Hvala M, Crnković B (1978) Discussion of the paper on the occurrence of the mineral trona ($Na_2CO_3.NaHCO_3*2H_2O$) in concrete deterioration products. Cement Concr Res 8:251–253. doi:10.1016/0008-8846(78)90014-5

Hammecker C (1995) The importance of the petrophysical properties and external factors in the stone decay on monuments. Pure Appl Geophys 145:337–361. doi:10.1007/BF00880275

Haneef SJ, Dickinson C, Johnson JB, Thompson GE, Wood GC (1992) Simulation of the degradation of coupled stones by artificial acid rain. Stud Conserv 37:105–112

Happell JD, Östlund G, Mason AS (2004) A history of atmospheric tritium gas (HT) 1950–2002. Tellus B 56:183–193. doi:10.1111/j.1600-0889.2004.00103.x

Hartog P, McKenzie P (2004) The effects of alkaline solutions on limestone. Discover Stone 3:34–49

Hawkins AB, Pinches GM (1987) Cause and significance of heave at Liandough Hospital, Cardiff – a case history of ground floor heave due to gypsum growth. Q J Eng Geol Hydrogeol 20:41–57. doi:10.1144/GSL.QJEG.1987.020.01.05

Haynie FH, Spence JW (1984) Air pollution damage to exterior household paints. J Air Pollut Control Assoc 34:941–944

Heaton THE (1987) $^{15}N/^{14}N$ ratios of nitrate and ammonium in rain in Pretoria, South Africa. Atmos Environ 21:843–852. doi:10.1016/0004-6981(87)90080-1

Heaton THE (1990) $^{15}N/^{14}N$ ratios of NOx from Vehicle engines and coal-fired power stations. Tellus B 42:304–307. doi:10.1034/j.1600-0889.1990.00007.x-i1

Heinemeier J, Jungner H, Lindroos A, Ringbom A, von Konow T, Rud N (1997) AMS $^{14}C$ dating of lime mortar. Nucl Instrum Methods in Phys Res B 123:487–495. doi:10.1016/S0168-583X(96)00705-7

Heiss K, Skalli M, Zallmanzig J (1991) Diagnosis study of the St. Marienkirche, Lippstadt, a mediaeval greensandstone monument, for conservatory planning. In: Baer NS, Sabbioni C, Sors AI (eds) Science, technology and European cultural heritage. Butterworth Heinemann, Oxford, pp 939–942

Herman DC, Maier RM (2000) Consequences of biogeochemical cycles gone wild. In: Maier RM, Pepper IL, Gerba CP (eds) Environmental microbiology. Academic Press, London, pp 347–361.

Hicks BB (1986) Measuring dry deposition: a re-assessment of the state of the art. Water Air Soil Pollut 30:75–90. doi:10.1007/BF00305177

Hildemann LM, Markowski GR, Cass G (1991) Chemical composition of emissions from urban sources of fine organic aerosol. Environ Sci Technol 25:744–759. doi:10.1021/es00016a021

Hizem N, Ben Fredj A, Ghedira L (2005) Determination of natural radioactivity in building materials used in tunisian dwellings by gamma ray spectrometry. Radiat Prot Dosimetry 114:533–537. doi:10.1093/rpd/nch489

Ho KF, Lee SC, Yu JC, Zou SC, Fung K (2002) Carbonaceous characteristics of atmospheric particulate matter in Hong Kong. Sci Total Environ 300:59–67. doi:10.1016/S0048-9697(02)00281-4

Hoke E (1978) Investigations of weathering crusts on Salzsburg stone monuments. Stud Conserv 23:118–126

Hon DN-S, Feist WC (1993) Interaction of sulfur dioxide and nitric oxide with photoirradiated wood surfaces. Wood Fiber Sci 25:136–141

Honeyborne DB (1998) Weathering and decay of masonry. In: Ashurst J, Dimes FG (eds) Conservation of building and decorative stone, part I. Butterworth-Heinemann, Oxford, pp 153–178

Hosono T, Uchida E, Suda C, Ueno A, Nakagawa T (2006) Salt weathering of sandstone at the Angkor monuments, Cambodia: identification of the origins of salts using sulfur and strontium isotopes. J Archaeol Sci 33:1541–1551. doi:10.1016/j.jas.2006.01.018

Huff AK, Thiemens MH (1998) $^{17}O/^{16}O$ and $^{18}O/^{16}O$ isotope measurements of atmospheric carbon monoxide and its sources. Geophys Res Lett 25:3509–3512. doi:10.1029/98GL02603

Iden IK, Hagelia P (2003) C, O and S isotopic signatures in concrete which have suffered thaumasite formation and limited thaumasite form of sulfate attack. Cement Concr Composites 25:839–846. doi:10.1016/S0958-9465(03)00110-0

Inkpen R (2003) The whole building and patterns of degradation. In: Brimblecombe P (ed) The effects of air pollution on the built environment. Imperial College Press, London, pp 393–422

Irving PM (1991) Acidic deposition: state of science and technology, vol III. Terrestrial materials, health and visibility effects. US National Acid Precipitation Assessment Program, Washington DC, 16-25-24-129

Itano Y, Bandow H, Takenaka N, Saitoh Y, Asayama A, Fukuyama J (2007) Impact of NOx reduction on long-term ozone trends in an urban atmosphere. Sci Total Environ 379:46–55. doi:10.1016/j.scitotenv.2007.01.079

Janhäll S, Olofsson KFG, Anderson PU, Pettersson JBC, Hallquist M (2009) Evolution of the urban aerosol during winter temperature inversion episodes. Atmos Environ 40:5355–5366. doi:10.1016/j.atmosenv.2006.04.051

Jenkin ME (2004a) Analysis of sources and portioning of oxidant in the UK – Part 1: the NOx-dependence of annual mean concentrations of nitrogen dioxide and ozone. Atmos Environ 38:5117–5129. doi:10.1016/j.atmosenv.2004.05.056

Jenkin ME (2004b) Analysis of sources and portioning of oxidant in the UK – Part2: contributions of nitrogen dioxide emissions and background ozone at a kerbside location in London. Atmos Environ 38:5131–5138. doi:10.1016/j.atmosenv.2004.05.055

Jenkin ME, Clemitshaw KC (2000) Ozone and other secondary photochemical pollutants: chemical processes governing their formation in the planetary boundary layer. Atmos Environ 34:2499–2527. doi:10.1016/S1352-2310(99)00478-1

Jenkin ME, Utembe SR, Derwent RG (2008) Modelling the impact of elevated primary $NO_2$ and HONO emissions on regional scale oxidant formation in the UK. Atmos Environ 42:323–336. doi:10.1016/j.atmosenv.2007.09.021

Johansson IG, Lindquist O, Mangio RE (1988) Corrosion of calcareous stones in humid air containing $SO_2$ and $NO_2$. Durab Build Mater 5:439–449

Johnson JB, Haneef SJ, Hepburn BJ, Hutchison GE, Thompson GE, Wood GC (1990) Laboratory exposure systems to simulate atmospheric degradation of building stone under dry and wet deposition conditions. Atmos Environ 24A:2585–2592. doi:10.1016/0960-1686(90)90136-B

Jones MS, O'Brien PF, Haneef SJ, Thompson GE, Wood GC, Copper TP (1996) A study of decay occurring in Leinster Granite, House No. 9, Trinity College. In: Riederer J (ed) Proceedings of the 8th international congress on deterioration of conservation of stone. Möller Druck und Verlag, Berlin, pp 211–221

Jones A, Haywood J, Boucher O, Kravitz B, Robock A (2010) Geoengineering by stratospheric $SO_2$ injection: results from the Met Office HadGEM2 climate model and comparison with the Goddard Institute for Space Studies Model. Atmos Chem Phys Discuss 10:7421–7434. doi:10.5194/acpd-10-7421-2010

Jouzel J, Hoffmann G, Koster RD, Masson V (2000) Water isotopes in precipitation: data/model comparison for present-day and past climates. Q Sci Rev 19:363–379. doi:10.1016/S0277-3791(99)00069-4

Judeikis HS, Stewart TB (1976) Laboratory measurement of SO2 deposition velocities on selected building materials and soils. Atmos Environ 10:769–776. doi:10.1016/0004-6981(76)90078-0

Kato S, Akimoto H, Bräunlich M, Röckmann T, Brenninkmeijer CAM (1999) Measurements of stable carbon and oxygen isotopic compositions of CO in automobile exhausts and ambient air from semi-urban Mainz, Germany. Geochem J 33:73–77

Kawamura K, Kaplan IR (1987) Motor exhaust emissions as a primary source of dicarboxylic acids in Los Angeles ambient air. Environ Sci Technol 21:105–110. doi:10.1021/es00155a014

Kendall C (1998) Tracing nitrogen sources and cycling in catchments. In: Kendall C, McDonnell JJ (eds) Isotope tracers in catchment hydrology. Elsevier, Amsterdam, pp 519–576

Kendall C, Sklash MG, Bullen TD (1995) Isotope tracers of water and solute sources in catchments, in solute modelling in catchment systems. Wiley, New York, pp 261–303

Kleinman LI (1991) Seasonal dependence of boundary layer peroxide concentration: the low and high NOx regimes. J Geophys Res 96:20721–20733. doi:10.1029/91JD02040

Klemm W, Siedel H (2002) Evaluation of the origin of sulphate compounds in building stone by sulphur isotope ratio. In: Siegesmund S, Weiss T, Vollbrecht A (eds) Natural stone weathering phenomena, conservation strategies and case studies. The Geological Society, London, pp 419–429. doi:10.1144/GSL.SP.2002.205.01.30

Koleske JV (ed) (1995) Paint and coating testing manual: 14th edition of the Gardner-Sward handbook, ASTM manual series no 17. ASTM, Philadelphia, 925 p

Kozlowski R, Magiera J, Weber J, Haber J (1990) Decay and conservation of Pinczow porous limestone. I. Lithology and weathering. Stud Conserv 35:205–221

Krouse HR (1980) Sulphur isotopes in our environment. In: Fritz P, Fontes JC (eds) Handbook of environmental geochemistry. Elsevier, Amsterdam, pp 435–471

Krupa SV (2002) Sampling and physico-chemical analysis of precipitation: a review. Environ Pollut 120:565–594. doi:10.1016/S0269-7491(02)00165-3

Kumar R, Rani A, Maharaj KK, Srivastava SS (2005) Atmospheric dry deposition to marble and red stone. J Atmos Chem 50:243–261. doi:10.1007/s10874-005-4633-5

Kunicki-Goldfinger J (2008) Unstable historic glass: symptoms, causes, mechanisms and conservation. Rev Conserv 9:47–60

Laurenzi-Tabasso M, Marabelli M (1992) Il degrado dei Monumenti in Roma in Rapporto all'inquinamento atmosferico. BetaGamma, Viterbo

Lea FM (1970) The chemistry of cement and concrete. E. Arnold Publishers Ltd, London

Lefèvre RA, Chabas A, Munier I, Lombardo T (2005) Modern and ancient glass in the polluted atmosphere. What is the prevailing phenomenon? Leaching, corrosion, soiling, encrusting…? In: van Grieken R, Janssens K (eds) Cultural heritage conservation and environmental impact

assessment by Non-destructive testing and micro-analysis. Taylor & Francis Group plc, London, pp 77–90
Leighton PA (1961) Photochemistry of air pollution. Academic, New York
Leissner J (2003) The effect of air pollution on glass. In: Brimblecombe P (ed) The effects of air pollution on the built environment. Imperial College Press, London, pp 249–265
Lerner DN, Barret MH (1996) Urban groundwater issues in the United Kingdom. Hydrogeol J 4(1):80–89. doi:10.1007/s100400050096
Létolle R (1980) Nitrogen-15 in the natural environment. In: Fritz P, Fontes JC (eds) Handbook of environmental geochemistry. Elsevier, Amsterdam, pp 407–733
Létolle R, Leroy P, Gegout P (1988) Fractionnements isotopiques cinétiques du $^{13}C$ dans le système $CO_2$–$CaCO_3$. C R Acad Sci Paris Serie 2(306):799–802
Létolle R, Gegout P, Rafai N, Revertegat E (1992) Stable isotopes of carbon and oxygen for the study of carbonation/decarbonation processes in concrete. Cement Concr Res 22:235–240. doi:10.1016/0008-8846(92)90061-Y
Levin I, Hesshaimer V (2000) Radiocarbon – a unique tracer of global carbon cycle dynamics. Radiocarbon 42:69–80
Lewin SZ (1982) The mechanism of masonry decay through crystallization. In: Conservation of historic stone buildings and monuments. National Academy Press, Washington, DC, pp 120–144. Available online at http://www.nap.edu/openbook.php?record_id=514&page=120. Accessed Aug 2010
Leysen L, Roekensand E, Van Grieken R (1989) Air pollution induced chemicals decay of a sandy-limestone (Cathedral in Belgium). Sci Total Environ 78:263–287. doi:10.1016/0048-9697(89)90038-7
Lipfert FW (1989) Atmospheric damage to calcareous stones: comparison and reconciliation of recent experimental findings. Atmos Environ 23:415–429. doi:10.1016/0004-6981(89)90587-8
Livingstone RA (1992) Graphical methods of examining the effects of acid rain and sulfur dioxide on carbonate stones. In: Delgado Rodrigues J, Henriques F, Telmo Jeremias F (eds) Proceedings of the 7th international congress on deterioration and conservation of stone. Laboratório Nacional de Engenharia Civil, Lisbon, pp 375–386
Lockhart LB (1962) Natural radioactive isotopes in the atmosphere at Kodiak and Wales, Alaska. Tellus 14:350–355. doi:10.1111/j.2153-3490.1962.tb01347.x
Lombardo T, Ionescu A, Chabas A, Lefèvre R-A, Ausset P, Candau Y (2010) Dose-response function for the soiling of silica-soda-lime glass due to dry deposition. Sci Total Environ 408:976–984. doi:10.1016/j.scitotenv.2009.10.040
Longinelli A, Bartelloni M (1978) Atmospheric pollution in Venice, Italy, as indicated by isotope analyses. Water Air Soil Pollut 10:335–341. doi:10.1007/BF00285061
Lopez-Arce P, Doehne E, Martin W, Pinchin S (2008) Magnesium sulfate salts and historic building materials: experimental simulation of limestone flaking by relative humidity cycling and crystallization of salts. Materiales de Construcción 58:125–142. doi:10.3989/mc.2008.v58.i289-290.77
Lovelock J (2008) A geophysiologist's thoughts on geoengineering. Philos Trans R Soc A 366:3883–3890. doi:10.1098/rsta.2008.0135
Luckat S (1976) Stone deterioration at the Cologne Cathedral due to air pollution. In: Proceedings of the international symposium on the conservation of stone, Bologna (Rossi-Manaresi R (ed). Centro C. Gnudi per la Conservazione delle sculture all'aperto, Bologna, pp 37–43
Lunde G, Gether J, Gjos N, Lande May Berit S (1977) Organic micropollutants in precipitation in Norway. Atmos Environ 11:1007–1977. doi:10.1016/0004-6981(77)90229-3
Maas S, Scheifler R, Benslama M, Crini N, Lucot E, Brahmia Z, Benyacoub S, Giraudoux P (2010) Spatial distribution of heavy metal concentrations in urban, suburban and agricultural soils in a Mediterranean city of Algeria. Environ Pollut 158:2294–2301. doi:10.1016/j.envpol.2010.02.001
MacLeod G, Hall AJ, Fallick AE (1990) An applied mineralogical investigation of concrete degradation in a major concrete road bridge. Mineralogical Magazine 54:637–44
MacLeod G, Fallick AE, Hall AJ (1991) The mechanism of carbonate growth on concrete structures, as elucidated by carbon and oxygen isotope analyses. Chem Geol 86:335–343. doi:10.1016/0168-9622(91)90015-O

Magee AW, Bull PA, Goudie AS (1988) Before rock decay: chemicals weathering of constituent grains by salts. In: Marinos PGAND, Koukis GC (eds) Engineering geology of ancient works, monuments and historical sites. Balkema, Rotterdam, pp 779–786

Manning M (1988) Corrosion of building materials due to atmospheric pollution in the united kingdom. In: Mellanby K (ed) Air pollution, acid rain and the environment. The Watt Committee on Energy, London, pp 37–66

Maravelaki-Kalaitzaki P (2005) Black crusts and patinas on Pentelic marble from the Parthenon and Erechtheum (Acropolis, Athens): characterization and origin. Anal Chim Acta 532:187–198. doi:10.1016/j.aca.2004.10.065

Maravelaki-Kalaitzaki P, Biscontin G (1999) Origin, characteristics and morphology of weathering crusts on Istria stone in Venice. Atmos Environ 33:1699–1709. doi:10.1016/S1352-2310(98)00263-5

Marchesini L (1969) Comportamento dei marmi e della pietre a Venezia. In: Roxxi-Manaresi R, Riccòmini E (eds) La conservazione delle sculture all'aperto. Rapporti della Soprintendenza alle Gallerie di Bologna, Bologna, pp 78–86

Martínez-Ramírez S, Puertas F, Blanco-Varela MT, Thompson GE (1997) Studies on degradation of lime mortars in atmospheric simulation chambers. Cement Concr Res 27:777–784. doi:10.1016/S0008-8846(97)00065-3

Martin P, Cabañas B, Villanueva F, Gallego MP, Colemar I, Salgado S (2010) Ozone and nitrogen dioxide levels monitored in an urban area (Ciudad Real) in central-southern Spain. Water Air Soil Pollut 208:305–316. doi:10.1007/s11270-009-0168–8

Mason AS (1977) Atmospheric HT and HTO, 4: estimations of atmospheric hydrogen residence time from interhemispheric tritium gas transport. J Geophys Res 82:5913–5916. doi:10.1029/JC082i037p05913

Matias JM, Alves CAS (2002) The influence of petrographic, architectural and environmental factors in decay patterns and durability of granite stones in Braga monuments (NW Portugal). In: Siegesmund S, Weiss T, Vollbrecht A (eds) Natural stone weathering phenomena, conservation strategies and case studies, The geological society, London special publication 205. The Geological Society, London, pp 273–281. doi:10.1144/GSL.SP.2002.205.01.20

Matsuoka N, Hirai E, Tagomori H, Momoshima N, Takashima Y (1994) Meteorological analysis of tritium concentrations in rain water collected in Fukuoka, Japan, from 1987–1991. Sci Total Environ 145:197–205. doi:10.1016/0048-9697(94)90114-7

Mazuret MA, Simoneit BR (1986) Organic components in bulk and wet-only precipitation. Crit Rev Environ Control 16:1–140. doi:10.1080/10643388609381740

McAlister JJ, Smith BJ, Torok A (2006) Element partitioning and potential mobility within surface dusts on buildings in a polluted urban environment, Budapest. Atmos Environ 40:6780–6790. doi:10.1016/j.atmosenv.2006.05.071

McBride JP, Moore RE, Witherspoon JP, Blanco RE (1978) Radiological impact of airborne effluents of coal and nuclear plants. Science 202:1045–1050. doi:10.1126/science.202.4372.1045

Mirwald PW, Kraus K, Wolff A (1988) Stone deterioration on the Cathedral of Cologne. In: Rossvall J, Aleby S (eds) Safeguarding our architectural heritage. Elsevier, Amsterdam, pp 365–386

Mook WG (1980) Carbon-14 in hydrogeological studies. In: Fritz P, Fontes JC (eds) Handbook of environmental geochemistry. Elsevier, Amsterdam, pp 49–73

Moore H (1977) The isotopic composition of ammonia, nitrogen dioxide and nitrate in the atmosphere. Atmos Environ 11:1239–1243. doi:10.1016/0004-6981(77)90102-0

Moreno F, Vilela SAG, Antunes ASG, Alves CAS (2006) Capillary-rising salt pollution and granitic stone erosive decay in the parish church of Torre de Moncorvo (NE Portugal)-implications for conservation strategy. J Cult Herit 7:56–66. doi:10.1016/j.culher.2005.10.006

Morillo E, Romero AS, Madrid L, Villaverde J, Maqueda C (2008) Characterization and sources of PAHs and potentially toxic metals in urban environments of Sevilla (southern Spain). Water Air Soil Pollut 187:41–51. doi:10.1007/s11270-007-9495-9

Morse JW, Mackenzie FT (1990) Geochemistry of sedimentary carbonates. Elsevier, Amsterdam

Mutch RD Jr, Mahony JD, Paquin PR, Cleary J (2007) Tritium in municipal solid waste leachate and gas: implications to treatment and disposal. HydroQual, Inc. Presentations at the water environment federation, specialty conferences, Rhode Island, July 29–Aug 1. Available online at http://www.hydroqual.com/publications/rdm_07_01_p.pdf. Accessed Aug 2010

Netterberg F, Bennet RA (2004) Blistering and cracking of airport runway surfacing due to salt crystallization. In: Proceedings of the 8th conference on asphalt pavements for Southern Africa (CAPSA'04), Sun City, paper 088. Available online at http://www.capsa-events.co.za/capsa04/Documents/088.pdf. Accessed Aug 2010

Neville A (2004) The confused world of sulfate attack on concrete. Cement Concr Res 34(8):1275–1296. doi:10.1016/j.cemconres.2004.04.004

Nord AG, Ericsson T (1993) Chemical analysis of thin black layers on building stone. Stud Conserv 38:25–35

Nord AG, Holenyi K (1999) Sulphur deposition and damage on limestone and sandstone in Stockholm City buildings. Water Air Soil Pollut 109:147–162. doi:10.1023/A:1005037617131

Nord AG, Tronner K, Boyce AJ (2001) Atmospheric bronze and copper corrosion as an environmental indicator. A study based on chemical and sulphur isotope data. Water Air Soil Pollut 127:193–204. doi:10.1023/A:1005254913598

Obika B, Freer-Hewish RJ, Fookes PG (1989) Soluble salt damage to thin bituminous road and runway surfaces. Q J Eng Geol Hydrogeol 22:59–73. doi:10.1144/GSL.QJEG.1989.022.01.05

Odler I (1998) Hydration, setting and hardening of Portland cement. In: Hewlett PC (ed) Lea's chemistry of cement and concrete. Elsevier, Oxford, pp 241–297

Oesch S, Faller M (1997) Environmental effects on materials: the effect of the air pollutants $SO_2$, $NO_2$, NO and $O_3$ on the corrosion of copper, zinc and aluminium. A short literature survey and results of laboratory exposures. Corrosion Sci 39:1505–1530. doi:10.1016/S0010-938X(97)00047-4

Ogawa M, Yoshida N (2005) Intramolecular distribution of stable nitrogen and oxygen isotopes of nitrous oxide emitted during coal combustión. Chemosphere 61:877–887. doi:10.1016/j.chemosphere.2005.04.096

Ollier C (1979) Weathering. Longman Group Ltd, London

Östman B, Voss A, Hughes A, Hovde PJ, Grexa O (2001) Durability of fire retardant treated wood products at humid and exterior conditions review of literature. Fire Mater 25:95–104. doi:10.1002/fam.758

Oyama T, Chigira M, Ohmura N, Watanabe Y (1996) Heave of house foundation by the chemical weathering of mudstone. In: Proceedings 4th international symposium on the geochemistry of the earth's surface, Yorkshire, pp 614–619

Pachiaudi C, Marechal J, Van Strydonck M, Dupas M, Dauchot-Dehon M (1986) Isotopic fractionation of carbon during co2 absorption by mortar. Radiocarbon 28:691–697

Parmar GR, Rao NN (2009) Emerging control technologies for volatile organic compounds. Crit Rev Environ Sci Technol 39:41–78. doi:10.1080/10643380701413658

Pavía S, Caro S (2006) Origin of films on Monumental Stone. Stud Conserv 51:177–188

Pawelczyk S, Pazdur A (2004) Carbon isotopic composition of tree rings as tool for bimonitoring $CO_2$ level. Radiocarbon 46:701–719

Pazdur A, Nakamura T, Pawelczyk S, Pawlyta J, Piotrowska N, Rakowski A, Sensula B, Szczepanek M (2007) Carbon isotopes in tree rings: climate and the Suess effect interferences in the last 400 years. Radiocarbon 49:775–788

Pearson FJ, Rightmire CT (1980) Sulphur and oxygen isotopes in aqueous sulphur compounds. In: Fritz P, Fontes JC (eds) Handbook of environmental geochemistry. Elsevier, Amsterdam, pp 227–257

Pearson JDM, Wells DM, Seller KJ, Bennerr A, Soares A, Woodall J, Ingrouille J (2000) Traffic exposure increases natural $\delta^{15}N$ and heavy metal concentrations in mosses. New Phytol 147:317–326. doi:10.1046/j.1469-8137.2000.00702.x

Penkett SA, Jones BMR, Brice KA, Eggleton AEJ (1979) The importance of atmospheric ozone and hydrogen peroxide in oxidising sulphur dioxide in cloud and rainwater. Atmos Environ 13:123–137. doi:10.1016/j.atmosenv.2007.10.065

Pierson WR, Brachaczek WW (1983) Particulate matter associated with vehicles on the road II. Aerosol Sci Technol 2:1–40. doi:10.1080/02786828308958610

Pilkey OH, Pilkey-Jarvis L (2007) Useless arithmetic: why environmental scientists can't predict the future. Columbia University Press, New York

Pio CA, Ramos MA, Duarte AC (1998) Atmospheric aerosol and soiling of external surfaces in an urban environment. Atmos Environ 32:1979–1989. doi:10.1016/S1352-2310(97)00507-4

Pleijel H, Klingberg J, Bäck E (2009) Characteristics of $NO_2$ pollution in the city of Gothenburg, South-West Sweden – Relation to $NO_x$ and $O_3$ levels, photochemistry and monitoring location. Water Air Soil Pollut: Focus 9:15–25. doi:10.1007/s11267-008-9201-y

Přikryl R, Svobodová J, Žák K, Hradil D (2004) Anthropogenic origin of salt efflorescences on sandstone sculptures (Charles bridge, Prague, Czech Republic) – mineralogical and stable isotope geochemistry evidence. Eur J Mineral 16:609–618. doi:10.1127/0935-1221/2004/0016-0609

Pryor SC, Barthelmie RJ (2000) Particle dry deposition to water surfaces: processes and consequences. Mar Pollut Bull 41:220–231. doi:10.1016/S0025-326X(00)00112-0

Puppala AJ, Saride S, Dermatas D, Al-Shamrani M, Chikyala V (2010) Forensic investigations to evaluate sulfate-induced heave attack on a tunnel shotcrete liner. J Mat Civ Eng 22:914–922. doi:10.1061/(ASCE)MT.1943-5533.0000087

Pye K, Mottershead DN (1995) Honeycomb weathering of Carboniferous sandstone in a sea wall at Weston-super-Mare, UK. Q J Eng Geol Hydrogeol 28:333–347. doi:10.1144/GSL.QJEGH.1995.028.P4.03

Pye K, Schiavon N (1989) Cause of sulphate attack on concrete, render and stone indicated by sulphur isotope ratios. Nature 342:663–664. doi:10.1038/342663a0

Quintana B, Álvarez-Iglesias P, Santamaría R, Rubio B, Pérez-Arlucea M (2006) Low-level gamma spectrometry for pollution assessment in San Simón Bay (Vigo, Spain). J Phys: Conf Ser 41:400–407. doi:10.1088/1742-6596/41/1/04

Rafai N, Letolle R, Blanc P, Person A, Gegout P (1991) Isotope geochemistry ($^{13}C$, $^{18}O$) of carbonation processes in concrete. Cem Concr Res 21:368–377. doi:10.1016/0008-8846(91)90018-D

Rafai N, Letolle R, Blanc P, Gegout P, Revertegat E (1992) Carbonation–decarbonation of concretes studied by the way of carbon and oxygen stable isotopes. Cem Concr Res 22:882–890. doi:10.1016/0008-8846(92)90112-9

Rahn T, Kitchen N, Eiler J (2002) D/H ratios of atmospheric $H_2$ in urban air: results using new methods for analysis of nano-molar $H_2$ samples. Geochim Cosmochim Acta 66:2475–2481. doi:10.1016/S0016-7037(02)00858-X

Rakowski AZ, Pawelczyk S, Pazdur A (2001) Changes of $^{14}C$ concentration in modern trees from Upper Silesia region, Poland. Radiocarbon 43:679–689

Rakowski AZ, Kuc T, Nakamura T, Pazdur A (2005) Radiocarbon concentration in urban area. Geochronometria 24:63–68

Reimann C, de Caritat P (2000) Intrinsic flaws of element enrichment factors (EFs) in environmental geochemistry. Environ Sci Technol 34:5084–5091. doi:10.1021/es001339o

Risacher F, Fritz B (2009) Origin of salts and brine evolution of Bolivian and Chilean salars. Aquatic Geochem 15:123–157. doi:10.1007/s10498-008-9056-x

Robbins NS, Kinniburgh DG, Bird MJ (1997) Generation of acid groundwater beneath City Road, London. In: Hawkins AB (ed) Ground chemistry: implications for construction. Balkema, Rotterdam, pp 225–232

Roberge PR (2000) Handbook of corrosion engineering. McGraw-Hill, New York

Roberge PR, Klassen RD, Haberecht PW (2002) Atmospheric corrosivity modeling – a review. Mater Design 23:321–330. doi:10.1016/S0261-3069(01)00051-6

Robert M, Bernabe E, Bromblet Ph, Jaunet AM, Verges BV, Penven MJ (1996) Identification of two alteration microsystems chemical and physical, causing granite and kersantite degradation in Brittany (France). In: Vicente MA, Delgado RJ, Acevedo J (eds) Degradation and conservation of granitic rocks in monument. Protection and conservation of the European cultural heritage research report no. 5. European Commission, Brussels, pp 67–71

Rodriguez-Navarro C; Doehne E (1999) Salt weathering: influence of evaporation rate, supersaturation and crystallization pattern. Earth Surf Process Landforms 24:191–209. doi:10.1002/(SICI)1096-9837(199903)24:3<191::AID-ESP942>3.0.CO;2-G

Rodriguez-Navarro C, Hansen E, Sebastian E, Ginell WS (1997a) The role of clays in the decay of ancient Egyptian limestone sculptures. JAIC 36:151–163. Available online at http://cool.conservation-us.org/jaic/articles/jaic36-02-005.html. Accessed Aug 2010

Rodríguez-Navarro C, Sebastián E, Rodríguez Gallego M (1997b) An urban model for dolomite precipitation: authigenic dolomite on weathered building stones. Sediment Geol 109(1–2):1–11. doi:10.1016/S0037-0738(96)00041-3

Roorda-Knape MC, Janssen NAH, de-Hartog J, van-Vliet PHN, H Harssena, Brunekreef B (1999) Traffic related air pollution in city districts near motorways. Sci Total Environ 235:339–341. doi:10.1016/S0048-9697(99)00217-X

Rösch H, Schwarz HJ (1993) Damage to frescoes caused by sulphate bearing salts: where does the sulphur come from? Stud Conserv 38:224–230

Rossi-Manaresi R (1991) Alterazione chimica e degrado di litotipi carbonatici nei monumenti. In: Zezza F (ed) Weathering and air pollution, First course. Community of Mediterranean universities, university school of monument conservation. Mario Adda editore, Bari, pp 139–145

Rozanski K, Araguas-Araguas L, Gonfiantini R (1993) Isotopic patterns in modern global precipitation. In: Swart PK, Lohmann KC, McKenzie J, Savin S (eds) In climate change in continental isotopic records, Geophysical monograph 78. American Geophysical Union, Washington, DC, pp 1–36

Rudd HJ (1995) Emissions of volatile organic compounds from stationary sources in the United Kingdom: speciation. AEA technology report, AEA/CS/REMA-029. AEA Technology, Oxfordshire

Ruijgrok W, Davidson CI, Nicholson KW (1995) Dry deposition of particles. Tellus 47B:587–601

Sabbioni C (1995) Contribution of atmospheric deposition to the formation of damage layers. Sci Total Environ 167:49–55. doi:10.1016/0048-9697(95)04568-L

Sabbioni C (2003) Mechanisms of air pollution damage to stone. In: Brimblecombe P (ed) The effects of air pollution on the built environment. Imperial College Press, London, pp 63–106

Sabbioni C, Zappia G, Ghedini N, Gobbi G (1996) Carbon due to atmospheric deposition on stone monuments and historical buildings. In: Riederer J (ed) Proceedings of the 8th international congress on deterioration of conservation of stone. Möller Druck und Verlag, Berlin, pp 333–340

Sabbioni C, Zappia G, Ghedini N, Gobbi G, Favoni O (1998) Black crusts on ancient mortars. Atmos Environ 32:215–223. doi:10.1016/S1352-2310(97)00259-8

Sabbioni C, Zappia G, Riontino C, Clanco-Varela MT, Aguilera J, Puertas F, Van Balen K, Toumbakari EE (2001) Atmospheric deterioration of ancient and modern hydraulic mortars. Atmos Environ 35:539–548. doi:10.1016/S1352-2310(00)00310-1

Sabbioni C, Ghedini N, Bonazza A (2003) Organic anions in damage layers on monuments and buildings. Atmos Environ 37:1261–1269. doi:10.1016/S1352-2310(02)01025-7

Saiz-Jimenez C (1993) Deposition of airborne organic pollutants on historic buildings. Atmos Environ 27B:77–85. doi:10.1016/0957-1272(93)90047-A

Saiz-Jimenez C (1995) Deposition of anthropogenic compounds on monuments and their effects on airborne microorganism. Aerobiologia 11:161–175. doi:10.1007/BF02450035

Sandberg D, Söderström O (2006) Crack formation due to weathering of radial and tangential sections of pine and spruce. Wood Mater Sci Eng 1:12–20. doi:10.1080/17480270600644407

Sanjurjo-Sánchez J, Vidal-Romani JR, Alves CAS (2004) Characterisation typology and origin of patinas in the church of Las Capuchinas, A Coruña (NW Spain). In: 6th international symposium on conservation of monuments in the Mediterranean Basin, Lisbon, 4–10 Apr, pp 202–206

Sanjurjo-Sánchez J, Vidal Romaní JR, Fernández-Mosquera D, Alves CA (2008) Study of origin and composition of coatings in a monument built with granitic rocks, by SEM, XRD, XRF and DTA-TGA. X-Ray Spectrom 37:346–354. doi:10.1002/xrs.1019

Sanjurjo-Sánchez J, Vidal-Romani JR, Alves CAS, Fernández-Mosquera D (2009) Origin of gypsum-rich coatings on historic buildings. Water Air Soil Pollut 204:53–68. doi:10.1007/s11270-009-0025-9

Sanjurjo-Sánchez J, Vidal-Romaní JR, Alves CAS (2011) Deposition of particles on gypsum-rich coatings of Heritage buildings in urban and rural environments. Construct Build Mater 25:813–822. doi:10.1016/j.conbuildmat.2010.07.001

Santhanam M, Cohen MD, Olek J (2001) Sulfate attack research – whither now? Cement Concr Res 31:845–851. doi:10.1016/S0008-8846(01)00510-5

Santhanam M, Cohen MD, Olek J (2002) Mechanism of sulfate attack: a fresh look. Part 1: summary of experimental results. Cement Concr Res 32:915–921. doi:10.1016/S0008-8846(02)00724-X

Saunders D (2000) Pollution and the national gallery. Natl Gallery Tech Bull 21:77–94

Savin SM (1980) Oxygen and hydrogen isotope effects in low-temperature mineral-water interactions. In: Fritz P, Fontes JC (eds) Handbook of environmental isotope geochemistry, vol 1. Elsevier, Amsterdam, pp 283–327

Schauer JJ, Kleeman MJ, Cass GR, Simoneit BR (1999) Measurements of emissions from air pollution sources. 2. $C_1$ through $C_{30}$ organic compounds from medium duty diesel trucks. Environ Sci Technol 33:1578–1587. doi:10.1021/es980081n

Schiavon N (1993) Microfabrics of weathered granite in urban monuments. In: Thiel MJ (ed) Conservation of stone and other materials I. E&FN Spon. Chapman & Hal, London, pp 271–278

Schiavon N (1996) Soiling of urban granite 1: microfabrics and mineralogical aspects. In; Vicente MA, Delgado RJ, Acevedo J (eds) Degradation and conservation of granitic rocks in monument. Protection and conservation of the European cultural heritage research report no. 5. European Commission, Brussels, pp 307–312

Schiavon N, Chiavari G, Fabbri D, Schiavon G (1994) Microscopical and chemical analysis of black patinas on granite. In: III international symposium on the conservation of monuments in the Mediterranean Basin, Venice, pp 93–99

Schleicher N, Recio-Hernández C (2010) Source identification of sulphate forming salts on sandstones from monuments in Salamanca, Spain-A stable isotope approach. Environ Sci Pollut Res Int 17:770–778. doi:10.1007/s11356-009-0196-3

Seal RR, Aalpers CN, Rye RO (2000) Stable isotope systematics of sulfate minerals. In: Aalpers CN, Jambor JL, Nordstrom DK (eds) Sulfate minerals. Crystallography, geochemistry and environmental significance. Mineralogical Society of America, Washington, DC, pp 541–602

Seinfeld JH (1986) Atmospheric chemistry and physics of air pollution. Wiley Interscience, New York

Seinfeld JH, Pandis SN (1997) Atmospheric chemistry and physics. Wiley, New York

Seymour RB (1988) Determination of chemical susceptibility, engineering plastics, vol 2, Engineered materials handbook. ASM International, Metals park, pp 571–574

Shestoperov S (1988) Road and building materials, vol 1. Mir Publishers, Moscow

Siedel H (2000) Evaluation of the environmental influence on sulphate salt formation at monuments in Dresden Germany by sulphur isotope measurements. In: 9th international congress on deterioration and conservation of stone, Venice, pp 401–409

Sillman S (1999) The relationship between ozone, NOx and hydrocarbons in urban and polluted rural environments. Atmos Environ 33:1821–1845. doi:10.1016/S1352-2310(98)00345-8

Sillman S, Samson FJ (1995) Impact of temperature on oxidant photochemistry in urban, polluted rural and remote environments. J Geophys Res 100:11497–11508. doi:10.1029/94JD02146

Simão J, Ruiz-Agudo E, Rodriguez-Navarro C (2006) Effects of particulate matter from gasoline and diesel vehicle exhaust emissions on silicate stones sulfation. Atmos Environ 40:6905–6917. doi:10.1016/j.atmosenv.2006.06.016

Simoneit BR, Mazuret MA (1981) Air pollution: the organic components. Crit Rev Environ Control 11:219–275. doi:10.1080/10643388109381689

Simpson W, TenWolde A (1999) Physical properties and moisture relations of wood. Wood handbook: wood as an engineering material. General technical report FPL; GTR-113. USDA Forest Service, Forest Products Laboratory 1999, Madison, pp 3.1–3.24

Sims I, Brown B (1998) Concrete aggregates. In: Hewlett PC (ed) Lea's chemistry of cement and concrete. Elsevier, Oxford, pp 907–1015

Singh H, Kumar PK (2009) Analysis of coal ash for race elements and their geo-environmental implications. Water Air Soil Pollut 198:87–94. doi:10.1007/s11270-008-9828-3

Skoulikidis T, Papakonstantinou-Ziotis P (1981) Mechanism of sulphation by atmospheric $SO_2$ of limestone's and marbles of the ancient monuments and statues. I. Observations in situ (Acropolis) and laboratory measurements. Br Corrosion J 16:63–69

Skoulikidis Th, Kalifatidou E, Tsakona K, Evangelatou M (1996) Salt spray tests on untreated and treated marble and stones. In Zezza F (ed) Origin, mechanisms, and effects of salts on degradation of monuments in marine and continental environments. Protection and conservation of the European cultural heritage research report no. 4. European Commission, Brussels, pp 87–98

Smith MR (ed) (1999) Stone: building stone, rock fill and armourstone in construction, Engineering geology special publications 16. Geological Society, London

Smith BJ, Magee RW, Whalley WB (1994) Breakdown patterns of quartz sandstone in a polluted urban environment, Belfast, Northern Ireland. In: Robinson ARD, Williams RBG (eds) Rock weathering and landform evolution. Wiley, Chichester, pp 131–150

Smith BJ, Turkington AV, Warke PA, Basheer PAM, McAlister JJ, Meneely J, Curran JM (2002) Modelling the rapid retreat of building sandstones: a case study from a polluted maritime environment. In: Siegesmund GS, Vollbrecht A, Weiss T (eds) Natural stone, weathering phenomena, conservation strategies and case studies, Special publications 205. Geological Society, London, pp 347–362. doi:10.1144/GSL.SP.2002.205.01.25

Smith BJ, Warke PA, Curran JM (2004) Implications of climate change and increased 'time-of-wetness' for the soiling and decay of sandstone structures in Belfast, Northern Ireland. In: Přikryl R (ed) Dimension stone 2004:new perspectives for a traditional building material: proceedings of the international conference on dimension stone 2004, Prague, Czech republic. Taylor & Francis Group, London, pp 9–14

Song XH, Polissar AV, Hopke PK (2001) Sources of the particle composition in the northeastern US. Atmos Environ 35:5277–5286. doi:10.1016/S1352-2310(01)00338-7

Sorantin H, Steger F (1984) Natural radioactivity of building materials in Austria. Radiat Prot Dosim 7:59–61

Steiger M (2003) Salts and crusts. In: Brimblecombe P (ed) The effects of air pollution on the built environment. Imperial College Press, London, pp 133–181

Steiger W, Dannecker W (1994) Determination of wet and dry deposition of atmospheric pollutants on building stones by field exposure experiments. In: III international symposium on the conservation of monuments in the Mediterranean Basin, Venice, pp 171–177

Steiger M, Wolf F, Dannecker W (1993) Deposition and enrichment of atmospheric pollutants on building stones as determined by field exposure experiments. In: Thiel J-M (ed) Conservation of stone and other materials. E & FN Spon, London, pp 35–42

Stevens CM, Krout L, Walling D, Venters A, Engelkemeir A, Ross LE (1972) The carbon isotopic composition of atmospheric carbon monoxide. Earth Planet Sci Lett 16:147–165. doi:10.1016/0012-821X(72)90183-5

Stevenson FJ, Cole MA (1999) Cycles of soil: carbon, nitrogen, phosphorus, sulfur, micronutrients. Wiley, New York

Stocksiefen W (1982) Soiling of natural stone facade slabs by materials formed by weathering. Fliesen Platten 32:44

Stojanovska Z, Nedelkovski D, Ristova M (2010) Natural radioactivity and human exposure by raw materials and end product from cement industry used as building materials. Radiat Meas 45:969–972. doi:10.1016/j.radmeas.2010.06.023

Storemyr P (2004) Weathering of soapstone in a historical perspective. Mater Charact 53:191–207. doi:10.1016/j.matchar.2004.06.015

Sulovsky P, Gregerova M, Pospisil P (1996) Mineralogical study of stone decay in Charles Bridge, Prague. In: Riederer J (ed) Proceedings of the 8th international congress on deterioration of conservation of stone. Möller Druck und Verlag, Berlin, pp 29–36

Taseiko OV, Mikhailuta SV, Pitt A, Lezhenin AA, Zakharov YV (2009) Air pollution dispersion within urban street canyons. Atmos Environ 43:245–252. doi:10.1016/j.atmosenv.2008.09.076

Tidblad V, Kucera J (2003) Air pollution damage to metals. In: Brimblecombe P (ed) Air pollution damage to metals. Imperial College Press, London, pp 227–247

Tjandraatmadja G, Diaper C (2006) Sources of critical contaminants. In: Domestic wastewater – A literature review. CSIRO and the SmartWater Fund, Australia, 78 p. Available online at http://www.smartwater.com.au/projects/round3/csiro/Documents/CSIRO_LiteratureReview_CriticalContaminants.pdf. Accessed Aug 2010

Toniolo L, Zerbi CM, Bugini R (2009) Black layers on historical architecture. Environ Sci Pollut Res 16:218–226. doi:10.1007/s11356-008-0046-8

Topal T, Sözmen B (2003) Deterioration mechanisms of tuffs in Midas monument. Eng Geol 68(3–4):201–223. doi:10.1016/S0013-7952(02)00228-4

Torfs K, Van-Grieken R (1997) Chemical relations between atmospheric aerosols, deposition and stone decay layers on historic buildings at the Mediterranean coast. Atmos Environ 31(15):2179–2192. doi:10.1016/S1352-2310(97)00038-1

Torfs K, Buzek F, Van-Grieken R (1996) Use of stable isotopes to study the deterioration of building stones. In: Riederer J (ed) Proceedings of the 8th international congress on deterioration of conservation of stone. Möller Druck und Verlag, Berlin, pp 783–789

Torfs KM, Van-Grieken RE, Buzek F (1997) Use of stable isotope measurements of evaluate the origin of sulfur in gypsum layers on limestone buildings. Environ Sci Technol 31:2650–2655. doi:10.1021/es970067v

Török A (2004) Gypsum-induced decay and weathering crusts on limestone buildings in urban environment of Budapest. In: 6th international symposium on the conservation of monuments in the Mediterranean Basin, Lisbon, pp 211–215

Torraca G (1988) Air pollution and the conservation of building materials. In: Rossvall J, Aleby S (eds) Air pollution and conservation. Safeguarding our architectural heritage. Elsevier, Amsterdam, pp 199–204

Townsend HE (ed) (2002) Outdoor atmospheric corrosion, ASTM International, STP 1421. ASTM, West Conshohocken, 385 p

Trethewey KR, Chamberlain J (1995) Corrosion for science and engineering. Longman Scientific and Technical, Harlow

Trudgill ST, Viles HA, Cooke RU, Inkpen RJ, Heathwaite AL, Houston J (1991) Trends in stone weathering and atmospheric pollution at St. Paul's Cathedral, London, 1980–1990. Atmos Environ 25A:2851–2853. doi:10.1016/0960-1686(91)90210-X

Trudgill ST, Viles HA, Inkpen R, Moses C, Gosling W, Yates T, Collier P, Smith DI, Cook RU (2001) Twenty-year weathering remeasurements at St Paul's Cathedral, London. Earth Surf Processes Landforms 26:1129–1142. doi:10.1002/esp. 260

Trujillano R, Iñigo AC, Rives V, Vicente MA (1996) Behaviour of three different types of granite under forced alteration. In: Vicente, MA, Delgado RJ, Acevedo, J (eds) Degradation and conservation of granitic rocks in monument. Protection and conservation of the European cultural heritage research report no. 5. European Commission, Brussels, pp 89–93

Tsunogai U, Yosuke Hachisu Y, Komatsu DD, Nakagawa F, Gamo T, Akiyama K (2003) An updated estimation of the stable carbon and oxygen isotopic compositions of automobile CO emissions. Atmos Environ 37:4901–4910. doi:10.1016/j.atmosenv.2003.08.008

Turkington AV, Paradise TR (2005) Sandstone weathering: a century of research and innovation. Geomorphology 67:229–253. doi:10.1016/j.geomorph.2004.09.028

Turpin BJ, Huntzicker JJ (1995) Identification of secondary organic aerosol episodes and quantification of primary and secondary organic aerosol concentrations during SCAQS. Atmos Environ 29:3527–3544. doi:10.1016/1352-2310(94)00276-Q

Turpin BJ, Huntzicker JJ, Larson SM, Cass GR (1991) Los Angeles summer midday particulate carbon: primary and secondary aerosol. Environ Sci Technol 25:1788–1793. doi:10.1021/es00022a017

Tzortzis M, Tsertos H, Christofides S, Christodoulides G (2003) Gamma radiation measurements and dose rates in commercially-used natural tiling rocks (granites). J Environ Radioact 70:223–235. doi:10.1016/S0265-931X(03)00106-1

Vallet JM, Gosselin C, Bromblet P, Rolland O, Vergès-Belmin V, Kloppmann W (2006) Origin of salts in stone monument degradation using sulphur and oxygen isotopes: first results of the Bourges cathedral (France). J Geochem Explor 88:358–362. doi:10.1016/j.gexplo.2005.08.075

Valls-del-Barrio S, Garcia Vallès M, Vendrell Saz M (2000) Mineralogical decay processes in stones: The case of the Lleida Cathedral. Europ J Miner 12:651–659.

Valls-del-Barrio S, Garcia-Valles M, Pradell T, Vendrell-Saz M (2002) The red-orange patina developed on a monumental dolostone. Eng Geol 63:31–38. doi:10.1016/S0013-7952(01)00066-7

van-der-Sloot HA (2000) Comparison of the characteristic leaching behavior of cements using standard (EN 196–1) cement mortar and an assessment of their long-term environmental behavior in construction products during service life and recycling. Cement Concr Res 30:1079–1096. doi:10.1016/S0008-8846(00)00287-8

Van-Grieken R, Torfs K (1996) Atmospheric aerosols and deposition near historic buildings: chemistry, sources, interrelations and relevance. In: Zezza F (ed) Origin, mechanisms and effects of salts on degradation of monuments in marine and coastal environments. European Commission Research Workshop, Bary, pp 23–35

Vazquez-Calvo C, Gómez Tubío B, Alvarez de Buergo M, Ortega Feliu I, Fort R, Respaldiza MA (2008) The use of a portable energy dispersive x-ray fluorescence spectrometer for the characterization of patinas form the architectural heritage of the Iberian peninsula. X-Ray Spectrom 37:399–409. doi:10.1002/xrs.1088. Available online at http://onlinelibrary.wiley.com/doi/10.1002/xrs.1088/abstract. this paper was published in 2008.

Veiga R, Sanches N, Anjos RM, Macario K, Bastos J, Iguatemy M, Aguiar JG, Santos AMA, Mosquera B, Carvalho C, Baptista Filho M, Umisedo NK (2006) Measurement of natural radioactivity in Brazilian beach sands. Radiat Meas 41:189–196. doi:10.1016/j.radmeas.2005.05.001

Verges-Belmin V (1994) Pseudomorphism of gypsum after calcite, a new textural feature accounting for the marble sulphation mechanism. Atmos Environ 28:295–304. doi:10.1016/1352-2310(94)90104-X

Vidal-Romaní JR, Bourne JA, Twidale CR, Campbell EM (2003) Siliceous cylindrical speleothems in granitoids in warm semiarid and humid climates. Zeitschirft für Geomorphol 47:417–437

Vilela SAG, Alves CAS, Hanson CMS, Rodrigues CII, Maia RN (2005) Estudo isotópico (d34S) de sulfatos em edificações de Braga (NW de Portugal).In: VIII Congresso de Geoquímica dos Países de Língua Portuguesa, Aveiro, pp 679–682

Viles HA (2002) Implications of future climate change for stone deterioration. In: Siegesmund S, Weiss T, Vollbrecht A (eds) Natural stone, weathering phenomena, conservation strategies and case studies, vol 205, Geological society, London, special publications. Geological Society, London, pp 407–418. doi:10.1144/GSL.SP.2002.205.01.29

Virgarzan R (2004) A review of sulfate ozone background levels and trends. Atmos Environ 38:3431–3442. doi:10.1016/j.atmosenv.2004.03.030

Vleugels G, Dewolfs R, Van Grieken R (1993) On the memory effect of limestone for air-pollution. Atmos Environ 27A:1931–1934. doi:10.1016/0960-1686(93)90298-D

Vollmer MK, Walter S, Bond SW, Soltic P, Röckmann T (2010) Molecular hydrogen ($H_2$) emissions and their isotopic signatures (H/D) from a motor vehicle: implications on atmospheric $H_2$. Atmos Chem Phys 10:5707–5718. doi:10.5194/acp-10-5707-2010

Wahlen M, Yoshinari T (1985) Oxygen isotope ratios in $N_2O$ from different environments. Nature 313:780–782. doi:10.1038/313780a0

Watanabe M, Handa T, Ichino T, Kuwaki N, Sakai J (2009) Characterization of patinas that formed on copper exposed in different environments for one month. Zairyo-to-Kankyo 58:143–157. doi:10.3323/jcorr.58.143

Watchman AL (1991) Age and composition of oxalate-rich crusts in the Northern Territory, Australia. Stud Conserv 36:24–32

Weaver ME (1991) Acid rain and air pollution vs. The buildings and outdoor sculptures of Montreal. APT Bull 23:13–19

Webster CR, Heymsfield AJ (2003) Water isotope ratios D/H, $^{18}O/^{16}O$, $^{17}O/^{16}O$ in and out of clouds map dehydration pathways. Science 302:1742–1745. doi:10.1126/science.1089496

Weldon DG (2009) Failure analysis of paints and coatings. Wiley, Chichester

Wendler E, Klemm DD, Snethlage R (1990) Contour scaling on building facades. Dependence on Stone type and environmental conditions. In: Advanced workshop "analytical methodologies of the investigation of damaged stone", Pavia

WHO (2000) Air quality guidelines for Europe, vol 2, European series, No 91. WHO Regional Publications, Copenhagen
Widory D (2007) Nitrogen isotopes: tracers of origin and processes affecting PM10 in the atmosphere of Paris. Atmos Environ 41:2382–2390. doi:10.1016/j.atmosenv.2006.11.009
Widory D, Javoy M (2003) The carbon isotope composition of atmospheric $CO_2$ in Paris. Earth Planet Sci Lett 215:289–298. doi:10.1016/S0012-821X(03)00397-2
Wik M, Renberg I (1996) Environmental records of carbonaceous fly-ash particles from fossil-fuel combustion. J Paleolimnol 15:193–206. doi:10.1007/BF00213040
Winkler EM (1994) Stone in architecture: properties, durability. Springer, Berlin
Wu Y, Hao J, Fu L, Wang Z, Tang U (2002) Vertical and horizontal profiles of airborne particulate matter near major roads in Macao, China. Atmos Environ 36:4907–4918. doi:doi:10.1016/S1352-2310(02)00467-3
Wüst RAJ, Schlüchter C (2000) The origin of soluble salts in rocks of the Thebes mountains, Egypt: the damage potential to ancient Egyptian wall art. J Archaeol Sci 27:1161–1172. doi:10.1006/jasc.1999.0550
Wypych G (1995) Handbook of material weathering. ChemTec Publishing, Toronto
Zannini A, Zambon L, Pagnin P (1991) Restoration of the "Vasca del Pincio" in Rome. In: Baer NS, Sabbioni C, Sors AI (eds) Science, technology and European cultural heritage. Butterworth Heinemann, Oxford, pp 745–746
Zezza F, Macrì F (1995) Marine aerosol and stone decay. Sci Total Environ 167:123–143. doi:10.1016/0048-9697(95)04575-L

# Chapter 3
# Novel Bioreactors for Waste Gas Treatment

**Eldon R. Rene, María Montes, María C. Veiga, and Christian Kennes**

## Contents




E.R. Rene • M. Montes • M.C. Veiga • C. Kennes (✉)
Chemical Engineering Laboratory, Faculty of Sciences, University of La Coruña, Rua da Fraga, 10, E-15008 La Coruña, Spain
e-mail: Kennes@udc.es



E. Lichtfouse et al. (eds.), *Environmental Chemistry for a Sustainable World: Volume 2: Remediation of Air and Water Pollution*, DOI 10.1007/978-94-007-2439-6_3,

**Abstract** Air quality has deteriorated constantly due to the rapid proliferation of fine-chemical and manufacturing industries, in addition to synchronized urbanization. Air pollution is an issue of sensible concern due to the serious implications on health and environment. The contaminants present in industrial waste-gases should either be destroyed or removed from an exhaust vent before it is emitted into the atmosphere by a particular process industry. Traditional waste-gas treatment systems, based on physico-chemical processes for pollutant removal, exhibits high energy costs, expend more chemicals, and produces more wastes. These systems have proved to be economical only when the waste-stream contains high concentrations of pollutants, which can either be recovered or destroyed completely. The use of well-adapted microorganisms to decontaminante the pollutants present in waste-gases, by maintaining them in suitable bioreactor configurations, can eliminate the drawbacks of classical physico-chemical techniques.

In recent years, biological waste-gas treatment systems have been successfully used in a wide variety of process industries, particularly when the exhaust gas-stream contains low concentrations of pollutant at high gas-flow rates. Biological waste-gas treatment systems are versatile in handling a wide variety of air emissions, comprising both individual and mixture of pollutants, from petrochemical, semi-conductor, paint and varnish production facilities, pharmaceutical and paper and pulp industry, among others. The different bioprocesses that are commonly used for air pollution control are the biofilters, biotrickling filters and bioscrubbers. However, recent scientific advances and technical discoveries have led to the development of novel bioreactor configurations that have shown to be successful and promising for treating air toxics with high removal efficiencies. We reviewed the operational details and performance of novel bioreactor configurations such as continuous suspended-growth bioreactors, monolith bioreactors, two-liquid-phase biotrickling filters, air-lift bioreactors, fluidized-bed bioreactors, and foam-emulsion bioreactors. Though all these bioreactors have not been examined at the industrial level for operational/practical constraints, they have shown to effectively handle a wide range of hydrophilic and hydrophobic air pollutants, when investigated at the lab-scale level for steady and transient-state performance. Closely looking at their advantages, we observed that all these novel bioreactors were designed to overcome some operational limitations, *viz.*, flow resistance/pressure drop, oxygen depletion and mass transfer problems for sparingly soluble pollutants, frequently encountered in conventional biological treatment systems.

**Keywords** Bioprocesses • Novel bioreactors • Volatile organic compounds • Biodegradation • Elimination capacity • Removal efficiency • Air quality • Gas treatment • Ozonation • Photooxidation • Biofilters • BTEX • MTBE • VOC • Trichloroethylene • $SO_2$ • $CS_2$ • $NH_3$ • $H_2S$ • CO • HAP • Biofuel • Bioenergy • Bioconversion • Biogas • Biomass • Air pollution • Methane • Hydrogen sulphide • Waste gas • Siloxanes • Carbon dioxide • Microbial kinetics • Bioreactor modelling • Greenhouse gas • Natural gas

## List of Abbreviations

| | |
|---|---|
| AGB | Acid gas biofilter |
| BTEX | Benzene toluene, ethyl benzene and xylene |
| CAAA | Clean air act amendments |
| CSGR | Continuous suspended growth reactor |
| CSTB | Continuous stirred tank bioreactor |
| CSTR | Continuous stirred tank reactor |
| DMS | Dimethyl sulphide |
| EBRT | Empty bed residence time |
| EC | Elimination capacity |
| HAPs | Hazardous air pollutants |
| ILR | Inlet loading rate |
| MEK | Methyl ethyl ketone |
| MIBK | Methyl isobutyl ketone |
| MTBE | Methyl tert butyl ether |

| | |
|---|---|
| PEG | Poly ethylene glycol |
| POTW | Public owned treatment work |
| PVA | Poly vinyl alcohol |
| RE | Removal efficiency |
| TCE | Trichloro ethylene |
| USEPA | United States Environmental Protection Agency |
| VICs | Volatile inorganic compounds |
| VOCs | Volatile organic compounds |

## 3.1 Introduction

### *3.1.1 Background Information on Air Pollution*

Ambient air is the most important natural resource and its availability in adequate quantity and quality is essential to all forms of life. Though air pollution from natural sources is known, it is the anthropogenic sources particularly combustion which have attracted more attention. Ambient air pollutants may include gases or particulate matter, which are small particles of dust, smoke, ash, pollen, or other substances (Kennes and Veiga 2001). Many air pollutants have been found to be harmful to both the environment and human health. The ecosphere continues to be challenged by the increasing amount and variability of toxic contaminants that are emitted by industrial activities. Such environmental contaminants are becoming more widespread as the pace of industrial activity accelerates, especially in developing countries, and also as a consequence of the difficulties in restricting the emissions of industrial activities from cross national boundaries (Daugulis 2001). However, with the increasing development of synthetic chemicals, particularly the petrochemicals, a new class of air pollutants, the Hazardous Air Pollutants (HAPs) have become a matter of concern in recent years. In general, air pollution control strategies and regulations have focused on the acute effects of air pollutants. In the past 50 years, however advances in medical and environmental sciences have led to better understanding of other deleterious effects of air pollutants.

Among these, the environmental and health effects of Volatile Organic Compounds (VOCs), and Volatile Inorganic Compounds (VICs) have resulted in increasingly stringent national and regional regulations. Volatile Organic Compounds are defined as "any compound of carbon, excluding CO, $CO_2$, carbonic acid, metallic carbides or carbonates, which participates in atmospheric photochemical reactions" (Nuñez 1998). The inorganic gases are called volatile inorganic compounds, and this includes inorganic gases such as hydrogen sulphide ($H_2S$), sulphur dioxide ($SO_2$), carbon disulphide ($CS_2$), and ammonia ($NH_3$). The Clean Air Act Amendments of 1990 (CAAA-90), proposed by the United States Environmental Protection Agency (US-EPA), has identified a list of 188 Hazardous Air Pollutants (HAPs), which includes VOCs and VICs. This list includes, 82 VOCs that have a vapor pressure ranging from 0.1 to 308 mmHg at 25°C. Since then, several conferences such as New York (1997) and Kyoto (1998) highlighted the influence of anthropogenic VOC emissions on the environment and the climate imbalance (Darracq et al. 2010).

## *3.1.2 Waste-Gas Treatment Methods*

### 3.1.2.1 Physico-Chemical Methods

The most commonly used physico-chemical air pollution control processes for removing pollutants present in a waste-gas stream are broadly classified into two types: destructive and recovery based. These collectively include processes such as absorption, adsorption, thermal and catalytic combustion, condensation, ozonation, chlorination, photo-catalytic oxidation and membrane separation. In general, both the amount and type of gas-phase pollutants to be treated are likely to have an effect on system configuration and its applicability. The greater the variability of substances to be controlled, the greater the limitation placed upon the selection process (Kennes and Veiga 2001). The economic feasibility of a process is generally decided by the flow rate, concentration, composition and type of waste-gas mixture to be treated. However the reduction mechanism in these processes may reduce the quality of the product or may yield products that require further treatment before ultimate discharge into the atmosphere (Khan and Ghoshal 2000). The disadvantages of these physico-chemical methods are their high-energy costs, as in the case of incinerators, the use of chemicals, which can be costly to purchase or dispose of and require operational safety procedures, in the case of chemical scrubbers, and the production of waste products, such as spent chemicals or spent activated carbon (van Groenestijn and Kraakman 2005).

### 3.1.2.2 Biological Methods

Biodegradation of organic and inorganic gas-phase pollutants has always been a promising and comprehendible technique for waste-gas treatment. In a fixed-bed system, such as a biofilter, the contaminants present in the waste-gas are transferred to the interface of the gas and biofilm that is attached to the bed packing material, where they may be absorbed into the biofilm. The dissolved contaminant in the liquid-phase (biofilm phase) is then utilized by the active microbial consortium as carbon and energy sources, converting the pollutants to carbon dioxide, water and other end-products (Kennes and Thalasso 1998; Kennes and Veiga 2001; Kennes et al. 2009a, b). The major advantages of biological methods are that they are rather inexpensive, reliable and environmentally compatible. Furthermore the pollutants are not merely transferred from one phase to another; instead they are completely mineralized to simple end products such as carbon dioxide and water. The extent to which biological waste-gas purification can occur is primarily determined by the physical and chemical properties of the pollutant(s), microbial physiology and ecology and other suitable environmental conditions.

A biological treatment system extends a new unconventional choice for the treatment of polluted gases, although it has already sporadically been used since the 1920s to remove malodorous compounds from sewage treatment plants (Ottengraf and Diks 1991; van Groenestijn and Hesselink 1993). From the European perspective, Germany and The Netherlands were the first countries to conduct on-field trials and implement biological methods for waste-gas treatment in practical situations (Leson and Winer 1991; Dragt and van Ham 1992). In the past few decades biological treatment techniques have acquired appreciable interest as they have proven to be competitive alternatives to physico-chemical methods, particularly for organic vapors in air streams at low concentrations and high flow rates. In the 1980s, the biofiltration market grew rapidly, and the 1980s and 1990s were the golden era of research and development on biological waste-gas treatment in Europe (van Groenestijn and Kraakman 2005).

A novel bioreactor can be defined as a new reactor configuration that has unique and improved performance characteristics compared to the existing bioreactor designs, and that will play a vital role in decontaminating/mineralizing the pollutants present in waste-gases in an economic and eco-friendly manner. Recent developments in bioreactor design have attempted to address some of the limitations of existing bioreactors. However, further progress in innovative bioreactor design remains a high priority, among others for environmental bioreactors, for increasing oxygen and pollutant transfer from the gas-phase to the microorganisms (Dumont et al. 2006), or to reduce pressure drop and clogging problems in packed bed bioreactors.

## 3.2 Physico-Chemical Processes for Waste-Gas Treatment

A large number of physico-chemical based treatment options exist for industrial waste-gases, which are primarily chosen depending on the type and composition of the pollutant present in it. The most prominent methods, pertaining to industrial application, *viz.*, absorption, adsorption, condensation, ozonation and photo-catalytic oxidation, are discussed in this section.

### *3.2.1 Absorption*

Absorption is a diffusional mass transfer operation by which soluble gaseous pollutants are removed by direct dissolution in a solvent liquid. The driving force for mass transfer is the concentration difference of the solute between the gaseous and liquid-phases. This process for waste-gas treatment uses water or low volatile organic liquids as the absorbing liquid (Schnelle and Brown 2002). According to Chungsiriporn et al. (2005), the most commonly used reactor configurations are packed and plate columns, open spray chambers and towers, cyclonic spray chambers and combinations of spray and packed chambers. For VOC treatment in wet-scrubbers, oxidizing chemicals, such as potassium permanganate ($KMnO_4$), chlorine ($Cl_2$) and sodium hypochlorite (NaOCl) have been used. The addition of a strong oxidizing reagent, like sodium hypochlorite in the scrubbing has shown to act both as conveyors and as oxidizers for the treatment of gas-phase VOCs. Chungsiriporn et al. (2005) studied toluene removal by oxidation with sodium hypochlorite in a spray-type wet-scrubber. The process parameters studied were gas-phase toluene concentration, gas-flow rate and sodium hypochlorite solution flow rate, sodium hypochlorite concentration, and oxidant drop size. It was observed that, an increase in sodium hypochlorite solution concentration, flow rate, and toluene concentration with low gas-flow rate (*i.e.*, high residence time) and smaller drop size in the spray scrubber improved the overall toluene removal efficiency, and a maximum removal efficiency of 85% was reported.

An absorption system can be designed to handle a capacity of 1–5 $m^3/s$, and VOCs concentrations ranging from 500 to 5,000 $ppm_v$, with removal efficiencies (REs) as high as 98% (William and Lead 1997). The effectiveness of these equipments requires the selection of an appropriate scrubbing liquid. Inorganic gases such as $SO_2$, $NH_3$, $H_2S$ and water soluble VOCs have been removed by this technique (Calvert and Englund 1984). Absorption involves high investment, large area and high operation costs.

### 3.2.2 Adsorption

Adsorption is a surface phenomenon in which the gaseous contaminant mixture is brought into contact with a suitable solid adsorbent. This technique is widely used to de-pollute waste-gas streams containing odors, VOCs and other trace pollutants. Generally, this process is classified as physical adsorption or chemisorption based on the interactions between the adsorbate and adsorbent. For VOC control, physical adsorption has been found to play a more significant role than chemisorption. It is reported to occur when organic molecules are held on the surfaces and in the pores of the adsorbent by weak van der Waals forces of attraction, characterized by low heat of adsorption and by the fact that adsorption equilibrium is reversible (Ruthven 1984). The traditional and advanced adsorption techniques can be classified as follows; (i) temperature-swing adsorption, (ii) pressure-swing adsorption, (iii) tapered pressure-swing adsorption, (iv) layered bed pressure-swing adsorption, (v) thermally coupled pressure-swing adsorption, (vi) vacuum-swing adsorption, and (vii) simulated moving-bed adsorption. The process fundamentals and the dynamics of pollutant removal and recovery efficiencies achieved in the above-mentioned techniques are given elsewhere (Ghoshal and Manjare 2002).

In waste-gas treatment, the commonly used adsorbent is activated carbon (Kennes and Veiga 2001). The adsorption capacity of a typical activated carbon process for a given VOC is represented by an adsorption isotherm for the amount of VOC adsorbed at constant temperature. After achieving the breakthrough, the pollutant should be released (desorption) from the surface to reuse the carbon, which is commonly done

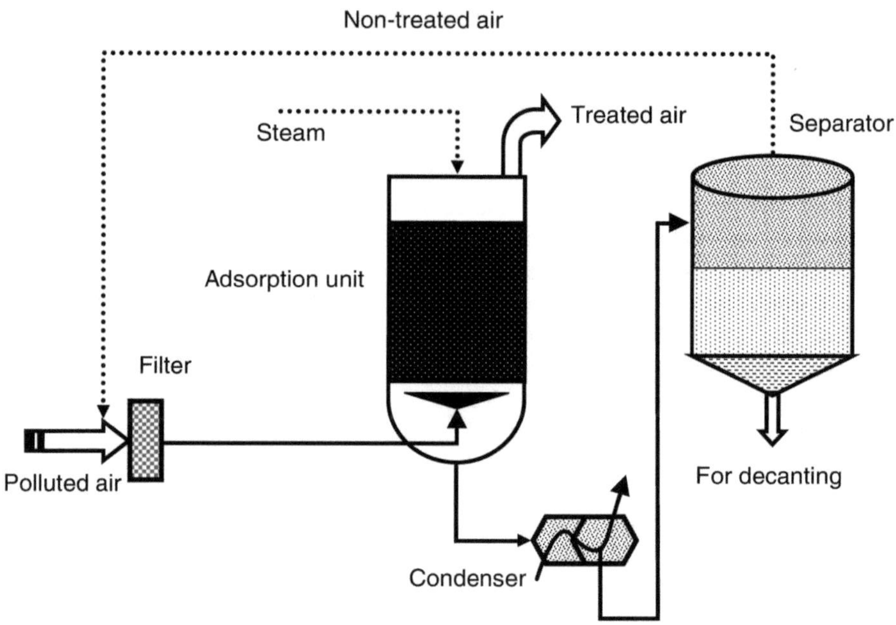


**Fig. 3.1** Typical activated carbon-based waste-gas purification system; steam is not fed continuously, but only for the regeneration cycle. For practical applications, chemically impregnated activated carbons are used to control inorganic pollutants such as hydrogen sulfide, mercury, and radon. This process is particularly effective for compounds (pollutants in gas-phase) that have higher molecular weight and high boiling points

with steam. A typical activated carbon based VOC removal plant (solvent recovery) is shown in Fig. 3.1. The waste-gas passes through a bed of activated carbon, where the pollutant is physically attracted and held to the surface of the carbon molecules. Clean air is then vented out to the atmosphere. When all the available surfaces of the carbon pores are occupied, it will not capture any additional pollutant, *i.e.,* the VOC present in the waste-gas. For recovering the solvent, for reuse, the most common procedure followed is desorption with steam. The hotter the carbon, the less solvent it can hold, so as the steam heats the carbon, solvent is released and flushed away by the steam. This mixture of steam and solvent is condensed by cooling and then separated in the simplest case by gravity decanting. Regeneration of carbon can be done on –or/off site, depending upon the process condition and local utilities. Moisture content, retention ability, pressure, concentration, type of adsorbent and desorption method used are some of the critical parameters that decides the efficiency of an adsorption process (Khan and Ghoshal 2000). Hydrophobic zeolites are now being considered as an alternative adsorbent since it has good thermal stability and hydrophobicity (Blocki 1993; Tsai et al. 1996). Many odorous organic compounds such as acetone, benzene, toluene, xylenes, trichloroethylene, hydrogen sulphide, mercaptans, sulphur dioxide, etc. have been successfully removed by this technique (Engleman 2000; Hamdi et al. 2004; Bagreev et al. 2005). Though adsorption based techniques have been implemented for some gas/vapor purification systems, it is still associated with several limitations, such as; (i) high initial costs, (ii) difficulty in maintenance of the complicated setup, (iii) high mechanical energy for pressurization, and (iv) desorbate recovered at very low purity.

### 3.2.3 Condensation

Condensation, refrigeration, and cryogenic systems are recovery based techniques that are used to remove the organic vapors, mostly VOCs, present in waste-gases, by condensing them onto a colder surface. Cryogenic condensation is a cost-effective technology for controlling emissions of VOCs from chemical processing facilities, only under certain conditions. These extremely cold conditions can be created by passing cold water through an indirect heat exchanger, by spraying cold liquid into an open chamber with the gas stream, by using a Freon –based refrigerant to create very cold coils, or by injecting cryogenic gases such as liquid nitrogen into the gas stream. The concentration of VOCs is reduced to the level equivalent to the vapor pressures of the compounds at the operating temperature. Condensation and refrigeration systems are usually used for high concentration, low gas-flow rate sources. Typical applications include gasoline loading terminals and chemical reaction vessels. Besides, they have proved to be more efficient for VOCs having a boiling point above 40°C, at waste-gas concentrations greater than 5,000 $ppm_v$. Low boiling-point pollutants can require extensive cooling or pressurization which could increase the operating costs (Khan and Ghoshal 2000). The removal efficiencies attainable with this approach depend strongly on the outlet gas temperature. For cold-water-based condensation systems, the outlet gas temperature is usually in the 4–10°C range, and the typical VOC removal efficiencies are in the 90–99% range depending on the vapor pressures of the specific compounds. For refrigerant and cryogenic systems, the removal efficiencies can be considerably above 99% due to the extremely low vapor pressures of essentially all VOC compounds at the very low operating temperatures of −50°C to less than −130°C. To avoid explosion hazards the condensing unit should be adequately enclosed and provided with inert gas blanketing that increases the overall equipment, operating and maintenance costs. This would require rigorous maintenance and skilled labours to handle the unit effectively.

### 3.2.4 Ozonation

Ozone is preferred for many oxidation processes owing to its high reactivity in both aqueous and gaseous phase. This is mainly attributed to the electronic configuration and resonance structure. Though at present there are many studies on the oxidation of pollutants in aqueous environment using direct ozonation (Corless et al. 1990; Xiong and Graham 1992), the studies in gas-phase are scarce (Shen and Ku 2002; Pengyi et al. 2003). Direct ozonation is not practiced in industrial situations owing to the high cost involved in producing ozone and possible transformation of the target pollutant into toxic intermediates.

### 3.2.5 Photo-Oxidation

Photo-oxidation is a promising method for the complete mineralization of organic compounds present in aqueous phase. In recent years, the application of this technology has been extended to waste-gas treatment, as demonstrated by Blystone et al. (1993) and Bhowmick and Semmens (1994). The application of ultraviolet (UV) photo-oxidation to remove gas-phase pollutants was studied in detail by Wang and Ray (2000). The following reasons were cited as the major driving force to use this technology for waste-gas treatment (Wang and Ray 2000); (i) lower UV absorption by air than water, (ii) higher mobility of dissociated species which prevents the reverse process of recombination of the radicals, (iii) higher UV-absorbance of the organics in the gas-phase compared to that in the liquid-phase, (iv) presence of excess oxygen in the gas-phase promotes oxidation by producing reactive species such as ozone, and (v) absence of scavengers such as bicarbonate and carbonate ions. In certain cases, UV oxidation process has been found to be more cost-effective than activated carbon-based adsorption process, as demonstrated by McGregor et al. (1988).

Recently, the photo-oxidation of VOCs, mainly maize derivatives, was investigated by Domeño et al. (2010), and the performance of photo-oxidation, chemical oxidation by hypo-chlorite and ozonation was tested. As shown in Fig. 3.2, the waste-gas from the industrial stack first enters a cyclonic heat exchanger, where some of the vapor from the gas condenses, small particulates are removed, and the temperature of the gas decreases. The gas from this stage then passes to the second section, where organic compounds contained in the gas are oxidized in a cylindrical chamber into which oxygen radicals and ozone, produced from the ambient air in the photo-oxidation module (at 254 and 185 nm), are continuously injected. It has been suggested that the ratio of emission to fresh air flow should be maintained at about 4:1. The following dimensions are suggested for these reactors; oxidation chamber–$3 \times 0.4 \times 0.4$ m, photo-oxidation module-$2 \times 0.7 \times 0.7$ m, with 64 lamps having a maximum power input of 1.4 kW. The ideal waste-gas contact time should be less than 10 s.

### 3.2.6 Photo-Catalytic Oxidation

The destruction of VOCs by photo-catalytic means is one of the emerging purification techniques for both wastewater and waste air, due to its versatility and high effectiveness at low temperatures. Heterogeneous photocatalysis is a process that is based on the photonic excitation of a catalyst (mostly $TiO_2$) at the appropriate wavelength of an incident UV light. In general, organic compounds have a redox potential

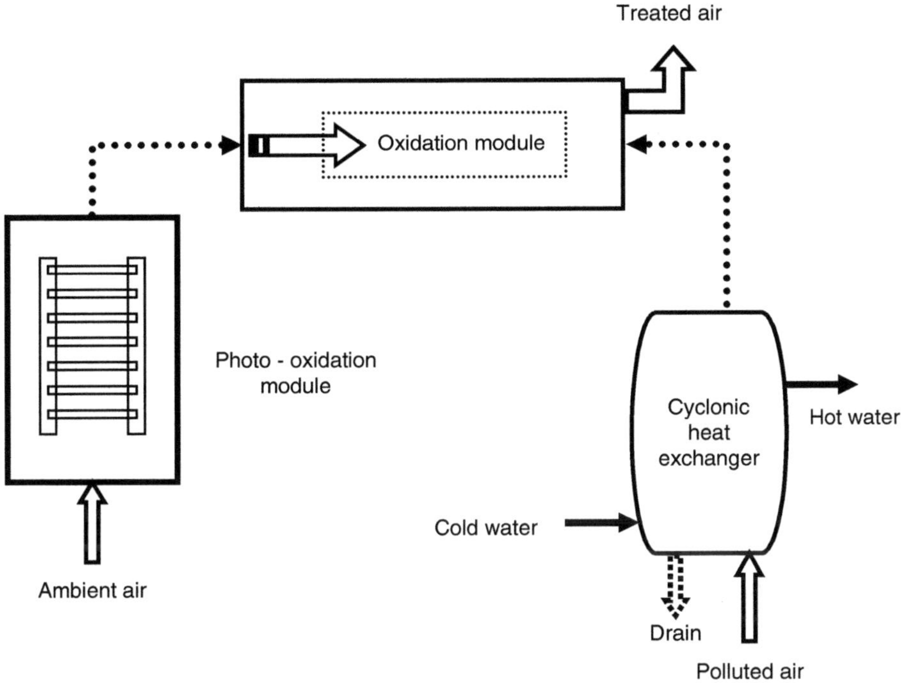


**Fig. 3.2** Schematic of photo-oxidation plant to treat volatile organic compounds and odors; photo-oxidation techniques would be more cost effective for the treatment of low levels of volatile organic compounds in waste-gases that arise from air stripping of organics from the aqueous phase

at higher energy than the valence band edge of common semiconductor oxides and therefore can act as an electron donor yielding a radical cation, which can further react with $H_2O$ or $O_2$. The chemistry occurring at the surface of a photo excited semiconductor is based on the radical formed from $O_2$, $H_2O$ and electron rich organic compounds. A general schematic of the mechanism of photocatalytic oxidation is illustrated in Fig. 3.3. Photocatalysis over $TiO_2$, a semiconductor oxide, is initiated by the absorption of photon with energy equal to or greater than the band gap of the semiconductor (3.2 e.V. for $TiO_2$), producing electron-hole ($e^-/h^+$) pairs. The photo-generated hole and electron pair move about freely on the crystal lattice, and they tend to recombine dissipating heat energy or reach the surface of catalyst to react with adsorbed species. If these charge carriers can be separated from each other, they can be used for oxidation or reduction of the gas-phase pollutants. The water molecules, *i.e.*, humidity, are reduced to hydroxyl radical ($OH^-$), and at the reducing site, the photo-generated electrons mainly react with molecular oxygen forming super oxide ($O_2^-$). This in turn undergoes successive reactions to form hydroxyl radicals. The hydroxyl radicals ($OH^-$) drive the chemical reaction by attacking oxidizable contaminants, producing a progressive breaking of molecules yielding $CO_2$, $H_2O$, mineral acids and other end-products. While the overall stoichiometry of most

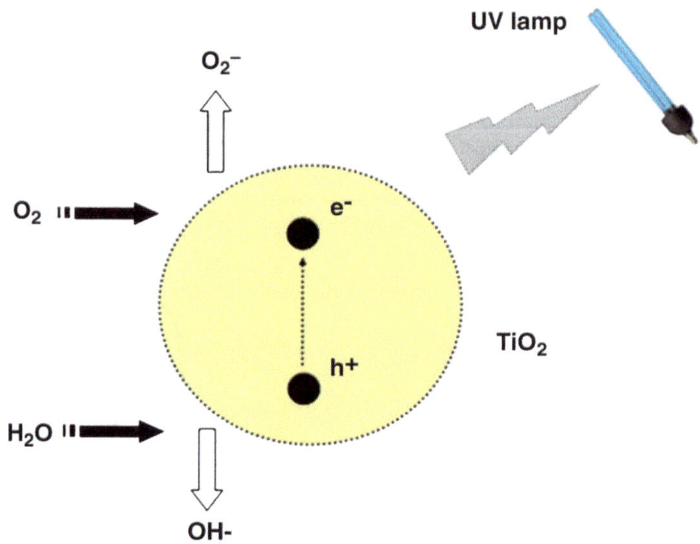


**Fig. 3.3** Schematic of the mechanism involved in photocatalytic oxidation; $TiO_2$ has been used as the catalyst to suit a wide range of industrial applications, including water treatment. In recent years, photocatalytic detoxification of volatile organic compounds is generally more efficient in the gas-phase than in aqueous phase

mineralization appears to be understood, details of the complex mechanism involved are still unclear. This mechanism primarily depends on the characteristics of irradiation, mass of photocatalyst and concentration of the reactants (Pichat 2003).

Most of the recent studies on the removal of VOCs have chosen $TiO_2$ as the catalyst for the following reasons: (i) inexpensive and easy availability (ii) operation at room temperature and pressure and (iii) effective degradation of VOCs into $CO_2$ and $H_2O$ (Lichtin and Sadeghi 1998; Jeong et al. 2004; Xie et al. 2004). Moreover recent studies have shown better removal of VOCs by a combined $O_3/TiO_2$/UV process, as excess ozone molecules could scavenge hydroxyl radicals produced from the excitation of $TiO_2$ by UV radiation (Shen and Ku 2002; Pengyi et al. 2003). Large scale installations for air purification by $TiO_2$ photocatalysis have been built by Trojan Technologies in North America and by the United Technologies of Connecticut (as reported by Pichat 2003). These systems were designed to strip hydrocarbons from the soil (or) groundwater and photo-catalytically irradiate (treat) them to acceptable limits. However because of the poor yield of photon utilization, the treatment of gaseous VOCs may require reactor sizes that are prohibitive in some cases. Though this technique has been reported to be simple, robust and flexible, its operating cost arises mainly from the UV lamps and the corresponding electrical consumptions. For these reasons, photocatalytic oxidation processes appear to be an insignificant option while designing/deciding appropriate VOC control techniques for process industries.

## 3.2.7 *Thermal and Catalytic Oxidation*

### 3.2.7.1 Thermal Oxidation

The process equipment essentially consists of a chambered unit, equipped with a propane or natural gas burner and a stack. The organic compounds present in the waste-gas are burned out completely with the evolution of heat. But unless the quantity of VOCs present is very large, additional energy is required to raise the VOC to the temperature required for complete destruction, which is supplied by an auxiliary fuel such as natural gas. Typical conditions for thermal oxidation are as follows; (i) 725–825°C in the combustion chamber and (ii) retention time of 0.5–1 s. The typical values for heat of combustion for benzene, toluene and xylene (BTX) are $4.1\times10^7$, $4.2\times10^7$, and $4.2\times10^7$ J/kg respectively. Flaring can be done to control a wide variety of flammable VOCs (Schnelle and Brown 2002), but cannot be used for dilute VOC streams, less than about 200 BTU/scf without supplemental fuel. A typical thermal oxidation system (fume incinerator) can handle a capacity of 0.02–5 $m^3$/min and VOC concentration ranging from 100 to 2,000 $ppm_v$.

Modern thermal oxidizers can achieve a RE of 90–98%, irrespective of the VOCs. In general, for thermal oxidizers to be economically acceptable, some form of heat recovery system must be used. The two basic types of heat exchangers namely recuperative or shell and tube type and the regenerative or ceramic bed type exchangers are commonly used (Donley and Lewandowski 1996). But this technique is a costly disposal method to treat low concentrations of organics contained in process exhausts (Khan and Ghoshal 2000). Some VOCs like halogenated and chlorinated VOCs would require additional downstream control equipments.

### 3.2.7.2 Catalytic Oxidation

Catalytic oxidation allows relatively lower process temperatures (200–500°C) with Pt, Pd, Rh, $CuO_2$, $MnO_2$ etc., as the catalyst depending on the gaseous pollutants present. Moreover complete mixing of $O_2$ and VOC is needed for chemical oxidation to occur, which is normally accomplished by ensuring high turbulence (large Reynolds number) within the oxidizer. Plasma Catalytic (Francke et al. 2000) and non-thermal plasma processing (Oda 2003) are the recently introduced developments in this field.

## 3.3 Biological Processes for Waste-Gas Treatment

The implementation and use of biological techniques for VOC removal at industrial scale is currently growing exponentially. The most commonly used biological waste-gas treatment techniques include bioreactor configurations such as the biotrickling filter, bioscrubber and biofilter. Though the mode of operation for all these configurations is very similar, they are distinguished by the behavior of the liquid-phase (continuously moving or stationary) and by the location of the microorganisms (freely dispersed or immobilized). However while choosing an appropriate treatment technique, focus is placed on the operational and control requirements needed to ensure an optimal chemical and physical environment for mass transfer and biodegradation of the pollutant in order to achieve high removal efficiencies (Kennes and Thalasso 1998; Waweru et al. 2000). Moreover microorganisms require a steady supply of organic material as a carbon source and $O_2$ as terminal electron acceptor for aerobic metabolism of the contaminant. Generally the pollutant is fed as the sole energy and carbon source to the microbes, where complete mineralization leads to the formation of $CO_2$, $H_2O$, biomass and some amount of heat. This is given by;

$$\underset{\text{(Substrate)}}{\text{Organic pollutant}} + \underset{\text{(Electron acceptor)}}{O_2} \xrightarrow{\text{Micro organisms}} \underbrace{CO_2 + H_2O}_{\text{Innocuous End products}} + \underset{\text{(Exothermicity)}}{\text{Heat}} + \underset{\text{(Residue)}}{\text{Biomass}} \quad (3.1)$$

Depending on the nature of the pollutant, other end products can also be produced. For example: during the biodegradation of dichloromethane, HCl may appear as one of the end-products. However this reaction (3.1) does not take place as simple as illustrated here, instead involves elaborate degradation pathways at controlled pH, temperature and other necessary conditions.

### 3.3.1 Performance Indicators

The performance of biological waste-gas treatment systems can be described with the parameters defined hereafter;
Volumetric Loading Rate:

$$VLR = \frac{Q}{V}, \left[\frac{m^3}{m^3.h}\right] \quad (3.2)$$

Mass Loading Rate:

$$MLR = \frac{Q.S_{in}}{V}, \left[\frac{g}{m^3.h}\right] \quad (3.3)$$

Elimination Capacity:

$$EC = \frac{Q.(S_{in} - S_{out})}{V}, \left[\frac{g}{m^3.h}\right] \quad (3.4)$$

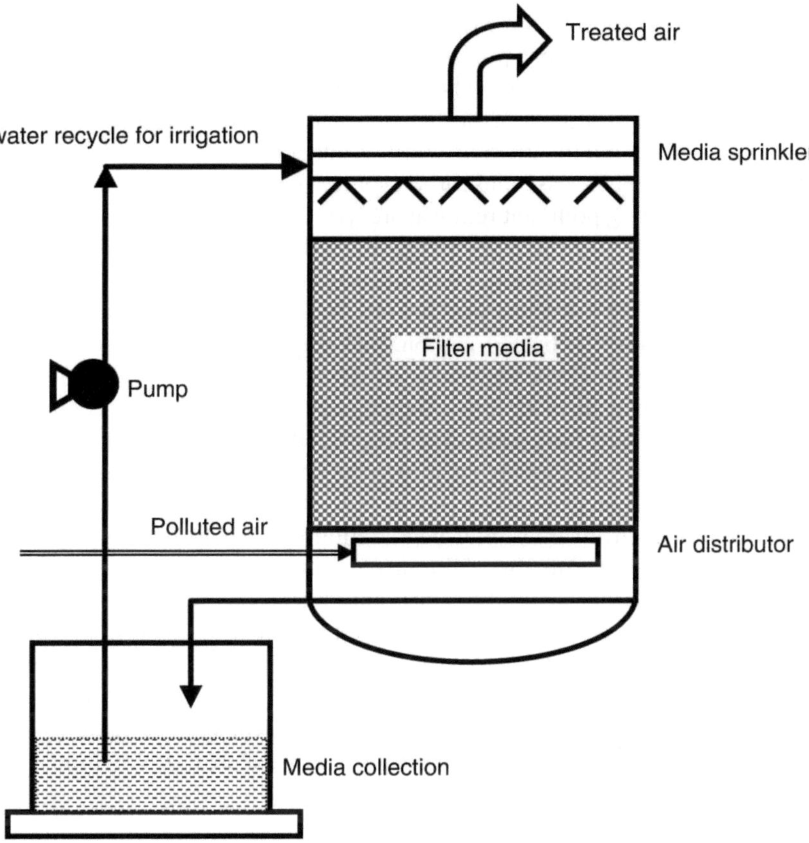


**Fig. 3.4** Schematic of a biotrickling filter; for industrial applications, it has been reported that an existing full-scale chemical scrubber was converted to a biological trickling filter for the removal of hydrogen sulphide ($H_2S$), at gas contact times as low as 1.6 s

Removal Efficiency:

$$RE = \frac{(S_{in} - S_{out})}{S_{in}} \times 100, [\%] \tag{3.5}$$

where, Q is the gas-flow rate ($m^3/h$), V is the volume of the filter bed ($m^3$) and $S_{in}$ and $S_{out}$ are, respectively, the inlet and outlet pollutant concentrations ($g/m^3$).

## 3.3.2 *Biotrickling Filters*

In biotrickling filters, the pollutant laden waste stream is passed over a microbial consortium immobilized on a support material having high surface area (Fig. 3.4).

The process of gas absorption, liquid-phase regeneration and subsequent degradation occurs simultaneously in one process apparatus. A continuous stream of re-circulating water containing essential nutrients for microbial growth is distributed over the filter bed; this water flows down as a thin film and wets the bio-layer. The waste-gas stream passes through the trickling biofilter, either co-or countercurrent to the liquid-phase and supplies the essential carbon source for microbes.

Factors affecting pollutant removal are; (i) composition and concentration of the waste-gas stream, (ii) structural configuration of the packing material, (iii) flow pattern, (iv) nutrient composition, (v) residence times, (vi) pH, and (vii) temperature (Chou and Wu 1999). Typical packing are made of inert materials such as ceramic or plastic structures, activated carbon, pall rings, celite or mixtures of materials (Pedersen and Arvin 1999; Shinabe et al. 1995; Weber and Hartmans 1996; Kennes and Thalasso 1998; Jin et al. 2005; Bailón et al. 2009; Rene et al. 2010a).

Biotrickling filters have been used successfully to treat compounds that produce acidic or alkaline metabolites such as halogenated hydrocarbons, $H_2S$, $NH_3$ and dichloromethane (Jin et al. 2005; Bailón et al. 2009; Rene et al. 2010b). The metabolites generated from microbial degradation can easily be removed by the re-circulating liquid stream. In addition, they are able to withstand high pollutant loadings due to higher internal biomass concentrations (Kennes et al. 2009a, b).

The main drawback of this system is the problem associated with mass transfer, especially for compounds that are hydrophobic. Still, they can be effective in the treatment of gaseous compounds with an air/water partition coefficient of less than 0.1 (Kennes and Thalasso 1998). The other drawbacks include channeling, and the accumulation of excess biomass, among others (Sorial et al. 1993; Smith et al. 1998; Kennes and Thalasso 1998).

### 3.3.3 Bioscrubbers

A typical bioscrubber consists of two units, as illustrated in Fig. 3.5. The first part is an absorption tower where pollutants are absorbed in a liquid-phase followed by biodegradation in a bioreactor. This bioreactor contains suspended activated sludge which is sufficiently aerated and much larger than the absorber. The effluent of this bioreactor is then recirculated over the absorption tower in a co-or counter-current mode to the flow of waste-gas. Microbial activity is enhanced by adding sufficient nutrients in the reactor. This makes them flexible to handle fluctuating loads of waste-gas streams (van Groenestijn and Hesselink 1993; van Groenestijn 2001; Kennes et al. 2009a). The main advantages of this technique are (i) better removal of reaction products by washing out (ii) no clogging problem and (iii) low occurrence of toxic metabolites in water phase.

Typical volatile compounds that have been removed in this system include; phenol, $NH_3$, methanol, isopropyl alcohol, acetone, heptane and $H_2S$ (Whaley et al. 1998; van Groenestijn 2001). However, this technique is only effective for pollutants having partition coefficient values less than 0.01 (Kok 1992). The mass transfer

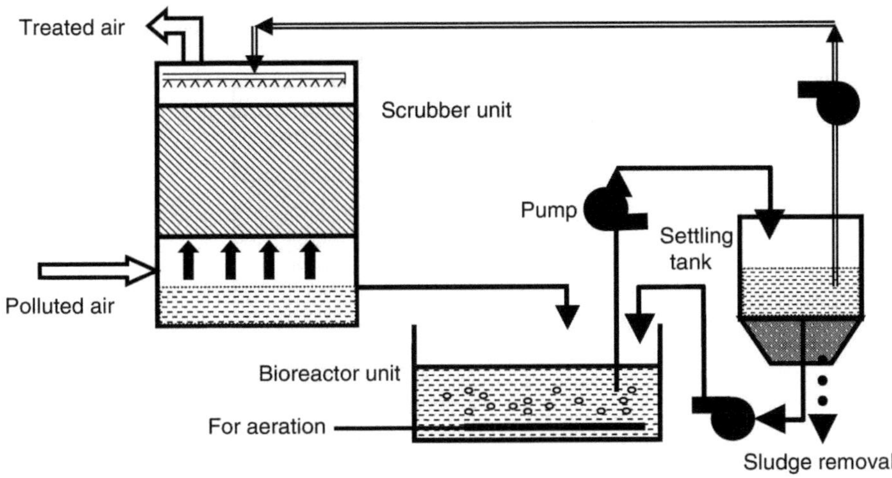


**Fig. 3.5** Schematic of a bioscrubber; when the polluted air enters the scrubber unit, the scrubbing liquid containing the microorganisms will solubilize the pollutants and the pollutants in liquid-phase would now be subsequently degraded by the microorganisms in the bioreactor unit (Kennes and Veiga 2001)

resistance of pollutant from gas-phase to water phase poses a major problem for VOCs like benzene, toluene, ethylbenzene and xylene (BTEX). The energy consumption is high due to continuous liquid recycle and extra aeration which increases the operating costs. Moreover there are sufficient chances for the slowest growing microbial community to get washed out resulting in lesser removal of the target contaminant.

A wide range of organic and inorganic compounds have been treated in bioscrubbers, reaching high removal efficiencies. The most successful removal occurs for low molecular weight and highly water soluble organic compounds. Organic compounds such as alcohols, aldehydes and some simple aromatics have demonstrated excellent biodegradability (Kennes and Thalasso 1998; Devinny et al. 1999; Kennes and Veiga 2001; Kennes et al. 2009a, b). Whaley et al. (1998) treated isopropyl alcohol, acetone and heptane in a pilot scale bioscrubber in a concentration range of 200–500 mg/m$^3$ and observed RE greater than 99%, 2 weeks after startup. Schippert (1986) investigated the removal of mixtures of butyl glycols, *n*-butanol, ethyl glycol, isobutanol, xylene and methyl isobutyl ketone (MIBK) and showed that xylene and MIBK were poorly removed (70%), compared to other VOCs (99%) in bioscrubbers. In general, the scrubbing liquid and waste-gas should be carefully selected, as they lose their cost efficiency for compounds having a Henry's coefficient ($H$)>0.01. Though bioscrubbing presents high potential for development because VOC transfer and biodegradation could be separately improved (Le Cloirec et al. 1999), they have not been used widely due to complex startup procedure, high operational cost and sludge generation.

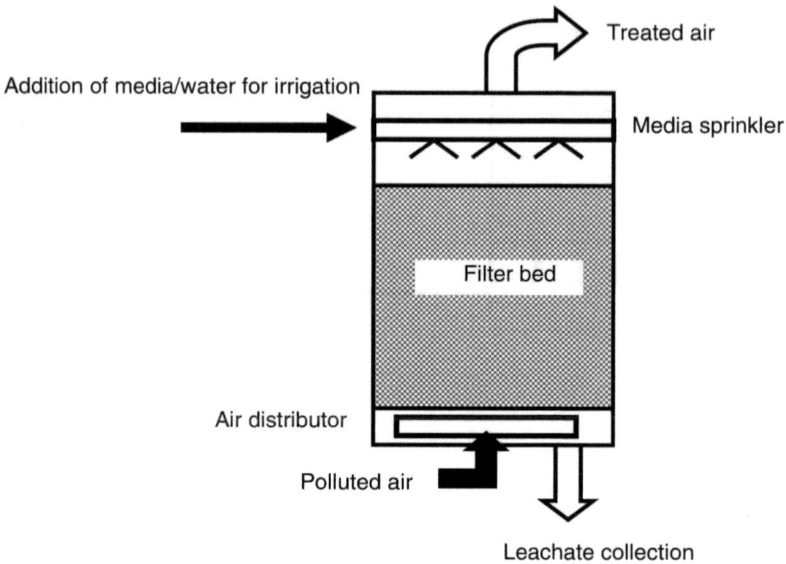


**Fig. 3.6** Schematic of a biofilter; peat and compost based biofilters have been used commercially for treating large volumes of air containing biodegradable volatile organic compounds at low concentrations (<200 $ppm_v$), but filter bed clogging, channelling and flow-maldistribution have often been reported

### *3.3.4 Biofilters*

Biofiltration can be done in aerobic/anaerobic bioreactors, in which a contaminated air stream is passed through a porous packed medium that supports a thriving population of microbes (Fig. 3.6). The principles governing this are similar to the common biofilm process. The contaminants of gas-phase (VOCs) are first absorbed from the air-phase to the water of the biofilm inter-phase. Then, this contaminant diffuses through the biofilm that holds the microorganisms. The microorganisms play a major role in converting them into $CO_2$, water and other end-products. On the contrary, most hydrogen sulphide ($H_2S$) degraders are autotrophic organisms, which has the ability to use carbon dioxide as its carbon source. However, in heterotrophic co-metabolic degradation processes, another carbon source may be required besides the pollutant itself. The microorganisms obtain sufficient energy from the oxidation of the chemical, while utilizing them as a primary substrate via specific enzymes. The degree of treatment depends on the type and composition of the waste-gas stream (solubility factor), activity of the microbes, pH, temperature, moisture content of the filter bed, nutrient and $O_2$ availability and other suitable conditions needed for biodegradation. The design of conventional biofilters have varied and improved over the past 50 years. Also a number of extensive reviews and studies about the technical aspects of biofiltration have been published in the past few years (Kennes and Thalasso 1998; Rene et al. 2004; Estévez et al. 2005; Mohammad et al. 2007; Jin et al. 2006a, 2008; Rene et al. 2009a; Kennes et al. 2009b).

**Table 3.1** List of individual pollutants treated in biofilter and the corresponding elimination capacity values, reported from literature for biofilters

| Pollutant | Packing material | Microorganism | $EC_{max}$ ($g/m^3.h$) | References |
|---|---|---|---|---|
| Dichloromethane | Compost, perlite, (50%v/v) | Activated sludge, *Pseudomonas putida* | >98%, 50 $ppm_v$ | Ergas et al. (1994) |
| Ethanol | Compost, polystyrene (1:1) | Mixed culture | 195 | Arulneyam and Swaminathan (2000) |
| Methanol | Wood chips | Mixed culture | 127 | Sagastume et al. (2001) |
| Methanol | Compost, polystyrene (6:4) | Mixed culture | 85 | Arulneyam and Swaminathan (2003) |
| Toluene | GAC, compost | Mixed culture | 97 | Hwang and Tang (1997) |
| Xylene | Conditioned peat balls | Activated consortium | 61 | Jorio et al. (2000) |
| Methanol | Lava rock | Activated sludge | 185 | Prado et al. (2005) |
| α-Pinene | Lava rock | *Ophiostoma stenoceras* | 143 | Jin et al. (2006a, b) |
| Styrene | Perlite | Mixed culture | 382 | Rene et al. (2009a) |
| Toluene | Perlite | *Exophiala oligosperma* and *Paecliomyces variotii* | 164 | Estévez et al. (2005) |
| Styrene | Perlite | *Sporothrix variecibatus* | 336 | Rene et al. (2010a) |

*EC* elimination capacity, *GAC* granular activated carbon, *v/v* volume to volume ratio

A wide variety of both organic and inorganic compounds such as alcohols, aldehydes, ketones, hydrocarbons, carboxylic acids, $H_2S$ and $NH_3$ have been treated in biofilters (Leson and Winer 1991; Jin et al. 2005; Prado et al. 2008; Kennes et al. 2009a). Table 3.1 shows typical examples of gas-phase pollutants treated in biofilters and their corresponding elimination capacity values.

The main advantages of using biofiltration over other conventional control methods are their low capital costs, low operating costs, use of less chemicals and low energy requirements. Biofiltration units can be designed to physically fit into any industrial setting. It can be also designed as an open field with the piping and delivery system placed underground. In addition, biofilters can be designed with stacked beds to minimize space requirements and multiple units can be run in parallel. Biofiltration is versatile enough to treat odors, toxic compounds and VOCs. The treatment efficiencies of these contaminants are above 90% for low concentrations of contaminants (<1,000 $ppm_v$). Different media, microbes and operating conditions can be used to tailor a biofilter system for many emission points.

## 3.4 Novel Bioreactors for Waste-Gas Treatment

Though conventional bioreactors configurations have been used in practice to solve a wide range of environmental problems, *i.e.,* for waste-gas treatment, they are still prone to operational problems. Five major operating problems; moisture, nutrient, pH control, headloss/clogging, and response to transient loadings are universal issues that have been well reported in the literature (Schroeder 2002). Besides, in conventional bioreactor configurations, fluctuating pollutant loads in waste-gas streams, especially of VOCs with low water solubility, can often not be satisfactorily removed. Concentration peaks leave the reactor virtually untreated, while periods without VOCs in the waste-gas lead to starvation of the bacteria (Studer and von Rohr 2008). Recently, several novel bioreactor configurations have been proposed by researchers, though most of them are still at their initial developmental stage. These reactors have been designed to overcome some operational limitations frequently encountered in conventional biological treatment systems, *viz.,* flow resistance/pressure drop and oxygen depletion in biofilters, and mass transfer limitations for sparingly soluble pollutants in the case of biotrickling filters and bioscrubbers. This section describes different emerging bioreactor configurations that have been tested for their performance using waste-gas streams of various characteristics; continuous suspended-growth bioreactors, monolith bioreactor, two-liquid-phase biotrickling filters, air-lift bioreactor, fluidized-bed bioreactor, two-stage bioreactors, and foam-emulsion bioreactor.

### *3.4.1 One- and Two-Liquid Phase Suspended-Growth Bioreactors*

Biological air pollution control in a suspended-growth reactor is one alternative design sometimes overlooked for the treatment of waste-gases. Generally, the air contaminants, such as odors and-or VOCs, are co-degraded with contaminants dissolved in inlet wastewater (Kennes and Veiga 2001). The continuous suspended-growth bioreactor (CSGB) has advantages for culturing microorganisms because it offers stable environmental conditions, which is important since biological systems can be very sensitive to transient process variations. Once the optimum conditions are determined, they can be maintained on a long-term basis in a carefully controlled bioreactor.

#### 3.4.1.1 Principle of Pollutant Removal in Continuous Suspended-Growth Reactor

Bioreactors are vessels in which a controlled environment is provided with the aim of improving the activity or performance of biocatalysts (Daugulis 2001). The operating principles of the suspended-growth bioreactors are relatively simple, and can

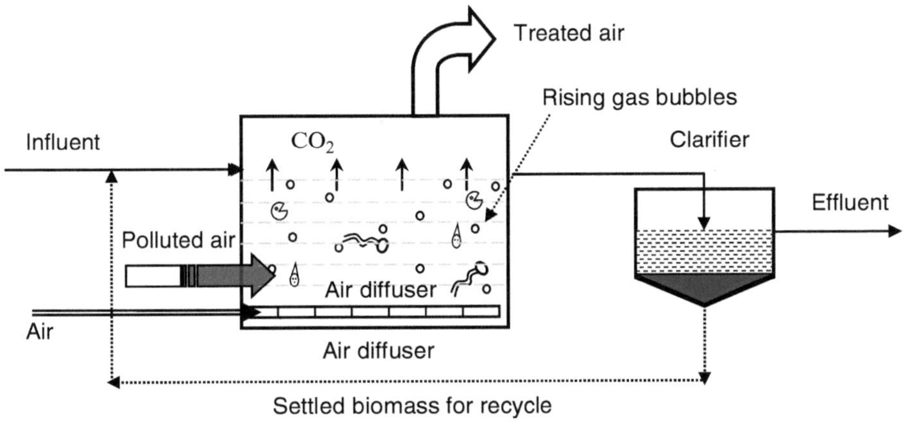


**Fig. 3.7** Schematic of a continuous suspended-growth bioreactor; air is supplied to the aeration tank from a compressor. The polluted air containing the volatile organic compounds are first absorbed in the water phase, and then degraded by the microorganisms. The mixed liquor from the aeration tank is then sent to the clarifier unit for removing the sludge

be compared to that of a continuous stirred-tank reactor (CSTR) used in chemical unit processes. As shown in Fig. 3.7, the medium and inoculum are introduced into a tank. The pollutants present in the waste-gas serve as the principal carbon and energy source for microbial growth. The polluted air, sparged through nozzles from the bottom of the suspended-growth bioreactor, is then transferred to the aqueous phase where it is biodegraded by the biocatalyst growing in suspension. In this process, the pollutant is biodegraded and converted into $CO_2$, $H_2O$ and mineral products. For environmental engineering applications, a suspended-growth bioreactor is a system that presents the following properties; (i) there is inflow and outflow of matter, (ii) chemical reactions occur within a well-defined system boundary, (iii) the system is in steady-state, *i.e.*, zero-accumulation rate, and (iv) the different species present in the system are homogeneously distributed.

#### 3.4.1.2 Factors Influencing Suspended-Growth Bioreactor Performance

Biomass Concentration

Biomass concentration and its activity can significantly affect the biodegradation rate of the target contaminant(s), and it has been suggested to maintain high biomass concentrations within the reactor, under all operating conditions. The microorganisms present, their specific growth rates, and prevailing nutrient conditions control the entire biodegradation process. Taking in account that such degradation is caused by suspended microorganisms in an agitated mixture, it is important to control the growth conditions in this kind of systems (Daugulis 1997). The microbial species could also affect to the optimum operating conditions in the continuous suspended-

**Table 3.2** Typical examples of pollutants treated in continuous stirred tank bioreactors

| Pollutant | Microorganism | EC ($g/m^3.h$) | RE (%) | References |
|---|---|---|---|---|
| Hexane | *Pseudomonas aeruginosa* | 45 | 19 | Muñoz et al. (2006) |
| Hexane | *Fusarium solani* | 50 | 28 | Arriaga et al. (2006) |
| Toluene | Microbial consortium | 326 | 95 | Daugulis and Boudreau (2008) |
| Toluene | Microbial consortium | <30 | >96 | Neal and Loehr (2000) |
| Dichloromethane | *Hyphomicrobium* KDM2, KDM4 | 110 | 90 | Bailón et al. (2009) |
| α-Pinene | *Pseudomonas fluorescens* NCIMB 11671 | 19 | 63 | Muñoz et al. (2008) |
| Methane | Microbial consortium | 75 | 34 | Rocha-Rios et al. (2009) |

*EC* elimination capacity, *RE* removal efficiency

**Table 3.3** Treatment efficiencies reported for aromatic organic compounds in suspended growth bioreactors (Adapted from Bielefeldt 2001)

| Pollutant | MLR (mg/l.h) | SLR (g/g VS.d) | EBRT, min | Inlet concentration (mg/l) | Removal efficiency (%) |
|---|---|---|---|---|---|
| Toluene | 11.6 | 0.47 | 10.5 | 2.1 | 99.4 |
| | 27 | 0.43 | 6.7 | 3.0 | 98.6 |
| Benzene | 58.9 | 4.02 | 10 | 9.7 | 96 |
| Ethylbenzene | 37.2 | 2.38 | 5 | 3.1 | 97 |
| *o* – Xylene | 34.9 | 1.05 | 5 | 2.9 | 89 |
| BTEX | 14.9 | 0.27 | 8.7 | 0.6 | 98.2 |

Typical operating temperatures ranged from 20°C to 25°C
*MLR* mass loading rate, *SLR* specific pollutant loading rate, *EBRT* empty bed residence time, *BTEX* benzene, toluene, ethylbenzene and xylene, *VS* volatile solids

growth bioreactor. Moreover, each microorganism requires different conditions for their growth. For example, as demonstrated by Powell and Hill (2008), an organism with a larger respiratory coefficient, and hence oxygen demand, will require more mixing to help transfer oxygen to the cells.

For the inoculation of suspended-growth bioreactors, wastewater sludge and pure or mixed cultures of bacterial or fungal populations are preferred. Table 3.2, shows the different gas-phase pollutants tested and the corresponding maximum EC values reported in the literature for continuous stirred-tank bioreactors (CSTB), simulating suspended-growth systems used for waste-gas treatment. Table 3.3 shows the typical organic pollutants treated in a continuous suspended-growth bioreactor, and their corresponding removal efficiencies. Nielsen et al. (2005) had shown earlier that, for a pure strain of bacteria treating VOCs, a constant biomass concentration will eventually be established due to cellular maintenance requirements, which are responsible for all of the substrate consumed. This principle can be extended to mixed microbial consortia present in the reactor, after an initial growth period (Daugulis and Boudreau 2008).

### Solubility of the Pollutant

Since all biological mechanisms take place in the presence of an aqueous-phase, the effectiveness of the processes for treatment of hydrophobic compounds is relatively poor. Therefore, there are many works based on the addition of an organic-phase with more affinity for the target compound than water. The continuous presence of heavy organic solvent in biological air treatment systems can also be advantageous for the removal of hydrophilic compounds from air stream, because they can reduce the level of inhibitory effect caused by high concentrations of organic pollutants to the existing microbial community (Fazaelipoor 2007). The most widely used organic phases are dodecane, hexadecane, and silicone oil. However, silicone oil seems to be the only solvent that did not present any problem so far, in terms of biodegradability or biocompatibility (Muñoz et al. 2007). Previously, the high performance generally reached in such two liquid-phase suspended-growth bioreactor has been compared to more conventional systems (Kennes et al. 2009a). Nevertheless, silicone oil has certain disadvantages such as its relatively high cost and because its recovery may increase process costs. Recently, it has been proposed that solid polymers could be used as an alternative organic-phase, for the degradation of hydrophobic compounds (Daugulis and Boudreau 2008).

### Suspended-Growth Bioreactor Response to Transient Operations

A general profile of transient-state operation in the form of medium and high instantaneous over loads and the corresponding response of any biological system in terms of removal efficiency are illustrated in Fig. 3.8. Only few experimental data are available in the literature that highlights the effect of transient or high loading conditions in a suspended-growth bioreactor.

Previously, for gas-phase toluene removal in a continuous stirred-tank bioreactor, when a sudden shock-load was imparted from 343 g/m$^3$.h to approximately 6,000 g/m$^3$.h, in a single-phase system, the removal efficiency of the system decreased from 97% to about 52% during the 60 min shock-load, decreasing further by another 14% for a brief period of time (20 min), reflecting the increased stripping of accumulated toluene once the step increase has been terminated (Daugulis and Boudreau 2008). Bailón et al. (2009) conducted a series of shock-loading experiments, *viz.*, as low-medium, medium-high, and low-high overloads and observed the performance changes during the treatment of gas-phase dichloromethane. At the highest conditions, when the load was increased from 70 to 500 g/m$^3$.h, the removal efficiency of the system dropped from 95% to minimum values of 50% at the end of 6 h shock-load. Besides, the performance during shock-loads were compared to those of a biotrickling filter, operated under similar conditions with gas-phase, and the continuous stirred-tank bioreactor clearly outperformed the biotrickling filter, at all loading conditions. For shock-loads tested in the presence of the second organic-phase by different researchers, an enhancement in the performance arising from the presence of a second-phase was observed. Anew, the instantaneous removal efficiency

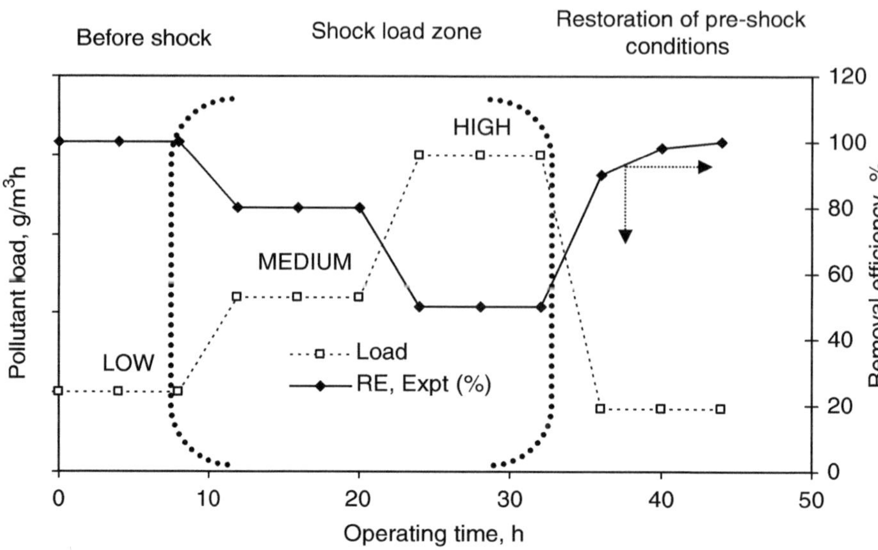


**Fig. 3.8** Schematic of low, medium and high over-loads in waste-gas treatment systems; sometimes sharp peak loads can occur almost instantaneously, leading to a decline in bioreactor performance (removal efficiency)

recovery times, after re-instating pre-shock conditions, also show a clear trend with the one-liquid-phase system recovering more rapidly, as a result of little pollutant absorption (Boudreau and Daugulis 2006).

### *3.4.2 Monolith Bioreactors*

A monolith reactor (Fig. 3.9) is a reactor in which the biomass is attached on a plastic, ceramic or metal structure with uniform parallel channels separated by thin walls. Commercially available monolith supports can be tailored to meet the needs of a relatively inexpensive, light weight, inert, bioreactor packing that provides a high specific surface area, *i.e.*, surface to volume ratio, to greatly improvise the mass transfer rate of pollutant (Jin et al. 2006b, 2008; Rene et al. 2010c).

#### 3.4.2.1 Flow Pattern in Monolith Channels

The operating mode depends on the size of the straight parallel channels. In large channels, the fluid trickles downwards along the channel walls and the gas travels co- or counter-currently through the channel in the core. In smaller channels, the dominant flow pattern is a segmented slug flow or bubble train flow of elongated bubbles and slugs, commonly referred to as Taylor flow in the literature (Kreutzer et al. 2005; Jin et al. 2006b). Theoretically, the gas and liquid would ideally follow near plug flow conditions as they travel with uniform velocities, as separated

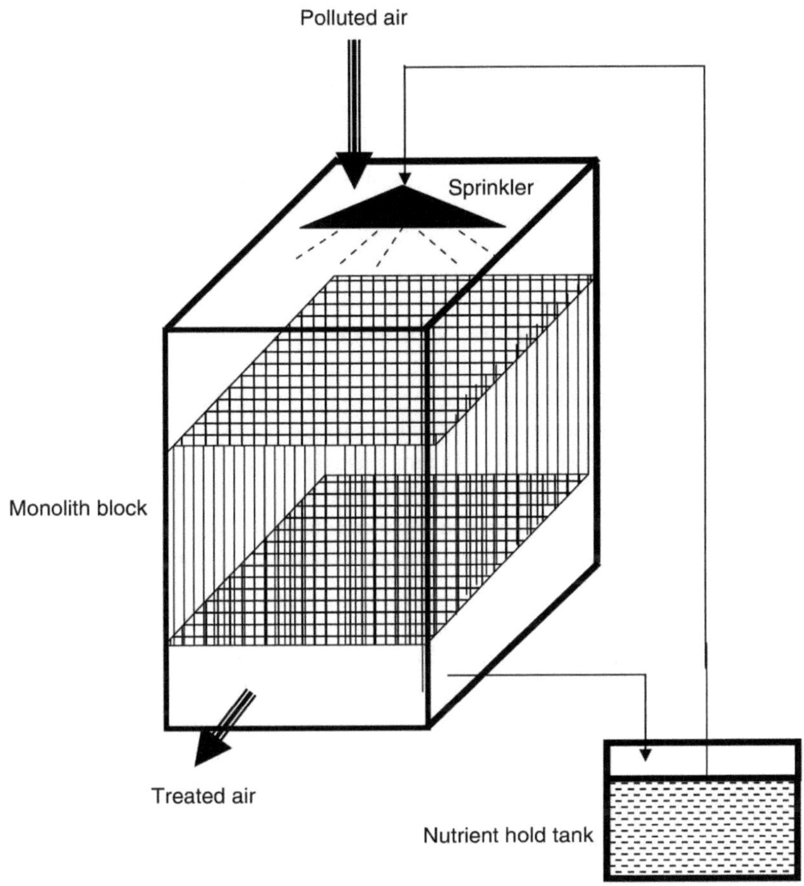


**Fig. 3.9** Schematic of a monolith bioreactor; excess biomass/biomass clogging can be removed/avoided by increasing the liquid trickling rate by fourfold. This bioreactor can also be operated by intermittently supplying the trickling medium, once sufficient biomass is attached (Rene et al. 2010c)

packages within a monolith channel. The wall is wetted by a thin liquid film, while the gas pollutant is easily transported through the film allowing higher mass transfer rates. The bio-catalytically active biofilm layer remains at the wall when the liquid slug passes by. Inside the liquid slug itself, a recirculation pattern is observed. This recirculation enhances transfer of gas from the caps of the bubble to the biocatalyst with higher removal by the attached biomass.

#### 3.4.2.2 Advantages of Monolith Support

In environmental applications, the monolith reactor is widely used as medium for non-biological catalyst support such as the cleaning of automobile exhaust gases and industrial off gases due to the following advantages; (i) simple reactor configuration and energy efficient operation due to low pressure drop, (ii) high external surface area and high mechanical strength, (iii) high interfacial mass transfer rates, (iv) better liquid distribution at low liquid flow rates, and (v) scaling up promises to be relatively easy compared to other reactor types (Jin et al. 2006b; Rene et al. 2010c). However, a continuous supply of nutritive aqueous phase is a prerequisite for optimal bioreactor performance with inert support materials. These nutrients are not naturally available in inert filter beds, and the aqueous nutrient solution acts as a main agent for oxygen and substrate transport from the gas to the biofilm (Kennes and Veiga 2002).

#### 3.4.2.3 Treatment of Gas-Phase Volatile Organic Compounds

The schematic of a monolith bioreactor used for waste-gas treatment is shown in Fig. 3.9. Monolith bioreactors have only been tested recently for the removal of both hydrophilic and hydrophobic VOCs from waste-gases (Jin et al. 2006b, 2008; Fang and Govind 2007; Rene et al. 2010c). Jin et al. (2008) used a yeast-dominant enriched culture, which could tolerate low-pH, to inoculate a monolith reactor for treating methanol polluted air. More than 80% removal, for methanol loads greater than 250 g/m$^3$.h, suggests the extent of mineralization possible for hydrophilic VOCs in such a monolith reactor, a load that is often not treatable in biofilters and biotrickling filters, at such low-pH. Fang and Govind (2007) tested an activated carbon coated monolith reactor having porous channel walls, made of porous cordierite to support biomass, under diffusive-and convective flow conditions to treat gas-phase toluene at low concentrations. High removal during the first 2 weeks of operation was attributed to adsorption of toluene by activated carbon on the monolith and the difference in removal efficiencies were a result of the different flow patterns inside the monolith modules. Though the maximum EC were low in both the flow types, the convective flow type gave higher EC (7 g/m$^3$.h) than the diffusive flow type configuration (3 g/m$^3$.h). Rene et al. (2010c) investigated styrene removal in a monolith bioreactor, inoculated with the fungus *Sporothrix variecibatus*, at different concentrations, ranging between 0.07 and 2.5 g/m$^3$, and at two different flow rates corresponding to empty bed residence times (EBRTs) of 77 and 19 s respectively. A maximum EC of 67.4 g/m$^3$.h was observed at an inlet styrene load of 73.5 g/m$^3$.h.

#### 3.4.2.4 Effect of Liquid Trickling Rate on the Removal Performance

For achieving high removal efficiencies for gas-phase pollutants in monolith bioreactors, the operational mode, *i.e.,* flow of pollutant and the trickling liquid, has to be in the Taylor flow regime. In this regime, the gas and liquid move through the channels as separate packages, ensuring plug flow behavior. Because of these properties of Taylor flow in capillaries, a high gas-liquid mass transfer rate is usually obtained. Jin et al. (2006b) studied the effect of the mode of feeding of the liquid-phase, first in trickling mode and later without trickling. It was observed that, the removal of gas-phase toluene in the trickling mode was lower than without trickling phase, and this observation seemed to have contradicted the theory that Taylor flow could enhance the mass transfer from the gas-phase to the liquid-phase. Jin et al. (2006b) further enunciated the reason for this observation, as follows; (i) when no liquid is recirculated in the reactor, the gas flow is uniform, and the liquid inside the channel forms a very thin liquid film, then the resistance between the gas-phase and the liquid-phase is low, (ii) in the trickling mode, although the Taylor flow generated by the liquid flow could enhance the mass transfer rates, the liquid was not uniformly distributed in the monolith. This could have possibly caused non-homogenous mass transfer in the different channels, and (iii) the hydrophobic nature of toluene would also explain its better removal in the absence of a continuous trickling aqueous phase.

#### 3.4.2.5 Response of Monolith Bioreactors to Transient-Loads

Since biological waste-gas systems include the presence of active microorganisms, they are sensitive to sudden variations of flow rate or concentration or a combination of both at once, *i.e.,* load fluctuations (Fig. 3.8). Shock-load tests have been mentioned in only one monolith bioreactor report, on gas-phase styrene removal (Rene et al. 2010c). Rene et al. (2010c) conducted low-medium and medium-high overload tests, at an EBRT of 77 s and 19 s, respectively. It was reported that, a sudden increase in the styrene load from 10 to 85 g/m$^3$.h decreased the RE from nearly 100–61% during medium shock-loads, and during this step increase, the EC increased from an original low value of 10 g/m$^3$.h to 55 g/m$^3$.h, and then remained almost constant during the 6-h medium shock-load. Anew, during high shock-load tests, *i.e.,* from 56 to 186 g/m$^3$.h, the removal efficiency dropped suddenly, to nearly 30% and then remained almost constant at such value during the shock-load period of 36 h. Furthermore, the EC values remained basically unchanged during the shock-load compared to its value before the shock-load. In both these shock-loads, the response of the monolith bioreactor was fast as seen from the immediate decrease in removal profile at high loads, and, when pre-shock steady state conditions were restored, the recovery time of the fungus-inoculated monolith bioreactor to achieve original performance was almost instantaneous, in less than 1 h (Rene et al. 2010c).

#### 3.4.2.6 Pressure Drop

Pressure drop is directly related to the accumulation of biomass in the narrow channels and on the surface of the monolith, and is responsible for part of the energy consumption during the operation of a monolith bioreactor. The pressure drop values usually start to increase, when new bacterial/fungal cells are formed in the monolith channels. Clogging of monolith bioreactors inoculated with fungi is somewhat faster than in reactors where bacteria are dominant populations (Kennes and Veiga 2004; Rene et al. 2010c). For fungal bioreactors, clogging occurs as a result of excess biomass growth and fluffy biofilm formation, which is controlled by phosphate limitation in fungi (Singh 2006). Rene et al. (2010c) showed that the pressure drop values increased to a maximum value of 7.2 cm $H_2O$ in a monolith bioreactor inoculated with the fungus *Sporothrix variecibatus*, during the biotreatment of gas-phase styrene. Similar observations were also reported in a previous study with methanol in monolith bioreactor and it was postulated that regular washing, by temporarily increasing the flow rate of the trickling liquid for a short period, of the monolith would remove the excess biomass and will allow for a long-term, stable operation (Jin et al. 2008).

### *3.4.3 Two-Liquid-Phase Biotrickling Filters*

For waste-gases containing hydrophobic compounds (low solubility in water), the efficiency of biological treatments is limited by the poor mass transfer of the pollutant from the gas-phase to the aqueous phase (Dumont et al. 2006). Under such conditions, the addition of a second non-aqueous liquid-phase with the already re-circulating nutrient media is a novel concept for use in biotrickling filters (Fig. 3.4). The first report on the application of two-phases for waste-gas treatment appeared around 1992, as a process improvement strategy for the treatment of hydrophobic contaminants in bioscrubbers (Cesario et al. 1992; Poppe and Schippert 1992). By doing so, high mass transfer rates can be achieved, compared to conventional biofilters and biotrickling filters, in the case of hydrophobic pollutants like α-pinene, hexane or toluene (Spigno et al. 2003; Arriaga et al. 2006; Muñoz et al. 2008; Bailón et al. 2009; Quijano et al. 2009a, b).

#### 3.4.3.1 Typical Organic Solvents Used and Advantages of Silicone Oil

In order to overcome mass transfer constraints, the addition of a water-immiscible organic solvent, such as silicone oil, oleyl alcohol, decane, n-hexadecane, dodecane, diethyl sebacate, 1-octadecane, isopropyl myristate, 2-undecanone, has been suggested (Daugulis 1997, 2001; Kennes and Veiga 2001; Muñoz et al. 2006; Mahanty et al. 2008; Fazaelipoor 2009; Bailón et al. 2009; Kennes et al. 2009a; Montes et al. 2009, 2010). Silicone oil, commercially known as polydimethylsiloxane $[(H_3C)[SiO(CH_3)_2]_nSi(CH_3)]$, has shown to serve the intended purpose by its addition in waste-gas treatment systems. 100% polydimethylsiloxane (PDMS) has unlimited useful life, very less volatility, high resistance to oxidation and exceptional stability to chemical attack. Silicone oil serves as the ideal candidate for use as an organic phase, in biological waste-gas treatment applications, due to the following advantages; (i) safe to use, (ii) high stability, (iii) non-toxic to the microorganisms, (iv) allows high pollutant solubility, which enables the treatment of high pollutant concentrations, (v) increases the specific interfacial area (gaseous) available for microorganisms to utilize the pollutant, by reducing the surface tension of water, (vi) controls the delivery pattern of the pollutant to the aqueous phase, (vii) enhances oxygen mass transfer rates and (viii) positive impact on downstream operations, such as product recovery (Daugulis 1997, 2001; Kennes and Veiga 2001; Bailón et al. 2009; Montes et al. 2009; Darracq et al. 2010; Mahanty et al. 2010). It is important to add only low amounts of organic phase to the system, since that additional phase will also result in an additional cost. A recent study has shown that the addition of only 5% silicone oil to the recirculating medium can easily increase the performance of mesophilic and thermophilic biotrickling filters for the removal of α-pinene in gas-phase (Montes et al. 2010).

#### 3.4.3.2 Mechanism of Pollutant Removal in the Presence of an Aqueous Organic Phase

The organic solvent, *i.e.*, the non-aqueous liquid-phase, when mixed with the recirculating medium forms a water-solvent emulsion and is continuously trickled over the filter bed containing the immobilized biocatalysts. The water-solvent emulsion tends to absorb the gas-phase pollutants and keeps them in the organic phase. The lower pollutant concentration in the aqueous phase resulting from this process is the driving force for pollutant diffusion from the organic to the aqueous phase (Dumont et al. 2006). During this process, complete biological regeneration of the liquid is also possible (Kennes and Veiga 2001). Furthermore, the presence of the second organic phase has shown to enhance oxygen mass transfer from the gas-phase to the microorganisms in order to improve the degradation of the pollutant.

#### 3.4.3.3 Performance of Two-Liquid-Phase Biotrickling Filters

Van Groenestijn and Lake (1999) treated hexane vapors in two biotrickling filters, connected in series as a two-stage bioreactor, inoculated with activated sludge from a wastewater treatment plant. In that study, a mixture of silicone oil and nutrient medium (1:1 v/v) was trickled over the top, and it was observed that, at a gas-flow rate of 120 l/h and hexane concentration of 10 $g/m^3$, at loading rates ranging between 83 and 97 $g/m^3.h$, removal efficiencies as high as 89% were attained. In another study, the stability of a biotrickling filter for treating gas-phase dichloromethane was evaluated for long-term operation with and without the addition of 10% (v/v) silicone oil to the nutrient medium (Bailón et al. 2009). The biotrickling filter was inoculated with pure strains of *Hyphomicrobium* KDM2 and KDM4, and the authors observed that, when adding 10% silicone oil to the biotrickling filter, the maximum EC increased by 25%, reaching 200 $g/m^3.h$ compared to the 160 $g/m^3.h$ achieved for the same biotrickling filter without silicone oil. There are however few reports in which silicone oil has been either added or coated onto the packing material for use in biofilters, as an alternate strategy to improve its performance. Budwill and Coleman (1997) coated peat granules with silicone oil for hexane vapor removal in a biofilter and observed a maximum EC of 16 $g/m^3.h$ with 66% RE, which is almost two times higher than the EC achieved in a control biofilter without silicone oil.

Fazaelipoor and Shojaosadati (2002) used enriched hydrocarbon degrading bacteria in two perlite biofilters, one serving as the control, while the other was coated with silicone oil ($v_{oil}/v_{perlite}=0.2$), to treat hydrocarbon vapors. The nutrient medium was intermittently recirculated, 10 min/2 h, for both the reactors to provide sufficient moisture for the filter bed. The results indicated that the performance of the silicone oil coated biofilter was only slightly better (20.4 g/m$^3$.h) than the non-coated biofilter (18.8 g/m$^3$.h). Djeribi et al. (2005) immersed lava rock grains into silicone oil for 1 h and used that lava rock, containing the adsorbed silicone oil, as the packing material in a biotrickling filter, as an alternative biphasic system strategy to treat gas-phase styrene. The inoculum for the biotrickling filter was an isolated consortium, containing different strains of *Pseudomonas spp.*, and *Achromobacter spp.*, previously obtained from a two-phase continuous stirred-tank bioreactor treating styrene. The results from that study show that, for a loading rate of 2,980 g/m$^3$.h, the RE was > 97% with a maximum EC of 2,900 g/m$^3$.h. A comparison between the performance of one-and two-liquid-phase biotrickling filters is shown in Table 3.4.

There is a concern to apply this technology in practical situations, despite the high performances reported in the literature for short-term laboratory experiments. Long term stability is crucial as the non-aqueous phase might partly be lost during continuous operations (Muñoz et al. 2007).

### 3.4.4 *Air-Lift Bioreactors*

Air-lift reactors were originally designed by chemical engineers for enhancing mass and heat transfer efficiency and to produce pharmaceuticals, polymers, and chemical feed stocks, among others (Bielefeldt 2001). This reactor configuration offers many advantages and is being applied in the field of wastewater engineering. The advantages include: complete mixing of species, simple mechanical design, low shear rate, high capacity, the absence of mechanical agitators, versatility for scale-up, and low power consumption for agitation and oxygenation (Chisti 1989). Additionally, when compared with similar reactor configurations like stirred tanks or bubble columns, shear stress can be relatively constant and mild throughout an air-lift bioreactor, a condition that favours stable microbial growth (Kanai et al. 1996).The application of this bioreactor configuration for waste-gas treatment has only been tested in laboratory conditions and was first reported in the 1990s (Ensely and Kurisko 1994; Ritchie and Hill 1995).

#### 3.4.4.1 Operation of Air-Lift Bioreactors

The schematic of a concentric, internal-loop air-lift bioreactor is shown in Fig. 3.10. Waste-gas is introduced into the air-lift bioreactor, through a sparger located at the central section, for the purpose of supplying oxygen to the bacteria and to provide mixing. The inner draft tube improves circulation and oxygen transfer and equalizes

**Table 3.4** Performance of trickling filter with one-and two-liquid phases for gas-phase VOC removal (Adapted from Kennes et al. 2009b)

| Pollutant | Packing material | ILR ($g/m^3.h$) | $EC_{max}$($g/m^3.h$) | References |
|---|---|---|---|---|
| Dichloromethane | Lava rock[a] | 360 | 160 | Bailón et al. (2009) |
| | Lava rock + silicone oil (10%)[a] | 465 | 195 | Bailón et al. (2009) |
| Hexane | Polyamide structure wire mat + silicone oil (10%) | 100 | 80 | van Groenestijn and Lake (1999) |
| | Fungal biofilter – granular expanded clay | 300 | 150 | Spigno et al. (2003) |
| | Fungal biofilter – perlite + silicone oil (10%)[b] | 180 | 165 | Arriaga et al. (2006) |
| | Fungal biofilter – perlite[b] | 164 | 110 | Arriaga et al. (2006) |
| | Peat biofilter + 20% silicone oil[c] | 24 | 16 | Budwill and Coleman (1997) |
| Hydrocarbon mixture | Perlite + silicone oil (20%)[d] | 31.1 | 20.4 | Fazaelipoor and Shojaosadati (2002) |
| Styrene | Lava rock + silicone oil (30%)[e] | 555<br>2,980[f] | 537<br>2,900[f] | Djeribi et al. (2005) |

*ILR* inlet loading rate, *EC* elimination capacity
Microorganism used in the BTFs are given here:
[a]*Hyphomicrobium sp.*, KDM2 and KDM4
[b]*Fusarium solani*
[c]The peat based biofilter was soaked with 20% silicone oil (w/w)
[d]Inoculated with 1 l of hydrocarbon degrading culture
[e]Four dominant strains, *viz., Psuedomonas putida* (C1), *Achromobacter sp.,* (C2), *Achromobacter sp.,* (C3) and *Pseudomonas aeruginosa* (C4) were used in the experiments
[f]ILR and EC obtained when experiments were conducted by progressively escalating gas-phase styrene concentrations, up to 6.6 $g/m^3$

shear forces in the reactor. The hydrostatic pressure difference due to gas sparging in the middle section causes a circulating motion in different parts within the reactor. The direction of waste-gas movement is upward in the light section and downward in heavier section, resulting in a well controlled fluid circulation pattern. These sections are named riser and down comer, respectively. Adequate headspace is provided above the riser and down comer to allow the treated gas to escape out. In some conventional air-lift bioreactors, the inner tube serves as the down comer, and the annular space between the two tubes serves as the riser. The biomass is usually dispersed in the liquid medium or immobilized onto suitable polymer matrix, such as polyvinyl alcohol (PVA)-sodium alginate gels (Vergara-Fernandez et al. 2008; Namgung et al. 2009).

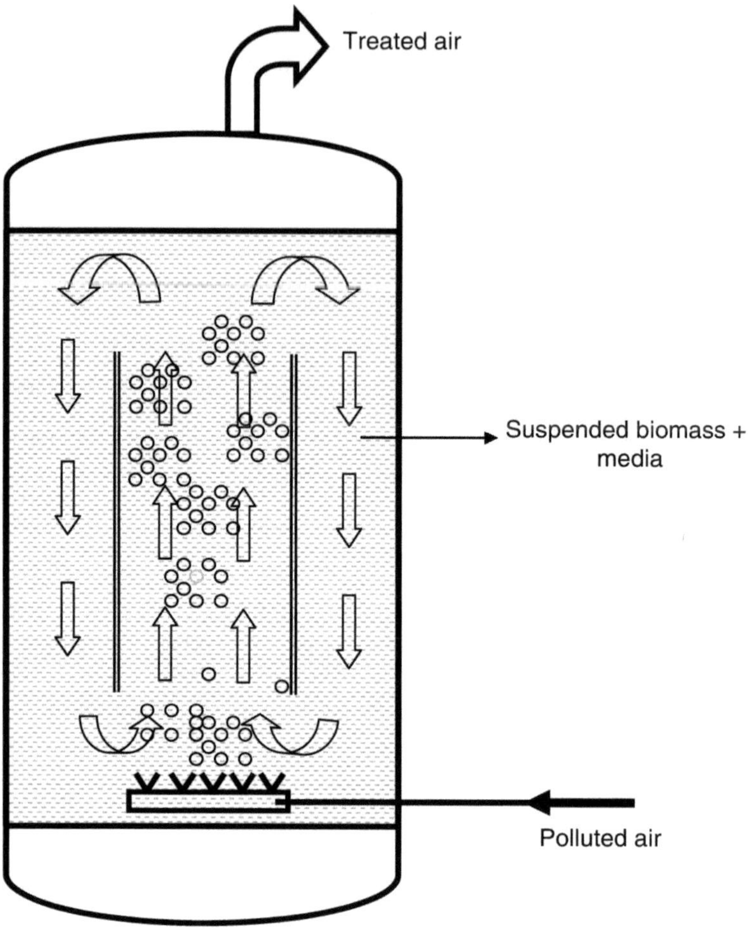


**Fig. 3.10** Schematic of an internal-loop air-lift bioreactor; it has been reported that an air-lift bioreactor requires less energy, and therefore, facilitate design scale-up, providing a potential practical replacement for traditional mechanically agitated configurations (Littlejohns and Daugulis 2009)

#### 3.4.4.2 Performance of Air-Lift Bioreactors for Waste-Gas Treatment

The application of air-lift bioreactors for waste-gas treatment is a relatively new technique, though this type of configuration has been extensively used in water treatment applications. Ritchie and Hill (1995) used a flat plate type sparger, in a circulating loop reactor (internal-loop) and reported >99% RE of phenol (<1 $g/m^3$) from contaminated air, for inlet loading rates up to 20 $g/m^3.h$. Khandan (1997) used a 3 m high air-lift bioreactor for the treatment of volatile hydrocarbon emissions (toluene) at a gas-flow rate of 1 l/min and at different temperatures between 20°C and 30°C. A reduction in the RE from nearly 100% to 70% was reported, when the gas-flow rate was increased to 10 l/min. The maximum EC of air-lift bioreactors appears to be a strong function of both the suspended biomass concentration and gas-flow rate. Wei et al. (1999) used a spinning-type sparger in an external loop air-lift bioreactor for the biodegradation of both *p*-cresol and ethanol from contaminated air. In batch experiments, having observed inhibition, the growth on both *p*-cresol by *Pseudomonas putida* and ethanol by *Acetobacter aceti*, respectively, were modeled using Haldane's type substrate inhibition kinetics. In the continuously operated bioreactor, it was observed that higher gas-flow rates resulted in higher build-up of *p*-cresol in the bioreactor (20 mg/l). Besides, for gas-phase *p*-cresol loading rates up to 8 $g/m^3.h$, inlet concentrations <0.6 $g/m^3$, a linear relationship was obtained between the rate of increase in biomass and the rate of *p*-cresol absorption. On the other hand, for ethanol removal using *Acetobacter aceti*, the bioreactor was able to handle high loading rates (220 $g/m^3.h$), but the microbial growth rate was much lower than those observed in batch experiments (0.05/h *vs.* 0.3/h in batch). Vergara-Fernandez et al. (2008) reported that an increase in biomass concentration in the reactor (V=2 l) will allow for greater elimination capacities. This was explained by the fact that a greater use of carbon source favors the mass transfer in the medium which also increases the pollutant solubility in the medium, produced by changes in the liquid-phase generated by microbial growth. For instance, when the average biomass concentration was 3,700 $g/m^3$, the EC was 203 $g/m^3.h$, and at a maximum biomass concentration of 8,000 $g/m^3$, the EC was 310 $g/m^3.h$. The same authors also reported that the gas-phase toluene RE decreased from 100% to 40% when the gas-flow rate was increased from 0.024 to 0.132 $m^3/h$. Namgung et al. (2009) tested an air-lift bioreactor, having suspended yeast-immobilized polymer media, for the biodegradation of a mixture of gas-phase toluene and methyl ethyl ketone (MEK). The yeast entrapment technique was adapted in this case, to achieve stable performance and high biodegradation rates. Besides, entrapment of microbes within a polymer matrix would minimize the impact received from unexpected shock-loads and help to maintain the microbial activity. Namgung et al. (2009) used the yeast strain *Candida tropicalis*, and a mixture of sodium alginate, polyethylene glycol (PEG) and powdered activated carbon to formulate the PEG-Alginate-Carbon-Yeast media. The EBRT was varied from 15 to 60 s and the maximum EC was reported as 70.4 and 56.4 $g/m^3.h$, for toluene and MEK, respectively. The authors attributed the high EC values to the adsorption capacity and biodegradation activity of the polymer media. The presence of this media also helped to dampen the fluctuations of loads, as well as interact with the active yeast cells in the liquid-phase for the biodegradation of VOCs. Nikakhtari and Hill (2006) designed an improvised version of the conventional air-lift bioreactor, by incorporating/placing a small packed bed near the riser section. The bioremediation of phenol polluted air was studied in that reactor using a pure strain of *Pseudomonas putida*. The authors have also reported a three-step procedure, using 0.2% polyethylenimine, to develop the biofilm on the packed bed (stainless steel mesh packing). Steady-state operation was achieved in less than 6 h, and the novel bioreactor was able to handle high phenol loads (33,120 $mg/m^3.h$), with 100% RE.

Jiangping et al. (2005) evaluated the performance of a gas-liquid-solid three-phase flow air-lift bioreactor for treating a contaminated air stream containing a mixture of ethyl acetate and ethanol. 100 g activated charcoal with the average diameter of 0.2 mm was mixed with activated sludge for biofilm formation onto the solid phase. The bioreactor was operated with a liquid flow rate of 100 mL/h, corresponding to 0.091/h of dilution rate. The maximum elimination capacities were 504 and 685 $g/m^3.h$, for ethyl acetate and ethanol in the mixture, respectively.

## 3.4.5 *Fluidized-Bed Bioreactor*

The fluidized-bed bioreactor (gas-solid) is a modified version of the conventional fluidized bed reactor used in chemical processes and other unit operations. These reactors can be thought upon as an attempt to improve the existing design of other biological systems such as biofilters (Clarke et al. 2008).

### 3.4.5.1 Operation of Fluidized-Bed Bioreactors

The schematic of a typical fluidized-bed bioreactor is shown in Fig. 3.11. The polluted air enters the reaction space with the help of a nozzle type distributor/sparger placed near the tapering section of the bioreactor. These bioreactors are generally constructed as hollow cylinders with perforated distribution plates, placed just above the sparger. An important design consideration, for biofilm development, is the selection of an appropriate particle type to use in the bed, and that the particle can be easily fluidized. Clarke et al. (2008) enunciated that the fluidized state is influenced by the particle properties such as size, size distribution, density, shape, and moisture content. Besides, the fluidization state can also affect the mass and heat transfer characteristics, and the conversion, *i.e.,* pollutant removal characteristics, in the bioreactor. Small particles in the size range of 0.15–0.3 mm can be used as the solid phase. Typical examples are; sand, carbon, fly ash, anthracite, glass, calcinated clay, etc. These fine particles can be easily fluidized by the upward flow of waste-gas entering the reactor. Poor fluidization behavior could lead to gas channeling and poor waste-gas treatment. In some cases, highly porous, fabricated media are also developed which allow biomass formation within the porous internal structure.

### 3.4.5.2 Performance of Fluidized-Bed Bioreactors for Waste-Gas Treatment

There are only a few reports pertaining to the use of fluidized-bed bioreactors for removing organic and inorganic vapor phase pollutants present in waste-gases. Wright and Raper (1998) investigated ammonia vapor removal in a three-phase sprouted bed bioreactor, consisting of larger particles than those used in typical fluidized bed systems to achieve the desired sprouting characteristics. Removal

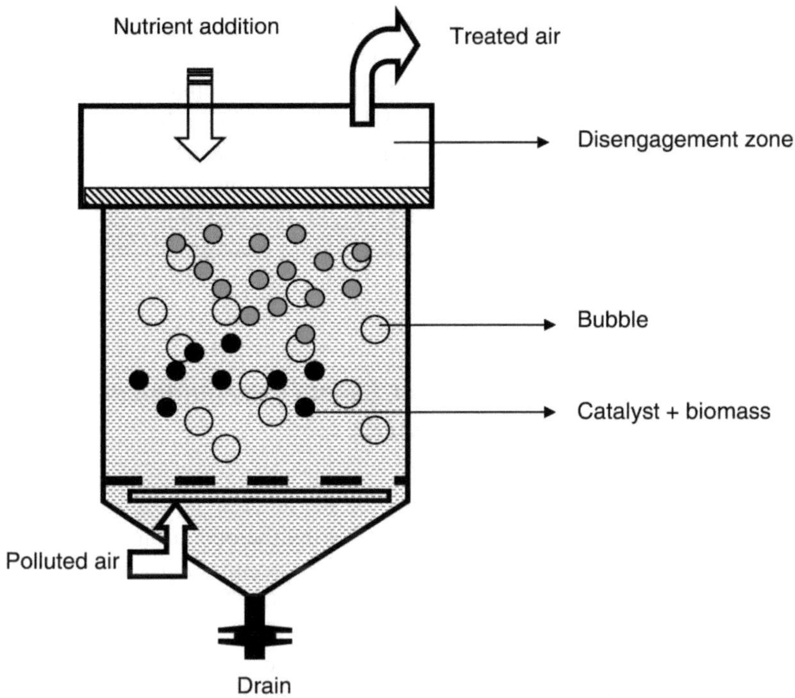


**Fig. 3.11** Schematic of fluidized-bed bioreactor; it has been shown that the performance (removal efficiency) in a fluidized bioreactor is better at the lowest velocity, when the bubbling bed fluidization regime predominates (Clarke et al. 2008)

efficiencies of ammonia in gas-phase varied from 0% to 40% for $NH_3$ loads varying between 500 and 650 g $NH_3/m^3$.h. The low removal efficiencies were attributed to the unevenly formed sprouting resulting from high wall adhesion and aggregation of the particles caused by the mineral solution. Leslous et al. (2004) reported the performance of a fluidized bed bioreactor for gas-phase ethanol biotreatment. Their study used moist scrap wood particles, and >80% RE was reported for ethanol loading rates <1,150 g/m$^3$.h. Clarke et al. (2007) used a gas-solid fluidized bed bioreactor for removing ethanol from polluted air, in a specially designed bioreactor packed with moist saw dust particles and glass spheres. The bioreactor, depending on the applied superficial gas-velocity, was designed to be operated in either packed or fluidized-mode. When the bioreactor was switched to fluidized mode, the sawdust and glass sphere mixture was maintained in a bubbling/slugging regime. The authors reported a maximum EC of 75 g/m$^3$ sawdust.h (during fluidized-bed mode) and 225 g/m$^3$.h during the packed bed-mode. The high EC value achieved during packed bed mode of operation was then attributed to the ability of the packing material to withstand the hydrophilic high ethanol loads, subsequently leading to high biomass growth. The low EC value in the case of the fluidized-bed bioreactor was speculated to be due to the inability of the microbial cells to attach to the rigid saw dust particles during fluidization. Besides, as a performance improvement strategy for the fluidized-bed bioreactor, it has been suggested to recycle the non-treated air from the exit to the inlet, resulting in an increase in the concentration of ethanol to the bottom of the bed. Clarke et al. (2008) used peat granules (Sauter mean diameter of 680 μm) and inoculated the fluidized-bed bioreactor with *Hansenula anomala,* for the biotreatment of gas-phase ethanol. In this study, the authors reported maximum ethanol ECs ranging between 1,150 and 1,520 g/m$^3$.h, when fluidized-bed bioremediation experiments were carried out at a superficial gas-velocity of 0.5, 0.75 and 1 m/s, respectively.

### *3.4.6 Two-Stage Reactors with at Least One Biological Step for Waste-Gas Treatment*

As the name implies, two-stage reactors exploit the inherent advantages of different processes used for waste-gas treatment, by adjusting the operation strategy as reactors connected in series or in one modified reactor configuration having different pollutant removal mechanisms (Kennes and Veiga 2001; Jin et al. 2008; Rene et al. 2009b, 2010b). Two-stage reactors are preferable for certain operating conditions and waste-gas characteristics, as given here; (i) when the waste-gas contains a mixture of inorganic and organic pollutants, and some of their degradation end-products are highly acidic, (ii) when it is desired to maintain different microbial species in the bioreactors for the removal of a mixture of pollutants, and (iii) for the treatment of exceedingly high loading rates of the pollutants, which could inhibit microbial activity in the biological reactor. Typical examples are shown in Fig. 3.12 (biotrickling filter followed by biofilter in one reactor configuration), Fig. 3.13 (adsorption followed by biological treatment) and Fig. 3.14 (advanced oxidation pre-treatment followed by biological treatment), respectively.

#### 3.4.6.1 Removal of Hydrogen Sulphide/Volatile Organic Compound Mixtures in Fixed-Bed Systems

Hydrogen sulphide ($H_2S$) and mixtures of VOCs are commonly present in waste-gases emitted from the head works and other facilities at public owned treatment works (POTWs). Cox et al. (2003) tested the removal of these pollutants in a biofilter and observed an EC of 13.8 g $H_2S$/m$^3$.h, while VOC removal was reportedly poor irrespective of the experimental conditions. About 25–35% of low concentrations of benzene, toluene and chlorobenzene was removed, while other chlorinated VOCs could not be removed in the biotrickling filter. The low VOC removal was attributed to the presence of inhibitory concentrations of sulphate in the recycle liquid and the possible accumulation of metabolites other than sulphate that inhibit VOC biodegradation. Under such condition and for situations like emissions from typical pulp and paper industries (as reported by Rene et al. 2009b, 2010b), a two-stage

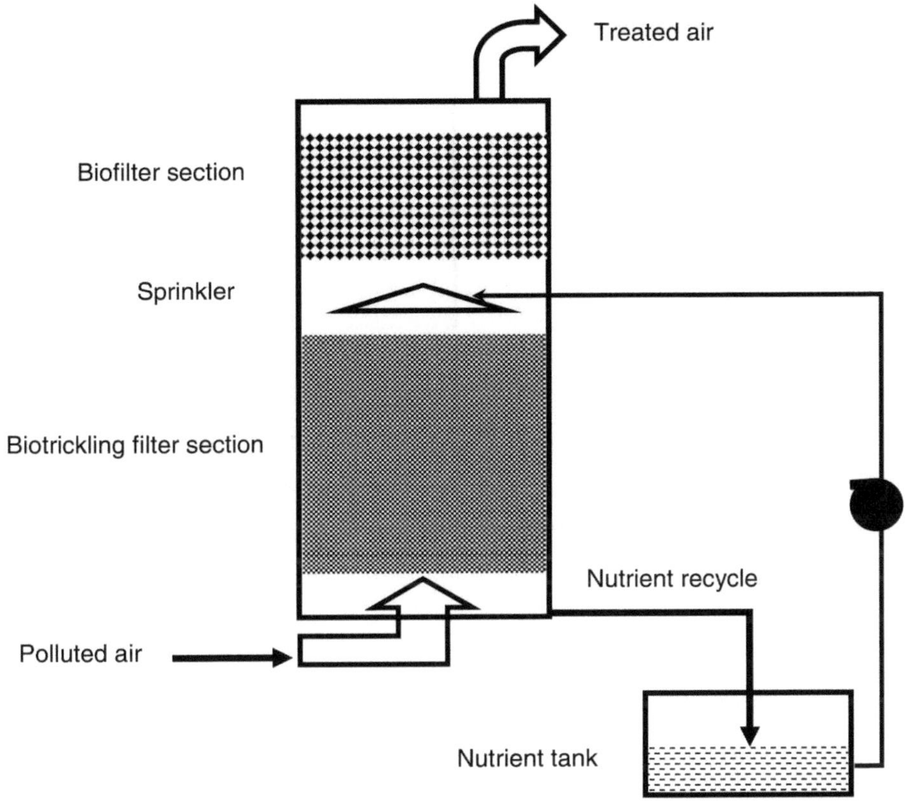


**Fig. 3.12** Schematic of a two-stage bioreactor (biotrickling filter + biofilter); used commonly for handling mixture of organic and inorganic gas-phase pollutants. Different microbial species can be independently maintained in this reactor configuration

bioreactor appears to be more promising and a practically feasible option for the co-treatment of $H_2S$ and VOCs and other gaseous-pollutant mixture, which has shown positive results in the recent past.

Chitwood et al. (1999) evaluated the feasibility of using a two-stage biofilter for the treatment of $H_2S$, air-toxics and smog precursors. The 1st stage acid-gas biofilter (AGB) packed with lava rock contained acidophilic autotrophic bacteria to remove $H_2S$, while the 2nd stage wood-chip biofilter removed other air-toxics that includes, methanol, acetone, methylene chloride, chloroform, toluene, xylene, ethyl benzene, methyl-tert-butyl ether (MTBE) and 2-methyl butane. However, they observed that the 1st stage AGB removed acetone and methanol completely, while other VOCs were intermittently removed depending on the concentrations, in addition to 99.6% removal of $H_2S$ at an ILR of 0.057 $g/m^3.h$.

Ruokojärvi et al. (2001) observed EC of 47.9 and 36.6 $g/m^3.h$ of $H_2S$ and dimethyl sulphide (DMS) in a two-stage biotrickling filter connected in series and inoculated

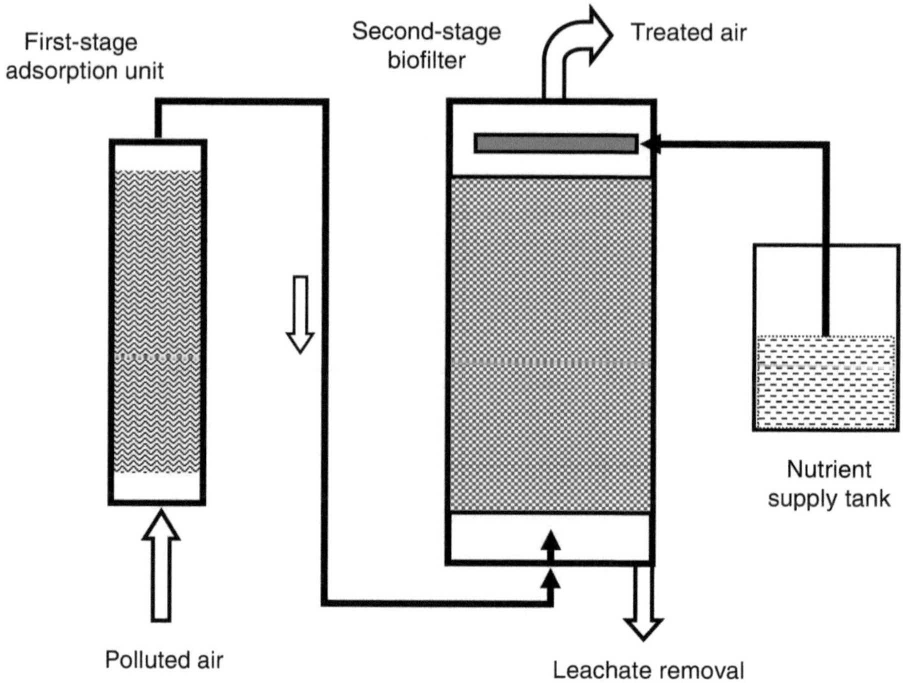


**Fig. 3.13** Schematic of a two-stage reactor (adsorption column + biofilter); used efficiently for flow equalization and resisting periodic shock loads. The first-stage adsorption unit would also release the pollutant, when required, to avoid microbial starvation in the next-stage biofilter

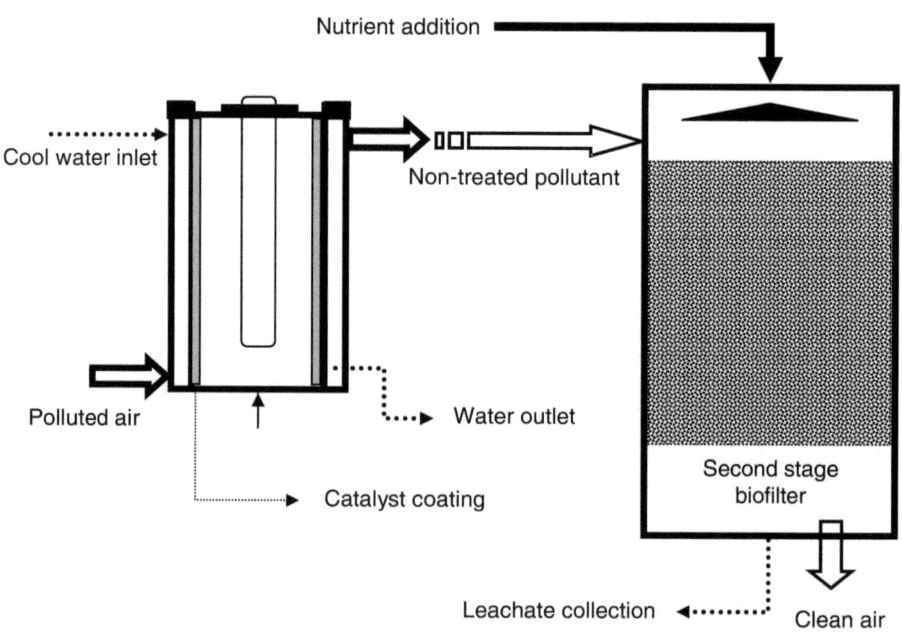


**Fig. 3.14** Schematic of a two-stage reactor (UV-photoreactor + biofilter); used efficiently for treating exceedingly high concentrations of gas-phase pollutant and during shock-loads

with a microbial consortium enriched from wastewater treatment plant sludge. Sercu et al. (2005) showed that high ECs could easily be achieved in a two-stage biotrickling filter for treating gas-phase DMS and $H_2S$. The 1st and 2nd stage biotrickling filters were inoculated with pure cultures of *Acidithiobacillus thiooxidans* and *Hyphomicrobium VS*, and the maximum EC achieved were 83 g $H_2S/m^3$.h in the 1st stage and 58 g DMS/$m^3$.h in the 2nd stage, that predominantly removed DMS. In yet another study, the feasibility of using sequential biofilters for $H_2S$ and a mixture of VOC vapors from wastewater treatment plant air was evaluated in laboratory-scale and field studies (Cox et al. 2003). At loading rate of 8.3 g $H_2S/m^3$.h and 33 g MTBE/$m^3$.h, near complete removal of these pollutants was noticed. However in field trials, the removal efficiency profiles followed the order, $H_2S$ > VOC > chlorinated VOC.

In a recent study, a hybrid-bioreactor system consisting of a low-pH biotrickling filter and a neutral pH biofilter contained within a single reactor column with different microbial populations and two different pH zones, was tested for removal of $H_2S$, methanol and α-pinene from polluted air. Though the hybrid reactor did not exhibit any synergistic effect in pollutant removal characteristics, even when fed with low concentrations of these pollutants, the results suggested the need for a two-stage system for achieving high ECs (Fig. 3.12). Low fungal growth in the biofilter section due to the only moderate tolerance of the fungus to acidification was considered as the major reason for such low EC values. The EC achieved from this study were 3.9, 4.3 and 1 g/$m^3$.h for $H_2S$, methanol and α-pinene respectively (Jin et al., non-published data). Rene et al. (2010b) evaluated the performance of a two-stage bioreactor (biotrickling filter followed by biofilter) to treat methanol, $H_2S$ and α-pinene from waste-gas, and reported that, hydrogen sulphide and methanol were removed in the first-stage, while α-pinene, was removed predominantly in the second-stage fungal biofilter. Besides, the first-stage biotrickling filter showed maximum elimination capacities of 45 g/$m^3$.h for hydrogen sulphide and 894 g/$m^3$.h for methanol, while the EC achieved for α-pinene removal in the fungal biofilter was about 140 g/$m^3$.h.

#### 3.4.6.2 Continuous Stirred Tank Bioreactor/Biotrickling Filter Combination

Manninen et al. (2003) studied the biodegradation of acetone, methanol, methyl ethyl ketone (MEK), naphthalene, α-pinene and toluene in a coupled bioreactor that consisted of a 1st stage liquid bioreactor and a 2nd stage biofilter. The coupled system yielded 97% overall VOC removal from initial start-up to shutdown. Lee (2003) used a two-stage continuous stirred-tank bioreactor/biotrickling filter for gas-phase treatment of trichloroethylene (TCE) by *Burkholderia cepacia*. The first stage continuous stirred-tank bioreactor was used for reactivation of the biomass, while the second-stage biotrickling filter was used to degrade TCE. This new operational design allowed the combined two-stage system to reach stable EC values during its 3 month operation, and an EC value of 28 mg TCE/l.d was reported. The addition of continuous stirred-tank bioreactor, *i.e.*, liquid-phase bioreactor, is also due to the greater feasibility of adding nutrients and neutralizing acidic metabolites formed during the biodegradation process.

#### 3.4.6.3 Adsorption Column/Bioreactor Combination

Shock-loads, in the form of sudden fluctuating loading rates, has always posed a major challenge for the design and operation of bioreactor systems. Short-term/temporary high loads can result in diminished treatment performance because of limitations in biological reaction capacity of the suspended/attached microorganisms, and the pollutant mass transfer rates (Kennes and Veiga 2001; Jin et al. 2007; Kennes et al. 2009a, b; Rene et al. 2010a, b; Nabatilan et al. 2010). On the other hand, low contaminant loading can lead to unexpected starvation conditions on the microorganisms present in the bioreactor, which eventually decreases cell activity and reactor performance during re-start. Weber and Hartmans (1995) used an activated carbon bed, prior to a biofilter, to minimize fluctuations in toluene loads. It was observed that the buffer capacity of the adsorbent depended highly on the desired concentration range of the contaminants entering the bioreactor and on the time available for desorption. For gas-phase toluene fluctuations up to 1,000 mg/$m^3$, the activated carbon bed was able to reduce it to a value of about 300 mg/$m^3$, which was then completely degraded in the biofilter. Similarly, Nabatilan et al. (2010) installed a column packed with granular activated carbon (GAC), prior to the biotreatment step, to achieve load equalization for the biological treatment (Fig. 3.13). The rationale for such a system was that during periods of high inlet concentrations, the GAC adsorbent can temporarily accumulate contaminants and then subsequently desorb contaminants during intervals when concentration in the waste-gas is lower. The results showed that, load equalization can be achieved in GAC columns which receive intermittent pollutant loading in combination with intermittent reduction of air flow rates during pollutant non-loading intervals.

#### 3.4.6.4 Ultraviolet Pre-treatment/Bioreactor Combination

The presence of hydrophobic VOCs in waste-gas, at high loading rates, would severely induce toxic and inhibitory effects for standalone bioreactors, such as biofilters, and biotrickling filters. Moussavi and Mohseni (2007) showed that UV photolysis can be used as the 1st stage system followed by a biofilter for the removal of toluene and *o*-xylene from polluted air. UV photolysis is an environmentally benign technology capable of oxidizing a wide range of contaminants using UV irradiation. However, the standalone use of UV photolysis for waste-gas treatment, depending on the chemical composition, is limited in application due to its tendency to produce water soluble by-products that are of environmental concern. Keeping this in mind, when a biofilter or a biotrickling filter is placed downstream, the by-products together with the non-treated air can be easily removed by the microbes present in the bioreactor. The use of UV/bioreactor combination (Fig. 3.14) has been reported in few studies and synergistic effects in pollutant removal have been observed (Koh et al. 2004; Mohseni and Zhao 2006). UV coupled biofilter receiving pretreated *o*-xylene was proven experimentally to exhibit greater EC values than those observed with standalone biofiltration (Mohseni and Zhao 2006). Moussavi and Mohseni (2007) showed that the two-stage system provided nearly 60% additional contaminant (toluene and *o*-xylene) removal compared to the sum of that offered by UV and a reference biofilter, and the maximum ECs in the coupled system were 45 and 50 $g/m^3.h$, respectively, for toluene and *o*-xylene. For practical purposes, it is advisable to use the UV pretreatment step only when it is desired. The UV lamps can be energized/turned-ON when the pollutant concentration is high, and during unexpected load fluctuations. Our own experience suggests, however, that the efficiency of using UV only is rather limited compared to a combined UV-$TiO_2$ system (non-published data).

Sometimes, depending on the waste-gas composition, bioscrubbers can also be combined with other treatment systems to improve performance. Hansen (1998) used a sequence of bioscrubber and an adsorption filter to remove mixtures of hydrophobic and hydrophilic VOCs. Standefer and Willingham (1998) used a combination of bioscrubbers to handle 60,000 $m^3$ gas/h containing nitrogen compounds, glycols, alcohols and aldehydes with a gas residence time less than 1 s. Their studies showed RE greater than 92% for readily soluble compounds, while RE ranged between 70% and 95% for other compounds. Beyreitz et al. (1989) used a biofilter and bioscrubber individually to treat emissions from a paint booth that contained six VOCs at a total loading rate of 70 $m^3/m^2.h$. The results indicated that 90% of the VOCs were removed in the biofilter, whereas only 80% could be removed in the bioscrubber. They further concluded that bioscrubbers are attractive for the removal of hydrophilic compounds whereas biofilters could be potentially put in use for hydrophobic compounds.

### 3.4.7 Foam-Emulsion Bioreactor

The mechanism of pollutant removal in a foam emulsion bioreactor (Fig. 3.15) is analogous to that of a two-liquid-phase reactor. The bioreactor consists of an emulsion of highly active pollutant degrading microorganisms and water-immiscible organic phase, which is made into foam by passing the waste-gas. The amount of organic phase is low and it uses a biocompatible surfactant for foam production (Shahna et al. 2010). There are only a few reports on foam-emulsion bioreactor for waste-gas treatment, and these results suggest that the foam-emulsion bioreactors could be an interesting alternative to conventional biological treatment systems, if high performance is desired (Kan and Deshusses 2003, 2005, 2006; Shahna et al. 2010). According to Kan and Deshusses (2005), the foam is continuously collapsed and the cells with the emulsion are reused after the desired level of biotreatment is achieved in the reactor. Besides, the foam-emulsion bioreactor gave high oxygen

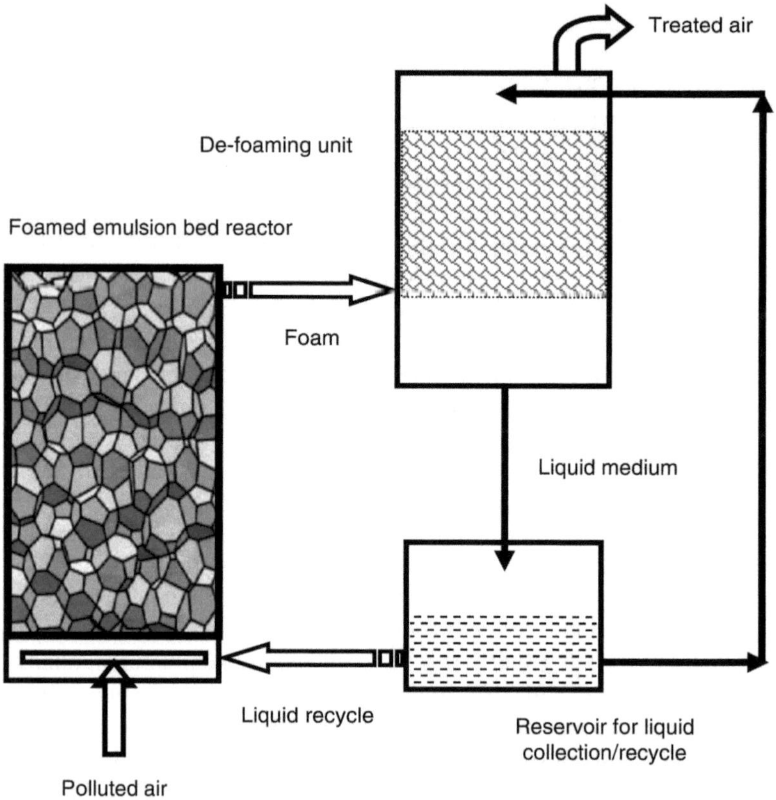


**Fig. 3.15** Schematic of foam-emulsion bioreactor; the process relies on the emulsion of an organic phase with a suspension of an actively growing culture of pollutant-degrading microorganisms, made into a foam with the air undergoing treatment (Kan and Deshusses 2003)

and pollutant mass transfer rates due to the large interfacial area between gas and liquid of the fine foam, and a high partitioning of pollutants into the organic phase.

For stable operation of the foam-emulsion bioreactor, foam stability, and cell activity are of prime importance. Besides, replacing and replenishing part of the culture would also be required to maintain high biodegradation activity and high pollutant removal in this reactor configuration. During short-term experiments (2–8 h), for gas-phase toluene removal, the foam-emulsion bioreactor reached an EC of 280 g/m$^3$.h, with 95% RE, at a gas residence time of 15 s, and still higher EC was reported (408 g/m$^3$.h) with 77% RE, when pure oxygen was added to the inlet air stream (Kan and Deshusses 2003). However, during long-term operations, the RE ranged from 89% to 94%, with EC value <226 g/m$^3$.h. For long-term bioreactor stability, it has been suggested to operate the bioreactor by replacing 20% of the culture with a concentrated nutrient solution, so as to maintain optimal cell growth, activity and to avoid nutrient limitations.

### 3.4.8 Other Novel Bioreactor Configurations for Volatile Organic Compound Treatment

Gas-phase benzene, toluene, ethylbenzene and xylene, commonly reported as BTEX, are usually present in emissions from petrochemical and pharmaceutical industries. The deleterious effects of these aromatics are well documented in the literature, and several studies have demonstrated their toxic and carcinogenic health effects. Shim et al. (2002) inoculated a novel fibrous-bed bioreactor with a co-culture of *Pseudomonas putida* and *Pseudomonas fluorescens*, and studied the kinetics of BTEX biodegradation in liquid-phase. The authors observed that the biodegradation rates of BTEX increased with increasing BTEX concentration and reactor loading rate. The maximum biodegradation rates for BTEX observed in that study were the highest for toluene (45 mg/l.h at a loading rate of 100 mg/l.h), followed by benzene, ethyl benzene and xylene isomers. The authors reported that the fibrous-bed bioreactor also had a stable long-term performance, maintaining its cell activity for efficient BTEX degradation without requiring additional nutrient source. The stable performance was later attributed to the high biomass concentrations and unique cell immobilization characteristics provided by the fibrous matrix, which allowed use of the reactor for continued regeneration, adaptation and selection of efficient BTEX degraders in the bioreactor environment.

Mudliar et al. (2008) described the operational characteristics of a novel single-stage rotating rope biofilter for the treatment of VOCs characterized by high volatility along with high water-solubility. The mechanism of VOC removal in this novel reactor combines both scrubbing and the biodegradation steps. A closed-type rotating biological reactor was modified for this purpose. The waste air containing the VOC was directly sparged through the water hold-up of the reactor and the water soluble VOCs were then absorbed in the aqueous phase. The VOC was degraded by the microbial consortium immobilized on the rope media. The following are the advantages of this reactor type, as reported by the authors; (i) high interfacial area per unit reactor liquid volume, (ii) high oxygen mass transfer rate, (iii) good microbial stability, and (iv) the ability to withstand high VOC loading rates.

## 3.5 Conclusions

The application of biological processes for waste-gas treatment has broadened due to the wide variety of cross-level research that has been carried out in recent years. Researchers have made successful attempts to study the mechanism of pollutant removal, fundamental microbial ecology in bioprocesses, the waste-gas flow pattern and other specific properties in detail, in different bioreactor configurations. This review shows that biological treatment systems can be more effective, reaching >90% removal efficiencies, compared to physico-chemical techniques for handling moderately high loading rates of volatile organic/inorganic compounds present in waste-gases. The novel bioreactors described in this review has shown to handle high loading rates of volatile organic compounds (example: gas-phase toluene, styrene) under both steady and transient-state operations, and can easily overcome operating problems like clogging, channeling, pH control and oxygen limitations experienced during biofilter and biotrickling filter operation. For waste-gas stream containing mixtures of organic and inorganic pollutants, two-stage bioreactors have shown to perform better, as they are able to operate independently, using specific microorganisms that target these pollutants individually. The performance of suspended-growth bioreactors for handling hydrophobic volatile organic compounds has improved tremendously in the last few years, thanks to the ability of the non-aqueous organic phase that has almost completely eliminated the commonly reported mass transfer limitations. Further research should be directed towards developing bioreactors that can provide high interfacial area with high mass transfer rates, robustness for scale-up, tolerance to high temperatures, and can perform at very low gas-residence times (<15 s).

**Acknowledgements** The authors thank the Spanish Ministry of Science and Innovation (Project: CTM2010-15796/TECNO), and European FEDER funds for financial support. Eldon R. Rene thanks the Spanish Ministry of Science and Innovation (MICINN), for his "Juan de La Cierva"–post doctoral contract. The PhD research contract of María Montes was funded through project CTM2007-62700/TECNO and from a fellowship offered by the University of La Coruña for doctoral students. We also thank M. Estefanía López, PhD student, for assisting us in depicting some of the high-quality figures provided in this chapter.

## References

Arriaga S, Muñoz R, Hernández S, Guieysse B, Revah S (2006) Gaseous hexane biodegradation by *Fusarium solani* in two liquid phase packed-bed and stirred-tank bioreactors. Environ Sci Technol 40:2390–2395

Arulneyam D, Swaminathan T (2000) Biodegradation of ethanol vapor in a biofilter. Bioproc Eng 22:63–67

Arulneyam D, Swaminathan T (2003) Biodegradation of methanol vapor in a biofilter. J Environ Sci 15:691–696

Bagreev A, Menendez JA, Dukhno I, Tarasenko Y, Bandosz TJ (2005) Oxidative adsorption of methyl mercaptan on nitrogen-enriched bituminous coal-based activated carbon. Carbon 43:208–210

Bailón L, Nikolausz M, Kästner M, Veiga MC, Kennes C (2009) Removal of dichloromethane from waste gases in one- and two-liquid-phase stirred tank bioreactors and biotrickling filters. Water Res 43:11–20

Beyreitz G, Hubner R, Saake M (1989) Biotechnologishe Behandlung losemittelhaltiger Abluft. Wasser Luft and Boden 9:62

Bhowmick M, Semmens MJ (1994) Batch studies on a close loop air stripping process. Water Res 28:2011–2019

Bielefeldt AR (2001) Activated sludge and suspended growth bioreactors. In: Kennes C, Veiga MC (eds) Bioreactors for waste gas treatment. Kluwer Academic Publisher, Dordrecht, pp 215–254

Blocki SW (1993) Hydrophobic zeolite adsorption. A proven advancement in solvent separation technology. Environ Prog 12:226–237

Blystone PG, Johnson MD, Haag WR, Daley PF (1993) Advanced ultraviolet flash lamps for the destruction of organic contaminants in air. In: Tedder DW, Pohland FG (eds) Emerging technologies in hazardous waste management III, vol 518, ACS Symposium Series. American Chemical Society, Washington, DC, p 380
Boudreau NG, Daugulis AJ (2006) Transient performance of two-phase partitioning bioreactors treating a toluene contaminated gas stream. Biotechnol Bioeng 94:448–457
Budwill K, Coleman RN (1997) Effect of silicone oil on biofiltration of *n*-hexane vapors. In: 11th forum for applied biotechnology, 62/4b, pp 1521–1528
Calvert S, Englund HM (1984) Handbook of air pollution technology. John Wiley and Sons, Inc, New York
Cesario MT, Beeftink HH, Tramper J (1992) Biological treatment of waste gases containing poorly-water soluble compounds. In: Dragt AJ, van Ham J (eds) Biotechniques for air pollution abatements and odour control policies. Elsevier Science Publishers, Amsterdam, pp 135–140
Chisti MY (1989) Airlift bioreactors. Elsevier, London
Chitwood DE, Devinny JS, Reynolds FE (1999) Evaluation of a two-stage biofilter for treatment of POTW waste air. Environ Prog 18:214–221
Chou MS, Wu FH (1999) Treatment of toluene in an air stream by a biotrickling filter packed with slags. J Air Waste Manage Assoc 49:386–398
Chungsiriporn J, Bunyakan C, Nikom R (2005) Treatment of toluene using wet scrubber with sodium hypochlorite oxidation reaction. In: Proceedings of the PSU-UNS international conference on engineering and environment-ICEE-2005, Novi Sad, Italy T11-3.1, pp 1–4
Clarke K, Hill GA, Pugsley TS (2007) Direct comparison of fluidized and packed bed bioreactors for bioremediation of an air pollutant. Intl J Chem React Eng V5(A11):1–12
Clarke K, Hill GA, Pugsley T (2008) Improved VOC bioremediation using a fluidized bed peat bioreactor. Proc Saf Environ Prot 86:283–290
Corless CE, Reynolds GL, Graham NJD, Perry R (1990) Ozonation of pyrene in aqueous solution. Water Res 24:1119–1123
Cox HHJ, Deshusses MA, Converse B, Shroeder ED, Vasooghi D, Samar P, Iranpour R (2003) Odor and VOC removal from wastewater treatment plant headwork's ventilation air using a biofilter. Water Environ Res 5:444–454
Darracq G, Couvert A, Couriol C, Amrane A, Thomas D, Dumont E, Andres Y, Le Cloirec P (2010) Silicone oil: an effective absorbent for the removal of hydrophobic volatile organic compounds. J Chem Technol Biotechnol 85:309–313
Daugulis AJ (1997) Partitioning bioreactors. Curr Opin Biotechnol 8:169–174
Daugulis AJ (2001) Two-phase partitioning bioreactors: a new technology platform for destroying xenobiotics. Trends Biotechnol 19:459–464
Daugulis AJ, Boudreau NG (2008) Solid-liquid two-phase partitioning bioreactors for the treatment of gas-phase volatile organic carbons (VOCs) by a microbial consortium. Biotechnol Lett 30:1583–1587
Devinny JS, Deshusses MA, Webster TS (1999) Biofiltration for air pollution control. Lewis Publisher, Boca Raton
Djeribi R, Dezenclos T, Pauss A, Lebeault J (2005) Removal of styrene from waste gas using a biological trickling filter. Eng Life Sci 5:450–457
Domeño C, Rodriguez-Lafuente A, Jm M, Bilbao R, Nerín C (2010) VOC removal and deodorization of effluent gases from an industrial plant by photo-oxidation, chemical oxidation and ozonization. Environ Sci Technol 44:2585–2591
Donley E, Lewandowski D (1996) Optimized design and operating parameters for minimizing emissions during VOC thermal oxidation. Metal Finish 94:52–58
Dragt AJ, van Ham J (1992) Biotechniques for air pollution abatement and odour control policies. Elsevier, Amsterdam
Dumont E, Andrès Y, Le Cloirec P (2006) Mass transfer coefficients of styrene and oxygen into silicone oil emulsions in a bubble reactor. Chem Eng Sci 61:5612–5619
Engleman VS (2000) Updates on choices of appropriate technology for control of VOC emissions. Metal Finish 98:433–445

Ensely BD, Kurisko PR (1994) A gas lift bioreactor for the removal of contaminants from the vapor phase. Appl Environ Microbiol 60:285–290
Ergas SJ, Kinney K, Fuller ME, Scow KM (1994) Characterization of a compost biofiltration system degrading dichloromethane. Biotechnol Bioeng 44:1048–1054
Estévez E, Veiga MC, Kennes C (2005) Biodegradation of toluene by the new fungal isolates *Paecilomyces variotii* and *Exophiala oligosperma*. J Ind Microbiol Biotechnol 32:33–37
Fang Y, Govind R (2007) A "convective" flow biofilter for the biofiltration of contaminants at sub-low concentrations. Remediation 18:95–107
Fazaelipoor MH (2007) A model for treating polluted air streams in a continuous two liquid phase stirred tank bioreactor. J Hazardous Materials 148:453–458
Fazaelipoor MH (2009) Analysis of a dual liquid phase biofilter for the removal of hydrophobic organic compounds from air streams. Chem Eng J 147:110–116
Fazaelipoor MH, Shojaosadati SA (2002) The effect of silicone oil on biofiltration of hydrophobic compounds. Environ Prog 21:221–224
Francke KP, Miessner H, Rudolph R (2000) Cleaning of air streams from organic pollutants by plasma–catalytic oxidation. Plasma Chem Plasma Proc 20:393–403
Ghoshal AK, Manjare SD (2002) Selection of appropriate adsorption technique for recovery of VOCs: an analysis. J Loss Prev Proc Ind 15:413–421
Hamdi B, Houari M, Hamoudi SA, Kessaïssia Z (2004) Adsorption of some volatile organic compounds on geomaterials. Desalination 166:449–455
Hansen N (1998) Odour's biological option. Water Qual Int Jul/Aug:12–14
Hwang S-J, Tang H-M (1997) Kinetic behavior of the toluene biofiltration process. J Air Waste Manage Assoc 47:664–673
Jeong J, Sekiguchi K, Sakamoto K (2004) Photochemical and photocatalytic degradation of gaseous toluene using short-wavelength UV irradiation with $TiO_2$ catalyst: comparison of three UV sources. Chemosphere 57:663–671
Jiangping W, Yu C, Dongyan C, Xiaoqiang J (2005) Removal of ethyl acetate in air streams using a gas-liquid-solid three phase flow airlift bioreactor. Biochem Eng J 24:135–139
Jin Y, Veiga MC, Kennes C (2005) Autotrophic deodorization of hydrogen sulphide in a biotrickling filter. J Chem Technol Biotechnol 80:998–1004
Jin Y, Veiga MC, Kennes C (2006a) Performance optimization of the fungal biodegradation of α-pinene in gas-phase biofilter. Proc Biochem 41:1722–1728
Jin Y, Veiga MC, Kennes C (2006b) Development of a novel monolith bioreactor for the treatment of VOC-polluted air. Environ Technol 27:1271–1277
Jin Y, Guo L, Veiga MC, Kennes C (2007) Fungal biofiltration of α–pinene: effects of temperature, relative humidity and transient loads. Biotechnol Bioeng 96:433–443
Jin Y, Veiga MC, Kennes C (2008) Removal of methanol from air in a low–pH trickling monolith bioreactor. Proc Biochem 43:925–931
Jorio H, Bibeau L, Viel G, Heitz M (2000) Effects of gas flow rate and inlet concentration on xylene vapors biofiltration performance. Chem Eng J 76:209–221
Kan E, Deshusses MA (2003) Development of foamed emulsion bioreactor for air pollution control. Biotechnol Bioeng 84:240–244
Kan E, Deshusses MA (2005) Continuous operation of foamed emulsion bioreactors treating toluene vapors. Biotechnol Bioeng 92:364–371
Kan E, Deshusses MA (2006) Scale-up and cost evaluation of a foamed emulsion bioreactor. Environ Technol 27:645–652
Kanai T, Uzumaki T, Kawase Y (1996) Simulation of airlift bioreactors: steady-state performance of continuous culture processes. Comput Chem Eng 20:1089–1099
Kennes C, Thalasso F (1998) Waste gas biotreatment technology. J Chem Technol Biotechnol 72:303–319
Kennes C, Veiga MC (2001) Bioreactors for waste gas treatment. Kluwer Academic Publishers, Dordrecht
Kennes C, Veiga MC (2002) Inert filter media for the biofiltration of waste gases-characteristics and biomass control. Rev Environ Sci Biotechnol 1:201–214

Kennes C, Veiga MC (2004) Fungal biocatalysts in the biofiltration of VOC polluted air. J Biotechnol 113:305–319

Kennes C, Rene ER, Veiga MC (2009a) Bioprocesses for air pollution control. J Chem Technol Biotechnol 84:1419–1436

Kennes C, Montes M, López ME, Veiga MC (2009b) Waste gas treatment in bioreactors: environmental engineering aspects. Can J Civil Eng 36:1887–1894

Khan FI, Ghoshal AK (2000) Removal of volatile organic compounds from polluted air. J Loss Prev Proc Ind 13:527–545

Khandan N (1997) Prototype development of a novel bioreactor configuration for treating VOC contaminated air streams. Technical completion report, institute for energy & the environment, New Mexico State University, Las Cruces

Koh LH, Kuhn D, Mohseni M, Allen DG (2004) Utilizing ultraviolet photooxidation as a pretreatment of volatile organic compounds upstream of a biological gas cleaning operation. J Chem Technol Biotechnol 79:619–625

Kok HJG (1992) Bioscrubbing of air contaminated with high concentrations of hydrocarbons. In: Dragt AJ, van Ham J (eds) Biotechniques for air pollution abatement and odour control policies. Elsevier Science Publishers, Amsterdam, pp 77–82

Kreutzer MT, Kapteijn F, Moulijn JA, Heiszwolf JJ (2005) Multiphase monolith reactors: chemical reaction engineering of segmented flow in microchannels. Chem Eng Sci 60:5895–5916

Le Cloirec P, Humeau P, Baleo JN (1999) Processes for biological purification of air loaded with volatile organic compounds. In: Proceedings of Biodepol, Rennes, France

Lee EY (2003) Continuous treatment of gas-phase trichloroethylene by *Burkholderia cepacia* G4 in a two-stage continuous stirred tank reactor/trickling biofilter system. J Biosci Bioeng 96:572–574

Leslous A, Delebarre A, Pre P, Warlus S, Zhang N (2004) Characterization and selection of materials for air biofiltration in fluidized beds. Int J Chem Reactor Eng V2(A20):1–19

Leson G, Winer AM (1991) Biofiltration: an innovative air pollution control technology for VOC emissions. J Air Waste Manage Assoc 41:1045–1054

Lichtin NN, Sadeghi M (1998) Oxidative photocatalytic degradation of benzene vapor over $TiO_2$. J Photochem Photobiol A Chem 113:81–88

Littlejohns JV, Daugulis AJ (2009) A two-phase partitioning airlift bioreactor for the treatment of BTEX contaminated gases. Biotechnol Bioeng 103:1077–1086

Mahanty B, Pakshirajan K, Dasu VV (2008) Biodegradation of pyrene by *Mycobacterium frederiksbergense* in a two-phase partitioning bioreactor system. Bioresour Technol 99:2694–2698

Mahanty B, Pakshirajan K, Dasu VV (2010) A two liquid phase partitioning bioreactor system for the biodegradation of pyrene: comparative evaluation and cost-benefit analysis. J Chem Technol Biotechnol 85:349–355

Manninen MR, Niemi BA, Kleinheinz GT (2003) Use of a coupled biological system to treat a chemically complex air stream. Arch Environ Contam Toxicol 45:1–10

McGregor FR, Piscaer PJ, Aieta EM (1988) Economics of treating waste gases from an air stripping tower using photochemically generated ozone. Ozone Sci Eng 10:339–351

Mohammad BT, Veiga MC, Kennes C (2007) Mesophilic and thermophilic biotreatment of BTEX-polluted air in reactors. Biotechnol Bioeng 97:1423–1438

Mohseni M, Zhao JL (2006) Coupling ultraviolet photolysis and biofiltration for enhanced degradation of aromatic air pollutants. J Chem Technol Biotechnol 81:146–151

Montes M, Rene ER, Veiga MC, Kennes C (2009) $\alpha$ - pinene removal from air in one- and two - liquid phase thermophilic and mesophilic biotrickling filters. In: Proceedings of the 3rd international congress on biotechniques for air pollution control, Delft, the Netherlands, pp 34–37

Montes M, Veiga MC, Kennes C (2010) Two-liquid-phase mesophilic and thermophilic biotrickling filters for the biodegradation of $\alpha$-pinene. Bioresour Technol 101:9493–9499

Moussavi G, Mohseni M (2007) Using UV pretreatment to enhance biofiltration of mixtures of aromatic VOCs. J Hazard Mater 144:59–66

Mudliar SN, Padoley KV, Bhatt P, Kumar SM, Lokhande SK, Pandey RA, Vaidya AN (2008) Pyridine biodegradation in a novel rotating rope bioreactor. Bioresour Technol 99:1044–1051

Muñoz R, Arriaga S, Hernández S, Guieysse B, Revah S (2006) Enhanced hexane biodegradation in a two phase partitioning bioreactor: overcoming pollutant transport limitations. Process Biochem 41:1614–1619
Muñoz R, Villaverde S, Guieysse B, Revah S (2007) Two-phase partitioning bioreactors for treatment of volatile organic compounds. Biotechnol Adv 25:410–422
Muñoz R, Chambaud M, Bordel S, Villaverde S (2008) A systematic selection of the non-aqueous phase in a bacterial two liquid phase bioreactor treating α-pinene. Appl Microbiol Biotechnol 79:33–41
Nabatilan MM, Harhad A, Wolenski PR, Moe WM (2010) Activated carbon load equalization of transient concentrations of gas-phase toluene: effect of gas flow rate during pollutant non-loading intervals. Chem Eng J 157:339–347
Namgung H-K, Kang JH, Hwang S-J, Song JH (2009) Removal of a mixture of VOCs using immobilized yeast strain *Candida tropicalis* in an airlift bioreactor. In: Proceedings of the 3rd international congress on biotechniques for air pollution control, Delft, the Netherlands, pp 63–67
Neal AB, Loehr RC (2000) Use of biofilters and suspended-growth reactors to treat VOCs. Waste Manag 20:59–68
Nielsen DR, Daugulis AJ, McLellan PJ (2005) Quantifying maintenance requirements from the steady-state operation of a two-phase partitioning bioscrubber. Biotechnol Bioeng 90:248–258
Nikakhtari H, Hill GA (2006) Continuous bioremediation of phenol-polluted air in an external airlift bioreactor with a packed bed. J Chem Technol Biotechnol 81:1029–1038
Nuñez C (1998) VOCs: sources, definitions and considerations for recovery. In: EPA, volatile organic compounds (VOC) recovery seminar, 16–17 Sept, Cincinnati, OH, pp 3–4
Oda T (2003) Non-thermal plasma processing for environmental protection: decomposition of dilute VOCs in air. J Electrost 57:293–311
Ottengraf SPP, Diks RMM (1991) Promising technique - process technology of biotechniques. LUCHT 4:135–144
Pedersen AR, Arvin E (1999) The function of a toluene degrading bacterial community in a waste gas trickling filter. Water Sci Technol 39:131–137
Pengyi Z, Fuyan L, Gang Y, Qing C, Wanpeng Z (2003) A comparative study on decomposition of gaseous toluene by $O_3$/UV, $TiO_2$/UV and $O_3$/$TiO_2$/UV. J Photochem Photobiol A Chem 156: 189–194
Pichat P (2003) Photocatalytic degradation of pollutants in water and air: basic concepts and applications. In: Tarr MA (ed) Chemical degradation methods for wastes and pollutants: environmental and industrial applications. Marcel Dekker, New York
Poppe W, Schippert E (1992) Das KCH-biosolv-ver-fahren in kombination mit einem biowascher herkomlicher art-eine verfahrensentwicklung zur abluftreinigung fur wasserlosliche und schwer wasserlosliche schadstoffe. In: Dragt AJ, van Ham J (eds) Biotechniques for air pollution abatements and odour control policies. Elsevier Science Publishers, Amsterdam, pp 71–76
Powell EE, Hill GA (2008) Optimization of continuously stirred tank bioreactor design for cost minimization: effect of microbial species and operating conditions. Int J Chem React Eng 6(A20):1–24
Prado OJ, Veiga MC, Kennes C (2005) Treatment of gas-phase methanol in conventional biofilters packed with lava rock. Water Res 39:2385–2393
Prado OJ, Veiga MC, Kennes C (2008) Removal of formaldehyde, methanol, dimethylether and carbon monoxide from waste gases of synthetic resin-producing industries. Chemosphere 70:1357–1365
Quijano G, Revah S, Gutierrez-Rojas M, Flores-Cotera LB, Thalasso F (2009a) Oxygen transfer in three-phase airlift and stirred tank reactors using silicone oil as transfer vector. Proc Biochem 44:619–624
Quijano G, Hernandez M, Thalasso F, Muñoz R, Villaverde S (2009b) Two - phase partitioning bioreactors in environmental biotechnology. Appl Microbiol Biotechnol 84:829–846
Rene ER, Arulneyam D, Swaminathan T (2004) Biofiltration. In: Pandey A (ed) Concise encyclopedia of bioresource technology. Haworth Press, New York, pp 31–39

Rene ER, Veiga MC, Kennes C (2009a) Performance of a biofilter for the removal of high concentrations of styrene under steady and non-steady state conditions. J Hazard Mater 168:282–290

Rene ER, Jin Y, Veiga MC, Kennes C (2009b) Two-stage gas-phase bioreactor for the combined removal of hydrogen sulphide, methanol and α-pinene. Environ Technol 30:1261–1272

Rene ER, Veiga MC, Kennes C (2010a) Biodegradation of gas-phase styrene using the fungus *Sporothrix variecibatus*: impact of pollutant load and transient operation. Chemosphere 79: 221–227

Rene ER, López ME, Veiga MC, Kennes C (2010b) Steady- and transient-state operation of a two-stage bioreactor for the treatment of a gaseous mixture of hydrogen sulphide, methanol and α-pinene. J Chem Technol Biotechnol 85:336–348

Rene ER, Lopez ME, Veiga MC, Kennes C (2010c) Performance of a fungal monolith bioreactor for the removal of styrene from polluted air. Bioresour Technol 101:2608–2615

Ritchie BJ, Hill GA (1995) Biodegradation of phenol polluted air using an external loop airlift bioreactor. J Chem Technol Biotechnol 62:339–344

Rocha-Rios J, Bordel S, Hernández S, Revah S (2009) Methane degradations in two-phase partition bioreactors. Chem Eng J 152:289–292

Ruokojärvi A, Ruuskanen J, Martikainen PJ, Olkkonen M (2001) Oxidation of gas mixtures containing dimethyl sulfide, hydrogen sulfide, and methanethiol using a two–stage biotrickling filter. J Air Waste Manage Assoc 51:11–16

Ruthven DM (1984) Principles of adsorption processes. John Wiley, New York

Sagastume FM, Sleep BE, Allen DG (2001) Effects of biomass growth on gas pressure drop in biofilters. J Environ Eng 127:388–396

Schippert E (1986) Biologische Abluftreinigung. Umwelt 8:527–529

Schnelle KB, Brown CA (2002) Absorption for HAP and VOC control. In: Air pollution control technology handbook. CRC Press, Boca Raton

Schroeder ED (2002) Trends in application of gas-phase bioreactors. Rev Environ Sci Biotechnol 1:65–74

Sercu B, Nuñez D, van Langenhove H, Aroca G, Verstraete W (2005) Operational and microbiological aspects of a bioaugmented two-stage biotrickling filter removing hydrogen sulfide and dimethyl sulphide. Biotechnol Bioeng 90:259–269

Shahna FG, Golbabaei F, Hamedi J, Mahjub H, Darabi HR, Shahtaheri SJ (2010) Treatment of benzene, toluene and xylene contaminated air in a bioactive foam emulsion reactor. Chin J Chem Eng 18:113–121

Shen YS, Ku Y (2002) Decomposition of gas-phase trichloroethene by the UV/$TiO_2$ process in the presence of ozone. Chemosphere 46:101–107

Shim H, Shin E-B, Yang S-T (2002) A continuous fibrous-bed bioreactor for BTEX biodegradation by a co-culture of *Pseudomonas putida* and *Pseudomonas fluorescens*. Adv Environ Res 7:203–216

Shinabe K, Oketani S, Ochi T, Matsumura M (1995) Characteristics of hydrogen sulphide removal by *Thiobacillus thio-oxidans* KSI isolated from a carrier-packed biological deodorization system. J Ferment Bioeng 80:592–598

Singh H (2006) Mycoremediation: fungal bioremediation. John Wiley & Sons Inc, Hoboken

Smith FL, Sorial GA, Suidan MT, Pandit A, Biswas P, Brenner RC (1998) Evaluation of trickle bed air biofilter performance as a function of inlet VOC concentration and loading, and biomass control. J Air Waste Manage Assoc 48:627–636

Sorial GA, Smith FL, Smith PJ, Suidan MT, Biswas P, Brenner RC (1993) Evaluation of biofilter media for treatment of air streams containing VOCs. In: Proceedings of water environment federation 66th annual conference, vol 10, Facility Operations, pp 429–439

Spigno G, Pagella C, Fumi MD, Molteni R, De Faveri DM (2003) VOCs removal from waste gases: gas-phase bioreactor for the abatement of hexane by *Aspergillus niger*. Chem Eng Sci 58:739–746

Standefer and Willingham (1998) Industrial application of biofilter technologies. In: Proceedings of the USC-TRG conference on biofiltration, Octoober 22–23, 1998, Los Angeles, California, pp 1–8

Studer M, von Rohr PR (2008) Novel membrane bioreactor: able to cope with fluctuating loads, poorly water soluble VOCs, and biomass accumulation. Biotechnol Bioeng 99:38–48
Tsai WT, Chang CY, Lee HC (1996) Adsorption of organic vapors containing N, N-dimethylformamide on activated carbon and hydrophobic zeolite. J Environ Sci Health A 31:995–1008
van Groenestijn JW (2001) Bioscrubbers. In: Kennes C, Veiga MC (eds) Bioreactors for waste gas treatment. Kluwer Academic Publisher, Dordrecht, pp 133–162
van Groenestijn JW, Hesselink PGM (1993) Biotechniques for air pollution control. Biodegradation 4:283–301
van Groenestijn JW, Kraakman NJR (2005) Recent developments in biological waste gas purification in Europe. Chem Eng J 113:85–91
van Groenestijn JW, Lake ME (1999) Elimination of alkanes from off-gases using biotrickling filters containing two liquid phases. Environ Prog 18:151–155
Vergara-Fernandez AO, Quiroz EF, Aroca GE, Pulido NAA (2008) Biological treatment of contaminated air with toluene in an airlift reactor. E J Biotechnol 11:1–7
Wang JH, Ray MB (2000) Application of ultraviolet photooxidation to remove organic pollutants in the gas phase. Sep Purif Technol 19:11–20
Waweru M, Herrygers V, Van Langenhove H, Verstraete W (2000) Process engineering of waste gas purification. In: Klein J, Winter J (eds) Biotechnology, 2nd edn. Wiley-VCH Verlag GmbH, Weinheim
Weber FJ, Hartmans S (1995) Use of activated carbon as a buffer in biofiltration of waste gases with fluctuating concentrations of toluene. Appl Microbiol Biotechnol 43:365–369
Weber FJ, Hartmans S (1996) Prevention of clogging in a biological trickle-bed reactor removing toluene from contaminated air. Biotechnol Bioeng 50:91–97
Wei VQ, Hill GA, Macdonald DG (1999) Bioremediation of contaminated air using an external-loop airlift bioreactor. Can J Chem Eng 77:955–962
Whaley MB, Monroig P, Villarubia E (1998) Allergen-isopropyl alcohol emissions removal using unique biological system. Environ Prog 17:256–259
William JC, Lead PE (1997) VOC control strategies in plant design. In: Chemical Processing: Project Engineering Annual, 44
Wright PC, Raper JA (1998) Investigation into the viability of a liquid-film three-phase sprouted bed biofilter. J Chem Technol Biotechnol 73:281–291
Xie C, Xu Z, Yang Q, Li N, Zhao D, Wang D, Du Y (2004) Comparative studies of heterogeneous photocatalytic oxidation of heptane and toluene on pure titania, titania–silica mixed oxides and sulfated titania. J Mol Cat Chem 217:193–201
Xiong F, Graham NJD (1992) Rate constants for herbicide degradation by ozone. Ozone Sci Eng 14:283–301

# Chapter 4
# Extraction Procedures for Organic Pollutants Determination in Water

**N. Etxebarria, P. Navarro, A. Prieto, M. Olivares, A. Usobiaga, L.A. Fernández, and Olatz Zuloaga**

## Contents




N. Etxebarria • M. Olivares • A. Usobiaga • L.A. Fernández • O. Zuloaga (✉)
Department of Analytical Chemistry, University of the Basque Country, PK 644, 48080 Bilbao, Spain
e-mail: olatz.zuloaga@ehu.es

P. Navarro
Laboratoire de Chimie Analytique Bio-Inorganique et Environnement, UMR 5254 CNRS-UPPA, 2 Avenue Président P. Angot, 64053 Pau, France

A. Prieto
Department of Analytical Chemistry, Helmholtz Centre for Environmental Research – UFZ, Permoserstraße 15, D-04318 Leipzig, Germany



E. Lichtfouse et al. (eds.), *Environmental Chemistry for a Sustainable World: Volume 2: Remediation of Air and Water Pollution*, DOI 10.1007/978-94-007-2439-6_4,

**Abstract** The European Water Framework Directive 2000/60 is one of the key legislations introduced for many years in water management and protection fields. The model of water management established by this directive is the river basin. There are a variety of indicators to define the good ecological and chemical status, including a list of priority and emerging pollutants to monitor. Compounds in the directive list are organic pollutants, e.g. hydrocarbons, organochlorine compounds, organic solvents, pesticides and chlorophenol, toxic metals (Hg, Ni, Cd and Pb) and one organometallic compound (tributyltin). Other chemical compounds such as pharmaceuticals, hormones and endocrine disruptor have also become important emerging contaminants due to their presence in environmental waters, the threat for drinking water sources and the concern about their possible estrogenic and other effects. For these reasons, some of them have been included as emerging pollutants in the directive. Although this directive does not specify the analytical methods and procedures to be used, the results should come to an agreement no matter which method is used and regardless in which lab the analysis is carried out. Therefore, a large effort to validate and harmonize analytical methods is very required to assure the quality of the results.

Consequently, the implementation of robust analytical methods is still a challenging requirement. Among the different steps of the analytical procedure sample preparation is not only the a major source of uncertainties but also where the risk of artifacts and contamination is the highest. Different extraction techniques for the analysis of organic pollutants in water samples are being developed and optimized in order to automate and miniaturize the extraction step, using low volumes of solvents or even solventless procedures in order to pre-concentrate the analytes in the final acceptor phase.

We review recent strategies for the analysis of organic pollutants in environmental water samples as alternative choices against the most classical liquid-liquid extraction or solid-phase extraction. Revised extraction techniques are classified in two main groups: (i) sorptive extractions and (ii) liquid-liquid microextraction. Among sorptive extractions, solid-phase microextraction, stir-bar sorptive extraction and microextraction by packed sorbent are described. Cloud point extraction, single drop microextraction, membrane liquid-phase microextraction and dispersive liquid-liquid microextraction are included in the group of liquid-liquid microextraction techniques. Among membrane liquid-phase microextraction the use of both porous and nonporous membranes is described. During the description of liquid-liquid microextraction techniques the use of ionic liquids is also mentioned due to the high number of applications found for both single drop microextraction and dispersive liquid-liquid microextraction. We discuss the basics, advantages, drawbacks and applications of the techniques for the analysis of organic pollutants in environmental water samples, including river, estuarine water, seawater or wastewater.

**Keywords** Water framework directive • Organic contaminants • Water samples • Extraction techniques • Solid-phase microextraction • Stir-bar sorptive extraction • Microextraction by packed sorbent • Cloud point extraction • Single drop microextraction • Membrane liquid-phase microextraction • Dispersive liquid-liquid microextraction • Ionic liquids

## List of Abbreviations

### Chemicals and Compounds

| Acronym Meaning | Chemical |
|---|---|
| [BMPL][PF3] | 1-methylpyrrolidinium tris(pentafluoroethyl)trifluorophosphate |
| [C4MIM][PF3] | 1-hexyl-3-methylimidazolium tris(pentafluoroethyl) trifluorophosphate |
| [C6MIM][PF6] | 1-hexyl-3-methylimidazolium hexafluorophosphate |
| [C8MIM][PF6] | 1-octyl-3-methylimidazolium hexafluorophosphate |
| [MC8IM][PF6] | 1-methyl-3-octylimidazolium hexafluorophosphate |
| [PH3T][PF3] | trihexyl(tetradecyl)phosphonium tris(pentafluoroethyl)trifluorsphosphate |
| 4nP | 4-nonylphenol |
| APs | alkylphenols |
| BPA | bisphenol-A |
| BTEX | benzene, toluene, ethylbenzene and xylenes |
| CAR | carboxen |
| ClP | chlorophenols |
| CW | carbowax |
| DBT | dibutyltin |
| DiEtHg | diethylmercury |
| DVB | divinylbenzene |
| ECDs | endrocrine disrupting chemicals |
| EME | electromembrane extraction |
| EtHg | ethylmercury |
| MBT | monobutyltin |
| MeHg | methylmercury |
| NPEs | nonylphenol ethoxylates |
| NPs | nonylphenols |
| OCP | organochlorine pesticide |
| OPP | organophosphorus pesticide |
| PA | polyacrilate |
| PAH | polycyclic aromatic hydrocarbon |
| PBBs | polybrominated biphenyls |
| PBDE | polybrominated byphenylether |
| PBDE | polybrominated diphenyl ethers |

| | |
|---|---|
| PC | polycarbonate |
| PCB | polychlorinated byphenyl |
| PCDD | polychlorinated dibenzodioxins |
| PDMS | polidimethylsiloxane |
| PEEK | polyetheretherketone |
| PEs | phtalate esters |
| POP | persistent organic pollutants |
| PP | polypropylene |
| PS | polystyrene |
| PTFE | polytetrafluoroethylene |
| PVDF | polyvinylidene difluoride |
| SVOC | semi-volatile organic compounds |
| TBAB | tetrabutylammonium bromide |
| TBT | tributyltin |
| TD | thermal desorption |
| TMAH | tetrabutylammonium hydroxide |
| TMSPA | 3-(trimethoxysilylpropyl) amine |
| TPhT | triphenyltin |
| TPR | template resin |
| VD-EPMS | vinylpyridine-ethylene dimethacrylate |
| VOC | volatile organic compounds |

**Techniques**

| | |
|---|---|
| CE-UV | capillary electrophoresis-ultraviolet detector |
| CF | continuous flow |
| CP | cloud-point |
| CPE | cloud point extraction |
| CRC | critical micellar concentration |
| CSM | cone shape |
| DAD | diode array detector |
| DD | drop-to-drop |
| DI-SBSE | direct immersion – SBSE |
| DI-SDME | direct immersion-single drop microextraction |
| dSPME | dual phase SPME |
| DI-SPME | direct immersion – SPME |
| DSD | directly suspended droplet |
| FTD | flame thermionic detection |
| GC | gas chromatography |
| GC/TOF-MS | gas chromatography/high-resolution time-of-flight mass spectrometry |
| GC-AED | gas chromatography with atomic emission detection |
| GC-ECD | gas chromatography-electron capture detector |
| GC-FID | gas chromatography-flame ionization detector |
| GC-FPD | gas chromatography flame photometric detection |
| GC-ICP-MS | gas chromatography inductively coupled plasma mass spectrometry |

| | |
|---|---|
| GC-MS | gas chromatography-mass spectrometry |
| GC-MS/MS | gas chromatography tandem mass spectrometry |
| HF | hollow fiber |
| HPLC-DAD | high-pressure liquid chromatography with diode array detector |
| HPLC-PAD | high performance liquid chromatography-photodiode array detector |
| HPLC-UV | high performance liquid chromatography-ultraviolet detector |
| HPLC-UV-ED | high performance liquid chromatography-ultraviolet detector-electrochemical detector |
| HS | headspace |
| HS-LPME | headspace-liquid phase microextraction |
| HS-SDME | headspace-single drop microextraction |
| HSSE | headspace sorptive extraction |
| IL | ionic liquids |
| IL-DI-SDME | ionic liquid-direct immersion-single drop microextraction |
| IL-HS-SDME | ionic liquid-headspace-single drop microextraction |
| IPPC | integrated Pollution Prevention and Control |
| IT-SPME | in tube solid phase microextraction |
| LC-FLD | liquid chromatography lateral flow devices |
| LC-MS/MS | liquid chromatography-tandem mass spectrometry |
| LD | liquid desorption |
| LDPE | low-density polyethylene |
| LLE | liquid-liquid extraction |
| LLL | liquid-liquid-liquid |
| LPME | liquid phase microextraction |
| LVI | large volume injector |
| LVI-GC-MS | large volume injection gas chromatography mass spectrometry |
| MASE | membrane-assisted solvent extraction |
| MEKC-UV | micellar electrokinetic chromatography-ultraviolet detector |
| MEPS | microextraction in packed sorbents |
| MEPS-BIN | microextraction by packed sorbent-barrel insert and needle (BIN) device |
| MESI | membrane extraction with sorbent interface |
| MIP | molecular imprinted polymers |
| MMLLE | microporous membrane liquid–liquid extraction |
| MW-HS-SPME | microwave assisted headspace- solid phase microextraction |
| PTV | programmable temperature vaporisers |
| REACH | registration, evaluation, restriction and authorisation of chemicals |
| RP-LC | reverse phase liquid chromatography |
| SBSE | stir bar sorptive extraction |
| SCX | strong cation exchanger |
| SDME | single-drop microextraction |
| SLM | supported liquid membrane |
| SPME | solid phase micro-extraction |
| SPME-MC | solid phase microextraction-micellar desorption |

| | |
|---|---|
| TD-GC-MS | thermal desorption-gas chromatography-mass spectrometry |
| TDU | thermal desorption unit |
| WFD | water framework irective |

## 4.1 Introduction

Pollution of marine and freshwaters is a challenging issue that mobilizes a vast amount of scientific, technical and legal resources, not only to understand and mitigate the ecological integrity or health status, but also to harmonize the policy responses (Rapport et al. 1998). As a matter of fact, since the fact that ecosystems are stressed by human activities is broadly accepted, developed societies are becoming aware of the effects and consequences of the presence of xenobiotics in many environmental compartments (Reid et al. 2005).

From the ecotoxicological point of view, the management of the environmental health issues requires the integration of several key actions that go from a deeper knowledge of toxicity mechanisms and biological responses to the standardization of the biological and chemical procedures (Lepom et al. 2009; Hansen 2007).

Some of the policy responses that have been promoted in the European Union are the Drinking Water Directive (1998/83/EC), the Water Framework Directive (2000/60/EC) and the Marine Strategy Framework Directive (2008/58/EC) or even the Integrated Pollution Prevention and Control Directive (IPPC 2008/1/EC), in addition to the REACH regulation (1907/2006) and all the supporting background documents such as the list of priority pollutants (2455/2001/EC) or the environmental quality standards in the field of water policy (2008/105/EC).

Generally speaking, the objectives of environmental analysis depend on the specific task considered, including regulatory requirements, routine monitoring, scientific research, etc., which involves the development of appropriate analytical methods for fast and sensitive detection of priority substances (Coquery et al. 2005). Anyhow, we should be aware of the uniqueness of most analyses due to the complexity of many sample matrices, the unpredictability of the interferences, or the low or very low concentrations of many contaminants (Roose and Brinkman 2005). These tight requirements have fostered unsurpassable analytical marks and the coining of the term "emerging contaminants" to mean unregulated chemicals. However, one drawback of this uniqueness of many tailor-made methods is that many compounds are overlooked or even ignored (Erickson 2002).

Within the last decades several studies have revealed that organic trace substances suspected for ecological and health risks are increasingly detected in natural or treated water bodies (Farré et al. 2008). These emerging contaminants, which have not been considered historically as pollutants, have now also become prominent agents of concern to environmental scientists. For screening and monitoring programs with high sample throughput, automated, cost-effective and user-friendly strategies for water analysis are needed. Nevertheless, legal issues such as the Water Framework Directive, with challenging analytical requirements, are under scientific query, among other reasons, due to the lack of knowledge on the fate and behaviour of contaminants in the environment (Vignati et al. 2009).

The analytical methodology includes a tiered scheme of procedures, going from sampling to instrumental analysis. From the point of view of method development the two key steps are the sample preparation and the instrumental analysis. In fact, the determination of organic trace compounds in environmental aqueous matrices requires an extraction and pre-concentration step prior to the chromatographic separation and detection. There are many recent reviews dealing with these methods and especially the chromatographic separation and the mass spectrometric determination of priority and emerging contaminants (Richardson 2009, 2010).

In this work we want to focus on the current development in the extraction processes of priority and some emerging contaminants from aqueous media. It is well-known that extraction methods are intended to either preconcentrate the analytes or to eliminate or simplify the sample matrix, getting rid of some of the matrix components (Mitra 2003). Some other minor applications include the adequation of the solvent to a given instrumental method or the purification of the analyte for further characterizations. Nowadays, the extraction methods include the use of sorptive extraction systems such as solid phase micro-extraction (SPME), stir bar sorptive extraction (SBSE) or microextraction by packed sorbents (MEPS) and liquid phase microextraction (LPME). Most of those extraction methods will be reviewed paying special attention to the key aspects of method development and to their applications in environmental water analysis.

## 4.2 Sorptive Extraction

Sample preparation techniques based upon sorptive extraction, such as solid-phase micro-extraction (SPME), stir-bar sorptive extraction (SBSE) and microextraction by packed sorbent (MEPS) have proven to be innovative and environmentally friendly alternatives to more solvent consuming techniques, such as, liquid-liquid extraction (LLE) or solid-phase extraction (SPE).

### *4.2.1 Solid-Phase Micro-Extraction (SPME) and Stir-Bar Sorptive Extraction (SBSE)*

The principles and applications of SPME and SBSE have been reviewed previously (Baltussen et al. 2002; Prieto et al. 2010a). Briefly, in SPME and SBSE the analytes are extracted from the aqueous matrix into a non-miscible liquid or solid polymer, mainly liquid polydimethylsiloxane (PDMS). PDMS is a very well-known stationary phase in gas chromatography (GC), thermally stable in a broad temperature range (−20°C to 320°C) and with remarkable diffusion properties. The major

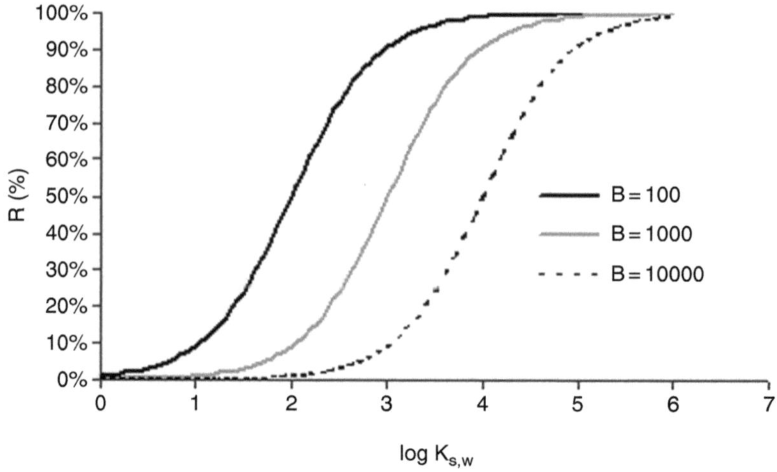


**Fig. 4.1** Theoretical recovery versus log $K_{polymer,w}$ and water phase volume during sorptive extraction

difference between SPME and SBSE relies on the amount of polymeric phase, which determines the extraction efficiency. However, one significant drawback of the increased fiber capacity is the loss of selectivity, since not only the analytes but also most of the interferences are exhaustively extracted (Lord and Pawliszyn 2000). In the case of SPME, the volume of the fibre is usually 0.5 μL, while for SBSE the smallest stir-bar has roughly 24 μL. This means that for the commonest extraction setup, i.e. 10 mL of aqueous solution, the phase ratio ($B=V_{sample}/V_{polymer}$) ranges from 400 (SBSE) to 20,000 (SPME). Therefore, while SPME is a micro-extraction technique, the recoveries are much higher when SBSE is used (see Fig. 4.1).

The sorptive extraction is controlled by the partition coefficient of the solute between the aqueous and sorbent phases ($K_{s,w}$). Recent studies have correlated this coefficient with the octanol-water distribution ($K_{o,w}$) when the sorbent phase is polydimethylsiloxane (PDMS), providing a good indication of the capabilities of extracting a given analyte (David and Sandra 2007; Kawaguchi et al. 2006b).

With SPME and SBSE there are two experimental work regimes according to the sampling time: equilibrium and kinetic. In the first case, if long extraction times are assured, the partitioning equilibrium between sample matrix and extraction phase is reached. In the second case, on the contrary, with very short extraction times a linear uptake is observed (Wells 2003). In the first approach, convection conditions do not affect the amount extracted and the extraction is limited the partition equilibria between the sample matrix and the sorptive phase. In the second approach, the extraction is interrupted prior to equilibrium and it is necessary to control the convection/agitation and the timing of the extraction must be constant in order to guarantee repeatable results (Lord and Pawliszyn 2000).

When these two extraction techniques are compared one of the clearest differences is the availability of coatings that can be found. In the case of SPME there are

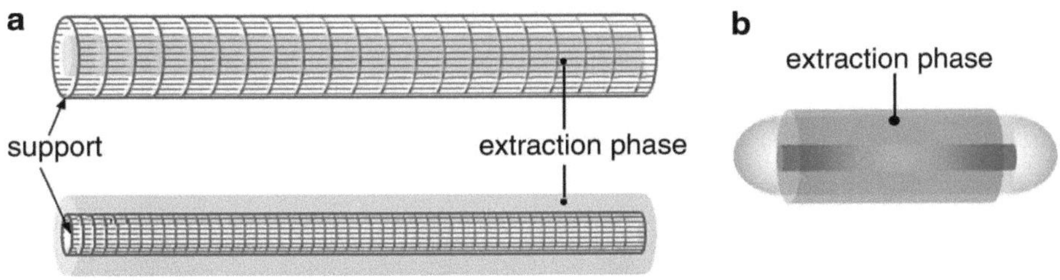


**Fig. 4.2** Solid phase micro-extraction (SPME) coating on the internal surface of the fibre or in the outer surface (**a**) and a stir-bar (**b**)

several coatings commercially available, such as PDMS, PDMS/divinylbenzene (DVB), polyacrilate (PA), carboxen/PDMS, carbowax(CW)/DVB, CW/template resin and DVB/carboxen/PDMS (Supelco 2004), which allow a higher selectivity of SPME. There are several factors, however, that must be taken into consideration when selecting SPME sorbents: concentration levels, detection limits, physical characteristics of the analytes and the resistance to high-temperatures, extreme pHs, salts or other additives (Wells 2003).

On the contrary, one of the major drawbacks of SBSE is that only non-polar PDMS coated stir-bars are commercially available so far. PDMS provides low recoveries for polar analytes and often in situ derivatization (see Fig. 4.1) is necessary. In-house coatings try to solve this problem by preparing coatings with different polarities. Sol-gel technology, monolithic materials, among others, have been reported as different approaches to solve this problem (Liu et al. 2004a, 2005b, 2007; Huang et al. 2010; Bagheri et al. 2010; Guan et al. 2007; Wang et al. 2008; Yao et al. 2008; Bagheri et al. 2008).

In the case of SPME, the coating can be placed on the outer surface of a fibre or on the internal surface of a capillary tube as shown in Fig. 4.2a (Wells 2003). In SBSE the stir-bar (so-called "twister") is coated with a polymer layer (see Fig. 4.2b).

#### 4.2.1.1 Method Development and Applications

SPME and SBSE consist of two principal steps: extraction and desorption. During the extraction step the polymer-coated fibre or stir-bar is put in contact with the solutes by direct immersion (DI-SPME or DI-SBSE) or by headspace sampling (HS-SPME or HSSE). In the direct immersion mode (see Fig. 4.3) the coated fibre or stir-bar is put in direct contact with the aqueous sample under controlled extraction conditions, while headspace extraction is performed by suspending the coated fibre or stir-bar in the headspace of the vial, which allows a static contact of the polymeric phase with the vapour phase of the aqueous matrix (see Fig. 4.3). The DI mode is significantly influenced by the sample matrix (Ho et al. 2008), while the HS mode can prevent the matrix effects and allows modification of the matrix without damaging the polymeric phase (Lord and Pawliszyn 2000).

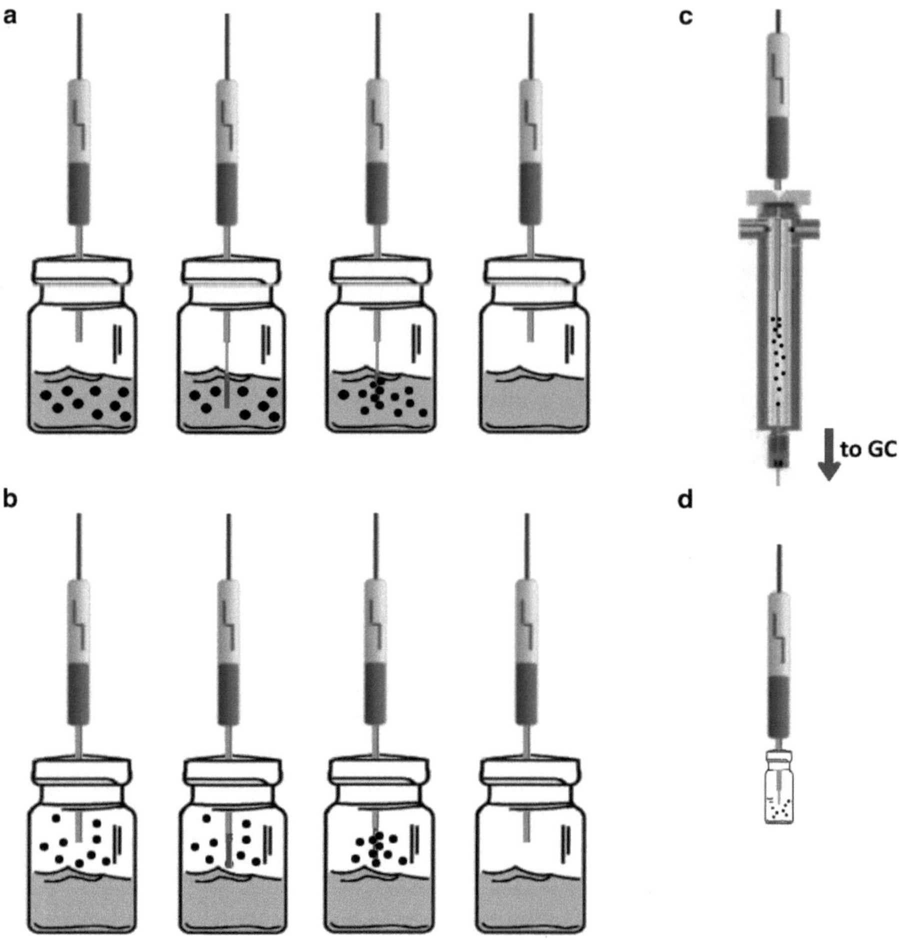


**Fig. 4.3** Extraction steps in Solid phase micro-extraction (SPME): extraction in the immersed mode (**a**), extraction in the headspace mode (**b**), thermal desorption (**c**) and chemical desorption (**d**)

The desorption step is another drawback of SBSE since the coated stir-bar cannot be directly desorbed in a simple spilt/splitless injection port of a gas chromatograph. Hence, the extraction step is followed by a thermal (TD) or liquid desorption (LD) previous to the electrophoretic or chromatographic separation and detection. Most applications involve the use of TD, which does not require the use of organic solvents, in order to introduce quantitatively thermally stable volatile and semi-volatile analytes to the gas chromatographic (GC) system (Baltussen et al. 2002). TD occurs at temperatures in the 150–300°C range and guarantees that the whole amount of analyte extracted is introduced to the chromatographic system, improving method sensitivity. TD of SPME fibres is accomplished in a simple split/splitless injector, whereas a thermal desorption unit (TDU) is necessary in the case of TD of stir-bars. A TDU consists of two programmable temperature vaporizers (PTV).

While the first PTV is heated in order to desorb the solutes, the second PTV is kept cool (−150°C to 40°C) in order to cryofocus the desorbed analytes before entering the GC column, since TD of stir-bars can take up to 15 min.

The use of LD is the alternative to those thermally labile analytes and is generally coupled to liquid chromatography (LC), capillary electrophoresis (CE) or even GC if large volume injection (LVI) is used. In LD mode, the use of organic stripping solvents (such as acetonitrile, methanol or mixtures with water or aqueous buffers) is required to accomplish the chemical desorption. Due to the polarity of most of the stripping solvents used, LD mode is mainly useful for non-volatile and thermo-labile compounds with an intermediate polarity (log $K_{o,w} \approx 2$).

The goal of the sorptive-extraction method development relies on the optimization of the factors affecting both the extraction and desorption steps as it has been already reviewed (Prieto et al. 2010a).

Regarding the extraction, all the parameters that could affect the partition of the analytes between the two phases (i.e. sample volume, nature and volume of the sorptive phase, boundary layer, pH, salt concentration or the presence of phase modifiers such as dissolved organic matter) would modify the extracted amount of analyte and therefore the recovery (Yang et al. 2008; Fattahi et al. 2007; Alzaga et al. 2003a; Hernandez et al. 2007; Lin et al. 2003). Additionally, the temperature is an important factor to control in the SPME and SBSE since it affects the partition coefficient and the time needed to reach the equilibrium (Farajzadeh and Hatami 2002; Lord and Pawliszyn 2000). Therefore, the most studied variables are extraction time, pH adjustment, addition of an inert salt, addition of an organic modifier, stirring rate, extraction temperature, sample volume and the nature and volume of the polymeric phase (Prieto et al. 2010a; Vazquez et al. 2008).

Regarding the desorption step, desorption time, desorption temperature and cryofocusing temperature (in the case of SBSE) are the most studied variables in the TD mode. For LD, on the contrary, stripping solvent nature, desorption time and desorption volume are the most frequently evaluated variables.

Both SPME and SBSE have been widely used for the extraction and pre-concentration of organic pollutants. Tables 4.1 and 4.2 summarize the applications found in the literature for the last decade.

#### 4.2.1.2 Derivatization

Derivatization of the analytes is a common option to increase the chromatographic sensibility of organic polar compounds (Lord and Pawliszyn 2000; Dean 1998). During sorptive extractions such as SPME and SBSE, derivatization can be performed in different ways (see Fig. 4.4). In *in-situ* derivatization, the derivatization reagent is added to the sample, analytes are derivatized and then the sorptive extraction is carried out (in the DI or HS modes) (Olejniczak and Staniewski 2007; Pan et al. 1997; Prieto et al. 2010a). Another option is *on-polymer* derivatization and, in this case, derivatization occurs on the polymer, which is put in direct contact with the derivatization reagent before or after the extraction process (Alzaga et al. 2003a;

**Table 4.1** SPME application to the determination of priority or emerging pollutants in water samples during the period 2000–2010

| Analyte | Matrix | Coating material | Extraction mode | Analysis technique | LOD | Apparent recovery (%) | Precision (%) | Reference |
|---|---|---|---|---|---|---|---|---|
| ClP | Landfill leachate | PA | HS-SPME | GC-MS | fg level | 83–114 | 4–14 | Ho et al. (2008) |
| ClP | Landfill leachate | | MW-HS-SPME | GC-ECD | 0.1–2.0 μg/L | >93 | <8 | Wei and Jen (2002) |
| 4-t-octylphenol, simazine, terbuthylazine, chlorpyrifos, terbumeton, and terbutryne | Environmental waters | CW/DVB | DI-SPME | GC/TOF-MS | | | | Hernandez et al. (2007) |
| N-contg. arom. heterocyclic compds., anilines, PEs, and OPPs | Environmental water | DVB/CAR/PDMS | DI-SPME | GC-MS | 6.1–21.8 ng/L | 92.8–103.5 | 4.2–8.2 | Gupta et al. (2008) |
| Halogenated aromatic hydrocarbons, aromatic hydrocarbons, aromatic ketones, OPPss | Drinking water | PDMS/DVB | DI-SPME | GC-FID | | | 4–18 | Guillot et al. (2006) |
| OCPs | Water | PDMS | HS-SPME | GC-AED | 0.050–0.097 μg/L | 96–103 | 3–8 | Ratola et al. (2006) |
| OCPs | Aqueous samples | PMDS/DVB | DI-SPME | GC-MS | | | <20 | Zambonin et al. (2002) |
| OPPs | River water | Sol-gel-based amino fiber | DI-SPME | GC-MS | 0.05–1 ng/L | 80–115 | <10.5 | Bagheri et al. (2010) |
| OPPs | Aqueous samples | PA | DI-SPME | GC-MS | 14–62 ng/L | | 2.5–4.6 | Singh et al. (2003) |
| PAHs | Scrubber water | PDMS/DVB | MW-HS-SPME | GC-FID | 0.35–7.53 μg/L | >88 | <17 | Wei and Jen (2007) |
| PBDE | Water | PDMS | DI-SPME | GC-ECD | 0.48–1.94 ng/L | | | Yu et al. (2008) |
| PCBs | River water | PDMS | MW-HS-SPME | GC-ECD | 0.27–1.34 ng/L | | 8–16 | Shu et al. (2003) |

| | | | | | | | | |
|---|---|---|---|---|---|---|---|---|
| Pesticides, POPs, and degradation products | Groudwater | | DI-SPME | GC-ECD<br>GC-MS | | | | Goncalves et al. (2007) |
| Phenols | Seawater | Poly(phthalazine) ether sulfone ketone | HS-SPME | GC | 20 μg/L | 94.8–117.3 | < 16 | Yao et al. (2008) |
| Phenols | River water and wastewater | Oxidized multi-walled carbon nanotubes | DI-SPME | HPLC-UV | 0.25–3.67 μg/L | | 1.5–12.95 | Liu et al. (2007) |
| Phenols | Seawater | Single-walled carbon nanotubes | DI-SPME | HPLC-UV | 0.9–3.8 ng/L1 | 87.5–102 | 0.7–3.2 | Li et al. (2010) |
| Phenols and ClPs | Tap water | TMSPA/ OH-PDMS | HS-SPME | GC-MS | 0.02–0.05 μg/L | >90 | 6.8–10 | Bagheri et al. (2008) |
| Phenylurea herbicides | Lake water | PDMS/DVB or CW/TPR | DI-SPME | HPLC-UV | 0.5–5.1 μg/L | | 1.0–5.9 | Lin et al. (2003) |
| Halogenated aromatic compounds, PCBs | Sediment porewater | PDMS | DI-SPME | GC-ECD | | | | Mayer et al. (2000) |
| Pyrethroids | Groundwater | PDMS/DVB | DI-SPME | HPLC-PIF-FD | 0.03–0.075 μg/L | 92–109 | 2–9 | Vazquez et al. (2008) |
| Halogenated aromatic compounds | River water | PDMS/DVB | MW-HS-SPME | GC-ECD | 0.5–270 ng/L | 94–107 | 3–14 | Huang et al. (2007) |
| Dimethyl, Dibutyl, Tributyl, | Water | DVB/CAR/PDMS | DI-SPME | GC-ICP-MS | 0.04–5 μg/L | 75–99 | 15 | Bianchi et al. (2010) |
| Organotins | Seawater | PDMS | HS-SPME | GC-MS | 0.4–4.6 ng/L | – | 12 | Chou and Lee (2005) |
| Organomercury Hg(II) | Seawater | PDMS | HS-SPME | GC-MS | 0.1 μg/L | – | 3–11 | Chou and Lee (2005), Carro et al. (2002) |
| Organometallic compounds | River and Seawater | DVB/CAR/PDMS | HS-SPME | GC-MS/MS | 4–33 ng/L | >80% | 4.8–21 | Beceiro-Gonzalez |

(continued)

**Table 4.1** (continued)

| Analyte | Matrix | Coating material | Extraction mode | Analysis technique | LOD | Apparent recovery (%) | Precision (%) | Reference |
|---|---|---|---|---|---|---|---|---|
| Organometallic compounds | River and Seawater | DVB/CAR/PDMS, PDMS | HS-SPME | GC-MS | 0.4–16.8 ng/L | 93–116 | 3–20 | Centineo et al. (2004) |
| Pharmaceutical products | Seawater, sewage water and ground water | CW/TPR, PDMS, PDMS/DVB, PA, CAR/PDMS | SPME-MD | LD-DAD | 0.1–12 μg/L | 80–113 | 3–17 | Torres Padron et al. (2009) |
| Non-steroial acidic anti-inflammatory drugs | Sewage water | PDMS, CAR/PDMS, PDMS/DVB, PA, CW/DVB | DI-SPME | GC-MS | 12–40 ng/L | | 4–8 | Rodriguez et al. (2004) |
| Pharmaceutical products | Wastewater | PA, PDMS/DVB, CW/TRP | dSPME | LC-MS | 0.005–0.050 μg/L | 93–108 | 8–12 | Unceta et al. (2010) |
| Estrogens | Surface water and waste water | PEEK | It-SPME | LC-MS/MS | | 86–107 | 0.9–8.8 | Mitani et al. (2005) |
| Estrogens | Sewage water | PA, PDMS/DVB, CAR/PDMS, CW/DVB | DI-SPME | GC-MS/MS | 0.2–3 ng/L | 80–120 | 1–12 | Carpinteiro et al. (2004) |
| Nonylphenol, nonylphenol ethoxylates and their acidic metabolites | River water | PDMS, PA, CW/DVB, DVB/CAR/PDMS | Derivatization-HS-SPME | GC-MS | 0.02–1.5 μg/L | | 11–25 | Diaz et al. (2002) |
| Estrogens | River water, wastewater | PA | DI-SPME | HPLC-UV-ED | UV:0.3–1.1 μg/L ED:0.06–0.08 μg/L | | UV: 16–31 ED: 12–25 | Penalver et al. (2002) |
| Estrogens and anabolic steroids | Urine | PA. CW/DVB | DI-SPME | GC-MS | 40 ng/L | | | Okeyo and Snow (1998) |

**Table 4.2** SBSE application to the determination of priority or emerging pollutants in water samples during the period 2000–2010

| Analyte | Sample | Method | Phase/Dimension | Desorption | Analysis technique | LODs | Precision (%) | Aparent recovery (%) | Reference |
|---|---|---|---|---|---|---|---|---|---|
| 17-β-estradiol | River water (10 mL) | SBSE | PDMS (10 mm×0.5 mm) | TD | GC-MS | 0.5 ng/L | 0.8–1.4 | – | Kawaguchi et al. (2006c) |
| 24 Priority Substances (WFD2000/60/EC) | Estuarine water (100 mL) | SBSE | PDMS (20 mm) | TD | GC-MS | 0.1–7.3 ng/L | 2.9–11.8 | 91.1–104.3 | Huertas et al. (2007) |
| 35 Priority SVOCs (WFD2000/60/EC) | Surface water (100 mL) | SBSE | PDMS (20 mm×0.5 mm) | TD | GC-MS | 0.04–10.7 ng/L | 0.8–18.9 | – | Leon et al. (2006) |
| OCPs | | | | | | | | | |
| OPPs | | | | | | | | | |
| Triazines | | | | | | | | | |
| PAHs | | | | | | | | | |
| APs | River water (2 mL) | SBSE | PDMS (10 mm×0.5 mm) | TD | GC-MS | 0.2–10 ng/L | 3.6–14.8 | 93.1–98.6 | Kawaguchi et al. (2005) |
| APs | Water (10 mL) | SBSE | PDMS (10 mm×0.5 mm) | TD | GC-MS | 0.1–3.2 | 1.6–11 | – | Nakamura and Daishima (2004) |
| ECDs (4nP,BPA,estogens) | River water | SBSE | PDMS (10 mm×0.5 mm) | TD | GC-MS | 1.2–2.6 ng/L | 2–9 | 77–95 | Magi et al. (2010) |
| ECDs (herbicides, pesticides, PAHs, PCBs, biocides, PEs and APs) | Water (30 mL) | SBSE | PDMS (20 mm×0.5 mm) | LD | LVI-GC-MS | – | 0.1–18.4 | – | Serodio and Nogueira (2004) |
| EDCs (steroid hormones) | Waste water (50 mL) | SBSE | VP-EDMA | LD | HPLC-DAD | 90–280 ng/L | 0.5–4.37 | 24.8–90.25 | Huang et al. (2009) |

(continued)

**Table 4.2** (continued)

| Analyte | Sample | Method | Phase/Dimension | Desorption | Analysis technique | LODs | Precision (%) | Apparent recovery (%) | Reference |
|---|---|---|---|---|---|---|---|---|---|
| EtHg,MeHg,Hg(II), DiEtHg | Drinking water (10 mL) | SBSE | PDMS (10 mm × 0.5 mm) | TD | GC-MS | 0.01–0.2 ng/L | 2.6–11.5 | 94.2–101.3 | Ito et al. (2008) |
| MeHg, butyltin | Water (10 mL) | HSSE | PDMS (20 mm × 0.5 mm) | TD | GC-MS | 0.4–5 ng/L | 4–17 | 99.8–120 | Prieto et al. (2008a) |
| MeHg, Hg (II) | River water (10 mL) | SBSE | PDMS (10 mm × 0.5 mm) | TD | GC-MS | 2–5 ng/L | 7–9 | – | Ito et al. (2009) |
| OPPs | Water (10 mL) | SBSE | PDMS (10 mm × 0.5 mm) | TD | GC-FPD | – | – | – | MacNamara et al. (2009) |
| OPPs, OCPs | Water (10 mL) | SBSE | PDMS (10 mm × 0.5 mm) | TD | GC-AED | 0.8–15.4 ng/L | 2.5–15 | – | Guan et al. (2008) |
| Organotin | Water (30 mL) | SBSE | PDMS (10 mm × 0.5 mm) | TD | GC-ICP-MS | 0.1 pg/L | 12 | – | Vercauteren et al. (2001) |
| PAHs | Rainfall water (10 mL) | SBSE | PDMS (10 mm × 1 mm) | LD | LC-FLD | 0.2–1.5 ng/L | 5.3–12.8 | 87.8–120.6 | Bourdat-Deschamps et al. (2007) |
| PAHs | Seawater (200 mL) | SBSE | PDMS (20 mm × 0.5 mm) | TD | GC-MS | 0.14–1,200 ng/L | 1–48 | – | Roy et al. (2005) |
| PAHs | Drinking water (40 mL) | SBSE | PDMS (10 mm × 0.5 mm) | LD | HPLC-FLD | 0.5–7.3 ng/L | 4–9 | 100 | Garcia-Falcon et al. (2004) |
| PAHs | Water (10 mL) | SBSE | PDMS (10 mm × 0.5 mm) | TD | GC-MS | 0.2–2.0 ng/L | 3–15 | 83–100 | Kolahgar et al. (2002) |
| PAHs<br>PCBs<br>PBDEs<br>NPs<br>OCPs | Seawater (100 mL) | SBSE | PDMS (20 mm × 1 mm) | TD | GC-MS | 0.011–2.5 ng/L | 2–24 | 86–118 | Sanchez-Avila et al. (2010) |

| | | | | | | | | | |
|---|---|---|---|---|---|---|---|---|---|
| PAHs<br>PCBs<br>PEs<br>NPs | Seawater (20 mL) | SBSE | PDMS (10 mm × 0.5 mm) | TD | GC-MS | 0.05–3.3 ng/L | 4–20 | – | Prieto et al. (2007) |
| PAHs<br>PCBs<br>PEs<br>NPs<br>PBDEs | Estuary and seawater | SBSE | PDMS (10 mm × 0.5 mm) | TD | GC-MS | 0.03–20.4 ng/L | 3–19 | 80–120 | Prieto et al. (2008b) |
| PBDEs | Waste water (3 mL) | SBSE | PDMS (20 mm × 0.5 mm) | LD | LVI-GC-MS | 0.3–203.4 ng/L | <12.1 | 65.6–116.9 | Serodio et al. (2007) |
| PBDEs | Water (100 mL) | SBSE | PDMS (20 mm × 0.5 mm) | TD | GC-MS | 0.4–9.4 ng/L | 0.7–5.3 | 94–105 | Llorca-Porcel et al. (2006) |
| PCB | Water (8 mL) | SBSE | PDMS (10 mm × 0.5 mm) | TD | GC-MS | 0.05–0.15 ng/L | 3.3–10.6 | 90–100 | Popp et al. (2005) |
| PCB | Groundwater (100 mL) | SBSE | PDMS (10 mm × 1.5 mm) | TD | GC-MS | 0.3–1.8 ng/L | 2.6–6.8 | – | Montero et al. (2004) |
| PEs | Drinking water (30 mL) | SBSE | PDMS (20 mm × 0.5 mm) | LD | LVI-GC-MS | 3–40 ng/L | 2.2–14.8 | – | Serodio and Nogueira (2006) |
| OCPs<br>OPPs<br>Carbamates<br>Pyrethroids | Water (5 mL) | SBSE | PDMS (10 mm × 0.5 mm) | TD (2 stir-bars simulta-neously) | GC-MS | 2.1–74 ng/L | 1.4–13 | 89–113 | Ochiai et al. (2008) |
| OCPs<br>OPPs<br>Carbamates<br>Pyrethroids | River water (10 mL) | HSSE | PDMS (10 mm × 0.5 mm) | TD | GC-MS | 0.2–20 ng/L | 1.4–20.2 | 80–90 | Nakamura and Daishima (2005) |

(continued)

**Table 4.2** (continued)

| Analyte | Sample | Method | Phase/Dimension | Desorption | Analysis technique | LODs | Precision (%) | Aparent recovery (%) | Reference |
|---|---|---|---|---|---|---|---|---|---|
| OCPs<br>OPPs<br>Triazines | Underground and superficial water (10 mL) | Sequential SBSE | PDMS (10 mm × 0.5 mm) | TD | GC-MS | – | 6.7–24.9 | – | Sampedro et al. (2009) |
| Phenols | Water (10 mL) | SBSE | PDMS (10 mm × 0.5 mm) | TD | GC-MS | 0.1–0.4 μg/L | 6–27 | – | Montero et al. (2005) |
| Phenols | Lake water and sea water (50 mL) | SBSE | VP-EDMA | LD | HPLC-DAD | 62–380 ng/L | 1.3–5 | – | Huang et al. (2008) |

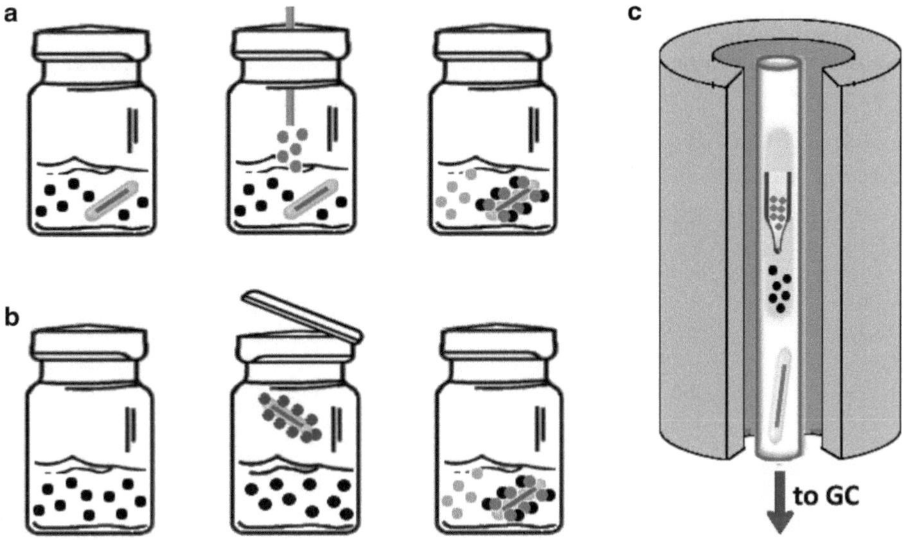


**Fig. 4.4** Derivatization modes in sorptive extraction: in situ (**a**), on-polymer (**b**) and in-tube (SBSE) (**c**)

Meyers and Almirall 2005; Lee et al. 2007; Prieto et al. 2010a). Finally, derivatization can be carried out directly in the injection port of a GC during the desorption step (Alzaga et al. 2003b; Alzaga and Bayona 2004). In the case of SBSE, this latter derivatization mode is known as in-tube derivatization since the stir-bar and the derivatization reagent are loaded onto the desorption tube that is introduced in the TDU (Prieto et al. 2010a).

### *4.2.2 Microextraction by Packed Sorbent (MEPS)*

MEPS is a recently developed technique that was introduced by Abdel-Rehim (Abdel-Rehim 2003, 2004; Abdel-Rehim et al. 2004) in the field of sample preparation. MEPS can combine sample processing and the extraction and injection steps in a fully automated way as an at-line sampling/injecting device to GC or LC (Jagerdeo and Abdel-Rehim 2009; Matysik and Matysik 2009; Morales-Cid et al. 2009; Schurek et al. 2008; Said et al. 2008). MEPS is not only a miniaturized format of SPE which is able to handle sample volumes as small as 10 μL. This technique integrates the sorbent directly into the syringe but not in a separate column as in commercial SPE. Thus, a fully automated method using MEPS is a promising approach to reduce time and labor effort in sample preparation and analysis (Abdel-Rehim et al. 2005, 2006a, b, c, 2007, 2008; El-Beqqali and Abdel-Rehim 2007, a, b). The MEPS technique has been used to extract a wide range of analytes in different samples (biological, environmental, food and beverages) but mainly for the extraction

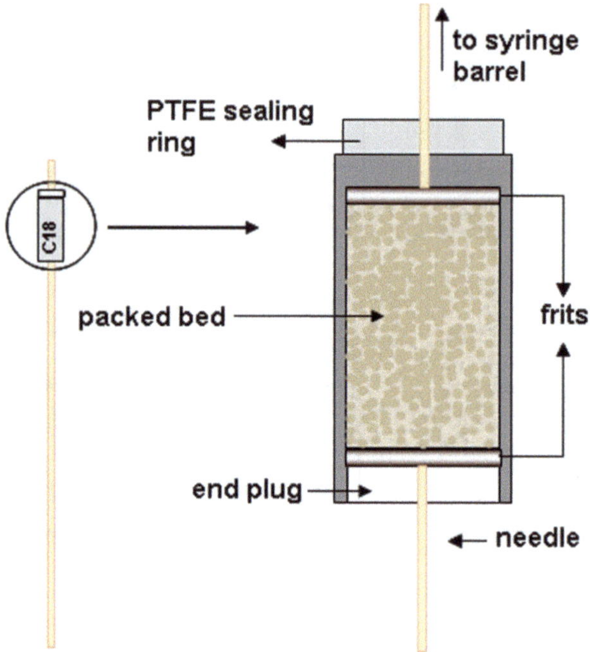


**Fig. 4.5** Scheme for a micro-extraction by packed sorbent (MEPS) syringe

of drugs and metabolites from biological matrices (plasma, blood, urine) (Abdel-Rehim 2004; Abdel-Rehim et al. 2004, 2006a, b; Jagerdeo and Abdel-Rehim 2009; Matysik and Matysik 2009).

In MEPS approximately 2 mg of the sorbent is thermo packed inside a syringe (100–250 μL) as a plug or between the barrel and the needle as a cartridge (Fig. 4.5). Sample extraction and enrichment takes place on the sorbent bed. The bed can be packed or coated to provide selective and suitable sampling conditions. Different sorbent materials such as silica based (C2, C8 and C18), carbon, polystyrene-divinylbenzene copolymer (PS-DVB), strong cation exchanger (SCX) or molecular imprinted polymers (MIPs) sorbents have been used. Thus, C2–C18 phases are suitable for lipophilic compounds and, for polar compounds such as acidic and basic analytes, polymeric phases (PS-DVB, SCX) could be more selective.

Obviously, analytes can be extracted using different approaches but a brief scheme of the main extraction steps is presented in Fig. 4.6. First, the sorbent is conditioned (elution solvent and water could be used) and the sample drawn through the sorbent by an autosampler. The extraction can be performed using two configurations, which differ in sample volumes. Sample vials with volumes of 2 and 10 mL allow the extraction of different sample amounts despite the insertion depth of the MEPS syringe into the vial is limited. Thus, an 800-μL sample can be extracted from a 2-mL standard vial. On the other hand, the use of 10-mL vials enables to extract up to 2-mL of the sample.

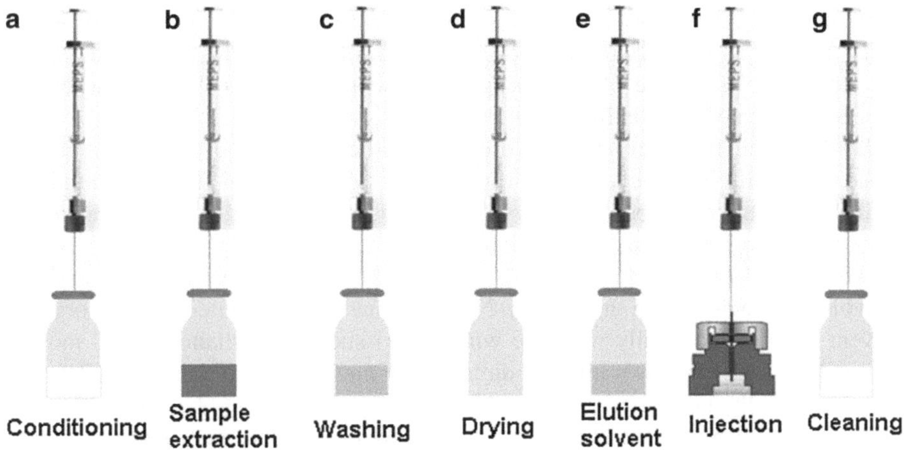


**Fig. 4.6** Steps in MEPS extraction: (**a**) conditioning of the MEPS syringe, (**b**) sample is loaded into the MEPS syringe, (**c**) sample is washed, (**d**) the MEPS syringe is dried, (**e**) analytes are eluted in the optimum solvent, (**f**) the extract is injected in a GC injection port and (**g**) MEPS syringe is washed previous to the next extraction. MEPS: micro-extraction by packed sorbent, GC: gas chromatography

Two possible extraction modes can also be used for the extraction of the sample by means of the way of sample loading: multiple draw-eject cycle mode and, alternatively, an extract-discard mode has also been described (Moeder et al. 2010). In the last mode, each sample aliquot is pumped only once through the MEPS-BIN (Microextraction by packed sorbent-barrel insert and needle (BIN) device) and the extracted sample portion is discarded into waste before the next aliquot from the sample is pumped. The solid phase is, then, washed to remove other interfering materials and dried drawing and pressing air through the sorbent at a chosen rate (empty vials can be used). Subsequently, the elution is carried out and the solvent injected directly into instrument's injector. The process can be fully automated without any modification of the instruments. After the extraction process, some wash cycles, each with the elution solvent, are used to clean the phase to avoid carryover.

#### 4.2.2.1 Critical Steps in the Microextraction by Packed Sorbent Extraction

Besides sample load, elution/injection volume, drying step and carry over effects should be evaluated, as these have been found as critical steps for the MEPs extraction. In this sense, in a first approach, an environmental method to determine polycyclic aromatic hydrocarbons (PAHs) in water samples was developed (El-Beqqali et al. 2006). The method used a draw-eject procedure for the extraction of the sample. On the contrary, at the extract-discard mode other applications can be found (Moeder et al. 2010) in which both extraction modes were evaluated. Therefore, they found too many draw-eject cycles (about 170 for one analysis) were needed in order to reach an extraction yield >80% for most of the analytes but this demands the mechanical sensitive of the syringe plunger and shortens the lifetime of the MEPS syringe (Abdel-Rehim 2003).

Besides, the fill/eject speed is crucial for the recovery of analytes. In this sense, high fill/eject speed values do not able adsorption of the compounds to MEPS and could cause poor recovery results and irreproducibility due to the bubble formation during the extraction/elution steps.

After the adsorption of compounds, the solid phase is washed and dried in order to avoid problems during the elution. Next, the analytes are eluted with an organic solvent (partially o totally miscible with water) such as methanol or LC mobile phase (10–50 μL) and then injected directly into the instrument's injector. The process can be fully automated. To reuse MEPS, the sorbent should be washed several times (3–15) with both the elution solvent and water solutions in order to eliminate carryover. This cleaning procedure removes most of the residual analytes and matrix components (Moeder et al. 2010; El-Beqqali et al. 2006).

Not only the possibility of bubble formation during the loading sample and/or elution steps, but also the no possibility of on-line coupling with both liquid or gas chromatographic systems are some drawbacks of MEPS technique. The speed of plunger movement is also crucial for the recovery of analytes, but mainly in the case of off-line MEPS. Too high speed of movement does not enable adsorption of the analyte to MEPS and leads to poor recovery results, as mentioned before.

#### 4.2.2.2 Applications and Future Perspectives

MEPS have been applied for the extraction of many drugs from biological samples but only a few applications can be found in the literature in the environmental field. The different contaminants extracted by MEPS from aqueous samples and the main characteristics and results obtained in the developed analytical methods are summarized in Table 4.3.

A higher selectivity could be obtained using a molecularly imprinted polymer (MIP) material. The superiority of MIPs as selective SPE sorbents for environmental analysis of different compounds has been demonstrated but, however, the use of MIPs as selective packing material for MEPS has not been developed yet in environmental applications. A critical point of the use of MIPs is their low capacity. Therefore, some features still need improvement. Currently, the MIPs used for SPE are typically prepared by a noncovalent imprinting technique, which gives quite low yields of specific binding sites. This results in low sample load capacity and high nonspecific binding. In this sense, the main advances expected in development of new MIP synthesis methods are the improvement of their capacity and selectivity (Nováková and Vlčková 2009). These tasks could be the basis for further studies on the application of MIPs as sorbents for online-MEPS procedures.

**Table 4.3** MEPS application to the determination of priority or emerging pollutants in water samples during the period 2000–2010

| Analyte | Matrix | Sorbent | Analysis technique | LODs | Apparent recovery (%) | Precision (%) | Reference |
|---|---|---|---|---|---|---|---|
| PAHs | Water | C8 | MEPS-GC-MS | 1–5 ng/L | 84–121 | 6–14 | El-Beqqali et al. (2006) |
| PAHs, PCBs, PEs, NPs, BPA, MeEE2, EE2 | Wastewater and snow samples | C18 | MEPS-LVI-GC-MS | 0.2–266 ng/L | 78–117 | 2.4–19 | Prieto et al. (2010b) |
| PBDEs | Wastewater | C18 | MEPS-GC-TOFMS | 0.5–9.5 ng/L | 72–108 | 2–12 | Schurek et al. (2008) |
| UV filters and polycyclic musks | Water | C18 and C8 | MEPS-LVI-GC-MS | 34–96 ng/L | 46–114 | 4–11% | Moeder et al. (2010) |

## 4.3 Liquid-Liquid Microextraction

Liquid-phase microextraction (LPME) is a simple liquid-liquid extraction (LLE) which uses only a few μL of acceptor phase in order to pre-concentrate the target analytes. Actually, the principles of LLE and the miniaturized nature of SPME are combined in LPME. Four main categories of LPME are found: (i) cloud-point extraction (CPE), (ii) single-drop microextraction (SDME), (iii) membrane liquid-phase microextraction and (iv) dispersive liquid-liquid microextraction (DLLME) (Sarafraz-Yazdi and Amiri 2010; Lambropoulou et al. 2007; Pena-Pereira et al. 2010).

### *4.3.1 Cloud Point Extraction*

Cloud point extraction (CPE) is based on the phase separation behaviour exhibited by aqueous solutions of certain surfactant micelles. Surfactants or surface-active agents are amphiphilic molecules that contain a polar hydrophilic head and a non-polar hydrophobic tail. While the head is an ionic or strongly polar group, in most cases, the tail is a linear or branched hydrocarbon chain with different carbon atoms that can be linear or branched and that can also contain aromatic rings. Above a minimum concentration, named critical micellar concentration (CRC), the hydrophobic tails tend to form aggregates called micelles. These uniquely organized structures have the capacity to solubilize solutes of different type of character and nature since the interaction between the analyte and the micelle can be electrostatic, hydrophobic or a combination of both (Xie et al. 2010; Sosa Ferrera et al. 2004; Rubio and Perez-Bendito 2003; Carabias-Martinez et al. 2000; Tani et al. 1997; Khammas 2009).

When a micellar solution of a surfactant is heated above a certain temperature (known as cloud-point, CP), the solution becomes turbid. Above the CP, the single isotropic micellar phase separates into two isotropic phases. One of the phases is rich in surfactant ("surfactant-rich phase") and is separated from the bulk aqueous solution, in which the concentration of surfactant is close to the CRC. Under optimal extraction conditions, the small volume surfactant-rich phase will contain the target analytes. The small volume of the surfactant rich phase allows the one-step extraction and pre-concentration of organic compounds prior to gas or liquid chromatographic analysis (Xie et al. 2010; Sosa Ferrera et al. 2004; Rubio and Perez-Bendito 2003; Carabias-Martinez et al. 2000; Tani et al. 1997).

The CP temperature depends on the structure of the surfactant and on its concentration and can be modified after the addition of salts, alkalis, acids, polymers, urea and other surfactants. The ionic strength not only alters the CP temperature but also facilitates the separation of the two phases since the density of the aqueous phase changes. Centrifugation can be used to speed up the separation of the two phases (Sosa Ferrera et al. 2004).

The CPE steps are the following (see Fig. 4.7): (i) addition of surfactant to the aqueous solution, (ii) alteration of temperature to separate the surfactant-rich phase

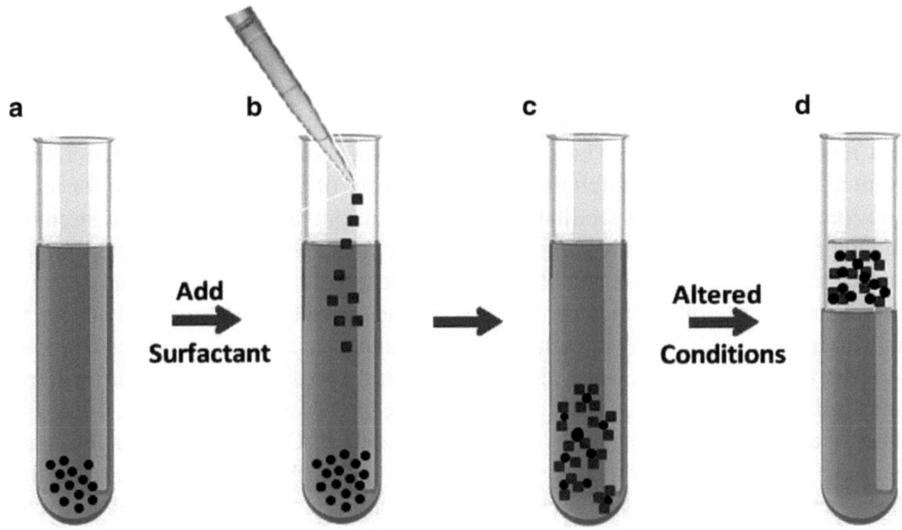


**Fig. 4.7** Scheme of the steps in cloud point extraction (CPE): (**a**) the initial aqueous solution containing the organic pollutants, (**b**) the addition of the aqueous micellar solution, (**c**) formation of the water insoluble chelates and (**d**) separation of both phases. The organic chelates are concentrated in a small volume of the surfactant-rich phase and separated from the aqueous phase

and the bulk aqueous phase and (iii) direct introduction of an aliquot of the surfactant-rich phase into the separation system or after a conditioning step (Sosa Ferrera et al. 2004). Depending on the density of the surfactant-rich phase, it can be either at the bottom or at the top. Actually, one drawback of CPE is that the surfactant-rich phase where the analytes are pre-concentrated is so viscous that it can only be injected into the instrument for analysis after dilution in aqueous or organic solvent (Xie et al. 2010).

Surfactants are classified as non-ionic, zwitterionic, cationic and anionic, although, up to now, the applications of cationic surfactants in CPE are scarce (Xie et al. 2010).

In most cases CPE is coupled with reverse-phase high performance liquid chromatography (RP-LC) since the surfactant-rich phase obtained during the extraction process is compatible with the mobile phases used in RP-LC. However, the coupling of CPE to GC is not direct, up to date, since the viscous and low-volatility surfactant-rich phase can block GC columns. Therefore, an extra step is introduced between the CPE step and GC analysis, which consists either on a back extraction into an organic solvent compatible with GC or on a normal-phase solid phase extraction clean-up. Therefore, it could be concluded that the coupling of CPE and GC is not straigthtforward and more complicated analysis procedures are necessary. Finally, although the use of surfactants in CE is a normal practice, the use of CPE as sample pre-concentration before CE is not extensive (Xie et al. 2010; Khammas 2009; Carabias-Martinez et al. 2003).

Among the variables studied during CPE the surfactant type and concentration, ionic strength, pH and equilibration temperature and time can be mentioned. It has been observed that the smaller the surfactant concentration, the higher the concentration factor; however, when the volume of the surfactant is very small accuracy and reproducibility suffer. Thus, a balance between surfactant concentration and volume is required. Ionic strength alters the density of the aqueous phase for most non-ionic surfactants and facilitates phase separation. Besides, the ionic strength can alter the CP temperature of non-ionic surfactants and contribute to the salting-out salting-in effect of the analytes. In the case of pH, its effect depends on the characteristics of both surfactant and analyte. The neutral form of the analytes interacts more strongly with the micellar aggregates and, thus, the pH should be adjusted in order to ensure that the analytes are in a neutral form before CPE. For zwitterionic surfactants pH has no significant effect, except at very acidic media, while for anionic and cationic surfactants pH has a great effect. Optimal equilibration temperature occurs when temperature is 15–20°C above the CP temperature of the surfactant. At lower temperatures, the phase separation does not occur, whereas higher temperatures may lead to the degradation of the analyte. Finally, the kinetics of extraction is very rapid and 10–20 min are sufficient for good extraction in most of the cases. Besides, centrifugation can accelerate equilibration time (Xie et al. 2010; Sosa Ferrera et al. 2004; Carabias-Martinez et al. 2003). In contrast with other extraction techniques where the addition of organic solvents such as methanol is studied in order to avoid the adsorption of analytes onto the extraction vessel walls, this is not necessary in CPE since the use of micelles avoids these adverse phenomena.

Applications of CPE to the determination of priority or emerging pollutants are listed in Table 4.4. Most applications are related to the analysis of PAHs, although applications to phenols, chlorophenols, polybrominated diphenyl ethers (PBDEs), polychlorinated dibenzodioxins (PCDDs), organophosphorus pesticides (OPPs), estrogens and hydrazine pesticides can also be found.

### *4.3.2 Single Drop Microextraction*

SDME is based on the preconcentration of target analytes in a small volume (0.5–3 μL) of organic solvent, aqueous solutions or ionic liquid (IL). In this sense, SDME is considered to be the basic liquid-liquid microextraction technique (Sarafraz-Yazdi and Amiri 2010; Nerin et al. 2009; Psillakis and Kalogerakis 2002; Xu et al. 2007; Lambropoulou et al. 2007).

Equilibration times, which range from seconds to hours, can happen in SDME, depending on the target analyte. In any case, regardless the sampling regime (i.e. equilibrium or kinetic), the extraction process is not usually exhaustive. In those cases in which the amount of extracted compound is negligible, several extractions can be performed on the same sample, as mentioned before for sorptive extraction (Jeannot et al. 2010).

**Table 4.4** CPE application to the determination of priority or emerging pollutants in water samples during the period 2000–2010

| Analyte | Matrix | Surfactant | Analysis technique | LOD | Apparent recovery (%) | Precision (%) | Reference |
|---|---|---|---|---|---|---|---|
| ClP | River water | 0.5% Triton X-114 | HPLC-UV | 1.7–5.0 μg/L | 40–100 | 4–10 | Seronero et al. (2000) |
| Estrogens | Water | o.25% Triton X-114 | HPLC-UV | 0.23–5 μg/L | 81.2–99.5 | 8.1–12 | Wang et al. (2006) |
| Hydrazine | Drinking water<br>River water | DAB | UV | 0.08 μg/L | 98.5–105 | 1.22 | Zarei (2007) |
| OPPs | Wastewater<br>Groundwater | POLE Genapol X-080 | HPLC-UV | 0.61–2.6 μg/L | 62.6–84.5 | 0.5–3.2 | Padron Sanz et al. (2004) |
| PAHs | Water | 1.0%PEG/PPG-18/18 dimethicone 1.0% PEG-12 dimethicone | HPLC-UV | | 93.6–100 | | Yao et al. (2007) |
| PAHs | Aqueous solution | 1.0% PEG/PPG-18/18 dimethicone | HPLC-UV | | 69–100 | | Yao and Yang (2008b) |
| PAHs | Seawater | 1.0% POLE | HPLC-FLD | 1.0150 ng/L | 32.6–115 | <10.4 | Pino et al. (2002) |
| PAHs | Water | 2% Tergitol TMN-6 | HPLC-UV | | 79.3–92.6 | | Yao and Yang (2007) |
| PAHs | Water | 10% Tergitol TMN-6 | HPLC-UV | | | | Yao and Yang (2008a) |
| PAHs | Water | 5% Triton X-114 | HPLC-FLD | 0.6–1.8 ng/L | | | Pongpiachan (2009) |
| PAHs | Aqueous solution | Tergitol 15-S-5 LE-203, Brij 30 | HPLC-UV | | 80–100 | | Hung and Chen (2007) |
| PAHs | Aqueous solution | 3% Tergitol 15-S-7 | HPLC-FLD | | 80–96 | | Bai et al. (2001) |
| PAHs | Aqueous Solution | 1% Tergitol 15-S-5 | HPLC-FLD | | | | Li et al. (2004) |
| PAHs | Aqueous solution | Tergitol 15-S-9 Neodol 25–7 Tergitol 15-S-7 | HPLC-UV | 0.1 mg/L | | | Hung et al. (2007) |
| PAHs | Seawater | 1.0% POLE/Brij 30 | HPLC-FLD | 32.4–231 ng/L | 72–99 | <8.07 | Delgado et al. (2004) |
| PAHs | Aqueous solution | Triton X-114 SDSA | GC-FID | 0.9–9.9 μg/L | 92–105 | | Sikalos and Paleologos (2005) |
| PBDEs | Water | Triton X-114 | GC-MS | 1–2 ng/L | >99.9 | ≤8.5 | Fontana et al. (2009) |

(continued)

**Table 4.4** (continued)

| Analyte | Matrix | Surfactant | Analysis technique | LOD | Apparent recovery (%) | Precision (%) | Reference |
|---|---|---|---|---|---|---|---|
| PCDDs | Water | 5% POLE | HPLC-UV | 0.05–12.80 μg/L | – | – | Padron Sanz et al. (2002) |
| PEs | Effluent water | 0.25% Triton X-114 | HPLC-UV | 1.0–3.8 μg/L | 86–103 | 1.9–3.9 | Wang et al. (2007) |
| Phenol | Wastewater | Monodisperse nonionic surfactants<br>Polydisperse nonionic surfactants | CE-UV | – | – | – | Taechangam et al. (2009) |
| Phenolic compounds | Rainwater<br>River water | PONPE 10 | MEKC-UV | 0.10–0.20 μg/L | >89.5 | | Stege et al. (2009) |
| Phenols | Seawater/depurated waste water | $C_{12}E_6/C_{12}E_9/C_{12}E_{10}$ | HPLC-UV | 0.6–3.5 μg/L | – | – | Santana et al. (2004) |
| Triazine herbicides | Aqueous solution | Triton X-114 | CE-UV | – | – | – | Carabias-Martinez et al. (2003) |

#### 4.3.2.1 Single Drop Microextraction Modes

So far, up to seven different modes of SDME are found in the literature (DI, HS, continuous flow (CF), drop-to-drop (DD), directly suspended droplet (DSD), liquid-liquid-liquid (LLL) and a combination of LLL and DSD). However, DI-SDME and HS-SDME are the most frequently found in the bibliography (38% and 41%, respectively), since they do not require extra-equipment (i.e., pump in CF) and are applied to the widest range of analytes (Sarafraz-Yazdi and Amiri 2010; Xu et al. 2007; Lambropoulou et al. 2007; Jeannot et al. 2010).

In DI-SDME the solvent drop is in direct contact with the aqueous solution and, therefore, the extraction solvent must be immiscible in water. Therefore, in DI-SDME the acceptor solvent is limited to non-polar or slightly polar organic solvents and ILs. In this sense, DI-SDME is limited to semi-volatile non-polar or slightly polar compounds, since volatile compounds are better extracted in the HS mode. The most used solvents are n-hexane and toluene followed by, in many cases, GC analysis. When LC is used after DI-SDME, solvent exchange from non-polar to polar is necessary, unless an IL was used for pre-concentration (Jeannot et al. 2010).

In HS-SDME the acceptor solvent is placed in the headspace of the extraction vial and, in this sense, there are no restrictions on the nature of the extraction solvent. However, solvents with a high vapor pressure (mostly preferred for GC analysis) evaporate too quickly during HS-SDME and, thus, low vapor pressure solvents are preferred. In this sense, the vapor pressure of the solvent is the primary consideration in order to study the best acceptor phase rather than the extraction efficiency. The organic solvents mainly used are 1-octanol, hexadecane, dodecane and decane, but polar solvents, including ILs, aqueous solutions and even pure water, can be used. Therefore, HS-SDME is applicable to both polar and non-polar compounds, volatile or semi-volatile (Xu et al. 2007; Lambropoulou et al. 2007; Jeannot et al. 2010). Additionally, non-volatile matrix interferences are reduced (Xu et al. 2007; Jeannot et al. 2010), as mentioned for HS-SPME and HSSE. GC is also the preferred separation technique in HS-SDME.

The first works on SDME were carried out by Liu and Dasgupta (1996), where a micro-drop of organic immiscible solvent was suspended in a larger aqueous drop, and at the same time by Jeannot and Cantwell (1996), who suspended a micro-drop at the end of a Teflon rod. The problem with these two approaches was that the extraction and injection were performed separately. Thus, Jeannot and Cantwell suggested the use of a microsyringe to suspend the drop (Jeannot and Cantwell 1997).

Both DI-SDME and HS-SDME can be performed in the dynamic mode and, in this case, not only the sample is stirred but also the extracting solvent is in motion in order to improve mass transfer and, thus, extraction efficiency (He and Lee 1997). The dynamic mode can, as well, be performed with both unexposed and exposed drops. In the unexposed or in-syringe mode (see Fig. 4.8), the extraction solvent with a small volume of the sample (1–3 μL) is withdrawn into the syringe needle, held for a time and then the sample is expelled. This cycle is repeated from 30 to 90 times. The in-syringe mode should be termed dynamic LPME since a drop configuration is not involved. In the exposed-drop, the drop is exposed to the sample, withdrawn into

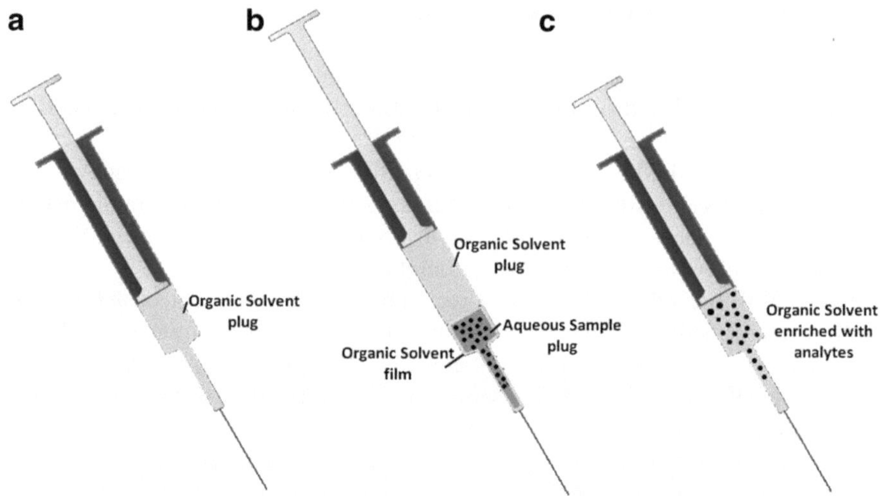


**Fig. 4.8** Dynamic liquid phase micro-extraction (LPME): (**a**) the acceptor solvent within the syringe plug, (**b**) small volume of the sample is withdrawn into the syringe and held for a time and (**c**) the acceptor phase is enriched with the analytes and the sample expelled

the needle for a specified time and expelled out to the needle tip again (Xu et al. 2007; Lambropoulou et al. 2007; Jeannot et al. 2010). During dynamic HS-LPME vapor pressure problems are overcome and the selection of organic solvent is more flexible than previously mentioned for static HS-SDME (Xu et al. 2007).

### 4.3.2.2 Ionic Liquids

As previously mentioned, ILs are found among the acceptor phases used in SDME. ILs are low-temperature melting salts that form liquids composed entirely of ions. Although the term IL was firstly used to include all types of thermally stable organic and inorganic melts, the emphasis has changed to low-melting point, organic, air stable salts with wide temperature-stable liquid ranges. The main types of room temperature ILs are alkylammonium, tetraalkylammonium, tetraalkylphosphonium, 1,3-dialkylimidazolium, and N-alkylpyridinium salts formed with weak nucleophilic anions such as bis(trifluoromethylsulfonyl)imide, hexafluorophosphate, tetrafluoroborate, perfluoroalkylsulfonate, etc. According to some authors ILs can be considered as environmentally friendly (Poole and Poole 2010; Aguilera-Herrador et al. 2008b). The most direct form of analysis of ILs extracts is LC but the coupling to GC is more troublesome since ILs should not enter the capillary column used in GC. For the coupling of IL-SDME with GC different approaches have been used. For instance, Aguilera-Herrador et al. (Aguilera-Herrador et al. 2008b) developed an interface which consisted in an injection zone, a removable unit and a transfer line for the direct coupling of ionic liquid based SDME (IL-SDME). Zhao et al. (2009a) used a homemade tube that was introduced in a glass injector insert of a GC for the IL-HS-SDME of chlorobenzene derivatives. Finally, Chisvert et al. (2009) used a commercially available TDU for TD-GC-mass spectrometry (TD-GC-MS) analysis of chlrobenzenes in water samples. Some applications of ILs in SDME are included in Table 4.5.

#### 4.3.2.3 Single Drop Microextraction (SDME) Development and Applications

During SDME development the most frequently studied variables are common to other extraction techniques such as inert salt addition, pH adjustment, agitation of the sample, extraction solvent nature, extraction time or volumes of donor and acceptor phase solutions (Psillakis and Kalogerakis 2002; Lambropoulou and Albanis 2007). High stirring rates improve mass transfer. However, in DI- SDME high stirring rates cause droplet instability (Psillakis and Kalogerakis 2002; Xu et al. 2007). Organic drop volume is a characteristic variable during SDME optimization. The use of a large organic drop results in an increase of the extraction efficiency, whereas larger drops are more difficult to manipulate and less reliable. Besides, large injection volumes cause band broadening in capillary GC. In this sense, approx. 1 μL organic drop volume is commonly used (Psillakis and Kalogerakis 2002). Extraction temperature is studied in HS-SDME in order to accelerate the mass transfer from the sample to the headspace. However, high temperatures decrease the organic solvent-headspace distribution coefficient and, therefore, a consensus is needed. In more sophisticated systems HS-SDME can be performed with solvent cooling (Yamini et al. 2004).

During SDME it is advisable to wash the syringe several times with the extraction solvent in order to eliminate air, since the presence of air bubbles can change the rate of extraction and give raise to non-reproducible results. Besides, precision is improved when the drop is placed in a reproducible and stable position (Psillakis and Kalogerakis 2002).

During SDME development it should be considered that there are two types of GC syringes, those where the plunger of the syringe is a wire inside the needle itself, and those where the plunger is a wire inside the glass barrel of the syringe. In the latter case, there is an extra or dead volume of solvent contained within the needle that is always retracted and not taken into account. This dead volume should be taken into account when conducting extraction rate experiments where the calibration standards are directly prepared into the solvent (Psillakis and Kalogerakis 2002).

Applications of SDME to priority or emerging pollutants are included in Table 4.5.

### *4.3.3 Membrane Liquid-Phase Microextraction*

Membrane liquid-phase microextraction was introduced years ago as a simple and inexpensive alternative to traditional liquid-liquid extraction (LLE) methodology

**Table 4.5** SDME application to the determination of priority or emerging pollutants in water samples during the period 2000–2010

| Analyte | Matrix | Mode | Extraction solvent | Analysis technique | LOD | Apparent recovery (%) | Precision (%) | Reference |
|---|---|---|---|---|---|---|---|---|
| APs | Tap, river, waste water | IL-LPME | $[C_6MIM][PF_6]$ | HPLC-FLD | 0.3–0.7 μg/L | 90–113 | 3.2–7.8 | Liu et al. (2004b) |
| APs, BPA | Seawater | DI-SDME | Decanol | HPLC-UV | 4–32 μg/L | 96–118 | 3.8–11 | Lopez-Darias et al. (2010) |
| BPA | River water | LPME | Toluene | GC-MS | 2 ng/L | 98.3–104.1 | 3.2–8.9 | Kawaguchi et al. (2006a) |
| BTEX | River, drinking, well, tap water | IL-HS-SDME | $[MC_8IM][PF_6]$ $[C_4MIM][PF_6]$ | GC-MS | 22–91 ng/L | 88.9–103.1 | 3.00–5.25 | Aguilera-Herrador et al. (2008c) |
| BTEX | Water | HS-SDME | n-Hexadecane | GC-FID | 0.72-5.0 μg/L | | 2.7–5.9 | Przyjazny and Kokosa (2002) |
| Carbamate pesticides OPPs | Lake water | DI-SDME | Toluene | GC-MS | 0.02–0.50 μg/L | 85.9–108.3 | 1.4–13.1 | Chen et al. (2009) |
| Chlorinated anilines | Tap, river and waste water | IL-HS-LPME | $[C_4MIM][PF_6]$ | HPLC-PAD | 0.5–1.0 μg/L | 81.9–99.6 | 5–7 | Peng et al. (2005) |
| Chlorobenzenes | Tap, river and wastewater | IL-HS-SDME | $[C_4MIM][PF_6]$ | HPLC-PDA | 0.102–0.203 μg/L | 60.8–120.6 | 1.6–5.1 | Vidal et al. (2007) |
| Chlorobenzenes | Wastewater | IL-HS-SDME | $[C_8MIM][PF_6]$ | GC-FID | 0.1–0.5 μg/L | 88.9–107.9 | 7.73–12.39 | Zhao et al. (2009a) |
| Chlorobenzenes | Tap, waste and sea water | HS-SDME | Dodecane | GC-ECD | 0.1–3.0 μg/L | 101.9–106.2 | 2.5–12.6 | Khajeh et al. (2006) |
| Chlorobenzenes | Tap, river effluent water | IL-HS-SDME | $[C_6MIM][PF_6]$ | TD-GC-MS | 2–4 μg/L | 90–115 | 3–17 | Chisvert et al. (2009) |
| ClP | Wastewater reservoir groundwater | DI-SDME | Decanoic acid and TMAH | HPLC-UV | 0.5–1 μg/L | 79–106 | 4.3–5.6 | Lopez-Jimenez et al. (2008) |

| | | | | | | | | |
|---|---|---|---|---|---|---|---|---|
| Dichloromethane, p-xylene, n-undecane, trichlormethane, carbon tetrachloride, | Aqueous solution | IL-DI-SDME | $[C_4MIM][PF_6]$ | GC-MS | 5.6–15.6 µg/L | | 3.4–4.4 | Aguilera-Herrador et al. (2008b) |
| Haloacetic acids | River water | DI-SDME | n-Octanol | GC-MS | 0.1–1.2 µg/L | 82.5–97.6 | 5.1–8.5 | Saraji and Mirmahdieh (2009) |
| OCPs | Wastewater | DI-SDME | Toluene | GC-MS | 22–101 ng/L | 43–103 | 1–15 | Cortada et al. (2009b) |
| OCPs | Natural and tap water | Static SDME | Hexane | GC-ECD | 5–200 ng/L | 83.3–98.3 | 3.2–10.7 | Zhao and Lee (2001) |
| OPPs | Tap water surface water | DI-SDME | Toluene | GC-MS | 10–73 ng/L | 57–102 | 11–19 | Lambropoulou et al. (2004) |
| OPPs | Lake water | Static SDME Cycle flow SDME | Toluene | GC-FPD | 0.21–0.56 ng/L | 90.7–106.5 | 1.7–10 | Xiao et al. (2006) |
| OPPs | Farm, well and river water | DI-SDME | $CCl_4$ | GC-FPD | 2-20 ng/L | 90–103 | 1.1–8.6 | Ahmadi et al. (2006) |
| OPPs and pyrethroid pesticides | River water | DI-SDME | Toluene | GC-FID | 0.3–3.0 µg/L | 73–104 | 7.7–18.8 | Pinheiro Anselmo and de Andrade (2009) |
| PAHs | Lake water | HS-SDME | β-cyclodextrin | HPLC-FLD | 4–247 ng/L | 95–111 | <10 | Wu et al. (2008) |
| PAHs | Tap, river and creek water | IL-DI-SDME | $[C_4MIM][PF_3]$, $[PH_3T]$ $[PF_3]$, [BMPL] $[PF_3]$ | HPLC-FLD | 0.04–1326.8 µg/L | 77–113 | 1.5–9.8 | Yao et al. (2009) |
| PAHs | Tap, waste, spring, pool, well water | HS-SDME | 1-butanol | GC-FID | 4-41 µg/L | 77–123 | 0.7–19.5 | Shariati-Feizabadi et al. (2003) |

(continued)

**Table 4.5** (continued)

| Analyte | Matrix | Mode | Extraction solvent | Analysis technique | LOD | Apparent recovery (%) | Precision (%) | Reference |
|---|---|---|---|---|---|---|---|---|
| Parabens cosmetic products | River water | DI-SDME | Hexyl acetate | GC-MS | 1–15 μg/L | 72.6–99.4 | 8.1–13 | Saraji and Mirmahdieh (2009) |
| PBDE | Tap and lake water | DI-SDME | Toluene | HPLC-UV | 0.7 μg/L | 91.5–103.8 | 4.4 | Li et al. (2007) |
| Phenol, ClP | River water | DI-SDME | Butyl acetate | GC-MS | 5–21 ng/L | 50–134 | 1.4–10.2 | Bagheri et al. (2004) |
| Phenols | River water | DI-SDME in-syringe derivatisation | Hexyl acetate | GC-MS | 4–61 ng/L | 92–144 | 4.8–12 | Saraji and Bakhshi (2005) |
| Phenols | Surface water | SDME-ion pair transfer derivatisation | TBAB | GC-MS | 0.48–1.15 μg/L | 85.7–121.4 | 3.9–7.6 | Fiamegos et al. (2008) |
| Phenols | Lake water wastewater | IL-HS-SDME | $[C_4MIM][PF_6]$ $[C_8MIM][PF_6]$ | GC-FID | 0.1–0.4 μg/L | 81–111 | 3.6–9.5 | Zhao et al. (2008) |
| TBT, TPhT | Water, wastewater | IL-HS-SDME | $[C_4MIM][PF_6]$ | HPLC-FLD | 0.62–0.95 μg/L | 86.9–92.1 | 7.8–8.3 | Sheikhloie et al. (2009) |
| Trihalomethanes | River, drirnking, tap, well and swimming pool water | IL-DI-SDME | $[C_6MIM][PF_6]$ $[C_8MIM][PF_6]$ | GC-MS | 0.5–0.9 μg/L | 91.6–101.7 | 3.1–11.0 | Aguilera-Herrador et al. (2008a) |
| Trihalomethanes | Well and tap water | HS-LPME | 1-octanol | GC-ECD | 0.15–0.40 μg/L | 101.0–109 | 8.2–11.3 | Zhao et al. (2004) |
| UV filters | Surface water | IL-DI-SDME | $[C_8MIM][PF_6]$ $[C_6MIM][PF_6]$ | HPLC-UV | 0.20–10.0 μg/L | 92–110 | 2.8–7.9 | Vidal et al. (2010) |
| α–and β- endosulfan | Tap water Surface water | DI-SDME | Isooctane | GC-ECD | 0.1–0.9 μg/L | 90–100 | 10.2–17.7 | Lopez-Blanco et al. (2003) |

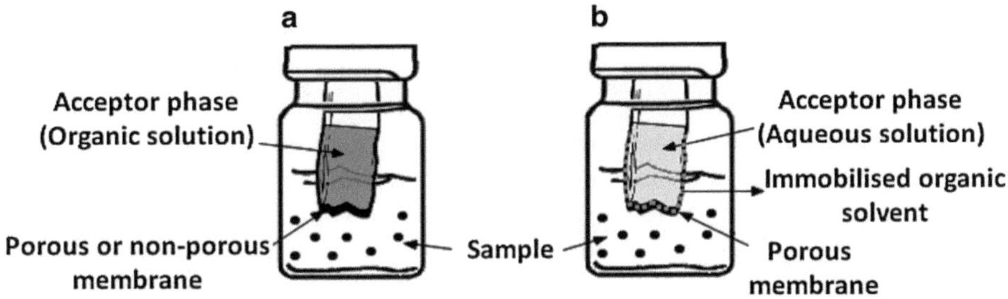


**Fig. 4.9** Technical set-up for membrane based liquid phase micro-extraction (LPME): (**a**) 2-phase configuration, used for both porous (*MMLLE*) and non-porous membranes (*MASE*), and (**b**) 3-phase configuration for (SLM) extraction. SLM: supported liquid membrane

and its development is still ongoing. The extraction takes place between the aqueous sample (donor phase) and a microvolume of acceptor phase, protected by a membrane that avoids the mixture of the two phases and acts as a selective barrier between the phases (Psillakis and Kalogerakis 2003b). The main advantages over conventional LLE are the avoidance of emulsion formation, the lack of the phase separation step and the use of modules with a high surface-area-to-volume ratio (Moreno Cordero et al. 2000).

There are two main categories depending on the nature of the membrane: porous or non-porous membrane techniques.

#### 4.3.3.1 Porous Membranes

When using porous membranes two techniques can be distinguished: (i) supported liquid membrane (SLM) extraction and (ii) microporous membrane liquid–liquid extraction (MMLLE).

SLM (see Fig. 4.9b) are three-phase extraction systems with an organic phase immobilized in a porous hydrophobic membrane separating two aqueous phases (Zorita et al. 2007). The analyte is first extracted into an organic solvent that impregnates the walls of the membrane, and then back-extracted into an aqueous acceptor solution adjusted to an adequate pH, depending on the acidic properties of the analytes. These three-phase systems have been shown to provide high selectivity, clean extracts, and to facilitate trace analysis by removing interferences and increasing the concentration of analytes to measurable levels (Jönsson and Mathiasson 1999).

MMLLE is a two-phase membrane extraction technique with one aqueous sample and an organic acceptor phase inside the microporous membrane (see Fig. 4.9a), where the same organic liquid is immobilized in the membrane pores. SLM is mainly used for the analysis of acidic or basic polar compounds that are easily protonated and often have low log $K_{ow}$ (Yamini et al. 2006; Kou et al. 2004), while MMLLE best suits the extraction of neutral and/or more hydrophobic organic compounds with high partition coefficients to the organic phase (Jönsson and Mathiasson 2000; Fontanals et al. 2006).

The most commonly used porous hydrophobic membranes are polypropylene (PP), polytetrafluoroethylene (PTFE) and polyvinylidene difluoride (PVDF) (Jönsson and Mathiasson 2000). PP membranes are much highly desirable as they provide unique suitability and performance for analyte extraction in terms of high porosity that can enhance mass transfer, compatibility and stability when used with wide range of organic solvents (Barri and Jönsson 2008). The main advantages of microporous membrane-based extraction techniques are high analyte capacity (collection in a liquid instead of adsorption), nearly total avoidance of organic solvent, easy handling and low analysis costs (Fernandez Laespada et al. 2001).

Several configurations have been applied, like flat sheet, spiral wound and hollow fiber (either rod-like or U-shaped) (Jönsson and Mathiasson 2001). Hollow fiber based SLM, also termed as hollow fiber liquid phase microextraction (HF-LPME), is the most popular and stable (He et al. 2004). It can be performed in both SLM and MLLE techniques and in two modes, static and dynamic. In the static mode, the HF is supported by a guiding tube into the sample, while in the dynamic mode, a conventional microsyringe with the HF attached to its needle is connected to a syringe pump to perform the extraction, as mentioned before for dynamic SDME (Basheer and Lee 2004). Dynamic extraction was claimed to provide better extraction efficiency and improved reproducibility when compared to the static mode (Sarafraz-Yazdi and Amiri 2010).

While enrichment, clean-up and low solvent consumption are the major advantages of the membrane based LPME and relatively long extraction times is perhaps the major disadvantage, due to passive diffusion as transport mechanism. However, the kinetics of the membrane was recently improved by application of an electrical potential difference over the SLM. This system was called electromembrane extraction (EME). One electrode is placed in the donor solution and the second one in the acceptor phase, promoting electrokinetic migration of the ionized analytes across the SLM. The use of these electrical potentials leads to shorter extraction time and to mass transfer strongly dependent on the compound, i.e. highly selective extraction to a large extent controlled by the extraction time (Pedersen-Bjergaard and Rasmussen 2008; Kjelsen et al. 2008; Gjelstad et al. 2007).

#### 4.3.3.2 Non-porous Membranes

On the other hand, membrane-assisted solvent extraction (MASE) or membrane extraction with sorbent interface (MESI) utilizes non-porous membranes. MASE, composed of three-phase aqueous–polymeric–organic system, involves no organic solvent deliberately immobilized in the polymeric material, but instead the organic solvent constitutes the acceptor side of the membrane (see Fig. 4.9a). The main difference between MASE and microporous membrane extraction techniques is that the membrane used in MASE is a low-density polyethylene (LDPE), dense PP, PDMS silicone rubbers and asymmetric composite polymeric membranes that are composed of a thin layer of silicone and another layer of polycarbonate (PC) or a relatively thick support layer of porous PP. By using nonporous polymeric membranes, the analyte extraction rate (permeability) is governed by a solution-diffusion mechanism that highly depends on the analyte solubility and diffusivity into the membrane material. Normally, nonpolar solvents are used, like heptane, hexane and cyclohexane (Hauser and Popp 2001b). After extraction, the organic solvent is usually collected for LVI into the GC instrument (Schellin and Popp 2003; Schellin and Popp 2006; Schellin and Popp 2005; Hauser et al. 2004). MASE has been proven to be an extremely simple, low cost and virtually solvent-free sample preparation technique, which provides a high degree of selectivity and enrichment by additionally eliminating the possibility of carryover between runs (Almeda et al. 2007; Marsin Sanagi et al. 2007).

Another application of non-porous membranes is MESI, an aqueous–polymeric–gaseous system (Luo and Pawliszyn 2000), operated in HS or DI sampling modes. The analytes from the aqueous sample diffuse through the nonporous polymeric membrane into the gaseous flowing stream on the other side of the membrane. The MESI technique uses also membranes made of silicone rubbers (PDMS) (Matz et al. 1999; Guo and Mitra 2000), PDMS–PC copolymer (Yu and Pawliszyn 2004) or Nafion (Shen and Pawliszyn 2001). MESI has limited applicability: analysis of volatile or semi-volatile organic compounds (VOCs and SVOCs) in aqueous and mainly air samples, due to the high permeability of this class of compounds and the unique selectivity provided by the silicone membranes. One of the limitations of MESI is that the technique is not applicable for extraction of polar and nonpolar, nonvolatile compounds, owing to permeation and trapping problems in the membrane system (Barri and Jönsson 2008).

In general, the main advantage of liquid homogeneous membranes as compared with polymeric membranes is the greater transport velocity through them, due to the great diffusivity of species in a liquid medium. Additionally, in these membranes it is easier to incorporate carriers to selectively increase the permeability of certain species, giving rise to facilitated or coupled transport processes. However, the membrane lifetime is usually longer for polymeric membranes, due to solvent leakage out of the liquid membranes, which becomes more pronounced as the polarity of the solvent increases; n-undecane and di-n-hexylether have shown the best stabilities (Moreno Cordero et al. 2000). Two other general advantages are ease of automation and amenability to coupling to analytical instruments on-line or in-line. In contrast, disadvantages inherent in the nature of the membrane system are that only some analyte classes can be processed simultaneously and a number of optimisation experiments must be often performed before practical problems can be addressed. Typically low stability of membranes and the long times required to extract analytes present at low concentrations are other alleged disadvantages (Arce et al. 2009).

#### 4.3.3.3 Method Development and Applications

Most of the variables affecting membrane liquid-phase microextraction are common to other pre-concentration techniques and include volume of the donor phase, pH of the aqueous solution, addition of inert salts to modify the ionic strength, extraction temperature or extraction time. Other variables such as the nature of the membrane or the nature and volume of the extraction solvent are more characteristic of these membrane-based extraction procedures (Psillakis and Kalogerakis 2003a).

From a technical point of view, membrane based liquid extraction techniques enable a high degree of flexibility, providing compatibility with GC, HPLC, CE and MS. They have been often applied to the analysis of several priority and emerging pollutants in water samples (Table 4.6).

## *4.3.4 Dispersive Liquid-Liquid Phase Microextraction*

DLLME was developed in 2006 by Rezaee et al. as a novel technique for the extraction of compounds from aqueous samples (Rezaee et al. 2006). As mentioned in the introduction of LPME techniques, DLLME avoids the high consumption of organic solvents involved in classical extraction techniques (Herrera-Herrera et al. 2010).

This technique, which allows the simultaneous extraction and concentration of analytes is based on the use of a ternary solvent system consisting of an aqueous phase, the sample from which the analytes are intended to be removed, and a mixture of two organic solvents: one which is water-miscible and acts as a dispersing agent and the other which is water-immiscible but miscible with the dispersing agent and of higher density, the extractant agent.

The dispersant-extractant mixture is contacted with the aqueous phase within a conical tube and the formation of an emulsion is readily observed. This maximizes the interface between phases, thus promoting the rapid transition of the analyte between the sample and the extractant.

After centrifugation a separation into two phases is obtained:

- On the one hand, the extracting agent settles at the bottom of the conical tube due to its greater density. This phase contains the extracted analytes and is known as the sedimented phase.
- On the other hand, the aqueous sample in which the dispersing agent is dissolved.

Figure 4.10 shows a series of images corresponding to the different stages of DLLME.

### 4.3.4.1 Method Development and Applications

Some of the characteristic variables in DLLME are mentioned below:

- Extracting solvent selection
  The extracting agent must satisfy a number of prerequisites: it should have a density and low water solubility to enhance the deposit of the sedimented phase at the bottom of the conical vial by centrifugation; it should be able to extract the compounds of interest and a good chromatographic behavior, because it will

**Table 4.6** Membrane based liquid extraction application to the determination of priority or emerging pollutants in water samples during the period 2000–2010

| Analytes | Matrix | Technique | Membrane material | Analysis technique | LOD | Apparent recovery (%) | Precision (%) | Reference |
|---|---|---|---|---|---|---|---|---|
| 17β-estradiol | Tap | HF-MMLLE | PP | GC-MS | 16 ng/L | – | 7.3 | Zorita et al. (2008, 2009) |
| Estrone | Sewage | | | LC-MS/MS | 3 ng/L | | 10.3 | |
| 17β-ethynylestradiol | | | | | 10 ng/L | | 8.9 | |
| Alkylphenols | Sea | HF-MMLLE | PP | GC-MS | 5–11 ng/L | 83.1–103.4 | 9.2–13.9 | Basheer and Lee (2004) |
| Chlorophenols | | | | | 15 ng/L | 75.4–116.2 | 5.9–9.2 | |
| Bisphenol-A | | | | | 14 ng/L | 103.5–119.7 | 10.9 | |
| Caffeine | River | MASE | PE | GC-MS | 25 ng/L | 37 (n=4) | 4 | Einsle et al. (2006), Chimuka et al. (2010) |
| Carbamate Pesticides | Distilled, tap waste | HF-MMLLE | PP | GC-MS | 0.2–0.8 ng/L | >83.04 | 4.9–7.8 | Zhang and Lee (2006) |
| Chlorophenols | Surface | SLM | Teflon | ECD | 10 μg/L | 90–112 | 2.8 | Almeda et al. (2007) |
| Chlorophenols | River<br>Leachate | HF-SLM | PP | HPLC-DAD | 0.1–5 μg/L | – | 2–12 | Liu et al. (2005a) |
| Chlorophenols | Sea | EME-SLM | PP | HPLC-UV | 100 ng/L | 6.8 | 74 | Lee et al. (2009) |
| Dicarboximide fungicides | Distilled, underground, river, sea, lake | HF-MMLE | PP | GC-ECD | 0.01 ng/L | >98 | 3.6–6.2 | Lambropoulou and Albanis (2007) |
| Dicarboximide fungicides | Natural and reagent water | MMLLE | PE | HPLC-UV | 1 ng/L | – | 2.7–3.9 | Sandahl et al. (2000) |
| Dinitrophenols | River<br>Sea<br>Lake | HF-SLM | PP | HPLC-UV | 6–8 ng/L | 12–43 | 3.1–16.6 | Berhanu et al. (2006b) |

(continued)

**Table 4.6** (continued)

| Analytes | Matrix | Technique | Membrane material | Analysis technique | LOD | Apparent recovery (%) | Precision (%) | Reference |
|---|---|---|---|---|---|---|---|---|
| Dinitrophenols | River<br>Leachate | HF-SLM | PP | HPLC-DAD | 95–506 ng/L | – | 1.7–3.6 | Lezamiz and Jönsson (2007) |
| Hexaconazole, Procymidone, Quinalphos, Vinclozolin | – | CSM-MMLLE | Nylon | microLC-UV/Vis | 1.1–1.9 μg/L | 83–108 | 6.3–7.5 | Marsin Sanagi et al. (2007) |
| Ibuprofen | Waste | HF-SLM | PP | LC-MS/MS | 14 ng/L | 86 | 30–32 | Quintana et al. (2004) |
| Ibuprofen | Waste | EME-SLM | PP | GC-MS | 130 ng/L | 13 | 20 | Basheer et al. (2010) |
| MBT, DBT, TBT | Industrially polluted | HF-MMLLE | PP | GC-FID | 0.5–1.5 μg/L | 47–91 | – | Cukrowska et al. (2004) |
| TPhT | Sea<br>River<br>Dam | | | | | | | |
| OCPs | Water<br>Sea | HF-MMLLE | Porous PP | GC-MS | 0.001–0.008 ng/L | | 4.7–10.6 | Basheer et al. (2004) |
| OCPs | Rain | HF-MMLLE | PP | GC-MS | 0.013–0.047 ng/L | >74.9 | 1.6–14 | Basheer et al. (2003) |
| OCPs | Sea | HF-MMLLE | PP | GC-MS | 0.013–0.047 ng/L | >81.6 | 1.6–14 | Basheer et al. (2002) |
| OPPs | River | HF-MMLLE | PP | GC-FTD | 0.005–0.042 ng/L | >89 | 4.3–11.1 | Lambropoulou and Albanis (2005) |
| OPPs | Lake | HF-MMLLE | PP | GC-MS | 0.006–0.2 ng/L | >83.1 | 3.5–8 | Chen and Huang (2006) |
| OPPs | Waste | MASE | Dense PP | LC-MS/MS | 1–25 ng/L | 5–98 | 1.5–27 | Quintana and Reemtsma (2006) |

| | | | | | | | | |
|---|---|---|---|---|---|---|---|---|
| PAHs | Rain | HF-MMLLE | PP | GC-MS | 2–47 ng/L | 70.5–114.3 | 4.2–10.6 | Basheer et al. (2003) |
| OCPs | | | | | 13–59 ng/L | 74.9–97.2 | 1.6–14.0 | |
| PAHs | Surface | MASE | Dense PP | LVI-GC-MS | 3–4 ng/L | 65–92 | 10–18 | Rodil et al. (2007) |
| PAHs | Estuary | | | | 0.2–14.4 ng/L | 81–120 | 3–19 | Prieto et al. (2008b) |
| PCBs | Sea | MASE | LDPE | LVI-GC-MS | 0.6–31.6 ng/L | 81–120 | | |
| NPs and NPEs | | | | | 307–317 ng/L | 102–121 | | |
| PEs | | | | | 29.8–252 ng/L | 81–117 | | |
| PBDEs | | | | | 0.1–5.3 ng/L | 92–121 | | |
| PBBs | | | | | 0.4–1.7 ng/L | 88–121 | | |
| PBDEs | Tap | HF-MMLLE | PP | GC-MS | 0.2–0.9 ng/L | 96.1–108.6 | 1.5–8.0 | Fontanals et al. (2006) |
| | River | | | | | 86.1–110.4 | 4.0–13.9 | |
| | Leachate | | | | | 85.4–106.7 | 2.4–8.2 | |
| PCBs | River | MASE | Dense PP | LVI-GC-MS | 2–10 ng/L | 88–100 | 3.7–12.2 | Schellin and Popp (2003) |
| Pesticides | Natural | HF-MMLLE | PP | LC-MS/MS | 26–81 ng/L | – | 5.6–11.8 | Trtic-Petrovic et al. (2010) |
| | Tap | | | | | | | |
| Phenolic Compounds (DNP, DNOC, Dinoseb, Dinoterb) | Reagent, river, lake, sea | HF-SLM | – | HPLC-UV | 0.006–0.008 ng/L | >52 | 2.4–16.6 | Berhanu et al. (2006a) |
| Phenols | Ground | MASE | Dense PP | LVI-GC-MS | 9–595 ng/L | 10.7–98.4 | 6.4–18.3 | Schellin and Popp (2005) |
| Phenoxy herbicides | Water | HF-MMLLE | PP | HPLC-UV | 0.1–0.5 ng/L | >94 | 7.6–7.5 | Wu et al. (2005) |
| Phthalate esters | Mineral | HF-SLM | PP | GC-MS | 5–100 ng/L | – | 4–19 | Psillakis and Kalogerakis (2003b) |
| Quinalphos | Underground | HF-MMLLE | Nylon | GC-ECD | 3–7 ng/L | 96.1–99.3 | 2.6–4.3 | Marsin Sanagi et al. (2010) |
| Methidathion | Farm | | | | | 98.6–101.5 | 2.8–4.9 | |
| Hexaconazole | | | | | | | | |

(continued)

**Table 4.6** (continued)

| Analytes | Matrix | Technique | Membrane material | Analysis technique | LOD | Apparent recovery (%) | Precision (%) | Reference |
|---|---|---|---|---|---|---|---|---|
| NPs | Sewage | HF-MMLLE | PP | GC-MS | 100 ng/L | 20.1 | 11.0 | Müller et al. (2003) |
| Bisphenol-A | | | | | 300 ng/L | 23.6 | 21.1 | |
| 17α-ethinylestradiol | | | | | 20 ng/L | 33.8 | 5.3 | |
| Triazine herbicides | Deionized | HF-MMLLE | Porous PP | GC-MS | 0.007–0.063 ng/L | >90.5 | 0.8–4.4 | Shen and Lee (2002) |
| Triazine herbicides | Waste | MASE | Dense PP | LVI-GC-MS | 10 ng/L | 30–125 | 7.0–9.1 | Hauser et al. (2004) |
| OCPs | | | | | 2–10 ng/L | 40–80 | 4.6–10.9 | |
| OPPs | | | | | 5–50 ng/L | 17–48 | 6–24.2 | |
| Triclosan | Tap<br>Surface | HF-MMLLE | PP | GC-MS | 20 ng/L | 83.6<br>114.1 | 6.9 | Zhao et al. (2007, 2010) |
| VOCs | River | MASE | Dense PP | GC-ECD | 5–50 ng/L | 43–96 | 4.3–9.7 | Schellin and Popp (2006) |
| VOCs | Surface<br>Underground | MESI | PDMS coated fused silica fiber | GC-FID | 100–1,000 ng/L | 3.8–6.2 | – | Luo and Pawliszyn (2000) |
| VOCs | Surface<br>Underground | MESI | Silicone | GC-FID | 100–5,000 ng/L | 2.5–12.8 | – | Hauser and Popp (2001a) |
| VOCs | Tap | MESI | Silicone poplycarbonate | GC-MS | 60 ng/L | – | – | Segal et al. (2000) |

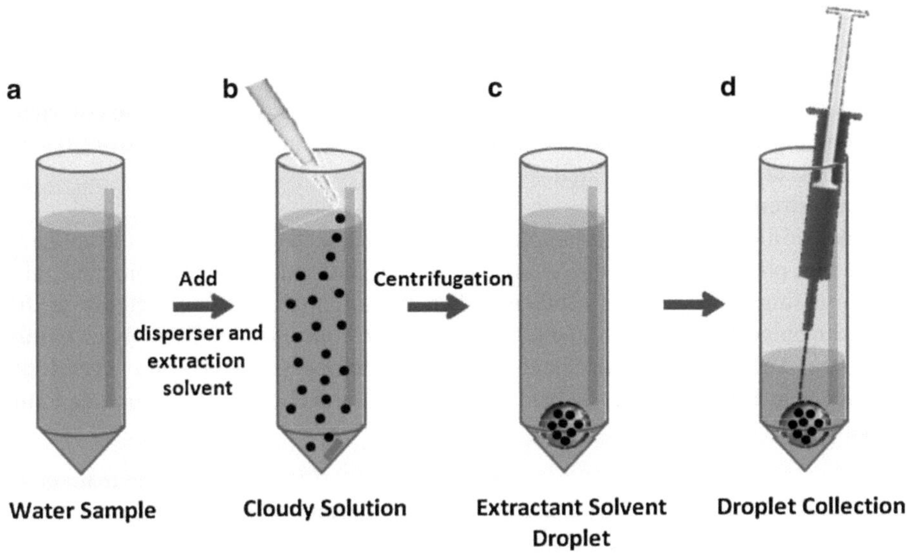


**Fig. 4.10** Steps in dispersive liquid-liquid phase micro-extraction (DLLME): (**a**) water sample, (**b**) addition of the disperser and extractant and formation of the cloudy solution, (**c**) extractant solvent droplet formation and (**d**) collection of the extractant solvent droplet

be usually injected directly into a GC system. The solvents most commonly used as extractants are halogenated hydrocarbons such as chloroform, carbon tetrachloride, trichloroethane, chlorobenzene, etc., although there are applications where carbon disulfide (Rahnama Kozani et al. 2007) and ILs (Zhou and Ye 2008) have been used. Typically, these solvents are used when the determination method is LC. The amount of extractant used is usually less than 200 μL, thus obtaining a very small drop of sedimented phase and achieving high enrichment factors.

- Dispersing solvent selection
  The solvent used as a dispersant should be miscible with the extracting agent and the aqueous sample, leading to the formation of an emulsion when the dispersant-extractant mixture is added over the sample. This phenomenon causes a large increase of the contact surface between the sample and the extractant, favoring the passage of the analytes into the organic phase and increasing, therefore, the extraction efficiency. The most commonly used solvents for this purpose are ethanol, methanol, acetone, acetonitrile and tetrahydrofuran.
- Extracting and dispersing volumes
  The amount of extracting agent used has a great influence on the enrichment factor. An increase in the added volume causes an increase in the size of the sedimented phase, leading to a decreased concentration of the analyte in this phase. This reduction results in a decrease in the enrichment factor as the initial concentration remains constant. The optimum quantity will be the one that generates a high enrichment factor and that is sufficiently high to allow an easy manipulation of the sedimented phase. Quantities used are usually in the 20–100 μL range.

Moreover, the amount of dispersing agent affects mainly the formation of the emulsion. The greater the dispersion degree, the better the contact between phases will be and the greater the extraction efficiency. This volume may also affect, to a lesser extent, the size of the sedimented phase, so that both contributions are to be taken into account. For 10 mL samples, volumes from 0.5 to 1.5 mL are normally selected.

- Ionic strength
  The addition of salt to the sample tends to decrease the solubility of the analytes in this phase, favoring its passage to the organic phase (extractant) and thus increasing the extraction efficiency. Yet, this often leads to an increase in the sedimented phase volume, which decreases the concentration of analytes in this phase. Therefore, the enrichment factor will be negatively affected. Therefore, special attention needs to be paid to both effects and choose the appropriate ionic strength.

Other variables such as the pH of the donor phase affect in the same manner as for the rest of the pre-concentration techniques mentioned above. On the other hand, it has been observed that the time for centrifugation, although a necessary step, is not significant and most authors use values of less than 5 min. Finally, the extraction time is considered as the lapse between the injection of the extractant-dispersant mixture on the sample and the centrifugation. This parameter has little influence on the extraction efficiency because the transfer of the analytes between the sample and the extractant occurs immediately after the dispersion and phase equilibrium is reached quickly (Montes et al. 2009a). This feature is one of the most important advantages of DLLME over other microextraction techniques.

DLLME is a relatively recent technique. After a first work in which the presence of PAHs in water samples was identified in 2006 (Rezaee et al. 2006), the same group continued with the study of other families of compounds such as organophosphorus pesticides (Berijani et al. 2006), trihalomethanes (Rahnama Kozani et al. 2007), chlorobenzenes (Kozani et al. 2007) and polychlorinated biphenyls (Rezaei et al. 2008).

However, due to its operational simplicity, low cost and, above all, because it is a good alternative to conventional liquid extraction for the analysis of several families of compounds in aqueous samples, with a negligible solvent consumption, it has become a very useful tool and there are many applications that have been published in the last 4 years.

Table 4.7 summarizes some of the characteristics of the application of DLLME to environmental aqueous samples with special focus on the analysis of priority and emerging organic pollutants.

#### 4.3.4.2 Modifications and Enhancements

Like many good analytical methods, DLLME has also seen adaptations and improvements. Among these, some of the most significant consist on the elimination of the

**Table 4.7** DLLME application to the determination of priority or emerging pollutants in water samples during the period 2000–2010

| Analyte | Matrix | Extractant | Dispersant | Analysis technique | LOD | Precision (%) | Apparent Recovery (%) | Reference |
|---|---|---|---|---|---|---|---|---|
| Amide herbicides | Water | Various | Acetone | GC-MS | 2–6 ng/L | 2.6–8.7 | | Zhao et al. (2009b) |
| Atrazine, Simazine | Water | CCl4 | Methanol | HPLC-UV | 0.04–0.1 μg/L | 4.6–6.4 | 61–91 | Zhou et al. (2009b) |
| Benomyl | Water | | | HPLC-FLD | 3.3 μg/L | | | Farhadi et al. (2009) |
| Carbamate pesticides | Water | Acetonitrile | Acetonitrile | GC-MS | 0.001–0.5 μg/L | 2.3–6.8 | 94–104 | Chen et al. (2010) |
| Carbamate pesticides | Water | Trichloromethane | Acetonitrile | HPLC-DAD | 0.1–0.5 μg/L | 3.5–8.7 | 86–97 | Liu et al. (2009c) |
| Carbamate pesticides | Water | Chlorobenzene | Acetonitrile | HPLC-UV | 0.1–0.9 μg/L | 1.8–4.6 | 74–94 | He et al. (2009b) |
| Carbendazim, Thiabendazole | Water, soil | Chloroform | Tetrahydrofuran | HPLC-FLD | 0.5–1 μg/L | 3.5–6.8 | 51–71 | Wu et al. (2009) |
| DDT | Water | CCl4 | Acetonitrile | HPLC-UV | 0.32–0.51 μg/L | 2.80–7.50 | 86–120 | Zhou et al. (2009a) |
| Dimethaclon | Water | CCl4 | Acetone | GC-ECD | 1 μg/L | 4.3–7.6 | 81 | Ruan et al. (2010) |
| Heterocyclic insecticides (fipronil, chlorfenapyr, buprofezin, and hexythiazox) | Water | Ionic liquids | Various | HPLC-DAD | 0.53–1.28 μg/L | 4.5–10.7 | 79–110 | Liu et al. (2009b) |
| Hydroxylated Benzophenones (UV filters) | Sea water | Chloroform | Acetone | GC-MS | 32–50 ng/L | 6 | 82–222 | Tarazona et al. (2010) |

(continued)

**Table 4.7** (continued)

| Analyte | Matrix | Extractant | Dispersant | Analysis technique | LOD | Precision (%) | Apparent Recovery (%) | Reference |
|---|---|---|---|---|---|---|---|---|
| Metacrate | Water | Dichloromethane | Methanol | HPLC-UV | 1 μg/L | 5.7 | 97–102 | Xia et al. (2008) |
| Methomyl | Water | Tetrachloroethane | Methanol | HPLC-UV | 3–5,000 μg/L | 2.6 | | Wei et al. (2007) |
| OCPs | Water | Hexadecane | Acetonitrile | GC-ECD | 0.01–0.1 μg/L | 5.8–8.8 | 83–102 | Leong and Huang (2008) |
| OCPs | Water | Tetrachloroethylene | Acetone | GC-MS | 1–25 ng/L | 5–15 | 70–120 | Cortada et al. (2009a) |
| OCPs | Water | Various | Various | GC-MS | 0.4–2.5 ng/L | 4–10 | 55–119 | Tsai et al. (2009) |
| OPPs | Water | Cyclohexane | Acetone | GC-FID, GC-MS | 3–4 μg/L (GC-FID), 3 ng/L (GC-MS) | 5.3–7.8 | 68–105 | Farajzadeh et al. (2009) |
| OPPs | Water | Chlorobenzene | Acetone | GC-FPD | 3–20 ng/L | 1.2–5.6 | 84–125 | Berijani et al. (2006) |
| OPPs | Water | [C8MIM][PF6] | Methanol | HPLC | 0.1–5 μg/L | 4.7 | 87–118 | He et al. (2009a) |
| OSPs | Water, soil, beverages | Various | Various | GC-FPD | 0.21–3.05 μg/L | 0.6–11.9 | 78–117 | Xiong and Hu (2008) |
| Parabens | Water | Octanol | Acetone | GC-FID | 5–15 ng/L | 2–3 | 25–72 | Farajzadeh et al. (2010a) |
| Phorate | Water | Various | Various | GC-FID | 1 ng/L | 6.7 | 104 | Xie et al. (2008) |
| Phtalates and polycyclic musks | Water | CCl4 | MeOH | GC-MS | 8–63 ng/L | 6–12 | 60–93 | Panagiotou et al. (2009) |
| Profenofos | Water | Trichloroethane | Acetonitrile | HPLC-UV | 2 mg/L | 1.4–6.1 | 87–118 | Shen et al. (2010) |
| Pyretroid pesticides | Water | Chlorobenzene | Acetone | GC-ECD | 0.04–0.1 μg/L | 3.1–7.2 | 76–116 | Zang et al. (2008) |
| Triazine herbicides | Water | Chlorobenzene | Acetone | GC-MS | 0.021–0.12 μg/L | 1.36–8.67 | 24–116 | Nagaraju and Huang (2007) |

| | | | | | | | | |
|---|---|---|---|---|---|---|---|---|
| Triazole pesticides | Water | 1,1,2,2-tetrachloro-ethane | methanol | GC-FID, GC-MS | 0.53–24 μg/L | <5.2 | | Farajzadeh et al. (2010b) |
| Triazophos, Carbaryl | Water, fruit juice | Tetrachloroethane | Acetonitile | HPLC-FLD | 12–16 ng/L | 1.4–2.7 | 80–118 | Fu et al. (2009) |
| Triclosan, Triclocarbam, Methyl-triclosan | Water | Various | Various | UHPLC-UV | 45–236 ng/L | 2–22 | 64–121 | Guo et al. (2009) |
| Carbofuran, clomazone, tebuconazole | Water | Various | Various | LC-MS/MS | (LOQ) 0.02 μg/L | 1.9–9.1 | 63–120 | Caldas et al. (2010) |

dispersing phase and promoting analyte phase transfer by means of ultrasounds (Regueiro et al. 2008), the solidification of the organic phase and its remotion with a spatula (Sobhi et al. 2008), the use of derivatization steps simultaneously with the extraction (Montes et al. 2009b; Chiang and Huang 2008), the use of liquid chromatography as analytical technique instead of the more common gas chromatography (Li et al. 2008; Melwanki and Fuh 2008), the use of ionic liquids (IL) (Zang et al. 2008; Fan et al. 2008; Pena et al. 2009) or the combination with other concentration and purification steps such as solid phase extraction (SPE) (Fattahi et al. 2007b; Liu et al. 2009a; Montes et al. 2009a).

## 4.4 Conclusion

As has been reviewed, there are several interesting alternatives to deal with the extraction of target analytes before taking them to the detection systems. Since the development of analytical strategies emphasizes the holistic view of the analytical process, from the sample treatment method to the quantification method, more than one fit-for-purpose method can be designed according to the requirements of the analysis, the instrumental facilities of the lab or the skills of the staff, to mention few of them.

Most of the extraction techniques can be seen as sophisticated variations of classical extraction procedures (i.e. liquid-liquid or solid-liquid extractions) and, on the whole, they offer a large variety of features to make them the chosen techniques for a given problem. Compared to the classical procedures, most of the methods provide added values in terms of throughput, efficiency, selectivity, scale, etc. just because they can be easily automated, or we can use them in microanalysis.

One of the most remarkable trends of these techniques is the miniaturization of the extraction methods. This fact makes possible the (micro)-extraction of samples and in many cases the automatization of the processes. When the later feature is not accomplished, i.e. all the operations are done by hand, one of the most extended problems is the lack of repeatability. On the contrary, when the automatization has been implemented the problems arise with the technical requirements of the robotic devices (home-made solutions vs. commercial accessories; proprietary arrangements, etc.) and the maintenance of the equipment.

One of the consequences of those trends is the life expectancy of the extraction techniques, especially the newest ones. As soon as the technique is matured and the applications are well established, as it can be the case of SPME or SBSE it is a question of time to get robotic set-ups extended to most of the brands, and to incorporate some of the well-established methods to the official ones. On the contrary, when the technique is too recent or the applications are not widely used, as it might be the MEPS or the membrane based extractions, it is hard to foresee the development of robotic systems or the incorporation to the official methods.

Nevertheless, leaving aside the specific requirements of the official methods (ruggedness, robustness, well validated methods, etc.), the development of new extraction methods and new applications in fields such as environment and toxicology, pharmacology and clinical analysis or food chemistry is still a challenge for researchers.

## References

Abdel-Rehim M (2003) In: Gazette CP (ed) AstraZeneca application, World Intelectual Property Organisation, London, UK, p 77

Abdel-Rehim M (2004) New trend in sample preparation: on-line microextraction in packed syringe for liquid and gas chromatography applications I. Determination of local anaesthetics in human plasma samples using gas chromatography-mass spectrometry. J Chromatogr B Analyt Technol Biomed Life Sci 801(2):317–321

Abdel-Rehim M, Altun Z, Blomberg L (2004) Microextraction in packed syringe (MEPS) for liquid and gas chromatographic applications. Part II – determination of ropivacaine and its metabolites in human plasma samples using MEPS with liquid chromatography/tandem mass spectrometry. J Mass Spectrom 39(12):1488–1493

Abdel-Rehim M, Skansen P, Vita M, Hassan Z, Blomberg L, Hassan M (2005) Microextraction in packed syringe/liquid chromatography/electrospray tandem mass spectrometry for quantification of olomoucine in human plasma samples. Anal Chim Acta 539(1–2):35–39

Abdel-Rehim M, Askemark Y, Norsten-Hoeoeg C, Pettersson K, Halldin M (2006a) Quantification of 4-OH-2,6-xylidine and its conjugates in human urine samples utilising microextraction in packed syringe on-line with liquid chromatography and electrospray tandem mass spectrometry (MEPS-LC-MS/MS). J Liquid Chromatogr Relat Technol 29(16):2413–2424

Abdel-Rehim M, Dahlgren M, Blomberg L (2006b) Quantification of ropivacaine and its major metabolites in human urine samples utilizing microextraction in a packed syringe automated with liquid chromatography-tandem mass spectrometry (MEPS-LC-MS/MS). J Sep Sci 29(11):1658–1661

Abdel-Rehim M, Dahlgren M, Blomberg L, Claude S, Tabacchi R (2006c) Microextraction in packed syringe (MEPS) utilizing methylcyanopropyl-silarylene as coating polymer for extraction of drugs in biological samples. J Liquid Chromatogr RelatTechnol 29(17): 537–2544

Abdel-Rehim M, Hassan Z, Skansem P, Hassan M (2007) Simultaneous determination of busulphan in plasma samples by liquid chromatography-electrospray ionization mass spectrometry utilizing microextraction in packed syringe (MEPS) as on-line sample preparation method. J Liquid Chromatogr Relat Technol 30(20):3029–3041

Abdel-Rehim M, Andersson A, Breitholtz-Emanuelsson A, Sandberg-Staell M, Brunfelter K, Pettersson K, Norsten-Hoeoeg C (2008) MEPS as a rapid sample preparation method to handle unstable compounds in a complex matrix: determination of AZD3409 in plasma samples utilizing MEPS-LC-MS-MS. J Chromatogr Sci 46(6):518–523

Aguilera-Herrador E, Lucena R, Cardenas S, Valcarcel M (2008a) Determination of trihalomethanes in waters by ionic liquid-based single drop microextraction/gas chromatographic/mass spectrometry. J Chromatogr A 1209(1–2):76–82

Aguilera-Herrador E, Lucena R, Cardenas S, Valcarcel M (2008b) Direct coupling of ionic liquid based single-drop microextraction and GC/MS. Anal Chem 80(3):793–800

Aguilera-Herrador E, Lucena R, Cardenas S, Valcarcel M (2008c) Ionic liquid-based single-drop microextraction/gas chromatographic/mass spectrometric determination of benzene, toluene, ethylbenzene and xylene isomers in waters. J Chromatogr A 1201(1):106–111

Ahmadi F, Assadi Y, Hosseini SMRM, Rezaee M (2006) Determination of organophosphorus pesticides in water samples by single drop microextraction and gas chromatography-flame photometric detector. J Chromatogr A 1101(1–2):307–312

Almeda S, Nozal L, Arce L, Valcárcel M (2007) Direct determination of chlorophenols present in liquid samples by using a supported liquid membrane coupled in-line with capillary electrophoresis equipment. Anal Chim Acta 587(1):97–103

Alzaga R, Bayona JM (2004) Determination of perfluorocarboxylic acids in aqueous matrices by ion-pair solid-phase microextraction-in-port derivatization-gas chromatography-negative ion chemical ionization mass spectrometry. J Chromatogr A 1042(1–2):155–162

Alzaga R, Pena A, Bayona JM (2003a) Determination of phthalic monoesters in aqueous and urine samples by solid-phase microextraction-diazomethane on-fibre derivatization-gas chromatography-mass spectrometry. J Sep Sci 26(1/2):87–96

Alzaga R, Pena A, Ortiz L, Bayona JM (2003b) Determination of linear alkylbenzensulfonates in aqueous matrices by ion-pair solid-phase microextraction-in-port derivatization-gas chromatography-mass spectrometry. J Chromatogr A 999(1–2):51–60
Arce L, Nozal L, Simonet BM, Valcárcel M, Ríos A (2009) Liquid-phase microextraction techniques for simplifying sample treatment in capillary electrophoresis. TrAC Trends Anal Chem 28(7):842–853
Bagheri H, Saber A, Mousavi SR (2004) Immersed solvent microextraction of phenol and chlorophenols from water samples followed by gas chromatography–mass spectrometry. J Chromatogr A 1046(1–2):27–33
Bagheri H, Babanezhad E, Khalilian F (2008) A novel sol-gel-based amino-functionalized fiber for headspace solid-phase microextraction of phenol and chlorophenols from environmental samples. Anal Chim Acta 616(1):49–55
Bagheri H, Ayazi Z, Babanezhad E (2010) A sol-gel-based amino functionalized fiber for immersed solid-phase microextraction of organophosphorus pesticides from environmental samples. Microchem J 94(1):1–6
Bai D, Li J, Chen SB, Chen BH (2001) A novel cloud-point extraction process for preconcentrating selected polycyclic aromatic hydrocarbons in aqueous solution. Environ Sci Technol 35(19):3936–3940
Baltussen E, Cramers CA, Sandra PJF (2002) Sorptive sample preparation – a review. Anal Bioanal Chem 373(1–2):3–22
Barri T, Jönsson J (2008) Advances and developments in membrane extraction for gas chromatography: techniques and applications. J Chromatogr A 1186(1–2):16–38
Basheer C, Lee HK (2004) Analysis of endocrine disrupting alkylphenols, chlorophenols and bisphenol-A using hollow fiber-protected liquid-phase microextraction coupled with injection port-derivatization gas chromatography–mass spectrometry. J Chromatogr A 1057(1–2):163–169
Basheer C, Lee HK, Obbard JP (2002) Determination of organochlorine pesticides in seawater using liquid-phase hollow fiber membrane microextraction and gas chromatography-mass spectrometry. J Chromatogr A 968(1–2):191–199
Basheer C, Balasubramanian R, Lee HK (2003) Determination of organic micropollutants in rainwater using hollow fiber membrane/liquid-phase microextraction combined with gas chromatography–mass spectrometry. J Chromatogr A 1016(1):11–20
Basheer C, Suresh V, Renu R, Lee HK (2004) Development and application of polymer-coated hollow fiber membrane microextraction to the determination of organochlorine pesticides in water. J Chromatogr A 1033(2):213–220
Basheer C, Lee J, Pedersen-Bjergaard S, Rasmussen KE, Lee HK (2010) Simultaneous extraction of acidic and basic drugs at neutral sample pH: a novel electro-mediated microextraction approach. J Chromatogr A 1217(43):6661–6667
Beceiro-Gonzalez E, Guimaraes A, Alpendurada MF (2009) Optimization of a headspace-solid-phase micro-extraction method for simultaneous determination of organometallic compounds of mercury, lead and tin in water by gas chromatography-tandem mass spectrometry. J Chromatogr A 1216(29):5563–5569
Berhanu T, Liu J, Romero R, Megersa N, Joensson JA (2006a) Determination of trace levels of dinitrophenolic compounds in environmental water samples using hollow fiber supported liquid membrane extraction and high performance liquid chromatography. J Chromatogr A 1103(1):1–8
Berhanu T, Liu J, Romero R, Megersa N, Jönsson JÅ (2006b) Determination of trace levels of dinitrophenolic compounds in environmental water samples using hollow fiber supported liquid membrane extraction and high performance liquid chromatography. J Chromatogr A 1103(1):1–8
Berijani S, Assadi Y, Anbia M, Milani Hosseini M, Aghaee E (2006) Dispersive liquid-liquid microextraction combined with gas chromatography-flame photometric detection. J Chromatogr A 1123(1):1–9
Bianchi F, Careri M, Maffini M, Mangia A, Mucchino C (2010) A fast and effective routine method based on SPME and GC/ICP-MS for the monitoring of organotin compounds in surface and sea water. Curr Anal Chem 6(3):223–227

Bourdat-Deschamps M, Daudin J, Barriuso E (2007) An experimental design approach to optimise the determination of polycyclic aromatic hydrocarbons from rainfall water using stir bar sorptive extraction and high performance liquid chromatography-fluorescence detection. J Chromatogr A 1167(2):143–153
Caldas SS, Costa FP, Primel EG (2010) Validation of method for determination of different classes of pesticides in aqueous samples by dispersive liquid-liquid microextraction with liquid chromatography-tandem mass spectrometric detection. Anal Chim Acta 665(1):55–62
Carabias-Martinez R, Rodriguez-Gonzalo E, Moreno-Cordero B, Perez-Pavon JL, Garcia-Pinto C, Fernandez Laespada E (2000) Surfactant cloud point extraction and preconcentration of organic compounds prior to chromatography and capillary electrophoresis. J Chromatogr A 902(1):251–265
Carabias-Martinez R, Rodriguez-Gonzalo E, Dominguez-Alvarez J, Garcia Pinto C, Hernandez-Mendez J (2003) Prediction of the behaviour of organic pollutants using cloudpoint extraction. J Chromatogr A 1005(1–2):23–34
Carpinteiro J, Quintana JB, Rodriguez I, Carro AM, Lorenzo RA, Cela R (2004) Applicability of solid-phase microextraction followed by on-fiber silylation for the determination of estrogens in water samples by gas chromatography-tandem mass spectrometry. J Chromatogr A 1056(1–2):179–185
Carro AM, Neira I, Rodil R, Lorenzo RA (2002) Speciation of mercury compounds by gas chromatography with atomic emission detection. Simultaneous optimization of a headspace solid-phase microextraction and derivatization procedure by use of chemometric techniques. Chromatographia 56(11/12):733–738
Centineo G, Blanco Gonzalez E, Sanz-Medel A (2004) Multielemental speciation analysis of organometallic compounds of mercury, lead and tin in natural water samples by headspace-solid phase microextraction followed by gas chromatography-mass spectrometry. J Chromatogr A 1034(1–2):191–197
Chen P, Huang S (2006) Determination of ethoprop, diazinon, disulfoton and fenthion using dynamic hollow fiber-protected liquid-phase microextraction coupled with gas chromatography-mass spectrometry. Talanta 69(3):669–675
Chen H, Chen R, Feng R, Li S (2009) Simultaneous analysis of carbamate and organophosphorus pesticides in water by single-drop microextraction coupled with GC-MS. Chromatographia 70(1/2):165–172
Chen H, Chen R, Li S (2010) Low-density extraction solvent-based solvent terminated dispersive liquid-liquid microextraction combined with gas chromatography-tandem mass spectrometry for the determination of carbamate pesticides in water samples. J Chromatogr A 1217(8):1244–1248
Chiang J, Huang S (2008) Simultaneous derivatization and extraction of anilines in waste water with dispersive liquid-liquid microextraction followed by gas chromatography-mass spectrometric detection. Talanta 75(1):70–75
Chimuka L, Msagati TAM, Cukrowska E, Tutu H (2010) Critical parameters in a supported liquid membrane extraction technique for ionizable organic compounds with a stagnant acceptor phase. J Chromatogr A 1217(16):2318–2325
Chisvert A, Roman IP, Vidal L, Canals A (2009) Simple and commercial readily-available approach for the direct use of ionic liquid-based single-drop microextraction prior to gas chromatography. Determination of chlorobenzens in real water samples as model analytical application. J Chromatogr A 1216(9):1290–1295
Chou C, Lee M (2005) Determination of organotin compounds in water by headspace solid phase microextraction with gas chromatography-mass spectrometry. J Chromatogr A 1064(1):1–8
Coquery M, Morin A, Bécue A, Lepot B (2005) Priority substances of the european water framework directive: analytical challenges in monitoring water quality. TrAC Trends Anal Chem 24(2):117–127
Cortada C, Vidal L, Pastor R, Santiago N, Canals A (2009a) Determination of organochlorine pesticides in water samples by dispersive liquid-liquid microextraction coupled to gas chromatography-mass spectrometry. Anal Chim Acta 649(2):218–221
Cortada C, Vidal L, Tejada S, Romo A, Canals A (2009b) Determination of organochlorine pesticides in complex matrices by single-drop microextraction coupled to gas chromatography-mass spectrometry. Anal Chim Acta 638(1):29–35

Cukrowska E, Chimuka L, Nsengimana H, Kwaramba V (2004) Application of supported liquid membrane probe for extraction and preconcentration of organotin compounds from environmental water samples. Anal Chim Acta 523(1):141–147

David F, Sandra P (2007) Stir bar sorptive extraction for trace analysis. J Chromatogr A 1152(1–2):54–69

Dean JR (1998) Extraction methods for environmental analysis. Wiley, Chinchester

Delgado B, Pino V, Ayala JH, González V, Afonso AM (2004) Nonionic surfactant mixtures: a new cloud-point extraction approach for the determination of PAHs in seawater using HPLC with fluorimetric detection. Anal Chim Acta 518(1–2):165–172

Diaz A, Ventura F, Galceran MT (2002) Simultaneous determination of estrogenic short ethoxy chain nonylphenols and their acidic metabolites in water by an in-sample derivatization/solid-phase microextraction method. Anal Chem 74(15):3869–3876

Einsle T, Paschke H, Bruns K, Schrader S, Popp P, Moeder M (2006) Membrane-assisted liquid–liquid extraction coupled with gas chromatography–mass spectrometry for determination of selected polycyclic musk compounds and drugs in water samples. J Chromatogr A 1124(1–2):196–204

El-Beqqali A, Abdel-Rehim M (2007) Quantitative analysis of methadone in human urine samples by microextraction in packed syringe-gas chromatography-mass spectrometry (MEPS-GC-MS). J Sep Sci 30(15):2501–2505

El-Beqqali A, Kussak A, Abdel-Rehim M (2006) Fast and sensitive environmental analysis utilizing microextraction in packed syringe online with gas chromatography-mass spectrometry. J Chromatogr A 1114(2):234–238

El-Beqqali A, Kussak A, Abdel-Rehim M (2007a) Determination of dopamine and serotonin in human urine samples utilizing microextraction online with liquid chromatography/electrospray tandem mass spectrometry. J Sep Sci 30(3):421–424

El-Beqqali A, Kussak A, Blomberg L, Abdel-Rehim M (2007b) Microextraction in packed syringe/liquid chromatography/electrospray tandem mass spectrometry for quantification of acebutolol and metoprolol in human plasma and urine samples. J Liquid Chromatogr Relat Technol 30(4):575–586

Erickson BE (2002) Analyzing the ignored environmental contaminants. Environ Sci Technol 36(7):140A–145A

Fan YC, Hu ZL, Chen ML, Tu CS, Zhu Y (2008) Ionic liquid based dispersive liquid-liquid microextraction of aromatic amines in water samples. Chin Chem Lett 19(8):985–987

Farajzadeh M, Hatami M (2002) Analysis of n-alkanes at sub microgram per liter level after direct solid phase microextraction from aqueous samples. Anal Sci 18(11):1221–1225

Farajzadeh MA, Seyedi SE, Shalamzari MS, Bamorowat M (2009) Dispersive liquid-liquid microextraction using extraction solvent lighter than water. J Sep Sci 32(18):3191–3200

Farajzadeh MA, Djozan D, Bakhtiyari RF (2010a) Use of a capillary tube for collecting an extraction solvent lighter than water after dispersive liquid-liquid microextraction and its application in the determination of parabens in different samples by gas chromatography-flame ionization detection. Talanta 81(4–5):1360–1367

Farajzadeh MA, Djozan D, Nouri N, Bamorowat M, Shalamzari MS (2010b) Coupling stir bar sorptive extraction-dispersive liquid-liquid microextraction for preconcentration of triazole pesticides from aqueous samples followed by GC-FID and GC-MS determinations. J Sep Sci 33(12):1816–1828

Farhadi K, Farajzadeh MA, Matin AA (2009) Liquid chromatographic determination of benomyl in water samples after dispersive liquid-liquid microextraction. J Sep Sci 32(14):2442–2447

Farré Ml, Pérez S, Kantiani L, Barceló D (2008) Fate and toxicity of emerging pollutants, their metabolites and transformation products in the aquatic environment. TrAC Trends Anal Chem 27(11):991–1007

Fattahi N, Samadi S, Assadi Y, Hosseini MRM (2007) Solid-phase extraction combined with dispersive liquid-liquid microextraction-ultra preconcentration of chlorophenols in aqueous samples. J Chromatogr A 1169(1–2):63–69

Fernandez Laespada ME, Calvo Seronero L, Perez Pavon JL, Garcia Pinto C, Moreno Cordero B (2001) Comparative study of silicone and microporous membranes for the separation of some phenols prior to high performance liquid chromatography analysis. J Sep Sci 24(7):526–532

Fiamegos YC, Kefala A, Stalikas CD (2008) Ion-pair single-drop microextraction versus phase-transfer catalytic extraction for the gas chromatographic determination of phenols as tosylated derivatives. J Chromatogr A 1190(1–2):44–51

Fontana AR, Silva MF, Martínez LD, Wuilloud RG, Altamirano JC (2009) Determination of polybrominated diphenyl ethers in water and soil samples by cloud point extraction-ultrasound-assisted back-extraction-gas chromatography–mass spectrometry. J Chromatogr A 1216(20):4339–4346

Fontanals N, Barri T, Bergström S, Jönsson J (2006) Determination of polybrominated diphenyl ethers at trace levels in environmental waters using hollow-fiber microporous membrane liquid–liquid extraction and gas chromatography–mass spectrometry. J Chromatogr A 1133(1–2):41–48

Fu L, Liu X, Hu J, Zhao X, Wang H, Huang C, Wang X (2009) Determination of two pesticides in soils by dispersive liquid-liquid microextraction combined with LC-fluorescence detection. Chromatographia 70(11/12):1697–1701

Garcia-Falcon MS, Cancho-Grande B, Simal-Gandara J (2004) Stirring bar sorptive extraction in the determination of PAHs in drinking waters. Water Res 38(7):1679–1684

Gjelstad A, Andersen TM, Rasmussen KE, Pedersen-Bjergaard S (2007) Microextraction across supported liquid membranes forced by pH gradients and electrical fields. J Chromatogr A 1157(1–2):38–45

Goncalves CM, Silva Joaquim CGEd, Alpendurada MF (2007) Evaluation of the pesticide contamination of groundwater sampled over two years from a vulnerable zone in Portugal. J Agric Food Chem 55(15):6227–6235

Guan W, Xu F, Liu W, Zhao J, Guan Y (2007) A new poly(phthalazine ether sulfone ketone)-coated fiber for solid-phase microextraction to determine nitroaromatic explosives in aqueous samples. J Chromatogr A 1147(1):59–65

Guan W, Wang Y, Xu F, Guan Y (2008) Poly(phthalazine ether sulfone ketone) as novel stationary phase for stir bar sorptive extraction of organochlorine compounds and organophosphorus pesticides. J Chromatogr A 1177(1):28–35

Guillot S, Kelly MT, Fenet H, Larroque M (2006) Evaluation of solid-phase microextraction as an alternative to the official method for the analysis of organic micro-pollutants in drinking water. J Chromatogr A 1101(1–2):46–52

Guo X, Mitra S (2000) On-line membrane extraction liquid chromatography for monitoring semi-volatile organics in aqueous matrices. J Chromatogr A 904(2):189–196

Guo J, Li X, Cao X, Li Y, Wang X, Xu X (2009) Determination of triclosan, triclocarban and methyl-triclosan in aqueous samples by dispersive liquid-liquid microextraction combined with rapid liquid chromatography. J Chromatogr A 1216(15):3038–3043

Gupta M, Pillai AKKV, Jain A, Verma KK (2008) Coupled in-tube and on-fibre solid-phase microextractions for cleanup and preconcentration of organic micropollutants from aqueous samples and analysis by gas chromatography-mass spectrometry. Anal Chim Acta 618(1):61–69

Hansen P (2007) Risk assessment of emerging contaminants in aquatic systems. TrAC Trends Anal Chem 26(11):1095–1099

Hauser B, Popp P (2001a) Combining membrane extraction with mobile gas chromatography for the field analysis of volatile organic compounds in contaminated waters. J Chromatogr A 909(1):3–12

Hauser B, Popp P (2001b) Membrane-assisted solvent extraction of organochlorine compounds in combination with large-volume injection/gas chromatography-electron capture detection. J Sep Sci 24(7):551–560

Hauser B, Schellin M, Popp P (2004) Membrane-assisted solvent extraction of triazines, organochlorine, and organophosphorus compounds in complex samples combined with large-volume injection-gas chromatography/mass spectrometric detection. Anal Chem 76:6029–6038

He Y, Lee HK (1997) Liquid-phase microextraction in a single drop of organic solvent by using a conventional microsyringe. Anal Chem 69(22):4634–4640

He T, Versteeg LAM, Mulder MHV, Wessling M (2004) Composite hollow fiber membranes for organic solvent-based liquid–liquid extraction. J Membr Sci 234(1–2):1–10

He L, Luo X, Xie H, Wang C, Jiang X, Lu K (2009a) Ionic liquid-based dispersive liquid-liquid microextraction followed high-performance liquid chromatography for the determination of organophosphorus pesticides in water sample. Anal Chim Acta 655(1–2):52–59

He L, Wang C, Sun Y, Luo X, Zhang J, Lu K (2009b) Dispersive liquid-liquid microextraction followed by high-performance liquid chromatography for the determination of three carbamate pesticides in water samples. Int J Environ Anal Chem 89(6):439–448

Hernandez F, Portoles T, Pitarch E, Lopez FJ (2007) Target and nontarget screening of organic micropollutants in water by solid-phase microextraction combined with gas chromatography/high-resolution time-of-flight mass spectrometry. Anal Chem 79(24):9494–9504

Herrera-Herrera AV, Asensio-Ramos M, Hernández-Borges J, Rodríguez-Delgado MÁ (2010) Dispersive liquid-liquid microextraction for determination of organic analytes. TrAC Trends Anal Chem 29(7):728–751

Ho H, Lee R, Lee M (2008) Purge-assisted headspace solid-phase microextraction combined with gas chromatography-mass spectrometry for determination of chlorophenols in aqueous samples. J Chromatogr A 1213(2):245–248

Hongxia Yu, Guanyong Su, Yijun Yu, Hongling Liu, Hanhua Lin, Qunsheng Lin (2008) Solid phase micro-extraction analytical method suitable for detection of organic pollutants in water. Patent Application Country: Application: CN; Patent Country: CN Patent 101,303,278, filed 2008, and issued 1112; Patent Application Date: 16 May 2008

Huang Y, Yang Y, Shu YY (2007) Analysis of semi-volatile organic compounds in aqueous samples by microwave-assisted headspace solid-phase microextraction coupled with gas chromatography-electron capture detection. J Chromatogr A 1140(1–2):35–43

Huang X, Qiu N, Yuan D (2008) Direct enrichment of phenols in lake and sea water by stir bar sorptive extraction based on poly(vinylpyridine-ethylene dimethacrylate) monolithic material and liquid chromatographic analysis. J Chromatogr A 1194(1):134–138

Huang X, Lin J, Yuan D, Hu R (2009) Determination of steroid sex hormones in wastewater by stir bar sorptive extraction based on poly(vinylpyridine-ethylene dimethacrylate) monolithic material and liquid chromatographic analysis. J Chromatogr A 1216(16):3508–3511

Huang X, Qiu N, Yuan D, Lin Q (2010) Preparation of a mixed stir bar for sorptive extraction based on monolithic material for the extraction of quinolones from wastewater. J Chromatogr A 1217(16):2667–2673

Huertas C, Morillo J, Usero J, Gracia-Manarillo I (2007) Validation of stir bar sorptive extraction for the determination of 24 priority substances from the European water framework directive in estuarine and sea water. Talanta 72(3):1149–1156

Hung K, Chen B (2007) Application of L3 sponge phase in extraction of polycyclic aromatic hydrocarbons. AICHE J 53(6):1450–1459

Hung K, Chen B, Yu LE (2007) Cloud-point extraction of selected polycyclic aromatic hydrocarbons by nonionic surfactants. Sep Purif Technol 57(1):1–10

Ito R, Kawaguchi M, Sakui N, Honda H, Okanouchi N, Saito K, Nakazawa H (2008) Mercury speciation and analysis in drinking water by stir bar sorptive extraction with in situ propyl derivatization and thermal desorption-gas chromatography-mass spectrometry. J Chromatogr A 1209(1–2):267–270

Ito R, Kawaguchi M, Sakui N, Okanouchi N, Saito K, Seto Y, Nakazawa H (2009) Stir bar sorptive extraction with in situ derivatization and thermal desorption-gas chromatography-mass spectrometry for trace analysis of methylmercury and mercury(II) in water sample. Talanta 77(4):1295–1298

Jagerdeo E, Abdel-Rehim M (2009) Screening of cocaine and its metabolites in human urine samples by direct analysis in real-time source coupled to time-of-flight mass spectrometry after online preconcentration utilizing microextraction by packed sorbent. J Am Soc Mass Spectrom 20(5):891–899

Jeannot MA, Cantwell FF (1996) Solvent microextraction into a single drop. Anal Chem 68(13):2236–2240

Jeannot MA, Cantwell FF (1997) Mass transfer characteristics of solvent extraction into a single drop at the tip of a syringe needle. Anal Chem 69(2):235–239

Jeannot MA, Przyjazny A, Kokosa JM (2010) Single drop microextraction—development, applications and future trends. J Chromatogr A 1217(16):2326–2336

Jönsson JÅ, Mathiasson L (1999) Liquid membrane extraction in analytical sample preparation: I. Principles. TrAC Trends Anal Chem 18(5):318–325

Jönsson JÅ, Mathiasson L (2000) Membrane-based techniques for sample enrichment. J Chromatogr A 902(1):205–225

Jönsson JÅ, Mathiasson L (2001) Membrane extraction in analytical chemistry. J Sep Sci 24(7):495–507

Kawaguchi M, Sakui N, Okanouchi N, Ito R, Saito K, Nakazawa H (2005) Stir bar sorptive extraction and trace analysis of alkylphenols in water samples by thermal desorption with in tube silylation and gas chromatography-mass spectrometry. J Chromatogr A 1062(1):23–29

Kawaguchi M, Ito R, Endo N, Okanouchi N, Sakui N, Saito K, Nakazawa H (2006a) Liquid phase microextraction with in situ derivatization for measurement of bisphenol A in river water sample by gas chromatography–mass spectrometry. J Chromatogr A 1110(1–2):1–5

Kawaguchi M, Ito R, Saito K, Nakazawa H (2006b) Novel stir bar sorptive extraction methods for environmental and biomedical analysis. J Pharm Biomed Anal 40(3):500–508

Kawaguchi M, Ito R, Sakui N, Okanouchi N, Saito K, Nakazawa H (2006c) Dual derivatization-stir bar sorptive extraction-thermal desorption-gas chromatography-mass spectrometry for determination of 17beta -estradiol in water sample. J Chromatogr A 1105(1–2):140–147

Khajeh M, Yamini Y, Hassan J (2006) Trace analysis of chlorobenzenes in water samples using headspace solvent microextraction and gas chromatography/electron capture detection. Talanta 69(5):1088–1094

Khammas ZA (2009) Recent trends for separation and preconcentration in metal ions and organic compounds analysis after cloud-point methodology: developments and analytical applications – a review. Eurasian J Anal Chem 4(1):1–35

Kjelsen IJØ, Gjelstad A, Rasmussen KE, Pedersen-Bjergaard S (2008) Low-voltage electromembrane extraction of basic drugs from biological samples. J Chromatogr A 1180(1–2):1–9

Kolahgar B, Hoffmann A, Heiden AC (2002) Application of stir bar sorptive extraction to the determination of polycyclic aromatic hydrocarbons in aqueous samples. J Chromatogr A 963(1–2):225–230

Kou D, Wang X, Mitra S (2004) Supported liquid membrane microextraction with high-performance liquid chromatography–UV detection for monitoring trace haloacetic acids in water. J Chromatogr A 1055(1–2):63–69

Kozani RR, Assadi Y, Shemirani F, Hosseini MM, Jamali MR (2007) Part-per-trillion determination of chlorobenzenes in water using dispersive liquid-liquid microextraction combined gas chromatography-electron capture detection. Talanta 72(2):387–393

Lambropoulou DA, Albanis TA (2005) Application of hollow fiber liquid phase microextraction for the determination of insecticides in water. J Chromatogr A 1072(1):55–61

Lambropoulou DA, Albanis TA (2007) Liquid-phase micro-extraction techniques in pesticide residue analysis. J Biochem Biophys Methods 70(2):195–228

Lambropoulou DA, Psillakis E, Albanis TA, Kalogerakis N (2004) Single-drop microextraction for the analysis of organophosphorus insecticides in water. Anal Chim Acta 516(1–2):205–211

Lambropoulou DA, Konstantinou IK, Albanis TA (2007) Recent developments in headspace microextraction techniques for the analysis of environmental contaminants in different matrices. J Chromatogr A 1152(1–2):70–96

Lee M, Chiu T, Dou J (2007) Determination of 1,3-dichloro-2-propanol and 3-chloro-1,2-propandiol in soy sauce by headspace derivatization solid-phase microextraction combined with gas chromatography-mass spectrometry. Anal Chim Acta 591(2):167–172

Lee J, Khalilian F, Bagheri H, Lee HK (2009) Optimization of some experimental parameters in the electro membrane extraction of chlorophenols from seawater. J Chromatogr A 1216(45):7687–7693

Leon VM, Llorca-Porcel J, Alvarez B, Cobollo MA, Munoz S, Valor I (2006) Analysis of 35 priority semivolatile compounds in water by stir bar sorptive extraction-thermal desorption-gas chromatography-mass spectrometry. Anal Chim Acta 558(1–2):261–266
Leong M, Huang S (2008) Dispersive liquid-liquid microextraction method based on solidification of floating organic drop combined with gas chromatography with electron-capture or mass spectrometry detection. J Chromatogr A 1211(1–2):8–12
Lepom P, Brown B, Hanke G, Loos R, Quevauviller P, Wollgast J (2009) Needs for reliable analytical methods for monitoring chemical pollutants in surface water under the European water framework directive. J Chromatogr A 1216(3):302–315
Lezamiz J, Jönsson JÅ (2007) Development of a simple hollow fibre supported liquid membrane extraction method to extract and preconcentrate dinitrophenols in environmental samples at ng L−1 level by liquid chromatography. J Chromatogr A 1152(1–2):226–233
Li J, Bai R, Chen B (2004) Preconcentration of phenanthrene from aqueous solution by a slightly hydrophobic nonionic surfactant. Langmuir 20(14):6068–6070
Li Y, Wei G, Wang X (2007) Determination of decabromodiphenyl ether in water samples by single-drop microextraction and RP-HPLC. J Sep Sci 30(16):2698–2702
Li Y, Hu J, Liu X, Fu L, Zhang X, Wang X (2008) Dispersive liquid-liquid microextraction followed by reversed phase HPLC for the determination of decabrominated diphenyl ether in natural water. J Sep Sci 31(13):2371–2376
Li Q, Liu Y, Wang Z, Chen W, Zhang Q (2010) Application of microwave-assisted extraction and its combination with other technologies on the determination of organic pollutants. Huaxue Tongbao 73(2):134–140
Lin H, Sung Y, Huang S (2003) Solid-phase microextraction coupled with high-performance liquid chromatography for the determination of phenylurea herbicides in aqueous samples. J Chromatogr A 1012(1):57–66
Liu H, Dasgupta PK (1996) Analytical chemistry in a drop. Solvent extraction in a microdrop. Anal Chem 68(11):1817–1821
Liu W, Wang H, Guan Y (2004a) Preparation of stir bars for sorptive extraction using sol–gel technology. J Chromatogr A 1045(1–2):15–22
Liu J, Chi Y, Jiang G, Tai C, Peng J, Hu J (2004b) Ionic liquid-based liquid-phase microextraction, a new sample enrichment procedure for liquid chromatography. J Chromatogr A 1026(1–2):143–147
Liu J, Jönsson JA, Mayer P (2005a) Equilibrium sampling through membranes of freely dissolved chlorophenols in water samples with hollow fiber supported liquid membrane. Anal Chem 77(15):4800–4809
Liu W, Hu Y, Zhao J, Xu Y, Guan Y (2005b) Determination of organophosphorus pesticides in cucumber and potato by stir bar sorptive extraction. J Chromatogr A 1095(1–2):1–7
Liu X, Ji Y, Zhang Y, Zhang H, Liu M (2007) Oxidized multiwalled carbon nanotubes as a novel solid-phase microextraction fiber for determination of phenols in aqueous samples. J Chromatogr A 1165(1–2):10–17
Liu X, Li J, Zhao Z, Zhang W, Lin K, Huang C, Wang X (2009a) Solid-phase extraction combined with dispersive liquid-liquid microextraction for the determination for polybrominated diphenyl ethers in different environmental matrices. J Chromatogr A 1216(12):2220–2226
Liu Y, Zhao E, Zhu W, Gao H, Zhou Z (2009b) Determination of four heterocyclic insecticides by ionic liquid dispersive liquid-liquid microextraction in water samples. J Chromatogr A 1216(6):885–891
Liu ZM, Zang XH, Liu WH, Wang C, Wang Z (2009c) Novel method for the determination of five carbamate pesticides in water samples by dispersive liquid-liquid microextraction combined with high performance liquid chromatography. Chin Chem Lett 20(2):213–216
Llorca-Porcel J, Martinez-Sanchez G, Alvarez B, Cobollo MA, Valor I (2006) Analysis of nine polybrominated diphenyl ethers in water samples by means of stir bar sorptive extraction-thermal desorption-gas chromatography-mass spectrometry. Anal Chim Acta 569(1–2):113–118
Lopez-Blanco MC, Blanco-Cid S, Cancho-Grande B, Simal-Gandara J (2003) Application of single-drop microextraction and comparison with solid-phase microextraction and solid-phase

extraction for the determination of alpha – and beta -endosulfan in water samples by gas chromatography-electron-capture detection. J Chromatogr A 984(2):245–252
Lopez-Darias J, German-Hernandez M, Pino V, Afonso AM (2010) Dispersive liquid-liquid microextraction versus single-drop microextraction for the determination of several endocrine-disrupting phenols from seawaters. Talanta 80(5):1611–1618
Lopez-Jimenez FJ, Rubio S, Perez-Bendito D (2008) Single-drop coacervative microextraction of organic compounds prior to liquid chromatography. Theoretical and practical considerations. J Chromatogr A 1195(1–2):25–33
Lord H, Pawliszyn J (2000) Evolution of solid-phase microextraction technology. J Chromatogr A 885:153–193
Luo YZ, Pawliszyn J (2000) Membrane extraction with a sorbent interface for headspace monitoring of aqueous samples using a cap sampling device. Anal Chem 72(5):1058–1063
MacNamara K, Leardi R, McGuigan F (2009) Comprehensive investigation and optimization of the main experimental variables in stir-bar sorptive extraction (SBSE)-thermal desorption-capillary gas chromatography (TD-CGC). Anal Chim Acta 636(2):190–197
Magi E, Di Marina C, Liscio C (2010) Passive sampling and stir bar sorptive extraction for the determination of endocrine-disrupting compounds in water by GC-MS. Anal Bioanal Chem 397(3):1335–1345
Marsin Sanagi M, See HH, Ibrahim WAW, Abu Naim A (2007) Determination of pesticides in water by cone-shaped membrane protected liquid phase microextraction prior to micro-liquid chromatography. J Chromatogr A 1152(1–2):215–219
Marsin Sanagi M, Abidin NAZ, Ibrahim WAW, Aboul-Enein HY (2010) Application of double-phase liquid phase microextraction in the determination of partition coefficients and analysis of pesticides in water samples. Chromatographia 71(5/6):461–467
Matysik S, Matysik F (2009) Microextraction by packed sorbent coupled with gas chromatography – mass spectrometry: application to the determination of metabolites of monoterpenes in small volumes of human urine. Microchim Acta 166(1):109–114
Matz G, Kibelka G, Dahl J, Lennemann F (1999) Experimental study on solvent-less sample preparation methods: membrane extraction with a sorbent interface, thermal membrane desorption application and purge-and-trap. J Chromatogr A 830(2):365–376
Mayer P, Vaes WHJ, Wijnker F, Legierse KCHM, Kraaij R, Tolls J, Hermens JLM (2000) Sensing dissolved sediment pore water concentrations of persistent and bioaccumulative pollutants using disposable solid-phase microextraction fibers. Environ Sci Technol 34(24):5177–5183
Melwanki MB, Fuh M (2008) Dispersive liquid-liquid microextraction combined with semi-automated in-syringe back extraction as a new approach for the sample preparation of ionizable organic compounds prior to liquid chromatography. J Chromatogr A 1198–1199:1–6
Meyers JE, Almirall JR (2005) Analysis of gamma-hydroxybutyric acid (GHB) in spiked water and beverage samples using solid phase microextraction (SPME) on fiber derivatization/gas chromatography-mass spectrometry (GC/MS). J Forensic Sci 50(1):31–36
Mitani K, Fujioka M, Kataoka H (2005) Fully automated analysis of estrogens in environmental waters by in-tube solid-phase microextraction coupled with liquid chromatography-tandem mass spectrometry. J Chromatogr A 1081(2):218–224
Mitra S (ed) (2003) Sample preparation techniques in analytical chemistry. Wiley, Hoboken
Moeder M, Schrader S, Winkler U, Rodil R (2010) At-line microextraction by packed sorbent-gas chromatography–mass spectrometry for the determination of UV filter and polycyclic musk compounds in water samples. J Chromatogr A 1217(17):2925–2932
Montero L, Popp P, Paschke A, Pawliszyn J (2004) Polydimethylsiloxane rod extraction, a novel technique for the determination of organic micropollutants in water samples by thermal desorption-capillary gas chromatography-mass spectrometry. J Chromatogr A 1025(1): 17–26
Montero L, Conradi S, Weiss H, Popp P (2005) Determination of phenols in lake and ground water samples by stir bar sorptive extraction-thermal desorption-gas chromatography-mass spectrometry. J Chromatogr A 1071(1–2):163–169

Montes R, Rodriguez I, Ramil M, Rubi E, Cela R (2009a) Solid-phase extraction followed by dispersive liquid-liquid microextraction for the sensitive determination of selected fungicides in wine. J Chromatogr A 1216(29):5459–5466

Montes R, Rodriguez I, Rubi E, Cela R (2009b) Dispersive liquid-liquid microextraction applied to the simultaneous derivatization and concentration of triclosan and methyltriclosan in water samples. J Chromatogr A 1216(2):205–210

Morales-Cid G, Gebefugi I, Kanawati B, Harir M, Hertkorn N, Rosselló-Mora R, Schmitt-Kopplin P (2009) Automated microextraction sample preparation coupled on-line to FT-ICR-MS: application to desalting and concentration of river and marine dissolved organic matter. Anal Bioanal Chem 395(3):797–807

Moreno Cordero B, Pérez Pavón JL, García Pinto C, Fernández Laespada ME, Carabias Martínez R, Rodríguez Gonzalo E (2000) Analytical applications of membrane extraction in chromatography and electrophoresis. J Chromatogr A 902(1):195–204

Müller S, Möder M, Schrader S, Popp P (2003) Semi-automated hollow-fibre membrane extraction, a novel enrichment technique for the determination of biologically active compounds in water samples. J Chromatogr A 985(1–2):99–106

Nagaraju D, Huang S (2007) Determination of triazine herbicides in aqueous samples by dispersive liquid-liquid microextraction with gas chromatography-ion trap mass spectrometry. J Chromatogr A 1161(1–2):89–97

Nakamura S, Daishima S (2004) Simultaneous determination of alkylphenols and bisphenol A in river water by stir bar sorptive extraction with in situ acetylation and thermal desorption-gas chromatography-mass spectrometry. J Chromatogr A 1038(1–2):291–294

Nakamura S, Daishima S (2005) Simultaneous determination of 64 pesticides in river water by stir bar sorptive extraction and thermal desorption-gas chromatography-mass spectrometry. Anal Bioanal Chem 382(1):99–107

Nerin C, Salafranca J, Aznar M, Batlle R (2009) Critical review on recent developments in solventless techniques for extraction of analytes. Anal Bioanal Chem 393(3):809–833

Nováková L, Vlčková H (2009) A review of current trends and advances in modern bio-analytical methods: chromatography and sample preparation. Anal Chim Acta 656(1–2):8–35

Ochiai N, Sasamoto K, Kanda H, Pfannkoch E (2008) Sequential stir bar sorptive extraction for uniform enrichment of trace amounts of organic pollutants in water samples. J Chromatogr A 1200(1):72–79

Okeyo PD, Snow NH (1998) Analysis of estrogens and anabolic steroids by SPME with on-fiber derivatization and GC/MS. J Microcolumn Sep 10(7):551–556

Olejniczak J, Staniewski J (2007) Enrichment of phenols from water with in-situ derivatization by in-tube solid phase microextraction-solvent desorption prior to off-line gas chromatographic determination with large-volume injection. Anal Chim Acta 588(1):64–72

Padron Sanz C, Sosa Ferrera Z, Santana Rodriguez JJ (2002) Extraction and preconcentration of polychlorinated dibenzo-p-dioxins using the cloud-point methodology. Application to their determination in water samples by high-performance liquid chromatography. Anal Chim Acta 470(2):205–214

Padron Sanz C, Halko R, Sosa Ferrera Z, Santana Rodriguez JJ (2004) Micellar extraction of organophosphorus pesticides and their determination by liquid chromatography. Anal Chim Acta 524(1–2):265–270

Pan L, Chong JM, Pawliszyn J (1997) Determination of amines in air and water using derivatization combined with solid-phase microextraction. J Chromatogr A 773(1 + 2):249–260

Panagiotou AN, Sakkas VA, Albanis TA (2009) Application of chemometric assisted dispersive liquid-liquid microextraction to the determination of personal care products in natural waters. Anal Chim Acta 649(2):135–140

Pedersen-Bjergaard S, Rasmussen KE (2008) Liquid-phase microextraction with porous hollow fibers, a miniaturized and highly flexible format for liquid–liquid extraction. J Chromatogr A 1184(1–2):132–142

Pena MT, Casais MC, Mejuto MC, Cela R (2009) Development of an ionic liquid based dispersive liquid-liquid microextraction method for the analysis of polycyclic aromatic hydrocarbons in water samples. J Chromatogr A 1216(36):6356–6364

Penalver A, Pocurull E, Borrull F, Marce RM (2002) Method based on solid-phase microextraction-high-performance liquid chromatography with UV and electrochemical detection to determine estrogenic compounds in water samples. J Chromatogr A 964(1–2):153–160

Pena-Pereira F, Lavilla I, Bendicho C (2010) Liquid-phase microextraction techniques within the framework of green chemistry. TrAC Trends Anal Chem 29(7):617–628

Peng J, Liu J, Jiang G, Tai C, Huang M (2005) Ionic liquid for high temperature headspace liquid-phase microextraction of chlorinated anilines in environmental water samples. J Chromatogr A 1072(1):3–6

Pinheiro Anselmo dS, de Andrade JB (2009) Development, validation and application of a SDME/GC-FID methodology for the multiresidue determination of organophosphate and pyrethroid pesticides in water. Talanta 79(5):1354–1359

Pino V, Ayala JH, Afonso AM, González V (2002) Determination of polycyclic aromatic hydrocarbons in seawater by high-performance liquid chromatography with fluorescence detection following micelle-mediated preconcentration. J Chromatogr A 949(1–2):291–299

Pongpiachan S (2009) Application of cloud point extraction for the determination of pyrene in natural water. Southeast Asian J Trop Med Public Health 40(2):392–400

Poole CF, Poole SK (2010) Extraction of organic compounds with room temperature ionic liquids. J Chromatogr A 1217(16):2268–2286

Popp P, Keil P, Montero L, Rueckert M (2005) Optimized method for the determination of 25 polychlorinated biphenyls in water samples using stir bar sorptive extraction followed by thermodesorption-gas chromatography/mass spectrometry. J Chromatogr A 1071(1–2):155–162

Prieto A, Zuloaga O, Usobiaga A, Etxebarria N, Fernandez LA (2007) Development of a stir bar sorptive extraction and thermal desorption-gas chromatography-mass spectrometry method for the simultaneous determination of several persistent organic pollutants in water samples. J Chromatogr A 1174(1–2):40–49

Prieto A, Zuloaga O, Usobiaga A, Etxebarria N, Fernandez LA, Marcic C, de Diego A (2008a) Simultaneous speciation of methylmercury and butyltin species in environmental samples by headspace-stir bar sorptive extraction-thermal desorption-gas chromatography-mass spectrometry. J Chromatogr A 1185(1):130–138

Prieto A, Telleria O, Etxebarria N, Fernandez LA, Usobiaga A, Zuloaga O (2008b) Simultaneous preconcentration of a wide variety of organic pollutants in water samples: comparison of stir bar sorptive extraction and membrane-assisted solvent extraction. J Chromatogr A 1214(1–2):1–10

Prieto A, Basauri O, Rodil R, Usobiaga A, Fernández LA, Etxebarria N, Zuloaga O (2010a) Stir-bar sorptive extraction: a view on method optimisation, novel applications, limitations and potential solutions. J Chromatogr A 1217(16):2642–2666

Prieto A, Schrader S, Moeder M (2010b) Determination of organic priority pollutants and emerging compounds in wastewater and snow samples using multiresidue protocols on the basis of microextraction by packed sorbents coupled to large volume injection gas chromatography–mass spectrometry analysis. J Chromatogr A 1217(38):6002–6011

Przyjazny A, Kokosa JM (2002) Analytical characteristics of the determination of benzene, toluene, ethylbenzene and xylenes in water by headspace solvent microextraction. J Chromatogr A 977(2):143–153

Psillakis E, Kalogerakis N (2002) Developments in single-drop microextraction. TrAC Trends Anal Chem 21(1):53–63

Psillakis E, Kalogerakis N (2003a) Developments in liquid-phase microextraction. TrAC Trends Anal Chem 22(9):565–574

Psillakis E, Kalogerakis N (2003b) Hollow-fibre liquid-phase microextraction of phthalate esters from water. J Chromatogr A 999(1–2):145–153

Quintana JB, Reemtsma T (2006) Potential of membrane-assisted solvent extraction for the determination of phosphoric acid triesters in wastewater samples by liquid chromatography–tandem mass spectrometry. J Chromatogr A 1124(1–2):22–28

Quintana JB, Rodil R, Reemtsma T (2004) Suitability of hollow fibre liquid-phase microextraction for the determination of acidic pharmaceuticals in wastewater by liquid chromatography–electrospray tandem mass spectrometry without matrix effects. J Chromatogr A 1061(1):19–26

Rahnama Kozani R, Assadi Y, Shemirani F, Milani Hosseini MR, Jamali MR (2007) Determination of trihalomethanes in drinking water by dispersive liquid-liquid microextraction then gas chromatography with electron-capture detection. Chromatographia 66(1/2):81–86
Rapport DJ, Gaudet CL, Constanza R, Epstein PR, Levins R (eds) (1998) Ecosystem health: principles and practice. Wiley-Blackwell, Chichester
Ratola N, Santos L, Herbert P, Alves A (2006) Uncertainty associated to the analysis of organochlorine pesticides in water by solid-phase microextraction/gas chromatography-electron capture detection-evaluation using two different approaches. Anal Chim Acta 573–574:202–208
Regueiro J, Llompart M, Garcia-Jares C, Garcia-Monteagudo JC, Cela R (2008) Ultrasound-assisted emulsification-microextraction of emergent contaminants and pesticides in environmental waters. J Chromatogr A 1190(1–2):27–38
Reid WR, Mooney HA, Cropper A, Capistrano D, Carpenter SR, Chopra K, Dasgupta P et al (2005) Report of the millennium ecosystem assessment. Ecosystems and human well-being: synthesis. Island Press, Washington, DC
Rezaee M, Assadi Y, Milani Hosseini M, Aghaee E, Ahmadi F, Berijani S (2006) Determination of organic compounds in water using dispersive liquid-liquid microextraction. J Chromatogr A 1116(1–2):1–9
Rezaei F, Bidari A, Birjandi AP, Milani Hosseini MR, Assadi Y (2008) Development of a dispersive liquid-liquid microextraction method for the determination of polychlorinated biphenyls in water. J Hazard Mater 158(2–3):621–627
Richardson SD (2009) Water analysis: emerging contaminants and current issues. Anal Chem 81(12):4645–4677
Richardson SD (2010) Environmental mass spectrometry: emerging contaminants and current issues. Anal Chem 82(12):4742–4774
Rodil R, Schellin M, Popp P (2007) Analysis of polycyclic aromatic hydrocarbons in water and beverages using membrane-assisted solvent extraction in combination with large volume injection–gas chromatography–mass spectrometric detection. J Chromatogr A 1163(1–2):288–297
Rodriguez I, Carpinteiro J, Quintana JB, Carro AM, Lorenzo RA, Cela R (2004) Solid-phase microextraction with on-fiber derivatization for the analysis of anti-inflammatory drugs in water samples. J Chromatogr A 1024(1–2):1–8
Roose P, Brinkman UAT (2005) Monitoring organic microcontaminants in the marine environment: principles, programmes and progress. TrAC Trends Anal Chem 24(11):897–926
Roy G, Vuillemin R, Guyomarch J (2005) On-site determination of polynuclear aromatic hydrocarbons in seawater by stir bar sorptive extraction (SBSE) and thermal desorption GC-MS. Talanta 66(3):540–546
Ruan S, Yang Y, Huang H (2010) Analysis of dimethachlon by ultrasonic-extraction with dispersive liquid-liquid microextraction coupled with gas chromatography. Fenxi Shiyanshi 29(1):111–114
Rubio S, Perez-Bendito D (2003) Supramolecular assemblies for extracting organic compounds. TrAC Trends Anal Chem 22(7):470–485
Said R, Hassan Z, Hassan M, Abdel-Rehim M (2008) Rapid and sensitive method for determination of cyclophosphamide in patients plasma samples utilizing microextraction by packed sorbent online with liquid chromatography-tandem mass spectrometry (MEPS-LC-MS/MS). J Liquid Chromatogr Relat Technol 31(5):683–694
Sampedro MC, Goicolea MA, Unceta N, Sanchez-Ortega A, Barrio RJ (2009) Sequential stir bar extraction, thermal desorption and retention time locked GC-MS for determination of pesticides in water. J Sep Sci 32(20):3449–3456
Sanchez-Avila J, Quintana J, Ventura F, Tauler R, Duarte CM, Lacorte S (2010) Stir bar sorptive extraction-thermal desorption-gas chromatography-mass spectrometry: an effective tool for determining persistent organic pollutants and nonylphenol in coastal waters in compliance with existing directives. Mar Pollut Bull 60(1):103–112
Sandahl M, Ulfsson E, Mathiasson L (2000) Automated determination of vinclozolin at the ppb level in aqueous samples by a combination of microporous membrane liquid-liquid extraction and adsorption chromatography. Anal Chim Acta 424(1):1–5

Santana CM, Ferrera ZS, Rodriguez JJS (2004) Extraction and determination of phenolic derivatives in water samples by using polyoxyethylene surfactants and liquid chromatography with photodiode array detection. J AOAC Int 87(1):166–171
Sarafraz-Yazdi A, Amiri A (2010) Liquid-phase microextraction. TrAC Trends Anal Chem 29(1):1–14
Saraji M, Bakhshi M (2005) Determination of phenols in water samples by single-drop microextraction followed by in-syringe derivatization and gas chromatography-mass spectrometric detection. J Chromatogr A 1098(1–2):30–36
Saraji M, Mirmahdieh S (2009) Single-drop microextraction followed by in-syringe derivatization and GC-MS detection for the determination of parabens in water and cosmetic products. J Sep Sci 32(7):988–995
Schellin M, Popp P (2003) Membrane-assisted solvent extraction of polychlorinated biphenyls in river water and other matrices combined with large volume injection–gas chromatography–mass spectrometric detection. J Chromatogr A 1020(2):153–160
Schellin M, Popp P (2005) Membrane-assisted solvent extraction of seven phenols combined with large volume injection–gas chromatography–mass spectrometric detection. J Chromatogr A 1072(1):37–43
Schellin M, Popp P (2006) Miniaturized membrane-assisted solvent extraction combined with gas chromatography/electron-capture detection applied to the analysis of volatile organic compounds. J Chromatogr A 1103(2):211–218
Schurek J, Pulkrabova J, Hajslova J, Lahoutifard N (2008) Application potential of microextraction in packed syringe coupled with gas chromatography time-of-flight mass spectrometry in analysis of brominated flame retardants in waste water. Organohalogen Compounds 70:2143–2146
Segal A, Górecki T, Mussche P, Lips J, Pawliszyn J (2000) Development of membrane extraction with a sorbent interface–micro gas chromatography system for field analysis. J Chromatogr A 873(1):13–27
Serodio P, Nogueira JMF (2004) Multi-residue screening of endocrine disrupters chemicals in water samples by stir bar sorptive extraction-liquid desorption-capillary gas chromatography-mass spectrometry detection. Anal Chim Acta 517(1–2):21–32
Serodio P, Nogueira JMF (2006) Considerations on ultra-trace analysis of phthalates in drinking water. Water Res 40(13):2572–2582
Serodio P, Cabral MS, Nogueira JMF (2007) Use of experimental design in the optimization of stir bar sorptive extraction for the determination of polybrominated diphenyl ethers in environmental matrices. J Chromatogr A 1141(2):259–270
Seronero LC, Fernandez Laespada ME, Perez Pavon JL, Moreno Cordero B (2000) Cloud point preconcentration of rather polar compounds: application to the high-performance liquid chromatographic determination of priority pollutant chlorophenols. J Chromatogr A 897(1+2):171–176
Shariati-Feizabadi S, Yamini Y, Bahramifar N (2003) Headspace solvent microextraction and gas chromatographic determination of some polycyclic aromatic hydrocarbons in water samples. Anal Chim Acta 489(1):21–31
Sheikhloie H, Saber-Tehrani M, Abrumand-Azar P, Waqif-Husain S (2009) Analysis of tributyltin and triphenyltin in water by ionic liquid-headspace single-drop microextraction then HPLC with fluorimetric detection. Acta Chromatogr 21(4):577–589
Shen G, Lee HK (2002) Hollow fiber-protected liquid-phase microextraction of triazine herbicides. Anal Chem 74(3):648–654
Shen Y, Pawliszyn J (2001) Catalytic reductions in membrane extraction as sample preparation for gas chromatographic analysis. J Sep Sci 24(7):623–626
Shen Y, Sun F, Dong J (2010) Determination of profenofos in water using dispersive liquid-liquid microextraction-high performance liquid chromatography. Fenxi Huaxue 38(4):551–554
Shu YY, Wang SS, Tardif M, Huang Y (2003) Analysis of polychlorinated biphenyls in aqueous samples by microwave-assisted headspace solid-phase microextraction. J Chromatogr A 1008(1):1–12

Sikalos TI, Paleologos EK (2005) Cloud point extraction coupled with microwave or ultrasonic assisted back extraction as a preconcentration step prior to gas chromatography. Anal Chem 77(8):2544–2549

Singh V, Jain A, Verma KK (2003) Solid phase microextraction of residues of organophosphorus pesticides from aqueous samples for their determination by gas chromatography-mass spectrometry. Indian J Chem Sect A Inorg Bio-Inorg Phys Theor Anal Chem 42A(12):2993–2999

Sobhi HR, Yamini Y, Esrafili A, Abadi RHHB (2008) Suitable conditions for liquid-phase microextraction using solidification of a floating drop for extraction of fat-soluble vitamins established using an orthogonal array experimental design. J Chromatogr A 1196–1197:28–32

Sosa Ferrera Z, Padron Sanz C, Mahugo Santana C, Santana Rodriguez JJ (2004) The use of micellar systems in the extraction and pre-concentration of organic pollutants in environmental samples. TrAC Trends Anal Chem 23(7):469–479

Stege PW, Sombra LL, Messina GA, Martinez LD, Silva MF (2009) Environmental monitoring of phenolic pollutants in water by cloud point extraction prior to micellar electrokinetic chromatography. Anal Bioanal Chem 394(2):567–573

Supelco (2004) Manufacturer data sheet. Supelco, Bellefonte, USA

Taechangam P, Scamehorn JF, Osuwan S, Rirksomboon T (2009) Effect of nonionic surfactant molecular structure on cloud point extraction of phenol from wastewater. Colloids Surf A Physicochem Eng Aspects 347(1–3):200–209

Tani H, Kamidate T, Watanabe H (1997) Micelle-mediated extraction. J Chromatogr A 780(1–2):229–241

Tarazona I, Chisvert A, Leon Z, Salvador A (2010) Determination of hydroxylated benzophenone UV filters in sea water samples by dispersive liquid-liquid microextraction followed by gas chromatography-mass spectrometry. J Chromatogr A 1217(29):4771–4778

Torres Padron ME, Sosa Ferrera Z, Santana Rodriguez JJ (2009) Coupling of solid-phase microextraction with micellar desorption and high performance liquid chromatography for the determination of pharmaceutical residues in environmental liquid samples. Biomed Chromatogr 23(11):1175–1185

Trtic-Petrovic T, Dordevic J, Dujakovic N, Kumric K, Vasiljevic T, Lausevic M (2010) Determination of selected pesticides in environmental water by employing liquid-phase microextraction and liquid chromatography-tandem mass spectrometry. Anal Bioanal Chem 397(6):2233–2243

Tsai W, Chuang H, Chen H, Huang J, Chen H, Cheng S, Huang T (2009) Application of dispersive liquid-liquid microextraction and dispersive micro-solid-phase extraction for the determination of quinolones in swine muscle by high-performance liquid chromatography with diode-array detection. Anal Chim Acta 656(1–2):56–62

Unceta N, Sampedro MC, Bakar NKA, Gomez-Caballero A, Goicolea MA, Barrio RJ (2010) Multi-residue analysis of pharmaceutical compounds in wastewaters by dual solid-phase microextraction coupled to liquid chromatography electrospray ionization ion trap mass spectrometry. J Chromatogr A 1217(20):3392–3399

Vazquez PP, Mughari AR, Galera MM (2008) Application of solid-phase microextraction for determination of pyrethroids in groundwater using liquid chromatography with post-column photochemically induced fluorimetry derivatization and fluorescence detection. J Chromatogr A 1188(2):61–68

Vercauteren J, Peres C, Devos C, Sandra P, Vanhaecke F, Moens L (2001) Stir bar sorptive extraction for the determination of ppq-level traces of organotin compounds in environmental samples with thermal desorption-capillary gas chromatography-ICP mass spectrometry. Anal Chem 73(7):1509–1514

Vidal L, Psillakis E, Domini CE, Grane N, Marken F, Canals A (2007) An ionic liquid as a solvent for headspace single drop microextraction of chlorobenzenes from water samples. Anal Chim Acta 584(1):189–195

Vidal L, Chisvert A, Canals A, Salvador A (2010) Ionic liquid-based single-drop microextraction followed by liquid chromatography-ultraviolet spectrophotometry detection to determine typical UV filters in surface water samples. Talanta 81(1–2):549–555

Vignati DAL, Valsecchi S, Polesello S, Patrolecco L, Dominik J (2009) Pollutant partitioning for monitoring surface waters. TrAC Trends Anal Chem 28(2):159–169

Wang L, Cai Y, He B, Yuan C, Shen D, Shao J, Jiang G (2006) Determination of estrogens in water by HPLC–UV using cloud point extraction. Talanta 70(1):47–51

Wang L, Jiang G, Cai Y, He B, Wang Y, Shen D (2007) Cloud point extraction coupled with HPLC-UV for the determination of phthalate esters in environmental water samples. J Environ Sci Beijing China 19(7):874–878

Wang Y, Li Y, Feng J, Sun C (2008) Polyaniline-based fiber for headspace solid-phase microextraction of substituted benzenes determination in aqueous samples. Anal Chim Acta 619(2):202–208

Wei M, Jen J (2002) Determination of aqueous chlorophenols by microwave-assisted headspace solid-phase microextraction and gas chromatography. Chromatographia 55(11/12):701–706

Wei M, Jen J (2007) Determination of polycyclic aromatic hydrocarbons in aqueous samples by microwave assisted headspace solid-phase microextraction and gas chromatography/flame ionization detection. Talanta 72(4):1269–1274

Wei G, Li Y, Wang X (2007) Application of dispersive liquid-liquid microextraction combined with high-performance liquid chromatography for the determination of methomyl in natural waters. J Sep Sci 30(18):3262–3267

Wells M (2003) In: Mitra S (ed) Sample preparation techniques in analytical chemistry. Wiley, Hoboken, New Jersey, USA

Wu J, Ee KH, Lee HK (2005) Automated dynamic liquid-liquid-liquid microextraction followed by high-performance liquid chromatography-ultraviolet detection for the determination of phenoxy acid herbicides in environmental waters. J Chromatogr A 1082(2):121–127

Wu Y, Xia L, Chen R, Hu B (2008) Headspace single drop microextraction combined with HPLC for the determination of trace polycyclic aromatic hydrocarbons in environmental samples. Talanta 74(4):470–477

Wu Q, Li Y, Wang C, Liu Z, Zang X, Zhou X, Wang Z (2009) Dispersive liquid-liquid microextraction combined with high performance liquid chromatography-fluorescence detection for the determination of carbendazim and thiabendazole in environmental samples. Anal Chim Acta 638(2):139–145

Xia J, Xiang B, Zhang W (2008) Determination of metacrate in water samples using dispersive liquid-liquid microextraction and HPLC with the aid of response surface methodology and experimental design. Anal Chim Acta 625(1):28–34

Xiao Q, Hu B, Yu C, Xia L, Jiang Z (2006) Optimization of a single-drop microextraction procedure for the determination of organophosphorus pesticides in water and fruit juice with gas chromatography-flame photometric detection. Talanta 69(4):848–855

Xie H, He L, Wu X, Fan L, Lu K, Wang M, Meng D (2008) Determination of phorate in water using dispersive liquid-liquid microextraction coupled with gas chromatography. Fenxi Huaxue 36(11):1543–1546

Xie S, Paau MC, Li CF, Xiao D, Choi MMF (2010) Separation and preconcentration of persistent organic pollutants by cloud point extraction. J Chromatogr A 1217(16):2306–2317

Xiong J, Hu B (2008) Comparison of hollow fiber liquid phase microextraction and dispersive liquid-liquid microextraction for the determination of organosulfur pesticides in environmental and beverage samples by gas chromatography with flame photometric detection. J Chromatogr A 1193(1–2):7–18

Xu L, Basheer C, Lee HK (2007) Developments in single-drop microextraction. J Chromatogr A 1152(1–2):184–192

Yamini Y, Hojjati M, Haji-Hosseini M, Shamsipur M (2004) Headspace solvent microextraction: a new method applied to the preconcentration of 2-butoxyethanol from aqueous solutions into a single microdrop. Talanta 62(2):265–270

Yamini Y, Reimann CT, Vatanara A, Jönsson JÅ (2006) Extraction and preconcentration of salbutamol and terbutaline from aqueous samples using hollow fiber supported liquid membrane containing anionic carrier. J Chromatogr A 1124(1–2):57–67

Yang Z, Maruya KA, Greenstein D, Tsukada D, Zeng EY (2008) Experimental verification of a model describing solid phase microextraction (SPME) of freely dissolved organic pollutants in sediment porewater. Chemosphere 72(10):1435–1440

Yao B, Yang L (2007) Ultrasonic assisted cloud point extraction of polyaromatic hydrocarbons. Sep Sci Technol 42(8):1843–1858

Yao B, Yang L (2008a) Pilot-scale ultrasonic assisted cloud point extraction of polycyclic aromatic hydrocarbons from polluted water. Sep Sci Technol 43(6):1564–1580
Yao B, Yang L (2008b) Stirring-assisted cloud-point extraction of polycyclic aromatic hydrocarbons. Ind Eng Chem Res 47(11):3949–3956
Yao B, Yang L, Hu Q, Akita S (2007) Cloud point extraction of polycyclic aromatic hydrocarbons in aqueous solution with silicone surfactants. Chin J Chem Eng 15(4):468–473
Yao G, Guan W, Xu F, Wang H, Guan Y (2008) Analysis of phenolic compounds in aqueous samples by gas chromatography coupled with headspace solid-phase microextraction using poly(phthalazine ether sulfone ketone) coated fiber. Se Pu 26(5):590–594
Yao C, Pitner WR, Anderson JL (2009) Ionic liquids containing the tris(pentafluoroethyl)trifluorophosphate anion: a new class of highly selective and ultra hydrophobic solvents for the extraction of polycyclic aromatic hydrocarbons using single drop microextraction. Anal Chem 81(12):5054–5063
Yu Y, Pawliszyn J (2004) On-line monitoring of breath by membrane extraction with sorbent interface coupled with CO2 sensor. J Chromatogr A 1056(1–2):35–41
Yu H, Guanyong S, Yijun Y, Hongling L, Hanhua L, Qunsheng L (2008) Solid phase micro-extraction analytical method suitable for detection of organic pollutants in water, Patent 20080516
Zambonin CG, Aresta A, Nilsson T (2002) Analysis of organochlorine pesticides by solid-phase microextraction followed by gas chromatography-mass spectrometry. Int J Environ Anal Chem 82(10):651–657
Zang X, Wang C, Gao S, Zhou X, Wang Z (2008) Analysis of pyrethroid pesticides in water samples by dispersive liquid-liquid microextraction coupled with gas chromatography. Fenxi Huaxue 36(6):765–769
Zarei AR (2007) Cloud point formation based on mixed micelle in the presence of electrolyte for extraction, preconcentration, and spectrophotometric determination of trace amounts of hydrazine in water and biological samples. Anal Biochem 369(2):161–167
Zhang J, Lee HK (2006) Application of liquid-phase microextraction and on-column derivatization combined with gas chromatography-mass spectrometry to the determination of carbamate pesticides. J Chromatogr A 1117(1):31–37
Zhao L, Lee HK (2001) Application of static liquid-phase microextraction to the analysis of organochlorine pesticides in water. J Chromatogr A 919(2):381–388
Zhao R, Lao W, Xu X (2004) Headspace liquid-phase microextraction of trihalomethanes in drinking water and their gas chromatographic determination. Talanta 62(4):751–756
Zhao R, Yuan J, Li H, Wang X, Jiang T, Lin J (2007) Nonequilibrium hollow-fiber liquid-phase microextraction with in situ derivatization for the measurement of triclosan in aqueous samples by gas chromatography-mass spectrometry. Anal Bioanal Chem 387(8):2911–2915
Zhao F, Li J, Zeng B (2008) Coupling of ionic liquid-based headspace single-drop microextraction with GC for sensitive detection of phenols. J Sep Sci 31(16–17):3045–3049
Zhao F, Lu S, Du W, Zeng B (2009a) Ionic liquid-based headspace single-drop microextraction coupled to gas chromatography for the determination of chlorobenzene derivatives. Microchim Acta 165(1–2):29–33
Zhao R, Diao C, Chen Q, Wang X (2009b) Sensitive determination of amide herbicides in environmental water samples by a combination of solid-phase extraction and dispersive liquid-liquid microextraction prior to GC-MS. J Sep Sci 32(7):1069–1074
Zhao R, Wang X, Sun J, Wang S, Yuan J, Wang X (2010) Trace determination of triclosan and triclocarban in environmental water samples with ionic liquid dispersive liquid-phase microextraction prior to HPLC-ESI-MS-MS. Anal Bioanal Chem 397(4):1627–1633
Zhou Q, Ye C (2008) Ionic liquid for improved single-drop microextraction of aromatic amines in water samples. Microchim Acta 162(1–2):153–159
Zhou Q, Pang L, Xiao J (2009a) Trace determination of dichlorodiphenyltrichloroethane and its main metabolites in environmental water samples with dispersive liquid-liquid microextraction in combination with high performance liquid chromatography and ultraviolet detector. J Chromatogr A 1216(39):6680–6684

Zhou Q, Pang L, Xie G, Xiao J, Bai H (2009b) Determination of atrazine and simazine in environmental water samples by dispersive liquid-liquid microextraction with high performance liquid chromatography. Anal Sci 25(1):73–76

Zorita S, Barri T, Mathiasson L (2007) A novel hollow-fibre microporous membrane liquid–liquid extraction for determination of free 4-isobutylacetophenone concentration at ultra trace level in environmental aqueous samples. J Chromatogr A 1157(1–2):30–37

Zorita S, Hallgren P, Mathiasson L (2008) Steroid hormone determination in water using an environmentally friendly membrane based extraction technique. J Chromatogr A 1192(1):1–8

Zorita S, Mårtensson L, Mathiasson L (2009) Occurrence and removal of pharmaceuticals in a municipal sewage treatment system in the south of Sweden. Sci Total Environ 407(8):2760–2770

# Chapter 5
# Oxidation of Amino Acids, Peptides, and Proteins by Chlorine Dioxide. Implications for Water Treatment

**Virender K. Sharma and Mary Sohn**

## Contents



**Abstract** Amino acids, proteins, and peptides occur ubiquitously in natural waters. Although these nitrogen compounds are not directly toxic, harmful byproducts can be formed by reactions with chemical oxidants during water treatment processes. Chlorination and chloramination of water that contains amino acids and other naturally occurring organic matter indeed produces toxic substances, often referred to as disinfection byproducts. The main advantage of using $ClO_2$ over other chlorine containing disinfectants is that the formation of harmful trihalomethanes (THMs), such as chloroform, during water treatment processes is minimized. Because $ClO_2$ is an effective and promising alternative to other chlorine containing disinfectants, the chemistry of $ClO_2$ interactions between amino acids, proteins, and peptides should be thoroughly understood, to ensure the safety of potable water supplies.

This chapter presents an overview of the aqueous chemistry of $ClO_2$ and its reactivity with amino acids, peptides, and proteins. The kinetics and products of the reactions are reviewed. Only a few amino acids have been reported to be reactive

V.K. Sharma (✉) • M. Sohn
Chemistry Department, Florida Institute of Technology,
150 West University Boulevard, Melbourne, FL, 32901, USA
e-mail: vsharma@fit.edu

E. Lichtfouse et al. (eds.), *Environmental Chemistry for a Sustainable World: Volume 2: Remediation of Air and Water Pollution*, DOI 10.1007/978-94-007-2439-6_5,

with $ClO_2$, and have been found to obey first-order kinetics with respect to the concentration of each reactant. The second-order rate constants vary from $10^{-2}$ to $10^7$ $M^{-1}$ $s^{-1}$ with an order of reactivity: cysteine > tyrosine > tryptophan > histidine > proline. Cysteine sulfonic acid and cystine are products of the oxidation of cysteine by $ClO_2$ in acidic and basic media, respectively. In reactions of histidine, tryptophan, and tyrosine with $ClO_2$, different products are obtained depending on the molar ratios of $ClO_2$ and the individual amino acid. Products also vary depending on whether oxygen is present in the reaction mixture. With excess molar amounts of $ClO_2$ relative to amino acids, low molecular weight compounds are obtained. The oxidation of bovine serum albumin and glucose-6-phosphate dehydrogenase (G6PD) by $ClO_2$ suggests the denaturing of proteins through attack on tryptophan and tyrosine residues. This information is especially important with respect to an understanding of the inactivation of microbes by $ClO_2$.

**Keywords** Kinetics • Products • Degradation • Amino acids • Proteins • Peptides • Chlorine dioxide • Disinfection • Chlorine dioxide • Disinfection byproducts • Oxidation kinetics • Trihalomethanes • Water treatment • NDMA (N-nitrosodimethylamine) • Glutathione • Protein denturation • Cysteine • Tyrosine • Tryptophan

## List of Abbreviations

| | |
|---|---|
| BSA | bovine serum albumin |
| DBPs | disinfection byproducts |
| DOC | dissolved organic carbon |
| G6PD | glucose-6-phosphate dehydrogenase |
| NOM | natural organic matter |
| PBS | phosphate buffer saline |
| THMs | trihalomethanes |

## 5.1 Introduction

### *5.1.1 Amino Acids, Peptides and Proteins Found in Natural Waters*

Amino acids are the building blocks of peptides and proteins and are characterized by a carboxylic acid group and an amino group, both bonded to the same carbon atom, referred to as the α-carbon (Abdelmoez et al. 2007). Amino acids are linked by peptide bonds to form either short chain peptides or longer chain polypeptides while the term protein is usually reserved for naturally occurring polypeptides that have definite three-dimensional structures under physiological conditions.

There are only 20 amino acids commonly found in proteins and all except for glycine, which has two hydrogen atoms bound to the α-carbon, are L-isomers (Creighton 1993). Proteins make up the majority of the dry mass of cells and play a major role in many biological processes (Alberts et al. 1994).

In industry, amino acids have many uses and applications including taste enhancement, animal feed, and medicinal uses and may therefore be found in industrial, commercial and domestic effluents. Amino acids are also constituents of nitrogen containing organic compounds, such as humic substances and proteins which contribute significantly to the organic matter found in surface waters, wastewaters, and biosolids (Bader and Hoigne 1982; Simat and Steinhart 1998; Higgins et al. 2008).

Amino acids are ubiquitous environmental constituents and have been found in diverse water sources (Thurman 1985; Chinn and Barrett 1999, 2000). In natural waters, amino acids typically account for 1–3% of the dissolved organic carbon (DOC) and the higher the level of DOC, the higher the concentration of amino acids which may occur as free amino acids, amino acids found as part of humic substances, or as bound amino acids e.g. part of polypeptides, proteins or other nitrogenous organic compounds (Thurman 1985). In surface waters, free amino acids have been found in concentrations ranging from ng to $\mu g\ L^{-1}$, while peptides and proteins were found at $mg\ L^{-1}$ levels (Lytle and Perdue 1981; Thurman 1985; Duguet et al. 1988; Chinn and Barrett 1999, 2000). The reported concentrations of total dissolved amino acids in shallow and deep groundwater as 20–100 $\mu g\ L^{-1}$ and 121–367 $\mu g\ L^{-1}$, respectively, with glycine, alanine, serine and aspartic acid dominating both samples (Spitzy 1982). The lowest average amino concentrations were determined in seawater and groundwater, followed in oligotrophic lakes, rivers, eutrophic lakes and marshes, while those waters with highest DOC levels, i.e. interstitial waters, had the highest levels (Thurman 1985).

Amino acids are expressed as $RCH(NH_3^+)COOH$, where based on the nature of the R groups, they may be classified as aliphatic, aromatic, or sulfur-containing amino acids. Most of the free amino acids usually undergo ionization of the carboxylate and amino groups, which may be described as follows:

$$RCH\left(NH_3^+\right)COOH \Leftrightarrow H^+ + RCH\left(NH_3^+\right)COO^- \qquad pK_{a1} \approx 2 \tag{5.1}$$

$$RCH\left(NH_3^+\right)COO^- \Leftrightarrow H^+ + RCH\left(NH_2\right)COO^- \qquad pK_{a2} \approx 8.7-10.5 \tag{5.2}$$

Algal cells also contain a wide range of organic nitrogen compounds including proteins, polysaccharides, and amino sugars (Fimmen et al. 2008). Extracellular enzymatic activity of these compounds produces amino acids and peptides, which are of great concern because of their contribution to the formation of disinfection byproducts (DBPs) (Lee and Westerhoff 2009). Chlorination or chloramination of water that contains amino acids and humic substances have been shown to produce toxic cyanogen chloride (Stelmaszynska and Zgliczynski 1978; Pattison and Davies 2001; Karpel vel Leitner et al. 2002; Boreen et al. 2008; Huang et al. 2009). Chlorination of aspartic acid, tyrosine and tryptophan has been shown to form dichloroacetonitrile (Trehy and Bieber 1981; Bieber and Trehy 1983; Trehy et al. 1986).

Dissolved organic nitrogen substances can also react with chlorine to give organo-chloramines (Shang et al. 2005; Ramos et al. 2008).

Peptides in water are of greater concern than free amino acids because chlorinated products of peptides are relatively more stable than chlorinated derivatives of amino acids (Margerum et al. 1994; Yang and Shang 2004; Abdelmoez et al. 2007; Na and Olson 2007). For example, the half-life of chloroalanylphenylalanine was determined to be 111 h at pH 7.0 and 23°C (Na and Olson 2007). Cyanobacteria produce various cyclic peptides in addition to toxic peptides such as microcystins (Stewart et al. 2008), and their treatment with free chlorine and inorganic chloramines produce DBPs. Chlorine dioxide ($ClO_2$) is an alternate oxidant that has attracted considerable attention (Sussman and Ward 1979; Aieta and Berg 1986; Lykins and Griese 1986; Graham et al. 1989; Myers 1990; White 1992; Narkis et al. 1995; Sutton et al. 2002).

### *5.1.2 $ClO_2$ Aqueous Chemistry*

Chlorine dioxide ($ClO_2$) in solution has been extensively used for the purification and disinfection of drinking water, sterilization of medical devices, sanitization of food products, bleaching of paper, and the biocidal treatment of cooling water (Tsai et al. 1995; Kim et al. 1999; Gordon and Rosenblatt 2005). The use of gaseous $ClO_2$ has been shown to be effective in treating mold affected buildings and in deactivating *Bacillus anthracis* spores in governmental and commercial buildings (Southwell 2003; Buttner et al. 2004; Wilson et al. 2005). The main advantage of using $ClO_2$ over other chlorine containing disinfectants is that it controls the formation of harmful trihalomethanes (THMs) when reacting with natural organic matter (NOM) (McVeigh et al. 1985; Lykins and Griese 1986; Gordon and Bubnis 1995; Long et al. 1997). For example, chloroform formation is significantly reduced when water is treated with $ClO_2$ as compared to $Cl_2$ (Fig. 5.1). Other halogenated organic byproducts are also greatly reduced (Richardson et al. 1994; Li et al. 1996; Richardson et al. 2003). $ClO_2$ can also degrade precursors of N-nitrosodimethylamine in water (Lee et al. 2007). The formation of nitrosodimethylamine from precursors in synthetic and natural waters was found to be significantly reduced by $ClO_2$ treatment (Lee et al. 2007).

Chlorine dioxide ($ClO_2$) is a yellowish-green gas (formula weight=67.46). In aqueous solution, it can be dissolved up to 20 g $L^{-1}$. Since chlorine dioxide exists as an unstable gas, it has to be produced on-site by use of a mechanical generator. Chlorine dioxide is most commonly generated by reacting sodium chlorite with $Cl_2$ gas (Eq. 5.3) and by mixing sodium chlorite with sodium hypochlorite and hydrochloric acid (Eq. 5.4)

$$2\,NaClO_2 + Cl_2 \rightarrow 2\,ClO_2 + 2NaCl \tag{5.3}$$

$$2\,NaClO_2 + NaOCl + 2\,HCl \rightarrow 2\,ClO_2 + 3\,NaCl + H_2O \tag{5.4}$$

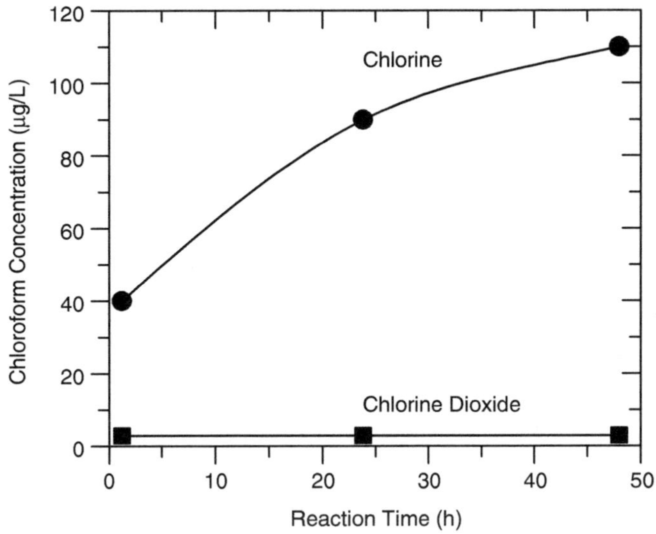


**Fig. 5.1** Chloroform production by chlorine and chlorine dioxide reaction with 5 mg $L^{-1}$ humic acid (Redrawn from Symons et al. 1981)

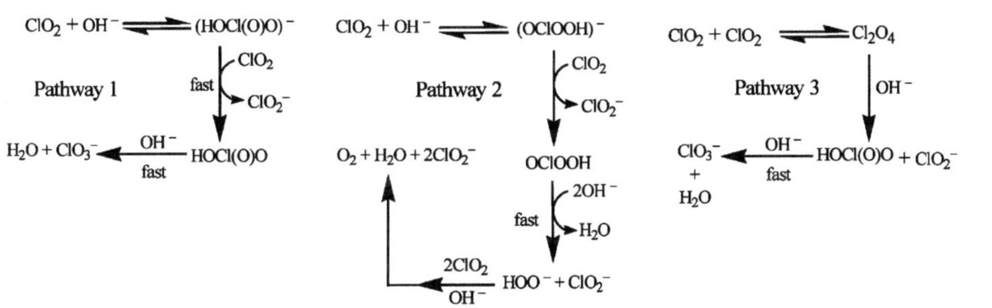


**Fig. 5.2** Pathways for decomposition of chlorine dioxide in basic solution (Reproduced from Odeh et al. 2002 with the permission of American Chemical Society)

In neutral aqueous solution, $ClO_2$ decomposes slowly while its decay is accelerated in basic solution (Odeh et al. 2002). Kinetic studies have shown three concurrent pathways (Fig. 5.2). Pathway 1 is first-order with respect to the concentration of $ClO_2$ and is a disproportionation reaction. Pathway 2 is also first-order in ($ClO_2$) and forms chlorite ($ClO_2^-$) while pathway 3 is second order in ($ClO_2$) and produces equal amounts of $ClO_2^-$ and chlorate ($ClO_3^-$). An adduct of $ClO_2$ with $OH^-$ ion has been suggested as an intermediate for pathways 1 and 2. In pathway 2, $OH^-$ attacks the oxygen atom of $Cl_2$ to result in peroxide intermediates, which finally yields $ClO_2^-$ and $O_2$ as products (Fig. 5.2). At millimolar or higher levels of $ClO_2$, disproportionation reactions dominate to form equal amounts of $ClO_2^-$ and $ClO_3^-$. However, at micromolar levels of $ClO_2$, the ratio of $ClO_2^-$ to $ClO_3^-$ formation increases significantly and pathway 3 becomes important. In drinking water treatment, 50–70% of $ClO_2$ is generally reduced to $ClO_2^-$ and 30% is transformed to $ClO_3^-$ and $Cl^-$ (Werdehoff and Singer 1987; USEPA 1999). Formation of $ClO_2^-$ and $ClO_3^-$ may cause hemolytic anemia and other health effects; hence dosage levels of $ClO_2$ in treating water are kept low. In the United States, dosages ranging from 1.0 to 1.4 mg $L^{-1}$ are used mainly for the preoxidation of surface water (Gates 1998). Higher dosages of $ClO_2$ (>1.4 mg $L^{-1}$) are likely to produce chlorite levels that exceed the United States Environmental Protection Agency standard of 1 mg $L^{-1}$.

## 5.2 Kinetics and Product Studies of $ClO_2$/Amino Acids, Peptides, and Proteins

Chlorine dioxide is a stable free radical and a selective powerful oxidant (Troitskaya et al. 1958).

$$ClO_2 + 4H^+ + 5e^- = Cl^- + 2H_2O \; E^o = 0.936 \text{ V} \quad (5.5)$$

The oxidations carried out by $ClO_2$ usually involve a single electron transfer (Tratnyek and Hoigne 1994). The reactivity of $ClO_2$ with inorganic and organic compounds obeys first-order kinetics with respect to $ClO_2$ and is also first-order with respect to the compound (Hoigne and Bader 1994). The second-order rate constants vary over a wide range ($10^{-5}$–$10^5$ $M^{-1}$ $s^{-1}$). The reactivity of $ClO_2$ with Fe(II), $O_3$, and $H_2O_2$ is high (Hoigne and Bader 1994; Wang et al. 2004).

Under water treatment conditions, aromatic hydrocarbons, carbohydrates, and molecules containing primary and secondary amines, aldehydes, and acetone are un-reactive, however, phenolic and tertiary amino group-containing compounds are reactive with $ClO_2$. The reactivity of $ClO_2$ with these compounds is governed by the pH; phenoxide ion and neutral species of the amine are much more reactive than either the neutral phenol or the protonated amine (Hoigne and Bader 1994; Tratnyek and Hoigne 1994).

The reactivity of amino acids and some peptides with $ClO_2$ have been examined by several workers (Kennaugh 1957; Masschelein and Rice 1979; Noss et al. 1983; Tan et al. 1987). The decay of chlorine dioxide was monitored in the reactions, which followed pseudo-first-order kinetics (Tan et al. 1987). Of the 21 amino acids and three peptides (L-aspartyl-L-phenylalanine methyl ester (aspartame), L-glycyl-L-tryptphan and L-tryptophylglycine), only few were reactive with $ClO_2$ at pH 6.0 (Tan et al. 1987). Among amino acids and peptides, the reactivity of cysteine, tryptophan, tyrosine, L-glycyl-L-tryptphan and L-tryptophylglycine were rapid while histidine, praline, and hydroxyproline had measurable rates. Other amino acids and aspartame did not show reactivity with $ClO_2$; similar to the results of a previous study carried out at pH 7.0 (Noss et al. 1986). The calculated second-order rate constants for the reactions are given in Table 5.1.

Mutagenic studies with the *Salmonella* microsome assay were conducted on the products of the reactions between $ClO_2$ with reacted amino acids and three peptides (Tan et al. 1987). The results suggested that the reaction products of tyrosine,

**Table 5.1** Reaction rate constants for consumption of chlorine dioxide by amino acids and peptides

| Compound | pH | $k$, $M^{-1}$ $s^{-1}$ | Reference |
|---|---|---|---|
| Glycine | 8.0 | $<10^{-5}$ | Hoigne and Bader (1994) |
| Proline | 6.0 | $3.4\times10^{-2}$ | Tan et al. (1987) |
| Hydroxyproline | 6.0 | $6.9\times10^{-2}$ | Tan et al. (1987) |
| Histidine | 6.0 | $5.4\times10^{-2}$ | Tan et al. (1987) |
| Cysteine | 7.0 | $1.0\times10^{7}$ | Ison et al. (2006) |
| Tyrosine | 7.0 | $1.8\times10^{5}$ | Ison et al. (2006) |
| *N*-acetyltyrosine | 6.2 | $3.2\times10^{4}$ | Napolitano et al. (2005) |
| Tryptophan | 7.0 | $3.4\times10^{4}$ | Stewart et al. (2008) |
| Glutathione | 5.9 | $5.8\times10^{2}$ | Ison et al. (2006) |

hydroxyproline and three peptides exerted mutagenic activity against both tester strains of TA98 and TA100 in the presence and absence of rat-liver S9 mix. However, reaction products of the reactions of $ClO_2$ with cysteine, praline and tryptophan (filtrate and precipitate) were determined to be non-mutagenic to both tester strains either in the presence or absence of the S9 mix.

Recently, a stopped-flow technique was applied to determine rate constants of the oxidation of cysteine, tryptophan, and tyrosine by $ClO_2$ as a function of pH (Darkwa et al. 2004; Napolitano et al. 2005; Ison et al. 2006; Stewart et al. 2008). These amino acids form chemical moieties of proteins such as the virus f2 (Noss et al. 1986). The reactions of these amino acids with $ClO_2$ were first-order with both reactants and reported rate constants at pH 7.0 are provided in Table 5.1. Amino acids containing sulfur or aromatic rings were determined to be the most reactive towards $ClO_2$. Cysteine with an –SH group, the most nucleophilic group studied, was the most reactive amino acid with $ClO_2$ (Ison et al. 2006). The oxidation of cysteine by $ClO_2$ has been examined in detail by determining the stoichiometry and products of the reaction (Darkwa et al. 2004; Ison et al. 2006). The reaction stoichiometry ([$ClO_2$]:[Cys]) depended on the pH, which were found to be 1:0.9 and 1:3.7 in acidic and basic media, respectively. At acidic pH, cysteine sulfonic acid was produced while cystine was obtained at alkaline pH. The balanced equations were given as:

$$\begin{aligned}&\text{Acidic pH}: 6ClO_2 + 5H_2NCH(COOH)CH_2SH + 3H_2O \rightarrow \\ &5Cl^- + 5H_2NCH(COOH)CH_2SO_3H + 6H^+\end{aligned} \tag{5.6}$$

$$\begin{aligned}&\text{Basic pH}: 2ClO_2 + 10H_2NCH(COOH)CH_2SH + 3H_2O \rightarrow \\ &2Cl^- + 5H_2NCH(COOH)CH_2SSH_2C(HOOC)HCNH_2 + 2H^+ + 4H_2O\end{aligned} \tag{5.7}$$

The proposed scheme explaining the products of Eqs. 5.6 and 5.7 are shown in Fig. 5.3. The reactive species of cysteine is the thiolate ion and the rate-determine step is proposed to involve an electron abstraction from thiolate ion by $ClO_2$ to give cysteinyl radical. The radical rapidly reacts with another $ClO_2$ molecule to form a cysteinyl-$ClO_2$ adduct, which disproportionates by two pH-dependent pathways to yield products of the reactions given in Eqs. 5.6 and 5.7. The relative reactivity of

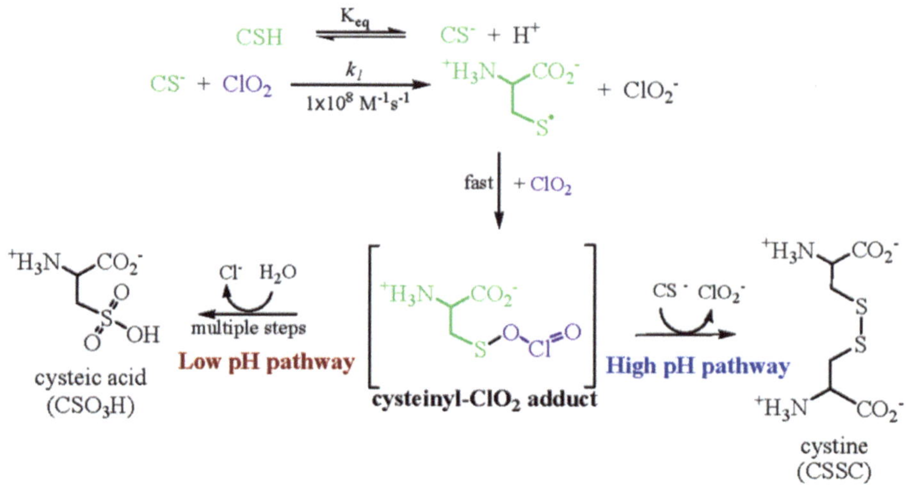


**Fig. 5.3** Proposed mechanism for the initial reaction between $ClO_2$ and subsequent decay of the cysteinyl-$ClO_2$ adduct (Reproduced from Ison et al. (2006) with the permission of American Chemical Society)

glutathione was found to be similar to that of cysteine and was proposed to go through similar steps in reaction with $ClO_2$ (Ison et al. 2006).

In the oxidation of tryptophan and tyrosine by $ClO_2$, similar initial steps to give adducts were demonstrated (Napolitano et al. 2005; Stewart et al. 2008). Adducts were short lived and their decay gave several products. One of the products in the oxidation of tryptophan by $ClO_2$ was identified as *N*-formylalkylnurenine (Stewart et al. 2008), while $ClO_2$ oxidation of tyrosine was found to produce dopaquinone (Napolitano et al. 2005). In a more recent study, a detailed product analysis was carried out for the reaction of $ClO_2$ with tryptophan, tyrosine, and histidine at molar ratios ranging of 0.25–4 in the presence and absence of oxygen (Navalon et al. 2009). The distribution of products varied with molar ratios and the presence of oxygen. For example, the product distribution observed in the reaction of tryptophan with $ClO_2$ is given in Table 5.2.

With an excess of $ClO_2$, the most abundant products were oxalic and fumaric acid in conjunction with minor amounts of 2-aminobenzoic acid, N-formylanthranilic acid, and 2-(2-oxoindolin-3-ylidene) acetic acid. These products of two or four carbon atoms suggest extensive C-C bond breaking occurrence in the tryptophan by $ClO_2$ to yield low molecular weight compounds. However, adding a stoichiometric amount of $ClO_2$ in the absence of oxygen produced products, which were structurally similar to the starting compounds. The product distribution given in Table 5.2 indicates that there are two preferential sites on tryptophan at which attack of $ClO_2$ could take place (Fig. 5.4).

One possible site of attack is on the nitrogen atom of the amino acid, yielding hydroxylamines, oximes, and imines as commonly observed in the oxidation of nitrogeneous organic compounds (Smith and March 2007). Further oxidation of

**Table 5.2** Product distribution observed for the reaction of $ClO_2$ with tryptophan at 25°C under argon or oxygen atmosphere using several reagent molar ratios

| | | |
|---|---|---|
| Tryptophane/$ClO_2$ molar ratio 0.25, under oxygen atmosphere, pH from 6 to 4 | | |
| HOOC–COOH 10% | 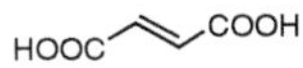 1% | 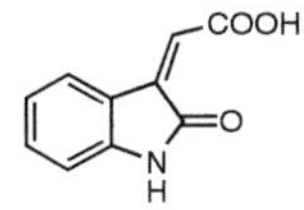 3% |
| 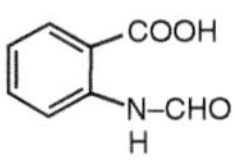 3% | 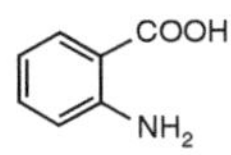 2% | |
| Tryptophane/$ClO_2$ molar ratio 0.5, under oxygen atmosphere, pH from 6 to 4 | | |
| 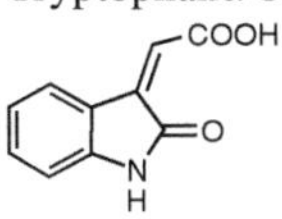 4% | 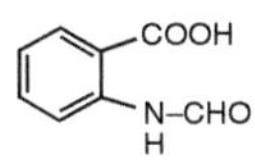 < 1% | 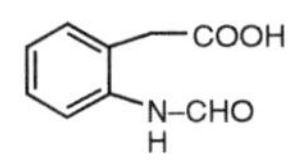 1%, |
| Tryptophane/$ClO_2$ molar ratio 1, under argon atmosphere, pH from 6 to 4 | | |
| 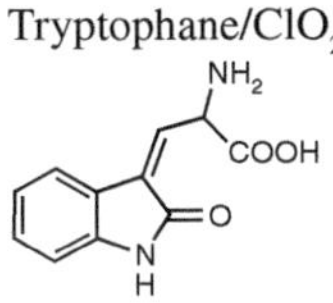 5% | 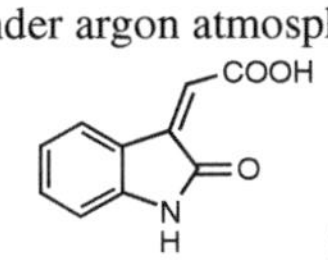 3% | 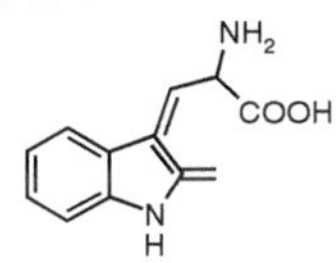 3% |
| Mixture of stereoisomers | | |
| 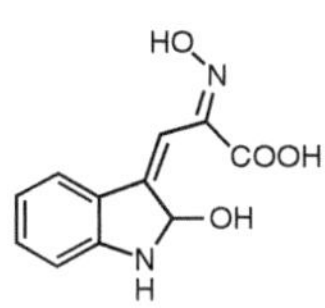 15% | 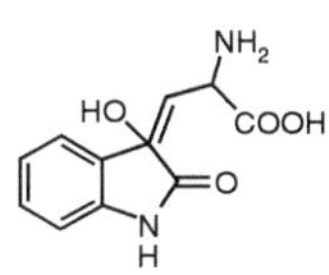 4% | |
| Mixture of stereoisomers | | |
| Tryptophane/$ClO_2$ molar ratio 0.5, under oxygen atmosphere, pH buffered 7.5 | | |
| 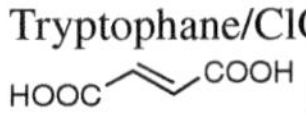 2% | 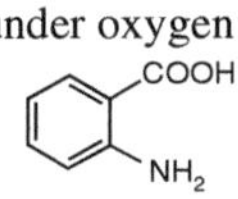 <1% | 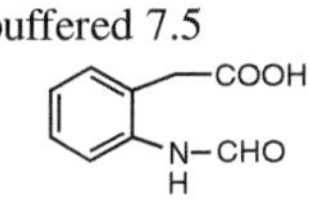 4% |
| 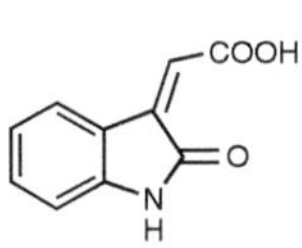 < 0.5% | 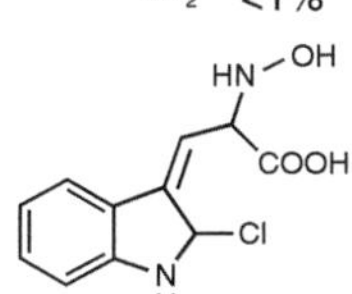 < 0.5% | 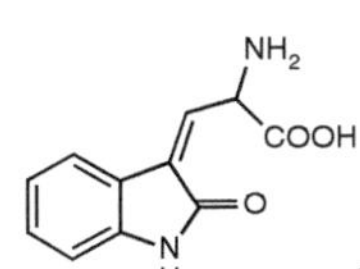 < 0.5% |
| 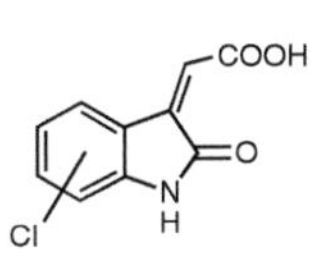 < 0.5% | 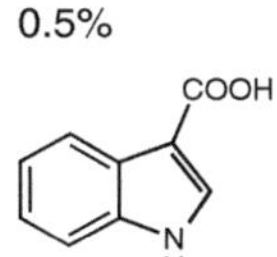 < 0.5% | 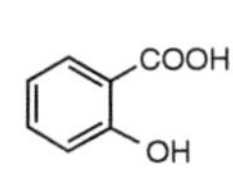 < 0.5% |

The product percentage has been estimated using nitrobenzene as external GC standard and is given with respect to initial moles of tryptophan (Reproduced from Navalon et al. 2009)

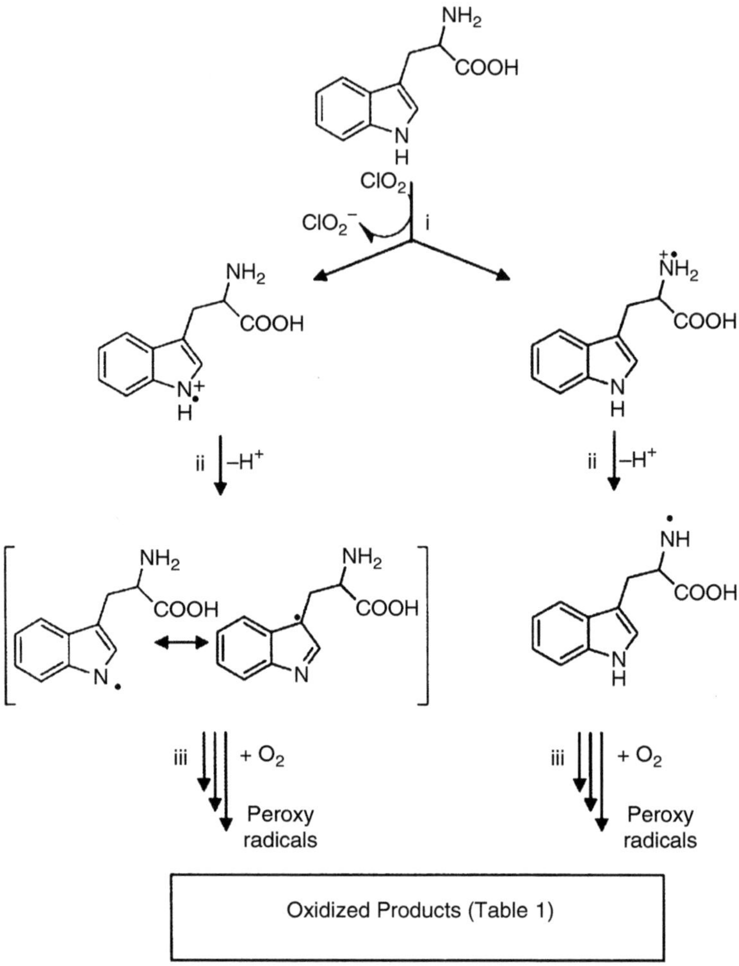


**Fig. 5.4** Proposed reaction scheme for the reaction of tryptophan with $ClO_2$ (Reproduced from Navalon et al. (2009) with the permission of Elsevier, Inc)

these products will ultimately form the observed oxidized products (Navalon et al. 2009). Another possibility is the attack on the two and three positions of the indole ring to cause initiation of hydroxylation of the ring, with subsequent formation of a carbonyl group. This may lead to ring opening; similar to what occurs with the oxidation of indole compounds (Katritzky 2002). It can be concluded that $ClO_2$ attacks initially either at an electron donating aromatic structure or at the amino acid nitrogen, to yield unstable primary products, which continue to degrade to simple, small carbon chains through a series of hydroxylation, ring opening, decarboxylation, and carbonyl formation steps (Navalon et al. 2009).

A study on the denaturing of proteins by $ClO_2$ has been performed using bovine serum albumin (BSA) and glucose-6-phosphate dehydrogenase (G6PD) of baker's yeast (*Saccharomyces cerevisiae*) (Ogata 2007). The denaturation of proteins by

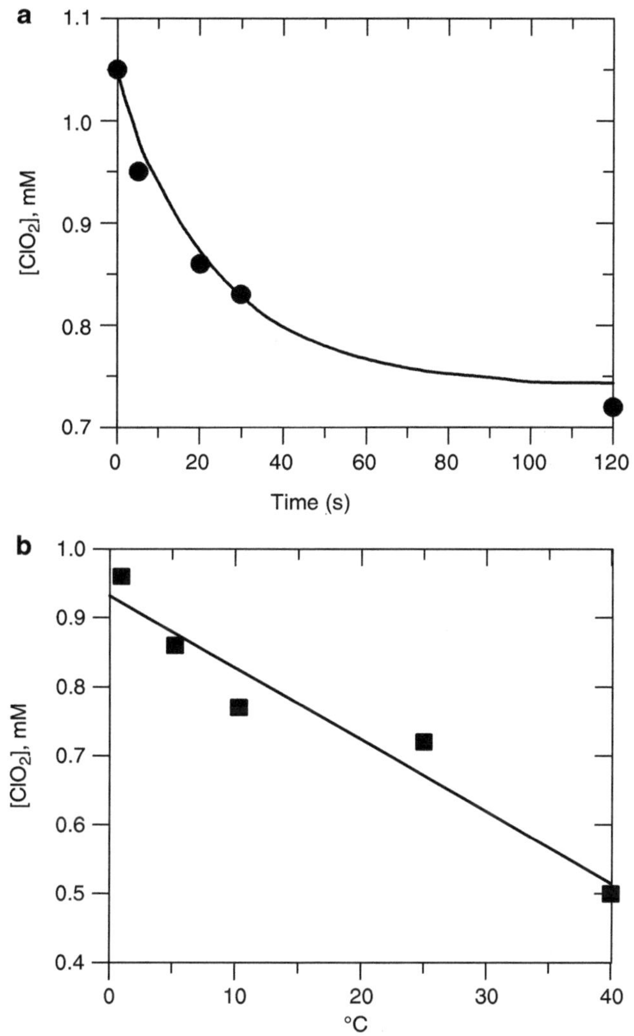


**Fig. 5.5** (**a**) Consumption of $ClO_2$ by the reaction with protein. $ClO_2$ was mixed at 25°C with 0.5 mg/mL bovine serum albumin (BSA). The amount of $ClO_2$ remaining in the reaction mixture was measured by iodometric titration. The curve was obtained by fitting the data to a one-phase exponential decay curve. (**b**) $ClO_2$ concentration remaining after a 2 min reaction as a function of temperature. Each point represents a mean of two experiments (Redrawn from Ogata (2007) with the permission of American Chemical Society)

$ClO_2$ has been shown to include a decrease in the α-helical content as well as in the transition temperature and endothermic transition enthalpy of heat-induced unfolding (Ogata 2007). This was demonstrated experimentally by quantifying the consumption of $ClO_2$ in the reaction of BSA and G6PD (Figs. 5.5a and 5.6a). The concentration of $ClO_2$ decreased exponentially and this decrease was temperature-dependent (Figs. 5.5b and 5.6b).

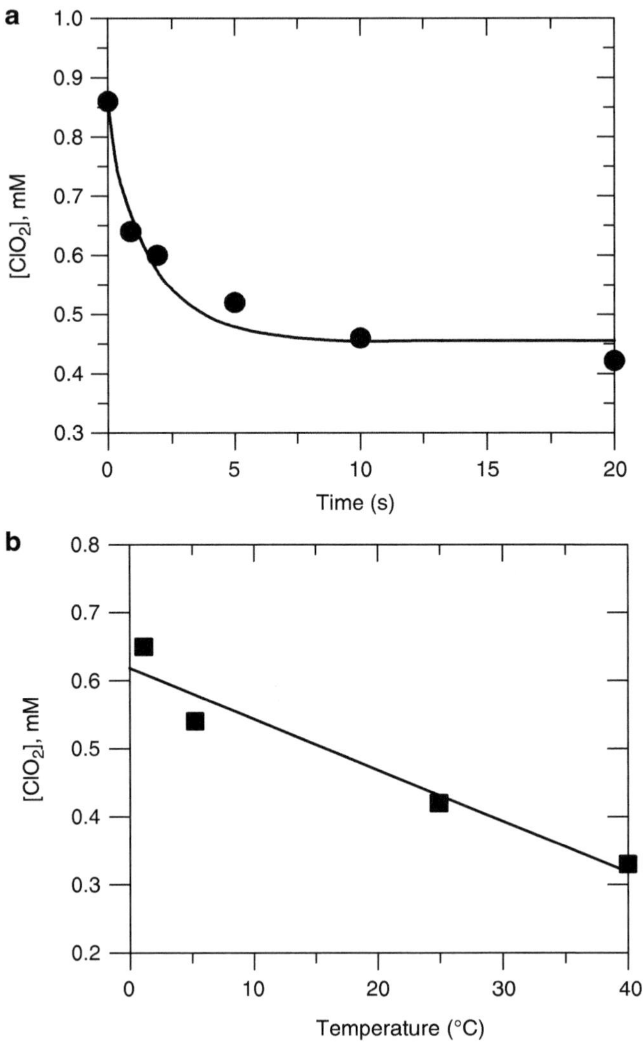


**Fig. 5.6** (**a**) Consumption of $ClO_2$ by the reaction with protein. $ClO_2$ was mixed at 25°C with 0.5 mg/mL glucose-6-phosphate dehydrogenase (G6PD). The amount of $ClO_2$ remaining in the reaction mixture was measured by iodometric titration. The curve was obtained by fitting the data to a one-phase exponential decay curve. (**b**) $ClO_2$ concentration remaining after a 2 min reaction as a function of temperature. Each point represents a mean of two experiments (Redrawn from Ogata (2007) with the permission of American Chemical Society)

The denaturing of proteins by $ClO_2$ was confirmed by the absorbance and fluorescence spectra of proteins with or without $ClO_2$ (Figs. 5.7 and 5.8). Spectral maxima in control experiments (without $ClO_2$ treatment) at 277 nm were observed, which were attributed to aromatic acids, which in this case were tryptophan

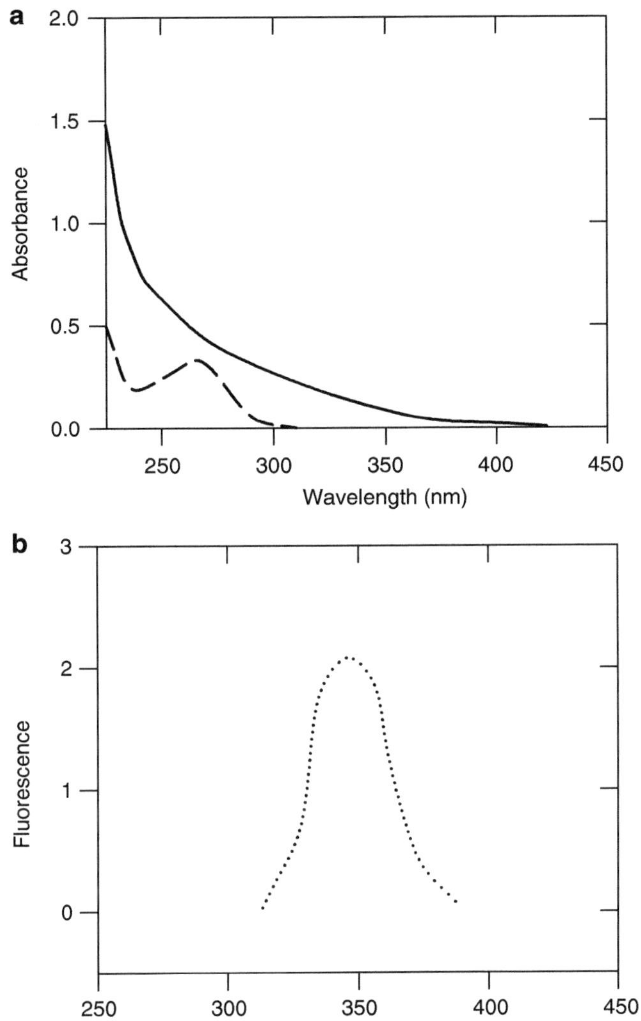


**Fig. 5.7** Absorption (**a**) and fluorescence (**b**) spectra of protein treated with or without $ClO_2$. The absorption and fluorescence (inset, excitation at 280 nm) spectra of BSA. 0.5 mg/mL treated with 1 mM $ClO_2$ in PBS at 25°C for 2 min are shown (*solid lines*). The $ClO_2$ reaction was terminated by 2 mM $Na_2S_2O_3$. As controls, spectra of protein treated with 1 mM $ClO_2$ that had been premixed with 2 mM $Na_2S_2O_3$ are shown (*dotted lines*). Each absorption spectrum was obtained after subtracting that of a mixture of 1 mM $ClO_2$ and 2 mM $Na_2S_2O_3$ as a background. The fluorescence intensity is shown in an arbitrary unit. Experiments were done twice independently, and similar results were obtained (Redrawn from Ogata (2007) with the permission of American Chemical Society)

($\varepsilon_{277nm} = 5,500\ M^{-1}\ cm^{-1}$) and tyrosine ($\varepsilon_{277\ nm} = 1,500\ M^{-1}\ cm^{-1}$). No peaks were observed at 277 nm for $ClO_2$ treated BSA and G6PD proteins (Figs. 5.7 and 5.8), which suggests that residues of tryptophan and tyrosine were destroyed in the treatment process. This was again supported by the fluorescence spectra of $ClO_2$-treated

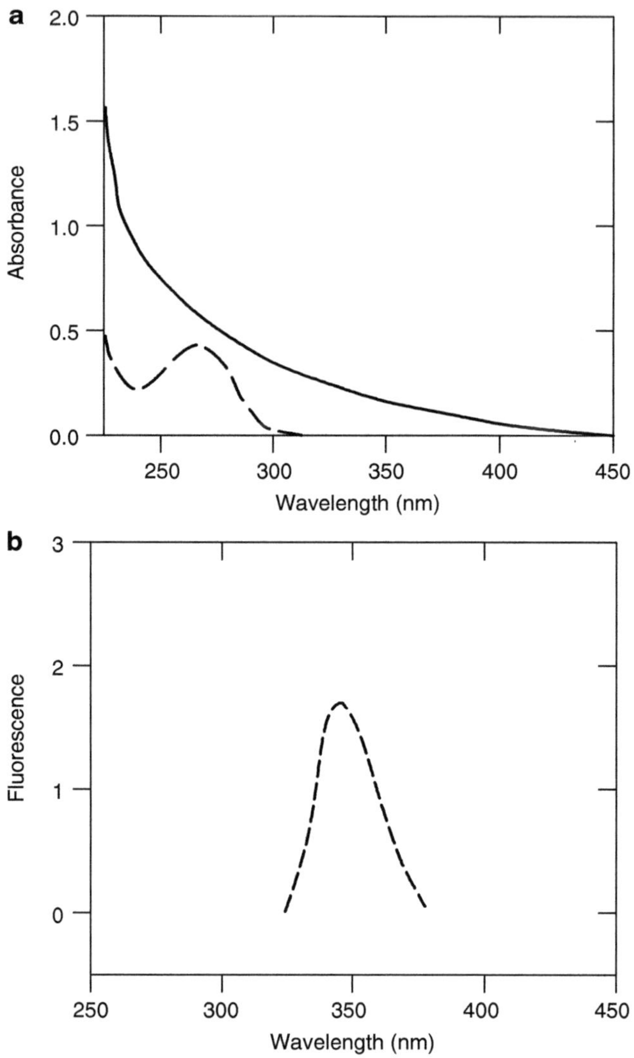


**Fig. 5.8** Absorption (**a**) and fluorescence (**b**) spectra of protein treated with or without $ClO_2$. The absorption and fluorescence (inset, excitation at 280 nm) spectra of G6PD. Each 0.5 mg/mL treated with 1 mM $ClO_2$ in PBS at 25°C for 2 min are shown (*solid lines*). The $ClO_2$ reaction was terminated by 2 mM $Na_2S_2O_3$. As controls, spectra of protein treated with 1 mM $ClO_2$ that had been premixed with 2 mM $Na_2S_2O_3$ are shown (*dotted lines*). Each absorption spectrum was obtained after subtracting that of a mixture of 1 mM $ClO_2$ and 2 mM $Na_2S_2O_3$ as a background. The fluorescence intensity is shown in an arbitrary unit. Experiments were done twice independently, and similar results were obtained (Redrawn from Ogata (2007) with the permission of American Chemical Society)

BSA and G6PD proteins excited at 280 nm. As shown in the inset of Fig. 5.6 (dotted line), proteins showed a characteristic fluorescence emission peak at ~350 nm, which disappeared completely after $ClO_2$ treatment (solid line). The analysis of products demonstrated that the tyrosine residue was transformed into *N*-formylalkylnurenine while the tyrosine residue converted to 3,4-dihydroxyphenylalanine and 2,4,5-trihydroxyphenylalanine after treatment with $ClO_2$ (Ogata 2007).

## 5.3 Conclusion

Among the amino acids tested for reactivity towards $ClO_2$, cysteine, tyrosine and tryptophan displayed the most rapid reactions with $ClO_2$. Cystein having a sulfhydryl group had the highest reactivity with $ClO_2$. The results of the oxidation products of the reaction of $ClO_2$ with histidine, tryptophan, and tyrosine provide insight on the reaction mechanism by which $ClO_2$ oxidizes amino acids present in natural waters. These amino acids are residues of peptides, proteins, and microbes and hence $ClO_2$ is highly effective in degrading these molecules/species. Studies on the oxidative modifications of tryptophan and tyrosine residues in BSA and G6PD provide information on factors involved in the denaturation of proteins by $ClO_2$. Noteworthy results in the literature have demonstrated the potential of $ClO_2$ in reducing the formation of THMs and NDMA. However, other DBPs may form from the oxidative reactions of amino acid containing compounds with $ClO_2$, suggesting a clear need for the evaluation of reaction products and intermediates for reaction of different peptides and proteins with $ClO_2$.

## References

Abdelmoez W, Nakahasi T, Yoshida H (2007) Amino acid transformation and decomposition in saturated subcritical water conditions. Ind Eng Chem Res 46:5286–5294

Aieta EM, Berg JD (1986) A review of chlorine dioxide in drinking water treatment. J Am Water Works Assoc 78:62–72

Alberts B, Bray D, Lewis J, Raff M, Roberts K, Watson JD (1994) Molecular biology of the cell, 3rd edn. Garland Publishing, Inc, New York, p 1294

Bader H, Hoigne J (1982) Determination of ozone in water by the indigo method; a submitted standard method. Ozone Sci Eng 4:169–176

Bieber TI, Trehy MJ (1983) Dihaloacetonitriles in chlorinated natural waters. Water Chlorination Environ Impact Health Eff 4:85–96

Boreen AL, Edhlund BL, Cotner JB, McNeill K (2008) Indirect photodegradation of dissolved free amino acids: the contribution of singlet oxygen and the differential reactivity of DOM from various sources. Environ Sci Technol 42:5492–5498

Buttner MP, Cruz P, Stetzenback LD, Klima-Comba AK, Stevens VL, Cronin TD (2004) Determination of the efficacy of two building decontamination strategies by surface sampling with culture and quantitative PCR analysis. Appl Environ Microbiol 70:4740–4747

Chinn R, Barrett, S (1999) Occurrence of amino acids in drinking water sources. Book of abstracts, ACS National Meeting 217

Chinn R, Barrett S (2000) Occurrence of amino acids in two drinking water sources. ACS Symp Ser 761:96–108

Creighton TE (1993) Proteins structures and molecular properties, 2nd edn. W. H. Freeman and Company, New York, pp 507

Darkwa J, Olojo R, Chikwana E, Simoyi RH (2004) Antioxidant chemistry: oxidation of L-cysteine and its metabolites by chlorite and chlorine dioxide. J Phys Chem A 108:5576–5587

Duguet JP, Anselme C, Mallevialle J (1988) Identification of organonitrogen compounds in waters and their changes in systems for potable water treatment. Water Supply 6:253–263

Fimmen RL, Trouts TD, Richter DD Jr, Vasudevan D (2008) Improved speciation of dissolved organic nitrogen in natural water: amide hydrolysis with fluorescence derivatization. J Environ Sci 20:1273–1280

Gates D (1998) The chlorine dioxide handbook. American Water Works Assocation, Denver

Gordon G, Bubnis B (1995). Chlorine dioxide chemistry issues. Proceedings of the Third International Symposium on Chlorine Dioxide: Drinking water, Process Water And Wastewater Issues, New Orleans, LA.

Gordon G, Rosenblatt A (2005) Chlorine dioxide: the current state of the art. Ozone Sci Eng 27:203–207

Graham NJD, Reynolds G, Buckley D, Perry R, Croll B (1989) Laboratory simulation disinfection regimes for trihalomethane control. J Instn Water Environ Manage 3:604–611

Higgins MJ, Adams G, Chen Y, Erdal Z, Forbes RH Jr, Glindermann D, Hargreaves JR, McEwen D, Murthy SN, Novak JT, Witherspoon J (2008) Role of protein, amino acids, and enzyme activity on odor production from anaerobically digested and dewatered biosolids. Water Environ Res 80:127–135

Hoigne J, Bader H (1994) Kinetics of reactions of chlorine dioxide (OClO) in water – I. rate constants for inorganic and organic compounds. Water Res 28:45–55

Huang J, Graham N, Templeton MR, Zhang Y, Collins C, Nieuwenhuijsen M (2009) A comparison of the role of two blue-green algae in THM and HAA formation. Water Res 43:3009–3018

Ison A, Odeh IN, Margerum DW (2006) Kinetics and mechanisms of chlorine dioxide and chlorite oxidations of cysteine and glutathione. Inorg Chem 45:8768–8775

Katritzky AR (2002) Advances in heterocyclic chemistry. Elsevier Science, San Diego, p 257

Kennaugh J (1957) Action of diaphanol on athropod cuticles. Nature 180:238

Kim JM, Huang T, Marshall MR, Wei C (1999) Chlorine dioxide treatment of seafoods to reduce bacterial loads. Food Sci 64:1089–1093

Lee W, Westerhoff P (2009) Formation of organic chloramines during water disinfection – chlorination versus chloramination. Water Res 43:2233–2239

Lee C, Schmidt C, Yoon J, Gunten UV (2007) Oxidation of N-nitrosodimethylamine (NDMA) precursors with ozone and chlorine dioxide: kinetics and effect on NDMA formation potential. Environ Sci Technol 41:2056–2063

Leitner NKV, Berger P, Legube B (2002) Oxidation of amino groups by hydroxyl radicals in relation to the oxidation degree of the alpha-carbon. Environ Sci Technol 36:3083–3089

Li JW, Yu Z, Cai X, Goa M, Chao F (1996) Trihalomethanes formation in water treated with chlorine dioxide. Water Res 30:2371–2376

Long BW, Miller RF, Rosenblatt AA (1997) Pilot study: the use of high-purity, chlorine-free chlorine dioxide to minimize trihalomethane formation. Chem Oxid 6:126–133

Lykins BW, Griese MH (1986) Using chlorine dioxide for trihalomethane control. J Am Water Works Assoc 78:88–93

Lytle CR, Perdue EM (1981) Free, proteinaceous, and humic-bound amino acids in river water containing high concentrations of aquatic humus. Environ Sci Technol 15:224–228

Margerum DW, Schurter LM, Hobson J, Moore EE (1994) Water chlorination chemistry: nonmetal redox kinetics of chloramine and nitrite ion. Environ Sci Technol 28:331–337

Masschelein WJ, Rice RG (1979) Chlorine dioxide: chemistry and environmental impact of oxychlorine compounds. Ann Harbor Science Publishers, Ann Harbor, pp 190

McVeigh JR, Lee, RG, Kelleher DL (1985) Chlorine dioxide use in pretreatment for control of trihalomethane generation, pp 825–839

Myers AG (1990) Evaluating alternative disinfectants for THM control in small systems. J Am Water Works Assoc 82:77–84
Na C, Olson TM (2007) Relative reactivity of amino acids with chlorine in mixtures. Environ Sci Technol 41:3220–3225
Napolitano MJ, Green BJ, Nicoson JS, Margerum DW (2005) Chlorine dioxide oxidations of tyrosine, N-acetyltyrosine, and dopa. Chem Res Toxicol 18:501–508
Narkis N, Katz A, Orshansky F, Kott Y, Friedland Y (1995) Disinfection of effluents by combination of chlorine dioxide and chlorine. Water Sci Technol 31:105–114
Navalon S, Alvaro M, Garcia H (2009) Chlorine dioxide reaction with selected amino acids in water. J Hazard Mater 164:1089–1097
Noss CI, Dennis WH, Olivieri VP (1983) Reactivity of chlorine dioxide with nucleic acids and proteins. Water Chlorination Environ Impact Health Eff 4:1077–1086
Noss CI, Hauchman FS, Olivieri VP (1986) Chlorine dioxide reactivity with proteins. Water Res 20:351–356
Odeh IN, Francisco JS, Margerum DW (2002) New pathways for chlorine dioxide decomposition in basic solution. Inorg Chem 41:6500–6506
Ogata N (2007) Denaturation of protein by chlorine dioxide: oxidative modification of tryptophan and tyrosine residues. Biochemistry 46:4898–4911
Pattison DI, Davies MJ (2001) Absolute rate constants for the reaction of hypochlorous acid with protein side chains and peptide bonds. Chem Res Toxicol 14:1453–1464
Ramos DR, Garcia MV, Canle LM, Santaballa JA, Furtmuller PG, Obinger C (2008) Myeloperoxidase-catalyzed chlorination: the quest for the active species. J Inorg Biochem 102:1300–1311
Richardson SD, Thruston AD Jr, Collette TW, Schenck Patterson K, Lykins BW Jr, Majetich G, Zhang Y (1994) Multispectral identification of chlorine dioxide disinfection byproducts in drinking water. Environ Sci Technol 28:592–599
Richardson SD, Thruston ADJ, Rav-Acha C, Groisman L, Popilevsky I, Juraev O, Glezer V, McKague AB, Plewa MJ, Wagner ED (2003) Tribromopyrrole, brominated acids, and other disinfection byproducts produced by disinfection of drinking water rich in bromide. Environ Sci Technol 37:3782–3793
Shang C, Qi Y, Xie L, Liu W, Yang X (2005) Kinetics of cyanogen chloride destruction by chemical reduction methods. Water Res 39:2114–2124
Simat TJ, Steinhart H (1998) Oxidation of free tryptophan and tryptophan residues in peptides and proteins. J Agric Food Chem 46:490–498
Smith MB, March J (2007) March's advanced organic chemistry: reactions, mechanisms, and structure. Wiley, Hoboken, p 2357
Southwell KL (2003) Chlorine dioxide dry fumigation in special collection libraries – a case study. Libr Arch Secur 18:39–49
Spitzy A (1982) Amino acids and sugars in deep and shallow groundwater from the hamburg region. Mitteilungen aus dem Geologisch-Palaeontologischen Institut der Universitaet Hamburg 52:743–748
Stelmaszynska T, Zgliczynski JM (1978) N-(2-oxoacyl)amino acids and nitriles as final products of dipeptide chlorination mediated by the Myeloperoxidase/H2O2/Cl- system. Eur J Biochem 92:301–308
Stewart DJ, Napolitano MJ, Bakhmutova-Albert EV, Margerum DW (2008) Kinetics and mechanisms of chlorine dioxide oxidation of tryptophan. Inorg Chem 47:1639–1647
Sussman S, Ward WJ (1979) Chlorine dioxide is an attractive (water) treatment alternative. Water Sew Works 126:120–121
Sutton KM, Elefritz R, Milligan J (2002) THM control in wastewater effluent with chlorine dioxide as a supplementary oxidant. In: Proceedings of the Water Environment Federation, Water Environment Federation, Dallas, pp 152–165
Symons JM, Love OT Jr, Carswell JK (1981) Treatment techniques for controlling trihalomethanes in drinking water. U.S. Environmental Protection Agency, Cincinnati, EPA 600/2-81-156, September.
Tan H, Wheeler WB, Wei C (1987) Reaction of chlorine dioxide with amino acids and peptides: kinetics and mutagenicity studies. Mutat Res 188:259–266

Thurman EM (1985) Organic geochemistry of natural waters. Martinus Nijhoff/Dr. W Junk Publishers, Dordrecht, p 512

Tratnyek PG, Hoigne J (1994) Kinetics of reactions of chlorine dioxide (OClO) in water – II. quantitative structure-activity relationships for phenolic compounds. Water Res 28:57–66

Trehy ML, Bieber TI (1981) Detection, identification, and quantitative analysis of dihaloacetonitriles in chlorinated natural waters In: Advances in the Identification and Analysis of Organic Pollutants in Water Vol 2, L.H. Keith (ed.) Ann Harbor, Michigan. Ann Harbor Science Publishers, Inc. p941–975.

Trehy ML, Yost RA, Miles CJ (1986) Chlorination byproducts of amino acids in natural waters. Environ Sci Technol 20(11):1117–1122

Troitskaya NV, Mishchenko KP, Flis IE (1958) The $ClO_2$ + e.dblharw.ClO2- equilibrium in aqueous solutions at various temperatures. Russ J Phys Chem 33:1614–1617

Tsai L, Higby R, Schade J (1995) Disinfection of poultry chiller water with chlorine dioxide: consumption and byproduct formation. J Agric Food Chem 43:2768–2773

U.S. Environmental Protection Agency, Office of Water (1999) Alternative disinfectants and oxidants guidance manual. EPA 815-R-99-014

Wang L, Odeh IN, Margerum DW (2004) Chlorine dioxide reduction by aqueous iron(II) through outer-sphere and inner-sphere electron-transfer pathways. Inorg Chem 43:7545–7551

Werdehoff KS, Singer PC (1987) Chlorine dioxide effects on THMFP, TOXFP and the formation of inorganic by-products. J Am Water Works Assoc 79:107–113

White GC (1992) Handbook of chlorination and alternative disinfectants, 3rd edn. Wiley, New York, pp 150

Wilson SC, Wu C, Andriychuck LA, Martin JM, Brasel TL, Jumper CA, Straus DC (2005) Effect of chlorine dioxide gas on fungi and mycotoxins associated with sick building syndrome. Appl Environ Microbiol 71:5399–5402

Yang X, Shang C (2004) Chlorination byproduct formation in the presence of humic acid, model nitrogenous organic compounds, ammonia, and bromide. Environ Sci Technol 38:4995–5001

# Chapter 6
# Biodegradation of Azo Dyes from Wastewater

**Pijush Kanti Mondal and Brajesh Chauhan**

## Contents



---

P.K. Mondal (✉)
Environmental Research Laboratory, Department of Applied Chemistry,
Faculty of Engineering and Technology, Aligarh Muslim University,
Aligarh 202 002, UP, India
e-mail: pijushamu@gmail.com

B. Chauhan
Environmental Engineering Section, Department of Civil Engineering,
Faculty of Engineering and Technology, Aligarh Muslim University,
Aligarh 202 002, UP, India

E. Lichtfouse et al. (eds.), *Environmental Chemistry for a Sustainable World: Volume 2: Remediation of Air and Water Pollution*, DOI 10.1007/978-94-007-2439-6_6,

**Abstract** Dyes are synthetic aromatic compounds with various functional groups. Colored dye effluents are highly toxic to the aquatic biota. They alter the symbiotic process by disturbing the natural equilibrium through reduced photosynthetic activity due to the coloration of the water in streams. The control of dyes containing water pollution has thus become of increasing importance in recent years. Amongst the synthetic dyes, azo dyes are the largest group, and account for more than half of the annually produced amounts of dyes. In recent years, several treatments have been proposed for efficient azo dye removal, most of them presenting some limitations such as poor capacity, generate waste products, incomplete mineralization or high operating cost. Nowadays combined anaerobic–aerobic treatments are considered as the most effective route. Removal of azo dyes is based on anaerobic-aerobic treatment for the reductive cleavage of the azo linkages of dyes and for the degradation of the aromatic amines from azo dye cleavage respectively. This chapter reviews the current technologies available for decolourisation of azo dyes and it suggests effective and economically attractive alternatives.

**Keywords** Azo dyes • Biodegradation • Bioreactors • Anaerobic–aerobic • Aromatic amines • Anaerobic • Aerobic • Biomass concentration • Aromatic amines • Toxicity • Wastewater treatment

## 6.1 Introduction

The last couple of years has seen a massive expansion in the chemicals industry; however, with this progress came the concomitant pollution of the environment with a significant range of xenobiotics (dye). Dye compounds form one such category of novel environmental contaminants and are used through a large number of industrial processes, most notably the textile, leather, plastics, food, pharmaceuticals and for manufacturing paints industries. Azo dyes represent the largest and most important group of dyes (Dafale et al. 2008a) mainly due to the simple synthesis. During manufacturing and usage of azo dyes an estimated amount of 10–15% is released into the environment. The release of these compounds into the environment is undesirable, not only because of their color but also because many azo dyes and their breakdown products are toxic, mutagenic, carcinogenic (Dos Santos et al. 2005) and negative influence on the living organisms.

Azo dyes can be reduced by azo reeducates in liver cells, skin surface micro flora, kidney cells and intestinal bacteria and thus aromatic amines (aryl amines) are released. Walker (1970) pointed out that the metabolism of azo dyes in mammal (dog) occurred when the result showed the product of azo dye (sulphanilic acid) was found in the urine of dogs after they consumed Orange I. Over the years epidemiological studies have shown the evidence that long-term occupational exposure to certain aromatic amines such as benzidine, 4-aminobiphenyl and 2-naphthylamine, that are used in dye industries increases the risk of developing cancer. To avoid the bad effect of azo group the concentration (mg/kg) of each harmful aromatic amine in dye products should not exceed 150 mg/kg. Although some azo group consisting compound have some bad effect but it is very important group for colour formation. Most of the colour component are consist of azo group. It's proper utilization can avoid it's bad effect. For this reason as a environmentalist we should know details about it.

Different physical, chemical and biological techniques can be applied to remove dyes from wastewater (Cooper 1993; Southern 1995; Vandevivere et al. 1998; Hao et al. 2000; Robinson et al. 2001). Each technique has its technical and economical limitations. Most physico-chemical dye removal methods have drawbacks because they are expensive, have limited versatility, are greatly interfered by other wastewater constituents, and/or generate waste products that must be handled. Alternatively, biological treatment may present a relatively inexpensive way to remove dyes from wastewater.

Biodegradation is defined as biologically mediated breakdown of chemical compounds. When biodegradation is complete the process is called mineralization i.e. the total breakdown of organic molecules into water, carbon dioxide and/or any other inorganic end products (Bennett and Faison 1997). Biodegradation of the dye by certain groups of fungi during dye removal has been extensively demonstrated. It has been demonstrated that under certain environmental conditions various microorganisms are able to transform azo dyes to noncolored products or even completely mineralize them (Stolz 2001; Xu et al. 2007). These synthetic compounds are resistant to aerobic degradation by bacteria. The strong electron-withdrawing character of the azo group stabilizes these aromatic pollutants against conversions by oxygenases (Kulla et al. 1983; Pagga and Brown 1986; Haug et al. 1991). In studies with more than 100 azo dyes tested in aerobic activated sludge systems, only a few were actually biodegraded (Shaul et al. 1991). Due to their recalcitrance in aerobic environments, the azo dyes eventually end up in anaerobic sediments, shallow aquifers, and groundwaters (Baughman and Perenich 1988) or in the gastrointestinal tracts of higher animals (Chung et al. 1992).

Under anaerobic conditions, azo dyes are readily cleaved via a four-electron reduction at the azo linkage generating aromatic amines (Brown and Laboureur 1983; Brown and Hamburger 1987). Biodegradation of dyes has been the subject of a large number of research papers and several review articles have been published (Walker 1970; Levine 1991; Chung and Cerniglia 1992; Bumpus 1995; Banat et al. 1996; Delee et al. 1998; McMullan et al. 2001; Stolz 2001; Pearce et al. 2003; Forgacs et al. 2004). Especially anaerobic azo dye reduction has been thoroughly investigated and most researchers agree that it is a non-specific and presumably extracellular process, in which reducing equivalents from either biological or chemical source are transferred to the dye.

The bacterial metabolism of azo dyes is initiated in most cases by a reductive cleavage of the azo bond, which results in the formation of colorless aromatic amines. Anaerobic reduction of the azo dyes is relatively easy to achieve, but the products have been found to be biorecalcitrant under anaerobic conditions (Stolz 2001). This observation coupled with the fact that many aromatic amines are completely degraded under aerobic conditions has led to the proposal that anaerobic–aerobic systems might be effective in achieving the complete biodegradation of azo dyes (Supaka et al. 2004).

Azo dyes are generally persistent under aerobic conditions (Isik and Sponza 2004c). However, under anaerobic conditions, they undergo facile reductive fission, yielding colorless aromatic amine compounds (Stolz 2001), that in turn generally require aerobic conditions for their biodegradation (Sponza and Isik 2002a; Ekici et al. 2001). Azo dyes require an anaerobic and an aerobic phase for their complete biodegradation. Combined anaerobic aerobic treatment is therefore the most logical strategy for the complete removal of azo dyes from wastewater.

A lot of research papers have reported the results of combined anaerobic–aerobic bioreactor treatment of azo dye containing wastewaters. This review article summarizes the results of those research studies and discusses the feasibility of combined anaerobic–aerobic treatment for the complete removal of azo dyes from wastewater.

## 6.2 Importance of Biological Degradation

Different physical, chemical and biological techniques can be applied to remove azo dyes from wastewater. These techniques include adsorption, coagulation-flocculation, reverse osmosis, oxidation, Photodegradation, membrane filtration and microbial degradation Each technique has its technical and economical limitations (Table 6.1). Most physico-chemical dye removal methods have drawbacks because they are expensive, have limited versatility, are greatly interfered by other wastewater

**Table 6.1** Current physical, chemical and biological technology for color removal

| Various methods | Advantages | Disadvantages |
|---|---|---|
| Adsorption | Good removal of a wide variety of dyes | Adsorbent requires regeneration or disposal. Expensive capital investment, long residence times, low adsorption capacity, frequent and expensive regeneration |
| Membrane technologies | Removes all dye types and efficiency is better than other | Concentrated sludge production and very expensive |
| Advanced oxidation/ photodegradation | Rapid process | High energy costs and unwanted byproduct released |
| Coagulation/flocculation, reverse osmosis | Economically feasible | High sludge production rate |
| Biodegradation | It is inexpensive and more environmental friendly. Good removal of a wide variety of dyes. Efficient COD and BOD reduction | Long residence times, may require nutrients, very large aeration tanks, lagoons, land areas, many toxic compounds not removed, variable color removal |
| Electrolysis | Not totally mineralized | High power source is required |

constituents, and/or generate waste products that must be handled. Alternatively, biological treatment may present an environmental-friendly methods that are becoming increasingly efficient and cost-effective for the remediation of azo dye-contaminated wastewater.

Microorganisms have the ability to adapt to a variety of environmental conditions. Their versatility makes them useful for various biotechnological applications. One such application is the use of microorganisms for removal of azo dyes from the environment, so-called bioremediation. Biodegradation is the process by which organic substances (like azo dyes) are broken down by other living organisms. Biological azo dye removal techniques are based on microbial biotransformation of dyes. As azo dyes are designed to be stable and long-lasting colorants, they are usually not easily biodegraded. Azo dyes can be degraded aerobically, with oxygen, or anaerobically, without oxygen. A term related to biodegradation is biomineralisation, in which azo dyes are converted into minerals.

Depending upon the number of azo-groups present they are called as monazo, disazo, trisazo, tetrakisazo and polyazo dyes. Polyazo, tetrakisazo and trisazo dyes are very difficult to biodegradation because they are very stable in nature.

## 6.3 Biodegradation of Azo Dyes by Combined Anaerobic-Aerobic Bioreactor

The combined anaerobic and aerobic degradation has been studied for the conversion of azo dyes by numerous researchers (Supaka et al. 2004; Ong et al. 2005; O'Neill et al. 2000b; Rajaguru et al. 2000). However, in most cases clear evidence for a complete biodegradation was not found, mainly due to a lack of proof for mineralization of the aromatic amines. In most experiments only the decolorization and organic load were measured while additional examinations, like the mineralization of the aromatic amines, were omitted. Only few references are available in which there exists real proof of mineralization of an azo dye by a bacterial co-culture under sequential anaerobic/aerobic batch conditions (Albuquerque et al. 2005). An overview of the research papers reviewed is presented in Table 6.2. A distinction was made between the different approaches used to obtain a combined anaerobic–aerobic reactor system.

## 6.4 Anaerobic Degradation of Azo Dyes

Various color removal efficiency listed in Table 6.2 show that removal of color is mainly associated with the anaerobic stage, whereas further decolorization in the aerobic stage is usually limited to a few extra percents. In one of the reactor studies it was shown that the color removal by a two-stage anaerobic–aerobic treatment process was 70% higher than that of a one-stage aerobic treatment process (Minke and Rott 2002).

**Table 6.2** Studies on biological degradation of azo dye-containing wastewater

| Treatment | Findings | Reference |
|---|---|---|
| Anaerobic–aerobic treatment of a dye wastewater by combination of RBC with activated sludge | Anaerobic rotating disc reactor, 7–8 h HRT, efficiency (an) 60–85%. Aerobic rotating disc reactor was used, 4–5 h HRT. Real wastewater (Textile dye wastewater with PVA and LAS as main COD) | Zaoyan et al. (1992) |
| Transformation of azo dye acid yellow 7 by wastewater biofilms | Rotating annular drum, total time 0.16–3 h, acid yellow 7 conc. 0–22 mg/l, synthetic wastewater, peptone, efficiency (an)18–97% | Harmer and Bishop (1992) |
| Anaerobic/aerobic treatment of selected azo dyes in wastewater | Anaerobic fluidized bed reactor, 5–40 mg/l, meat extract and peptone substrate are used, efficiency (an) 20–90% | Seshadri et al. (1994) |
| Study of dye waste treatment using anaerobic–aerobic process | Real wastewater, textile dye wastewater with PVA and LAS as main COD, efficiency (an) 90–95% | Jianrong et al. (1994) |
| Aerobic biodegradation of azo dyes in biofilms | Rotating annular drum, synthetic wastewater, total time 2 h, acid yellow 8, meat extract and peptone, yeast extract and trout chow substrate are used, efficiency(an) 20–90% | Jiang and Bishop (1994) |
| Two stage anaerobic/aerobic treatment of sulfonated azo dyes | Anaerobic fluidized bed reactor, 31 h HRT, swisher aerobic reactor, peptone, YE, trout chow, efficiency(an) ~62%, recovery <1%, detection in MS | FitzGerald and Bishop (1995) |
| Biological treatment of dye wastewaters using an anaerobic–oxic system | UASB, 6–10 h HRT, Aerobic tank, 6.5 h HRT, dye manufacturing factory, mixed with simulated municipal wastewater, efficiency(an) 70–80%, efficiency (a)10–20% | An et al. (1996) |
| Purification of wastewaters containing azo dyes | Anaerobic rotating disc reactor , 31 h HRT, 600 mg/l aerobic rotating disc reactor, 7.5 h HRT, acetate and yeast extract, efficiency(an) 70%, DOC analysis | Sosath et al. (1997) |
| Combined biological and chemical treatment of textile dye-house wastewater using rotating disc reactors | Anaerobic rotating disc reactor , 15 h HRT, 530 mg/l aerobic rotating disc reactor, 7.5 h HRT, acetate and yeast extract, efficiency(an) 90–95%, efficiency (a)69–83%, detection in HPLC | Sosath and Libra (1997) |
| GAC-amended UASB reactor for the stable treatment of toxic textile wastewater | UASB , 24–48 h HRT, aerobic tank reactor, efficiency(an) 80–95%, efficiency (a)69–83%, colored textile wastewater | Kuai et al. (1998) |

| | | |
|---|---|---|
| Anaerobic and aerobic treatment of a simulated textile effluent | Inclined tubular digester, 34–84 h HRT, 1,500 mg/l aerobic tank, acetate and starch, efficiency (an) 62%, detection in HPLC | O'Neill et al. (1999a) |
| Colour in textile effluents—sources, measurement, discharge consents and simulation: a review | UASB, 24–48 h HRT, 1,500 mg/l aerobic tank, acetate and starch, efficiency (an) 78%, detection in HPLC | O'Neill et al. (1999b) |
| Biomineralisation of azo dyes and their breakdown products in anaerobic-aerobic hybrid and UASB reactors | UASB with aerated upper part, total time 1–100 h. Synthetic wastewater, direct yellow-26 conc. 300 mg/l, ethanol co-substrate, efficiency (an) 40–70% | Kalyuzhnyi and Sklyar (2000) |
| Effect of chemical structures of reactive dyes on color removal by an anaerobic–aerobic process | 18 h anaerobic cycle, 5 h aerobic cycle, reactive black 5, glucose and acetate, efficiency (an) 58–63% | Luangdilok and Panswad (2000) |
| Reactive textile dye colour removal in a sequencing batch reactor | 13 h anaerobic cycle, 8 h aerobic cycle, reactive violet 5 conc. 60–100 mg/l, starch, efficiency (an) 30–90% | Lourenco et al. (2000) |
| Biodegradation of azo dyes in a sequential anaerobic–aerobic system | Anaerobic filter, 36 h HRT, aerobic tank 36 h HRT, acid yellow 10 conc. 5–100 mg/l, glucose, efficiency (an) 90–100% recovery 25–50%, efficiency (a) 100% 2 | Rajaguru et al. (2000) |
| Anaerobic–aerobic biotreatment of simulated textile effluent containing varied ratios of starch and azo dye | UASB, 24 h HRT, aerobic tank, 19 h HRT, reactive red 141 conc. 150–750 mg/l, starch and acetate, efficiency (an) 38–59% | O'Neill et al. (2000a) |
| Azo-dye degradation in an anaerobic–aerobic treatment system operating on simulated textile effluent | UASB, 24 h HRT, aerobic tank, 19 h HRT, reactive red 141 conc. 450 mg/l, starch and acetate, efficiency (an) 64% aerobic colour removal-11% | O'Neill et al. (2000b) |
| Degradation of azo dye Mordant Yellow 10 in a sequential anaerobic and bioaugmented aerobic bioreactor | UASB, 25 h HRT, aerobic tank, 10 h HRT, metanil yellow-10 conc. 25–150 mg/l, ethanol, efficiency (an) 100%. Aromatic amine recovery and removal-100% | Tan et al. (1999) |
| Biodegradation of Disperse Blue 79 using sequenced anaerobic/aerobic biofilters | Anaerobic filter, aerobic filter, 12–24 h HRT, 25–150 mg/l glucose and acetate, efficiency (an) 100%, 65% aromatic amine are degraded in aerobic reactor | Cruz and Buitron (2001) |
| Biological degradation of dye-containing wastewater | Anaerobic rotating disc reactor , 15 h HRT, 530 mg/l aerobic rotating disc reactor, 7.5 h HRT, acetate and yeast extract, efficiency(an) 90–95%, efficiency (a)69–83%, detection in HPLC, MS and LCMS | Sarsour et al. (2001) |

(continued)

**Table 6.2** (continued)

| Treatment | Findings | Reference |
|---|---|---|
| Effect of some operational parameters on textile dye biodegradation in a sequential batch reactor | 9–12 h anaerobic cycle, 8–12 h aerobic cycle, reactive violet 5 conc. 60–100 mg/l, starch, efficiency (an) 30–90% | Lourenco et al. (2001) |
| Decolorisation of azo-reactive dye by polyphosphate and glycogen-accumulating organisms in an anaerobic–aerobic sequencing batch reactor | 18 h anaerobic cycle, 5 h aerobic cycle, reactive black 5 conc. 10 mg/l, acetate glucose, efficiency (an) 68–72% | Panswad et al. (2001a) |
| Integrated and sequential anaerobic/aerobic biodegradation of azo dyes | EGSB with oxygenation of recycled effluent, 36–43 h total time required, metanil yellow 10 conc. 59–65 mg/l, ethanol, efficiency (an)100%, detection in HPLC | Tan et al. (1999) and Tan (2001) |
| Anaerobic/aerobic treatment of coloured textile effluents using sequencing batch reactors | 18.5 h anaerobic cycle, 0.5 h aerobic cycle, synthetic wastewater, reactive black 5 conc. 533 mg/l, starch, PVA, CMC, efficiency (an) 86–96% | Shaw et al. (2002) |
| Investigations to anaerobic pre-treatment of highly colored textile industry wastewater | Anaerobic filter, 26–90 h HRT, aerobic biodegradability test, real wastewater, starch and PVA, efficiency (an) 89–92% aerobic colour removal 1–7% | Minke and Rott (2002) |
| Decolorization and azo dye degradation by anaerobic/aerobic sequential process | UASB, 3–30 h HRT, aerobic tank, 10–30 h HRT, reactive black 5 conc. 100 mg/l, glucose, efficiency (an) 82–98% | Sponza and Isik (2002a) |
| Ultimate azo dye degradation in anaerobic/aerobic sequential processes | UASB, 3–30 h HRT, aerobic tank, 10–108 h HRT, reactive black 5 conc. 100 mg/l, glucose, efficiency (an) 82–98%. Detection in UV-visible spectrophotometer | Sponza and Isik (2002b) |
| Attempts to the decolorization and mineralization of the azo dye C.I. Reactive Black 5 | Anaerobic rotating disc, 31 h HRT, aerobic rotating disc, 7.5 h HRT, reactive black 5 conc. 600 mg/l, acetate and yeast extract, efficiency(an) 95%. Aromatic amine recovery 83%, removal-100% | Wiesmann et al. (2002) |
| Aromatic amine degradation in a UASB/CSTR sequential system treating Congo Red dye | UASB, 2.6–26 h HRT, aerobic tank, 10–102 h HRT, direct red 28 conc. 100–4,000 mg/l, glucose, efficiency(an) 97%. Aromatic amine recovery 95%, removal-100% | Isik and Sponza (2003) |
| Decolorization of simulated textile wastewater in an anaerobic–aerobic sequential treatment system | Anaerobic filter, 12–72 h HRT, 50–400 mg/l, aerobic tank, 10 h HRT, reactive red 195, molasses, efficiency (an) 60–100%, aerobic colour removal 15% | Kapdan et al. (2003) |

| | | |
|---|---|---|
| Analysis of secondary metabolite fate during anaerobic–aerobic azo dye biodegradation in a sequential batch reactor | 9–13 h anaerobic cycle, 8–12 h aerobic cycle, reactive violet 5 conc. 100 mg/l, starch, efficiency (an) 90% | Lourenco et al. (2003) |
| The toxicity of textile reactive azo dyes after hydrolysis and decolourisation | Baffled reactor with anaerobic and aerobic compartments, total time 48 + 18 h, reactive black 5, starch and PVA, efficiency 88% | Gottlieb et al. (2003) |
| Two stage biological treatment of a diazo reactive textile dye and the fate of the dye metabolites | Anaerobic rotating disc, 15 h HRT, aerobic rotating disc, 7.5 h HRT, reactive black 5, acetate and yeast extract, efficiency (an)100%, aerobic colour removal 35% | Libra et al. (2004) |
| Anaerobic/aerobic sequential treatment of a cotton textile mill wastewater | UASB, 30 h HRT, aerobic tank, 108 h HRT, cotton mill waste water, efficiency (an) 46–55%, aerobic colour removal-25%, aromatic amines removal 35–90% | Isik and Sponza (2004a) |
| Decolorization of azo dyes under batch anaerobic and sequential anaerobic/aerobic conditions | UASB, 3–30 h HRT, aerobic tank, 10–108 h HRT, reactive black 5 conc. 100 mg/l, glucose, efficiency (an) 98%, aerobic colour removal-20% | Isik and Sponza (2004b) |
| C.I. Direct Black 38 azo dye through decolorization in an anaerobic/aerobic sequential reactor system | UASB, 86.4 h HRT, aerobic tank, 432 h HRT, direct black 38 conc. 3,200 mg/l, glucose, efficiency (an) 81%, aerobic colour removal-13%, aromatic amines recovery 74% and remove 81% | Isik and Sponza (2004c) |
| Biological sulphate reduction and redox mediator effects on azo dye decolourisation in anaerobic–aerobic sequencing batch reactors | 10.5 h anaerobic cycle, 10 h aerobic cycle, 24 h total time, reactive violet 5 conc. 100 mg/l, starch, efficiency (an) 90–99% detection in HPLC, MS, LCMS | Albuquerque et al. (2005) |
| Evaluation of an integrated anaerobic/aerobic SBR system for the treatment of wool dyeing effluents | 0–12 h anaerobic cycle, 8–12 h aerobic cycle, 24 h total time, wool dyeing effluent with azo and anthraquinone dyes, additional colour removal as percentage of the influent colour, detection in UV spectrophotometry | Cabral Goncalves et al. (2005) |
| Reactor performances and fate of aromatic amines through decolorization of Direct Black 38 dye under anaerobic/aerobic sequentials | UASB, 15–16.5 h HRT, aerobic tank, 55–60 h HRT, direct black 38 conc. 100–3,200 mg/l, glucose, efficiency (an) 80–100%, aromatic amines recovery 85–95% and remove 50% | Sponza and Isik (2005) |

(continued)

**Table 6.2** (continued)

| Treatment | Findings | Reference |
|---|---|---|
| Application of anaerobic–aerobic sequential treatment system to real textile wastewater for color and COD removal | Anaerobic filter, 12–72 h HRT, aerobic tank, 10 h HRT, textile wastewater with added glucose and nutrients. efficiency (an) 60–85%, aerobic colour removal 10% | Kapdan and Alparslan (2005) |
| Decolorization of azo dye (Orange II) in a sequential UASB–SBR system | UASB, 24 h HRT, sequential batch reactor, 21.5–24 h HRT, acid yellow 7 conc. 60–300 mg/l, glucose and peptone, efficiency (an) 60–97% | Ong et al. (2005) |
| Degradation of acid orange 7 by a controlled anaerobic–aerobic sequencing batch reactor | Acid orange 7, glucose as the co-substrate in a molar relationship 1:40 (substrate/co-substrate), the efficiency of removal of total organic carbon was between 85% and 90%. 85% efficiency during the anaerobic stage | Buitron et al. (2006) |
| An integrated process for the treatment of CETP wastewater using coagulation, anaerobic and aerobic process | Mixture of waste coming from 525 small-scale industries, color removal of 74% and COD reduction of 75% at a concentration of 10 and 4 g/l, culture Pseudomonas aeroginosa is used | Moosvi and Datta (2007) |
| Reduction and partial degradation mechanisms of azo dye amaranth by Shewanella decolorationis S12 | Naphthylaminesulfonic azo dye amaranth, S. decolorationis strain S12 was used, this is the first demonstration of complete biological reduction and partial mineralization of naphthylaminesulfonic azo dye by S.12 | Pandey et al. 2007 |
| Anaerobic/aerobic treatment of a simulated textile wastewater | UASB, 24 h HRT, sequential batch reactor, 19.17 h HRT, mixing of various dyes of total concentration, CSTR reactor glucose and acetic acid, efficiency (an) 97% | Isik and Sponza (2008) |
| Effect of cycle time on biodegradation of azo dye in sequencing batch reactor | Anaerobic–aerobic sequencing batch reactor (SBR), Remazol brilliant violet 5R, The percentage reductions in color by the anaerobic stage of the SBR were at 72%, 89% and 86% for the 24-h, 12-h and 6-h cycle times, respectively | Cinara et al. (2008) |
| Kinetic study approach of remazol black-B use for the development of two-stage anoxic–oxic reactor for biodegradation of azo dyes by activated consortium | Remazol black-B, The activation energy ($E_a$) and frequency factor ($K_0$) based on the Arrhenius equation was calculated as 11.67 kcal $mol^{-1}$ and $1.57 \times 107$ mg lgMLSS$^{-1}$ $h^{-1}$, respectively. Colour removal of 84% and COD removal 80% was achieved | Dafale et al. (2008a, b) |

| | | |
|---|---|---|
| Assessment of the biodegradability of a monosulfonated azo dye and aromatic amines | (Acid Orange 7) by anaerobic biomass (colour removal of 96%) was achieved in the first sample port of the reactor profiles, COD degradation rate (4.1–2.5 g COD removed $l^{-1}$ $d^{-1}$ for 60 and 300 mg $l^{-1}$ of dye, respectively), suggesting inhibitory effects | Carvalho et al. (2007) |
| Anaerobic decolorization bacteria for the treatment of azo dye in a sequential anaerobic and aerobic MBR | Anaerobic sequencing batch reactor (SBR) combined with an aerobic membrane bioreactor, Reactive black 5, 36 h HRT, efficiency 99% | You and Teng (2009) |
| Biodegradation potential of pure and mixed bacterial cultures for removal of 4-nitroaniline from textile dye wastewater | Acid Red-88, reactive black-5, direct red-81, and disperse orange-3, efficiency 100 μmol/l of 4-nitroaniline within 72 h three bacteria identified as *Acinetobacter sp.*, *Citrobacter freundii* were isolated from enrichment cultures of activated sludge | Khalid et al. (2009) |
| The effect of oxygen on anaerobic color removal of azo dye in a sequencing batch reactor | Remazol brilliant violet 5R, anaerobic-aerobic sequencing batch reactor was used, 96%. anaerobic color removal efficiency, catechol 2,3-dioxygenase and catechol 1,2-dioxygenase, are break down product | Cinar et al. (2009) |
| Hybrid UASFB-aerobic bioreactor for biodegradation of acid yellow-36 in wastewater | Acid yellow-36, UASFB, 24 h HRT, sequential batch reactor, sodium acetate as a co substrate, efficiency (an) 97% | Ahmad et al. (2010) |
| Biodegradation of Reactive blue 13 in a two-stage anaerobic/aerobic fluidized beds system with a *Pseudomonas* sp. isolate | Azo textile dye reactive blue 13, *Pseudomonas* sp. strain L1, an overall color removal of 83.2% and COD removal of 90.7% was achieved at pH 7. A residence time of 70 h and a glucose concentration of 2 g/l, HRT=70 h and $C_{glucose}$=2,000 mg/l. Oxygen was contributing to blocking the azo bond cleavage | Lin et al. (2010) |

These results are in agreement with previously published data on recalcitrance of azo dyes in aerobic sludge environments (Pagga and Brown 1986; Shaul et al. 1991; Ganesh et al. 1994; Ekici et al. 2001). It is reasonable to assume that anaerobic color removal is mainly due to azo dye reduction. The extent of azo dye color removal achieved in the anaerobic stages of the studies listed in Table 6.1 was mostly higher than 70% and in several cases almost 100%. Ranges of anaerobic color removal efficiencies refer to different circumstances tested.

### 6.4.1 Dye Structure

Different color removal efficiencies were investigated according to dye structure. Normally monoazo, disazo dyes were shown better efficiency in synthetic wastewater treatment. In real wastewater treatment the efficiency was slightly low because of various unwanted compound are there. The total mineralization of dyes was dependent on dye structure if the reduced aromatic amine are unsulphonated then aerobic biodegradation are efficient (Tan 2001).

### 6.4.2 Hydraulic Retention Time

The effect of HRT was positively related to the degradation efficiency. Table 6.2 shows that the steady state results obtained at a gradually rising HRT. The overall degradation of dye and co-substrate was enhanced as HRT increased at a certain level, but the enhancement became lower immediately if the HRT increased at a high rate (Isik and Sponza 2004b; Albuquerque et al. 2005). It depends upon system to system.

### 6.4.3 Biomass Concentration

Biomass concentration was measured in terms of volatile suspended solid (VSS). The sludge was quantified gravimetrically by weighing the oven-dried samples at 105°C for 24 h. Oven-dried solid samples were scrapped out from the supports and ignited at 550°C for 2 h to estimate the volatile solid (VSS) content. Several studies reported that when biomass concentration (g/l) decreases then solid retention time will decrease and total degradation efficiency falls (Lourenco et al. 2000). Systems with a higher biomass retention capacity (e.g. upflow anaerobic reactors) might therefore be better suited for azo dye decolorization than systems with a lower biomass retention capacity (e.g. SBR systems). The higher Reactive Black 5 color removal efficiency of an upflow anaerobic sludge blanket (UASB) reactor

(Sponza and Isik 2002a) in comparison to those of the anaerobic phases of SBR systems operated with similar dye concentrations and reaction times (Luangdilok and Panswad 2000; Panswad and Luangdilok 2000; Lourenco et al. 2001; Panswad et al. 2001a) supports this hypothesis.

### 6.4.4 Alternative Electron Acceptors

Biodegradation of azo dyes is an oxidation–reduction reaction in which the dye acts as an electron acceptor. The presence of an alternative electron acceptor may compete with the azo dye for reducing equivalents. The presence of nitrate, a normal constituent of textile processing wastewater, was shown to slow down decolorization (Lourenco et al. 2003; Panswad et al. 2001b). Those results are agreement with previously published data from batch experiments to azo dye decolorization in the presence of nitrate (Carliell et al. 1995) and nitrite (Wuhrmann et al. 1980). In contrast, the presence of sulfate, also a normal constituent of textile-processing wastewater, was not found to significantly affect dye decolorization (Panswad and Luangdilok 2000; Albuquerque et al. 2005), which is in agreement with previously published data from batch experiments to azo dye decolorization in the presence of sulfate (Carliell et al. 1995; Van der Zee et al. 2003a). As sulfate can be biologically reduced to sulfide, a well-known bulk reductant of azo dyes, it has been suggested that its presence may rather stimulate than competitively suppress azo dye decolorization. According to Van der Zee et al. (2003a) Chemical reduction of the dyes due to sulphide was greatly stimulated of which sulfate influenced dye reduction in accordance to biogenic sulfide formation from sulfate reduction. This is also supported by Kim et al. (2008) where addition of sulfide led to an increase in the microbial decolorization of more than 9% in comparison with the case where no sulfide was added.

### 6.4.5 Primary Substrate Concentration

The presence of an electron donor is a prerequisite for azo dye reduction. In theory, the required amount of electron-donating primary substrate is low, four reducing equivalents per azo linkage, i.e. 32 mg COD per mmol monoazo dye. Competition for reducing equivalents by other reactions will increase the required amount of primary substrate. Moreover, the rather slow process of azo dye reduction may benefit kinetically by a higher primary substrate concentration. In the studies listed, the electron-donating primary substrate (biodegradable COD) was usually in large excess of the stoichiometric quantity needed for azo dye reduction. Only in one of the studies Cruz and Buitron (2001) has the reactor been operated for a while under clear substoichiometric conditions, when the reactor influent contained no other electron donor than the dye's dispersing agent. Consequently, the reactor's anaerobic color removal efficiency dropped. Substoichiometric conditions might also explain the low color removal efficiency before raising the level of primary substrate in another reactor system (Sosath and Libra 1997). Two other studies report sub-optimal primary substrate levels in large excess of the stoichiometric level. Kapdan et al. (2003), investigating the decolorization of Reactive Red 195 at a concentration of 100 mg/l (0.1–0.2 mM, assuming that the dye's unknown molecular weight lies between 500 and 1,000 g/mol), report declining color removal efficiencies at substrate concentrations below 3,000 mg/l, which is in large excess of the theoretically necessary 3.2–6.4 mg/l. Likewise, O'Neill et al. (2000a) reported improved color removal efficiencies when doubling the primary substrate concentration, even though the original primary substrate concentration was already 60–300 times in excess of the stoichiometric amount.

### *6.4.6 Primary Substrate Type*

The electron-donating primary substrates used in the reactor studies varied from simple substrates like acetate, ethanol and glucose to more complex ones, including relevant constituents of textile-processing wastewaters like starch, polyvinylalchol (PVA) and carboxymethyl cellulose (CMC). In all cases, azo dye decolorization occurred, which suggests that the process is relatively non-specific with respect to its electron donor. Yet, some primary substrates may be better suitable for delivering reducing equivalents to azo dyes, either because of the substrate itself or because of the microorganisms involved. Three of the studies listed in Table 6.2 report substrate effects. In one of those SBR studies, the anaerobic removal of AO7 color, which was very poor when a starch derivative was used as main COD source, greatly raised when lactate was added to the reactor influent (Albuquerque et al. 2005). Another SBR study reported considerably higher anaerobic color removal efficiencies with acetate and nutrient broth than with glucose, an effect that was subscribed to a better azo dye-reducing capacity of polyphosphate-accumulating microorganisms versus that of glycogen-accumulating microorganisms (Panswad et al. 2001a). In a consequent study by the same research group, the substrate effect, now comparing acetate and glucose both in the presence of nutrient broth, was less pronounced (Panswad et al. 2001b). Therefore, the substrate effect may have been at least partly due to (compounds present in the) nutrient broth and does not necessarily relate completely to the microbial composition of the biomass.

### *6.4.7 Dye Concentration*

Dye concentration is an another factor that may play a vital role in the color removal process As seen in Table 6.2, large variations in dye concentrations have been applied in the reactor studies. In several cases, the applied dye concentrations largely exceed the 10–250 mg/l range of normal concentrations in dye house effluents (O'Neill et al. 1999b). High dye concentrations may negatively affect the anaerobic color removal efficiency, either by exceeding the reactor's biological azo dye reduction capacity or by causing toxicity to the anaerobic biomass. Investigations with different dye concentrations usually reported higher net color removal efficiencies at lower dye concentrations (Seshadri et al. 1994; Luangdilok and Panswad 2000; O'Neill et al. 2000a; Rajaguru et al. 2000; Cruz and Buitron 2001; Kapdan and Oztekin 2003; Sponza and Isik 2005), even though the amount of dye reduced per unit of time may increase with increasing dye concentrations, e.g. as reported by Cruz and Buitron (2001).

### 6.4.8 Dye Toxicity

Toxicity to microorganisms affecting the transforming is obviously an important factor governing the fate and behaviour of a chemical compound in the natural environment. Dye toxicity is defined as the % inhibition of microbial growth at a particular substrate concentration. However, genotoxicity is defined as the gene mutation, chromosomal alteration or DNA damage caused by the substrate. In case of azo dyes, polyazo dyes is most toxic than others due to steric, electronic and hydrodrophobic effects of the substituents.

## 6.5 Aromatic Amines Formation Due to Azo Dye Reduction

The stoichiometric reduction of an azo bond yielding aromatic amines requires two reducing equivalents, which were typically supplied by an electron-donating co substrate. Many reports indicate that electron-donating co substrates are required for azo reduction by bacteria (Van der Zee and Villaverde 2005). It is also likely that co substrates could act as donors of reducing equivalents to specific azo reductases. As has been demonstrated in many research papers, e.g. those summarized by Field et al. (1995) azo dye decolorization due to anaerobic reduction results in the formation of anaerobic amines. In several of the reactor studies listed in Table 6.2, attention has been paid to aromatic amine formation, all of them providing evidence for the formation of aromatic amines under anaerobic conditions, indicating azo dye reduction.

## 6.6 Aerobic Biodegradation of Aromatic Amines

Azo dye reduction leads to the formation of aromatic amines. Aromatic compounds possess large negative resonance energy, resulting in thermodynamic stability. Aromatic amines are generally not degraded and accumulate under anaerobic conditions Field et al. (1995), with the exception of a few aromatic amines characterized by the presence of hydroxyl and/or carboxyl groups (Razo-Flores et al. 1996). Microorganisms, particularly bacteria, have evolved enzyme systems that degrade the benzene structure under aerobic and anoxic conditions (Schink et al. 2000). The aromatic amines are susceptible to mineralization in the periphery of the biofilm, if the right bacteria are present. Mineralization of the aromatic amines by aerobic bacteria and aerobic sludge in treatment plants is more common and, therefore, aerobic conditions are preferable to degrade the accumulated aromatic amine. However, it should be noted that some aromatic amines are readily autoxidized in the presence of oxygen to humic like oligomeric and polymeric structures. The results taken as a whole suggest that many of the aromatic amines from anaerobic cleavage of azo dyes were removed in the consequent aerobic stage. However, some aromatic amines may not be removed.

## 6.7 Recent Advances

Recently researchers have taken up various methods which can enhance the process of dye removal by wild-type of bioreactor or by used various isolated species. Combining compound analysis with toxicity measurements will deliver insight in the size, composition, and potential harm of the recalcitrant fraction, thereby providing the information needed to judge whether discharge can be allowed or whether tertiary treatment (e.g. advanced oxidation, adsorption, coagulation) will have to be applied.

## 6.8 Discussion

Recent literature reviewed in this paper indicates that although a large number of lab-scale studies have been conducted on decolourization of synthetic dye solutions through combined anaerobic-aerobic biodegradation, there is a need to generate relative performance data on industrial effluents. According to the concept of combined anaerobic–aerobic treatment, azo dyes should be removed from the water phase by (anaerobic) reduction followed by (aerobic) oxidation of the dyes' constituent aromatic amines. The anaerobic–aerobic reactor studies reviewed in this paper show that a generally high extent of color removal can be obtained, and several studies further more provide evidence for removal of aromatic amines. Combined anaerobic–aerobic treatment therefore holds promise as a method to completely remove azo dyes from wastewater. However, the results of the reactor studies reveal some possible limitations, both with respect to azo dye reduction and with respect to the fate of aromatic amines. Nevertheless, more research should also be conducted on the probable routes for the safe and eco-friendly way of handling the residues and byproducts obtained after decolourization process.

## References

Ahmad R, Mondal PK, Usmani SQ (2010) Hybrid UASFB-aerobic bioreactor for biodegradation of acid yellow-36 in wastewater. Bioresour Technol 101(10):3787–3790

Albuquerque MGE, Lopes AT, Serralheiro ML, Novais JM, Pinheiro HM (2005) Biological sulphate reduction and redox mediator effects on azo dye decolourisation inanaerobic–aerobic sequencing batch reactors. Enzyme Microb Technol 36(5–6):790–799

An H, Qian Y, Gu XS, Tang WZ (1996) Biological treatment of dye wastewaters using an anaerobic–oxic system. Chemosphere 33(12):2533–2542

Banat IM, Nigam P, Singh D, Marchant R (1996) Microbial decolorization of textile-dye-containing effluents: a review. Bioresour Technol 58(3):217–227

Baughman GL, Perenich TA (1988) Fate of dyes in aqueous system: solubility and partitioning of hydrophobic dyes and related compounds. Environ Toxicol Chem 1988(7):183–199

Bennett JW, Faison BD (1997) Use of fungi in biodegradation. In: Environmental Microbiology, ASM Press, Washington.

Brown D, Hamburger B (1987) The degradation of dyestuffs: part III-investigations of their ultimative biodegradability. Chemosphere 16:1539–1553

Brown D, Laboureur P (1983) The degradation of dyestuffs, Part 1, primary biodegradation under anaerobic conditions. Chemosphere 12(3):397–404

Buitron G, Martinez KM, Vargas A (2006) Degradation of acid orange 7 by a controlled anaerobic–aerobic sequencing batch reactor. Water Sci Technol 54(2):187–192

Bumpus JA (1995) Microbial degradation of azo dyes. Prog Ind Microbiol 32:157–176

Cabral Goncalves I, Penha S, Matos M, Santos AR, Franco F, Pinheiro HM (2005) Evaluation of an integrated anaerobic/aerobic SBR system for the treatment of wool dyeing effluents. Biodegradation 16(1):81–89

Carliell CM, Barclay SJ, Naidoo N, Buckley CA, Mulholland DA, Senior E (1995) Microbial decolourisation of a reactive azo dye under anaerobic conditions. Water SA 21(1):61–69

Carvalho C, Fernandes A, Lopes A (2007) Electrochemical degradation applied to the metabolites of acid orange 7 anaerobic biotreatment. Chemosphere 67:1316–1324

Chung KT, Cerniglia CE (1992) Mutagenicity of azo dyes: structure–activity relationships. Mutat Res 277(3):201–220

Chung KT, Stevens SE Jr, Cerniglia CE (1992) The reduction of azo dyes by the intestinal microflora. Crit Rev Microbiol 18:175–190

Cinar O, Yasara S, Kertmena M, Demirozb K, OzguYigit N, Kitis M (2008) Effect of cycle time on biodegradation of azo dye in sequencing batch reactor. Process Saf Environ Prot 86:455–460

Cinar O, Demiroz K, Kanat G, Uysal Y, Yaman C (2009) The effect of oxygen on anaerobic color removal of azo dye in a sequencing batch reactor. Clean Soil Air Water 37(8):657–662

Cooper P (1993) Removing colour from dyehouse waste waters—a critical review of technology available. J Soc Dyers Colour 109:97–100

Cruz A, Buitron G (2001) Biodegradation of disperse blue 79 using sequenced anaerobic/aerobic biofilters. Water Sci Technol 44(4):159–166

Dafale N, Rao NN, Meshram SU, Wate SR (2008a) Decolorization of azo dyes and simulated dye bath wastewater using acclimatized microbial consortium biostimulation and halo tolerance. Bioresour Technol 99(7):2552–2558

Dafale N, Watea S, Meshramb S, Nandy T (2008b) Kinetic study approach of remazol black-B use for the development of two-stage anoxic–oxic reactor for decolorization/biodegradation of azo dyes by activated bacterial consortium. J Hazard Mater 159:319–328

Delee W, O'Neill C, Hawkes FR, Pinheiro HM (1998) Anaerobic treatment of textile effluents: a review. J Chem Technol Biotechnol 73(4):323–335

Dos Santos AB, Traverse J, Cervantes FJ, van Lier JB (2005) Enhancing the electron transfer capacity and subsequent color removal in bioreactors by applying thermophilic anaerobic treatment and redox mediators. Biotechnol Bioeng 89(1):42–52

Ekici P, Leupold G, Parlar H (2001) Degradability of selected azo dye metabolites in activated sludge systems. Chemosphere 44:721–728

Field JA, Stams AJM, Kato M, Schraa G (1995) Enhanced biodegradation of aromatic pollutants in cocultures of anaerobic and aerobic bacterial consortia. Antonie van Leeuwenhoek Int J Gen Mol Microbiol 67:47–77

FitzGerald SW, Bishop PL (1995) Two stage anaerobic/aerobic treatment of sulfonated azo dyes. J Environ Sci Health A Toxic/Hazard Subst Environ Eng 30(6):1251–1276

Forgacs E, Cserhati T, Oros G (2004) Removal of synthetic dyes from wastewaters: a review. Environ Int 30(7):953–971

Ganesh R, Boardman GD, Michelsen D (1994) Fate of azo dyes in sludges. Water Res 28(6):1367–1376

Gottlieb A, Shaw C, Smith A, Wheatley A, Forsythe S (2003) The toxicity of textile reactive azo dyes after hydrolysis and decolourisation. J Biotechnol 101(1):49–56

Hao OJ, Kim H, Chang PC (2000) Decolorization of wastewater. Crit Rev Environ Sci Technol 30(4):449–505

Harmer C, Bishop P (1992) Transformation of azo dye AO-7 by wastewater biofilms. Water Sci Technol 26(3/4):627–636

Haug W, Schmidt A, Nortermann B, Hempel DC, Stolz A, Knackmuss HJ (1991) Mineralization of the sulfonated azo dye mordant yellow 3 by a 6-aminonaphthalene-2-sulfonate-degrading bacterial consortium. Appl Environ Microbiol 57:3144–3149

Isik M, Sponza DT (2003) Aromatic amine degradation in a UASB/CSTR sequential system treating Congo Red dye. J Environ Sci Health A Toxic/Hazard Subst Environ Eng 38(10):2301–2315

Isik M, Sponza DT (2004a) Anaerobic/aerobic sequential treatment of a cotton textile mill wastewater. J Chem Technol Biotechnol 79(11):1268–1274

Isik M, Sponza DT (2004b) Decolorization of azo dyes under batch anaerobic and sequential anaerobic/aerobic conditions. J Environ Sci Health A Toxic/Hazard Subst Environ Eng 39(4):1107–1127

Isik M, Sponza DT (2004c) Monitoring of toxicity and intermediates of C.I. Direct Black 38 azo dye through decolorization in an anaerobic/aerobic sequential reactor system. J Hazard Mater 114(1–3):29–39

Isik M, Sponza DT (2008) Anaerobic/aerobic treatment of a simulated textile wastewater. J Sep Purif Technol 60:64–72

Jianrong Z, Yanru Y, Huren A, Yi Q (1994) A study of dyewaste treatment using anaerobic–aerobic process. In: Proceedings of the seventh international symposium on anaerobic digestion, Cape Town, pp 360–363

Jiang H, Bishop PL (1994) Aerobic biodegradation of azo dyes in biofilms. Water Sci Technol 29(10–11):525–530

Kalyuzhnyi S, Sklyar V (2000) Biomineralisation of azo dyes and their breakdown products in anaerobic–aerobic hybrid and UASB reactors. Water Sci Technol 41(12):23–30

Kapdan IK, Alparslan S (2005) Application of anaerobic– aerobic sequential treatment system to real textile waste- water for color and COD removal. Enzyme Microb Technol 36(2–3): 273–279

Kapdan IK, Oztekin R (2003) Decolorization of textile dyestuff Reactive Orange 16 in fed-batch reactor under anaerobic condition. Enzyme Microb Technol 33(2–3):231–235

Kapdan IK, Tekol M, Sengul F (2003) Decolorization of simulated textile wastewater in an anaerobic–aerobic sequential treatment system. Process Biochem 38(7):1031–1037

Khalid A, Arshad M, Crowley DE (2009) Biodegradation potential of pure and mixed bacterial cultures for removal of 4-nitroaniline from textile dye wastewater. Water Res 43:1110–1116

Kim SY, Young AJ, Woo KB (2008) The effects of reductant and carbon source on the microbial decolorization of azo dyes in an anaerobic sludge process. Dyes Pigment 76:256–263

Kuai L, De Vreese I, Vandevivere P, Verstraete W (1998) GAC-amended UASB reactor for the stable treatment of toxic textile wastewater. Environ Technol 19(11):1111–1117

Kulla HG, Klausener F, Meyer U, Ludeke B, Leisinger T (1983) Interference of aromatic sulfo groups in microbial degradation of the azo dyes orange I and orange II Arch. Microbiology 135:1–7
Levine WG (1991) Metabolism of azo dyes: implication for detoxification and activation. Drug Metab Rev 23(3–4):253–309
Libra JA, Borchert M, Vigelahn L, Storm T (2004) Two stage biological treatment of a diazo reactive textile dye and the fate of the dye metabolites. Chemosphere 56(2):167–180
Lin J, Zhang X, Li Z, Lei L (2010) Biodegradation of Reactive blue 13 in a two-stage anaerobic/aerobic fluidized beds system with a *Pseudomonas* sp. Isolate. Bioresour Technol 101(1):34–40
Lourenco ND, Novais JM, Pinheiro HM (2000) Reactive textile dye colour removal in a sequencing batch reactor. Water Sci Technol 42(5–6):321–328
Lourenco ND, Novais JM, Pinheiro HM (2001) Effect of some operational parameters on textile dye biodegradation in a sequential batch reactor. J Biotechnol 89(2–3):163–174
Lourenco ND, Novais JM, Pinheiro HM (2003) Analysis of secondary metabolite fate during anaerobic–aerobic azo dye biodegradation in a sequential batch reactor. Environ Technol 24(6):679–686
Luangdilok W, Panswad T (2000) Effect of chemical structures of reactive dyes on color removal by an anaerobic–aerobic process. Water Sci Technol 42(3–4):377–382
McMullan G, Meehan C, Conneely A, Nirby N, Robinson T, Nigam P, Banat IM, Marchant R, Smyth WF (2001) Mini review: microbial decolourisation and degradation of textile dyes. Appl Microbiol Biotechnol 56(1–2):81–87
Minke R, Rott U (2002) Untersuchungen zur innerbetrieblichen anaeroben Vorbehandlung stark farbiger Abwasser der Textilveredelungsindustrie (Investigations to anaerobic pre-treatment of highly colored textile industry wastewater). Wasser Abwasser 143(4):320–328
Moosvi S, Datta M (2007) An integrated process for the treatment of CETP wastewater using coagulation, anaerobic and aerobic process. Bioresour Technol 98:3384–3392
O'Neill C, Hawkes FR, Esteves SRR, Hawkes DL, Wilcox SJ (1999a) Anaerobic and aerobic treatment of a simulated textile effluent. J Chem Technol Biotechnol 74(10):993–999
O'Neill C, Hawkes FR, Hawkes DL, Lourenco ND, Pinheiro HM, Delee W (1999b) Colour in textile effluents—sources, measurement, discharge consents and simulation: a review. J Chem Technol Biotechnol 74(11):1009–1018
O'Neill C, Hawkes FR, Hawkes DW, Esteves S, Wilcox SJ (2000a) Anaerobic–aerobic biotreatment of simulated textile effluent containing varied ratios of starch and azo dye. Water Res 34(8):2355–2361
O'Neill C, Lopez A, Esteves S, Hawkes FR, Hawkes DL, Wilcox S (2000b) Azo-dye degradation in an anaerobic–aerobic treatment system operating on simulated textile effluent. Appl Microbiol Biotechnol 53(2):249–254
Ong SA, Toorisaka E, Hirata M, Hano T (2005) Decolorization of azo dye (Orange II) in a sequential UASB–SBR system. Sep Purif Technol 42(3):297–302
Pagga U, Brown D (1986) The degradation of dyestuffs: part II. Behaviour of dyestuffs in aerobic biodegradation tests. Chemosphere 15(4):479–491
Pandey A, Singh P, Iyengar L (2007) Bacterial decolorization and degradation of azo dyes. Int Biodeterior Biodegrad 59:73–84
Panswad T, Luangdilok W (2000) Decolorization of reactive dyes with different molecular structures under different environmental conditions. Water Res 34(17):4177–4184
Panswad T, Iamsamer K, Anotai J (2001a) Decolorisation of azo-reactive dye by polyphosphate and glycogen-accumulating organisms in an anaerobic–aerobic sequencing batch reactor. Bioresour Technol 76:151–159
Panswad T, Techovanich A, Anotai J (2001b) Comparison of dye wastewater treatment by normal and anoxic anaerobic/aerobic SBR activated sludge processes. Water Sci Technol 43(2):355–362
Pearce CI, Lloyd JR, Guthrie JT (2003) The removal of colour from textile wastewater using whole bacterial cells: a review. Dyes Pigment 58(3):179–196

Rajaguru P, Kalaiselvi K, Palanivel M, Subburam V (2000) Biodegradation of azo dyes in a sequential anaerobic– aerobic system. Appl Microbiol Biotechnol 54:268–273

Razo-Flores E, Donlon BA, Field JA, Lettinga G (1996) Biodegradability of N-substituted aromatics and alkylphenols under methanogenic conditions using granular sludge. Water Sci Technol 33(3):47–57

Robinson T, McMullan G, Marchant R, Nigam P (2001) Remediation of dyes in textile effluent: a critical review on current treatment technologies with a proposed alternative. Bioresour Technol 77(3):247–255

Sarsour J, Janitza J, Gahr F (2001) Biological degradation of dye-containing wastewater (Biologische Abbau farbstoff haltiger Abwasser). Wasser Luft Boden 6:44–46

Schink B, Philipp B, Muller J (2000) Anaerobic degradation of phenolic compounds. Review Naturwissenschaften 87:12–23

Seshadri S, Bishop PL, Agha AM (1994) Anaerobic/aerobic treatment of selected azo dyes in wastewater. Waste Manage 14(2):127–137

Shaul GM, Holdsworth TJ, Dempsey CR, Dostal KA (1991) Fate of water soluble azo dyes in the activated sludge process. Chemosphere 22:107–119

Shaw CB, Carliell CM, Wheatley AD (2002) Anaerobic/aerobic treatment of coloured textile effluents using sequencing batch reactors. Water Res 36(8):1993–2001

Sosath F, Libra JA (1997) Purification of wastewaters containing azo dyes. Acta Hydrochim Hydrobiol 25(5):259–264

Sosath F, Libra J, Wiesmann U (1997) Combined biological and chemical treatment of textile dyehouse wastewater using rotating disc reactors. In: Kornmuller A (ed) Treatment of wastewaters from textile processing. Technische Universitat Berlin Universitatsbibliothek, Berlin, pp 229–243, http://www.ecoprocess.de/pdf/SFB1997Sosath.pdf

Southern TG (1995) Technical solutions to the colour problem: a critical review. In: Cooper P (ed) Colour in dyehouse effluent. Society of Dyers and Colourists, Bradford, pp 73–91

Sponza DT, Isik M (2002a) Decolorization and azo dye degradation by anaerobic/aerobic sequential process. Enzyme Microb Technol 31(1–2):102–110

Sponza DT, Isik M (2002b) Ultimate azo dye degradation in anaerobic/aerobic sequential processes. Water Sci Technol 45(12):271–278

Sponza DT, Isik M (2005) Reactor performances and fate of aromatic amines through decolorization of Direct Black 38 dye under anaerobic/aerobic sequentials. Process Biochem 40(1):35–44

Stolz A (2001) Basic and applied aspects in the microbial degradation of azo dyes. Appl Microbiol Biotechnol 56(1–2):69–80

Supaka N, Juntongjin K, Damronglerd S, Delia M-L, Strehaiano P (2004) Microbial decolorization of reactive azo dyes in a sequential anaerobic–aerobic system. Chem Eng J 99(2):169–176

Tan NCG (2001) Integrated and sequential anaerobic/aerobic biodegradation of azo dyes. PhD thesis, Agro technology and Food Sciences, Sub-department of Environmental Technology, Wageningen University, Wageningen

Tan NCG, Opsteeg JL, Lettinga G, Field JA (1999) Integrated anaerobic/aerobic EGSB bioreactor for azo dye degradation. In: Fass R, Flashner Y, Reuveny S (eds) Bioremediation of nitroaromatics and haloaromatic compounds. Battelle, Columbus/Richland, pp 253–258

Van der Zee FP, Villaverde S (2005) Combined anaerobic–aerobic treatment of azo dyes—a short review of bioreactor studies. Water Res 39:1425–1440

Van der Zee FP, Bisschops IAE, Blanchard VG, Bouwman RHM, Lettinga G, Field JA (2003a) The contribution of biotic and abiotic processes during azo dye reduction in anaerobic sludge. Water Res 37(13):3098–3109

Vandevivere PC, Bianchi R, Verstraete W (1998) Treatment and reuse of wastewater from the textile wet-processing industry: review of emerging technologies. J Chem Technol Biotechnol 72(4):289–302

Walker R (1970) The metabolism of azo compounds: a review of the literature. Food Cosmet Toxicol 8(6):659–676

Wiesmann U, Sosath F, Borchert M, Riedel G, Breithaupt T, Mohey El-Dein A, Libra J (2002) Attempts to the decolorization and mineralization of the azo dye C.I. Reactive Black 5. Wasser Abwasser 143(4):329–336

Wuhrmann K, Mechsner K, Kappeler T (1980) Investigation on rate-determining factors in the microbial reduction of azo dyes. Eur J Appl Microbiol Biotechnol 9:325–338

Xu M, Guo J, Sun G (2007) Biodegradation of textile azo dye by Shewanella decolorationis S12 under microaerophilic conditions. Appl Microbiol Biotechnol 76:719–726

You SJ, Teng JY (2009) Anaerobic decolorization bacteria for the treatment of azo dye in a sequential anaerobic and aerobic membrane bioreactor. Journal of the Taiwan Institute of Chemical Engineers 40(5):500–504

Zaoyan Y, Ke S, Guangliang S, Fan Y, Jinshan D, Huanian M (1992) Anaerobic–aerobic treatment of a dye wastewater by combination of RBC with activated sludge. Water Sci Technol 26(9–11):2093–2096

# Chapter 7
# Air Quality Monitoring with Lichens in India. Heavy Metals and Polycyclic Aromatic Hydrocarbons

**Vertika Shukla and D.K. Upreti**

## Contents



**Abstract** Air quality monitoring is an integral part of an Environmental Impact Assessment Programme. It can be carried out either by direct air samplers or by employing biomonitoring plants. Monitoring of air pollution using living organisms provide information on the nature and quantity of pollutants at low cost. Lichens can be used as biomonitoring organism because they have slow growth rate and can survive longer than vascular plants and are highly dependent on atmospheric source

V. Shukla (✉)
Department of Environmental Science, Babasaheb Bhimrao Ambedkar (Central) University, Vidya Vihar, Raebareli Road, Lucknow 226 025, UP, India
e-mail: vertika_shukla@rediffmail.com

D.K. Upreti
Lichenology Laboratory, CSIR-National Botanical Research Institute, Rana Pratap Marg, Lucknow 226 001, India
e-mail: upretidk@rediffmail.com

E. Lichtfouse et al. (eds.), *Environmental Chemistry for a Sustainable World: Volume 2: Remediation of Air and Water Pollution*, DOI 10.1007/978-94-007-2439-6_7,

for nutrients. The feature that makes lichens a very useful group of plants in comparison to other vascular plants is that they do not shed their parts and remain a repository of facts. Lichen biomonitoring in a diverse and wide geographic area of India can provide a cost effective alternative to monitor the air quality of such a wide region. The data presented here is the consolidated information about the biomonitoring studies carried out in India till date, which will provide the base line data for future biomonitoring studies and prospects of utilizing lichens in air quality studies in India. The pollutants discussed are Pb, Zn, Ni, Cu, Hg, Cr and polycyclic aromatic hydrocarbons.

**Keywords** India • Air pollution • Anthropogenic origin • Environmental monitoring • Biomonitoring • Lichen • Lichen diversity • Heavy metal • Polycyclic aromatic hydrocarbons • Pyrogenic origin • Petrogenic origin • Physiological response

## 7.1 Introduction

Air pollution is a serious global problem especially in developing countries like India. In India air pollution is mainly due to unplanned developmental processes: growing urban settlements, increasing traffic, industrial development (Singh et al. 2007). Major input of contaminants in metropolitan cities of India, like Delhi, Kolkata and Mumbai, is due to vehicular activity, which ranges upto 70%, 52% and 30%, of the total pollution load in these cities, respectively. There has been an exponential increase in the number of vehicles since 1971 (Fig. 7.1). India has 23 major cities of over one million population and air pollution levels in these cities exceed the WHO standards in many of them (Gupta et al. 2002; Chauhan et al. 2010). A variety of

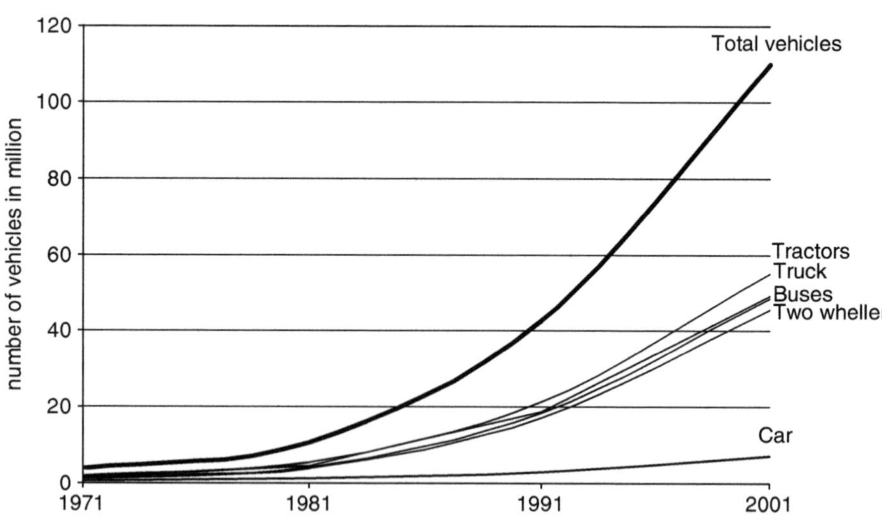


Fig. 7.1 Increase in number of vehicles in India between 1971 and 2001 (Chauhan et al. 2010)

**Table 7.1** Terminology used in various aspect of biomonitoring studies (Wittig 1993; Markert et al. 2003; Markert et al. 1997)

| Terminology | |
|---|---|
| 1. Bioindicator | Organisms that provide information on the environment or the quality of environmental changes |
| i. Reaction indicators | Organisms which have a sensitive reaction to air pollutants and which are used especially in studying the effects of pollutants on species composition, and on physiological and ecological functioning |
| ii. Accumulation indicators | Organisms that readily accumulate a range of pollutants and are therefore used especially when monitoring the amount of pollutants and their distribution |
| 2. Biomonitor | Organisms that provide quantitative information on the quality of the environment |
| i. Passive biomonitors | Organisms which occurs naturally in the study area |
| ii. Active biomonitors | Organisms are transplanted into the research area under controlled conditions for a specific period of time |

substances, which include greenhouse gases (including carbon dioxide and methane), sulphur dioxide and NOx gases, organic dust, and particulate matter having natural as well as anthropogenic origin, are the major source of air pollution. After the release, pollutants undergo physical, chemical, and photochemical transformations, which ultimately decide their fate and atmospheric concentrations. The major concern about the air pollutants is that it does not remain confined near the source of emission, but spread over distances (WHO 2000).

Two methodologies are known for air quality monitoring, (i) direct collection of airborne particulate matter, precipitation and total deposit using air samplers or (ii) employing biomonitors (bioaccumulators). The first approach is aimed at quantitative surveys of pollutants at different locations. It provides the temporal and spatial representation of measurements at different locations. Direct measurements on a large scale are extremely costly, therefore approach involving use of biomonitors is regarded a non-expensive and reliable means of air quality status assessment in a country or a region. Certain types of biological organisms like lichens provide a measure of integrated exposure over a certain amount of time (Markert et al. 2003). The term Bioindicator/biomonitor is used to refer to an organism, or a part of it, that depicts the occurrence of pollutants on the basis of specific symptoms, reactions, morphological changes or concentrations. There is considerable variation in the use of the terms bioindicator and biomonitor and related terminologies as in Table 7.1 (Wittig 1993; Markert et al. 2003; Markert et al. 1997).

High sensitivity towards changes in microclimatic condition in the ambient environment is an important aspect for its utility in assessment of air quality and its effects on lichen provides useful information on the state of health of the environment (Manning and Feder 1980). Since lichen sampling is relatively simple and no expensive technical equipments are needed, therefore air quality monitoring with the help of biomonitoring with lichens can be effectively utilized for air quality monitoring in India as the country has wide geographical area and numerous anthropogenic sources of air pollution.

## 7.2 Lichen as an Effective Biomonitoring Organism

Lichens are slow-growing symbiotic associations of fungi (mycobionts) and green algae or cyanobacteria (photobionts), which forms a common thallus which lacks roots or waxy cuticles and relies mainly on the atmospheric input of mineral nutrients. They have wide geographical distribution (Wolterbeek et al. 2003).

Tolerance to most of the heavy metals and the slow growth rate, are the main factors which makes lichens good indicators or/and monitors of metal pollution (Seaward 1974). Lichens have ability to bioaccumulate airborne substance to concentrations much higher in comparison to the ambient atmosphere, and the element contents of lichen thalli proved to be directly correlated with environmental levels (Sloof 1995; Bari et al. 2001; Shukla and Upreti 2007a, 2008a; Conti and Cecchetti 2001).

Lichens have a wide variety of growth forms. Main morpho-types are fruticose (thallus has a shrub-like appearance or is branched), foliose (thalli are 'leafy' with distinct upper and lower sides) and crustose (thallus forms a crust over the substrate). The fruticose lichens are the most sensitive towards pollution followed by foliose and crustose forms (Awasthi 2000), while foliose lichens are better accumulators in comparison with fruticose ones (Swinscow and Krog 1988; St. Clair et al. 2002a, b). In general foliose lichens are employed in the biomonitoring studies (Fig. 7.2) which may be due to the morphology of the foliose lichens that provides larger surface area for the absorption/adsorption of the pollutants. As in the case of *P. hispidula*, dense tuft of rhizinae present on the lower side of the thallus act as a reservoir for metals (Goyal and Seaward 1982). Selection of the species of lichens to be utilized as a biomonitor is depended on the morphology and anatomy of the particular species (Riga-Karandinos and Karandinos 1998).

### *7.2.1 Characteristic Features of Lichens Utilized in Biomonitoring Studies*

Lichens has been recognized and successfully utilized as biological indicators of air quality. They are among the most valuable and reliable biomonitors of atmospheric pollution. Primarily lichens were utilized to monitor gaseous pollution, namely sulphur ($SO_2$), nitrogen ($NO_x$, $NH_3$, $NO_3$, etc.) (Rao and LeBlanc 1967; Vestergaard et al. 1986). Lichens show high sensitivity towards sulphur dioxide because their efficient absorption systems result in rapid accumulation of sulphur when exposed to high levels of sulphur dioxide pollution (Wadleigh and Blake 1999). The algal partner (5–10% of total thallus structure) is most affected by the sulphur dioxide as chlorophyll is irreversibly converted to phaeophytin and thus photosynthesis is inhibited (Upreti 1994). Lichens also absorb sulphur dioxide dissolved in water (Hawksworth and Rose 1970). Excessive levels of pollutants in the atmosphere, especially $SO_2$, has detrimental effect on the physiology and morphology of sensitive species, causes extinction of the species, which ultimately results in changed lichen diversity pattern (Haffner et al. 2001; Purvis 2000).

**Fruticose Lichen**
Highly sensitive to slight alteration in the microclimatic conditions.
Extremely susceptible to elevation of $SO_2$ levels e.g. *Usnea longissima*, and other species of *Usnea*

**Foliose Lichen**
Sensitivity varies from species to species
Generally pollution intermediate to pollution tolerant species.
Most of the established bioindicator/ biomonitor species belong to this category. e.g. *Phaeophyscia hispidula, Hypogymnia physodes, Parmelia sulcata, Pseudoevernia furfuracea*

**Crustose Lichen**
Resistant
Some lichen of this category are Bioindicator e.g. *Caloplaca* sp., *Lecanora* achroa etc

**Fig. 7.2** Growth forms of lichen, their sensitivity and biomonitoring potential

Change in the community structure of epiphytic lichens is a sensitive indicator of the biological effects of air pollutants (Richardson 1991; Pinho et al. 2004; Paoli and Loppi 2008). Rise in the number of nitrophillous lichens especially members of the *Physciaceae* family (including *Phaeophyscia* and *Pyxine*) has been attributed to climate change and also to dry and dusty conditions in urban centers (van Herk et al. 2002; Loppi and Pirintos 2000). In addition to lichen diversity changes, lichens are being widely used for monitoring accumulation of heavy metals (Loppi et al. 1998; Garty et al. 2003; Shukla and Upreti 2007a).

Lichens are employed for air quality assessment in three ways viz., by identifying and mapping all lichen species in an area (LDV and IAP studies), transplanting healthy lichens into a polluted area and measuring thallus deterioration and sampling an individual species and measuring contaminants accumulated within the thallus (Pfeiffer and Barclay-Estrup 1992; Garty et al. 2003; Loppi and Frati 2006; Pinho et al. 2004; Shukla and Upreti 2007a; Shukla et al. 2010b).

Mechanisms involved to accumulate substances in lichens from their ambient atmosphere include, particulate trapping, ion exchange, extra cellular absorption, hydrolysis, and intracellular uptake (Nieboer et al. 1978).

## 7.3 World Scenario

Nylander (1866), first discussed that the absence of lichens in city of Paris due to air pollution originating from the surrounding buildings. Since then more than 2,000 references are available on the similar and increasingly extensive example of lichen changes around towns and industrial complexes all over the world.

Damage caused due to increasing air pollution (especially metal stress) has been estimated in transplanted lichens by estimating physiological parameters (Richardson 1992; Pirintsos et al. 2004; Haffner et al. 2001, Dzubaj et al. 2008; González and Pignata 1994; Garty et al. 1997; Baddeley et al. 1972). Pollution level of critically polluted area with sparse or no lichen (lichen desert zone) have been successfully assessed with lichen transplants studies (Garty et al. 2003). *Hypogymnia physodes* (pollution tolerant lichen) abundant in the Pacific Northwest, is being frequently transplanted worldwide for air quality monitoring (Brodo 1961; Holopainen 1984; Farkas et al. 1985; Gailey et al. 1985; Shirazi et al. 1996; McCune et al. 1996; Vestergaard et al. 1986; Pirintsos et al. 2006).

Most of the recent literature on lichens and air pollution are related to monitoring of metals concentration in lichens, its effects on lichen distribution or on ecophysiological parameters (Loppi and Bonini 2000; Poličnik et al. 2008; Thrower 1980; Cabral 2003; Loppi et al. 1997; Dzubaj et al. 2008). Quite few researches are available on biomonitoring of persistent organic pollutants in relation with lichens and mosses (Augusto et al. 2009; Guidotti et al. 2003; Ockenden et al. 1998; Muir et al. 1993; Shukla et al. 2010; Shukla and Upreti 2009). The use of lichens as indicators of air pollution has been well studied in Europe and northern America (Herzig et al. 1989; Conti and Cecchetti 2001; Loppi and Frati 2006; Thormann 2006; Pinho et al. 2004; Crespo et al. 2004); however, little is known about air pollution and its effects on lichen in India.

## 7.4 Biomonitoring Studies Carried Out in India Till Date

Indian landscape has wide geographical area ranging from tropical to alpine climate. It has rich lichen diversity with more than 2,300 species, reported from different regions of the country, belonging to 305 genera and 74 families. Out of these species reported about 520 species (22.5%) are endemic (Singh and Sinha 2010). There are

**Fig. 7.3** Map of India showing different cities where long term biomonitoring studies are being carried out and some of the lichen rich areas of the country.* J&K: Jammu & Kashmir, GHNP: Great Himalayan National Park, NBR: NandaDevi Biosphere Reserve, MPG: Milam & Pindari Glacier, PWS: Panch marhi Wildlife Sanctuary, AABR: Amarkantak-Achanakmaar Biosphere Reserve, S: Sikkim, NH: Nilgiri Hills, which have contributed to the rich lichen diversity of India.(* Source: Lichen rich area (Upreti and Nayaka 2008). *1*. DEHRADUN, *2*. HARIDWAR, *3*. RISHIKESH, *4*. SRINAGAR (G.), *5*. CHOPTA, *6*. BADRINATH, *7*. PITHORAGARH)

lichen hot spots like Great Himalayan National Park, Western Ghats, North-Eastern states and Nanda Devi Biosphere Reserve (Upreti and Nayaka 2008) but the urban areas have poor diversity, and only some tolerant species belonging to members of Physciaceae dominates the lichen flora dominates the lichen flora (Fig. 7.3) of urban center (Shukla 2007).

***Phaeophyscia hispidula*** **(Ach.) Moberg**

*Characteristic feature*: dense tuft of rhizinae on the lower surface which act as reservoir for pollutants especially Iron.
*Distribution in India:* has wide distribution. Common in the sub tropical and temperate climates. Has exclusive occurrence in highly polluted city centers.

***Pyxine subcinerea*** **Stirton**.

*Characteristic feature:* presence of lichexanthone in the upper cortex which acts as uv filter and plays major role in thermoprotection of this organism in extreme urban climate.
*Distribution in India:* wide distribution in the tropical region of the country.

***Dirinaria applanata*** **(Fée) D.D. Awasthi**

*Characteristic feature:* presence of different type of cuticle, whitish and flour like in appearance composed of superficial deposits most commonly calcium oxalate.
*Distribution in India:* commonly distributed in tropical areas.

**Fig. 7.4** Different lichen species widely used for biomonitoring studies in India. 1. *Phaeophyscia hispidula* (Ach.) Moberg, 2. *Pyxine subcinerea* Stirton.3. *Dirinaria applanata* (Fée) D.D. Awasthi

In tropical climate of India *Pyxine subcinerea*, *Phaeophyscia hispidula* and *Dirinaria applanata* (Fig. 7.4), common foliose member of lichen family *Physciaceae*, having well known pollution tolerant characters are the most suitable lichen taxa for carrying out Active (transplant) as well as Passive (organism taken from their natural habitat) biomonitoring.

In tropical India, *Shorea robusta* has been reported to be an excellent host for lichen growth (Satya et al. 2005). *Mangifera indica* is the other tropical tree having suitable habitat for luxuriant growth of many lichen taxa. In sub-temperate areas of India trees of *Melia sp.* Also provide suitable suitable for growth of lichen along the road side.

Systematic study on lichens was initiated in India by Awasthi (1957), who deciphered the lichen flora of India, Nepal and Sri Lanka (Awasthi 1988; Awasthi 1991; Awasthi 2007). Biomonitoring studies with lichens in India were initiated by Das in the year 1986. Since then several studies have been carried out in various parts of India.

### *7.4.1 Bioindicator Studies in India*

Foliose and fruticose epiphytic lichens are best suited for biomonitoring studies (Seaward 1993; van Dobben and ter Braak 1999). In urban environment, sulphur dioxide along with the $NO_X$ gases (resulting from vehicular emissions) has antagonistic effect on lichen community at relatively high doses of gases in the environment (Balaguer et al. 1997). Changes in the composition of lichen diversity including frequency, density and abundance provide first hand evidence on the alterations in air quality of an area due to air pollution or microclimatic changes (van Herk et al. 2002; Aptroot and van Herk 2007).

Since the last century there has been a considerable change/decline in the lichen biodiversity all around the world (Hauck 2009). It has been observed that rate of decline of lichen biodiversity in Himalayan region especially Garhwal Himalayas is quite faster (Upreti and Nayaka 2008). There is a considerable increase in the abundance of thermophilous and poleotolerant lichens in temperate climate of Garhwal Himalayas. Increase in frequency and abundance of thermophilous and poleotolerant lichen such as *Phaeophyscia hispidula* (Ach.) Moberg, *P. orbicularis (*Necker) Moberg, *Pyxine cocoes* (Swartz.) Nyl., *P. subcinerea* Stirton, *Candelaria concolor* (Dicks.) B. Stein, *Punctelia borreri* (Sm.) Krog, *Physcia biziana* (Massal.) Zahlbr and *Xanthoria elegans* (Link.) Th. Fr, *Physconia enteroxantha* (Nyl.) Poelt, indicate changed environmental conditions in Garhwal Himalayas (Shukla and Upreti 2010a; Shukla 2007).

Lichen flora of Kolkata revealed the exclusive occurrence of pollution resistant species, *Parmelia caperata* (= *Flavoparmelia caperata*) on the road side trees of Kolkata. The most probable reason for existence of resistant species was long range dispersal of pollutants (caused by nearby factories) with wind (Das et al. 1986).

Lichen flora of Garden city, Bangalore was explored by Nayaka et al. (2003). Significant change in the lichen diversity was observed in comparison to earlier study conducted 18 years earlier. There were only four species common between two studies. Air quality of Pune city in Maharashtra Province was assessed by distribution of lichens in the city. It was observed out of the 20 streets/sites of the Pune city surveyed, only 11 sites showed the presence of lichens Nayaka and Upreti (2005a).

Grid plotting technique has been utilized to map lichen diversity in Lucknow city. In this study distribution of each species was plotted in 1×1 km grid in all direction. The distribution data of lichens collected from all the four areas viz. North, East, West and South, provided four distinct zones, viz. Zone A with no lichen growth, was the area within the center of the city upto 5 km all around, Zone B- showed presence of some calcareous lichens mostly in the areas with old historical buildings, Zone C- had scare growth of few crustose and foliose lichen in the localities with scattered mango trees, Zone D- showed normal growth of different epiphytic lichen taxa together with same foliicolous (leaf inhabiting) lichens, an indication of a more or less pollution free environment (Saxena and Upreti 2004).

An earlier enumeration of lichens of Indian Botanical Garden collected by Kurz in 1865 and described by Nylander in 1867 was compared by Upreti et al. (2005). It is interesting to note that in the last more than 140 years the lichen flora of the area has been changed significantly as only three species out of 50 species (recorded earlier) were common between the two studies.

## 7.4.2 *Biomonitoring Studies in India*

### 7.4.2.1 Heavy Metals

Lichen morphology and physiology enable it to be a good reservoir of metals. Various heavy metals such as Pb, Zn, Ni, Cu, Hg, and Cr considered as toxic for many other living organisms, may be accumulated simultaneously in lichen thallus (Garty 1993). Accumulation of metals by lichen thalli is one of the extensively studied aspects in lichen biomonitoring (Nieboer et al. 1978).Vehicular activity is reported to be the main source of atmospheric Cr, Cu and Pb (Loppi et al. 1998; Tuba and Csintalan 1993).

In India, air quality of major cities in the country has been evaluated with the help of lichens. Vehicular pollution contributes major input of wide range of pollutants in the atmosphere including heavy metals (Chauhan et al. 2010). Biomonitoring studies are mainly concentrated on bioaccumulation of metallic content since in the urban centers vehicular and anthropogenic activity is the main contributing source of pollution. The negative correlation of concentration of Pb with the distance from the source of pollution has been well documented by Dubey et al. (1999).

Height plays an important role in accumulation in lichen, which has observed by transplant study using lichen, *Dirinaria consimilis* in two vertical positions in Lucknow city. Cr and Cu were reported to be in higher concentration at higher height while maximum zinc was detected at lower height (Bajpai et al. 2004). Commercial and industrial sites of Lucknow city were also surveyed for their metallic input to the environment by Mishra et al. (2003). Sample facing towards the source of pollution was reported to have higher concentration of metals than the leeward one. Saxena et al. (2007) analysed the effect of vehicular activity on accumulations in six common lichen species growing on *Mangifera indica* trees in Mango orchards surrounding in the north side of the Lucknow city. The study revealed the higher concentrations of Pb (3.3–15.6 μg $g^{-1}$) Zn (49–219 μg $g^{-1}$) Cu (10.2–66.6 μg $g^{-1}$) and Fe (1,748–19,374 μg $g^{-1}$).

Lichens have also been used for study of manganese emissions from mines. Magnasite mining area of Chandak in Pithoragarh, Uttarakhand was studied by Nayaka et al. (2005b), the concentration of Manganese (Mn) in different lichen taxa varied considerably. The concentration of Manganese varied between 40 and 50 μg $g^{-1}$ in different lichen species.

In order to study the bioaccumulation pattern and effect of common fungicide, Bordeaux mixture (mixture of copper sulphate and hydrated lime; $CuSO_4 + Ca(OH)_2$), on lichens. *Cryptothecia punctulata* was analysed which bioaccumulated Cu, Ca, and S as 575, 10, 21,000 μg $g^{-1}$ respectively (Nayaka et al. 2005c).

The different urban areas in the Garhwal Himalayas were studied for their lichen distribution and air quality assessment (inorganic as well as organic contaminants). The common foliose lichen *Pheophyscia hispidula* has been successfully employed for passive biomonitoring to observe the inorganic and organic pollutants in the ambient air of urban areas viz. capital city of DehraDun and Holy pilgrimage center Badrinath, Pauri, Srinagar (Garhwal) and Chopta-Tungnath. (Shukla 2007; Shukla and Upreti 2007a).

Madhya Pradesh, a central province of Republic of India is known for its historical monuments. These monuments were constructed by sedimentary rock, which act as suitable substratum for lichenization. *Lepraria lobificans* Nyl., leprose lichen growing naturally on monuments and buildings in the city Mandav in central India was utilized for passive monitoring of atmospheric metals. Samples collected from road site revealed the maximum concentration of Fe, Cd, Cr, Ni, and Zn. *L. lobificans* exhibited the higher accumulation of Fe than foliose and fruticose lichens (Bajpai et al. 2010a).

Arsenic in different growth forms of lichens growing on old monuments in the city of Mandav, Dhar district of Madhya Pradesh, was analyzed by Bajpai et al. (2009). Among the different growth forms, foliose lichens were found to accumulate higher amounts of arsenic followed by leprose form. It was observed that cyanolichens exhibit higher concentration of arsenic than the green photobionts-containing squammulose form.

#### 7.4.2.2 Physiological Response

Diversity of epiphytic lichens responds to air pollutants emitted by pollution source but it requires continuous exposure for a period, which results in the changes in species number and composition. It is the changes on physiological level which results in extinction or tolerance of the species in response to the level of pollution, which is reflected by the change in the species composition. Therefore monitoring physiological changes in lichen exposed to air pollution can help to detect early stress symptoms (Paoli and Loppi 2008).

In India, urban centers has poor diversity, some tolerant species dominate the lichen flora mostly dominated by members of *Physciaceae* family (Shukla 2007). In order to investigate the probable adaptation prevalent in the naturally and luxuriantly lichen growing in these centers. Stress physiology of *P. hispidula* was studied by Shukla and Upreti (2007b) in Pauri and Srinagar, Uttaranchal. Chl. a, Chlorophyll degradation ratio and protein revealed positive correlation with the increase in pollution.

Other poleotolerant species of *Physciaceae* family, *Pyxine subcinerea* and *Rinodina sophodes* and *Pyxine cocoes* from Srinagar (Garhwal), Kanpur city and Raebareily respectively, were evaluated for its physiological response to metallic stress (Shukla and Upreti 2008a; Satya and Upreti 2009; Bajpai et al. 2010).

#### 7.4.2.3 Polycyclic Aromatic Hydrocarbons

Polycyclic Aromatic Hydrocarbons lies in the category of human health hazardous compound having known carcinogenic and mutagenic properties. The main sources of PAHs are combustion processes and oil spillage. In India traditional usage of coal and wood for cooking and other purposes cause high input of PAH to the environment (Shukla et al. 2010b).

In order to quantify PAHs in Garhwal Himalayas, Shukla et al. (2010b) studied lichens collected from different sites from Garhwal Himalayas, Uttaranchal, which revealed pyrolytic origin while PAH concentration in lichen *Phaeophyscia hispidula* from DehraDun, capital city of Uttaranchal revealed mixed origin, a major characteristic of urban environment. The study established the utility of *P. hispidula* as an excellent biomonitoring organism in monitoring both PAH and metals in the temperate regions of India (Shukla and Upreti 2009).

In urban centers vehicular traffic has been recognized as significant sources of pollution. The ambient air in urban environment favours only nitrophilous lichens to flourish well especially, members of Physciaceae family. *P. hispidula* (a member of Physciaceae family) presents a large contact surface for atmospheric pollutants and is able to accumulate high amounts of heavy metals in polluted areas (Shukla and Upreti 2007a).

#### 7.4.2.4 Role of Lichen Substances (Chemical Response)

In India, few studies (Shukla 2007) have been conducted regarding role of lichen substances in response to air pollution. The morphology and physiology together with lichen substances of the *Pyxine subcinerea* and *P. hispidula* is known to play a key role in the thermo protection of cellular structure in urban environment. It has been found that chemically, in *P. subcinerea* main constituent is lichaxanthone and it tends to increase in urban sites in comparison to the control sites (Shukla and Upreti 2008b). High pigment concentration may be a complimentary feature for the high success of these species in the urban climate.

## 7.5 Advantage of Biomonitoring with Lichen in Indian Perspective

Lichens are cheap and effective indicators and monitors of atmospheric pollution because of their high sensitivity towards ambient microclimatic changes, which can be monitored on quantitative and qualitative scale. This gives them a considerable advantage over other biological means for environmental monitoring (Manning and Feder 1980). Lichen diversity studies provide preliminary information on the air quality status thus they can help to install analytical instruments at specific location for detailed analysis. Further, transplant studies is an alternative to installation of air samplers at busy polluted (lichen desert), data thus collected may be utilized for long term biomonitoring studies (Richardson 1991).

As lichens do not shed their parts and remain a repository of facts, which reduces the need for continuous chemical monitoring. Thus biomonitoring studies using lichens make it possible to recognize source and spatial dispersal of air pollution which is an important step in environment protection practice moreover any improvement in air quality can be tracked by reinvasion studies (Augusto et al. 2009; Conti and Cecchetti 2001; Hawksworth and Rose 1970).

India, a megadiversity region, having rich diversity of different plant groups including lichens, therefore long term biomonitoring studies data can provide first hand evidence about the air quality of an area Furthermore, some species of lichens like *Phaeophyscia, Dirinaria* and *Pyxine* can be best utilized for screening the accumulation pattern as they have vast geographical distribution used for passive biomonitoring while urban center can be monitored by transplanting lichen taken from the lichen rich areas. Therefore biological monitoring employing lichen is well suited in Indian perspective as it is cost effective and installation of analytical instruments in such a large region is not feasible. Advantage of the biomonitoring studies lie on the fact that it provides accurate air quality (environmental condition) without installation of costly air monitoring instruments.

Thus biomonitoring with lichens can play a pivotal role in the environmental conservation programme in India which provides reliable information on the present environmental condition and any improvement or deterioration in the air quality.

## 7.6 Conclusion

Owing to the vast geographical area of India, the number of studies so far conducted on lichens in relation to pollution monitoring is quite scarce. There is a need to carry out extensive biomonitoring studies with the help of lichens in different region of the country to know the level of prevailing air pollution (in the polluted area) and to study the effect of restoration measures by lichen reinvasion and long term biomonitoring studies. The present review provides the consolidated information available on biomonitoring studies carried out in India along with the established bioindicator species i.e. *Phaeophyscia hispidula*. The present information reported will act as a base line data which can be used in future for carrying out long term monitoring studies in the country.

**Acknowledgements** Author (V.S.) is thankful to the Vice Chancellor, Babasaheb Bhimrao Ambedkar (Central) University, Lucknow for providing Laboratory facilities. Thanks are due to Ms. Aparna Dwivedi for valuable help. This work has been catalyzed and supported by Scientific and Engineering Research Council, Department of Science & Technology, New Delhi (SR/FT/LS-028/2008).

## References

Aptroot A, van Herk CM (2007) Further evidence of the effects of global warming on lichens, particularly those with *Trentipohlia* phycobionts. Environ Pollut 146(2):293–298

Augusto S, Máguas C, Matos J, Pereira MJ, Soares A, Branquinho C (2009) Spatial modeling of PAHs in lichens for fingerprinting of multisource atmospheric pollution. Environ Sci Technol 43(20):7762–7769

Awasthi DD (1957) A new species of *Parmelia* from Kodaikanal, S. India. Curr Sci 26:123–124

Awasthi DD (1988) A key to the macrolichens of India and Nepal. J Hattori Bot Lab 65:207–303

Awasthi DD (1991) A key to the microlichens of India Nepal and Sri Lanka. Biblioth Lichenol 40:1–337

Awasthi DD (2000) A hand book of lichens. Bishan Singh Mahendra Pal Singh, Dehradun

Awasthi DD (2007) A compendium of the macrolichens from India, Nepal and Sri Lanka. Bishan Singh Mahendra Pal Singh, Dehradun

Baddeley MA, Ferry BW, Finegan EJ (1972) The effects of sulphur dioxide on lichen respiration. Lichenologist 5:283–291

Bajpai R, Upreti DK, Mishra SK (2004) Pollution monitoring with the help of lichen transplant technique at some residential sites of Lucknow city, Uttar Pradesh. J Environ Biol 25(5):191–195

Bajpai R, Upreti DK, Dwivedi SK (2009) Arsenic accumulation in lichens of Mandav monuments, Dhar district, Madhya Pradesh, India. Environ Monit Assess 159:437–442. doi:10.1007/s10661-008-0641-7

Bajpai R, Upreti DK, Dwivedi SK (2010a) Passive monitoring of atmospheric heavy metals in a historical city of central India by *Lepraria lobificans* Nyl. Environ Monit Assess 166:477–484. doi:10.1007/s10661-009-1016-4

Bajpai R, Upreti DK, Nayaka S, Kumari B (2010b) Biodiversity, bioaccumulation and physiological changes in lichens growing in the vicinity of coal based thermal power plant of Raebareli district, north India. J Hazard Mater 174:429–436

Balaguer L, Manrique E, Ascaso C (1997) Predictability of the combination effects of sulphur dioxide and nitrate on green algal lichen *Ramalina farinacea*. Can J Bot 75:1836–1842

Bari A, Rosso A, Minciardi MR, Troiani F, Piervittori R (2001) Analysis of heavy metals in atmospheric particulates in relation to their bioaccumulation in explanted *Pseudevernia furfuracea* thalli. Environ Monit Assess 69:205–220

Brodo IM (1961) Transplant experiments with corticolous lichens using a new technique. Ecology 42:838–841

Cabral JP (2003) Copper toxicity to five Parmelia lichens in vitro. Environ Exp Bot 49:237–250

Chauhan A, Pawar M, Kumar R, Joshi PC (2010) Ambient air quality status in Uttarakhand (India): a case study of Haridwar and Dehradun using air quality index. J Amer Sci 6(9):565–574

Conti ME, Cecchetti G (2001) Biological monitoring: lichens as bioindicator of air pollution assessment – a review. Environ Pollut 114:471–492

Crespo A, Divakar PK, Arguello A, Gasca C, Hawksworth DL (2004) Molecular studies on *Punctelia* species of the Iberian Peninsula, with an emphasis on specimens newly colonizing Madrid. Lichenol 36(5):299–308
Das G, Das AK, Das JN, Guo N, Majumdar R, Raj S (1986) Studies on the plant responses to air pollution, occurrence of lichen in relation to Calcutta city. Indian Biologists 17(2):26–29
Dubey AK, Pandey V, Upreti DK, Singh J (1999) Accumulation of lead by lichens growing in and around Faizabad, U.P., India. J Environ Biol 20(3):223–225
Dzubaj A, Bačkor M, Tomko J, Peli E, Tuba Z (2008) Tolerance of the lichen Xanthoria parietina (L.) Th. Fr. to metal stress. Ecotox Environ Safe 70:319–326. doi:10.1016/j.ecoenv.2007.04.002
Farkas E, Lokos L, Verseghy K (1985) Lichens as indicator of air pollution in the Budapest agglomeration. I Air pollution map based on floristic data and heavy metal concentration measurement. Acta Bot Hung 31:45–68
Gailey FAY, Smith GH, Rintoul LJ, Lloyd OL (1985) Metal deposition patterns in central Scotland, as determined by lichen transplants. Environ Monit Assess 5:291–309
Garty J (1993) Lichens as biomonitors of heavymetal pollution In: Markert, B. (eds.) Plants as biomonitors VCH Verlagsgesellschaft mbh, Germany, pp 193–263
Garty J, Kloog N, Wolfson R, Cohen Y, Karnieli A, Avni A (1997) The influence of air pollution on the concentration of mineral elements, on the spectral reflectance response and on the production of stress-ethylene in the lichen *Ramalina duriaei*. New Phytol 137:587–597
Garty J, Tomer S, Levin T, Lehra H (2003) Lichens as biomonitors around a coal-fired power station in Israel. Environ Res 91:186–198
González CM, Pignata ML (1994) The influence of air pollution on soluble proteins, chlorophyll degradation, MDA, sulphur and heavy metals in a transplanted lichen. Chem Ecol 9:105–113
Goyal R, Seaward MRD (1982) Metal uptake in terricolous lichens. II. Translocation in the thallus of Peltigera canina. New Phytol 90:85–98
Guidotti M, Stella D, Owczarek M, DeMarco A, De Simone C (2003) Lichens as polycyclic aromatic hydrocarbon bioaccumulators used in atmospheric pollution studies. J Chromatogr 985(1–2):185–190
Gupta HK, Gupta VB, Rao CVC, Gajghate DG, Hasan MZ (2002) Urban air quality and its management strategy for an metropolitan city of India. Bull Environ Contam Toxicol 68:347–354
Haffner E, Lomsky B, Hynek V, Hallgren JE, Batic F, Pfanz H (2001) Air pollution and lichen physiology. Physiological responses of different lichens in a transplant experiment following an $SO_2$-gradient. Water Air Soil Pollut 131:185–201
Hauck M (2009) Global warming and alternative causes of decline in arctic-alpine and boreal-montane lichens in north-western central Europe. Glob Change Biol 15:2653–2661
Hawksworth DL, Rose F (1970) Qualitative scale for estimating sulphur dioxide air pollution in England and Wales using epiphytic lichen. Nature 227:145–148
Herzig R, Liebendörfer L, Urech M, Ammann K (1989) Passive biomonitoring with lichens as a part of an integrated biological measuring system for monitoring air pollution in Switzerland. Int J Environ An Ch 35:43–57
Holopainen T (1984) Types and distribution of ultra structural symptoms in epiphytic lichens in several urban and industrial environments in Finland. Ann Bot Fennici 21:213–229
Loppi S, Bonini I (2000) Lichens and mosses as biomonitors of trace elements in areas with thermal springs and fumarole activity (Mt. Amiata, central Italy). Chemosphere 41:1333–1336
Loppi S, Frati L (2006) Lichen diversity and lichen transplants as monitors of air pollution in a rural area of central Italy. Environ Monit Assess 114:361–375. doi:10.1007/s10661-006-4937-1
Loppi S, Pirintsos SA (2000) Effect of dust on epiphytic lichen vegetation in the mediterranean area (Italy and Greece). Isr J Plant Sci 48:91–95
Loppi S, Nelli L, Ancora S, Bargagli R (1997) Passive monitoring of trace elements by means of tree leaves, epiphytic lichens and bark substrate. Environ Monit Assess 45:81–88
Loppi S, Pacioni G, Olivieri N, Di Giacomo F (1998) Accumulation of trace metals in the lichen *Evernia prunsteri* transplanted at biomonitoring sites in central Italy. Bryologist 101(3):451–454

Manning WJ, Feder WA (1980) Biomonitoring air pollutants with plants. Applied Science Publishers, London

Markert B, Oehlmann J, Roth M (1997) General aspects of heavy metal monitoring by plants and animals. In: Subramanian KS, Iyengar GV (eds) Environmental biomonitoring – exposure, assessment and specimen banking, vol 654, ACS symposium series. American Chemical Society, Washington D.C., pp 19–29

Markert BA, Breure AM, Zechmeister HG (2003) Definitions, strategies and principles for bioindication/biomonitoring of the environment. In: Markert BA, Breure AM, Zechmeister HG (eds) Bioindicators and biomonitors. Elsevier, Oxford, pp 3–39

McCune B, Derr CC, Muir PS, Shirazi M, Sillet SC, Peck JE, Dale WJ (1996) Pendents for measuring lichen growth. Lichenologist 28(2):161–169

Mishra SK, Upreti DK, Pandey V, Bajpai R (2003) Pollution monitoring with the help of lichens transplant technique in some commercial and industrial areas of Lucknow city. Polln Res 22(2):221–225

Muir DCG, Segstro MD, Welbourn PM, Toom D, Eisenreich SJ, Macdonald CR, Whelpdale DM (1993) Patterns of accumulation of airborne organochlorine contaminants in lichens from the upper great lakes region of Ontario. Environ Sci Technol 27:1201–1210

Nayaka S, Upreti DK (2005a) Lichen flora of Pune city (India) with reference to air pollution. In: Abstracts of the IIIrd international conference on plants and environmental pollution, NBRI, Lucknow, 28 Nov–2 Dec 2005

Nayaka S, Upreti DK, Gadgil M, Pandey V (2003) Distribution pattern and heavy metal accumulation in lichens of Bangalore city with special reference to Lalbagh garden. Curr Sci 84(5):674–680

Nayaka S, Upreti DK, Pandey V, Pant V (2005b) Manganese (Mn) in lichens growing on magnasite rocks in India. Bull Bri Lic Soc 97:66–68

Nayaka S, Singh PK, Upreti DK (2005c) Fungicidal elements accumulated in Cryptothecia punctata (Ascomycetes) lichens of an Arecanut orchard in South India. J Environ Biol 26(2):299–300

Nieboer E, Richardson DHS, Tomassini FD (1978) Mineral uptake and release by lichen: an overview. Bryologist 81:226–246

Nylander W (1866) Les lichens Jardin du Luxembourg. Bull Soc Bot France 13:364–372

Ockenden WA, Steinnes E, Parker C, Jones KC (1998) Observations on persistant organic pollutant in plants: implications for their use as passive air sampler and for POP cycling. Environ Sci Technol 32:2721–2726

Paoli L, Loppi S (2008) A biological method to monitor early effect of the air pollution. Environ Pollut 155:383–388. doi:10.1016/j.envpol.2007.11.004

Pfeiffer HN, Barclay-Estrup P (1992) The use of a single lichen species, Hypogymnia physodes as an indicator of air quality in northwestern Ontario. Bryologist 95:38–41

Pinho P, Augusto S, Branquinho C, Bio A, Pereira MJ, Soares A, Catarino F (2004) Mapping lichen diversity as a first step for air quality assessment. J Atmos Chem 49:377–389

Pirintsos SA, Kotzabasis K, Loppi S (2004) Polyamine production in lichens under metal pollution stress. J Atmos Chem 49:303–315

Pirintsos SA, Matsi T, Vokou D, Gaggi C, Loppi S (2006) Vertical distribution patterns of trace elements in an urban environment as reflected by their accumulation in lichen transplants. J Atmos Chem 54:121–131. doi:10.1007/s10874-006-9019-9

Poličnik H, Simončič P, Bati F (2008) Monitoring air quality with lichens: a comparison between mapping in forest sites and in open areas. Environ Pollut 151:395–400. doi:10.1016/j.envpol.2007.06.003

Purvis W (2000) Lichens. The Natural History Museum, London

Rao DN, LeBlanc F (1967) Influence of an iron sintering plant on corticolous epiphytes in Wawa, Ontario. Bryologist 70:141–157

Richardson DHS (1991) Lichens as biological indicators. Recent developments. In: Jeffery DW, Madden B (eds) Bioindicators and environmental management. Academic, London, pp 263–272

Richardson DHS (1992) Pollution monitoring with lichens, vol 19, Naturalist's handbooks. Richmond Publishing Co. Ltd, Slough, p 76

Riga-Karandinos NA, Karandinos GM (1998) Assessment of air pollution from a lignite power plant in the plain of megalopolis (Greece) using as a biomonitors three species of lichens; impact on some biochemical parameters of lichens. Sci Total Environ 215:167–183

Satya UDK (2009) Correlation among carbon, nitrogen, sulphur and physiological parameters of *Rinodina sophodes* found at Kanpur city, India. J Hazard Mater. doi:10.1016/j/jhazmat.2009.04.063

Satya, Upreti DK, Nayaka S (2005) *Shorea robusta* – an excellent host tree for lichen growth in India. Curr Sci 89(4):594–595

Saxena S, Upreti DK (2004) Lichen flora of Lucknow district with reference to air Pollution studies in the area. Ph.D. thesis, Lucknow University, Lucknow

Saxena S, Upreti DK, Sharma N (2007) Heavy metal accumulation in lichens growing in north side of Lucknow city. J Environ Biol 28(1):45–51

Seaward MRD (1974) Some observations on heavy metal toxicity and tolerance in lichens. Lichenologist 6:158–164

Seaward MRD (1993) Lichens and sulphur dioxide air pollution field studies. Environ Rev 1:73–91

Shirazi AM, Muir PS, McCune B (1996) Environmental factors influencing the distribution of lichen lobaria oregano and L. Pulmonaria. Bryologist 99(1):12–18

Shukla V (2007) Lichens as bioindicator of air pollution. Final technical report. Science and Society Division, Department of Science and Technology, New Delhi. Project No. SSD/SS/063/2003

Shukla V, Upreti DK (2007a) Heavy metal accumulation in *Phaeophyscia hispidula* en route to Badrinath, Uttaranchal, India. Environ Monit Assess 131:365–369. doi:10.1007/s10661-006-9481-5

Shukla V, Upreti DK (2007b) Physiological response of the lichen *Phaeophyscia hispidula* (Ach.) essl. To the urban environment of Pauri and Srinagar (Garhwal), Himalayas. Environ Pollut 150:295–299. doi:10.1016/j.envpol.2007.02.010

Shukla V, Upreti DK (2008a) Effect of metallic pollutants on the physiology of lichen, *Pyxine subcinerea* Stirton in Garhwal Himalayas. Environ Monit Assess 141:237–243. doi:10.1007/s10661-007-9891-z

Shukla V, Upreti DK (2008b) Role of lichen substances in the thermoprotcction of *Pyxine subcinerea* Stirton in extreme environmental conditions of Garhwal Himalayas. In: Abstracts of the 3rd J & K science congress, Jammu University, Jammu, 26–28th Feb 2008

Shukla V, Upreti DK (2009) Polycyclic aromatic hydrocarbon (PAH) accumulation in lichen, *Phaeophyscia hispidula* of DehraDun city, Garhwal Himalayas. Environ Monit Assess 149(1–4):1–7

Shukla V, Upreti DK (2010) Changing lichen diversity in and around urban settlements of Garhwal Himalayas due to increasing anthropogenic activities. Environ Monit Assess. doi:10.1007/s10661-010-1468-6

Shukla V, Upreti DK, Patel DK, Tripathi R (2010) Accumulation of polycyclic aromatic hydrocarbons in some lichens of Garhwal Himalayas, India. Int J Environ Waste Manag 5(1/2):104–113

Singh KP, Sinha GP (2010) Indian lichens an annotated checklist. Botanical Survey of India, Ministry of Environment and Forest, Kolkata

Singh AK, Gupta HK, Gupta K, Singh P, Gupta VB, Sharma RC (2007) A comparative study of air pollution in Indian cities. Bull Environ Contam Toxicol 78:411–416

Sloof JE (1995) Lichens as quantitative biomonitors for atmospheric trace-elements deposition, using transplants. Atmos Environ 29:11–20

St. Clair BS, St. Clair LL, Mangelson FN, Weber JD (2002a) Influence of growth form on the accumulation of airborne copper by lichens. Atmos Environ 36:5637–5644

St. Clair BS, St Clair LL, Weber JD, Mangelson FN, Eggett LD (2002b) Element accumulation patterns in foliose and fruticose lichens from rock and bark substrates in Arizona. Bryologist 105:415–421

Swinscow TDV, Krog H (1988) Macrolichens of East Africa. British Museum (Natural History), London

Thormann MN (2006) Lichens as indicators of forest health in Canada. Forest Chron 82(3):335–343

Thrower SL (1980) Air pollution and lichens in Hong kong. Lichenologist 12(3):305–311

Tuba Z, Csintalan Z (1993) Bioindication of road motor traffic caused heavy metal pollution by lichen transplants. In: Markert B (ed) Plants as biomonitors indicators for heavy metal in the terrestrial environment. VCH, Weinheim, pp 329–341

Upreti DK (1994) Lichens: the great benefactors. Appl Bot Abst 14(3):64–75

Upreti DK, Nayaka S (2008) Need for creation of lichen garden and sanctuaries in India. Curr Sci 94(8):976–978

Upreti DK, Nayaka S, Bajpai A (2005) Do lichens still grow in Kolkata city? Curr Sci 88(3):338–339

van Dobben HF, ter Braak CJF (1999) Ranking of epiphytic lichen sensitivity to air pollution using survey data: a comparison of indicator scales. Lichenologist 31(1):27–39

van Herk CM, Aptroot A, van Dobbin HF (2002) Longterm monitoring in the Netherlands suggests that lichen respond to global warming. Lichenologist 34:141–154

Vestergaard N, Stephansen U, Rasmussen L, Pilegaard K (1986) Airborne heavy metal pollution in the environment of a Danish steel plant. Water Air Soil Pollut 27:363–377

Wadleigh MA, Blake DM (1999) Tracing sources of atmospheric sulphur using epiphytic lichens. Environ Pollut 106:265–271

Wittig R (1993) General aspects of biomonitoring heavy metals by plants. In: Markert B (ed) Plants as biomonitors – indicators for heavy metals in the terrestrial environment. VHC, Weinheim, pp 3–27

Wolterbeek HTh, Garty J, Reis MA, Freitas MC (2003) Biomonitors in use: lichens and metal air pollution. In: Markert BA, Breure AM, Zechmeister HG (eds) Bioindicators and biomonitors. Elsevier, Oxford, pp 377–419

World Health Organization (WHO) (2000) Guideline for air quality. WHO Regional Publications, Geneva

# Part II
# Remediation

# Chapter 8
# Electrochemical Remediation Technologies for Waters Contaminated by Pharmaceutical Residues

**Enric Brillas and Ignasi Sirés**

## Contents



**Abstract** The presence of pharmaceutical micropollutants in the environment has become of major concern in the last decades. Many electrochemical technologies are currently available for the remediation of waters contaminated by refractory organic pollutants. They are mainly defined as eco-friendly water treatments since the main reagent involved is a clean species, the electron. Recent reviews have focused on the destruction of pharmaceutical residues by the application of methods

---

E. Brillas • I. Sirés (✉)
Laboratori d'Electroquímica dels Materials i del Medi Ambient,
Departament de Química Física, Facultat de Química, Universitat de Barcelona,
Martí i Franquès 1-11, Barcelona, 08028, Spain
e-mail: brillas@ub.edu; i.sires@ub.edu; isiresa@gmail.com

E. Lichtfouse et al. (eds.), *Environmental Chemistry for a Sustainable World: Volume 2: Remediation of Air and Water Pollution*, DOI 10.1007/978-94-007-2439-6_8,

like ozonation and advanced oxidation processes. Here, we present an overview on electrochemical methods devised for the removal of pharmaceutical residues from both synthetic solutions and real pharmaceutical wastewaters. The fundamentals and experimental set-ups involved in different technologies such as electrocoagulation, anodic oxidation, electro-oxidation with active chlorine, electro-Fenton, photoelectro-Fenton, and photoelectrocatalysis, among others, are discussed. Progress on the promising solar photoelectro-Fenton process devised and further developed in our laboratory is especially highlighted and documented. The destruction of the individual pharmaceuticals and the abatement of total organic carbon or reduction of chemical oxygen demand allow the comparison between the different methods. In some cases, the routes for the complete degradation of the initial pollutants are discussed.

**Keywords** Advanced oxidation processes • Anodic oxidation • Boron-doped diamond anode • Carboxylic acids • Electrochemical technologies • Electrocoagulation • Electro-Fenton • Electro-oxidation with active chlorine • Gas-diffusion electrode • Hydrogen peroxide • Hydroxyl radical • Organic pollutants • Pharmaceuticals • Photoelectrocatalysis • Photoelectro-Fenton • Reaction pathway • Solar-assisted processes • Total organic carbon (TOC) • Wastewater • Water treatment

## 8.1 Introduction

Since the 1970s, the impact of chemical contamination on the aquatic environment has mainly focused on persistent priority pollutants such as acutely toxic/carcinogenic pesticides, synthetic dyes, and industrial chemicals. This pollution considerably affects the feasibility of water reuse from effluents of sewage treatment plants (STPs), which is an essential task due to the severe global water problems. Recently, pharmaceuticals are receiving increasing attention as potential bioactive chemicals in the environment, being considered as emerging pollutants because they still remain unregulated or are currently undergoing a regularization process under the assessment of the corresponding governmental agencies, although the directives and legal frameworks are not yet set-up (Daughton and Ternes 1999; Esplugas et al. 2007). Actually, intensive research on this topic started about 15 years ago (Kümmerer 2009). In the literature, different terms such as 'drugs', 'pharmaceuticals', 'pharmaceutically active compounds (PhACs)', 'active pharmaceutical ingredients (APIs)', and 'pharmaceuticals and personal care products (PPCPs)' are routinely employed to refer to the same emerging issue. These compounds and their bioactive metabolites, all of them considered as 'pharmaceutical residues' in this review, are continuously introduced in the aquatic environment, where they can reach trace concentrations (i.e., micropollutants found in the ng $L^{-1}$ or μg $L^{-1}$ range) and become persistent. This affects the water quality and drinking water supplies and may constitute a potential risk for the ecosystems and the human and animal welfare in the long term (Klavarioti et al. 2009). In fact, it seems probable that most urban wastewater is contaminated with medicinal compounds (Jones et al. 2005). Table 8.1

**Table 8.1** Classification, use, and electrochemical treatment of common pharmaceuticals contained in synthetic solutions prepared using ultra-pure, distilled, or deionized water, as well as in real wastewaters

| Class | Specific compounds or groups | Use | Technology |
|---|---|---|---|
| *Synthetic aqueous solutions* | | | |
| Analgesics/Non-steroidal anti-inflammatories (NSAIDs) | Paracetamol (acetaminophen), salicylic acid[a], diclofenac, ibuprofen, ketoprofen, naproxen | NSAIDs are the most used and abused drugs in the world and have analgesic, antipyretic, and anti-inflammatory effects | 2,3,4,6 |
| Antibiotics | Sulfonamides, fluoroquinolones, tetracyclines | Used to treat fungal and internal bacterial infections in both, humans and animals | 2,3,4,6 |
| Antiepileptics | Carbamazepine | They stop, prevent, or control convulsions, partial seizures, and generalized tonic-chronic seizures. | 7 |
| Antihypertensives (β-blockers) | Atenolol, metoprolol, propranolol | Chemicals that allow the treatment of cardiovascular disorders by reducing the blood pressure in the arteries. | 2,3,4 |
| Antineoplastics | Cyclophosphamide, ifosfamide | They can treat solid tumors, lymphomas, leukemias, and are helpful in some autoimmune disorders such as rheumatoid arthritis | 2 |
| Antiseptics/Antimicrobials | Chlorophene, triclosan, triclocarban, chloroxylenol | Chemical agents that slow/stop the growth of micro-organisms (germs) on external surfaces such as the epidermis and help prevent infections. | 2,3,4 |
| Contraceptives/Steroidal hormones | 17α-ethinylestradiol, 17β-estradiol | Chemicals that are taken by mouth to inhibit normal fertility by acting on the hormonal system | 2 |
| Hypnotic/Anti-anxiety agents | Diazepam, riluzole | They relieve anxiety, nervousness, and tension associated with anxiety disorders | 3 |
| Blood lipid regulators | Clofibric acid[b] | They reduce the cholesterol and triglyceride levels in the blood | 2,3,4,7 |
| *Real wastewaters* | | | |
| Clinical/Hospital wastewaters | | | 2 |
| Pharmaceutical wastewaters | | | 1,2,5 |

Technology: (1) Electrocoagulation; (2) Electrochemical oxidation; (3) Electro-Fenton and anodic Fenton treatment; (4) UVA/Solar photoelectro-Fenton; (5) Internal micro-electrolysis; (6) Photoelectrocatalysis; (7) Corona discharge

[a]Metabolite of acetylsalicylic acid

[b]Active metabolite of lipid regulators (clofibrate, theofibrate, etofibrate)

collects several common pharmaceutical residues, whose treatment will be discussed and referenced along this chapter. They have been detected in the aquatic environment and are used in either human or in veterinary medicine, as well as in other activities.

Several publications have been devoted to review the occurrence and fate of pharmaceuticals in the environment (Daughton and Ternes 1999; Pauwels and Verstraete 2006; Khetan and Collins 2007; Kümmerer 2009; Rahman et al. 2009). Since thousands of tons of these chemicals are largely consumed worldwide, they are ubiquitous in soil and water. This kind of pollution arises from various routes, including emission from production sites, direct disposal of overplus drugs in households and hospitals, excretion from urine or faeces after drug administration to humans and animals, and water treatments in fish farms. Veterinary pharmaceuticals, for example, contaminate soil directly via manure and surface and ground waters by runoff from fields (Khetan and Collins 2007). Pharmaceuticals persist in the environment mainly because the elimination of such highly polar compounds in the STPs is incomplete, ranging between 60% and 90% even under the possible action of biodegradation, deconjugation, partitioning, and photodegradation steps (Jones et al. 2005). Once they enter in these plants, three main fates are possible: (1) the compound is ultimately mineralized to carbon dioxide, inorganic ions, and water, (2) the compound does not degrade because of its lipophilicity, and it is partially retained in the sedimentation sludge, and (3) the compound metabolizes to a more hydrophilic species, thus remaining soluble. The STP effluents then contain biorecalcitrant unmetabolized and metabolized pharmaceutical residues that are released in the receiving surface waters, mainly rivers (Klavarioti et al. 2009).

Pharmaceuticals released into the environment may impose toxicity potentially on any level of the biological hierarchy. Although the active ingredients of these compounds have been selected or designed because of their activity against organisms, their effects on living beings in the aquatic and terrestrial environment are not well documented (Kümmerer 2009). For example, it is usually accepted that some pharmaceuticals may cause long-term, irreversible changes to the micro-organisms genome, even at low contents, which therefore increases their resistance to them (Esplugas et al. 2007; Klavarioti et al. 2009). However, several studies on the effects of antibiotics and other drugs such as diclofenac, ibuprofen, and carbamazepine on *Daphnia*, algae, and bacteria have shown that chronic lowest observed effect concentrations (LOEC) under standard laboratory conditions are about two orders of magnitude higher than the maximum contents detected for STP effluents. It has also been found that the LOEC of diclofenac for fish toxicity is in the range of wastewater concentration, whereas the LOEC of the beta-blocker propranolol for zooplankton and benthic organisms is close to the maximum STP effluent concentrations (Khetan and Collins 2007; Kümmerer 2009). The question is certainly troublesome because the negative effects can remain unnoticed, being slowly accumulated until the evident changes are finally revealed (Daughton and Ternes 1999). Furthermore, this kind of pollutants often occur as complex mixtures, giving rise to 'drug cocktails' whose toxicity has been seldom predicted, as in the case of beta-blockers (Cleuvers 2005). More toxicological studies are then needed to clarify the possible effects of the large variety of pharmaceuticals found in the environment on living beings. On the other hand, some pharmaceuticals have also been classified as endocrine disrupting compounds (EDCs) and, consequently, their presence in the aquatic systems has risen considerable concern since these chemicals cause harmful effects on the human endocrine system (Esplugas et al. 2007; Klavarioti et al. 2009; Rahman et al. 2009).

The aforementioned considerations clearly reflect the need for the complete removal of pharmaceuticals and their metabolites from aquatic systems to avoid their potential toxicity and other possible dangerous health effects. These pollutants cannot be totally destroyed in the STPs using conventional techniques such as biological remediation and physicochemical treatments including coagulation, volatilization, adsorption, sedimentation, and filtration (Jones et al. 2005; Suárez et al. 2008; Rahman et al. 2009). In most STPs, disinfection of final effluents is mandatory and usually chlorination or UV irradiation are applied; however, both techniques exhibit low oxidation ability and the trace organic pollutants are refractory to their action (Khetan and Collins 2007; Suárez et al. 2008). For all these reasons, recent research has focused on the application of ozonation and advanced oxidation processes (AOPs) to the treatment of pharmaceutical residues and pharmaceutical wastewaters.

The AOPs are environmentally friendly chemical, photochemical, or electrochemical methods sharing the common feature of the in situ production of the hydroxyl radical ($^{\bullet}OH$) as their main oxidant. The $^{\bullet}OH$ radical is the second strongest oxidizing agent known after fluorine, having such a high standard reduction potential ($E^{\circ}(^{\bullet}OH/H_2O) = 2.8$ V vs. Standard Hydrogen Electrode, SHE) that it is able to non-selectively react with most organic pollutants via hydroxylation or dehydrogenation until their total mineralization. Several comprehensive and extensive reviews have reported the effective use of $O_3$, chemical AOPs such as $O_3/H_2O_2$, Fenton's reagent ($Fe^{2+}/H_2O_2$), ultrasound, and wet oxidation, and photochemical AOPs such as $O_3$/UV, $H_2O_2$/UV, photo-Fenton ($Fe^{2+}/H_2O_2$/UV), and $TiO_2$ photocatalysis; they have been used alone or combined with physicochemical or biological treatments, aiming at the remediation of synthetic and real wastewaters with drugs in general (Ikehata et al. 2006; Esplugas et al. 2007; Klavarioti et al. 2009) and with special attention for chemical and pharmaceutical industrial wastewaters (Bloecher 2007), urban wastewater (Suárez et al. 2008), and drinking water (Rahman et al. 2009). However, only some few papers have summarized some data on the performance of the electrochemical advanced oxidation processes (EAOPs) such as the anodic oxidation (AO), electro-Fenton (EF), and photoelectro-Fenton (PEF) to remove pharmaceuticals (Garrido et al. 2007; Brillas et al. 2008, 2009; Klavarioti et al. 2009).

Over the past 15 years, the electrochemical technologies have experienced a large development for their use as an alternative wastewater treatment technology (Chen 2004; Oturan and Brillas 2007; Brillas et al. 2009; Panizza and Cerisola 2009). Currently, they offer promising approaches to remove pollutants from industrial effluents. The main advantage of these technologies is their environmental compatibility because the main reagent, the electron, is a clean reagent. Other advantages include their versatility, high energy efficiency, amenability to automation, easy handling because of the simple equipment required, and safety because they operate under mild conditions (Anglada et al. 2009; Martínez-Huitle and Brillas 2009). Despite their potential application, most studies related to the electrochemical treatment of waters contaminated by pharmaceuticals have been published over the last 6 years.

Here, we present a general overview on the electrochemical remediation technologies that have been applied to the removal of pharmaceutical residues from both, synthetic solutions and real pharmaceutical wastewaters. Table 8.1 summarizes the treatments applied to the different classes of pharmaceuticals and types of solutions. Despite their complex nature, varied functionalities, and broad spectrum of physicochemical and biological properties, these molecules can be progressively degraded by the electrochemical processes that will be presented. The most conventional electrocoagulation process will be firstly considered, as well as its combination with other methods. The treatments by means of different EAOPs like the AO with different anodes, electro-oxidation with active chlorine, EF and related approaches, and PEF with UV or direct solar irradiation are further detailed, followed by the use of novel emerging technologies such as the internal micro-electrolysis, photoelectrocatalysis (PEC), and corona discharge. A global perspective on the fundamentals and experimental set-ups is offered, and lab and pilot plant experiments are examined and discussed.

## 8.2 Electrocoagulation

Coagulation is a traditional physicochemical treatment of phase separation for the decontamination of wastewaters containing colloidal and ionic species before discharge to the aquatic environment. Coagulating agents such as $Fe^{3+}$ or $Al^{3+}$ ions are usually added in the form of chloride salts for precipitation. Similar effects can be achieved by means of the electrochemical technology known as electrocoagulation (EC) (Chen 2004; Martínez-Huitle and Brillas 2009). In EC, a current is applied to dissolve Fe (or steel) or Al anodes immersed in the polluted water to release the corresponding metal ions that form different Fe(II) (and/or Fe(III)) or Al(III) species with hydroxide ion depending on the pH of the medium. These species act as coagulants or destabilizing agents that neutralize charges and separate colloids and ionic products from the wastewater by sedimentation, finally producing some sludge. Otherwise, the coagulated particles can also be separated by electroflotation when they are attached to the bubbles of $H_2$ gas evolved at the cathode, being transported to the solution surface where they can be withdrawn.

When EC is applied with an iron or steel anode, $Fe^{2+}$ is dissolved in the wastewater from the anodic oxidation of Fe by reaction (8.1) with $E^{o} = -0.44$ V vs. SHE:

$$Fe \rightarrow Fe^{2+} + 2e^{-} \quad (8.1)$$

whereas $H_2$ gas is generated at the cathode from direct reduction of proton in acidic medium by reaction (8.2) or from water reduction in alkaline medium, along with hydroxide ion, by reaction (8.3) with $E^o = -0.83$ V vs. SHE:

$$2H^+ + 2e^- \rightarrow H_{2(g)} \tag{8.2}$$

$$2H_2O + 2e^- \rightarrow 2OH^- + H_{2(g)} \tag{8.3}$$

$H^+$ consumption from reaction (8.2) or $OH^-$ production from reaction (8.3) causes an increase in pH during the electrolysis. The insoluble $Fe(OH)_2$ precipitates at pH > 5.5, then remaining in equilibrium with $Fe^{2+}$ up to pH 9.5 or with monomeric species such as $Fe(OH)^+$, $Fe(OH)_2$, and $Fe(OH)_3^-$ at higher pH values.

In the presence of $O_2$, dissolved $Fe^{2+}$ is oxidized to insoluble $Fe(OH)_3$ by reaction (8.4):

$$4Fe^{2+} 10H_2O + O_{2(g)} \rightarrow 4Fe\left(OH\right)_{3(s)} + 8H^+ \tag{8.4}$$

$Fe(OH)_3$ coagulates from pH > 1.0 and the insoluble flocs can be in equilibrium with soluble monomeric species like $Fe^{3+}$, $Fe(OH)^{2+}$, $Fe(OH)_2^+$, $Fe(OH)_3$, and $Fe(OH)_4^-$ as a function of pH. The insoluble $Fe(OH)_2$ and/or $Fe(OH)_3$ flocs retain colloids and ionic species and coagulate to form particles that are separated from the wastewater by sedimentation or electroflotation.

In the case of EC with Al, the anodic reaction leads to soluble $Al^{3+}$ by reaction (8.5) with $E^o = 1.66$ V vs. SHE:

$$Al \rightarrow Al^{3+} + 3e^- \tag{8.5}$$

$Al^{3+}$ is transformed into soluble monomeric species such as $Al(OH)^{2+}$, $Al(OH)_2^+$, and $Al(OH)_3$ in acid medium, and $Al(OH)_4^-$ in alkaline medium. These species are converted into insoluble $Al(OH)_3$ flocs by a complex precipitation kinetics. Freshly formed amorphous $Al(OH)_3$ flocs can adsorb ions and even soluble organic compounds and/or trap colloidal particles, which then coagulate to produce particles that precipitate, usually near neutral pH.

The EC method has been envisaged as a pre-treatment of wastewaters from pharmaceutical industries, although little work has been made in this way. A first paper by Deshpande et al. (2005) described the successful degradation of a pharmaceutical effluent using a cast iron anode with the removal of 72% of chemical oxygen demand (COD) and the improvement of the biochemical oxygen demand (BOD)/COD ratio from 0.18 to 0.3, which points out the enhancement of the wastewater biodegradability. A more comprehensive study was further reported by Deshpande et al. (2009) for the treatment of a high-strength wastewater from an industry manufacturing the drug ranitidine with 34 $g L^{-1}$ COD, 16 $g L^{-1}$ BOD, and 775 Pt-Co units color at pH 6.8–7.2. The EC experiments were performed with 150 mL of wastewaters in a batch tank reactor like that of Fig. 8.1, with a mechanical stirrer and equipped with Al/Al, mild steel/mild steel or stainless steel/stainless steel pairs. The electrodes exhibited 25 $cm^2$ geometric immersed area. Comparative electrolyses at 8 $mA\ cm^{-2}$

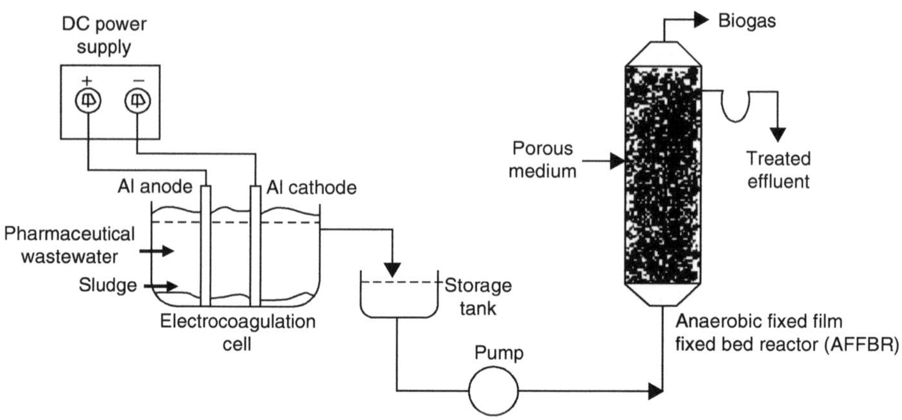


**Fig. 8.1** Experimental set-up of a bench-scale system for electrocoagulation combined with an anaerobic fixed film fixed bed reactor (AFFBR) for the treatment of a real pharmaceutical wastewater. The wastewater is electrolyzed in an undivided Al/Al cell and the liquid supernatant of the sludge formed is stored in a tank to be further pumped into the AFFBR packed with fibrous synthetic nylon pads for the attachment of biomass, which releases gas and the final treated effluent. Adapted from Deshpande et al. (2010)

revealed that COD could be reduced by 24% at 25 min, 25% at 120 min, and 33% at 120 min as maximum, respectively, attaining corresponding final BOD/COD values of 0.546, 0.519, and 0.538 and final color of 235, 325, and 490 Pt-Co units. These results indicated a slight enhancement of the biodegradability in all cases and a better efficiency using the Al/Al pair during shorter time, which allowed the least energy consumption (=95.83 Wh (kg COD)$^{-1}$) and the greatest anode efficiency (=5.76 kg COD A$^{-1}$ m$^{-2}$h$^{-1}$). Based on this behavior, Deshpande et al. (2010) recently considered the decontamination of the same pharmaceutical wastewater by combining an EC pre-treatment with the above Al/Al tank reactor and an anaerobic fixed film fixed bed reactor (AFFBR) packed with fibrous synthetic nylon pads for the attachment of biomass. Figure 8.1 shows a scheme of the coupled system, where the supernatant liquid of the EC cell was stored in a tank before being pumped into the AFFBR producing the treated effluent and biogas. As can be seen in Fig. 8.2, similar final COD (23–24%) and BOD (35–37%) removals were found by applying 4, 8, and 12 mA cm$^{-2}$ at the Al/Al cell, being required a shorter time and higher energy consumption with increasing current density due to the larger extent of the parasitic reactions, mainly the water electrolysis reaction and the heating due to the Joule effect. A study in the pH range 5–11 also demonstrated that the higher COD abatement was achieved at the initial pH of the wastewater (natural pH), with a final value of 7.5. Optimum removals of 80% COD and 85% BOD were obtained when the AFFBR was fed with the EC pre-treated effluent at a loading rate of 4.0 kg COD m$^{-3}$ day$^{-1}$ and retention time of 2 days, which were much higher than 68% COD and 72% BOD obtained for the received effluent not having undergone the EC pre-treatment. The yield of the biogas generated was 0.3 m$^3$ $CH_4$ (kg COD)$^{-1}$.

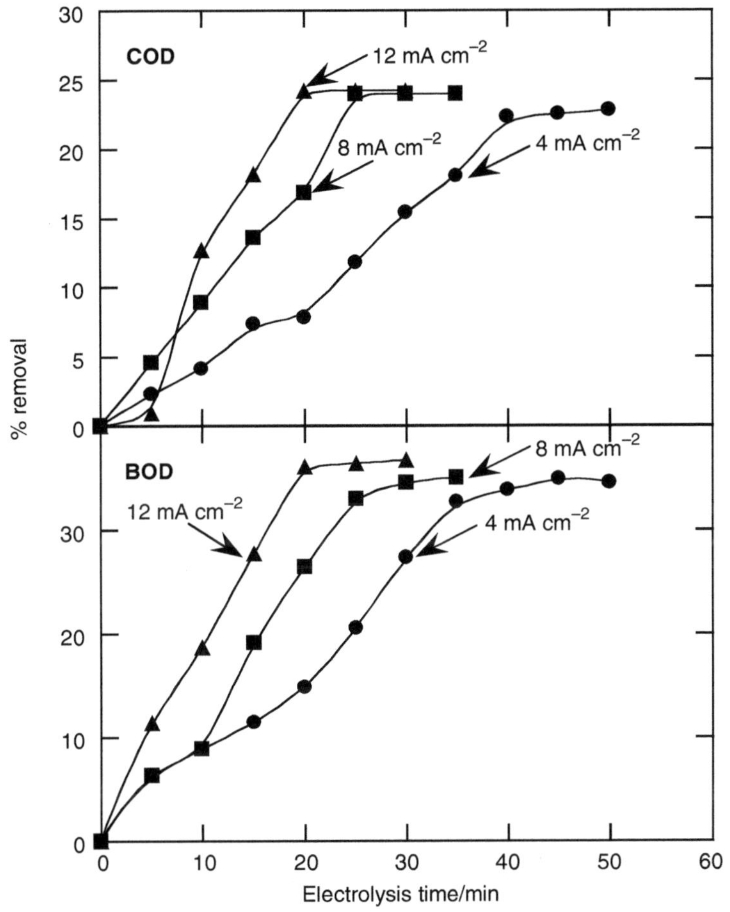


**Fig. 8.2** Effect of current density on the percentage of COD and BOD removal with the electrolysis time for the cell of Fig. 8.1 equipped with two parallel Al plates of 25 $cm^2$ area to treat 150 mL of a pharmaceutical wastewater with $COD_0 = 34.4\ g L^{-1}$ and $BOD_0 = 22\ g L^{-1}$ at pH 7.2. Final COD and BOD reductions are almost independent of current density. COD: Chemical oxygen demand, BOD: Biochemical oxygen demand. Adapted from Deshpande et al. (2010)

These results open the door to future research on the coupling of EC-biological systems with supply of a very small current to the EC pre-treatment to minimize energy consumption in view of the much longer time needed for the biological treatment.

The use of EC pre-treatment followed by $TiO_2$ photocatalysis has also been recently explored. The latter method is an AOP that involves the irradiation of anatase $TiO_2$ (nano)particles in colloidal suspension by UV photons of sufficient energy ($\lambda < 380$ nm) to promote an electron from the valence band to the conduction band ($e_{CB}^-$) with a band gap of 3.2 eV, thus generating a positively charged vacancy or hole ($h_{VB}^+$) from reaction (8.6). Heterogeneous $^{\bullet}OH$ can then formed from reaction between the photogenerated vacancy and adsorbed water by reaction (8.7). As a result, the organics can be directly oxidized by the hole or by the $^{\bullet}OH$. Other weaker reactive oxygen species (ROS) such as superoxide radical ion ($O_2^{-\bullet}$), hydroperoxyl radical ($HO_2^{\bullet}$), and $H_2O_2$, as well as homogeneous $^{\bullet}OH$, can be produced from the photoinjected electron by reactions (8.8)–(8.11) (Boroski et al. 2009; Martínez-Huitle and Brillas 2009).

$$TiO_2 + h\nu \rightarrow e^-_{CB} + h^+_{VB} \quad (8.6)$$

$$h^+_{VB} + H_2O \rightarrow {}^{\bullet}OH + H^+ \quad (8.7)$$

$$e^-_{CB} + O_2 \rightarrow O_2^{-\bullet} \quad (8.8)$$

$$O_2^{-\bullet} + H^+ \rightarrow HO_2^{\bullet} \quad (8.9)$$

$$2HO_2^{\bullet} \rightarrow H_2O_2 + O_2 \quad (8.10)$$

$$H_2O_2 + O_2^{-\bullet} \rightarrow {}^{\bullet}OH + OH^- + O_2 \quad (8.11)$$

The major loss in efficiency in $TiO_2$ photocatalysis is due to the recombination of the electrons promoted to the valence band, either with unreacted holes by reaction (8.12) or with adsorbed hydroxyl radical by reaction (8.13):

$$e^-_{CB} + h^+_{VB} \rightarrow TiO_2 + \text{heat} \quad (8.12)$$

$$e^-_{CB} + {}^{\bullet}OH \rightarrow OH^- \quad (8.13)$$

Note that reaction (8.13) considerably limits the efficiency of available hydroxyl radicals formation, which truly constitutes the major drawback of this process.

Boroski et al. (2009) studied the combination of EC and $TiO_2$ photocatalysis to degrade an effluent from a pharmaceutical and cosmetic industry with 1.8 $g L^{-1}$ COD and 25.5 FTU turbidity of pH 7.2. The EC pre-treatment was carried out with 500 mL of wastewater in a batch cylindrical reactor with two 31 $cm^2$ Fe electrodes as the anode and cathode under magnetic stirring at a constant current density of 76.3 $mA\ cm^{-2}$. The initial pH was adjusted in the range 4.0–10.0 and the best decontamination was attained starting from pH 6.0 up to pH 10.5 at 90 min, with reduction of 91% COD and 86% turbidity with 92% current efficiency. The EC pre-treatment then yielded a residual wastewater still containing about 160 $mg\ L^{-1}$ COD due to some remaining refractory organics. This effluent was further treated in a $UV/TiO_2/H_2O_2$ system composed of a borosilicate glass erlenmeyer of 300 mL in a photoreactor box of 80 cm × 50 cm with three 250 W mercury lamps on the top side at 15 cm of the sample. The optimum degradation was obtained operating with 0.25 $g L^{-1}$ $TiO_2$ and 10 mM $H_2O_2$ at pH 3.0 for 4 h, giving a final wastewater with near 50 $mg\ L^{-1}$ COD (97% COD removal), i.e., with a clear loss of organic matter compared with that of the pre-treated effluent. More research on EC-$TiO_2$

photocatalysis systems should be made to assess its potential viability at industrial level, because the use of mercury lamps may certainly be a major drawback considering the economic balance. Particular attention should be paid to an enhanced photocatalytic process, namely photoelectrocatalysis or photoelectrochemical treatment, where reactions (8.12)–(8.13) are inhibited or, at least, minimized (see Sect. 8.5.2).

## 8.3 Electrochemical Oxidation

### *8.3.1 Anodic Oxidation of Non-chloride Solutions*

The most popular electrochemical technique for wastewater remediation is the electrochemical oxidation, which is frequently called anodic oxidation when non-chloride solutions are treated. This procedure involves the oxidation of pollutants in an electrolytic cell by: (1) direct electron transfer to the anode and (2) indirect or mediated oxidation with heterogeneous ROS formed from water discharge at the anode, such as physically adsorbed "active oxygen" (physisorbed $^{\bullet}OH$) or chemisorbed "active oxygen", i.e., oxygen within the lattice of a metal oxide anode (MO). The existence of these species allowed the proposal of two approaches (Comninellis 1994; Marselli et al. 2003; Panizza and Cerisola 2009): (1) electrochemical conversion, where the refractory organics are selectively transformed into biodegradable compounds like carboxylic acids by chemisorbed "active oxygen", and (2) electrochemical combustion, where the organics are mineralized by physisorbed $^{\bullet}OH$.

The use of low cell voltages avoiding $O_2$ evolution only allows the direct electron transfer to the anode and frequently causes the loss of the anode activity because of the adsorption of some by-products. For wastewater treatment, high cell voltages are then applied to achieve the simultaneous oxidation of pollutants and water, maintaining the anode activity; but, the anode material has a strong influence on the selectivity and efficiency of the process, yielding either electrochemical conversion or electrochemical combustion depending on the case. Comninellis (1994) proposed a model to explain this different behavior of the electrodes assuming the existence of the so-called "active" and "non-active" anodes, respectively. Examples include Pt, $IrO_2$, and $RuO_2$ for the former and $PbO_2$, $SnO_2$, and boron-doped diamond (BDD) for the latter.

The model of Comninellis presupposes that in both kind of anodes, denoted as M, water is oxidized leading to the formation of physisorbed hydroxyl radical ($M(^{\bullet}OH)$) by reaction (8.14). In the case of "active" anodes, this radical interacts so strongly with their surface that it is transformed into the chemisorbed "active oxygen" or superoxide MO from reaction (8.15):

$$M + H_2O \rightarrow M(^{\bullet}OH) + H^+ + e^- \qquad (8.14)$$

$$M(^{\bullet}OH) \rightarrow MO + H^+ + e^- \tag{8.15}$$

Reaction (**8.15**) may occur when higher oxidation states are available for a metal oxide anode, above the thermodynamic standard potential for oxygen evolution ($E^o = 1.23$ V vs. SHE). The MO/M pair is a mediator in the oxidation of an organic compound (R) given by reaction (8.16), which competes with oxygen evolution via chemical decomposition of MO from reaction (8.17):

$$MO + R \rightarrow M + RO \tag{8.16}$$

$$MO \rightarrow M + \tfrac{1}{2}O_2 \tag{8.17}$$

Conversely, the surface of "non-active" anodes interacts so weakly with M($^{\bullet}$OH) that this radical reacts directly with the organics giving, for example, the following total mineralization reaction:

$$aM(^{\bullet}OH) + R \rightarrow M + mCO_2 + nH_2O + xH^+ + ye^- \tag{8.18}$$

where R is a hydrocarbon with $m$ carbon atoms and $a = (2m + n)$ oxygen atoms. Reaction (8.18) competes with side reactions of M($^{\bullet}$OH) such as its direct oxidation to $O_2$ from reaction (8.19) or its dimerization to hydrogen peroxide by reaction (8.20):

$$M(^{\bullet}OH) \rightarrow M + \tfrac{1}{2}O_2 + H^+ + e^- \tag{8.19}$$

$$2M(^{\bullet}OH) \rightarrow 2M + H_2O_2 \tag{8.20}$$

The destruction of the organic pollutant by direct electron transfer to the anode does not take place in a "non-active" anode such as BDD, because it only acts as an inert substrate and as a sink for the removal of electrons mediated by the M($^{\bullet}$OH). The BDD anode is the best "non-active electrode" verifying this behavior (Marselli et al. 2003; Panizza and Cerisola 2009) and is the most suitable anode for treating organics by AO. Owing to the electrogeneration of the hydroxyl radical as a mediator, the AO is considered as an EAOP.

However, when high cell voltages are applied, a weaker oxidant such as ozone can also be generated from water discharge at the anode surface ($E^o = 1.51$ V vs. SHE) by reaction (8.21):

$$3H_2O \rightarrow O_3 + 6H^+ + 6e^- \tag{8.21}$$

In AO, different ROS are then produced, such as heterogeneous $^{\bullet}$OH by reaction (8.14), $H_2O_2$ by reaction (8.20), and $O_3$ by reaction (8.21), although the physisorbed $^{\bullet}$OH is the strongest oxidant of pollutants. Moreover, when BDD is used, other weaker oxidizing species like peroxodisulfate, peroxodicarbonate, and peroxodiphosphate can also be competitively formed with ROS from the direct oxidation of the electrolyte ions, such as bisulfate by reaction (8.22), bicarbonate by reaction (8.23), and phosphate by reaction (8.24) (Panizza and Cerisola 2009):

$$2HSO_4^- \rightarrow S_2O_8^{2-} + 2H^+ + 2e^- \tag{8.22}$$

$$2HCO_3^- \rightarrow C_2O_6^{2-} + 2H^+ + 2e^- \tag{8.23}$$

$$2PO_4^{3-} \rightarrow P_2O_8^{4-} + 2e^- \tag{8.24}$$

### 8.3.2 *Electro-Oxidation with Active Chlorine*

A very different behavior is found when the wastewater contains chloride ions since the generated active chlorine species such as $Cl_2$, HClO and/or $ClO^-$, $ClO_2^-$, $ClO_3^-$ and $ClO_4^-$ can attack the organics in competition with ROS (Bergmann et al. 2009; Martínez-Huitle and Brillas 2009; Panizza and Cerisola 2009). This procedure is usually called electro-oxidation with active chlorine and is based on the direct oxidation of $Cl^-$ ion at the anode to yield soluble chlorine by reaction (8.25):

$$2Cl^- \rightarrow Cl_2(aq) + 2e^- \tag{8.25}$$

The electrogenerated $Cl_2$ can form bubbles if its local concentration exceeds its solubility, although it partly diffuses away from the anode and can react with chloride ion to form trichloride ion from reaction (8.26); otherwise, it is rapidly hydrolyzed and disproportionated to hypochlorous acid and chloride ion from reaction (8.27):

$$Cl_2(aq) + Cl^- \rightleftarrows Cl_3^- \tag{8.26}$$

$$Cl_2(aq) + H_2O \rightleftarrows HClO + Cl^- + H^+ \tag{8.27}$$

Hypochlorous acid is then in equilibrium with hypochlorite ion at $pK_a = 7.55$:

$$HClO \rightleftarrows ClO^- + H^+ \tag{8.28}$$

From the above equilibria, it is found that $Cl_3^-$ is formed in a very low concentration up to pH ca. 4, while the predominant species are $Cl_2$ until pH near 3, HClO in the pH range 3–8, and $ClO^-$ at pH > 8 (Martínez Huitle and Brillas 2009). The mediated oxidation with active chlorine species is expected to be faster in acid than in alkaline media because of the higher standard potential of $Cl_2$ ($E^o = 1.36$ V vs. SHE) and HClO ($E^o = 1.49$ V vs. SHE) compared to $ClO^-$ ($E^o = 0.89$ V vs. SHE). Nevertheless, only the chemical action of hypochlorite ion and generated ROS is commonly referred for organic mineralization since most electrolyses are performed with alkaline wastewaters. The concentration of hypochlorite ion, however, can be limited by its anodic oxidation to chlorite ion from reaction (8.29), followed by the successive oxidation to chlorate and perchlorate ions from reactions (8.30) and (8.31), respectively:

$$ClO^- + H_2O \rightarrow ClO_2^- + 2H^+ + 2e^- \tag{8.29}$$

$$ClO_2^- + H_2O \rightarrow ClO_3^- + 2H^+ + 2e^- \quad (8.30)$$

$$ClO_3^- + H_2O \rightarrow ClO_4^- + 2H^+ + 2e^- \quad (8.31)$$

It is worth mentioning that the formation and accumulation of toxic chloroderivatives, trihalomethanes, and chloramines is a possible major drawback of the electro-oxidation in the presence of active chlorine.

As a conclusion from Sects. 8.3.1 and 8.3.2, the complexity arising from both, the generation of ROS in the AO treatment of non-chloride effluents and chlorine active species when using chloride effluents, makes the degradation rate and current efficiency of these processes to be a function of different experimental parameters such as pH, temperature, stirring, substrate concentration, and current density, and their effects need to be studied to reach the best operative conditions of the applied method.

### *8.3.3 Treatment of Synthetic Solutions*

The mineralization ability of the AO and/or electro-oxidation with active chlorine has been tested for several synthetic solutions using a large variety of anodes and electrolytic systems. Table 8.2 collects the percentage of TOC decay determined for various common pharmaceuticals under selected conditions. An inspection of these data allows concluding that the use of a BDD anode yields a much higher mineralization degree compared to other "non-active" anodes such as $PbO_2$ and $SnO_2$, as well as to "active" anodes such as Pt and $RuO_2$, regardless of the presence/absence of chloride ion and the electrolytic system utilized.

Several authors have studied the characteristics and influence of the experimental parameters on the oxidation power of the AO process. For example, Sirés et al. (2006a) tested the degradation of clofibric acid solutions in 0.05 M $Na_2SO_4$ as the background electrolyte using a stirred tank reactor of 100 mL capacity. The tank reactor was a thermostated cylindrical cell equipped with a Pt or BDD anode and a stainless steel cathode. Figure 8.3 shows that the drug solutions are poorly decontaminated using a Pt anode leading to 36% TOC removal after consuming 21 Ah $L^{-1}$ during 7 h at 300 mA. In contrast, the same solutions undergo almost total mineralization with 98% TOC decay using a BDD anode since hydroxyl radicals formed at its surface (BDD($^{\bullet}$OH)) from reaction (8.14) have a much higher oxidation power than those produced on Pt (Pt($^{\bullet}$OH)). These results also prove that the oxidation ability of each anode is pH–independent. For the more potent BDD anode, Fig. 8.4 depicts the need of a higher specific charge as current increases from 100 to 450 mA; in contrast, the inverse trend is found for the degradation rate and the time required. This means that when current rises, a higher quantity of BDD($^{\bullet}$OH) is produced enhancing the oxidation of the organics; but, the relative proportion of available radicals for the electrodegradation decreases as current rises, due to the higher rate of the parasitic reactions (8.19)–(8.22) that waste more charge without valuable

**Table 8.2** Percentage of TOC decay obtained for the electro-oxidation of synthetic pharmaceutical solutions with or without active chlorine species under selected conditions

| Pharmaceutical | Anode | Solution | Electrolytic system | % TOC decay | Charge[b] / C $(mg_C)^{-1}$ | Ref. |
|---|---|---|---|---|---|---|
| Paracetamol | Pt | 100 mL of 157 mg $L^{-1}$ drug in 0.05 M $Na_2SO_4$, pH 3.0 | Stirred tank reactor, 300 mA for 360 min | 17 | 3,812 | Brillas et al. (2005) |
| | BDD | | | 98 | 661 | |
| | $Ti/IrO_2$ | 50 mL of 1 mM drug in 0.025 M $Na_2SO_4$, pH 7.8 | Divided plug flow reactor, liquid flow rate 1 mL $min^{-1}$, 200 mA for 210 min | 1 | 52,500 | Waterston et al. (2006) |
| | $Ti/SnO_2$ | | | 40 | 1,313 | |
| | BDD | | | 73 | 719 | |
| | $Ti/RuO_2$ | 100 mL of 1 mM drug in 0.1 M NaCl | Stirred tank reactor, 80 mA for 300 min | 28 | 536 | Boudreau et al. (2010a) |
| | BDD | | | 55 | 273 | |
| Diclofenac | BDD | 100 mL of 30 mg $L^{-1}$ drug in: | Stirred tank reactor with three electrodes, 4 V vs. SCE[a] for 240 min | | | Zhao et al. (2009a) |
| | | – 0.1 M $Na_2SO_4$ | | 83 | 171 | |
| | | – 0.1 M NaCl | | 72 | – | |
| | Pt | 100 mL of 175 mg $L^{-1}$ drug in 0.05 M $Na_2SO_4$, phosphate buffer pH 6.5 | Stirred tank reactor, 300 mA for 360 min | 46 | 1,409 | Brillas et al. (2010) |
| | BDD | | | 97 | 668 | |
| Ibuprofen | $Ti/Pt/PbO_2$ | 200 mL of 1.75 mM drug in 0.035 M $Na_2SO_4$ | Stirred tank reactor, 30 mA $cm^{-2}$ for 360 min | 75 | 16 | Ciríaco et al. (2009) |
| | BDD | | | 92 | 13 | |
| Sulfamethoxazole | BDD | 45 mL of 1 mM drug in | Stirred tank reactor, 20 mA for 300 min | | | Boudreau et al. (2010b) |
| | | – 0.1 M $NaClO_4$ | | 78 | 85 | |
| | | – 0.1 M NaCl | | 69 | 97 | |
| Enrofloxacin | BDD | 600 mL of 1.6 g $L^{-1}$ drug in 0.05 M $Na_2SO_4$, pH 3.0 | Undivided flow cell, liquid flow rate at 2.5 L $min^{-1}$, 3.9 A for 15 h | 83 | 423 | Guinea et al. (2009a) |
| 17α-ethinylestradiol | $Ti/SnO_2$ | 100 mL of 2.0 mg $L^{-1}$ drug in 0.2 M $Na_2SO_4$, pH 6.2 | Stirred tank reactor, 60 mA for 480 min | 79 | 13,671 | Feng et al. (2010) |
| 17β-estradiol | BDD | 250 mL of 0.5 mg $L^{-1}$ drug in 0.1 M $Na_2SO_4$, pH 6 | Stirred tank reactor, 350 mA for 270 min | 94 | 60,855 | Yoshihara and Murugananthan (2009) |
| Clofibric acid | Pt | 100 mL of 179 mg $L^{-1}$ drug in 0.05 M $Na_2SO_4$, pH 3.0 | Stirred tank reactor, 300 mA for 420 min | 36 | 2,100 | Sirés et al. (2006a) |
| | BDD | | | 98 | 771 | |

[a]Saturated calomel electrode

[b]Required charge for the treatment, in Coulombs per milligram of carbon mineralized

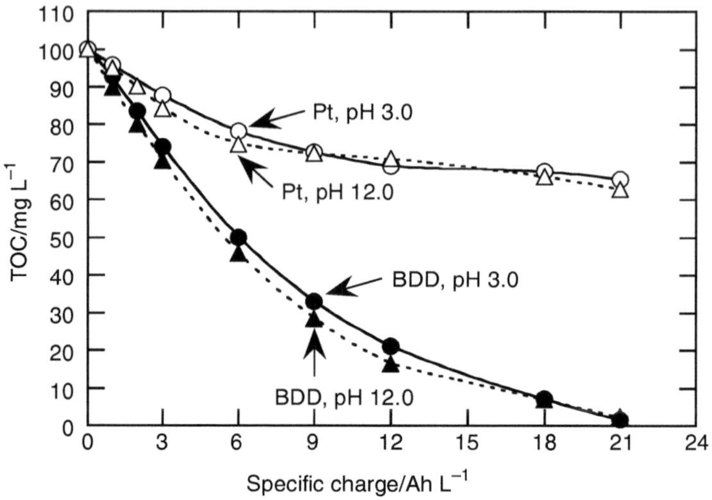


**Fig. 8.3** TOC removal versus specific charge for the anodic oxidation of 100 mL of 179 mg L$^{-1}$ clofibric acid solutions in 0.05 M $Na_2SO_4$ at 300 mA and 35.0°C using a stirred tank reactor with a Pt or BDD anode and a stainless steel cathode of 3 cm$^2$ area. Regardless of the anode used, TOC decays at a similar rate at pH acid and alkaline, but more rapidly with BDD attaining almost overall mineralization. Adapted from Sirés et al. (2006a). TOC: total organic carbon, BDD: boron-doped diamond

oxidation events. Figure 8.4 also shows that increasing temperature from 25°C to 45°C at 300 mA accelerates the mineralization process, which is due to the quicker transport of reactants to the BDD surface. The use of the BDD anode is very effective for any drug content since it can reduce 98% TOC of a solution 550 mg L$^{-1}$, i.e., near saturation, with consumption of 30 Ah L$^{-1}$ at 300 mA, as can be seen in Fig. 8.4. Less charge is proportionally consumed with rising drug content owing to the reaction of greater amounts of organics with BDD(•OH) and the smaller extent of the parasitic reactions (8.19)–(8.22). Therefore, under these experimental conditions at 300 mA, the process was controlled by the diffusion of the organic molecules towards the anode surface, which means that the system operated above the limiting current density.

Since clofibric acid is a phenolic drug, its concentration decay was determined by reversed-phase high performance liquid chromatography (HPLC) and, surprisingly, it was found that the drug was removed more rapidly using Pt than BDD. This behavior can be observed in Fig. 8.5 and is accounted for by the existence of more interactions favoring the particular reaction of clofibric acid with •OH, whereas BDD tends to favor the simultaneous oxidation of the initial pollutants and its intermediates.

For each anode, the same decay rate following a pseudo first-order kinetics is obtained at pH 3.0 and 12.0, confirming that the formation of oxidant Pt(•OH) or BDD(•OH) from reaction (8.14) is pH–independent.

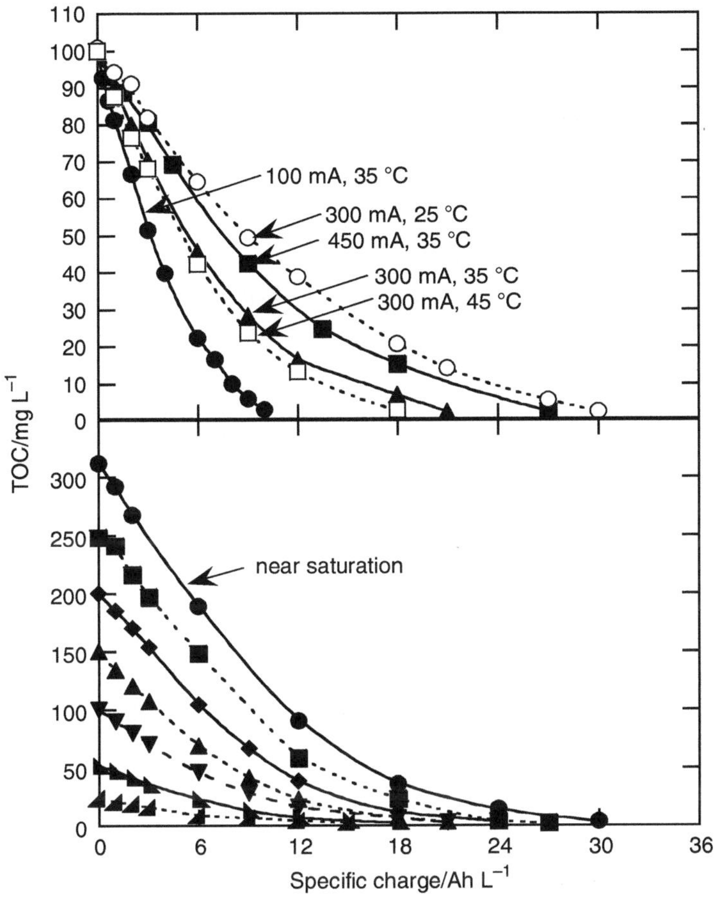


**Fig. 8.4** TOC abatement for the anodic oxidation with a BDD anode at pH 12.0 of a 179 mg L$^{-1}$ clofibric acid solution at different current values and temperatures (*upper*) and several drug concentrations at 300 mA and 35°C (*lower*). A higher charge is required at increasing current and at decreasing temperature. A quicker mineralization is achieved as the initial concentration increases. Adapted from Sirés et al. (2006a). TOC: total organic carbon, BDD: boron-doped diamond

Clofibric acid yields aromatics and carboxylic acids as by-products. Aromatic intermediates were determined by gas chromatography coupled to mass spectrometry (GC-MS) and quantified by reversed-phase HPLC. Generated carboxylic acids were detected by ion-exclusion HPLC. Figure 8.6 exemplifies the evolution of selected intermediates for a solution at 300 mA. Aromatics such as 4-chlorophenol, 4-chlorocatechol, and 1,4-benzoquinone are slowly accumulated and destroyed. In contrast, carboxylic acids such as 2-hydroxyisobutyric, maleic, and oxalic persist using a Pt anode, but are efficiently removed using a BDD anode. The high oxidation power of AO with BDD is then due to the destruction of the generated carboxylic acids. On the other hand, ion chromatograms revealed the release of chloride ion,

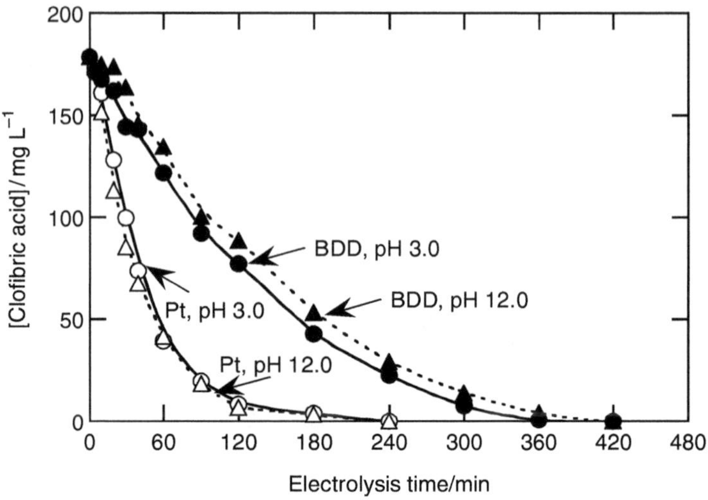


**Fig. 8.5** Clofibric acid concentration decay for the same experiments of Fig. 8.3. For each anode, the decay rate is equal in both media, but the drug disappears more quickly with Pt. Adapted from Sirés et al. (2006a)

which remained stable using Pt and was slowly oxidized to chlorine from reaction (8.25) at BDD. Based on these results, the reaction sequence of Fig. 8.7 was proposed for the mineralization of clofibric acid with $^{\bullet}OH$ formed at the Pt or BDD anode surface. The process starts with the breakage of the C(1)–O bond of the drug to yield 4-chlorophenol and 2-hydroxyisobutyric acid. Hydroxylation of 4-chlorophenol then gives 4-chlorocatechol, 4-chlororesorcinol, or hydroquinone, which are further transformed into 1,2,4-benzenetriol and 1,4-benzoquinone with loss of chloride ion. The destruction of these compounds yields tartronic, maleic, fumaric, and formic acids. The three former acids are oxidized to oxalic acid, which is also generated from the initially generated 2-hydroxyisobutyric acid via pyruvic acid. The ultimate carboxylic acids, oxalic and formic, are finally converted into $CO_2$.

Our group has reported a similar behavior for the AO degradation of two other phenolic compounds such as paracetamol (Brillas et al. 2005) and diclofenac (Brillas et al. 2010) under the above operative conditions using the same stirred undivided cell with a Pt or BDD anode. More than 97% TOC removal was achieved starting with up to 1 g L$^{-1}$ of paracetamol within the pH range 2.0–12.0 using a BDD anode, whereas the use of Pt comparatively yielded a very poor decontamination (see Table 8.2).

$NH_4^+$ and $NO_3^-$ ions were released in both cases. Hydroquinone and 1,4-benzoquinone were detected as the aromatic by-products, while oxalic and oxamic acid were identified as carboxylic acids that are completely removed at the end of the AO with BDD, but remain stable using Pt. In contrast, the mineralization of diclofenac with BDD was only effective in neutral phosphate buffer of pH 6.5 (see Table 8.2),

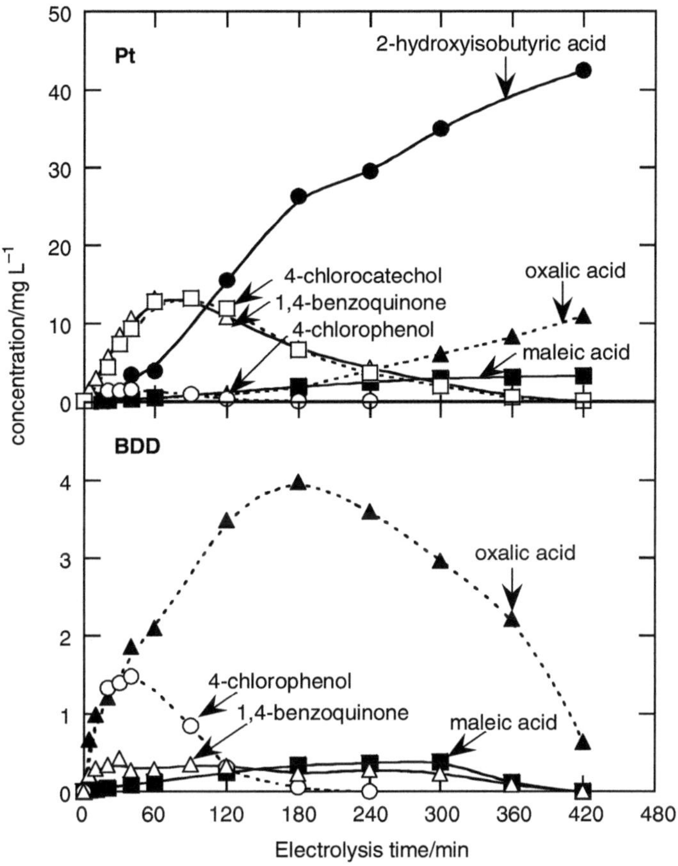


**Fig. 8.6** Evolution of the concentration of selected aromatics and carboxylic acids detected as intermediates during the anodic oxidation of a 179 mg L$^{-1}$ clofibric acid solution under the same conditions of Fig. 8.3 at pH 3.0. While aromatics are slowly accumulated and removed with both anodes, carboxylic acids are only destroyed using a BDD anode. Adapted from Sirés et al. (2006a). BDD: boron-doped diamond

because its protonated form precipitated at lower pH values. This drug was oxidized to 2-hydroxyphenylacetic acid, 2,5-dihydroxyphenylacetic acid, 2,6-dichloroaniline, and 2,6-dichlorohydroquinone, which were transformed into small carboxylic acids. For the AO with Pt, high amounts of malic, succinic, tartaric, and oxalic acids were accumulated in the bulk, whereas the most persistent oxalic and oxamic acids were finally destroyed with BDD. Inorganic ions such as $Cl^-$, $NH_4^+$, and $NO_3^-$ were released in all cases.

Recently, our group has explored the superiority of the oxidation of the veterinary fluoroquinolone antibiotic enrofloxacin by AO with a BDD anode compared with two other AOPs like ozonation and Fenton's reagent (Guinea et al. 2009a). The treatments were carried out with solutions containing 1.6 g L$^{-1}$ of this drug.

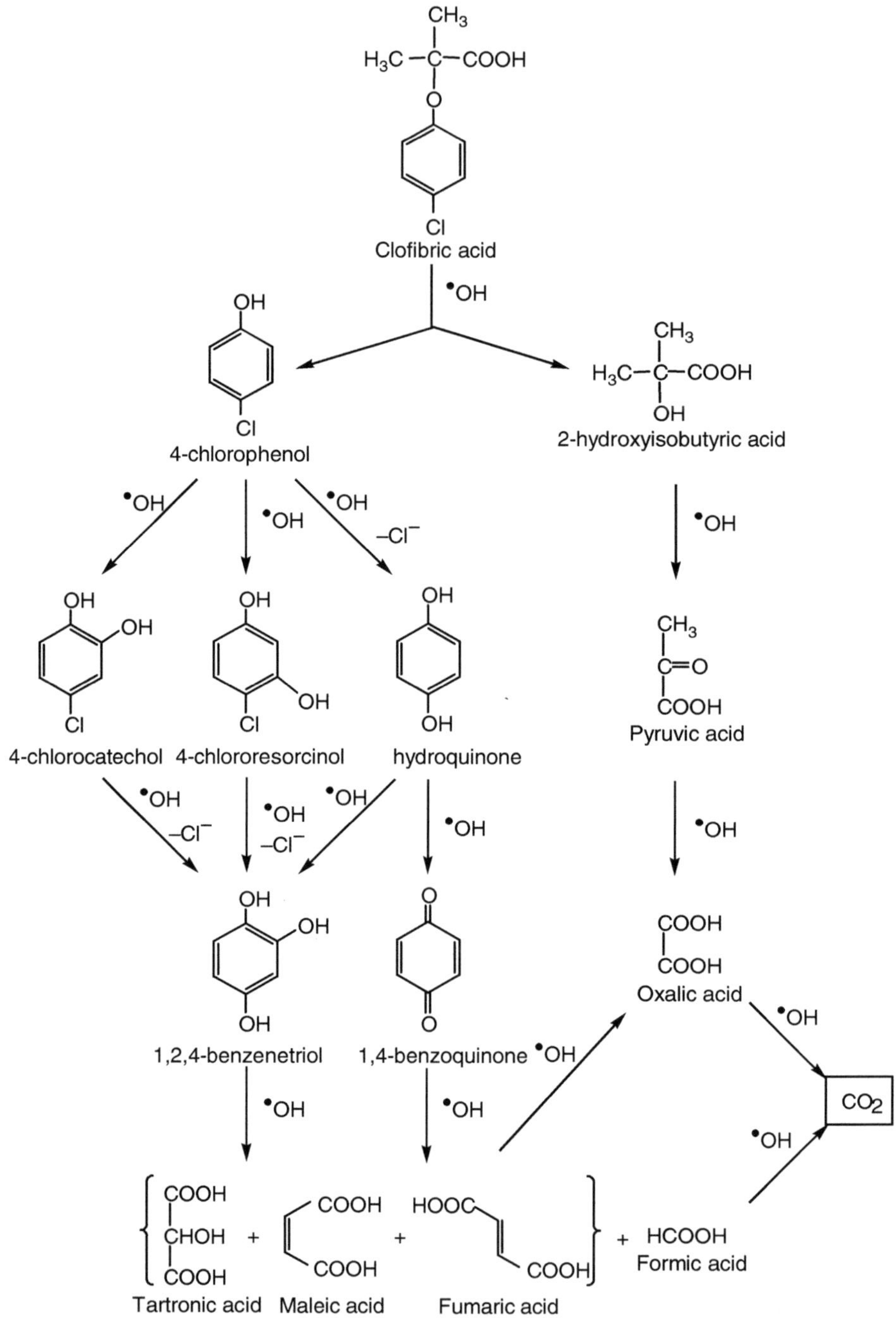


**Fig. 8.7** Reaction sequence proposed for the clofibric acid mineralization by anodic oxidation with a Pt or BDD anode. Initial attack of the hydroxyl radical yields 4-chlorophenol and 2-hydroxyisobutyric acid, which are finally transformed into oxalic and formic acids as the ultimate carboxylic acids. Adapted from Sirés et al. (2006a). BDD: boron-doped diamond

The electrochemical oxidation was performed under recirculation of 600 mL of the solution with 0.05 M $Na_2SO_4$ at pH 3.0 through a BDD/stainless steel flow cell of 78 $cm^2$ electrode area at 2.5 L $min^{-1}$. After 15 h at 3.9 A, 83% TOC was removed. At this time, about 50% of TOC removal was achieved using either ozonation at pH 12.0 or Fenton's reagent at pH 3.0, indicating that the AO is preferable to mineralize enrofloxacin. Based on this, an accurate study was further made about the influence of the nature of the BDD on its electrocatalytic behavior (Guinea et al. 2009b). High diamond/graphite ratios for this synthesized material were found essential for the complete mineralization of enrofloxacin to carbon dioxide thanks to the greater concentration of $^{\bullet}OH$ near the BDD surface. The degradation rate of the antibiotic was also enhanced when increasing the concentration of doping boron and decreasing the electrode thickness.

The decay kinetics of the sulfonamide antibiotic sulfamethoxazole at BDD in the anodic compartment of a plug-flow microreactor has also been investigated (Li et al. 2008). A change from pseudo zero-order to pseudo first-order reaction was observed after 75 min of electrolysis at 20 mA of 50 mL of 1 mM drug in 0.025 M $Na_2SO_4$ at pH 6–7 under recirculation at 1 mL $min^{-1}$. This behavior was interpreted on the basis that the kinetics varied from a pure current control (zero order) to a pure mass transport control (first order) as a result of the decrease in the sulfamethoxazole concentration, thus slowing down the process. The use of a wide range of current values and drug contents showed a partial kinetic order of 0.4–0.5 with respect to each parameter. The mineralization efficiency for sulfamethoxazole increased with increasing current and drug concentration, which are the most significant variables for its destruction by AO.

Ciríaco et al. (2009) compared the effectiveness of BDD with that of other "non-active" anodes like $PbO_2$ coated on Pt/Ti. For a 1.75 mM ibuprofen solution in 0.035 M $Na_2SO_4$ electrolyzed in a stirred tank reactor of 200 mL at 30 mA $cm^{-2}$ for 360 min, a higher TOC removal was found using BDD (92%) than a Pt/Ti/$PbO_2$ anode (75%). Operating at a lower current density (i.e., below the limiting current density), the performance of both electrodes was similar and so, at 10 mA $cm^{-2}$, a current control with efficiency near 100% was obtained in both cases. These results are indicative of a larger generation of hydroxyl radical at the BDD anode from reaction (8.14) as the applied current and cell voltage increase. On the other hand, Waterston et al. (2006) examined the comparative removal of paracetamol with "non-active" anodes like BDD and Ti/$SnO_2$ and the "active" Ti/$IrO_2$ anode using a divided plug-flow reactor. Figure. 8.8 shows that its decay rate agrees with a pseudo first-order kinetics decreasing in the order BDD > Ti/$SnO_2$ > Ti/$IrO_2$ when the drug was treated at 500 mA. This trend is due to the gradually smaller production of hydroxyl radicals. The "non-active" anodes mineralized the drug, being BDD better than Ti/$SnO_2$.

At 200 mA for 210 min, for example, 73% and 40% TOC removals with 26% and 14% current efficiency were obtained, respectively. In contrast, the "active" Ti/$IrO_2$ anode did not yield significant mineralization since the solution electrolyzed with this anode at 200 mA for 300 min gave 88% of 1,4-benzoquinone. From these findings, the electrochemical conversion of the drug via direct two-electron oxidation

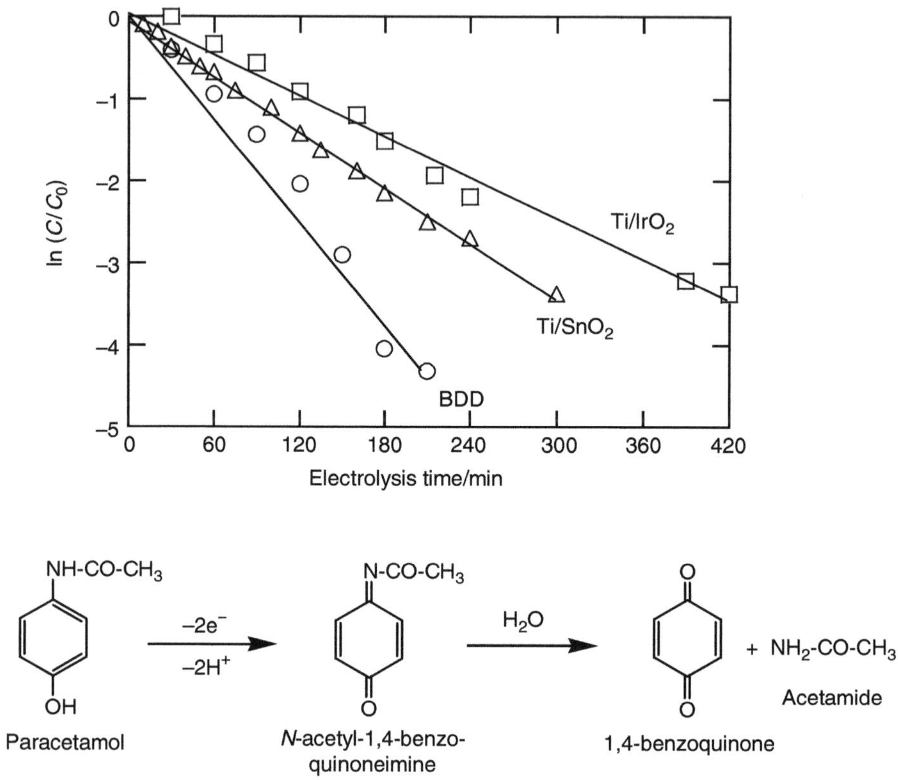


**Fig. 8.8** Comparative kinetic analysis for the decay of 1 mM paracetamol in 0.025 M $Na_2SO_4$ of pH 7.8 at three electrodes using a 50 mL solution recirculating through the anodic compartment of a divided plug-flow reactor at 1 mL min$^{-1}$ and applying 500 mA (*upper*). The BDD anode removed much more rapidly the drug than Ti/$SnO_2$ and Ti/$IrO_2$ anodes. Paracetamol is converted into 1,4-benzoquinone without further mineralization using Ti/$IrO_2$ (*down*). Adapted from Waterston et al. (2006). BDD: boron-doped diamond

to *N*-acetyl-1,4-benzoquinoneimine, followed by hydrolysis to 1,4-benzoquinone with loss of acetamide, was proposed, as shown in Fig. 8.8.

Feng et al. (2010) studied the degradation of the steroid estrogen 17α-ethinylestradiol, which is an estradiol derivative, with the "non-active" Ti/$SnO_2$ anode using 100 mL solutions in 0.2 M $Na_2SO_4$ in a cylinder cell with a stainless steel cathode at 60 mA and 25°C. For 10 mg L$^{-1}$ of the drug, for example, its concentration decayed 89% in 2 h and the solution TOC was reduced by 78% in 8 h, ending in a negligible estrogenic activity. The effectiveness of this treatment was related to the generation of hydroxyl radicals causing electrochemical combustion. In contrast, the AO treatment of tetracyclines (so-called because of the presence of four hydrocarbon rings) has been explored in chloride-free media with different "active" anodes to show the effectiveness of this method to promote the electrochemical conversion of the substrates. A very slow decay of tetracycline, without practical mineralization, was reported by Vedenyapina et al. (2008) when they electrolyzed 50 mg $L^{-1}$ of this antibiotic in 0.05 M $H_2SO_4$ in a divided cell equipped with smooth Pt plates as the anode and cathode by applying a current density as low as 5 mA $cm^{-2}$. Under these weak oxidation conditions and after 24 h of treatment, the aromatic rings of tetracycline were broken, thus losing its biological activity. Similarly, Zhang et al. (2009) showed the decay of more than 90% of 100 mg $L^{-1}$ of tetracycline in 0.1 M $Na_2SO_4$ of pH between 3.9 and 10.0 after 60 min of electrolysis in a stirred tank reactor of 200 mL with a 31.5 $cm^2$ Ti/$RuO_2$-$IrO_2$ anode at 47.6 mA $cm^{-2}$. This implies the supply of 300 C (mg tetracycline)$^{-1}$. It followed a pseudo first-order reaction and its rate increased with increasing current density, but it was not affected by pH and drug content. Addition of a hydroxyl radical scavenger such as *tert*-butanol did not vary the kinetics process, as expected if very small amount of hydroxyl radicals are produced at the Ti/$RuO_2$-$IrO_2$ anode from reaction (8.14) and tetracycline is then directly oxidized via electron transfer without $CO_2$ production. The same behavior was described by Rossi et al. (2009) for the degradation of 50 mL of 2 mM oxytetracycline hydrochloride in a phosphate buffer of pH 5.45 in a stirred tank reactor with a 2 $cm^2$ Ti/$RuO_2$ anode. The electrochemical conversion of this antibiotic was achieved in 120 min of treatment at 50 mA $cm^{-2}$ with complete loss of its antibacterial activity with respect to *Staphylococcus aureus*. Conversely, Weichgrebe et al. (2004) described the effective removal of drug residues like aspirin, tetracycline, and gentamicine with "active" anodes of $MnO_2$-C and Pt, although an increase in the biological availability of the treated solution was found with progressive oxidation of the drugs.

The degradative action of "active" anodes is strongly enhanced when chloride ions are present in the synthetic solution and generated active chlorine, alone or in combination with hydroxyl radicals, oxidizes the organic matter. This was well proven by Boudreau et al. (2010a) by comparing the destruction of paracetamol in $NaClO_4$ and NaCl media using BDD and Ti/$RuO_2$ anodes. Solutions of the pharmaceutical in 0.1 M of each electrolyte were electrolyzed in a stirred tank reactor at 80 mA. Figure 8.9 depicts that in 0.1 M $NaClO_4$, about 80% and 95% paracetamol is destroyed in 300 min by AO with Ti/$RuO_2$ and BDD, respectively, while in 0.1 M NaCl, it disappears in 30 min and at a similar rate with both anodes. Moreover, when the Ti/$RuO_2$ anode is used, no mineralization takes place in 0.1 M $NaClO_4$ and TOC is only reduced by 26% in 0.1 M NaCl. In contrast, the BDD anode causes 84% and 54% mineralization in such media, respectively. This behavior confirms the oxidation of paracetamol by direct electron transfer to Ti/$RuO_2$ in perchlorate medium only giving electrochemical combustion, whereas in chloride medium this anode generates active chlorine species like $Cl_2$, HClO, and/or $ClO^-$ from reactions (8.25)–(8.28) that rapidly destroy the drug and promote its partial mineralization. The oxidation with BDD in perchlorate solution takes place via hydroxyl radical and this oxidant mineralizes the drug in large extent, but in chloride medium the formation of active chlorine species predominates and they remove mainly the drug, at the same rate as in Ti/$RuO_2$. The higher mineralization degree found in 0.1 M NaCl for BDD indicates that generated hydroxyl radicals help to the conversion of

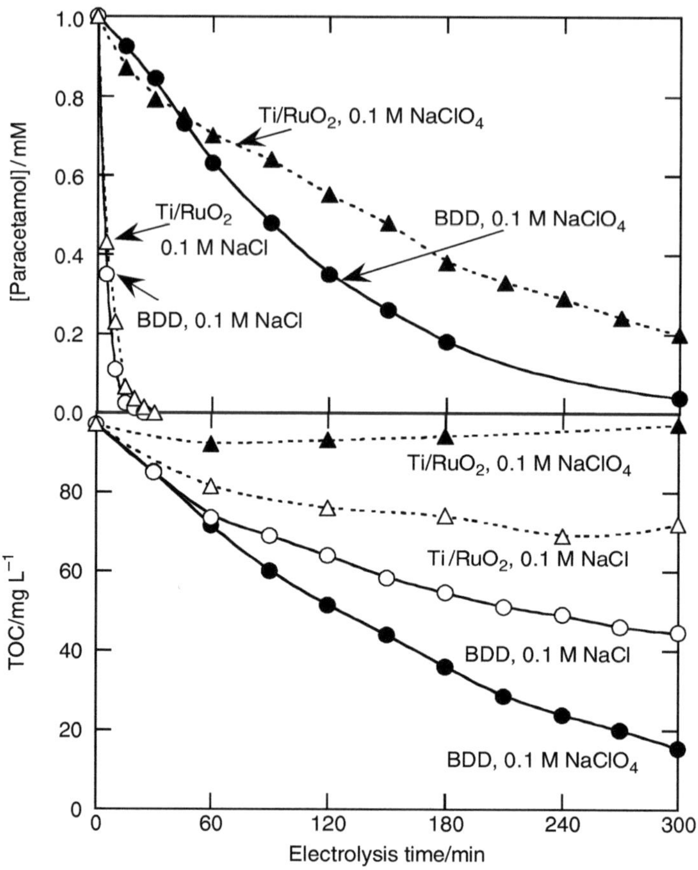


**Fig. 8.9** Decay of paracetamol (*upper*) and TOC (*down*) during the treatment of 100 mL of a 1 mM drug solution with different electrolytes in a stirred tank reactor at 80 mA. In 0.1 M $NaClO_4$, the AO process with hydroxyl radicals is much more effective for BDD than for $Ti/RuO_2$. In 0.1 M NaCl, active chlorine is also formed as oxidant, although $^{\bullet}OH$ generated at BDD enhance the mineralization degree. Adapted from Boudreau et al. (2010a). TOC: total organic carbon, BDD: boron-doped diamond. AO: advanced oxidation

chlorinated by-products into $CO_2$. The existence of these different oxidative ways including electro-oxidation with active chlorine enhanced by AO with hydroxyl radicals was confirmed by the same authors for the degradation of sulfamethoxazole at BDD (Boudreau et al. 2010b).

The formation of chloroderivatives using electro-oxidation with active chlorine has been demonstrated by Zhao et al. (2009a) for the mineralization of 100 mL of a 30 mg $L^{-1}$ diclofenac solution in a stirred tank reactor equipped with a 12 $cm^2$ BDD anode, a Pt wire as the cathode and a saturated calomel electrode (SCE) as the reference electrode. By applying a potential of 4 V to the anode in a 0.1 M $Na_2SO_4$ solution, 83% of TOC was reduced in 240 min. GC-MS analysis of the electrolyzed solutions at 120 min revealed the generation of intermediates such as benzoic acid, 2,6-dichlorobenzenamine, and 2,5-dihydroxybenzyl alcohol, which were further converted into smaller molecules such as hexanoic, butanedioic, malonic, 2-hydroxyl-propanoic, 2-hydroxyl-acetic, and oxalic acids that could be slowly mineralized to $CO_2$. In the presence of 0.1 M NaCl at pH 6.7, dichlorodiclofenac was determined as the main intermediate, probably generated by reaction of the initial substrate with HClO, and subsequent chlorinated by-products generated were slowly removed but the TOC reduction in 240 min decreased to 72%.

Similarly, Pauwels et al. (2006) reported 85% removal of 10 μg $L^{-1}$ of 17α-ethinylestradiol in tap water by combining the oxidation with hydroxyl radicals and active chlorine species originated in an undivided electrolytic device with stable $TiO_2$ electrodes. The estrogen 17β-estradiol has also been efficiently degraded at a BDD anode in the presence of different electrolytes (Murugananthan et al. 2007; Yoshihara and Murugananthan 2009). Electrolyses were performed in a stirred tank reactor using 250 mL solutions with 0.25–0.75 mg $L^{-1}$ of the drug in 0.1 M $Na_2SO_4$ and 0.1 M NaCl at pH 6 and 25°C. For a 0.5 mg $L^{-1}$ solution at 100 mA, 17β-estradiol disappeared in 40 min for the sulfate medium and only in 10 min for the chloride medium, indicating that HClO produced from chloride ion oxidation attacks more efficiently the drug than hydroxyl radical formed in 0.1 M $Na_2SO_4$. For the same solution in the latter conditions, 94% TOC reduction was obtained operating at 350 mA for 270 min, indicating that the AO treatment yields almost total mineralization of the drug. No information on TOC abatement under chloride conditions was given.

The abatement of ofloxacin and lincomycin, i.e., fluoroquinolone and lincosamide antibiotics, respectively, in 0.01 M $Na_2SO_4$ and 0.02 M NaCl at Pt, Ti/$IrO_2$-$Ta_2O_5$, graphite, and a 3D activated carbon pellets was examined by Carlesi Jara et al. (2007) using a bench-scaled divided flow cell with electrodes of 200 $cm^2$ and applying up to 40 mA $cm^{-2}$ under recirculation of 50 mg $L^{-1}$ solutions. The oxidation process followed a pseudo first-order kinetics in sulfate medium, being accelerated in the presence of chloride by the formation of hypochlorite ion. The best anode was the 3D electrode, followed by Pt. For the latter anode, for example, 85% of ofloxacin removal and 50% COD decay were determined after 240 min of electrolysis at 20 mA $cm^{-2}$ in 0.01 M $Na_2SO_4$. Based on these results, Carlesi Jara and Fino (2010) developed a complex model to assess the cost optimization of the current density applied to the electrochemical oxidation of pharmaceutical wastewaters considering the removal of 50 mg $L^{-1}$ of ofloxacin and ranitidine solutions in 0.01 M $Na_2SO_4$ at a Pt anode in a divided batch cell. They found that the optimum current density depended directly on the electrode cost and its service life and inversely on the electric current price, although it was very far from the diffusion-limiting value and hence, far from the maximum current efficiency since the model included all the phenomenology associated with rises in the electrical field across the cell and the thermal and fluodinamic effects associated with current increments.

### 8.3.4 Treatment of Real Wastewaters

The study of real pharmaceutical wastewaters by electrochemical oxidation is more difficult than that of synthetic wastewaters because they are composed of a complex mixture of compounds including ions such as $NH_4^+$, $Cl^-$, $NO_3^-$, and $SO_4^{2-}$. The indirect oxidation of organic and inorganic species with the generated $^{\bullet}OH$ and active chlorine is then expected using the potent BDD anode. This has been proven by Perez et al. (2010) when examining the electrochemical destruction of reverse osmosis concentrates generated in tertiary STPs of pH 7.4–8.1 containing 124–140 mg $L^{-1}$ COD and several pharmaceuticals with contents between 1.5 and 27 μg $L^{-1}$, as well as ca. 207 mg $L^{-1}$ $NH_4^+$, 630 mg $L^{-1}$ $Cl^-$, 12 mg $L^{-1}$ $NO_3^-$, and 584 mg $L^{-1}$ $SO_4^{2-}$. Volumes of 2 L of the effluent were electrolyzed in a flow plant with an undivided cell equipped with two circular BDD electrodes of 70 $cm^2$ area under recirculation at 10 L $min^{-1}$. Application of current densities of 5, 10, and 20 mA $cm^{-2}$ caused the complete decay of COD in 480, 360, and 300 min, respectively. For the two latter current densities, all the $NH_4^+$ ions were oxidized to $NO_3^-$ ions. Simultaneously, $Cl^-$ ion was partially oxidized and converted into active chlorine species from reactions (8.25)–(8.28) and $ClO_3^-$ ion from reactions (8.29) and (8.30). Moreover, all drugs were also completely destroyed at a similar rate for all current densities verifying a pseudo first-order kinetics because their diffusion towards the anode limited their reaction with generated oxidants.

The aforementioned findings demonstrate the complexity of the electrochemical oxidation of real wastewaters with BDD due to generation of potentially toxic inorganic by-products (e.g., chlorinated ions, $NO_3^-$, etc.). These parasitic processes can be diminished using less potent "active" anodes and applying directly the electro-oxidation with active chlorine. In this way, Babu et al. (2009) utilized the flow plant of Fig. 8.10 to decontaminate a pharmaceutical effluent mainly containing gentamicin and dexamethasone with 1.4 g $L^{-1}$ COD at pH 7.6. The system worked in continuous mode at 10–40 L $h^{-1}$ after the addition of 2–5 g $L^{-1}$ NaCl to the effluent. The excellent COD removal obtained under these conditions by oxidation with active chlorine species produced on the Ti/$RuO_2$ coated expanded-mesh anode can be observed in the same figure. The percentage of COD reduction did not vary for NaCl contents $\geq$3 g $L^{-1}$ and increased with increasing the retention time in the cell, i.e., with decreasing liquid flow rate, and with greater current density up to 40 mA $cm^{-2}$. Its maximum value of 86% was obtained at 10 L $h^{-1}$ for this current density. These results prove that the electro-oxidation with active chlorine is very effective and capable of increasing the quality of the treated wastewater to the reuse standard prescribed for pharmaceutical industries. The effectiveness of this procedure was also tested for the inactivation of antineoplastics like epirucibin hydrochloride, irinotecan hydrochloride, vincristine sulfate, mitomycin C, paclitaxel, methotrexate, and cisplatin in clinical wastewaters with about 100% removal of cytotoxicity, mutagenecity, and antibacterial activity (Hirose et al. 2005; Kobayashi et al. 2008).

Lazarova and Spendlingwimmer (2008) compared the removal of pharmaceuticals like diclofenac and ibuprofen in urine or yellow waters by membrane separation methods and electrochemical oxidation in a batch pilot plant with a reactor equipped with graphite electrodes. Nano-filtration allowed the separation of both pharmaceuticals from yellow waters, but with the disadvantage of losing part of urea, a valuable product as a nitrogen fertilizer. In contrast, the electrochemical oxidation was more useful because it destroyed the compounds without loss of urea. Finally, the higher

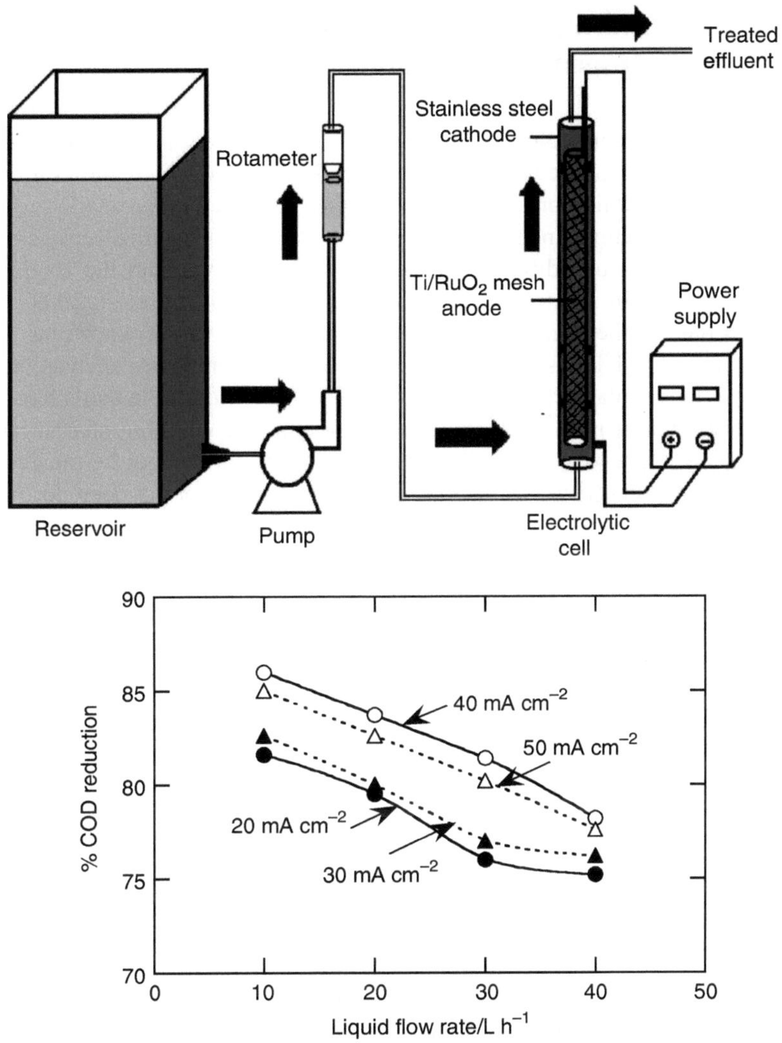


**Fig. 8.10** Schematic set-up of the bench-scaled continuous flow plant used for the treatment of a real pharmaceutical wastewater by electro-oxidation with active chlorine (*upper*) and percentage of COD reduction versus liquid flow rate at different current densities upon the addition of 3 $g L^{-1}$ NaCl (*down*). The wastewater with 1.4 $g L^{-1}$ COD was pumped at constant flow rate regulated by a rotameter from the reservoir to the cylindrical electrochemical cell of 4.5 L capacity with a cathode of 110 cm height × 7 cm diameter and an anode of 100 cm height × 5 cm diameter. The maximum COD reduction was achieved at 10 $L h^{-1}$ and 40 $mA cm^{-2}$. Adapted from Babu et al. (2009). COD: chemical oxygen demand

effectiveness of the AO with BDD compared to ozonation for removing various pharmaceutical residues from a SWT effluent has been checked by Menapace et al. (2008), who also proved that the consecutive application of both methods yielded 97–100% destruction of all compounds.

## 8.4 Indirect Electro-Oxidation in the Bulk with Electrogenerated Fenton's Reagent

The technologies discussed in this section are based on the main action of the $^{\bullet}OH$ generated in the bulk of the treated contaminated solutions, which allows minimizing the diffusional limitations inherent to the $^{\bullet}OH$ generated at the anode surface in the AO systems explained in the previous section. The EAOPs with electrogenerated Fenton's reagent include different processes, which derive from the original EF process introduced by the teams of Brillas and Oturan (Brillas et al. 2009). Here, attention is paid to the classical EF configurations and a variant known as anodic Fenton treatment (AFT), as well as to the photoassisted processes such as the PEF with UV or direct solar irradiation. As explained below, these methods have been applied to the treatment of synthetic aqueous solutions containing pharmaceutical residues, being especially relevant the intensive work carried out by our research team in the last years. However, actual wastewaters have not yet been decontaminated by these processes.

### *8.4.1 Electro-Fenton*

The fundamentals and main characteristics of the chemical and electrochemical technologies based on Fenton's reaction chemistry for water and soil treatment have been recently reviewed (Pignatello et al. 2006; Brillas et al. 2009). The generally accepted classical mechanism for the Fenton process in acidic medium is initiated by the formation of hydroxyl radical in accordance with the classical Fenton's reaction (8.32) between Fenton's reactants ($H_2O_2 + Fe^{2+}$) occurring in the bulk:

$$Fe^{2+} + H_2O_2 + H^+ \rightarrow Fe^{3+} + H_2O + {}^{\bullet}OH \qquad (8.32)$$

The technique becomes operative if the contaminated solution is at the optimum pH of 2.8–3.0, where it can be propagated by the catalytic behavior of the $Fe^{3+}/Fe^{2+}$ couple. Interestingly, only a small catalytic amount of $Fe^{2+}$ is required because this ion is regenerated from the so-called Fenton-like reaction (8.33), as well as by the reduction of $Fe^{3+}$ with several species such as the radical $HO_2^{\bullet}$, the organic radical $R^{\bullet}$, and/or the superoxide ion ($O_2^{\bullet-}$):

$$Fe^{3+} + H_2O_2 \rightarrow Fe^{2+} + HO_2^{\bullet} + H^+ \qquad (8.33)$$

The attack of $^{\bullet}OH$ to saturated (RH) or aromatic (ArH) organics gives dehydrogenated or hydroxylated derivatives. In contrast, reactions promoting the consumption of $^{\bullet}OH$ by the Fenton's reagent are major wasting reactions that decrease the oxidizing power of the Fenton systems. The catalytic activity of iron species is mainly determined by the solution pH and therefore, a strict pH control is needed.

From a conceptual standpoint, the use of the electrochemical alternative seems more appealing since it allows the continuous electrogeneration and/or regeneration of the reactants ($H_2O_2$ and $Fe^{2+}$), thus increasing the efficacy of the processes and enhancing their environmental compatibility. The EF process was the first method proposed among the EAOPs based on the cathodic electrogeneration of $H_2O_2$ and laid the foundation for a large variety of related processes. It is an emerging technology that has been successfully applied to the treatment of acidic aqueous solutions containing organic pollutants including pesticides, organic synthetic dyes, pharmaceuticals and personal care products, and a great deal of industrial pollutants (Brillas et al. 2009). The great interest for all these indirect electro-oxidation methods arises from the fact that reactions involved in the cathodic reduction of oxygen proceed at low potential and in homogeneous environment. The EF method is based on the continuous electrogeneration of $H_2O_2$ at a suitable cathode fed with pure $O_2$ or air by reaction (8.34):

$$O_2 + 2H^+ + 2e^- \rightarrow H_2O_2 \tag{8.34}$$

along with the addition of an iron catalyst to the treated solution to produce oxidant $^{\bullet}OH$ at the bulk via Fenton's reaction (8.32). In an undivided cell, the process is complicated by the simultaneous destruction of pollutants with ROS, mainly the heterogeneous $^{\bullet}OH$ like Pt($^{\bullet}OH$) or BDD($^{\bullet}OH$) formed at a Pt or BDD anode, respectively.

The EF technology uses three- and two-electrodes divided and undivided electrolytic cells in which $H_2O_2$ is continuously electrogenerated. Cathode materials such as graphite, carbon felt, and gas diffusion electrodes (GDE) are most typically employed. Anode materials such as graphite, Pt, metal oxides and BDD are usually preferred. Figure 8.11 shows two examples of stirred undivided tank reactors. They have been utilized for the treatment of 100–250 mL of synthetic pharmaceutical solutions by EF.

In EF and most related methods, soluble $Fe^{3+}$ can be cathodically reduced to $Fe^{2+}$ by reaction (8.35), with $E^o = 0.77$ V vs. SHE:

$$Fe^{3+} + e^- \rightarrow Fe^{2+} \tag{8.35}$$

The faster the regeneration of $Fe^{2+}$ by reaction (8.35), the quicker the production of $^{\bullet}OH$ from Fenton's reaction (8.32).

The efficiency of the oxidation process is a function of various operation parameters such as temperature, solution pH, $O_2$ feeding, stirring rate or liquid flow rate, electrolyte composition, applied potential or current, and catalyst and pollutant concentrations, some of which control the regeneration ability of $Fe^{2+}$ from $Fe^{3+}$ produced along the process.

Preliminary work dealing with the application of EF to the treatment of pharmaceutical residues was started by Oturan et al. (1992a, b) using a divided three-electrode cell with a toxic Hg pool under air bubbling They studied the polyhydroxylation of the NSAIDs ketoprofen and salicylic acid at pH 3.0 and applying a cathode potential $E_{cat} = -0.5$ V vs. SCE, with the addition of $Fe^{3+}$. The reactivity of ketoprofen was

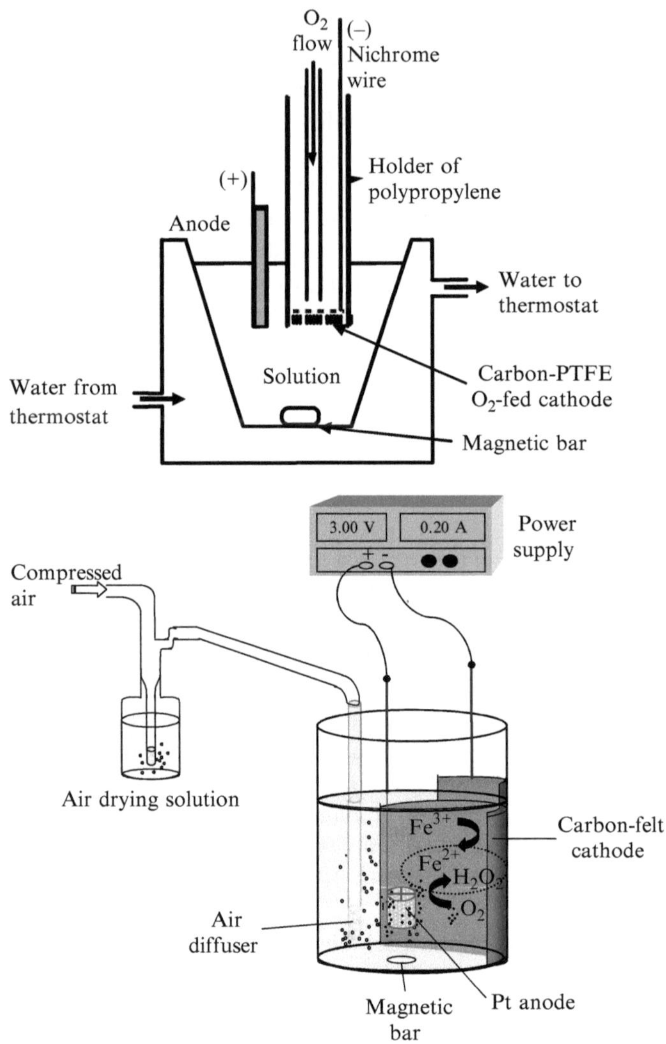


**Fig. 8.11** Schemes of bench-scaled, open, stirred two-electrode undivided tank reactors used for the electro-Fenton treatment of synthetic pharmaceutical solutions. $H_2O_2$ is electrogenerated either at a gas diffusion electrode (GDE) composed of a carbon-PTFE cloth directly fed with $O_2$ (*upper*) or a carbon-felt electrode surrounding the anode and with bubbling air through the solution (*down*). Adapted from Brillas et al. (2009)

found to be similar to that of salicylic acid. More environmentally friendly systems have been developed since then, using two-dimensional GDEs or three-dimensional carbon felts as the cathode instead of Hg, as shown in Fig. 8.11. Using a carbon-felt cathode in a divided three-electrode cell at $E_{cat}$=−0.5 V vs. SCE, Oturan et al. (1999a, b) identified four monohydroxylated derivatives of the neuroprotector drug riluzole upon the EF degradation at pH 2.0. These intermediates were identical to the natural metabolites and could be more easily oxidized than the starting drug.

More recently, the carbon-felt cathode has been applied to the decontamination of single and multicomponent solutions of the β-blockers atenolol, metoprolol, and propranolol (Sirés et al. 2010). Bulk electrolyses were performed at pH 3.0 and room temperature using a cell like that of Fig. 8.11 (down) containing 220 mL of drug solution in 0.05 M $Na_2SO_4$ under the application of a constant current in the range 30–300 mA. Continuous saturation of air at atmospheric pressure to produce $H_2O_2$ from reaction (8.34) was ensured by bubbling compressed air. The EF degradation of atenolol with a Pt anode and 0.2 mM $Fe^{2+}$ showed a higher performance than the AO treatment (i.e., in the absence of $Fe^{2+}$), with fast destruction of this drug following pseudo-first order kinetics and fast degradation. Solutions with 0.15 mM atenolol (25 mg $L^{-1}$ TOC) could be only partially mineralized by AO due to the low oxidation power of Pt and the formation of highly refractory by-products and so, only 15% TOC removal was achieved at 360 min, whereupon the TOC remained unchanged. In contrast, the complete mineralization of the solutions was ensured after 360 min at 300 mA by EF, due to the high amount of $^{\bullet}OH$ formed in the bulk. Under the latter conditions, atenolol disappeared in just 6 min. Multicomponent solutions with 0.15 mM of each β-blocker (total TOC = 85 mg $L^{-1}$) were electrolyzed under similar conditions and a high mineralization rate was observed up to the overall TOC removal at 480 min. The use of a gradient HPLC method allowed the simultaneous determination of the three pharmaceuticals. According to the absolute rate constants determined by the competition kinetics method, the order of reactivity of $^{\bullet}OH$ with each β-blocker increases in the order: atenolol ($1.42 \times 10^9$ L $mol^{-1}$ $s^{-1}$) < metoprolol ($2.07 \times 10^9$ L $mol^{-1}$ $s^{-1}$) < propranolol ($3.36 \times 10^9$ L $mol^{-1}$ $s^{-1}$). The structural differences account for their different reactivity.

Isarain-Chávez et al. (2010) also tested the use of carbon felt for the degradation of atenolol. Novel undivided, four-electrode combined cells containing a GDE and a carbon-felt cathode and equipped with two power supplies were proven to enhance the mineralization in EF because of the fast $Fe^{2+}$ regeneration at the carbon-felt cathode. Thus, solutions of 100 mL with 158 mg $L^{-1}$ atenolol (100 mg $L^{-1}$ TOC), 0.5 mM $Fe^{2+}$, and 0.05 M $Na_2SO_4$ were degraded at constant current (50 mA in the GDE cell and 12 mA in the carbon felt cell) at 35°C, yielding mineralization degrees of 81% and 90% for the combined Pt/GDE-Pt/carbon felt and BDD/GDE-Pt/carbon felt configurations, respectively. Atenolol disappeared in about 25 min decays in these combined four-electrode cells, following again a pseudo-first order kinetics, and it was observed that the destruction was four times quicker than in two-electrode Pt/GDE and BDD/GDE cells. The time course of the concentration of the aromatic and short-chain carboxylic acid intermediates demonstrated the progressive detoxification of the atenolol solutions during the application of the aforementioned EF treatments.

In the above works, a reaction sequence was proposed for atenolol mineralization. The initial formation of hydroxylated atenolol, *p*-hydroxyphenylacetamide, and hydroxylated *p*-hydroxyphenylacetamide as the aromatic by-products is illustrated in

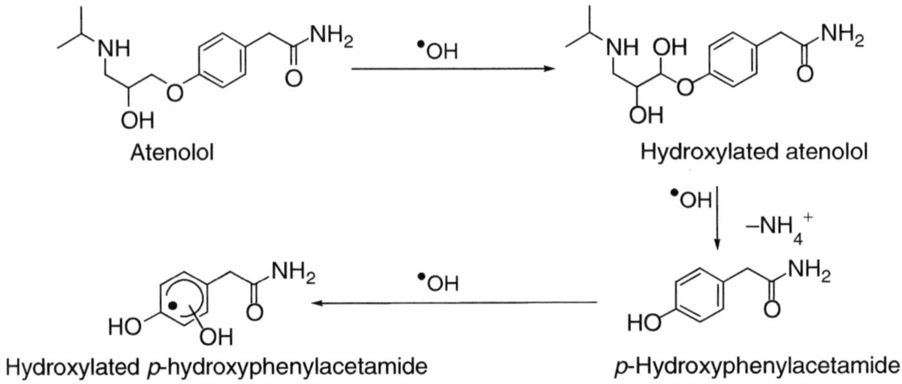


**Fig. 8.12** Initial reaction sequence proposed for the electro-Fenton (EF) degradation of atenolol using an undivided Pt/carbon felt cell. Atenolol is hydroxylated and deaminated to form the hydroxylated *p*-hydroxyphenylacetamide. Adapted from Sirés et al. (2010)

the main pathway of Fig. 8.12 (Sirés et al. 2010). The aryl moiety undergoes further oxidation and aliphatic carboxylic acids such as maleic, fumaric, tartaric, tartronic, glycolic, formic, and oxalic are formed upon cleavage of the benzenic rings (Isarain-Chávez et al. 2010), as can be seen in Fig. 8.13. Acetamide is also formed and oxidized to oxamic acid. Since $Fe^{2+}$ is largely regenerated at the carbon-felt cathode, the Fe(II)-oxalate and Fe(II)-oxamate complexes formed can be oxidized by ${}^{\bullet}OH$ and BDD(${}^{\bullet}OH$). In contrast, the competitively produced Fe(III)-oxalate and Fe(III)-oxamate complexes can only be removed with BDD(${}^{\bullet}OH$). It can be seen that almost 100% of the initial N content was accumulated predominantly as $NH_4^+$ ion.

A larger insight into the degradative behavior of pharmaceutical residues has been reported by Brillas' group, who selected several compounds to study the performance of undivided two-electrode cells like that of Fig. 8.11 (upper) containing a carbon-PTFE GDE as the cathode and different anode materials such as Pt and BDD. The experiments were always done by adding 0.2–2 mM $FeSO_4$ as the iron catalyst source to the starting solutions. Table 8.3 summarizes some selected results for the degradation of paracetamol, clofibric acid, and salicylic acid. Former studies were focused on exploring the effect of $Cu^{2+}$ as a co-catalyst in the EF treatment of paracetamol using a Pt/GDE cell (Sirés et al. 2004, 2006b), because a positive and synergistic effect had been previously observed for the EF treatment of other N-containing organic pollutants such as nitrobenzene (Brillas et al. 2004).Solutions of 100 mL containing up to 1 $g\,L^{-1}$ of this drug in 0.05 M $Na_2SO_4$ were electrolyzed in the pH range 2.0–6.0 at a constant current density between 33 and 150 mA $cm^{-2}$ at 35°C for 6 h. Table 8.3 shows that the oxidation ability of the EF system with 1 mM $Fe^{2+}$ and 1 mM $Cu^{2+}$ was greater than that with 1 mM $Fe^{2+}$ alone, which is much more evident at high current density. A 60% mineralization was obtained as maximum at 4 h using 1 mM $Fe^{2+}$ because of the formation of stable complexes of Fe(III) with the ultimate oxalic and oxamic acids formed that cannot be attacked by ${}^{\bullet}OH$ formed from Fenton's reaction (8.32).

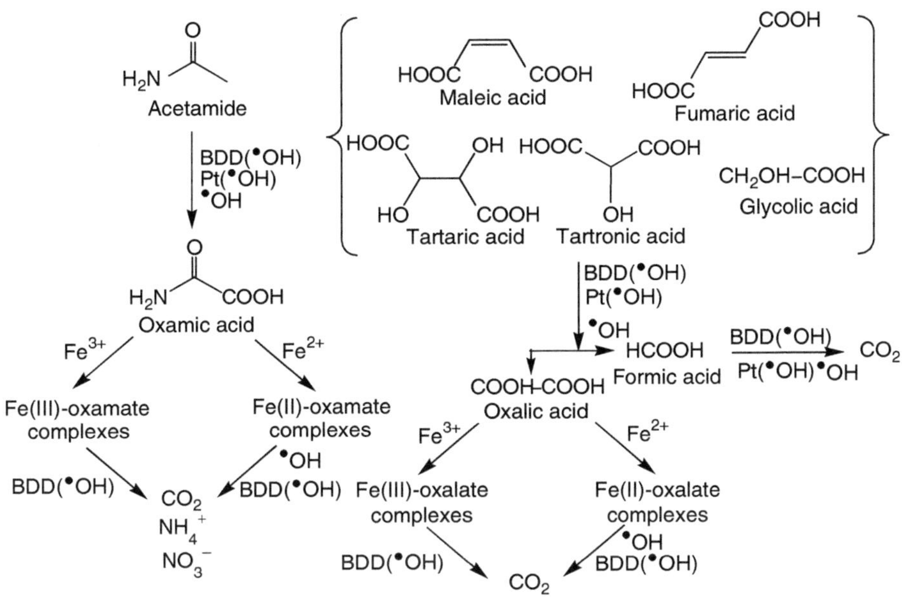


**Fig. 8.13** Reaction sequence for the destruction of carboxylic acids formed during the electro-Fenton (EF) degradation of atenolol in acidic aqueous solutions using combined Pt/GDE-Pt/carbon felt and BDD/GDE-Pt/carbon felt cells. The ultimate oxalic and oxamic acids form complexes with $Fe^{2+}$ and $Fe^{3+}$ ions that are oxidized with ˙OH in the bulk and/or with heterogeneous BDD(˙OH). Adapted from Isarain-Chávez et al. (2010)

In contrast, up to 81% TOC removal was achieved with the addition of $Cu^{2+}$ since the Cu(II) complexes are oxidized by ˙OH. Paracetamol was very rapidly destroyed, only requiring 6 min at 100 mA $cm^{-2}$ in both systems. The initial nitrogen was mainly transformed into $NH_4^+$ ions during all treatments. From the identification of intermediates by GC-MS and HPLC, it was concluded that the initial hydroxylation of paracetamol produces hydroquinone with a release of acetamide, which is converted into oxamic acid. The simultaneous opening of the benzenic ring of *p*-benzoquinone coming from hydroquinone degradation leads to oxalic acid via oxidation of generated carboxylic acids such as ketomalonic, maleic, and fumaric. The removal of final oxalic and oxamic acids was strongly affected by the action of catalysts.

Further work of Brillas' group followed a different strategy trying to enhance the performance of the EF process, focusing on the nature of the anode rather than on the addition of metal co-catalysts. The EF treatment of the blood lipid regulator metabolite clofibric acid and the NSAID salicylic acid was carried out using both, undivided Pt/GDE and BDD/GDE cells like that of Fig. 8.11 (upper) (Sirés et al. 2007a, b; Guinea et al. 2008). Comparative electrolyses were carried out with 100 mL of 179 mg $L^{-1}$ clofibric (100 mg $L^{-1}$ TOC) in 0.05 M $Na_2SO_4$ at pH 3.0, 100 mA $cm^{-2}$, and 35.0°C, with electrodes of 3 $cm^2$ area. As shown in Table 8.3, the

**Table 8.3** Time required for the total disappearance of various pharmaceutical residues ($t_{TD}$) and percentage of TOC removal after 3 or 4 h of treatment of 100 mL of synthetic aqueous solutions containing 100 mg $L^{-1}$ TOC in 0.05 M $Na_2SO_4$ at pH 3.0 and 35 °C by electro-Fenton (EF) and photoelectro-Fenton (PEF) processes using a stirred tank reactor with a Pt or BDD anode and a GDE cathode

| | Catalyst | | | Electrochemical process-cell | | | | | | | |
|---|---|---|---|---|---|---|---|---|---|---|---|
| | | | | EF-Pt/GDE | | EF-BDD/GDE | | PEF-Pt/GDE | | PEF-BDD/GDE | |
| Compound | $[Fe^{2+}]$ /mM | $[Cu^{2+}]$ /mM | $j$ /mA $cm^{-2}$ | $t_{TD}$ /min | % TOC removal | $t_{TD}$ /min | % TOC removal | $t_{TD}$ /min | % TOC removal | $t_{TD}$ /min | % TOC removal |
| Paracetamol[a] | 1 | – | 33 | | 55 | | | | 75 | | |
| | | | 100 | 6 | 59 | | | 6 | 79 | | |
| | | | 150 | | 60 | | | | 80 | | |
| | 1 | 1 | 33 | | 53 | | | | 65 | | |
| | | | 100 | 6 | 80 | | | 6 | 95 | | |
| | | | 150 | | 81 | | | | 96 | | |
| Clofibric acid[b] | 1 | – | 33 | | | 17 | 78 | 12 | 91 | | |
| | | | 100 | 7 | 73 | 7 | 93 | 7 | 92 | 7 | >96 |
| | | | 150 | | | 4 | 95 | 5 | 92 | | |
| Salicylic acid[c] | 0.5 | – | 33 | 30 | 57 | 30 | 73 | 30 | >96 | 30 | >96 |
| | | | 100 | | 62 | | 85 | | >96 | | >96 |
| | | | 150 | | 64 | | 90 | | >96 | | >96 |

Data from Sirés et al. (2006b, 2007a, b), Guinea et al. (2008)
Initial concentration of the pharmaceutical residues: [a]157 mg $L^{-1}$, [b]179 mg $L^{-1}$, and [c]165 mg $L^{-1}$
Percentage of TOC removal at: [a,b]4 h, [c]3 h

superiority of BDD compared with Pt as the anode was confirmed in EF with 1.0 mM $Fe^{2+}$ because 93% TOC removal could be achieved in the BDD/GDE cell after 4 h. A higher removal rate was obtained as the current density increased due to the greater production of BDD(•OH) at the anode and •OH in the bulk, reaching almost total mineralization after about 6 h. Such mineralization degree can not be achieved in EF with Pt, which then confirms the interest of the BDD/GDE system as an alternative oxidation technology. On the other hand, clofibric acid decays agreed with a pseudo first-order kinetics, with total disappearance in 17, 7, and 4 min at 33, 100, and 150 mA $cm^{-2}$. Similar trends for TOC and pollutant concentration decays were found for the treatment of 165 mg $L^{-1}$ salicylic acid (100 mg $L^{-1}$ TOC) (Guinea et al. 2008). Table 8.3 shows that only partial TOC abatement was achieved by EF, whereas the initial drug required 30 min to disappear. This time is higher than that found for paracetamol and clofibric acid (see Table 8.3), due to the formation of hardly oxidizable Fe(III)-salicylate complexes. Solutions of 100 mL of the biocide chloroxylenol (100 mg $L^{-1}$, i.e., 61.5 mg $L^{-1}$ TOC), the NSAID ibuprofen (41 mg $L^{-1}$, i.e., 31 mg $L^{-1}$ TOC), and the antibiotic enrofloxacin (158 mg $L^{-1}$, i.e., 100 mg $L^{-1}$ TOC) were also partially mineralized by EF with Pt/GDE and BDD/GDE cells (Skoumal et al. 2008, 2009; Guinea et al. 2010).

In a collaboration between Brillas and Oturan's groups, the oxidation ability of four EF systems with undivided cells, namely Pt/GDE, BDD/GDE, Pt/carbon felt, and BDD/carbon felt like those shown in Fig. 8.11 was comparatively studied by considering the degradation of the phenolic antimicrobials chlorophene, triclocarban, and triclosan with $Fe^{3+}$ as the catalyst (Sirés et al. 2007c, d). In the cells with the GDE, $H_2O_2$ was largely accumulated but $Fe^{3+}$ content remained unchanged, so that pollutant decay was enhanced by increasing the initial $Fe^{3+}$ concentration because this promoted faster $Fe^{2+}$ regeneration at the cathode with greater •OH production from Fenton's reaction (8.32). In contrast, when the carbon-felt cathode was used, $H_2O_2$ was electrogenerated in relatively small extent, whereas $Fe^{2+}$ was largely accumulated as a result of the quick $Fe^{3+}$ cathodic reduction, being required a $Fe^{3+}$ concentration as low as 0.2 mM to obtain the maximum •OH generation rate. Note that such results on the $Fe^{2+}$ regeneration totally agree with those discussed above for the treatment of atenolol. Under these conditions, absolute rate constant values of $1.00 \times 10^{10}$ and $5.49 \times 10^{9}$ $M^{-1}$ $s^{-1}$ were obtained for the decay reactions of chlorophene and triclosan, respectively. A poor mineralization degree was found for the Pt/GDE cell because of the difficult oxidation of final Fe(III)-oxalate with •OH. These complexes were completely destroyed using a BDD anode at a high current thanks to the great amount of BDD(•OH) generated on its surface. Complete mineralization was also achieved in both carbon felt cells with 0.2 mM $Fe^{3+}$ owing to the efficient oxidation of Fe(II)-oxalate complexes with •OH in the bulk. The highest oxidation power regarding total mineralization was attained for the BDD/carbon felt cell at a high current by the additional destruction of Fe(II)-oxalate complexes with BDD(•OH). The overall release of $Cl^-$ ion was confirmed for all treatments. Electrolyses in hydroorganic medium were found to be an interesting alternative for the study of the reaction pathway of low water-soluble pollutants such as triclocarban. Thus, primary intermediates such as 2,4-dichlorophenol, 4-chlorocatechol, chlorohydroquinone, and chloro-*p*-benzoquinone were identified for triclosan, whereas urea, hydroquinone, chlorohydroquinone, 1-chloro-4-nitrobenzene, and 1,2-dichloro-4-nitrobenzene were reported for triclocarban.

### 8.4.2 Anodic Fenton Treatment

The AFT is a variant of the EF process where the reagent $H_2O_2$ is not cathodically generated, but it is added to the solution. A divided cell is employed to avoid the negative effect of the formation of $OH^-$ at the cathode from water reduction, which raises pH and is then against the optimum conditions of $^{\bullet}OH$ production from Fenton's reaction (8.32). This process has been extensively developed for water remediation by Lemley's group, who have considered different bench-scale electrolytic systems (Brillas et al. 2009). The most recent one is referred to as "membrane AFT" since an ion exchange membrane is used to connect the anodic and cathodic half cells. The system contains an iron plate with an area of 20–30 $cm^2$ as anode and a graphite rod as cathode, treating small volumes of 90–350 ml of anolyte with low contents of organic pollutants and NaCl as the background electrolyte at pH 2.0–3.0 and a constant current of 0.01–0.24 A for short time periods of 2–40 min. The sacrificial iron anode allows the $Fe^{2+}$ electrogeneration from its anodic dissolution by reaction (8.1), whereas $H_2$ is produced from $H_2O$ reduction at the inert cathode by reaction (8.3). The continuous addition of $H_2O_2$ into the anode half-cell is provided by a peristaltic pump, and thus the organic matter is oxidized by the $^{\bullet}OH$ formed from Fenton's reaction (8.32). The controlled $Fe^{2+}$ electrogeneration along with the pH stability in the anode half-cell minimize the extent of the accumulation of $Fe(OH)_3$ sludge. In most trials, the fastest and overall removal of starting pollutants and their primary by-products is obtained with an $[Fe^{2+}]/[H_2O_2]$ ratio of 1:10, but a low mineralization degree is usually attained.

Although the former studies of Lemley's group using this process assessed the oxidation power of AFT to remove various pesticides, two agricultural antibiotics belonging to the sulfonamides group, namely sulfamethazine and sulfadiazine, have been recently degraded (Neafsey et al. 2010). Galvanostatic electrolyses were performed at room temperature employing a stirred tank reactor equipped with an anion exchange membrane, which separated the anodic half-cell containing 400 mL of the antibiotic solution from the cathodic half-cell containing 400 mL of 0.08 M NaCl. The AFT process was firstly optimized for the treatment of sulfamethazine by considering the effect of various experimental parameters. It was found that the optimal $[Fe^{2+}]/[H_2O_2]$ ratio was of 1:10 and the optimal $Fe^{2+}$ delivery rate was in the range 38.9–78 μM $min^{-1}$. Under these conditions, Fig. 8.14 shows that sulfamethazine disappears in 4–8 min from solutions with a concentration of 100 μM. Similarly, sulfadiazine was degraded within 6–8 min. Plausible degradation pathways were proposed for both antibiotics, involving the removal of the sulfonyl ($-SO_2$) group as the first step, along with the hydroxylation of the extrusion product. However, the mineralization of the antibiotics was not assessed.

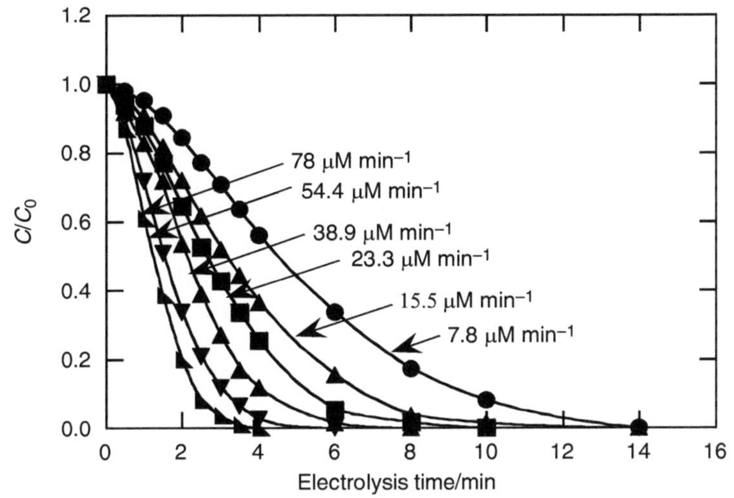


**Fig. 8.14** Normalized concentration destruction of 100 μM sulfamethazine by anodic Fenton treatment in the anodic compartment of a divided cell at various $Fe^{2+}$ delivery rates at the optimal $[Fe^{2+}]/[H_2O_2]$ ratio of 1:10. Overall removal of the antibiotic was achieved between 4 and 8 min operating at $Fe^{2+}$ delivery rates > 38.9 μM $min^{-1}$. Adapted from Neafsey et al. (2010)

### *8.4.3 Photoelectro-Fenton*

The PEF method has been proposed and extensively studied by Brillas' group using undivided, open Pt/GDE and BDD/GDE cells like those of Fig. 8.11 (upper). In this technique, the solution is treated under EF conditions and simultaneously irradiated with UVA light to accelerate the mineralization rate of organics. The $[Fe(OH)]^{2+}$ ion, which is the pre-eminent form of Fe(III) at pH 2.8–3.5, plays a key role when the dark EF process is photoassisted by UVA lamps as energy source. Hydroxyl radicals are produced from Fenton's reaction (8.32) as well, but the drawback of the large accumulation of Fe(III) species decelerating the treatment is avoided from the reductive photolysis of $[Fe(OH)]^{2+}$ by reaction (8.36), thus regenerating the $Fe^{2+}$ ion and producing additional $^{\bullet}OH$ (Pignatello et al. 2006):

$$\left[\mathrm{Fe}(\mathrm{OH})\right]^{2+} + h\nu \rightarrow \mathrm{Fe}^{2+} + {}^{\bullet}\mathrm{OH} \tag{8.36}$$

In addition, UVA light can induce the photodegradation of some oxidation by-products or their complexes with Fe(III) promoting $Fe^{2+}$ regeneration. This is the case for the photodecarboxylation of Fe(III)-carboxylate species, according to the general reaction (8.37), which turns into reaction (8.38) for oxalate complexes (Safarzadeh-Amiri et al. 1997):

$$\mathrm{Fe}(\mathrm{OOCR})^{2+} + h\nu \rightarrow \mathrm{Fe}^{2+} + \mathrm{CO}_2 + \mathrm{R}^{\bullet} \tag{8.37}$$

$$2Fe(C_2O_4)_n^{(3-2n)} + h\nu \rightarrow 2Fe^{2+} + (2n-1)C_2O_4^{\ 2-} + 2CO_2 \qquad (8.38)$$

First works focused on exploring the degradation of paracetamol using a Pt/GDE cell (Sirés et al. 2004, 2006b). Table 8.3 shows that PEF with a 6 W UVA light of $\lambda_{max}=360$ nm yields much better results than the corresponding EF treatments at any current density tested. Furthermore, the combination of 1 mM $Fe^{2+}$, 1 mM $Cu^{2+}$, and UVA light was the only one that led to the complete TOC removal, which was almost reached after 4 h at 100 mA $cm^{-2}$. This can be accounted for by the fact that the complexes of Fe(III) with oxalic and oxamic acids, which are not removed in dark conditions, can be photolyzed by UVA light giving almost total mineralization. Figure 8.15 shows that the synergistic effect of all the catalysts promotes the quickest decontamination because $^{\bullet}OH$ in the bulk can destroy the Cu(II)-oxalate and Cu(II)-oxamate complexes generated and UVA light can photodegrade the competitively formed Fe(III) complexes. Paracetamol was destroyed after just 6 min at 100 mA $cm^{-2}$, as also found in EF, which shows that UVA light does not photolyzes directly the drug.

Further work aimed at comparing the Pt/GDE and BDD/GDE cells for the treatment of clofibric acid by PEF (Sirés et al. 2007a, b). The superiority of PEF compared with the analogous EF treatments discussed in Sect. 8.4.1 is shown in Table 8.3, although the major differences were mainly given in the Pt/GDE cell. For example, at 100 mA $cm^{-2}$, up to 92% TOC removal was achieved by PEF, whereas only 73% was reached in EF. In the case of PEF, similar values of 91–92% were obtained at 4 h at any current density, but the total mineralization was attained at 7, 6, and 5.5 h at 33, 100, and 150 mA $cm^{-2}$, respectively (not shown), which can be explained by the quicker generation of $^{\bullet}OH$ at a higher applied current. The use of BDD led to a quicker degradation, with more than 96% TOC removal at 4 h at 100 mA $cm^{-2}$. On the other hand, assuming pseudo first-order kinetics for clofibric acid decays, a similar apparent rate constant ($k$) value of $(1.35 \pm 0.10) \times 10^{-2}$ $s^{-1}$ was obtained for all the EF and PEF treatments, a value much greater than that obtained in AO, thus confirming the higher reactivity of $^{\bullet}OH$ formed in the bulk to remove aromatics compared with that of Pt($^{\bullet}OH$) and BDD($^{\bullet}OH$) decay rate following a pseudo first-order kinetics.

The performance of the PEF process with Pt/GDE and BDD/GDE cells was also assessed for salicylic acid solutions (Guinea et al. 2008). As can be seen in Table 8.3, both systems led to the complete mineralization at 3 h even at the lowest current density, thus ameliorating the results provided by the EF systems. Salicylic acid disappeared in 30 min, as in the case of EF, indicating that the drug was not directly photolyzed by UVA light. Also, solutions containing 100 mL of 61.5 mg $L^{-1}$ of the antiseptic chloroxylenol were treated under similar conditions, and it was feasible to reach a 93% mineralization after 4 h by PEF with BDD at 33 mA $cm^{-2}$, only requiring 20 min for the destruction of the pharmaceutical (Skoumal et al. 2008).

The analysis of the treated solutions by GC-MS and HPLC allowed detecting intermediates such as: (1) 4-chlorophenol, 4-chlorocatechol, hydroquinone, *p*-benzoquinone and 2-hydroxyisobutyric, tartronic, maleic, fumaric, formic and oxalic acids for

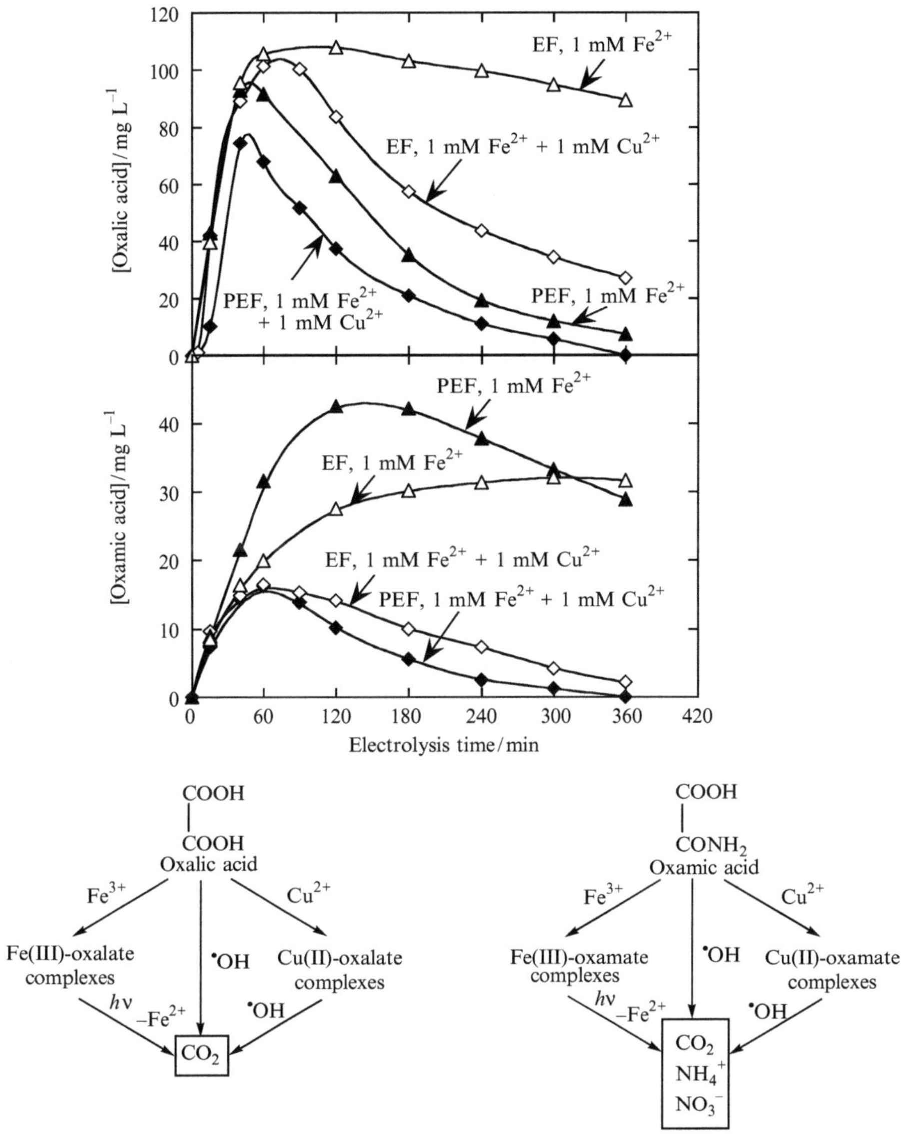


**Fig. 8.15** Evolution of the concentration of oxalic and oxamic acids during the degradation of 100 mL of 157 mg L$^{-1}$ paracetamol solutions at pH 3.0, 100 mA cm$^{-2}$ and 35 °C by electro-Fenton (EF) and photoelectron-Fenton (PEF) processes with 1 mM $Fe^{2+}$ and/or 1 mM $Cu^{2+}$ as catalysts using a Pt/GDE cell (*upper*) and reaction sequence proposed for the destruction of both acids (*down*). Fe(III)-oxalate and Fe(III)-oxamate complexes are photolyzed with UVA light. Cu(II)-oxalate and Cu(II)-oxamate complexes competitively produced are destroyed with hydroxyl radicals in the bulk. Adapted from Sirés et al. (2006b)

clofibric acid; (2) 2,3-dihydroxybenzoic, 2,5-dihydroxybenzoic, 2,6-dihydroxybenzoic, α-ketoglutaric, glycolic, glyoxylic, maleic, fumaric, malic, tartronic and oxalic acids for salicylic acid; and (3) 2,6-dimethylhydroquinone, 2,6-dimethyl-*p*-benzoquinone, 3,5-dimethyl-2-hydroxy-*p*-benzoquinone and maleic, malonic, pyruvic, acetic and oxalic acids for chloroxylenol. Hydroxylation followed by the generation of carboxylic acids is then the main degradative route of these aromatic pharmaceutical residues, as stated above for paracetamol and atenolol.

Ibuprofen, enrofloxacin, and atenolol have also been degraded by PEF with Pt and BDD (Skoumal et al. 2009; Guinea et al. 2010; Isarain-Chávez et al. 2010). Isarain-Chávez et al. (2010) reported that solutions of 100 mL with 158 mg $L^{-1}$ atenolol (100 mg $L^{-1}$ TOC), 0.5 mM $Fe^{2+}$, and 0.05 M $Na_2SO_4$ were degraded by PEF at constant current at 35°C, yielding the complete mineralization in 360 min for both, the combined Pt/GDE-Pt/carbon felt and BDD/GDE-Pt/carbon felt configurations. In fact, the two- and four-electrode cells that they proposed combining GDE and carbon-felt cathodes showed a similar oxidation power to reach the almost overall mineralization as a result of the quick photolysis of the iron complexes under UVA irradiation. Therefore, the potent, synergystic action of the UVA irradiation makes less significant the effect of such an innovative combination of both kinds of cathodes in PEF.

A clear disadvantage of the classical PEF process with UV irradiation is the excessive economical cost arising from the use of artificial light. Luckily, recent work has shown that the alternative use of sunlight ($\lambda > 300$ nm) as a free and renewable energy source in the so-called solar photoelectro-Fenton (SPEF) process is more viable for wastewater remediation. Brillas' group has reported that direct solar irradiation on solutions treated in small tank reactors as well as in flow plants accelerates the decay of organics in SPEF compared with PEF because of the higher production rate of $^{\bullet}OH$ by reaction (8.36) resulting from the greater photon intensity supplied to the system (Brillas et al. 2009). The SPEF method seems more beneficial for water remediation than PEF, although further insight is needed for its future application at an industrial scale.

Comparative SPEF treatments have been performed with stirred Pt/GDE and BDD/GDE cells in order to analyze the possible differences with the PEF treatments. Salicylic acid, ibuprofen, and enrofloxacin were the pharmaceutical residues under study (Guinea et al. 2008; Skoumal et al. 2009; Guinea et al. 2010). Figure 8.16 shows the evolution of the solution TOC over the electrolysis time for the SPEF treatments with 100 mL of contaminated solution in the cell directly exposed to solar radiation, with a mirror at the bottom to better collect the sun rays.

The experiments were made in Barcelona (latitude: 41°21' N, longitude: 2°10' E), being the direct average UV irradiation intensity close to 17 W $m^{-2}$. The mineralization of the three compounds by SPEF was very rapid. Salicylic acid was completely mineralized in only 120 min, following a similar trend for the Pt and BDD cells and with an energy cost of 10 and 20 kWh $m^{-3}$, respectively. For ibuprofen and enrofloxacin, the TOC abatement was slightly quicker using BDD than using Pt. While enrofloxacin was totally mineralized in 240 min with BDD, ibuprofen was rapidly destroyed up to 120 min, attaining 84–88% TOC removal, whereupon the

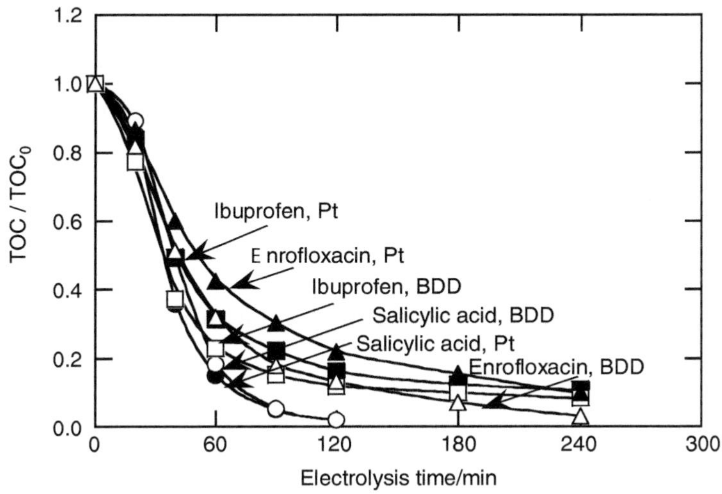


**Fig. 8.16** Normalized TOC abatement for the solar photoelectro-Fenton treatment of 100 mL of 164 mg $L^{-1}$ salicylic acid, 41 mg $L^{-1}$ ibuprofen, and 158 mg $L^{-1}$ enrofloxacin in 0.05 M $Na_2SO_4$ with 0.5 mM $Fe^{2+}$ of pH 3.0 using Pt/GDE and BDD/GDE cells at 33 mA $cm^{-2}$. The salicylic acid solution was more rapidly mineralized and the use of a BDD anode enhanced the degradation process compared with Pt because complexes of Fe(III) are oxidized with hydroxyl radicals formed at BDD and photolyzed under UVA irradiation. Adapted from Guinea et al. (2008, 2010) and Skoumal et al. (2009)

degradation became much slower finally reaching 89% and 92% TOC removal using Pt and BDD, respectively. A much longer elcctrolysis time is then required to destroy the refractory by-products and attain a high mineralization degree. Although not shown in Fig. 8.16, a much faster TOC decay compared with PEF was always observed, which is owed to: (1) the additional production of $^{\bullet}OH$ from photoreduction reaction (8.36) and (2) the quicker photolysis of Fe(III)-carboxylate complexes with the greater UV energy supplied by solar irradiation. The decay of drugs in the above solutions was always very rapid. For example, ibuprofen disappeared totally in only 20 min, somewhat quicker than 30 min needed for PEF destruction, whereas enrofloxacin removal required only 20 min in all PEF and SPEF treatments. Pseudo-first order kinetics was valid to describe all the concentration decays.

Finally, as a first step to show the viability of SPEF at industrial scale, a flow plant with filter-press Pt/GDE or BDD/GDE reactors coupled to a solar photoreactor was used for the treatment of 2.5 L of 158 mg $L^{-1}$ enrofloxacin solutions in 0.05 M $Na_2SO_4$, with 0.2 mM $Fe^{2+}$ as the optimized catalyst concentration (Guinea et al. 2010). The liquid was recirculated with a peristaltic pump at a liquid flow rate of 200 L $h^{-1}$ and its temperature was kept at 35°C by two heat exchangers. The GDE cathode was fed with compressed air and the exposed geometric area of all the electrodes was 20 $cm^2$ with an interelectrode gap of 1.2 cm. The solar photoreactor was a polycarbonate box of 600 mL irradiation volume, connected to the liquid outlet of the electrochemical cell, with a mirror at the bottom and tilted 30°. As a result, the use of BDD led to a much quicker TOC removal compared to Pt, and the decontamination was much more rapid than in the small cell. Using BDD, an 86% TOC removal was achieved after 300 min at 50 mA $cm^{-2}$, with an energy cost of 21.2 kWh $m^{-3}$. The mineralization of enrofloxacin was accompanied by the release of inorganic ions such as $F^-$ (100% of the initial F), $NH_4^+$ (54% of the initial N), and $NO_3^-$ (41% of the initial N), whose concentrations were practically independent of the anode used.

## 8.5 Other Related Electrochemical Technologies

The treatment of pharmaceuticals such as ethinylestradiol contained in urine by electrodialysis has led to positive results concerning the toxicity reduction, as demonstrated from the 99.7% of removal efficiency of estrogenicity (Escher et al. 2006). However, the efficacy of such filtration treatment for the removal of pharmaceutical residues is not as well documented as in the case of other processes like the internal micro-electrolysis, PEC, and corona discharge, as discussed below.

### *8.5.1 Internal Micro-electrolysis*

This method is a variant of EC described in Sect. 8.2, but without an external power supply to feed electricity to the electrodes. It is based on the use of galvanic cells formed between grains or particles of iron and carbon in the pharmaceutical effluent (Jin et al. 2002; Wang et al. 2009). The spontaneous Fe dissolution from anode reaction (8.1) and $H_2$ evolution from cathode reaction (8.2) or (8.3) are the main processes taking place during the internal micro-electrolysis. The $Fe(OH)_2$ and/or $Fe(OH)_3$ produced from reaction (8.4) can then remove the organic contaminants through adsorption onto, co-precipitation with, and enmeshment in their floccules. Furthermore, activated carbon can also adsorb some organic pollutants, especially the hydrophobic ones (Wang et al. 2009).

Jin et al. (2002) examined the degradation of 500 mL of pharmaceutical wastewaters with initial COD between 2.6 and 25 $g L^{-1}$ in a batch tank reactor under stirring by adding 80 g of iron-carbon grains in the pH range 2.3–9.1 during 120 min as maximum. They found that after neutralization and sedimentation of samples, more than 90% COD was removed using neutral and alkaline pH for internal micro-electrolysis during more than 20 min, indicating that this procedure can be very effective for the decontamination of pharmaceutical effluents. More recently, Wang et al. (2009) designed a pilot plant inspired in the combined process of Fig. 8.17 for treating a pharmaceutical effluent containing 10–20 $g L^{-1}$ COD with a high salinity of 61 $g L^{-1}$ at pH 9. The internal micro-electrolysis was initially carried out by adding 1.3 g of scrap iron and 40 g of carbon to 75 L of the raw wastewater at pH 2.0–3.0 in a tank under aeration for 5 h, followed by coagulation in a sedimentation tank at pH

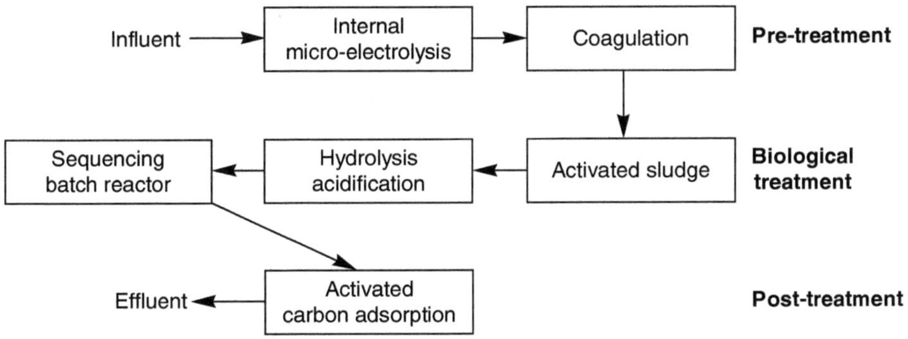


**Fig. 8.17** Flow chart for a combined treatment processes for the overall treatment of a real pharmaceutical wastewater. Internal micro-electrolysis followed by coagulation was utilized as a pre-treatment to remove COD and increase the biodegradability of the resulting wastewater for further biological treatment. Adapted from Wang et al. (2009)

7.0–8.0 for 30 min. The pre-treated effluent was then fed into an activated sludge reactor for 5 h and the resulting wastewater was treated by hydrolysis acidification at 38°C for 6 h before its introduction into sequencing batch reactor (SBR) of four steps for 8 h. The biologically treated effluent was finally post-treated in an activated carbon adsorption reactor for 1.5 h. In this process, iron-carbon internal micro-electrolysis followed by coagulation yielded 64–72% COD removal and increased threefolds the wastewater biodegradability. Thanks to the effectiveness of this pre-treatment, COD was reduced by 90% and 96% after the biological treatment and post-treatment, respectively.

## *8.5.2 Photoelectrocatalysis*

The electrochemical application of photocatalysis is the PEC method, which provides much higher efficiency for wastewater remediation. This photoassisted electrochemical method consists in the application of a constant bias anodic potential ($E_{anod}$) usually to a $TiO_2$-based thin film anode subjected to UV illumination. The electrochemical cells used in PEC are tank or flow reactors that permit the pass of UV light through a quartz glass to reach the exposed surface of the anode with the minimum loss of incident irradiation. Under these conditions, the photoinduced electrons are continuously extracted from the anode by an external electrical circuit to be injected into the cathode. This favors the production of a higher amount of holes from reaction (8.6) and heterogeneous $^{\bullet}OH$ from reaction (8.7), largely enhancing the oxidation compared with photocatalysis since reactions (8.8)–(8.13) are inhibited (Brillas et al. 2009).

Former studies on the viability of PEC for pharmaceutical destruction have been reported by Liu et al. (2009a, b), who treated synthetic tetracycline wastewaters

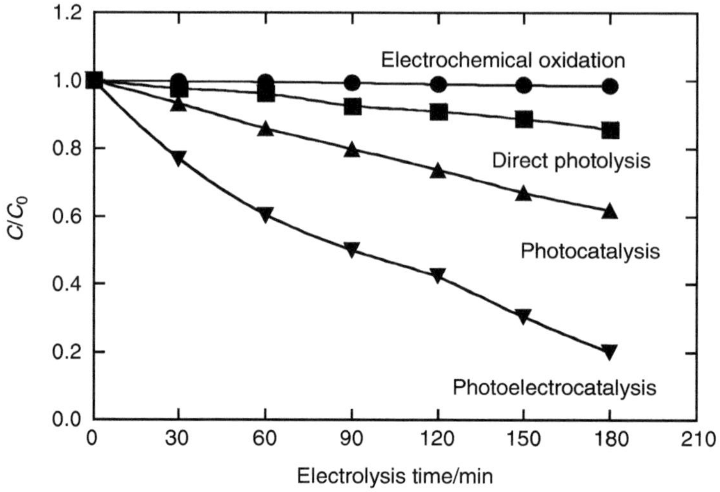


**Fig. 8.18** Normalized concentration removal for different treatments of 50 mg $L^{-1}$ tetracycline in 0.02 M $Na_2SO_4$ of pH 4.5 in a stirred rectangular quartz reactor of 25 mL capacity containing a $TiO_2$ anode, at a potential of 0.5 V vs. Ag/AgCl under 4 W UV irradiation at $\lambda_{max}$ = 254 nm, and a Pt cathode. The drug was much more rapidly removed for photoelectrocatalysis than for photocatalysis due to the suppression of recombination between the photogenerated electron/hole pairs by the external electric field. Adapted from Liu et al. (2009a)

using a $TiO_2$ nanotube arrays photoanode in a stirred undivided three-electrode rectangular quartz reactor. The cell was filled with 25 mL of 50 mg $L^{-1}$ drug solution in 0.02 M $Na_2SO_4$ of pH 4.5 and contained, apart from the anode, a Pt wire as the cathode and a Ag/AgCl reference electrode. The solution was illuminated with a 4 W UV lamp of $\lambda_{max}$ = 254 nm with an intensity of 2 mW $cm^{-2}$. By applying $E_{anod}$ = 0.5 V vs. SCE for 180 min, Fig. 8.18 depicts that tetracycline concentration is reduced by 81% for PEC, whereas it only undergoes a 2% decay using electrochemical oxidation "without light irradiation" due to the very low oxidation power of the $TiO_2$ anode. It can be observed that the drug content decays 16% for direct photolysis "without electrodes" and 38% for photocatalysis "without potential supply". The superiority of photoelectrocatalysis over photocatalysis can be mainly attributed to the suppression of the recombination process between the photogenerated electron/hole pairs from reaction (8.12) by the external electric field. After 3 h of PEC treatment, 81% of decolorization and 41% TOC removal were determined. The same group has reported the PEC treatment of ibuprofen and naproxen using an innovative porous $Bi_2MoO_6$ film deposited onto BDD (Zhao et al. 2009b). This photoanode of 11 $cm^2$ area was exposed to a 150 W Xe lamp to simulate visible light irradiation and it was immersed into 60 mL of the drug solution in 0.12 mg $L^{-1}$ $Na_2SO_4$ filling a cylindrical quartz cell that contained a Pt wire as the cathode and a SCE reference electrode.

For a solution with 10 mg $L^{-1}$ ibuprofen at $E_{anod}$ = 2 V vs. SCE, for example, 86% drug content decay and 72% TOC removal were found for the PEC process, which were much higher than the corresponding values of 64% and 42% obtained for electrochemical oxidation and 21% and 8% for photocatalysis. In this case, the electrochemical oxidation was effective by the ability of $Bi_2MoO_6$ coating to generate ROS at the great $E_{anod}$ applied. These results confirm the possibility of using PEC for pharmaceutical wastewaters treatment with sunlight as irradiation source, although more research is needed to demonstrate its viability in practice.

### *8.5.3 Corona Discharge*

Krause et al. (2009) have recently proposed the destruction of several pharmaceuticals by a pulse or continuous corona discharge generated by two barrier electrodes that causes the generation of ROS such as $^{\bullet}OH$, $^{\bullet}O$, $H_2O_2$, etc. Cells operating in batch with non-circulation or recirculation of the wastewater, with different catalytic cathodes, were employed. The corona discharge was produced with 25–35 kV voltage amplitude at 30 kHz frequency level and a total energy of 250 W. The drug concentration in 200 mL solutions with 0.1 mM of clofibric acid, the antiepilectic drug carbamazepine, and the constrast medium iopromide was rapidly decreased operating by both pulse and continuous mode with a BDD catalytic cathode, disappearing in about 30 min. The kinetics for the decay of these compounds became slower with other cathodes, following the order BDD > Ti > $IrO_2$ > Fe. This sequence was ascribed to the greater production of $O_2$ due to its lower oxidation potential, with the consequent loss of ROS to oxidize the pollutants. Although these findings evidence that this method may be able to remove pharmaceuticals in wastewaters, the high energy required prevents its use in practice.

## 8.6 Conclusion

The treatment of pharmaceutical residues by means of several AOPs is progressively receiving greater attention due to their efficacy and/or efficiency. Particularly, in the last decade the electrochemical technologies have proven to be an efficient, versatile alternative. The EAOPs, especially those based on the electrogenerated Fenton's reagent ($H_2O_2 + Fe^{2+}$) such as the EF and SPEF processes, have been the most applied electrochemical decontamination processes for the treatment of synthetic aqueous solutions containing pharmaceutical residues. These methods exhibit high oxidation ability due to the generation of $^{\bullet}OH$ in the bulk from the Fenton's reaction between cathodically electrogenerated $H_2O_2$ and added iron catalyst. The $^{\bullet}OH$ radical is able to lead very quickly to the cleavage of the aromatic rings as well as to the progressive oxidation of most of the aliphatic carboxylic acid by-products. However, the formation of some refractory carboxylic acids requires the use of UV radiation from artificial lamps or sunlight, and/or the use of high oxidation power anodes to take advantage of the anodic oxidation process based on the role of heterogeneous •OH radical. On the other hand, the real hospital and pharmaceutical wastewaters have been mostly treated by electrochemical oxidation with "non active" anodes such as BDD and electrocoagulation. Since these wastewaters usually contain $Cl^-$ ion, the electro-oxidation process usually proceeds via the formation of active chlorine species, thus being interesting the use of "active" anodes such as $Ti/RuO_2$. Other technologies such as the internal micro-electrolysis and PEC have been less documented for the treatment of these pollutants and need more research for its thorough understanding. In summary, the use of renewable energy sources to power the processes is priority to enhance the sustainability of all these electrochemical treatments. This can definitely increase the acceptance of the electrochemical technologies, as it is foreseen in the case of the SPEF. This latter method is probably the most promising technology among those explained in this chapter.

## References

Anglada A, Urtiaga A, Ortiz I (2009) Contributions of electrochemical oxidation to waste-water treatment: fundamentals and review of applications. J Chem Technol Biotechnol 84:1747–1755. doi:10.1002/jctb.2214

Babu BR, Venkatesan P, Kanimozhi R, Basha CA (2009) Removal of pharmaceuticals from wastewater by electrochemical oxidation using cylindrical flow reactor and optimization of treatment conditions. J Environ Sci Health A 44:985–994. doi:10.1080/10934520902996880

Bergmann MEH, Rollin J, Iourtchouk T (2009) The occurrence of perchlorate during drinking water electrolysis using BDD anodes. Electrochim Acta 54:2102–2107. doi:10.1016/j.electacta.2008.09.040

Bloecher C (2007) Elimination of micropollutants and hazardous substances at the source in the chemical and pharmaceutical industry. Water Sci Technol 56(12):119–123. doi:10.2166/wst.2007.820

Boroski M, Rodrigues AC, Garcia JC, Sampaio LC, Nozaki J, Hioka N (2009) Combined electrocoagulation and $TiO_2$ photoassisted treatment applied to wastewater effluents from pharmaceutical and cosmetic industries. J Hazard Mater 162:448–454. doi:10.1016/j.jhazmat.2008.05.062

Boudreau J, Bejan D, Bunce NJ (2010a) Competition between electrochemical advanced oxidation and electrochemical hypochlorination of acetaminophen at boron-doped diamond and ruthenium dioxide based anodes. Can J Chem 88:418–425. doi:10.1139/V10-017

Boudreau J, Bejan D, Li S, Bunce NJ (2010b) Competition between electrochemical advanced oxidation and electrochemical hypochlorination of sulfamethoxazole at a boron-doped diamond anode. Ind Eng Chem Res 49:2537–2542. doi:10.1021/ie900614d

Brillas E, Baños MA, Camps S, Arias C, Cabot PL, Garrido JA, Rodríguez RM (2004) Catalytic effect of $Fe^{2+}$, $Cu^{2+}$ and UVA light on the electrochemical degradation of nitrobenzene using an oxygen-diffusion cathode. New J Chem 28:314–322. doi:10.1039/b312445b

Brillas E, Sirés I, Arias C, Cabot PL, Centellas F, Rodríguez RM, Garrido JA (2005) Mineralization of paracetamol in aqueous medium by anodic oxidation with a boron-doped diamond electrode. Chemosphere 58:399–406. doi:10.1016/j.chemosphere.2004.09.028

Brillas E, Garrido JA, Rodríguez RM, Arias C, Cabot PL, Centellas F (2008) Wastewaters by electrochemical advanced oxidation processes using a BDD anode and electrogenerated $H_2O_2$ with Fe(II) and UVA light as catalysts. Port Electrochim Acta 26:15–46

Brillas E, Sirés I, Oturan MA (2009) Electro-Fenton process and related electrochemical technologies based on Fenton's reaction chemistry. Chem Rev 109:6570–6631. doi:10.1021/cr900136g

Brillas E, Garcia-Segura S, Skoumal M, Arias C (2010) Electrochemical incineration of diclofenac in neutral aqueous medium by anodic oxidation using Pt and boron-doped diamond anodes. Chemosphere 79:605–612. doi:10.1016/j.chemosphere.2010.03.004

Carlesi Jara C, Fino D (2010) Cost optimization of the current density for electroxidation wastewater processes. Chem Eng J 160:497–502. doi:10.1016/j.cej.2010.03.060

Carlesi Jara C, Fino D, Specchia V, Saracco G, Spinelli P (2007) Electrochemical removal of antibiotics from wastewaters. Appl Catal B- Environ 70:479–487. doi:10.1016/j.apcatb.2005.11.035

Chen G (2004) Electrochemical technologies in wastewater treatment. Sep Purif Technol 38:11–41. doi:10.1016/j.seppur.2003.10.006

Ciríaco L, Anjo C, Correia J, Pacheco M, Lopes A (2009) Electrochemical degradation of ibuprofen on Ti/Pt/$PbO_2$ and Si/BDD electrodes. Electrochim Acta 54:1464–1472. doi:10.1016/j.electacta.2008.09.022

Cleuvers M (2005) Initial risk assessment for three β-blockers found in the aquatic environment. Chemosphere 59:199–205. doi:10.1016/j.chemosphere.2004.11.090

Comninellis Ch (1994) Electrocatalysis in the electrochemical conversion/combustion of organic pollutants for waste water treatment. Electrochim Acta 39:1857–1862. doi:10.1016/0013-4686(94)85175-1

Daughton CG, Ternes TA (1999) Pharmaceuticals and personal care products in the environment: agents of subtle change? Environ Health Perspect 107(suppl 6):907–938. doi:10.2307/3434573

Deshpande A, Lokesh KS, Bejankiwar RS, Gowda TPH (2005) Electrochemical oxidation of pharmaceutical effluent using cast iron electrode. J Environ Sci Eng 47:21–24

Deshpande AM, Satyanarayan S, Ramakant S (2009) Electrochemical pretreatment of wastewater from bulk drug manufacturing industry. J Environ Eng 135:716–719. doi:10.1061/(ASCE)EE.1943-7870.0000029

Deshpande AM, Satyanarayan S, Ramakant S (2010) Treatment of high-strength pharmaceutical wastewater by electrocoagulation combined with anaerobic process. Water Sci Technol 61:463–472. doi:10.2166/wst.2010.831

Escher BI, Pronk W, Suter MJF, Maurer M (2006) Monitoring the removal efficiency of pharmaceuticals and hormones in different treatment processes of source-separated urine with bioassays. Environ Sci Technol 40:5095–5101. doi:10.1021/es060598w

Esplugas S, Bila DM, Krause LGT, Dezotti M (2007) Ozonation and advanced oxidation technologies to remove endocrine disrupting chemicals (EDCs) and pharmaceuticals and personal care products (PPCPs) in water effluents. J Hazard Mater 149:631–642. doi:10.1016/j.jhazmat.2007.07.073

Feng Y, Wang C, Liu J, Zhang Z (2010) Electrochemical degradation of 17-alpha-ethinylestradiol (EE2) and estrogenic activity changes. J Environ Monit 12:404–408. doi:10.1039/b923495k

Garrido JA, Brillas E, Cabot PL, Centellas F, Arias C, Rodríguez RM (2007) Mineralization of drugs in aqueous medium by advanced oxidation processes. Port Electrochim Acta 25:19–41

Guinea E, Arias C, Cabot PL, Garrido JA, Rodríguez RM, Centellas F, Brillas E (2008) Mineralization of salicylic acid in acidic aqueous medium by electrochemical advanced oxidation processes using platinum and boron-doped diamond as anode and cathodically generated hydrogen peroxide. Water Res 42:499–511. doi:10.1016/j.watres.2007.07.046

Guinea E, Brillas E, Centellas F, Cañizares P, Rodrigo MA, Sáez C (2009a) Oxidation of enrofloxacin with conductive-diamond electrochemical oxidation, ozonation and Fenton oxidation. A comparison. Water Res 43:2131–2138. doi:10.1016/j.watres.2009.02.025

Guinea E, Centellas F, Brillas E, Cañizares P, Sáez C, Rodrigo MA (2009b) Electrocatalytic properties of diamond in the oxidation of a persistent pollutant. Appl Catal B- Environ 89:645–650. doi:10.1016/j.apcatb.2009.01.028

Guinea E, Garrido JA, Rodríguez RM, Cabot PL, Arias C, Centellas F, Brillas E (2010) Degradation of the fluoroquinolone enrofloxacin by electrochemical advanced oxidation processes based on hydrogen peroxide electrogeneration. Electrochim Acta 55:2101–2115. doi:10.1016/j.electacta.2009.11.040

Hirose J, Kondo F, Nakano T, Kobayashi T, Hiro N, Ando Y, Takenaka H, Sano K (2005) Inactivation of antineoplastics in clinical wastewater by electrolysis. Chemosphere 60:1018–1024. doi:10.1016/j.chemosphere.2005.01.024

Ikehata K, Naghashkar NJ, El-Din MG (2006) Degradation of aqueous pharmaceuticals by ozonation and advanced oxidation processes: a review. Ozone: Sci Eng 28:353–414. doi:10.1080/01919510600985937

Isarain-Chávez E, Arias C, Cabot PL, Centellas F, Rodríguez RM, Garrido JA, Brillas E (2010) Mineralization of the drug β-blocker atenolol by electro-Fenton and photoelectro-Fenton using an air-diffusion cathode for $H_2O_2$ electrogeneration combined with a carbon-felt cathode for $Fe^{2+}$ regeneration. Appl Catal B- Environ 96:361–369. doi:10.1016/j.apcatb.2010.02.033

Jin Y, Zhang Y, Li W (2002) Experimental study on micro-electrolysis technology for pharmaceutical wastewater treatment. J Zhejiang Univ Sci 3:401–404. doi:10.1631/jzus.2002.0401

Jones OAH, Voulvoulis N, Lester JN (2005) Human pharmaceuticals in wastewater treatment processes. Crit Rev Environ Sci Technol 35:401–427. doi:10.1080/10643380590956966

Khetan SK, Collins TJ (2007) Human pharmaceuticals in the aquatic environment: a challenge to green chemistry. Chem Rev 107:2319–2364. doi:10.1021/cr020441w

Klavarioti M, Mantzavinos D, Kassinos D (2009) Removal of residual pharmaceuticals from aqueous systems by advanced oxidation processes. Environ Int 35:402–417. doi:10.1016/j.envint.2008.07.009

Kobayashi T, Hirose J, Sano K, Hiro N, Ijiri Y, Takiuchi H, Tamai H, Takenaka H, Tanaka K, Nakano T (2008) Evaluation of an electrolysis apparatus for inactivating antineoplastics in clinical wastewater. Chemosphere 72:659–665. doi:10.1016/j.chemosphere.2008.02.020

Krause H, Schweiger B, Schuhmacher J, Scholl S, Steinfeld U (2009) Degradation of the endocrine disrupting chemicals (EDCs) carbamazepine, clofibric acid, and iopromide by corona discharge over water. Chemosphere 75:163–168. doi:10.1016/j.chemosphere.2008.12.020

Kümmerer K (2009) The presence of pharmaceuticals in the environment due to human use - present knowledge and future challenges. J Environ Manage 90:2354–2366. doi:10.1016/j.jenvman.2009.01.023

Lazarova Z, Spendlingwimmer R (2008) Treatment of yellow water by membrane separations and advanced oxidation methods. Water Sci Technol 58:419–426. doi:10.2166/wst.2008.667

Li S, Bejan D, McDowell MS, Bunce NJ (2008) Mixed first and zero order kinetics in the electrooxidation of sulfamethoxazole at a boron-doped diamond (BDD) anode. J Appl Electrochem 38:151–159. doi:10.1007/s10800-007-9413-2

Liu Y, Gan X, Zhou B, Xiong B, Li J, Dong C, Bai J, Cai W (2009a) Photoelectrocatalytic degradation of tetracycline by highly effective $TiO_2$ nanopore arrays electrode. J Hazard Mater 171:678–683. doi:10.1016/j.jhazmat.2009.06.054

Liu Y, Zhou B, Li J, Gan X, Bai J, Cai W (2009b) Preparation of short, robust and highly ordered $TiO_2$ nanotube arrays and their applications as electrode. Appl Catal B- Environ 92:326–332. doi:10.1016/j.apcatb.2009.08.011

Marselli B, Garcia-Gomez J, Michaud PA, Rodrigo MA, Comninellis Ch (2003) Electrogeneration of hydroxyl radicals on boron-doped diamond electrodes. J Electrochem Soc 150:D79–D83. doi:10.1149/1.1553790

Martínez-Huitle CA, Brillas E (2009) Decontamination of wastewaters containing synthetic organic dyes by electrochemical methods: a general review. Appl Catal B- Environ 87:105–145. doi:10.1016/j.apcatb.2008.09.017

Menapace HM, Diaz N, Weiss S (2008) Electrochemical treatment of pharmaceutical wastewater by combining anodic oxidation with ozonation. J Environ Sci Health A 43:961–968. doi:10.1080/10934520801974558

Murugananthan M, Yoshihara S, Rakuma T, Uehara N, Shirakashi T (2007) Electrochemical degradation of 17β-estradiol (E2) at boron-doped diamond (Si/BDD) thin film electrode. Electrochim Acta 52:3242–3249. doi:10.1016/j.electacta.2006.09.073

Neafsey K, Zeng X, Lemley AT (2010) Degradation of sulfonamides in aqueous solution by membrane anodic Fenton treatment. J Agric Food Chem 58:1068–1076. doi:10.1021/jf904066a

Oturan MA, Brillas E (2007) Electrochemical advanced oxidation processes (EAOPs) for environmental applications. Port Electrochim Acta 25:1–18

Oturan MA, Pinson J, Bizot J, Deprez D, Terlain B (1992a) Reaction of inflammation inhibitors with chemically and electrochemically generated hydroxyl radicals. J Electroanal Chem 334:103–109. doi:10.1016/0022-0728(92)80563-J

Oturan MA, Pinson J, Deprez D, Terlain B (1992b) Polyhydroxylation of salicylic acid by electrochemically generated OH radicals. New J Chem 16:705–710

Oturan MA, Pinson J, Oturan N, Deprez D (1999a) Hydroxylation of aromatic drugs by the electro-Fenton method. Formation and identification of the metabolites of riluzole. New J Chem 23:793–794. doi:10.1039/A903791H

Oturan MA, Pinson J, Traikia M, Deprez D (1999b) The electrochemical oxidation of riluzole, a neuroprotective drug: comparison with the reaction with oxygen derived radicals. J Chem Soc Perkin Trans 2 3:619–622. doi:10.1039/a807354f

Panizza M, Cerisola G (2009) Direct and mediated anodic oxidation of organic pollutants. Chem Rev 109:6541–6569. doi:10.1021/cr9001319

Pauwels B, Verstraete W (2006) The treatment of hospital wastewater: an appraisal. J Water Health 4:405–416. doi:10.2166/wh.2006.025

Pauwels B, Deconinck S, Verstraete W (2006) Electrolytic removal of 17α-ethinylestradiol (EE2) in water streams. J Chem Technol Biotechnol 81:1338–1343. doi:10.1002/jctb.1515

Perez G, Fernandez-Alba AR, Urtiaga AM, Ortiz I (2010) Electro-oxidation of reverse osmosis concentrates generated in tertiary water treatment. Water Res 44:2763–2772. doi:10.1016/j.watres.2010.02.017

Pignatello JJ, Oliveros E, MacKay A (2006) Advanced oxidation processes for organic contaminant destruction based on the Fenton reaction and related chemistry. Crit Rev Environ Sci Technol 36:1–84. doi:10.1080/10643380500326564

Rahman MF, Yanful EK, Jasim SY (2009) Occurrences of endocrine disrupting compounds and pharmaceuticals in the aquatic environment and their removal from drinking water: challenges in the context of the developing world. Desalination 248:578–585. doi:10.1016/j.desal.2008.05.105

Rossi A, Alves VA, Da Silva LA, Oliveira MA, Assis DOS, Santos FA, De Miranda RRS (2009) Electrooxidation and inhibition of the antibacterial activity of oxytetracycline hydrochloride using a $RuO_2$ electrode. J Appl Electrochem 39:329–337. doi:10.1007/s10800-008-9676-2

Safarzadeh-Amiri A, Bolton JR, Cater SR (1997) Ferrioxalate-mediated photodegradation of organic pollutants in contaminated water. Water Res 31:787–798. doi:10.1016/S0043-1354(96)00373-9

Sirés I, Arias C, Cabot PL, Centellas F, Rodríguez RM, Garrido JA, Brillas E (2004) Paracetamol mineralization by advanced electrochemical oxidation processes for wastewater treatment. Environ Chem 1:26–28. doi:10.1071/EN04018

Sirés I, Cabot PL, Centellas F, Garrido JA, Rodríguez RM, Arias C, Brillas E (2006a) Electrochemical degradation of clofibric acid in water by anodic oxidation. Comparative study with platinum and boron-doped diamond electrodes. Electrochim Acta 52:75–85. doi:10.1016/j.electacta.2006.03.075

Sirés I, Garrido JA, Rodríguez RM, Cabot PL, Centellas F, Arias C, Brillas E (2006b) Electrochemical degradation of paracetamol from water by catalytic action of $Fe^{2+}$, $Cu^{2+}$, and UVA light on electrogenerated hydrogen peroxide. J Electrochem Soc 153:D1–D9. doi:10.1149/1.2130568

Sirés I, Arias C, Cabot PL, Centellas F, Garrido JA, Rodríguez RM, Brillas E (2007a) Degradation of clofibric acid in acidic aqueous medium by electro-Fenton and photoelectro-Fenton. Chemosphere 66:1660–1669. doi:10.1016/j.chemosphere.2006.07.039

Sirés I, Centellas F, Garrido JA, Rodríguez RM, Arias C, Cabot PL, Brillas E (2007b) Mineralization of clofibric acid by electrochemical advanced oxidation processes using a boron-doped diamond anode and $Fe^{2+}$ and UVA light as catalysts. Appl Catal B- Environ 72:373–381. doi:10.1016/j.apcatb.2006.12.002

Sirés I, Garrido JA, Rodríguez RM, Brillas E, Oturan N, Oturan MA (2007c) Catalytic behavior of the $Fe^{3+}/Fe^{2+}$ system in the electro-Fenton degradation of the antimicrobial chlorophene. Appl Catal B- Environ 72:382–394. doi:10.1016/j.apcatb.2006.11.016

Sirés I, Oturan N, Oturan MA, Rodríguez RM, Garrido JA, Brillas E (2007d) Electro-Fenton degradation of antimicrobials triclosan and triclocarban. Electrochim Acta 52:5493–5503. doi:10.1016/j.electacta.2007.03.011

Sirés I, Oturan N, Oturan MA (2010) Electrochemical degradation of β-blockers. Studies on single and multicomponent synthetic aqueous solutions. Water Res 44:3109–3120. doi:10.1016/j.watres.2010.03.005

Skoumal M, Arias C, Cabot PL, Centellas F, Garrido JA, Rodríguez RM, Brillas E (2008) Mineralization of the biocide chloroxylenol by electrochemical advanced oxidation processes. Chemosphere 71:1718–1729. doi:10.1016/j.chemosphere.2007.12.029

Skoumal M, Rodríguez RM, Cabot PL, Centellas F, Garrido JA, Arias C, Brillas E (2009) Electro-Fenton, UVA photoelectro-Fenton and solar photoelectro-Fenton degradation of the drug ibuprofen in acid aqueous medium using platinum and boron-doped diamond anodes. Electrochim Acta 54:2077–2085. doi:10.1016/j.electacta.2008.07.014

Suárez S, Carballa M, Omil F, Lema JM (2008) How are pharmaceutical and personal care products (PPCPs) removed from urban wastewaters? Rev Environ Sci Biotechnol 7:125–138. doi:10.1007/s11157-008-9130-2

Vedenyapina MD, Eremicheva Y, Pavlov VA, Vedenyapin AA (2008) Electrochemical degradation of tetracycline. Russ J Appl Chem 81:800–802. doi:10.1134/S1070427208050145

Wang K, Liu S, Zhang Q, He Y (2009) Pharmaceutical wastewater treatment by internal micro-electrolysis-coagulation, biological treatment and activated carbon sorption. Environ Technol 30:1469–1474. doi:10.1080/09593330903229164

Waterston K, Wang JW, Bejan D, Bunce NJ (2006) Electrochemical waste water treatment: electrooxidation of acetaminophen. J Appl Electrochem 36:227–232. doi:10.1007/s10800-005-9049-z

Weichgrebe D, Danilova E, Rosenwinkel KH, Vedenjapin AA, Baturova M (2004) Electrochemical oxidation of drug residues in water by the example of tetracycline, gentamicin and aspirin. Water Sci Technol 49(4):201–206

Yoshihara S, Murugananthan M (2009) Decomposition of various endocrine-disrupting chemicals at boron-doped diamond electrode. Electrochim Acta 54:2031–2038. doi:10.1016/j.electacta.2008.07.006

Zhang H, Liu F, Wu X, Zhang J, Zhang D (2009) Degradation of tetracycline in aqueous medium by electrochemical method. Asia Pac J Chem Eng 4:568–573. doi:10.1002/apj.286

Zhao X, Hou Y, Liu H, Qiang Z, Qu J (2009a) Electro-oxidation of diclofenac at boron doped diamond: kinetics and mechanism. Electrochim Acta 54:4172–4179. doi:10.1016/j.electacta.2009.02.059

Zhao X, Qu J, Liu H, Qiang Z, Liu R, Hu C (2009b) Photoelectrochemical degradation of anti-inflammatory pharmaceuticals at $Bi_2MoO_6$-boron-doped diamond hybrid electrode under visible light irradiation. Appl Catal B- Environ 91:539–545. doi:10.1016/j.apcatb.2009.06.025

# Chapter 9
# Biogas Technologies and Cleaning Techniques

**M. Estefanía López, Eldon R. Rene, María C. Veiga, and Christian Kennes**

## Contents



M.E. López • E.R. Rene • M.C. Veiga • C. Kennes (✉)
Chemical Engineering Laboratory, Faculty of Sciences,
University of La Coruña, Rua da Fraga, 10, La Coruña E-15008, Spain
e-mail: Kennes@udc.es

E. Lichtfouse et al. (eds.), *Environmental Chemistry for a Sustainable World: Volume 2: Remediation of Air and Water Pollution*, DOI 10.1007/978-94-007-2439-6_9,

**Abstract** The acclivitious rise in fuel costs and taxation burden have significantly triggered the scientific community to make substantial changes in the pattern of using renewable energy sources. Gases arising from anaerobic digesters and landfills are widely used as fuel to produce electricity, drive pumps and fire boilers. Biogas can be defined as a biofuel produced by a large number of anaerobic microbial species that inherently possess the capability to ferment organic matter under controlled temperature, moisture and pH to yield a high energy value fuel. Methane and carbon dioxide are the two predominant compounds present in biogas, along with trace quantities of gases such as water vapour, hydrogen sulfide ($H_2S$), halogenated hydrocarbons, siloxanes, ammonia, nitrogen, and oxygen. Biogas can be used as a fuel for generating heat and electricity and in feed boilers. However, biogas has to be cleaned and upgraded, in order to use it effectively as a fuel or to distribute it in the natural gas grid. Biogas cleaning not only increases the calorific value, but also reduces the risk of corrosion.

In this chapter, we outline the importance of biogas from their source, property and application view-points. Secondly, the various physico-chemical and biological methods commonly used for biogas cleaning were reviewed by highlighting the operational advantages and limitations in each case. We have also presented the different strategies adopted to control specific impurities present in biogas, *i.e.*, $CO_2$, siloxanes, $H_2S$ and other trace compounds.

**Keywords** Biogas composition • Biogas cleaning • Physico-chemical treatment • Biological treatment • Bioreactors • Hydrogen sulfide • Siloxanes • Cryogenic • $CO_2$ • $H_2S$ • Water scrubbing • Activated carbon • Filter • Dioxin • Furan • CFC • $NH_4$ • Bioprocess • Biodegradation • Biofilter • Bioscrubber • Biotrickling filter • Waste gas • Air pollution • Water pollution • Odour control • Greenhouse gases • Microbial kinetics • Reactor modelling • Photocatalysis • Absorption • Adsorption

## 9.1 Introduction

In recent years, renewable energy sources, in particular wind, solar, tidal, geothermal, and biomass, are gaining significant interest because they promise to be a feasible and eco-friendly alternative to the acclivitous problems related to future fossil fuels shortage and the difficulty of setting up hydro-electric and thermo-electric power stations. House (2007) explained the need for alternative energy sources as "rising energy prices and concerns about long-term sustainability have once again brought renewable energy sources to the forefront". Through sustained research efforts from practicing engineers, scientists and researchers from a wide variety of disciplines, several newer techniques have emerged for the production of biogas, from diversified domestic, industrial and agricultural sources.

Biogas can be defined as "an end-product originating from the anaerobic digestion of hydrocarbons, comprising mainly methane ($CH_4$), carbon dioxide ($CO_2$), and other components" (Walsh et al. 1988). To make the definition much simpler, biogas is the result of the complete microbial degradation of organic matter, under anaerobic conditions. This organic matter may be wastes from different sources, such as sewage sludge, landfills or industry.

Furthermore, there are several reasons; apart from non-fossil fuels energy production, for the utilization of anaerobic digestion. In nature, organic wastes will be degraded by the action of microorganisms, as it occurs in landfills, releasing a complex mixture of gases, mainly methane and carbon dioxide. Methane is a hazardous pollutant and has a greenhouse gas effect 23-times greater than carbon dioxide, (Petersson and Wellinger 2009). Therefore, collecting biomethane generated through anaerobic digestion, for subsequent combustion, will reduce the great environmental damage that can occur releasing those amounts of methane into the ambient atmosphere. It can also be argued that, after anaerobic digestion, the organic matter will be stabilized and the final digestate, rich in nutrients, can be used as a fertilizer.

However, biogas not only contains methane and carbon dioxide, but also other gases such as hydrogen sulfide, siloxanes, halogenated hydrocarbons, or ammonia that are produced during the anaerobic digestion step. These impurities can cause health and environmental problems, and can lead to corrosion and failure of the process equipments, pipelines, nozzles and connections, as they tend to accumulate/build-up. In addition, the presence of such impurities will reduce the final methane content in biogas, which in turn reduces its calorific value during combustion. Thus, it is clear from the above-mentioned reasons that, in order to improve the quality of biogas, for economic considerations and from an environmental perspective, biogas has to be upgraded, just like the pre-treatment steps undertaken for other fossil fuels, prior to its utilization.

## 9.2 Biogas Technology

As previously mentioned, many different organic sources may be used in biogas generation. Although most of the natural organic material, except lignin (Sleat and Mah 1987; Petersson and Wellinger 2009), can be used as a biogas precursor, the most important sources are those processes that are directly involved with waste management, mainly sewage sludge, landfills, agriculture wastes, manure, and industrial food wastes (Fig. 9.1).

The anaerobic digestion process, responsible for biogas production, is carried out in a closed reactor called a digester. Different designs of this reactor type have been studied and optimized to suit industrial operations, but the final configuration of the digester depends on many factors, *viz.*, characteristics and concentration of the raw organic feed, the scale of application and energy requirements, among others (Chynoweth 1987).

Inside the digester, a mixed consortium of both anaerobic and facultative microorganisms will carry out the transformation of organic compounds into methane. This process, illustrated in three stages, as hydrolysis, acid forming and methanogenesis, is shown in Fig. 9.2. In the first stage, complex insoluble polymers (proteins, lipids and carbohydrates), are converted to simple soluble compounds (aminoacids, fatty acids and monosaccharides). The transformation of these complex substrates is carried out by anaerobic or facultative hydrolytic bacteria. During the acid forming stage, both anaerobic and facultative bacteria will degrade the simple substrates, originated in the hydrolysis step, through a series of fermentative processes. The main products

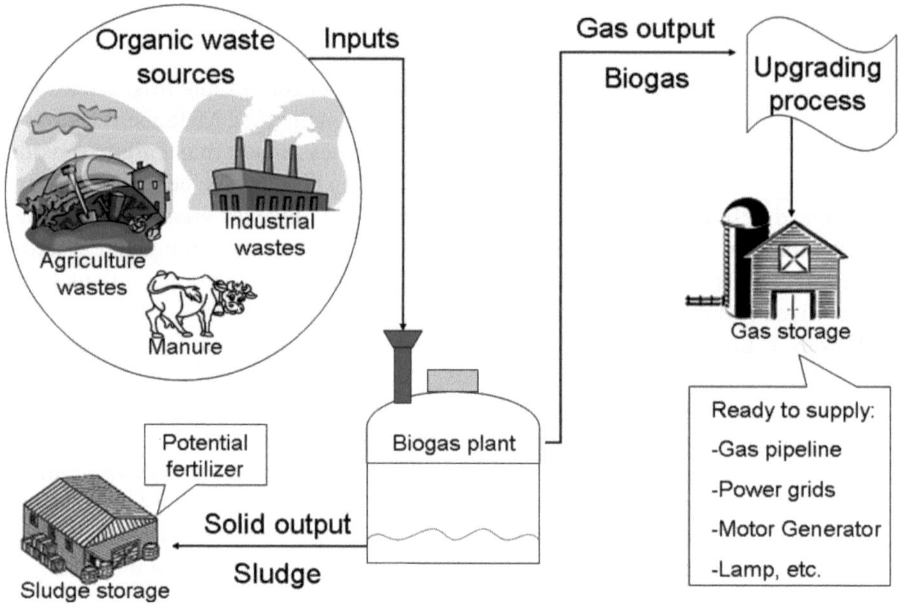


**Fig. 9.1** Schematic of biogas production, sludge storage and supply unit. After pre-conditioning the raw material (organic waste), microbial conversion of organic matter to methane and other useful end-product is achieved in the biogas plant, under well-optimized environmental conditions. This process is anaerobic and is quintessentially carried out by the action of various groups of anaerobic bacteria

of fermentation are organic acids, alcohols, carbon dioxide, hydrogen, and compounds originated from proteins degradation, such as sulfides and ammonia. However, not all the organic acids can be used as substrate by methanogenic bacteria, and hence they have to be degraded into other compounds like acetate (Gerardi 2003). Acetogenic microorganisms are involved in the production of acetate from organic acids and alcohols (Eq. 9.1). Acetate can also be produced from the reaction of carbon dioxide and hydrogen originated from previous steps (Eq. 9.2).

$$\text{Organic acids} + \text{alcohols} \rightarrow \text{Acetate} \quad (9.1)$$

$$2CO_2 + 4H_2 \rightarrow CH_3COOH + 2H_2O \quad (9.2)$$

Finally, the methanogenic bacteria will use these final products in order to generate methane, using different paths [Eqs. 9.3–9.5].

$$CH_3COOH \rightarrow CH_4 + CO_2 \quad (9.3)$$

$$2C_2H_5OH + CO_2 \rightarrow CH_4 + 2CH_3COOH \quad (9.4)$$

$$CO_2 + 4H_2 \rightarrow CH_4 + 2H_2O \quad (9.5)$$

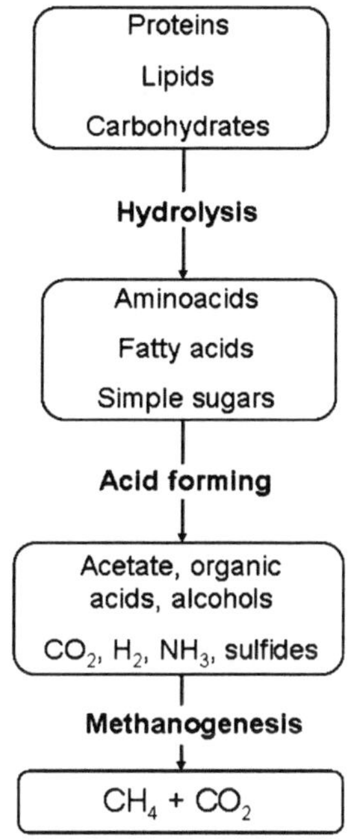


**Fig. 9.2** Different stages of the anaerobic digestion (Gerardi 2003). The transformation of these complex substrates to methane is carried out by anaerobic or facultative hydrolytic bacteria. The methane thus produced can be burned to produce both heat and electricity, usually with a reciprocating engine

### *9.2.1 Biogas Composition*

Biogas composition will depend on the type of waste source used. Biogas compounds are listed in Table 9.1, while Table 9.2 shows the composition of some important trace compounds, such as siloxanes.

As shown in Table 9.1, the methane content in raw biogas can vary significantly depending on the origin of the organic waste. This variation implies that the methane potential calorific value will also be different. Therefore, depending on the methane content, biogas will be used for different applications. For example, poultry farm waste produces high biogas yields (0.065–0.116 $m^3$ biogas $kg^{-1}$ dung), while pre-treated crop waste and water hyacinth have the potential to produce 0.037–0.045 $m^3$ biogas $kg^{-1}$, respectively. For raw biogas containing >75% methane, the heat value ranges from 20,100 to 28,900 kcal $Nm^{-3}$ (Harasimowicz et al. 2007). The calorific value has been reported to range from 22.5 to 25 MJ $m^{-3}$, assuming methane with about 35,800 KJ $m^{-3}$, which is roughly 6.25–10 kWh $m^{-3}$ (Jordão and Pessoa 1995; Salomon and Lora 2009).

### *9.2.2 Advantages and Applications of Biogas*

From an aesthetic and hygienic view-point, biogas technology contributes significantly to improvise living conditions and socio-economics in rural areas. It has been

**Table 9.1** Composition of biogas obtained from different sources

| Biogas | $CH_4$ (%) | $CO_2$ (%) | $O_2$ (%) | $N_2$ (%) | $H_2S$ (ppm) | $NH_3$ (ppm) | Reference |
|---|---|---|---|---|---|---|---|
| Landfill | 44.0 | 40.1 | 2.6 | 13.2 | n.a. | n.a. | Jaffrin et al. (2003) |
| Landfill | 47–57 | 37–41 | <1 | <1–17 | 36–115 | n.a. | Rasi et al. (2007) |
| Landfill | 35–65 | 15–50 | 0–5 | 5–40 | 0–100 | 5 | Petersson and Wellinger (2009) |
| Sewage digester | 58 | 33.9 | 0 | 8.1 | 24.1 | n.a. | Rasi et al. (2007) |
| Sewage digester | 59.6 | 39.1 | 0.2 | 0.9 | 2,897 | n.a. | Osorio and Torres (2009) |
| Municipal sludge | 60 | 33 | 0 | 1 | n.a. | n.a. | Favre et al. (2009) |
| Agriculture residues | 68 | 26 | 0 | 1 | n.a. | n.a. | Favre et al. (2009) |
| Farm biogas plant | 55–58 | 37–38 | <1 | <1–2 | b.d. | n.a. | Rasi et al. (2007) |
| From anaerobic digester | 53–70 | 15–50 | 0–5 | 0.2 | 0–10,000 | <100 | Persson et al. (2006) |

*n.a* not available, *b.d* below detectable limit

**Table 9.2** Composition of important trace compounds from different biogas sources

| Biogas | Siloxanes (mg $m^{-3}$) | Halogenated compounds (mg $m^{-3}$) | Sulfur compounds except $H_2S$ (mg $m^{-3}$) | Reference |
|---|---|---|---|---|
| Landfill | 7.84 | 194.9 | 22.9 | Jaffrin et al. (2003) |
| Landfill | 24.15 | n.a. | n.a. | Haberbauer (2005) |
| Sewage digester | 7.04 | n.a. | n.a. | Haberbauer (2005) |

*n.a* not available

reported that, in developing countries, by installing biogas plants, diseases like asthma, lung problems, and eye infections have considerably decreased in the same area when compared to the pre-biogas plant times. Salomon and Lora (2009) estimated the electricity generation potential of biogas from different sources of organic residues, *viz.,* sugar and alcohol industry residues (vinasse), urban solid and liquid wastes (garbage and sewage) and livestock wastes (bovine and swine manure), and described the advantage of biogas production as follows; "the bio-digestion process can reduce the polluting potential of organic residue discharges with high contents of biochemical oxygen demand (BOD), and at the same time, it can produce methane and a fertilizer agent as residue". Biogas can be used for the generation of electricity with the help of internal combustion engines, as a fuel in substitution of natural or liquefied petroleum gas (LPG), heat generation for incubators and coolers, and as an alternative fuel for gasoline engines (Roubaud and Favrat 2005; Purohit and Kandpal 2007).

## 9.3 Biogas Cleaning Techniques

The biogas produced from different sources using conventional or emerging biogas technologies may contain various impurities. Biogas, when upgraded efficiently, can be used for all applications designed for natural gas. However, in order to use biogas as a vehicle fuel it has to be enriched in methane, and this is primarily achieved by carbon dioxide removal which then enhances the energy value of the gas to give longer driving distances with a fixed gas storage volume. Hydrogen sulfide should also be removed before the gas can be used for generation of electricity or heat. The degree of treatment, however, depends on the desired gas utilization process. The presence of hydrogen sulfide in biogas can lead to serious problems on process equipments, due to its toxic and corrosive nature, besides leading to sulfur oxide emissions during combustion. From an environmental and health view-point, hydrogen sulfide is a flammable and highly toxic gas, heavier than air. Its odour threshold is about 0.00047 ppmv (Aroca et al. 2007). Upon inhalation, hydrogen sulfide reacts with enzymes in the bloodstream and inhibits cellular respiration resulting in pulmonary paralysis, sudden collapse, and death. Higher concentrations (200–300 ppm) may result in respiratory arrest leading to coma and unconsciousness (Syed et al. 2006).The removal of carbon dioxide and hydrogen sulfide, is usually recommended so as to avoid severe down-stream damage, increase process safety and increase the operational efficiency of the equipment/down-stream process. The different physico-chemical and biological methods used for biogas cleaning are discussed in the next section.

### 9.3.1 *Physico-Chemical Methods*

#### 9.3.1.1 Removal of Carbon Dioxide

Adsorption

*Pressure Swing Adsorption (PSA)*

Carbon dioxide can be removed from the biogas by adsorption on a surface, normally activated carbon, molecular sieves or zeolites. In pressure swing adsorption, adsorption takes place at elevated pressure and the material is regenerated by reducing the pressure and subsequent application of a light vacuum. However, this process requires a pre-treatment step in order to remove water vapour and hydrogen sulfide present in the inlet stream, *i.e.,* biogas. Hydrogen sulfide pose a severe operating problem during adsorption at elevated pressures, as it could easily damage the adsorbent at such high pressures, very fast, because its adsorption is irreversible in these types of materials. The presence of water can destroy the structure of the adsorbent. In order to solve this, a pre-treatment step, normally adsorption onto impregnated activated carbon followed by an increase in the temperature between 60°C and

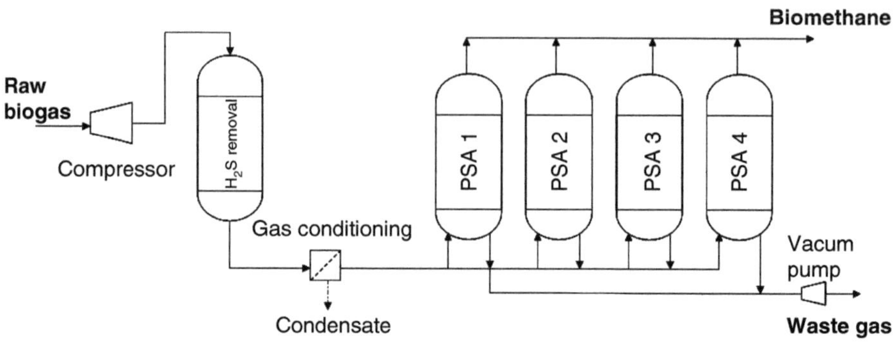


**Fig. 9.3** Typical schematic of a pressure swing adsorption unit. The four adsorber vessels (PSA1-PSA4) operate in an alternating cycle of adsorption, regeneration and pressure build-up. An optimized PSA-plant can achieve high enrichment of $CH_4$ (>97%) and low levels of air emissions and solid waste. The advantages of pressure swing adsorption can be summarized as follows (Schulte-Schulze Berndt 2005)

90°C can easily remove gas-phase hydrogen sulfide from biogas. After the desulfurization step, water is removed by condensation (Schulte-Schulze Berndt 2005).

The schematic of pressure swing adsorption equipment, illustrating the several stages of unit operation is shown in Fig. 9.3. The system consists of several vessels, often four, working in parallel, in an alternating cycle of adsorption, regeneration and pressure build-up. When the adsorbing material in one vessel becomes saturated, the biogas is sent to another vessel in which the adsorbing material has been already regenerated. The regeneration of the material is usually done in several steps. First, the pressure is reduced by linking the vessel with an already regenerated vessel. Then the pressure is again reduced to almost standard atmospheric values, and finally, the vessel is completely evacuated using a vacuum pump (Petersson and Wellinger 2009). The waste stream contains $N_2$, $O_2$, and $CO_2$, along with small quantities of $CH_4$, which can be re-circulated again into the inlet in order to recover that methane.

The methane-free outlet gas-stream leaving the vessel consists of mainly carbon dioxide, which can be released directly into the atmosphere or can be sent to further treatment. For instance, the outlet gas-stream can be linked to a generator and by doing so $CO_2$ emissions to the atmosphere can be avoided (Persson et al. 2006).

- High methane concentration (≥96%) in the treated biogas
- Co-removal of other impurities, such as halogenated and silicon compounds, and partial removal of oxygen and nitrogen
- No handling of water or chemicals, therefore no liquid waste stream
- Simple and automatic operation

However, this technique also has some disadvantages;

- Emission of waste gas stream of carbon dioxide and impurities
- The requirement of a pretreatment unit for desulfurization
- Energy requirements in compressors and heat exchangers

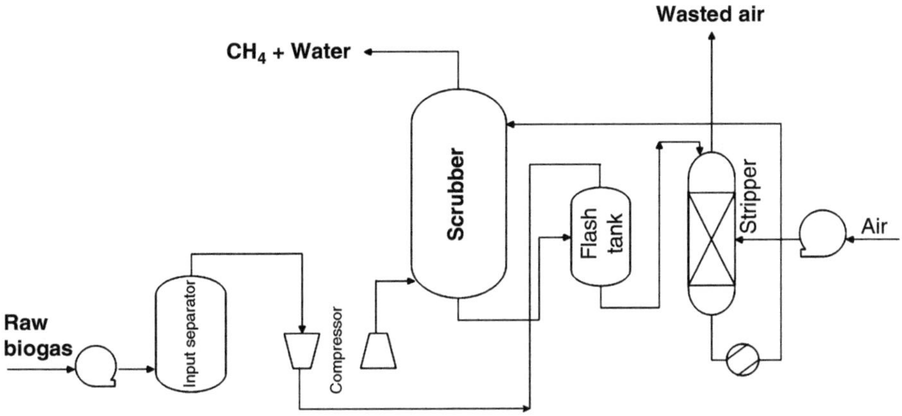


**Fig. 9.4** Schematic of a conventional water scrubber (*adapted from* de Hullu et al. 2008). In a typical scrubbing unit, water soluble components removed from the gas-phase are absorbed by the continuously spraying water-phase. By adapting proper pH control devices, and under well-optimized process conditions, high concentrations of volatile sulphur compounds, nitrogen and chlorine can be removed easily

## Absorption

In absorption techniques for biogas purification, the raw gas is put in contact with a non-volatile liquid phase, the purpose being the mass transfer of the contaminant from the gas-phase to the liquid phase (Kennes et al. 2001). The main idea of cleaning biogas through absorption is to transfer carbon dioxide from the gas-phase to the stationary liquid phase, for example water, in which that pollutant is more soluble than methane. The different absorption technologies use different types of absorbents, *viz.*, water scrubbing, organic physical scrubbing and chemical scrubbing.

### *Water Scrubbing*

Carbon dioxide, as well as hydrogen sulfide, are more soluble in water than methane, and in conventional scrubbing columns, these pollutants can easily be dissolved in aqueous phase, while the methane concentration in the gas-phase increases. In a typical water scrubber, biogas is compressed, in order to increase its solubility in water, and fed through the bottom of an absorption column, contacting a counter-flowing stream of water (Fig. 9.4). The column is filled with a packing material in order to increase the contact surface between the liquid and biogas. The outlet gas-stream at the top of the column is biogas enriched in methane and saturated water, whereas the liquid collected at the bottom contains the pollutants. The water leaving the column also contains some amounts of methane, which can once again be treated in order to recover it. The liquid outlet is transferred to a flash tank where the dissolved gases, containing mainly carbon dioxide and some methane, are released and transferred back to the biogas inlet stream. After the flash tank, the liquid stream is transferred to a stripper, where water is regenerated by desorption with a counter flow of air, into which carbon dioxide will be released. During the stripping/regeneration process, hydrogen sulfide reacts with the oxygen of air, forming elemental sulfur which can cause corrosion to the system. Moreover, the outlet gas stream contains traces of hydrogen sulfide that will be released into atmosphere. If the hydrogen sulfide concentration is high, it is more beneficial to remove it prior to the water absorption step (Persson et al. 2006).

There are two possible paths for the liquid stream; (1) In *single pass scrubbing*, the washing water is used only once. The advantage is that the inlet water is free from trace quantities of both $H_2S$ and $CO_2$. The disadvantage of this technique is that it requires a large amount of water and then, at the end of the process, it has to be treated in a waste water treatment plant. (2) In *regenerative absorption*, the washing water is regenerated by desorption after washing the biogas. Water can be recirculated in the system and thus significant reduction of the water quantity used is achieved compared to single pass scrubbing. Water scrubbing requires a large amount of water, so single pass scrubbing is neither very economic nor an eco-friendly option. Hence, in most of the cases regenerative absorption is preferred (de Hullu et al. 2008).

By using water scrubbers, methane concentrations greater than 97% can be achieved, although it produces two waste streams, one in the liquid and another in the gas-phase. Water scrubbing is the most common upgrading technique and plants are commercially available from several suppliers in a broad range of capacities (Petersson and Welinger 2009).

*Organic Physical Scrubbing*

Organic physical scrubbing is very similar to water scrubbing; with the difference that the carbon dioxide is absorbed in an organic solvent in which carbon dioxide and hydrogen sulfide are more soluble than in water. In this process, water and halogenated hydrocarbons, which are also present in biogas (for example-from landfills) are separated. However, a large amount of energy is needed to regenerate the organic solvent from hydrogen sulfide; thereby it is often better to separate the hydrogen sulfide prior to the absorption step. The most commonly used solvent for biogas absorption is polyethylene glycol, also known commercially as Selexol® and Genosorb® (Persson et al. 2006). The major difference between water and polyethylene glycol is that carbon dioxide and hydrogen sulfide are more soluble in the organic solvent, hence operational costs, as well as solvent regeneration and pumping requirements, will be reduced. Besides, water and halogenated hydrocarbons are also removed. The polyethylene glycol scrubbing medium is then recirculated into the system and regeneration is done by heating and/or depressurizing. Hydrogen sulfide, water, oxygen and nitrogen may be removed together with carbon dioxide using this technique.

*Chemical Scrubbing*

Chemical absorption involves the formation of reversible chemical bonds between the pollutants and the solvent. Regeneration of the solvent, therefore, involves breaking of these bonds and correspondingly, a relatively high energy input (Kapdi et al. 2004). In chemical scrubbers, carbon dioxide reacts chemically with the absorbent present in the liquid-phase. The absorbents used are alkanol amine solutions, like mono ethanol amine (MEA), di-methyl ethanol amine (DMEA) or tertiary amines. The reaction, for a primary amine, takes place between the solutions amino group and carbon dioxide according to the equation described herein (Kumar et al. 2002);

$$2RNH_2 + CO_2 \rightarrow RNHCOO^- + RNH_3^+ \qquad (9.6)$$

The enriched bio-methane stream will eventually be released at the top of the scrubber. As this chemical reaction is strongly selective, methane losses are usually low. The liquid containing carbon dioxide is regenerated in a subsequent step by heating, and, the clean solvent is recirculated into the system again. Using this technique, it is possible to achieve methane contents around 99% (Petersson and Wellinger 2009) (Fig. 9.5).

One disadvantage of the technique is that it can only treat one compound at a time, *i.e.,* for example: carbon dioxide. If hydrogen sulfide is present in the inlet gas, it will also be absorbed in the amine scrubber solution. Desorption of hydrogen sulfide is not as easy as for carbon dioxide and high temperatures have to be applied to recover it. Hence, prior to applying this technique for removing carbon dioxide, hydrogen sulfide should be removed.

Membrane Technique

The driving force behind the membrane separation process is the difference in partial pressure between different gases present in biogas. Certain molecules, due to the difference in particle size or affinity, can be transported through a membrane while others can't (de Hullu et al. 2008). This separation technique is highly dependent on the type of membrane used. Many different membranes are commercially available, with different specifications, in order to permit or avoid the transport of specific compounds. There are two basic systems for biogas purification with membranes: a *high pressure gas separation* with gas-phases on both sides of the membrane, and a *low-pressure gas liquid absorption separation,* where a liquid absorbs the molecules diffusing through the membrane (Wellinger and Lindberg 2000).

The inlet gas-stream passes through a filter in order to remove water, oil drops and aerosols, which could negatively affect the membrane performance. Additionally, hydrogen sulfide is usually removed in another stage by adsorption with activated carbon (Petersson and Wellinger 2009). The hydrogen sulfide free biogas is then subjected to further purification using gas permeable membranes. In the *high pressure gas separation*, the carbon dioxide will permeate through the membrane at a

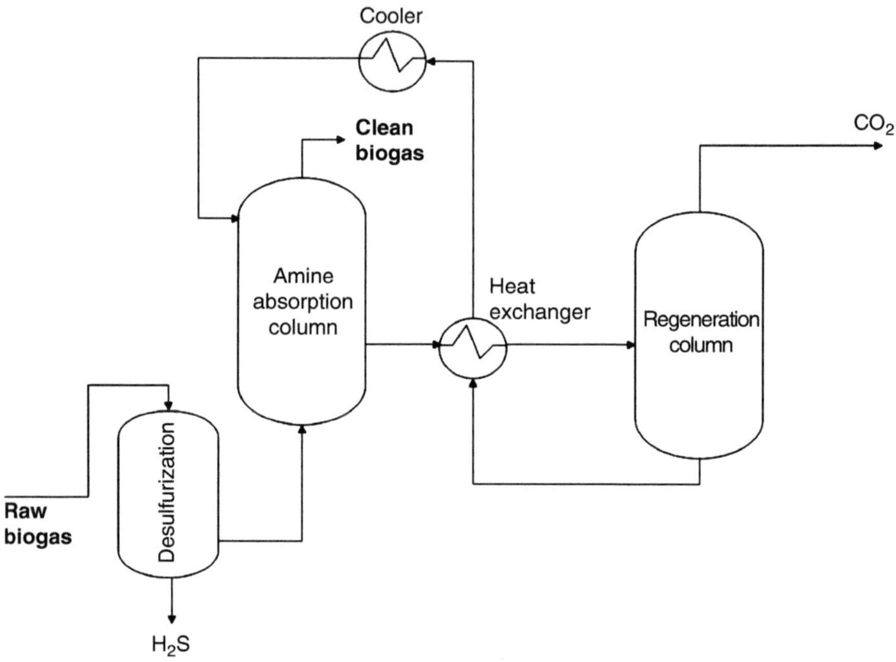


**Fig. 9.5** Schematic of a chemical scrubber (*adapted from* de Hullu et al. 2008). The contaminants are removed from the gas stream by means of a scrubbing liquid (chemical). The scrubbing unit is packed with chemically inert packings to increase mass-tranfer and to provide large wetted surface area. The wetted area provides better contact between the gas and the scrubbing liquid

faster rate than the other biogas components. Thus, the pressurized feed stream is separated into a carbon dioxide-rich, low pressure permeate stream on one side and a carbon dioxide-depleted high pressure $CH_4$ gas-stream on the other side (Fig. 9.6).

For biogas cleaning, membranes having high selectivity and $CO_2$ permeance capacity are usually preferred, in order to minimize $CH_4$ loss and achieve high methane purity. The membranes are also expected to operate at high pressure in the presence of water vapour. Conventional membranes for $CO_2/CH_4$ separation are dense polymeric membranes based on the solution-diffusion mechanism (Deng and Hägg 2010). Membranes used in this technique are made of materials that are permeable to carbon dioxide, water and ammonia. Hydrogen sulfide and oxygen permeate through the membrane to some extent while nitrogen and methane only pass to a very low extent.

In the *liquid absorption separation,* a microporous hydrophobic membrane is used in order to separate the gaseous phase impurities using a liquid phase. The molecules from the gas-stream, which are able to diffuse through the membrane, will be absorbed by a liquid stream that is kept on the other side of the membrane.

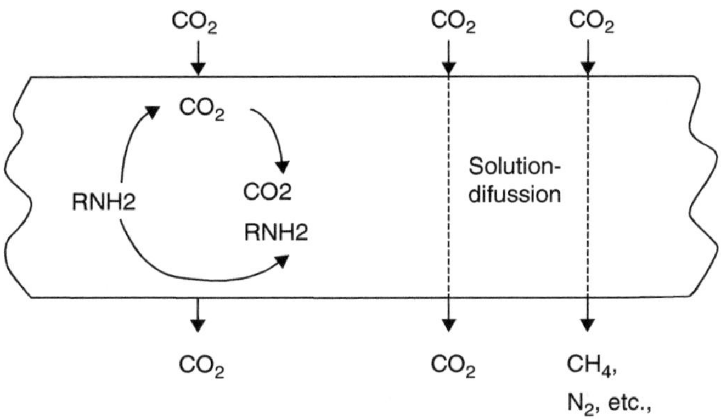


**Fig. 9.6** Schematic of $CO_2$ removal from biogas using membrane technique. Membrane separation techniques fall under two categories; *viz.*, high pressure gas separation, and gas-liquid adsorption. In the former case, selective separation of $H_2S$ and $CO_2$ from $CH_4$ can be achieved by continuously operating the system in three stages, yielding more than 96% pure $CH_4$

The removal of gaseous components using this technique is very efficient. The absorbent normally used is sodium hydroxide. The liquid sodium hydroxide waste stream containing saturated hydrogen sulfide can also be used for heavy metal removal from wastewater. Finally, carbon dioxide is removed by an amine solution, in an amine absorption column (see chemical scrubbing). The use of membrane permeation technique for upgrading biogas has been investigated thoroughly and is being used presently at the industrial level (Favre et al. 2009).

### Cryogenic Technique

Cryogenic technique is based on the differences of boiling and sublimation points between methane and the impurities, particularly for the separation of carbon dioxide and methane. Methane has a boiling point of −160°C at atmospheric pressure whereas carbon dioxide has a boiling point of −78°C (Persson et al. 2006). This implies that carbon dioxide can be separated from the biogas as a liquid by cooling the gas mixture at elevated pressure. Methane can be taken out in gas or liquid phase, depending on the system configuration. The basis of cryogenic separation technique is that biogas is compressed and then cooled followed by some expansion step. These phenomena make the carbon dioxide to condensate. After carbon dioxide has been removed as a liquid, the gas can be cooled further to condensate the methane. The separated carbon dioxide is clean and can be used elsewhere or sold.

To avoid freezing and other problems during the cryogenic compression-expansion process, water and hydrogen sulfide need to be removed from the raw biogas. Water and siloxanes are also removed during cooling of the gas. Cooling usually takes place in several steps in order to remove the different gases in the biogas individually and to optimize the energy recovery.

Cryogenization is an emerging technique and is still under experimentation. de Hullu et al. (2008) reported simulated data from Aspen Plus software package, operating at –90°C and 40 bars. The simulation showed removal efficiencies of 91%.

#### 9.3.1.2 Removal of Hydrogen Sulfide

Chemical Absorption

Chemical absorption, using sodium hydroxide (NaOH), is one of the oldest methods to remove gas-phase hydrogen sulfide. Sodium hydroxide reacts with hydrogen sulfide to form sodium sulfide or sodium hydrogen sulfide. The water absorption capacity is enhanced by sodium hydroxide, and due to this enhancement the volume of water required and pumping demands are reduced. The main disadvantage of the technique is that large quantities of water contaminated with sodium sulfide are produced (Wellinger and Lindberg 2000). Besides, dealing with the spent caustic solution needs high technical facilities and expertise. Because of this, sodium hydroxide washing is not very useful and economical as a biogas cleaning technique, except if the biogas contains high concentrations of hydrogen sulfide, and for the fact that hydrogen sulfide can be converted into a more stable or valuable product like sulfur (Petersson and Wellinger 2009).

There are also two other chemical absorption-based processes where hydrogen sulfide is used to produce iron oxide and elemental sulfur. In the first process, hydrogen sulfide reacts with iron hydroxide or iron oxide to form iron sulfide (FeS). The iron oxide containing bed is usually made of oxidized steel wool (rust coated), wood chips covered with iron oxide or pellets made of red mud or some waste product from aluminum production. The system normally consists of two reaction beds/stages, as shown in Fig. 9.7. The first stage is desulfurizing the biogas, and the second stage consists of regenerating the material with air. During the regeneration stage, iron sulfide is oxidized with air and iron oxide or hydroxide is recovered together with elemental sulfur (Petersson and Wellinger 2009).

The second process is absorption of hydrogen sulfide into iron-chelated solutions. This type of absorption provides a good selectivity for hydrogen sulfide, high removal efficiencies and low consumption of chemicals, because the iron-chelated solutions act as a pseudo-catalyst that can be regenerated.

The mechanism reaction for this process can be given as follows, (de Hullu et al. 2008):

$$S^{-2} + Fe^{+3} \rightarrow S + Fe^{+2} \tag{9.7}$$

$$Fe^{+2} + O_2\left(aq\right) \rightarrow Fe^{+3} + H_2O \tag{9.8}$$

EDTA is the most commonly used chelating agent. In this process, carried out at ambient temperatures, the sulfur produced can be removed easily from the slurry by sedimentation or filtration operations. Removal of $H_2S$ from the biogas is almost

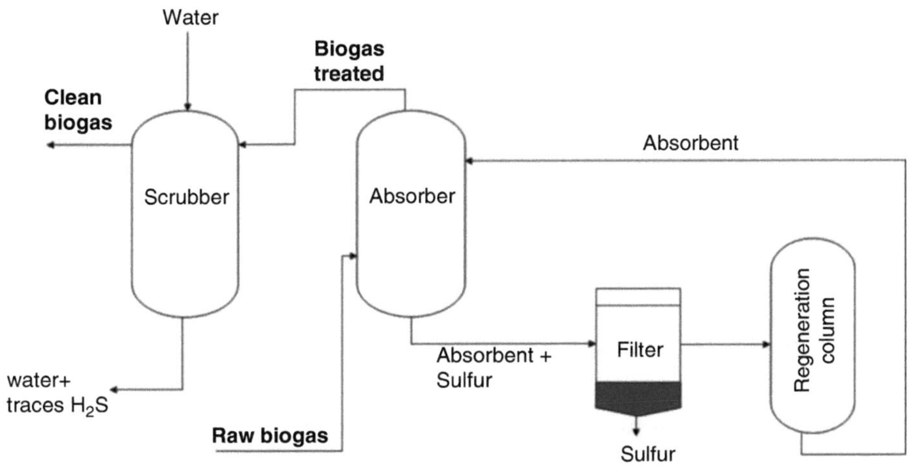


**Fig. 9.7** Schematic of a chemical absorption column for $H_2S$ removal from biogas (adapted from de Hullu et al. 2008). Desulfurizing steo is carried out in the first-stage, followed by material regeneration in the second-stage. Iron sulfide is oxidized with air in the second-stage and iron oxide or hydroxide is recovered together with elemental sulfur

complete with this technique. As only hydrogen sulfide can be removed by this technique, a secondary treatment step is required for removing carbon dioxide from the biogas.

### Impregnated Activated Carbon

Adsorption by activated carbon is a practically viable technique to remove hydrogen sulfide. The activated carbon is usually impregnated with caustic materials like sodium or potassium hydroxide or otherwise modified (Bandosz et al. 2000). In the presence of oxygen, hydrogen sulfide can be converted by caustic catalysis, to produce elemental sulfur and water. The optimal reaction pressures vary between 7 and 8 bars and typical operating temperatures range from 50°C to 70°C (Wellinger and Lindberg 2000). The sulfur produced is adsorbed by the activated carbon while remaining traces of hydrogen sulfide would react with the base and thus can be easily immobilized (Turk et al. 1992). If the biogas is previously washed in scrubbers, in order to remove carbon dioxide, then the air stream will be humidified during the scrubbing. After the washing step the air stream is blown through the activated carbon vessels. This presence of water will facilitate the reaction to sulfur conversion. When the activated carbon bed is saturated, it can be replaced for a fresh one, or regenerated by washing with water. Impregnated activated carbon is a common pretreatment step for hydrogen sulfide removal, as reported earlier in section "Adsorption" on pressure swing adsorption.

Precipitation

Hydrogen sulfide removal can be achieved inside the digester by adding iron salts like $FeCl_2$, $FeCl_3$ or $FeSO_4$. These compounds provide iron cations, *viz.*, $Fe^{2+}$ or $Fe^{3+}$, owing to which hydrogen sulfide will be transformed in the almost insoluble form, iron sulfide. The method is primarily used in digesters when the hydrogen sulfide concentrations in the biogas are greater than 1,000 ppmv (Petersson and Wellinger 2009).

## *9.3.2 Biological Techniques*

### 9.3.2.1 Biological Techniques for Carbon Dioxide Removal

Although very little has been reported about biological methods for the removal of carbon dioxide from biogas, recently some research groups have been investigating the possibility to use natural biodegradation processes such as photosynthesis or enzymatic catabolism, in biogas upgrading.

Converti et al. (2009) have reported a laboratory-scale apparatus for biogas production and purification using a two-stage biological system, *viz.,* anaerobic digestion of sewage sludge from a waste water treatment plant followed by carbon dioxide removal by the photosynthetic *Cyanobacteria sp., Arthrospira platensis* (anaerobic digester + photo-bioreactor). *Cyanobacteria sp.* as well as some microalgae, can remove simultaneously, organic and inorganic pollutants, some of these chemicals are very hazardous like phosphoric compounds. *A. Platensis sp.* presents the advantage of mixotrophic metabolism, to remove either carbon dioxide or organic pollutants. Mixed sludge was added daily into the digester by fed-batch pulse feeding mode of operation and retention times were varied in the range of 6.2–50 days (beyond mesophilic digester ranges). The biogas thus produced was transferred into a sulfuric acid solution and fed into the photo-bioreactor daily, by liquid displacement. *A. Platensis* was initially acclimated in a batch system and later cultivated in the photo-bioreactor that was made from a 1.0 L closed glass vessel, illuminated by a 40 W-fluorescent lamp. Temperature was maintained at the microorganism's optimum value of 30°C. Carbon dioxide was the only carbon source available for the growth of *cyanobacteria.* The results from that study showed almost complete carbon dioxide removal from biogas (99%) and, the removal rate showed a near linear relationship with the microbial growth. Besides, from the photosynthesis step, oxygen was also produced in a range of 10–24%.

A research group in Sweden has studied the use of the enzyme carboanhydrase for carbon dioxide removal from biogas. Carboanhydrase is an enzyme present in human blood that catalyzes the dissolution of carbon dioxide generated during cell metabolism. The dissolved carbon dioxide is transported as a carbonate to the lungs (Eq. 9.9), where the same enzyme catalyzes the reverse reaction where carbon dioxide and water is formed (Petersson and Wellinger 2009).

$$CO_2 + H_2O \leftrightarrow H^+ + HCO_3^- \quad (9.9)$$

According to their study, biogas could be purified up to a methane content of 99% (Mattiasson 2005).

#### 9.3.2.2 Biological Techniques for Hydrogen Sulfide Removal

Bioprocesses for odour control are not a new technology, and bioreactors were already used several decades ago in waste water treatment plants (Pomeroy 1957; Ottengraf and Diks 1992). However, recently they have further been developed and optimized as relatively new alternatives to non-biological processes, for waste-gas treatment (Kennes and Veiga 2001; Kennes et al. 2009a) and several new types of reactors, other than conventional biofilters, have recently been used as well. Biological treatment systems are based on the activity of microorganisms to decontaminate polluted air through a series of phenomenological steps like absorption, adsorption, diffusion and biodegradation (Rene et al. 2009). The pollutants are oxidized to innocuous end-products like carbon dioxide, water and biomass, by a thriving microbial population present in the bioreactor (Kennes and Veiga 2001). For biogas upgrading, the most common biological systems are biofilters, biotrickling filters and bioscrubbers.

##### Microbial Considerations

Biogas can be desulfurized through the action of microorganisms. The main best performing species are the chemotrophs, namely *Thiobacillus sp.* and *Sulfolobus sp.* In oxygen-rich environments, chemotrophs use oxygen as their electron acceptor, and hydrogen sulfide, thiosulfate or elemental sulfur as the electron donor. Under oxygen limited conditions, sulfur is the major end-product, which can be represented by the following equation (Kuenen 1975):

$$H_2S + CO_2 + \text{nutrients} + O_2 \rightarrow \text{cells} + \text{nutrients} + \text{S and/or } SO_4^{2-} \quad (9.10)$$

Oxygen is the key parameter to control the level of final oxidation (Rene et al. 2010), therefore in biogas treatment, the main end-product would be elemental sulfur, considering that biogas is anaerobic. Recently, several studies have reported the biological treatment of hydrogen sulfide using sulfur-utilizing chemolithoautotrophic denitrifiers. Among others, two species are well known, *Thiobacillus denitrificans* and *Thiomicrospira denitrificans*. These microorganisms can degrade hydrogen sulfide in the absence of oxygen, using nitrate as the electron acceptor. Sulfoxidation by these microorganisms can lead to the formation of elemental sulfur or sulfate, under both aerobic and anaerobic conditions (Beristain Cardoso et al. 2006).

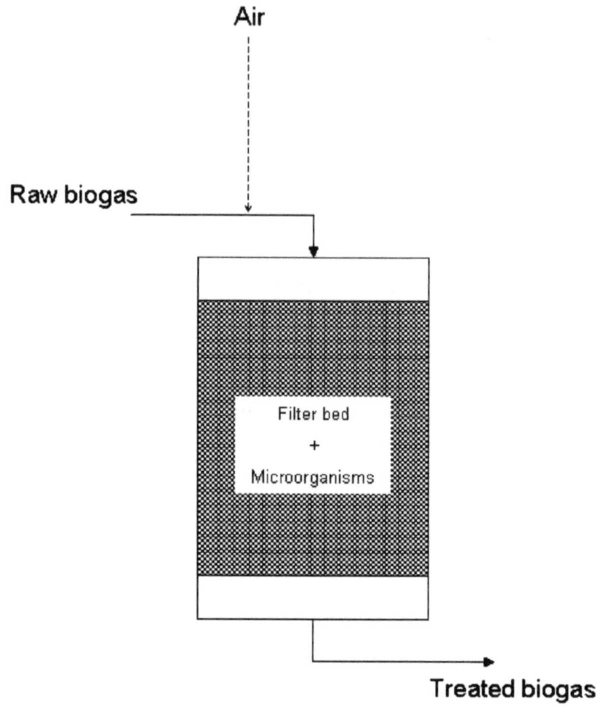


**Fig. 9.8** Schematic of a biofilter for $H_2S$ removal. A well-optimized biofilter provides the required ideal habitat for the growth of sulfide-oxidizing bacteria, to the exclusion of competing microbes which can normally predominate in such systems. Microorganisms play a vital role in oxidizing hydrogen sulfide to sulfuric acid

## Overview of Bioreactor Configurations

### *Biofilters*

A biofilter consists of a filter bed, with an organic packing like peat or compost or synthetic ones like plastic pall rings (Fig. 9.8). A nutrient solution comprising inorganic salts and trace elements is added periodically to the bioreactor to supply the nutrients required for microbial growth, and to maintain optimal moisture contents in the filter bed (usually 40–60%). The packed bed acts as a carrier medium for the biomass and the nutrient source (Kennes and Thalasso 1998). The raw biogas is allowed to pass through the filter bed, in an upflow or downflow mode, and hydrogen sulfide present in the biogas is degraded by the microorganisms.

One disadvantage is the drop in pH down to extremely low values, often $<3$, during the conversion of hydrogen sulfide into sulfur or sulfate, which then needs periodic adjustment of media pH, by the addition of NaOH (Jin et al. 2005).

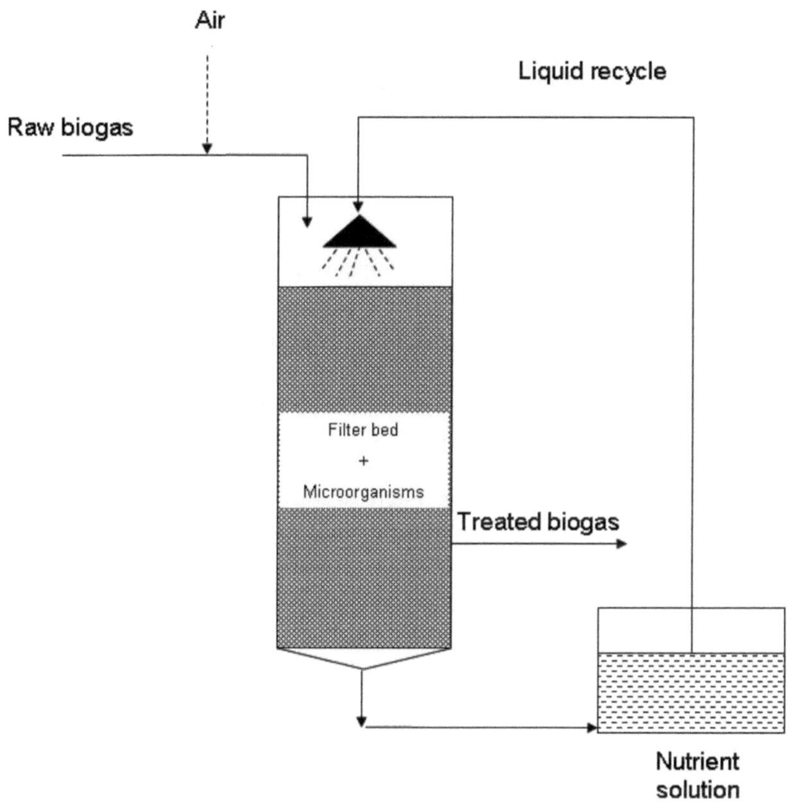


**Fig. 9.9** A typical schematic of biotrickling filter. In practice, pilot-scale or full-scale biotrickling filters are periodically exposed to many perturbations as a result of fluctuation or discontinuous shock loads. Industrial experiences have shown that $H_2S$ concentrations can be reduced from ›2,000 ppm to below 3 ppm

*Biotrickling Filters*

The main differences between conventional biofilters and biotrickling filters, is that in the later, the aqueous phase is continuously trickled over the packing and the bed is always made of inert materials, (Kennes et al. 2009b). A schematic of the biotrickling filter is shown in Fig. 9.9. Biotrickling filters are more complex than biofilters, yet they have been used in field situations to remove hydrogen sulfide from biogas. This is due to its better performance in handling acidic compounds like hydrogen sulfide, and its easiness in controlling the different physico-chemical operational parameters, *viz.*, pH, temperature and others.

Profactor Produktionsforschungs GmbH® recently designed a biotrickling filter, seeded with aerobic bacterial consortia, to remove hydrogen sulfide from biogas. The system did not pose any operational problem and was found to be effective during long-term operation, bringing down hydrogen sulfide concentrations from about 2,000 ppm to below 3 ppm (Trogisch and Baaske 2004; Ahrer et al. 2006).

Soreanu et al. (2008) developed a pilot-scale biotrickling filter for hydrogen sulfide removal from biogas, using *Thiobacillus denitrificans* as the predominant biocatalyst. Biogas produced from a pilot anaerobic digester was continuously fed to an anoxic biotrickling filter. Nitrate-rich waste water from a pilot-scale sequencing batch reactor was used as nutrient medium and the same effluent, rich in microorganisms, acted as an inoculum. The system was tested for around 5 months, operating with hydrogen sulfide concentrations between 1,000 and 4,000 ppmv, achieving removal efficiencies greater than 99%.

Another biotrickling filter inoculated also with *Thiobacillus denitrificans* as the dominant genus, was operated by Bailón (2007), for hydrogen sulfide removal from biogas. Aerobic *Thiobacillus sp.* were selected rather than anaerobic bacteria, due to several reasons, as low nutritional requirements, high hydrogen sulfide affinity or more economic value of the air compared to nitrate. Due to safety reasons, the inlet stream used in lab-scale assays was a mixture of nitrogen (65%), carbon dioxide (35%) and traces of hydrogen sulfide, instead biogas. The biotrickling filter consisted of three packed beds, stacked one above the other, and oxygen enriched nutrient medium was supplied to these beds by bubbling air through a bubble column that contained the nutrient medium. Reported removal efficiences were >99% at 1,000 ppm hydrogen sulfide concentrations. Even when working at 2,000 ppm, the removal never dropped below 93.5%.

*Bioscrubbers*

The schematic of a bioscrubber is shown in Fig. 9.10. Removal of hydrogen sulfide in a bioscrubber is achieved in two stages. The first-stage involves physical or chemical absorption of the pollutant in a liquid stream (normally water), followed by the biological treatment of this liquid in a separate bioreactor unit. Aerobic bioscrubbers have been reported in the literature to yield higher hydrogen sulfide removals than other biological processes. For anaerobic gases like biogas, similar systems can perform successfully taking into account the limitations of oxygen supply. Besides, compared to pure absorption-based techniques, like liquid redox systems (iron cheleates); bioscrubbers may serve as a good eco-friendly alternative (van Groenestijn 2001). As described previously, chemical absorption based on iron chelated solutions needs one step only to regenerate the liquid absorbent, while in the case of bioscrubbers; regeneration can easily be achieved by the activity of microorganisms like *Acidithiobacillus ferrooxidans*.

Mesa et al. (2002) used a two-stage bioscrubber to remove hydrogen sulfide from biogas. In the first step, the biogas was absorbed in a ferric sulphate solution, producing ferrous sulphate and elemental sulfur. The ferrous sulphate produced was regenerated by biodegradation using a low-pH tolerant bacterium, *Acidithiobacillus ferrooxidans*, in the next stage bioreactor. By working at such low-pH values, the reactor required no addition of any chelating agent.

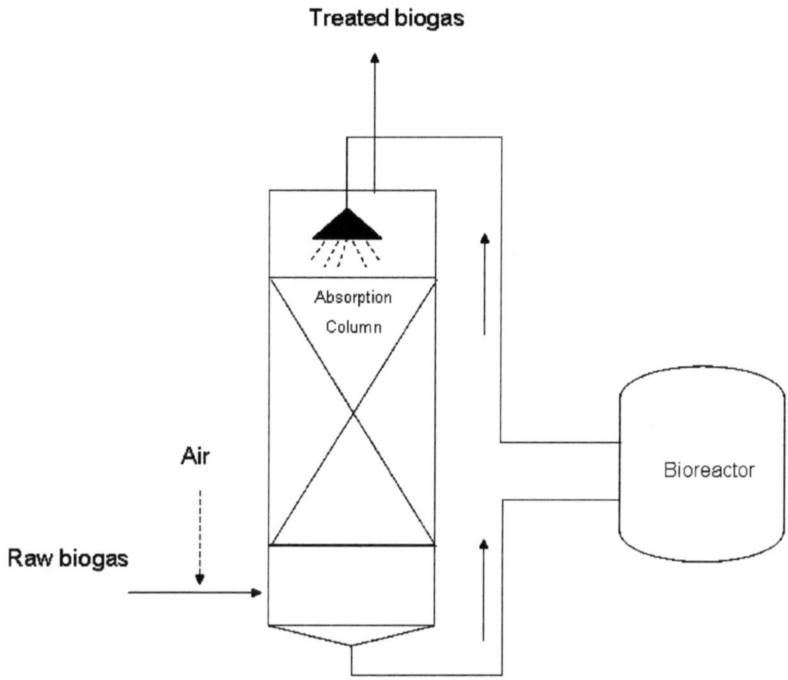


**Fig. 9.10** Schematic of a bioscrubbers (absorption tower+bioreactor). It consists of two units, the absorption column and the bioreactor. The process involves in recirculating a biologically active, nutrient-rich scrubbing solution over a packed tower, while polluted air is forced upward through the media bed. Biodegradation of pollutants in liquid-phase occurs in the bioreactor unit

## 9.4 Removal of Siloxanes from Biogas

Siloxanes are man-made organic products that belong to low molecular weight organosilicons. These compounds are used as a starting material for the formation of high molecular weight polyorganosiloxane that are present in many products of hygiene, health care and industrial products, as gums, shampoos, deodorants, etc. (Wheless and Jeffrey 2004). In biogas arising from sewage treatment plants, as well as landfills, low molecular weight siloxanes volatilize into the digester gas and landfill gas.

During biogas combustion, in order to generate power, through gas turbines, boilers or internal combustion engines, siloxanes are converted to silicone dioxide ($SiO_2$), which can deposit in biogas-fuelled power plant equipments. These deposits can cause serious damages to the process equipments, *viz.*, abrasion to the engine parts, or result in the build-up of layers, leading to severe damage of the heat conduction and lubrication systems. They can also poison the catalysts used in gas purification. Anew, such deposits can cause alterations in the geometrical structure

**Table 9.3** Selected physical properties of D3, D4, D5, L2, and Trimethylsilanol

| Compound | Boiling point (°C) | $P_v$ (mm Hg) at 25°C | Water solubility at 25°C (mg/l) | Log $K_{ow}$ |
|---|---|---|---|---|
| Trimethylsilanol | 69.5 | 16.0 | 35,000 | 1.14 |
| L2 | 106.9 | 42.2 | 0.93 | 4.2 |
| D3 | 135.2 | 8.6 | 1.56 | 3.85 |
| D4 | 175.7 | 0.99 | 0.056 | 4.45 |
| D5 | 211.2 | 0.174 | 0.017 | 5.2 |

Adapted from Schweigkofler and Niessner (2001)

of the combustion chamber, inducing higher emissions of carbon monoxide and formaldehyde (Ajhar et al. 2010). This not only increases the operation costs, but also severely affects the combustion efficiency, suggesting the need for an appropriate gas cleaning technique to remove siloxanes from biogas.

Low molecular weight siloxanes (D3, D4, D5, and L2) are sparingly soluble in water, have a density less than one and possess relatively high oil–water separation coefficient values (Table 9.3).

As evident from the structure of D3, D4, D5, and L2 illustrated in Fig. 9.11, siloxane compounds are formed by Si-O bonds, with hydrocarbon groups on the silicone atom. This type of bond is not usually present in nature, and for this reason, siloxanes are widely considered as inert with respect to natural degradation in the environment.

Two types of chemical structures are found in siloxanes, *i.e.*, cyclic and linear. Three compounds are most commonly present in biogas, *viz.*, octamethylcyclotetrasiloxane and declamethylciclotetrasiloxane, abbreviated as D4 and D5 and belonging to the cyclic type, while L2 (hexamethyldisiloxane) is of linear type, representing more than 90% of the total siloxane composition.

## 9.4.1 Siloxane Removal Technologies

### 9.4.1.1 Adsorption

Adsorption based techniques are the most efficient amongst all the siloxane removal technologies, in particular activated carbon-based adsorption (Wheless and Jeffrey 2004), although polymer beads, molecular sieve and silica gel have been tested for D5 and L2 removal (Schweigkofler and Niessner 2001). The adsorption capacity of activated carbon for volatile siloxane removal from biogas depends on the physico-chemical properties of activated carbon, BET surface area and impregnation or microporous volume of the activated carbon (Schweigkofler and Niessner 2001; Matsui and Imamura 2009; Finocchio et al. 2009). Competitive adsorption pattern has reported to occur at adsorption sites, when the biogas contains compounds such as non-volatile sulfur, and halogenated compounds, in addition to siloxanes

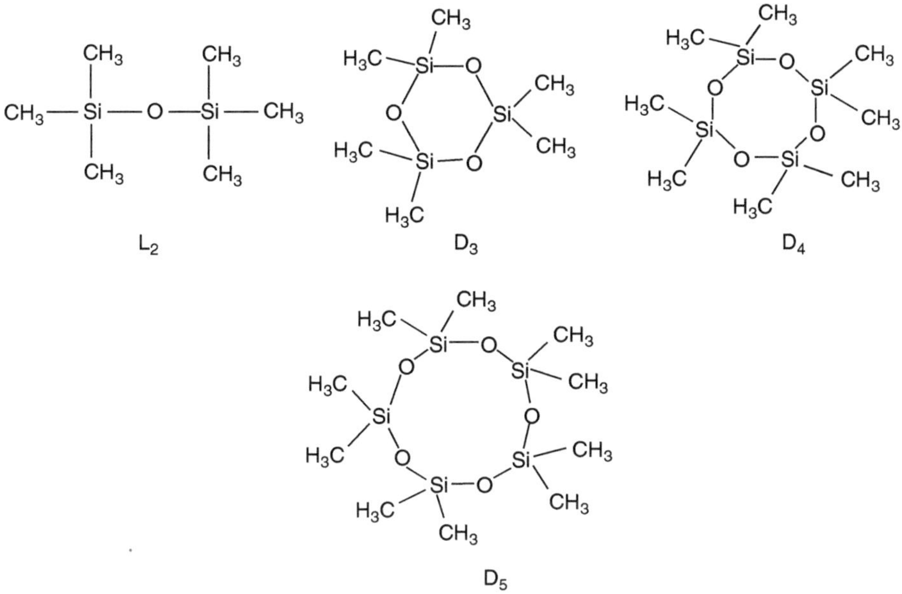


**Fig. 9.11** Chemical structure of L2, D3, D4, and D5. Siloxanes possess useful inherent properties; high compressibility, low flammability, low surface tension, and high thermal stability. Siloxanes can seriously damage the energy recovery equipment (heat exchangers, gas engines), thus reducing the economic benefit of using biogas (Dewil et al. 2006)

(Urban et al. 2009). Similar competitive behaviour can also occur between different types of siloxanes, for instance, L2 breakthrough is sooner than D5. Apart from that, the physical properties of siloxanes, *i.e.,* low water solubility, temperature and humidity, will also have a significant influence on siloxane removal in activated carbon beds (Ajhar et al. 2010).

In order to extend the life of activated carbon, it has been suggested to pre-treat biogas, so as to remove other odorous compounds present in the raw biogas. This is usually achieved by using a water scrubber for the removal of hydrophilic compounds or an absorber containing non-impregnated/impregnated activated carbon. Sufficient care should also be taken to ensure that the biogas is dry before subjecting it to activated carbon adsorption for siloxane removal.

Silica gel has also shown good potential for D5 and L2 removal from biogas. Schweigkofler and Niessner (2001) tested different types of adsorbents (activated carbon, polymer beads, molecular sieve and silica gel) for D5 and L2 removal, and it was observed that all the tested adsorbents exhibited large adsorption capacities for D5, and only silica gel and one of the activated carbon types did not get the breakthrough for L2. Desorption efficiencies of L2 and D5 were greater than 95% for silica gel, while around 74–83% of D5 and >95% of L2 was desorbed using activated carbon. Wheless and Jeffrey (2004) used silica gel and reported a 50% increase in siloxane loading capacity compared to activated carbon.

#### 9.4.1.2 Absorption

Although volatile siloxanes are reported as being highly non-reactive compounds, in the presence of strong acids or bases, generation of silicones is catalysed. Chemical absorption by a strong base is not very helpful because the reaction would produce carbonates, eventually leading to precipitation problems in the equipment. D5 and L2 absorption by strong sulfuric and nitric acid solutions was tested by Schweigkofler and Niessner (2001). High removal efficiencies (>95%) were achieved at 60°C. However, this technology is not of easy practical application in biogas plants, since strong acidic solutions are potentially dangerous for the health and the environment.

Selexol, a dimethyleter of polyethylene glycol, can be used in biogas processing plants to achieve high siloxane removals of >98% (Wheless and Jeffrey 2004). Several other solvents and mineral oils have also been evaluated for their absorptive potential to capture siloxanes, yielding less satisfactory results (Ajhar et al. 2010).

#### 9.4.1.3 Cryogenation

Deep chilling process, described earlier for carbon dioxide removal, can also be used for the removal of siloxanes. Ajhar et al. (2010) reported a theoretical model to predict siloxane concentrations of an ideal gas purified by deep chilling at 1 bar, and temperatures between 0°C and −70°C. According to the model, condensation of siloxanes depends strongly on their different physico-chemical properties. More volatile compounds are more difficult to condensate. Thus, L2, D3 and L3 cannot be removed to concentrations lower than those present in biogas. Temperatures between −50°C and −70°C are required in order to achieve acceptable removals of D5, L4 and D4.

There are also a few studies wherein siloxane removal has been achieved at moderately extreme temperatures. Rossol et al. (2003), achieved removal efficiencies of 80–90%, by deep chilling of digester gas, at–30°C. At the same operating temperature, Wheless and Jeffrey (2004), reported removals up to 50% from landfill gas, in laboratory-scale experiments and in practical situations. Schweigkofler and Niessner (2001), evaluated the effect of gas refrigeration on biogas originated from a waste disposal site and from a sewage treatment plant, at 5°C. The authors observed nearly 20–25% removal of low volatile compounds, while D5 was removed by 82–88% of its initial concentration. However, more than 88% of the initial siloxanes concentrations were found in both the dried landfill gas and sewage gas, after purification, which was later attributed to the siloxane washed out in the acidic condensate (Schweigkofler and Niessner 2001; Wheless and Jeffrey 2004; Ajhar et al. 2010).

The initial investment and operation costs for any cryogenation based technique is relatively high, compared to other cheaper methods such as activated carbon. Chilling is usually preferable and considered to be economically viable only for high flow rates and elevated siloxane loads (Urban et al. 2009).

#### 9.4.1.4 Membranes

The main reason for using membrane separation is to enrich the methane content in the biogas. Membranes that possess high siloxane/methane selectivity can be used to serve this purpose. Although membrane based removal techniques have been established very well for carbon dioxide removal, including industrial applications (Favre et al. 2009), there are only few published results showing the application of membranes for siloxane removal where removal efficiencies greater than 80% were obtained (Ajhar et al. 2010). The initial investments and operational costs have to be analyzed well before implementing this technique to suit a particular industrial facility. The selection of the membrane type is also crucial for long-term performance.

#### 9.4.1.5 Biodegradation of Siloxanes

As a synthetic compound, it was assumed that silicones were not able to be removed through microbial degradation in the environment. Despite this, recent studies have come to challenge this assumption. Graiver et al. (2003) reported PDMS biodegradation by *Arthobacter* and *Fusarium oxyspurum schlechtendahl*. Popat and Deshusses (2008), reported results from laboratory scale studies for D4 biodegradation by aerobic and anaerobic means in biotrickling filters. In the aerobic biotrickling filter, different empty residences times (EBRT) were tested, achieving D4 removals of 43% at an EBRT of 19.5 min. Even greater removals were reached at higher EBRT, in the range of 30–40 min. Only one EBRT of 4 min was reported in the anaerobic biotrickling, and the corresponding D4 removal was 15%. However, working at such high EBRT would involve too significant increases in the equipment and operation costs. In order to understand the mechanism of siloxanes biodegradation, the authors also estimated the D4 mass transfer coefficient between the liquid and the gas-phase. The results showed that the maximum mass transfer rate was markedly lower than the D4 load during the experiments, and it was concluded that mass transfer is a major rate-limiting step for the D4 removal performance in biotrickling filters.

Addition of a second, non-miscible organic phase was also tested in order to solve the mass transfer problems. The performance was carried in a biotrickling filter by the addition of oleyl alcohol. D4 removal appeared very high, 70%, during the first week, but after that it dropped down to 10%. The partition coefficients between air and the organic phase were also calculated and it was found that mass transfer was faster in oley alcohol than in water. Therefore, the initial high removals observed in the biotrickling filter with oley alcohol were attributed to the absorption of the D4 into the organic phase. Overall, the authors concluded that D4 microbial degradation is poor and even microorganisms with the potential to degrade siloxanes need time to adapt their metabolisms with D4 as the only carbon source.

In yet another recent study, Accettola et al. (2008) studied the biodegradation of D4 and D3 by activated sludge bacteria, with the dominant genus *Pseudomonas*, isolated from a waste water treatment plant. D4 biodegradation experiments were carried out in batch cultures for a period of more than 90 days. The results showed that some amount of D4 was degraded via hydrolysis to generate DMSD (dimethylsilanediol). This non-biological degradation was found to be higher in the presence of microorganisms than in the blanks. It was therefore suggested that biomass could act as a catalyst in the reaction. Accettola et al. (2008) also studied D3 biodegradation in an aerobic biotrickling filter and removal efficiencies varying between 10% and 20%, at an EBRT of 3.6 min, were reported.

#### 9.4.1.6 Other Techniques

There are several other techniques that have been tested only recently for siloxane removal; purification by catalyst (Urban et al. 2009), purification using per-oxidation agents (Appels et al. 2008) and ultra-filtration (Rossol et al. 2003).

## 9.5 Removal of Other Compounds

### *9.5.1 Halogenated Hydrocarbons*

Halogenated hydrocarbons, particularly chloro- and flouro-compounds, are often present in landfill gas. However, they are reported only rarely in the case of biogas from the digestion of sewage sludge or organic waste (Persson et al. 2006).

These compounds can be oxidized during biogas combustion, generating corrosive products that can cause corrosion in CHP engines, in the combustion chamber, valves, cylinder heads and other equipment. Moreover, during the combustion they can be the precursors for dioxines (PCDDs) and furanes (PCDFs) generation.

The most common fluorinated contaminants are the chlorofluorocarbons (CFCs), which were widely used as refrigerants. Although their use has been completely forbidden since the late 1980s, they are persistent compounds in landfills due to their slow volatilization from old waste. Cleaning technologies normally applied to remove these compounds, are the same as described for carbon dioxide removal.

A special technique based on adsorption, is to pass the biogas through pressurised tube exchangers filled with specific activated carbon. Small molecules like methane and carbon dioxide will pass through the tube, while big molecules will be adsorbed. The system consists of two stages, one vessel will adsorb the halogen compounds from the biogas and at the same time, the other vessel is used for regeneration, desorbing the pollutants by heating, (Wellinger and Lindberg 2000).

### *9.5.2 Removal of Water*

Biogas is saturated with water vapour after leaving the digester. Water can condensate in gas pipelines and lead to corrosion together with sulfur compounds. It can also affect other purification systems, as in the case of the adsorption columns packed with activated carbon, in which water can destabilize the structure of the material. A simple condensation process can be used to eliminate water, usually carried out by increasing the pressure or decreasing the temperature.

Adsorption based removal techniques include the use of silica, activated charcoal or molecular sieves. Regeneration of the materials is easily achieved by heating or decreasing the pressure. A simple absorption technology is the use of hygroscopic salts, (Petersson and wellinger 2009).

### *9.5.3 Removal of Ammonia*

Ammonia is formed as an end-product during the degradation of proteins in the digester. The amount of ammonia present in biogas will depend on the substrate's composition and the pH in the digester. During biogas upgradation, above all during the drying and in the adsorption process, ammonia is usually removed. Therefore a separate cleaning step is generally not needed, if the amount of ammonia is not very high. However, in some cases ammonia concentrations will be higher, like in biogas originated from animal manure, where it can reach 2,000 ppm (Guo et al. 2009). In such cases, other techniques are involved, like catalytic oxidation and biological degradation.

Guo et al. (2009), designed a packed bed N-TRAP system, impregnated with sulfuric acid, for removal of ammonia originated in biogas. The N-TRAP adsorbent was prepared by waste wood-shaving sand anaerobic digestion bio-waste. The authors reported adsorption capacities much higher than those achieved with activated carbon under the same conditions. Another advantage is that the presence of water does not appear to affect the system, contrary to activated carbon. Finally, the adsorbent saturated with ammonia, can be used as a fertilizer, providing nutrients as nitrogen and sulfur for plant growth.

Regarding biological techniques, co-treatment processes for the removal of hydrogen sulfide and ammonia as a mixture has been reported. However, inhibitions of ammonia degradation at high loadings of hydrogen sulfide were observed. Those effects can occur due to the accumulation of elemental sulfur and ammonium sulfate in the system (Kim et al. 2002; Malhautier et al. 2003).

Jiang et al. (2009) proposed a horizontal biotrickling filter packed with exhausted activated carbon, for the co-removal of hydrogen sulfide and ammonia. The biofilter was inoculated with both autotrophic sulfide oxidizing and nitrifying bacteria. Working in a range of 20–100 ppmv, the maximum removal efficiency achieved was 95% for hydrogen sulfide, at a gas residence time greater than 8 s. At a gas residence time greater than 4 s, a maximum removal of 98% was reached for ammonia.

### *9.5.4 Removal of Oxygen and Nitrogen*

Oxygen is normally not present in the digester. However, oxygen is consumed by facultative microorganisms if at all they are present in the digester. The presence of nitrogen in biogas is indicative of the presence of air in the system. If the gas is collected under pressure, like in the case of landfills, biogas can contain some air (Petersson and Wellinger 2009). Some amount of oxygen is tolerable in biogas, however, high quantities would pose serious problems since oxygen is the key of explosion risks. The presence of nitrogen is not desirable due to its diluting effect that will result in a decrease of the methane content.

Oxygen and nitrogen can be removed by adsorption with activated carbon or molecular sieves. Some full-scale systems for carbon dioxide removal, like PSA or membranes, as well as systems used for desulfurization, can also remove part of them (Persson et al. 2006).

Both compounds are difficult and expensive to remove; hence their presence should be avoided unless the biogas is used for CHPs or in boilers.

## 9.6 Conclusions

In this chapter, we reviewed and systematically discussed the merits and demerits of different physico-chemical and biological techniques for the removal of the undesirable compounds present in biogas. Choosing one technology or another, however requires a complex analysis involving different factors, as the capacity for treating different pollutants, *i.e.,* the undesirable component of biogas, and its cost-economics largely varies amongst different techniques. From an economic view-point, techniques that produce less waste streams, has less energy requirements, involves low investments, operational and maintanence costs are usually prefered. Keeping this in mind, it would be wise to choose wet-scrubbers for carbon dioxide removal. Water scrubbers are easy to operate and less expensive than other techniques for carbon dioxide removal due to less energy requirements and non-utilization of chemicals. Furthermore, with the help of an additional pretreatment step for desulfurization, they can also remove hydrogen sulfide and other volatile inorganic compounds present in biogas (ammonia). Chemical scrubbing with selexol®, is an attractive option due to its high performance capacity for the simultaneous removal of different compounds, including siloxanes. Several field trials have demonstrated that high pressure gas separation membranes and pressure swing adsorption can easily handle techniques that generate a liquid waste stream. However for the simultaneous removal of different compounds membranes are not found to be effective. On the contrary, energy requirements in pressure swing adsorption are higher due to the usage of several compressors. Regarding biological techniques, they are less energy consuming, cheap, reliable and eco-friendly alternatives (end products are innocuous), compared to physico-chemical techniques. Conventional bioreactor configurations have shown great potential in hydrogen sulfide removal under different operational conditions, both steady and transient-state, as well as for the co-treatment of hydrogen sulfide and ammonia. Bioreactors usually requires more time for start-up (several days to a few months), can experience clogging/channelling problems due to excess biomass growth, and pose operational risks if oxygen has to be supplied continuously. Research on developing bioreactors for carbon dioxide and siloxanes removal has just been initiated. Further research directions should be aimed at optimizing and combining the highly performing physico-chemical methods, that generates less waste streams, with a suitable biological method that can decontaminate maximum number of compounds possible. Besides, for practical situations, it is also desirable to develop a biological step for the simultaneous treatment of hydrogen sulfide and siloxane.

**Acknowledgements** The PhD research of M.E. López and our research on biofuels is being partly funded by the Spanish Ministry of Science and Innovation (Project: CTM2010-15796/TECNO). E. R. Rene would like to thank the Spanish Ministry of Science and Innovation (MICINN), for his post-doctoral ("Juan de La Cierva") contract.

## References

Accettola F, Guebitz GM, Schoeftner R (2008) Siloxane removal from biogas by biofiltration: biodegradation studies. Clean Technol Environ Policy 10:211–218

Ahrer W, Accettola F, Trogisch S, Schöftner R (2006) Upgrading of biogas for its usage in high-temperature fuel cells and gas grid injection. In: Lens P, Kennes C, Le Cloirec P, Deshusses MA (eds) Waste gas treatment for resource recovery, vol 28. IWA Publishing, London, pp 479–487

Ajhar M, Travesset M, Yuce S, Melin T (2010) Review: siloxane removal from landfill and digester gas-a technology overview. Bioresour Technol 101:2913–2923

Appels L, Baeyens J, Dewil R (2008) Siloxane removal from biosolids by peroxidation. Energy Convers Manage 49:2859–2864

Aroca GE, Urrutia H, Núñez D, Oyarzún P, Arancibia A, Guerrero K (2007) Comparison on the removal of hydrogen sulfide in biotrickling filters inoculated with *Thiobacillus thioparus* and *Acidithiobacillus thiooxidans*. J Biotechnol 10:515–520

Bailón L (2007) An innovate biotrickling filter for $H_2S$ removal from biogas. In: Kennes C, Veiga MC (eds) Proceedings of the 2nd international congress on biotechniques for air pollution control, Universidade da Coruña, A Coruña, 3–5 October 2007, pp 215–224

Bandosz TJ, Bagreev A, Adib F, Turk A (2000) Unmodified versus caustic-impregnated carbons for control of hydrogen sulfide emissions from sewage plants. Environ Sci Technol 34:1069–1074

Beristain Cardoso R, Sierra-Alvarez R, Rowlette P, Razo Flores E, Gómez J, Field JA (2006) Sulfide oxidation under chemolithoautotrophic denitrifying conditions. Biotechnol Bioeng 95:1148–1157

Chynoweth DP (1987) Overview. In: Chynoweth DP, Isaacson R (eds) Anaerobic digestion of biomass. Elsevier Applied Science Publishers, Ltd, New York, pp 1–14

Converti A, Oliveira RPS, Torres BR, Lodi A, Zilli M (2009) Biogas production and valorization by means of a two-step biological process. Bioresour Technol 100:5771–5776

de Hullu J, Maassen JIW, van Meel PA, Shazad S, Vaessen JMP, Bini L, Reijenga JC (2008) Comparing different biogas upgrading techniques. Project report commissioned by Dirkse Milieutechniek BV and submitted to the Technical University of Eindhoven

Deng L, Hägg M-B (2010) Techno-economic evaluation of biogas upgrading process using $CO_2$ facilitated transport membrane. Int J Greenh Gas Con 4:638–646

Dewil R, Appels L, Baeyens J (2006) Energy use of biogas hampered by the presence of siloxanes. Energy Convers Manage 47:1711–1722

Favre E, Bounaceur R, Roizard D (2009) Biogas, membranes and carbon dioxide capture. J Memb Sci 328:11–14

Finocchio E, Montanari T, Garuti G, Pistarino C, Federico F, Cugino M, Busca G (2009) Purification of biogases from siloxanes by adsorption: on the regenerability of activated carbon sorbents. Energ Fuel 23:4156–4159

Gerardi MH (2003) The microbiology of anaerobic digesters. Wiley, Hoboken

Graiver D, Farminer KW, Narayan R (2003) A review of the fate and effects of silicones in the environment. J Polym Environ 11:129–136

Guo XJ, Tak JK, Johnson RL (2009) Ammonia removal from air stream and biogas by a $H_2SO_4$ impregnated adsorbent originating from waste wood-shavings and biosolids. J Hazard Mater 166:372–376

Haberbauer M (2005) Biofuel quality for fuel cell applications. In: Lens P, Westermann P, Haberbauer M, Moreno A (eds) Biofuels for fuel cells. IWA Publishing, London, pp 403–413

Harasimowicz M, Orluk P, Zakrzewska-Trznadel G, Chmielewskia AG (2007) Application of polyimide membranes for biogas purification and enrichment. J Hazard Mater 144:698–702

House H (2007) Alternative energy sources-biogas production. In: Proceedings of the London swine conference-today's challenges - tomorrow's opportunities, London, Ontario, 3–4 April 2007, pp 119–128

Jaffrin A, Bentounes N, Joan AM, Makhlouf S (2003) Landfill biogas for heating greenhouses and providing carbon dioxide supplement for plant growth. Biosyst Eng 86:113–123

Jiang X, Yan R, Tay JH (2009) Simultaneous autotrophic biodegradation of $H_2S$ and $NH_3$ in a biotrickling filter. Chemosphere 75:1350–1355

Jin Y, Veiga MC, Kennes C (2005) Effects of pH, $CO_2$ and flow pattern on the autotrophic degradation of hydrogen sulfide in a biotrickling filter. Biotechnol Bioeng 92:462–471

Jordão EP, Pessoa CA (1995) Tratamento de esgotos domésticos, 3rd edn. ABES-Associação Brasileira de Engenharia Sanitária e Ambiental, Rio de Janeiro

Kapdi SS, Vijay VK, Rajesh SK, Prasad R (2004) Biogas scrubbing, compression and storage: perspective and prospectus in Indian context. Renew Energ 30:1195–1202

Kennes C, Thalasso F (1998) Waste gas biotreatment technology. J Chem Technol Biotechnol 72:303–319

Kennes C, Veiga MC (2001) Conventional biofilters. In: Kennes C, Veiga MC (eds) Bioreactors for waste gas treatment. Kluwer Academic Publisher, Dordrecht, pp 47–98

Kennes C, Veiga MC, Prado O (2001) Non-biological treatment technologies. In: Kennes C, Veiga MC (eds) Bioreactors for waste gas treatment. Kluwer Academic Publisher, Dordrecht, pp 17–46

Kennes C, Rene ER, Veiga MC (2009a) Bioprocesses for air pollution control. J Chem Technol Biotechnol 84:1419–1436

Kennes C, Montes M, López ME, Veiga MC (2009b) Waste gas treatment in bioreactors: environmental engineering aspects. Can J Civil Eng 36:1887–1894

Kim HS, Kim YJ, Chuung JS, Xie Q (2002) Long-term operation of a biofilter for simultaneous removal of $H_2S$ and $NH_3$. J Air Waste Manag 52:1389–1398

Kuenen JG (1975) Colourless sulphur bacteria and their role in the sulphur cycle. Plant Soil 43:49–76

Kumar PS, Hogendoorn JA, Feron PHM, Versteeg GF (2002) New absorption liquids for the removal of $CO_2$ from dilute gas streams using membrane contactors. Chem Eng Sci 57:1639–1651

Malhautier L, Gracian C, Roux JC, Fanlo JL, Le Cloirec P (2003) Biological treatment process of air loaded with an ammonia and hydrogen sulfide mixture. Chemosphere 50:145–153

Matsui T, Imamura S (2009) Removal of siloxane from digestion gas of sewage sludge. Bioresour Technol 101:529–532

Mattiasson B (2005) *Ekologisk lunga för biogasuppgradering.* Nationellt Samverkansprojekt Biogas i Fordon, 1–14

Mesa MM, Macías M, Cantero D (2002) Biological iron oxidation by Acidithiobacillus ferrooxidans. Chem Biochem Eng Q 16:69–73

Osorio F, Torres JC (2009) Biogas purification from anaerobic digestion in a wastewater treatment plant for biofuel production. Renew Energ 34:2164–2171

Ottengraf SPP, Diks R (1992) Process technology of biotechniques. In: Dragt AJ, van Ham J (eds) Biotechniques for air pollution abatement and odour control policies. Elsevier, Amsterdam, pp 17–32
Persson M, Jönsson O, Wellinger A (2006) Biogas upgrading to vehicle fuel standards and grid injection. In: Task 37 - Energy from biogas and landfill gas. IEA Bioenergy, pp 1–34
Petersson A, Wellinger A (2009) Biogas upgrading technologies-developments and innovations. In: Task 37 - Energy from biogas and landfill gas. IEA Bioenergy, pp 1–19
Pomeroy RD (1957) Deodorizing gas streams by the use of microbiological growths. US patent 2.793.096
Popat SC, Deshusses MA (2008) Biological removal of siloxanes from landfill and digester gases: opportunities and challenges. Environ Sci Technol 42:8510–8515
Purohit P, Kandpal TC (2007) Techno-economics of biogas-based water pumping in India: an attempt to internalize $CO_2$ emissions mitigation and other economic benefits. Renew Sust Energ Rev 11:1208–1226
Rasi S, Veijanen A, Rintala J (2007) Trace compounds of biogas from different biogas production plants. Energy 32:1375–1380
Rene ER, Jin Y, Veiga MC, Kennes C (2009) Two-stage gas-phase bioreactor for the combined removal of hydrogen sulphide, methanol and α-pinene. Environ Technol 30:1261–1272
Rene ER, López ME, Veiga MC, Kennes C (2010) Steady- and transient-state operation of a two-stage bioreactor for the treatment of a gaseous mixture of hydrogen sulphide, methanol and α-pinene. J Chem Technol Biotechnol 85:336–348
Rossol D, Schmelz KG, Hohmann R (2003) Siloxane im Faulgas. KA-Abwasser Abfall 8:8
Roubaud A, Favrat D (2005) Improving performances of a lean burn cogeneration biogas engine equipped with combustion prechambers. Fuel 84:2001–2007
Salomon KR, Lora EES (2009) Estimate of the electric energy generating potential for different sources of biogas in Brazil. Biomass Bioenerg 33:1101–1107
Schulte-Schulze Berndt A (2005) Biogas upgrading with pressure swing adsorption versus biogas reforming. In: Lens P, Westermann P, Haberbauer M, Moreno A (eds) Biofuels for fuel cells. IWA Publishing, London, pp 414–429
Schweigkofler M, Niessner R (2001) Removal of siloxanes in biogases. J Hazard Mater 83:183–196
Sleat R, Mah R (1987) Hydrolytic bacteria. In: Chynoweth DP, Isaacson R (eds) Anaerobic digestion of biomass. Elsevier Applied Science Publishers, Ltd, New York, pp 15–33
Soreanu G, Béland M, Falletta P, Edmonson K, Seto P (2008) Laboratory pilot scale study for $H_2S$ removal from biogas in an anoxic biotrickling filter. Water Sci Technol 57:201–207
Syed M, Soreanu G, Falletta P, Béland M (2006) Removal of hydrogen sulfide from gas streams using biological processes - a review. Can Biosyst Eng 48:1–14
Trogisch S, Baaske WE (2004) Biogas powered fuel cells-case studies for their implementation. Trauner Verlag publishers, Linz
Turk A, Sakalis E, Rago O, Karamitsos H (1992) Activated carbon systems for removal of light gases. Ann N Y Acad Sci 661:221–227
Urban W, Lohmann H, Gomez JIS (2009) Catalytically upgraded landfill gas as a cost-effective alternative for fuel cells. J Power Sources 193:359–366
van Groenestijn JW (2001) Bioscrubbers. In: Kennes C, Veiga MC (eds) Bioreactors for waste gas treatment. Kluwer Academic Publisher, Dordrecht, pp 133–162
Walsh JL, Smith MS, Ross CC, Harper SR (1988) Biogas utilization handbook. Georgia Tech Research Institute Publisher Atlanta, Georgia, p 133
Wellinger A, Lindberg A. (2000) Biogas upgrading and utilization. Task 24: Energy from biological conversion of organic waste. IEA Bioenergy, pp 1–19
Wheless E, Jeffrey P (2004). Siloxanes in landfill and digester gas update. In: 27th SWANA landfill Symposium, San Antonio, 22–25 March 2004

# Chapter 10
# Heavy Metals: Toxicity and Removal by Biosorption

**Ackmez Mudhoo, Vinod Kumar Garg, and Shaobin Wang**

## Contents



A. Mudhoo (✉)
Department of Chemical and Environmental Engineering, Faculty of Engineering,
University of Mauritius, Reduit, Mauritius
e-mail: ackmezchem@yahoo.co.uk

V.K. Garg
Department of Environmental Science and Engineering, Guru Jambheshwar University
of Science and Technology, Hisar 125001, Haryana, India

S. Wang
Department of Chemical Engineering, Curtin University of Technology, G.P.O. Box U1987,
Perth 6845, WA, Australia

E. Lichtfouse et al. (eds.), *Environmental Chemistry for a Sustainable World: Volume 2: Remediation of Air and Water Pollution*, DOI 10.1007/978-94-007-2439-6_10,

**Abstract** Industrialization, urbanization and various anthropogenic activities such as mining and agriculture have increased releases of toxic heavy metals into the natural environment such as soils, lakes, rivers, groundwaters and oceans. The release of heavy metals in biologically available forms alter both natural and man–made ecosystems. Although some heavy metal ions are essential micronutrients for plant metabolism, they become highly toxic when they occur at high concentrations in soils, groundwaters and waste streams. Moreover, heavy metals are not biodegradable and persist in the environment. Conventional methods for the removal of the heavy metals ions from contaminated wasters and wastewaters include chemical precipitation, electroflotation, ion exchange, reverse osmosis and adsorption onto activated carbon. Recently, pioneering research on biosorption of heavy metals has led to the identification of microbes that are extremely effective in bioconcentrating metals. Biosorption is the binding and concentration of an element from aqueous solutions by organic materials such as microbial biomass. The major advantages of biosorption over conventional treatment methods include low cost, high efficiency, minimization of chemical or biological sludge, regeneration of biosorbents and possible metal recovery. Due to their humic acid content, vermicomposts are a novel and effective variety of biosorbent for removing metallic ions such as Pb, Ni, Cr, Cd and V from wastewaters. The types of biosorbents surveyed in this chapter are fungal biomass, biomass of nonliving, dried brown marine algae, agricultural wastes and residues, composite chitosan biosorbent prepared by coating chitosan, cellulose based sorbents, and bacterial strains.

**Keywords** Biosorption • Heavy metals • Green chemistry • Langmuir • Freundlich Toxicity • Cadmium • Chromium • Copper • Nickel • Lead • Zinc • Vermicomposting

## 10.1 Introduction

Rapid industrialization and urbanization have resulted in elevated emission of toxic heavy metals entering the biosphere. Various anthropogenic activities such as mining and agriculture have polluted extensive areas throughout the world (Lu et al. 2004). The release of heavy metals in biologically available forms may damage or alter both natural and man–made ecosystems (Tyler et al. 1989). Although some heavy metals ions are essential micronutrients for plant metabolism but if present in excess concentration in soil, groundwater and some aqueous waste streams can become extremely toxic to the receiving living micro and macro environments. Authorities have been applying very stringent regulations for the treatment of industrial wastewaters recently because of the threatening level of the environmental pollution faced as a result of wastewater streams being discharged indiscriminately and which have unusually elevated concentrations of toxic organic pollutants and heavy metals ions.

Indeed, wastewaters from various industries, such as metal finishing, electroplating, plastics, pigments and mining, contain several heavy metals of health and environmental concern, such as cadmium, copper, chromium, zinc and nickel (Dang et al. 2009). Industrial wastewater containing heavy metals is a threat to the public health because of the accumulation of the heavy metals in the aquatic life which is transferred to human bodies through the food chain. All the more, nowadays, an increasing number of hazardous organic compounds together with variable levels of heavy metals ions are also being discharged into the environment (Aksu 2005). Most of the organic pollutants are degraded or detoxificated by physical, chemical and biological treatments before released into the environment. Although the biological treatments are a removal process for some organic compounds, their products of biodegradation may also be hazardous. Moreover, some non–degradable compounds like the heavy metals ions discharged into the environment along with the treated compounds can cause problems due to non–degradability, bioaccumulation, biomagnification and transport to long distances. As a result, some organic molecules and the heavy metals ions are not biodegradable and persist in the environment.

Conventional methods for the removal of the heavy metals ions from wastewaters include chemical precipitation, electroflotation, ion exchange, reverse osmosis and adsorption onto activated carbon (Cimino et al. 2005). But due to operational demerits, high cost of the treatment and the generation of toxic chemical sludges, some new technologies have been tried for a long time (Elouear et al. 2008). Among them less expensive non–conventional adsorbents like apple waste (Maranon and Sastre 1991), peanut hull carbon (Periasamy and Namasivayam 1995), agricultural wastes (Azab and Peterson 1989) and red mud (Apak et al. 1998) are being investigated for the removal of ions like the Cd and Ni ions. Sud et al. (2008) propose the use of agricultural waste materials as bioadsorbents of heavy metals as a low cost and highly efficient technology, because the functional groups present in agricultural waste biomass (acetamido, alcoholic, carbonyl, phenolic, amido, amino and sulphydryl groups) have affinity for heavy metals ions to form metal complexes or chelates that immobilize the contaminants through reactions of chemisorption, complexation, adsorption on surface, diffusion through pores and ion exchange. As a result, researchers and engineers, all alike, have been oriented toward the practical use of adsorbents for the treatment of wastewater polluted by heavy metals (Kocasoy and Güvener 2009).

Many agricultural wastes, including barks, manures, and composts, contain high levels of lignocellulosic materials. Harman et al. (2007) have hypothesized that the lignin fraction, which contains numerous reactive groups, would be highly effective in binding and removing heavy metals ions from contaminated water, and, further, that the absorptive capabilities of the materials would be strongly affected by the pH of the solution. A series of materials have been tested by Harman et al. (2007), and, at pH levels above about 5.5, these materials were highly effective in removing heavy metals ions, generally as large or larger than nickel, but ineffective in removal of lighter ions such as sodium or magnesium. Various barks were generally observed to be the most effective and were capable of removing more than 90% of iron, copper, or lead from solutions in simple shake flask experiments. Harman et al. (2007) also highlighted that materials that retain cellular structures and that have high lignin contents were highly effective with barks possessing these properties. At alkaline pH levels, many heavy metals ions precipitate, but three separate lines of evidence from the extensive study of Harman et al. (2007) indicate that ions were removed from aqueous solutions by absorption to barks rather than by precipitation. At acidic pH levels, they also were partially effective in removal of the oxyanion chromate. The study of Harman et al. (2007) hence underpinned that biosorption is becoming a promising alternative to replace or supplement the present removal processes of pollutants from wastewaters and other contaminated aqueous media.

Heavy metals ions have recently been of great and renewed concern because of the extreme toxicity and/or persistency in the natural waters and wastewaters. This comprehensive review gives an account of the toxicity of some selected heavy metals ions (Cd, Cr, Cu, Ni, Pb and Zn) and the performance of novel (*green*) biosorbents that have been tested for their removal by adsorption from contaminated waters. To the purpose of a balanced and comprehensive approach to the review on heavy metals ions removal by new biosorbents, this entry also describes the essentials of the vermicomposting process and the wide variety of research recently conducted to evaluate the potential of vermicomposts, fungal biomass, biomass of nonliving, dried brown marine algae, agricultural wastes and residues, composite chitosan biosorbent prepared by coating chitosan, cellulose based sorbents, and bacterial strains as effective heavy metals biosorbents.

## 10.2 Biosorption of Heavy Metals

Pollutants such as heavy metals, volatile organic compounds and dissolved solids are found in wastewaters. Heavy metal remediation of aqueous streams is of special concern due to recalcitrant and persistency of heavy metals in environment (Sud et al. 2008). They are removed on adsorbents such as activated carbon, clay and sediments in riverbeds and in suspension. Over the years, the role of adsorption in water/wastewater treatment has been critically investigated (Oke et al. 2008). Adsorption of heavy metals onto suspended particles had been studied as a model of transportation of metals in rivers and sea. The effects of chemical composition and particle size on adsorption by suspended particles had also been studied. Attention had mostly been on carbon as adsorption materials and indeed Erhan et al. (2004) had documented about 37 sources of carbon, which had been studied as adsorbents for the removal of water pollutants. Three problems associated with the use of carbon for the adsorption of pollutants in a regular volume of waters are: (1) its relatively high cost in making renovation of water by this means very expensive, (2) the fragility of present types of activated carbon makes the regeneration and reuse of activated carbon difficult, (3) activated carbon is virtually limited to the removal of non–polar materials (Schroeder 1977). Hence, conventional treatment technologies for the removal of these toxic heavy metals are not economical and further generate huge quantity of toxic chemical sludge, and research efforts have made that biosorption is now emerging as a potential and powerful alternative to the existing conventional technologies for the removal and/or recovery of metal ions from aqueous solutions.

Biosorption is the binding and concentration of adsorbate(s) from aqueous solutions (even very dilute ones) by certain types of inactive, dead, microbial biomass. The major advantages of biosorption over conventional treatment methods include: low cost, high efficiency, minimization of chemical or biological sludge, regeneration of biosorbents and possibility of metal recovery (Sud et al. 2008). Another powerful technology is adsorption of heavy metals by activated carbon for treating domestic and industrial wastewater. However the high cost of activated carbon and its loss during the regeneration restricts its application. Since the 1990s the adsorption of heavy metals ions by low cost renewable organic materials has gained momentum. Recently attention has been diverted towards the biomaterials which are byproducts or the wastes from large scale industrial operations and agricultural waste materials.

Hence, pioneering research on biosorption of heavy metals, intrinsically guided by the emerging concept of *Green Chemistry*, has led to the identification of a number of microbial biomass types that are extremely effective in concentrating metals. *Green Chemistry* (environmentally benign chemistry) is the utilization of set of principles that reduces or eliminates the use or generation of hazardous substances in the design, manufacture and application of chemical products (Kidwai and Mohan 2005). In practice, green chemistry is taken to cover a much broader range of issues than the definition suggests. As well as using and producing better chemicals with less waste, green chemistry also involves reducing other associated environmental impacts, including reduction in the amount of energy used in chemical processes (Kidwai and Mohan 2005).

*Green Chemistry* is not different from traditional chemistry in as much as it embraces the same creativity and innovation that has always been central to classical chemistry. However, there is a crucial difference in that historically synthetic chemists have not been seen to rank the environment very high in their priorities (Kidwai and Mohan 2005). But with an increase in consciousness for environmental protection, environmental pollution prevention, safer industrial ecology and cleaner production technologies, throughout the world, there is a heightened interest and almost a grand challenge for chemists to develop new products, processes and services that achieve necessary social, economical and environmental objectives. Since the types of chemicals and the types of transformations are very much varied in the chemical industry and chemical research worlds, so are the green chemistry solutions that have been proposed.

Some types of biomass are waste byproducts of large–scale industrial fermentations while other metal–binding biomass types can be readily harvested from the oceans. These biomass types can accumulate in excess of 25% of their dry weight in deposited heavy metals: Pb, Cd, U, Cu, Zn, Cr and others. Some biosorbents can bind and collect a wide range of heavy metals with no specific priority, whereas others are specific for certain types of metals. When choosing the biomass for metal biosorption experiments, its origin is a major factor to be considered. In general terms, biomass can come from industrial wastes which should be obtained free of charge, organisms that can be obtained easily in large amounts in nature (*e.g.*, bacteria, yeast, algae) or fast–growing organisms that are specifically cultivated or propagated for biosorption

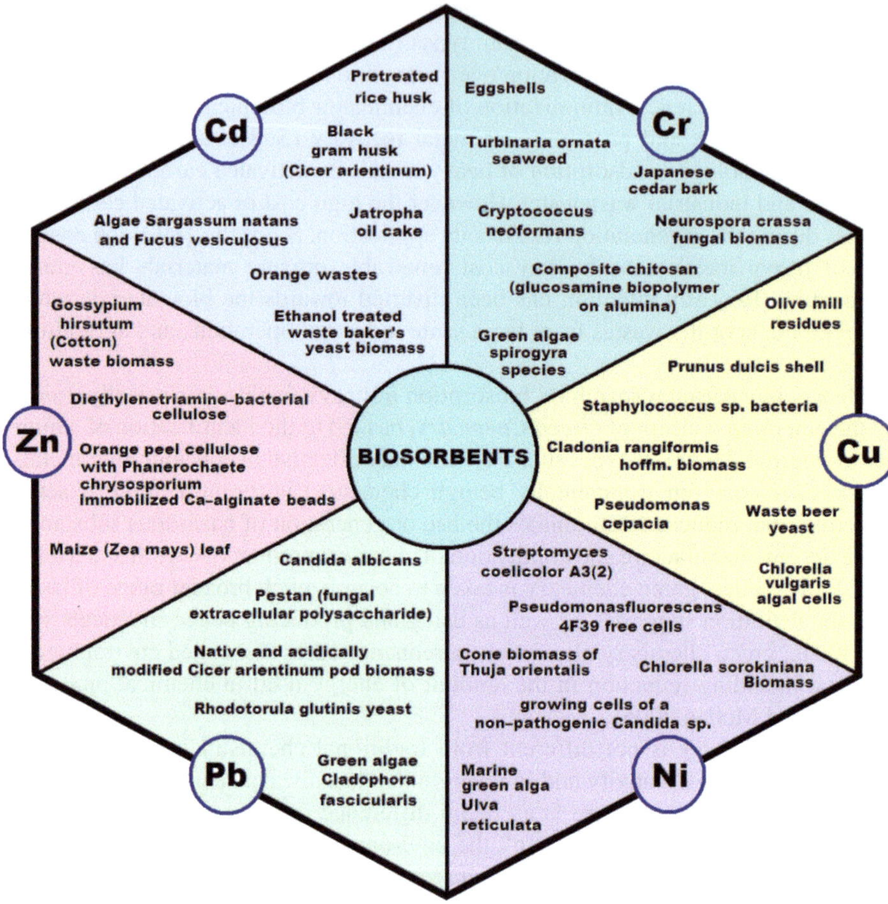


**Fig. 10.1** Biosorbents prepared from various biomass sources for removal by adsorption of heavy metals such as chromium, cadmium, zinc, lead, nickel and copper. The several biomass sources tested for their heavy metal uptake capacity can be grouped broadly in the following categories: organic wastes and residues (e.g. black gram husk, orange peel wastes and olive mill residues), algae (e.g. *turbinia ornata* and *ulva reticulata*), cellulose based sorbents; bacterial strains (e.g. *Cryptococcus neoformans*, *staphylococcus* sp. bacteria and *streptomyces coelicolor* A3(2)) and fungal biomass (e.g. *rhodotorula glutinis* yeast, waste beer yeast and ethanol treated waste baker's yeast)

purposes (crab shells, seaweeds). Figure 10.1 presents some of the biosorbents that have been recently tested and analysed for their metal ions removal capacity. Research on biosorption is revealing that it is sometimes a complex phenomenon where the metallic species could be deposited in the solid biosorbent through various sorption processes, such as ion exchange, complexation, chelation, microprecipitation and oxidation/reduction. The sections to follow highlight the pollution and toxicity characteristics of some selected heavy metals (Cd, Cr, Cu, Ni, Pb and Zn) and summarize in a representative manner the various biosorbents that have been studied for their removal from synthetic and/or natural contaminated aqueous media.

### 10.2.1 The Vermicomposting Process

Composting is a controlled self–heating, aerobic solid phase biodegradative process of organic materials (Ryckeboer et al. 2003). The process comprises mesophilic and thermophilic phases involving numerous microorganisms. In several successive steps, microbial communities degrade organic substrates into more stable, humified forms and inorganic products, generating heat as a metabolic waste product (Mason 2006; Mohee and Mudhoo 2005; Mudhoo and Mohee 2006,2007,2008; Nakasaki et al. 2005; Richard et al. 2006; Ryckeboer et al. 2003). During composting, compounds containing carbon and nitrogen are transformed through successive activities of different microbes to more stable organic matter, which chemically and biologically resembles humic substances. The rate and extent of these transformations depend on available substrates and the process variables used to control composting (Gajalakshmi and Abbasi 2008). Due to the complexity of substrates and intermediate products, microbial diversity and the succession of populations is a prerequisite to ensure complete biodegradation (Ryckeboer et al. 2003). Due to the dynamic process, both in time and space (microhabitats) (Das and Keener 1997; Ryckeboer et al. 2003; Taiwo and Oso 2004; Xi et al. 2005), which is reflected by constantly changing pH, humidity, oxygen partial pressure and temperature it is extremely difficult to detect, all the microorganisms involved. Research on composts is also relatively difficult because the process can hardly be simulated in the laboratory since all major gas and temperature fluxes are to a large extent determined by the physical extension of the system.

Vermicomposting is the term given to the process of conversion of biodegradable matter by earthworms into vermicompost (Garg et al. 2006; Tognetti et al. 2005). In the process, a major fraction of the nutrients contained in the organic matter is converted to more bioavailable forms. The first step in vermicomposting occurs when earthworms break the substrate down to small fragments as a prelude to ingesting the substrate (Gajalakshmi and Abbasi 2008). The earthworms possess a grinding gizzard that enables the mincing of the substrate. This increases the surface area of the substrate, facilitating microbial action. The substrate is then ingested and goes through a process of "digestion" brought about by numerous species of bacteria and enzymes present in the worm gut (Gajalakshmi and Abbasi 2008). During this process, important plant nutrients such as nitrogen, potassium, phosphorus, and calcium present in the feed material are converted into forms that are much more water–soluble and bioavailable to the plants than those in the parent substrate (Martín–Gil et al. 2008; Ravikumar et al. 2008). The earthworms derive their nourishment from the microorganisms that grow upon the substrate particles. At the same time, they promote further microbial activity in the residuals so that the faecal material that they produce is much more fragmented and microbially active than what was ingested (Gajalakshmi and Abbasi 2008). Worms can digest more than their own weight each day, and since the retention time of the waste in the earthworm is short, large quantities are passed through an average population of earthworms. In vermicomposting the earthworms take over the roles of both turning and maintaining aeration (Contreras–ramos et al. 2006) for the organics to be in an aerobic condition, thereby eliminating the need for mechanical or forced aeration.

In the digestive system of these worms, microorganisms are responsible for transforming some 40 (Pereira and Arruda 2003) organic species (proteins, nucleic acids, fats, carbohydrates) into a more stable product (vermicompost). This product presents a high cation exchange capacity (CEC), high humidity content, wide particle size distribution, high concentration of nutrients (Ca, Mg, Na, K, P, S, N) and a characteristic black colour due to the presence of humic substances (Atiyeh et al. 2002; Canellas et al. 2002; Contreras–ramos et al. 2005; Arancon et al. 2006).

Vermicomposting is not an entirely exothermic process and, unlike conventional composting processes (Mason 2006), does not lead to any perceptible rise in the vermireactor temperature (Gajalakshmi and Abbasi 2008). To ensure that the earthworms remain maximally active, the vermireactor should be kept at conditions of temperature and soil moisture as close to the given earthworm species' naturally preferred habitat as possible. Vermicompost from different sources such as municipal solid waste (Alves and Passoni 1997), pig manure (Atiyeh et al. 1999) and decomposed pods of green gram (Kamergam et al. 1999) promote plant growth. Vermicompost can also be obtained when cattle manure, together with soil, are used for earthworm diet. Earthworm species such as *Eisenia fetida*, *Lumbricus terrestris* and *Lumbricus rubellus* have been successfully used in composting processes. Although a great deal of research has been conducted on vermicomposting, Table 10.1 summarizes but a few selected studies that have been conducted to treat organic wastes/substrates by vermicomposting technology. The main substrates treated, the experimental conditions, the composting process performance and main conclusions have been jotted down. The reader is however invited to read though the works of Dominguez et al. (1997), Logsdon (1994), Nagavallemma et al. (2004), Sharma et al. (2005), Tognetti et al. (2005) and Wang et al. (2007) for more valuable (and technical) information on the vermicomposting process.

### *10.2.2 Heavy Metals Biosorption by Vermicomposts*

At present there are only a few studies regarding the treatment of wastewaters containing heavy metals by vermicompost. The efficiency of vermicompost for removing heavy metals from aqueous solutions or industrial effluents has not been studied in detail except to mention a few and namely Jordão et al. (2002), Pereira and Arruda (2003), Carrasquero Durán et al. (2006), Jadia and Fulekar (2008), Urdaneta et al. (2008) and Jordão et al. (2009). Those researchers found that metal concentrations in the purified effluents were below the maximum values established for waste discharges by the prevalent local standards. They also reported that the vermicompost residues obtained from the metal retention process could be applied as a fertilizer to agricultural lands. In their review of vermicomposting and vermicompost, Sharma et al. (2005) have given a comprehensive description of vermicompost, and the main their main conclusions are discussed in the following section.

**Table 10.1** Examples of vermicomposting of organic wastes

| Substrates | Experimental conditions | Composting performance | Main conclusions | Reference |
|---|---|---|---|---|
| Domestic (household) wastes excluding nonrecyclable items such as plastic, rubber, polythene bags, wood, cardboard and glass | Laboratory conditions for vermicomposting process using earthworms *P. excavatus* and *P. sansibaricus*. | Domestic waste processed by *P. sansibaricus* showed about 6% more total N and about 7% more available P at the end than by *P. excavatus*. As compared with the initial level organic C content as well as C:N ratio showed a considerable reduction that was noted higher in substrate with *P. sansibaricus* than those by *P. excavates*. Mean individual live weight, maximum individual growth rate (mg wt./worm/day), number of cocoons and reproduction rate (cocoon/worm/day) were higher in bedding with *P. sansibaricus* | Study clearly indicated that vermicomposting of domestic waste could be an effective technology to convert the negligible resource into some value–added products, e.g. vermicomposts and earthworms on a low–input basis | Suthar and Singh (2008) |
| Leaf litter ensuing from the trees of mango (*Mangifera indica*) | Earthworm *Eudrilus eugeniae* Kinberg used for process in vermireactors | Earhworms grew well in all reactors, increasing their zoomass by ~103% and producing ~157 offspring | Ability of the earthworms to survive, grow and breed in the vermireactors fed with composted mango tree leaves, and a rising trend in vermicast output inspite of the death of a few worms after 4 months of reactor operation, indicated the sustainability of this type of vermireactors. The studies also indicated that even better vermireactor efficiency could be possible by modifying the reactor geometry | Gajalakshmi et al. (2005) |

(continued)

**Table 10.1** continued

| Substrates | Experimental conditions | Composting performance | Main conclusions | Reference |
|---|---|---|---|---|
| Primary sewage sludge (PSS) amended with cow dung (CD) | Epigeic earthworm *Eisenia fetida* used in laboratory scale experiments for the evaluation of vermistabilization of PSS and CD mixtures after 15 weeks in terms of fertilizer quality of the products and; growth and reproduction of *Eisenia fetida* up to 11 weeks in different vermireactors | Maximum worm biomass was attained in 10% PSS + 90% CD mixture while, the worm growth rate was highest in 30% PSS + 70% CD feed mixture | Results indicated that PSS could be converted into good quality manure by vermicomposting if mixed in appropriate ratio (30–40%) with cow dung | Gupta and Garg (2008) |
| Livestock organic wastes – cow and goat manures | Earthworm, *Eisenia fetida*. Used and each treatment group consisted of six replicates and worm vermicasts were examined after 5 weeks | Earthworm biomass and reproductive performance, in terms of number of worms after 5 weeks of experiment, were higher in cattle manure than in goat manure. The cocoon production per worm in cattle manure was higher than in goat manure | Cattle manure provided a more nutritious and friendly environment to the earthworms than goat manure | Loh et al. (2005) |
| Textile mill sludge mixed with cow dung and/or agricultural residues | Earthworm *Eisenia fetida* used for the process for 11 weeks in the laboratory under controlled environmental conditions | Maximum growth and reproduction was obtained in 100% cow dung, but worms grew and reproduced favorably in 80% cow dung + 20% solid textile mill sludge and 70% cow dung + 30% solid textile mill sludge also | Vermicomposting can be an alternate technology for the management of textile mill sludge if mixed with cow dung in appropriate quantities | Kaushik and Garg (2004) |

| | | | | |
|---|---|---|---|---|
| Kitchen waste, agro–residues, institutional and industrial wastes including textile industry sludge and fibres | *Eisenia foetida* earthworm used for the vermicomposting | The increase in TKN for different feed substrates was found in the order: textile sludge > textile fibre = institutional waste > agro–residues > kitchen waste. Available Phosphorus increased 1.4–6.5-fold in different feed mixtures in comparison to control. Reduction in TOC was highest in agro–residues (threefold) followed by kitchen waste (2.2-fold), institutional waste (1.7-fold) and textile industrial wastes (sludge, 1.5-fold and fibre, 1.68-fold) in earthworm–inoculated pots than control. | Vermicomposting (using *E. foetida*) is a suitable technology for the decomposition of different types of organic wastes | Garg et al. (2006) |

Vermicompost is a 'peat–like' material with high porosity, aeration, drainage, water holding capacity and microbial activity (Ansari 2008; Zaller 2006). It contains most nutrients in plant available forms such as nitrates, phosphates, exchangeable calcium and soluble potassium and has large particular surface area that provides many microsites for microbial activity and for the strong retention of nutrients (Singh et al. 2008; Zaller 2006). The plant growth regulators and other plant growth influencing materials i.e. auxins, cytokinins and humic substances produced by microorganisms have been reported from vermicompost (Arancon et al. 2006; Bachman and Metzger 2008). The humic materials extracted from vermicomposts have been reported to produce auxin–like cell growth and nitrate metabolism of carrots (Muscolo et al. 1999). However humic substances can occur naturally in mature animal manure, sewage sludge or paper–mill sludge but their amount and rates of production are increased dramatically by vermicomposting (Sharma et al. 2005; Romero et al. 2007). Humic acid like components (HAL) were isolated by conventional procedures from various organic wastes including animal manures, a municipal solid refuse and a sewage sludge that were composted for 2–3 months with the earthworms *E. fetida* or *Lumbricus rubellus* by Hervas et al. (1989), Landgraf et al. (1998), Benítez et al. (2000) and Anastasi et al. (2005). Vermicompost HAL containing appreciable amounts of iron and copper inner sphere complexes of definite chemical and geometrical forms, similar to those found in humid acid from soil and other sources, can be considered adequate analogues of soil humic acid with respect to their metal complexion properties and behaviour (Sharma et al. 2005).

Vermicompost has been proposed as a biosorbent for removing metallic ions such as Pb, Ni, Cr and V from wastewaters by Urdaneta et al. (2008). They reported it as a promising alternative for decontamination taking advantage of the humic and fulvic substances interaction with metals. Humic and fulvic substances have a high capacity to interact with metal ions and organic and mineral substances, forming soluble or non soluble complexes with varied stability and physico–chemical properties. The involved processes are: mineral dissolution, formation of metal complexes soluble in water, formation of metal–ligand complexes soluble in water, adsorption/desorption processes (solid phase) of metallic ions, adsorption onto external mineral surfaces and adsorption onto inter–laminar mineral surfaces (Benedetti and Van Rinniburgh 1995; Gardea–Torresdey et al. 1996b). The polyvalent metals are strongly complexed with humic substances. The content and quality of the organic matter is correlated with the capacity to bond metals. Factors that affect the stability of the complex thus formed are the oxidation state of the metal, the coordination number and binding field effects of the ligand. The stability of the complex increases with the cation valence which generally diminishes with the ionic radius (Vinkler et al. 1976; Murray and Liner 1984; Pommery et al. 1988). The organic structure of humic acid is naturally oxidized, giving it a negative charge and is as shown in Fig. 10.2. Positive ions, attracted to broken bonds at the site of the oxidation, create sites for micronutrients to attach on. The oxidized sites on the molecule are normally saturated with potassium, which is readily exchanged for all major micronutrient ions in the soil. Table 10.2 presents a summary of studies that have addressed research on heavy metals removal/binding from solutions by humic acids.

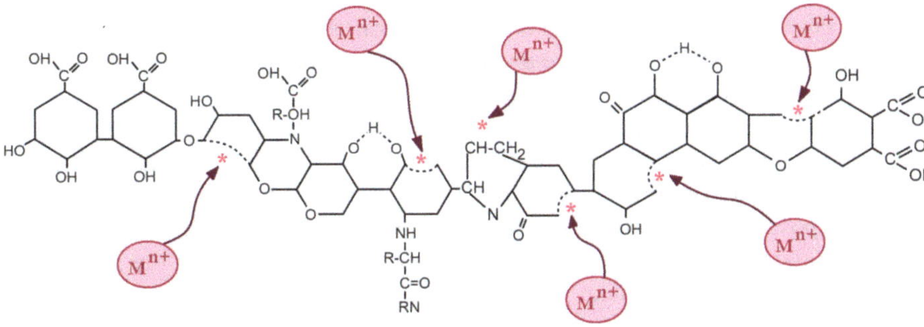

**Fig. 10.2** Oxidized humic acid molecule with metal ions **M**$^{n+}$ (M: Cd, Cr, Cu, Ni, Pb and/or Zn; n: 2, 3 or 6) attaching to oxidized binding site (shown with asterisk). Positive ions, attracted to broken bonds at the site of the oxidation, create sites for micronutrients and microflora to attach. Every oxidized humic acid molecule can potentially accommodate upto six metal ions. Courtesy of Brian Phelps (http://www.phelpstek.com/)

**Table 10.2** Reported studies on heavy metal ions adsorption by humic acids

| Heavy metal(s) | Reference |
|---|---|
| Cadmium, lead | Seki and Suzuki (1995) |
| Copper, cadmium, lead | Liu and Gonzalez (2000) |
| Copper, zinc, cobalt, cadmium | Spark et al. (1997) |
| Copper, cadmium | Arias et al. (2002) |
| Lead, zinc, copper | Gao et al. (1999) |
| Arsenic | Thanabalasingam and Pickering (1986) |
| Nickel | Green–Pedersen et al. (1997) |
| Copper, nickel | El–Eswed and Khalili (2006) |
| Nickel, zinc | Dries et al. (2005) |
| Copper, cadmium, lead | Gondar et al. (2006) |

The following discussions focus on the research advances that have been realized when using composts for the removal of some heavy metals. It is brought to the notice of the reader that our review of over 2,500 published research papers revealed that literature on heavy metals biosorption by compost is relatively scanty but the pool of results obtained so far support the suitability of compost as a novel and efficient heavy metals biosorbent. Jordão et al. (2002) reported that Cu, Zn and Ni retention by cattle manure vermicompost from electroplating wastes were close to 100%. They also reported that it was not necessary to correct the effluent pH during the treatment process to reach the levels recommended by Brazilian legislation for discharge into water courses; whilst Pereira and Arruda (2003) evaluated the potential of vermicompost to adsorb Cd from both synthetic solution and mineral water. For cadmium preconcentration procedures, this potential was confirmed, as high pre–concentration factors and recoveries were obtained by Pereira and Arruda (2003). This characteristic is of high analytical interest, as vermicompost is easily available at low cost. Related to adsorption behaviour, it is possible to predict that vermicompost may adsorb other metallic species as efficiently as $Cd^{2+}$. This affirmative was based by Pereira and Arruda (2003) on the structural properties of vermicompost, as this material demonstrates desirable characteristics (high CEC values), which can be applicable to different metals.

According to all these characteristics, the use of vermicompost for determination of environmental contaminants is feasible; mainly those related to determining the results of treatments of effluents with high initial metal concentrations. For environmental purposes, when complex matrices are considered, an increase in the vermicompost mass could be made to compensate possible differences in the adsorption properties due to the presence of concomitants. Vermicompost (humic material) samples were submitted to characterization by infrared spectroscopy, X–ray diffractometry, thermogravimetric analysis and electron microscopy. In addition, its CEC was determined. Pereira and Arruda (2003) designed factorial experiments in order to elucidate the interaction mechanisms between $Cd^{2+}$ and the adsorptive sites of vermicompost and the maximum adsorptive capacity for cadmium was evaluated by a Langmuir isotherm. The characterization revealed that vermicompost exhibited high CEC values (72.9 meq per 100 g), high surface area (porous material), chelating groups and a maximum adsorptive capacity for Cd (38.6 mg/g) at pH 5.0. This potential application was exploited in the analysis of synthetic and real (mineral water) samples with Cd concentrations of *ca* 5 μg/L, requiring a preconcentration step. Pereira and Arruda (2003) noted an enrichment factor of 100, and this was a major breakthrough in its own merit.

Similarly, Ulmanu et al. (2003) and Chen et al. (2005) have investigated the sorption of Cd ions onto two other types of composts. Ulmanu et al. (2003) found that of all the adsorbents studied, bentonite and compost presented the highest removal efficiencies, reaching 99% for copper when cadmium is also present, for initial solution concentrations of up to 100 mg/L, while Chen et al. (2005) found that the maximum uptake estimated with the Langmiur isotherm model were 833.33 mg/g for $Cd^{2+}$, 1,000.00 mg/g for $Pb^{2+}$ and 44.44 mg/g for $Cr^{3+}$, respectively, when spent mushroom compost of *Lentinus edodes* was used as a biosorbent for adsorbing cadmium, lead and chromium from solutions under batch conditions. All the results showed that vast potential sorption capacity was existed in the biomass (compost) for adsorbing these three kinds of metals studied.

The efficiency of the sulphate reducing bacteria–based *in situ* treatment of acid mine drainage is often limited by the low degradability of the current carbon sources, typically complex plant–derived materials (Gibert et al. 2005). In such non–sulphate–reducing conditions, field and laboratory experiences have shown that mechanisms other than sulphide precipitation should be considered in the metal removal, i.e. metal (oxy)hydroxides precipitation, co–precipitation with these precipitates, and sorption onto the organic matter. In this respect, Gibert et al. (2005) focused their research on the sorption of Zn and Cu sorption on vegetal compost and attempted to develop a general and simple model for the prediction of their distribution in organic–based passive remediation systems. The model considered two kinds of sorption sites and the existence of monodentate and bidentate metal–binding reactions, and it assumed that only free $M^{2+}$ species can sorb onto the compost surface. The acid–base properties of the compost were studied by means of potentiometric titrations in order to identify the nature of the involved surface functional groups and their density. The distribution coefficient ($K_D$) for both Zn and Cu were determined from batch experiments as a function of pH and metal concentration. The model of Gibert et al. (2005) yielded the predominant surface complexes at the experimental conditions, being $>SO_2Zn$ for Zn and $>SO_2HCu^+$ and $(>SO_2H)_2Cu$ for Cu, with log $K_M$ ($K_M$=complexation constant) values of −2.10, 3.36 and 4.65, respectively. The results presented by Gibert et al. (2005) further demonstrated that their proposed model provided a very good description of the sorption process of Zn and Cu onto the vegetal compost used in their experiments.

Still in 2005, Wei et al. used a kitchen waste compost to sorb Cr for various times from water containing either $Cr(NO_3)_3$ or $CrO_3$ in different concentrations. Scanning Electron Microscopy results showed that the composts had been partially oxidized by $Cr^{6+}$ during the sorption experiments while X–ray absorption near edge structure simulation suggested that about 54.1–61.0% Cr sorbed on the compost is in form of organic $Cr^{3+}$ through ionic exchange process with the rest being existent as $Cr(NO_3)_3$ in the $Cr^{3+}$ sorption case.

Some studies have also been conducted to assess the potential of vermicomposts to sorb heavy metals and the results have been positive. Carrasquero Durán et al. (2006) have studied the adsorption of $Pb^{2+}$ on vermicompost at 11°C, 30°C and 50°C by using Langmuir and Freundlich models. The maximum adsorption capacities were 116.3, 113.6 and 123.5 μg/g for each temperature, respectively. The differences in Fourier Transform Infrared Spectroscopy spectra of vermicompost at pH 3,8 and pH 7.0 in the region from 1,800 to 1,300/cm were interpreted on the basis of carboxyl acid ionization that reduced band intensity around 1,725/cm producing signals at 1,550 and 1,390/cm of carboxylate groups. Similar changes were detected at pH 3.8 when ionic lead was present suggesting that heavy metal complexation occurred throughout a cationic exchange reaction. Jadia and Fulekar (2008) conduced a greenhouse experiment to determine the phytotoxic effect of heavy metals such as Cd, Cu, Ni, Pb and Zn on the growth of Sunflower (*Helianthus annuus*): on the seed germination, root/shoot growth and uptake of metals in soil–vermicompost media. The selected metals were dosed at various concentrations ranging from 0, 5, 10, 20, 40 and 50 mg/L separately in soil–vermicompost media (3:1) in pot experiment. The seed germination, root and shoot growth were found significantly affected by these metals at higher concentration of 40 and 50 mg/L. However, the lower concentration of heavy metals ranging from 5 to 20 mg/L doses were observed to be stimulating the root and shoot length and increase biomass of the sunflower plant. The research study of the sunflower indicated the heavy metal uptake at the concentrations 5, 10, 20, 40 and 50 mg/L. Jadia and Fulekar (2008) concluded that vermicompost could be used to remediate metals–contaminated sites because it binds metals and increases uptake by providing nutrients such as sodium, magnesium, iron, zinc, manganese and copper which serve as a natural fertilizer giving high yield of biomass and microbial consortium helps the overall growth of the sunflower plant.

Urdaneta et al. (2008) used vermicompost as adsorbent substrate for removing Pb, Ni, V and Cr from wastewaters. In their work, after a preliminary physical and chemical characterization of the vermicompost, the optimal parameters for the heavy metal adsorption were obtained. A synthetic multi elemental solution of Pb, Cr and Ni and a solution of $NH_4VO_3$ for vanadium were evaluated. The highest adsorption and removal of the metals was observed for a vermicompost mass of 2 g per 500 mL using a particle size from 75 to 841 μm for Pb, Cr and Ni, and 841 till 1,192 μm for V, and the mean removal for each element was around 95% for Pb. Later, Carrasquero–Durán and Flores (2009) evaluated the immobilization of lead ions by a vermicompost with calcite added by adsorption isotherms and the results were explained on basis of the pH dependent surface charge and by IR spectroscopy. Their results showed maximum adsorption values between 113.6 mg/g (33°C) and 123.5 mg/g (50°C), and the differences in the IR spectra at pH 3.8 and 7.0 in the region from 1,800 to 1,300/cm, were interpreted on the basis of the carboxyl acid ionization, that reduced the band intensity around 1,725/cm, producing signals at 1,550 and 1,390/cm of carboxylate groups. Jordão et al. (2009) studied the adsorption of $Zn^{2+}$ from both synthetic solution and kaolin industry wastewater by cattle manure vermicompost. In synthetic solution, the maximum adsorption capacity of the vermicompost for $Zn^{2+}$ ions was 20.48 mg $g^{-1}$ at 25°C when the vermicompost dose was 1 g/10 mL and the initial adjusted pH was 2. The batch adsorption studies of $Zn^{2+}$ ion vermicompost using kaolin wastewater have shown the maximum adsorption capacity was 2.49 mg/g at pH 2. The small values of the constant related to the energy of adsorption (from 0.07 to 0.163 L/mg) indicated that $Zn^{2+}$ ions were binded strongly to the vermicompost. The values of the separation factor, $R_L$, which has been used to predict affinity between adsorbate and adsorbent were between 0 and 1, indicating that sorption was very favourable for $Zn^{2+}$ in synthetic solution and kaolin wastewater.

Lately, a couple of fresh studies on the use of composts for heavy metals adsorption have been reported by Kocasoy and Güvener (2009) and Boni and Sbaffoni (2009). Kocasoy and Güvener (2009) determined the retention capacity of compost for copper, zinc, nickel and chromium. For this purpose, experiments in batch–mixing reactors with initial metal concentrations ranging from 100 to 1,000 mg $L^{-1}$ were carried out. Kocasoy and Güvener (2009) observed that compost could repeatedly be used in metal sorption processes, and their experiments indicated that compost has high retention capacities for copper, zinc and nickel, but not for chromium. Thus, compost is a very good potential sorbent for copper, zinc and nickel and may find place in industrial applications. In the same strike, solid waste which is another source of significant environmental pollution will be reduced by being converted into a beneficial compost (Kocasoy and Güvener 2009).

Lastly, Boni and Sbaffoni (2009) have presented their results of a column reactor test, aiming at evaluating the performance of a biological permeable barrier made of low–cost waste materials, for $Cr^{6+}$ removal from contaminated groundwater. A 1:1 by volume mixture of green compost and siliceous gravel was tested as reactive medium in the experimental activity. A 10 mg/L $Cr^{6+}$ contaminated solution was used and the residual $Cr^{6+}$ concentration along the column height and in the outlet was determined in the water samples collected daily. Boni and Sbaffoni (2009) noted that the $Cr^{6+}$ removal efficiency was higher than 99% during the entire experimental activity. The influence of the biological activity on $Cr^{6+}$ removal efficiency was evaluated by varying the organic carbon and nitrogen dosages in the contaminated solution fed to the system. According to Boni and Sbaffoni (2009), the $Cr^{6+}$ removal was closely linked to the biological activity of the native biomass of compost. Two main processes involved were: adsorption on the organic–based matrix and reduction into $Cr^{3+}$ mediated by the anaerobic microbial metabolism of the bacteria residing in green compost. Boni and Sbaffoni (2009) concluded that by virtue of the proven efficiency and the low–cost, the reactive medium consisting of compos and siliceous gravel represents a suitable alternative to conventional approaches to chromium remediation.

## 10.3 Scientific Basis of Biosorption

Several important aspects of biosorption for heavy metal removal need to be considered when exploring this emerging bioremediation technique for optimization purposes. The underlying principles of biosorption for removal of metal ions, the kinetics of mass transfer during the process of biosorption of metal ions, the theory and models that can be used to describe the mass transfer process and the thermodynamics of biosorption of heavy metals onto biomass and the models which can be used to quantify metal–biomass interactions at equilibrium, all are key knowledge areas in biosorption science which have been hence so far relatively well presented, discussed and reviewed in the literature. The following subsections make a brief revisit of these key aspects. The following subsections make a brief revisit of these key aspects. However, the reader is earnestly directed to more comprehensive and extensive reviews by Davis et al. (2003), Figueira et al. (2000), Loukidou et al. (2004), Naja and Volesky (2006), Vijayaraghavan and Yun (2008) and Volesky (2001) where these scientific aspects of biosorption have been excellently reported.

### *10.3.1 Biosorption Experimental Procedures*

A biosorption process can be performed via several modes (Vijayaraghavan and Yun 2008); of which, batch and continuous modes of operation are frequently employed to conduct laboratory scale biosorption processes. Although most industrial applications prefer a continuous mode of operation, batch experiments have to be used to evaluate the required fundamental information, such as biosorbent efficiency, optimum experimental conditions, biosorption rate and possibility of biomass regeneration. The factors influencing bacterial batch biosorption are solution pH, temperature, ionic strength, biosorbent dosage, biosorbent size, initial solute concentration and agitation rate (Vijayaraghavan and Yun 2008). Of these, the solution pH usually plays a major role in biosorption, and seems to affect the solution chemistry of metals/dyes and the activity of the functional groups of the biomass. For metals, the pH strongly influences the speciation and biosorption availability of the metal ions. At higher solution pH, the solubility of metal complexes decreases sufficiently allowing precipitation, which may complicate the sorption process. The activity of binding sites can also be altered by adjustment of the pH (Vijayaraghavan and Yun 2008). Temperature seems to affect biosorption only to a lesser extent within the range from 20°C to 35°C. Higher temperatures usually enhance sorption due to the increased surface activity and kinetic energy of the solute (Vijayaraghavan and Yun 2008); however, physical damage to the biosorbent can be expected at higher temperatures. Due to the exothermic nature of some adsorption processes, an increase in temperature has been found to reduce the biosorption capacity of the biomass. Another important parameter in biosorption is the ionic strength, which influences the adsorption of solute to the biomass surface. The effect of ionic strength may be ascribed to the competition between ions, changes in the metal activity, or in the properties of the electrical double layer. When two phases, e.g. biomass surface and solute in aqueous solution are in contact, they are bound to be surrounded by an electrical double layer owing to electrostatic interaction. Thus, adsorption decreases with increase in ionic strength.

The dosage of a biosorbent strongly influences the extent of biosorption. In many instances, lower biosorbent dosages yield higher uptakes and lower percentage removal efficiencies. An increase in the biomass concentration generally increases the amount of solute biosorbed, due to the increased surface area of the biosorbent, which in turn increases the number of binding sites. The initial solute concentration seems to have impact on biosorption, with a higher concentration resulting in a high solute uptake (Vijayaraghavan and Yun 2008). This is because at lower initial solute concentrations, the ratio of the initial moles of solute to the available surface area is low; subsequently, the fractional sorption becomes independent of the initial concentration. However, at higher concentrations, the sites available for sorption become fewer compared to the moles of solute present and; hence, the removal of solute is strongly dependent upon the initial solute concentration. With appropriate agitation, mass transfer resistance can be minimized. When increasing the agitation rate, the diffusion rate of a solute from the bulk liquid to the liquid boundary layer surrounding particles becomes higher due to the enhanced turbulence and the decrease in the thickness of the liquid boundary layer. Hence, at higher agitation rates, the boundary layer becomes very thin, which usually enhances the rate at which a solute diffuse through the boundary layer (Vijayaraghavan and Yun 2008).

### *10.3.2 Batch Data Modeling and Isotherms*

Biosorption modeling can be performed in two general ways: empirical or mechanistic equations, which are able to explain, represent and predict the experimental behaviour (Vijayaraghavan and Yun 2008). The quality of a biosorbent is judged by how much sorbate it can attract and retain in an immobilized form. The solute uptake by a biosorbent can be calculated from the differences between the initial quantities of solute added to that contained in the supernatant, which is achieved using the following Eq. 10.1.

$$Q = \frac{V_0 C_0 - V_f C_f}{M} \tag{10.1}$$

where $Q$ is the solute uptake (mg/g); $C_0$ and $C_f$ the initial and equilibrium solute concentrations in solution (mg/L), respectively; $V_0$ and $V_f$ the initial and final solution volumes (L), respectively; and $M$ the mass of biosorbent (g). The sorption uptake can be expressed in different units depending on the purpose of the exercise: for example, milligrams of solute sorbed per gramof the (dry) biosorbentmaterial (the basis for engineering process–mass balance calculations), or mmol/g (when the stoichiometry and/or mechanism are to be considered) (Vijayaraghavan and Yun 2008). A biosorption isotherm, the plot of uptake ($Q$) versus the equilibrium solute concentration in the solution ($C_f$), is often used to evaluate the sorption performance. Isotherm curves can be evaluated by varying the initial solute concentrations, while fixing the environmental parameters, such as pH, temperature and ionic strength. In general, the uptake increases with increase in concentration, and will reach saturation at higher concentrations. In most biosorption studies, pH seems to be an important parameter for the evaluation of an isotherm (Vijayaraghavan and Yun 2008).

### *10.3.3 Empirical Modeling and Isotherms*

Empirical models are simple mathematical relationships, characterized by a limited number of adjustable parameters, which give a good description of the experimental behavior over a large range of operating conditions. Some frequently employed and well established empirical models involve two, three or even four parameters to model the isotherm data (Vijayaraghavan and Yun 2008). Although these conventional empirical models do not reflect the mechanisms of sorbate uptake, they are capable of reflecting the experimental curves. Also, in most cases, the assumptions from which these models were derived are not valid for biosorption. Despite this, conventional adsorption isotherm models are used with a high rate of success for replicating biosorption isotherm curves. Within the literature, the Langmuir and Freundlich models (two–parameter models) have been used to describe biosorption isotherm. The models are simple, well–established and have physical meaning and are easily interpretable, which are some of the important reasons for their frequent and extensive use (Vijayaraghavan and Yun 2008).

Some other two–parameter models widely used for describing biosorption isotherms include the Temkin isotherm, the Dubinin–Radushkevich model, the Redlich–Peterson model, the Sips model, the Khan model, the Radke–Prausnitz model and the Toth model. Of these three–parameter models, the Redlich–Peterson and Sips models have been used with most success.

The Langmuir model (Eq. 10.2) incorporates two easily interpretable constants: $Q_{max}$, which corresponds to the maximum achievable uptake by a system;

and $b_L$, which is related to the affinity between the sorbate and sorbent. The Langmuir constant "$Q_{max}$" is often used to compare the performance of biosorbents; while the other constant "$b_L$" characterizes the initial slope of the isotherm. Thus, for a good biosorbent, a high $Q_{max}$ and a steep initial isotherm slope (i.e., high $b_L$) are generally desirable. A typical Langmuir adsorption isotherm is shown in Fig. 10.3. The Freundlich isotherm (Eq. 10.3) was originally empirical in nature, but was later interpreted as the sorption to heterogeneous surfaces or surfaces supporting sites with various affinities. It is assumed that the stronger binding sites are initially occupied, with the binding strength decreasing with increasing degree of site occupation. It incorporates two constants: $K_F$, which corresponds to the binding capacity; and $n_F$, which characterize the affinity between the sorbent and sorbate.

$$Q = \frac{Q_{\max} b_L C_f}{1 + b_L C_f} \tag{10.2}$$

$$Q = K_F C_f^{\frac{1}{n_F}} \tag{10.3}$$

Equations 10.4–10.10 depict the Temkin, Dubnin–Radushkevich, Redlich–Peterson, Sips, Khan, Radke–Prausnitz and Toth isotherms. More details of these models may be obtained from Vijayaraghavan and Yun (2008). $b_{Te}$ is the Temkin constant related to the heat of sorption; $a_{Te}$ the Temkin isotherm constant; $R$ the gas constant (8.314 J/mol.K); $T$ the absolute temperature; $Q_D$ the Dubinin–Radushkevich model uptake capacity and $B_D$ the Dubinin–Radushkevich model constant.

$$Q = \frac{RT}{b_{Te}} \ln\left(a_{Te} C_f\right) \tag{10.4}$$

$$Q = Q_D e^{\left[-B_D\left(RT \ln\left\{1+\frac{1}{C_f}\right\}\right)^2\right]} \tag{10.5}$$

$$Q = \frac{K_{RP} C_f}{1 + a_{RP} C_f^{\beta_{RP}}} \tag{10.6}$$

$$Q = \frac{K_S C_f^{\beta_S}}{1 + a_S C_f^{\beta_s}} \tag{10.7}$$

$$Q = \frac{Q_{\max} b_K C_f}{\left(1 + b_K C_f\right)^{a_K}} \tag{10.8}$$

$$Q = \frac{a_R r_R C_f^{\beta_R}}{a_R + r_R C_f^{\beta_R - 1}} \tag{10.9}$$

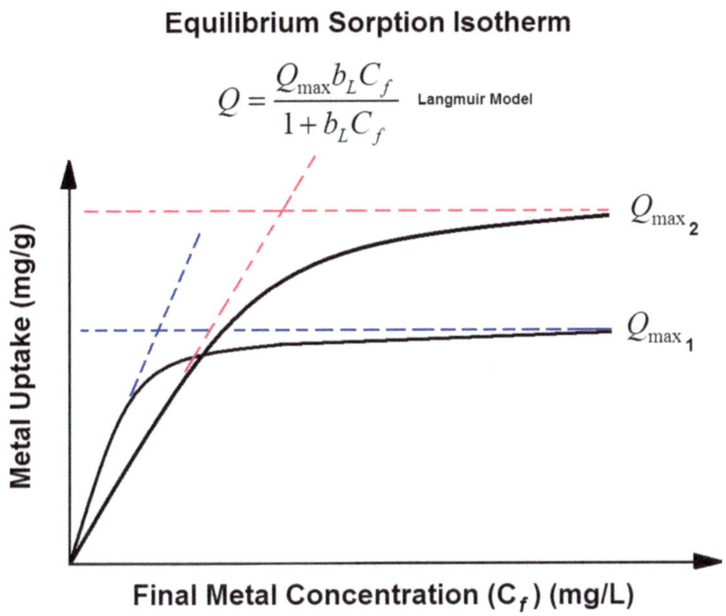


**Fig. 10.3** Biosorption–Langmuir isotherm relationship curves showing significance of $b_L$ with respect to $Q_{max}$. $b_L$ indicates the initial rise in slope and hence gives a very good indication of the sorbent affinity for the adsorbate. A steeper slope indicates a higher affinity of a sorbent for an adsorbate when several adsorption systems are being compared. Adapted from Volesky and Schiewer (1999)

$$Q = \frac{Q_{max} b_T C_f}{\left[1 + \left(b_T C_f\right)^{1/n_T}\right]^{n_T}} \tag{10.10}$$

$K_{RP}$ is the Redlich–Peterson model isotherm constant, $a_{RP}$ the Redlich–Peterson model constant; $\beta_{RP}$ the Redlich–Peterson model exponent; $K_S$ the Sips model isotherm constant; $a_S$ the Sips model constant; $\beta_S$ the Sips model exponent; $b_K$ the Khan model constant; $a_K$ the Khan model exponent; $a_R$ and $r_R$ are Radke–Prausnitz model constants; $\beta_R$ the Radke–Prausnitz model exponent; $b_T$ the Toth model constant and $n_T$ the Toth model exponent.

## 10.3.4 Mechanistic Models

Mechanistic models have been proposed to describe solute adsorption onto the surfaces of biomass. The development of a mechanistic model is usually based on preliminary biomass characterization, with the formulation of a set of hypothesized reactions between the sorbent sites and solutes, which also considers the particular solution chemistry of the solutes. Mechanistic models can often be characterized by the different degrees of complexity or accuracy in a system description to account for the surface heterogeneity and other factors that contribute to non–ideal adsorption phenomena. Mechanistic modeling of biosorption has been attempted in several investigations, with significant success.

### 10.3.5 Kinetic Modeling

Mathematical models that can describe the behaviour of a batch biosorption process operated under different experimental conditions are very useful for scale up studies or process optimization (Loukidou et al. 2004). A number of models with varying degrees of complexicity have been developed to describe the kinetics of metal biosorption in batch systems. According to the kinetic model selection criteria, several reaction–based and diffusion–based models were tested for the simulation of the obtained experimental data. The finally selected kinetic models will be those, which not only fit closely the data, but also represent reasonable sorption mechanisms (Loukidou et al. 2004).

Over 20 models have been reported in the literature, all of which have attempted to quantitatively describe the kinetic behavior during the adsorption process. Each adsorption kinetic model has its own limitations, which are derived according to specific experimental and theoretical assumptions. Even though they violate the fundamental assumptions, many adsorption models have been used to successfully test experimental biosorption data. Of these, pseudo–first and pseudo–second order models (Eqs. 10.11 and 10.12, respectively) have often been used to describe biosorption kinetic data.

$$Q_t = Q_e\left(1 - e^{-k_1 t}\right) \tag{10.11}$$

$$Q_t = Q_e\left(1 - \frac{1}{1 + Q_e k_2 t}\right) \tag{10.12}$$

where $Q_e$ is the amount of solute sorbed at equilibrium (mg/g); $Q_t$ the amount of solute sorbed at time $t$ (mg/g); $k_1$ the first order equilibrium rate constant ($min^{-1}$) and $k_2$ the second order equilibrium rate constant (g/mg min). In most published cases involving biosorption, the pseudo–first order equation was found to not fit well over the entire contact time range, but was generally applicable over the initial periods of the sorption process.

### 10.3.6 Main Mechanisms of Metal Biosorption

Different metal–binding mechanisms have been postulated to be active in biosorption metal uptake such as chemisorption by ion–exchange, complexation, coordination, chelation; physical adsorption and microprecipitation (Volesky 2001). There are also possible oxidation–reduction reactions taking place in the biosorbent. Due to the complexity of biomaterials and biosorbents, it is also plausible that at least some of these mechanisms are acting simultaneously to varying extents depending on the biosorbent composition, surface properties and functional chemical groups, and the solution environment (Volesky 2001). Biomass materials offer several molecular groups that are known to offer ion exchange sites, carboxyl, sulphate, phosphate, and amine, could be the main ones (Volesky 2001).

Ion–exchange is an important concept in biosorption, because it explains many of the observations made during heavy metal uptake experiments (Davis et al. 2003). It should be pointed out that the term ion–exchange does not explicitly identify the binding mechanism, rather it is used here as an umbrella term to describe the experimental observations (Davis et al. 2003). The precise binding mechanism(s) may range from physical (i.e. electrostatic or London–van der Waals forces) to chemical binding (i.e. ionic and covalent). The term sorption would refer to binding of a metal cation to a free site as opposed to one that was previously occupied by another cation. It is distinct from adsorption that defines binding in terms of a physical rather than chemical surface phenomenon. In the case of biosorption of heavy metals by biomass, the mechanisms can be viewed, in principle, as being extracellular, or occurring discretely at the cell wall. Intracellular sorption would normally imply bioaccumulation by a viable organism (Davis et al. 2003).

## 10.4 Heavy Metals Toxicity

The term heavy metals refers to metals and metalloids having densities greater than 5 g $cm^{-3}$ and is usually associated with pollution and toxicity although some of these elements (essential metals) are required by organisms at low concentrations (Adriano 2001). Heavy metals toxicity and the danger of their bioaccumulation in the food chain represent one of the major environmental and health problems of our modern society. Primary sources of pollution is from the burning of fossil fuels, mining and melting of metallic ferrous ores, municipal wastes, fertilizers, pesticides, and sewage sludge (Peng et al. 2006). The most common heavy metals contaminants are: cadmium (Cd), chromium (Cr), copper (Cu), lead (Pb), nickel (Ni) and zinc (Zn).

Metal–bearing effluents generally emanate from metallurgical industries, electroplating and metal finishing industries and hazardous waste sites (Elouear et al. 2008). For example, cadmium ($Cd^{2+}$) is a non–essential and a non–biodegradable metal ion which slowly accumulates in the body of living creatures, usually through food chain (Srivastava et al. 2006). Nickel ($Ni^{2+}$) is also toxic, especially to activated sludge bacteria. The presence of $Ni^{2+}$ has been reported to be detrimental to the operation of anaerobic digesters used in wastewater treatment plants (Patterson 1977). $Cd^{2+}$ and $Ni^{2+}$ ions are frequently encountered together in industrial wastewaters. Hence, it is essential to removal Cd and Ni from industrial wastewaters before transport and cycling into the nature environment. The soil has been traditionally the site for disposal for most of the heavy metals wastes which need to be treated (Aboulroos et al. 2006).

Unlike organic compounds, metals cannot be degraded, and their cleanup requires their immobilization and toxicity reduction or removal. In recent years, scientists and engineers have started to generate cost effective technologies which includes use of microorganisms/ biomass or live plants for cleaning of polluted areas (Kuzovkina et al. 2004). The fate of toxic metallic cations in the soil environment depends largely on the interactions of these metals with inorganic and organic surfaces. The extent to which a metallic cation interacts with these surfaces determines the concentration of metal in solution and, consequently, the potential for movement into groundwater or uptake by plants. A considerable amount of literature is available on adsorption or complexation of various heavy metals by soils (Mullen et al. 1989). Table 10.3 summarizes some of the studies conducted to assess the complexation of heavy metals by different soils.

### *10.4.1 Cadmium*

Cadmium (Cd) is one of the most toxic heavy metals and is considered non–essential for living organisms. Cadmium has been recognized for its negative effect on the environment where it accumulates throughout the food chain posing a serious threat to human health. Cadmium pollution has induced extremely severe effects on plants (Baszynski 1986). Cadmium , which is widely used and extremely toxic in relatively low dosages, is one of the principal heavy metals responsible for causing kidney damage, renal disorder, high blood pressure, bone fraction and destruction of red blood cells (Drasch 1993). Because of the toxicity and bioaccumulation, $Cd^{2+}$ has been considered as a priority pollutant by the US Environmental Protection Agency (Krishnan and Anirudhan 2003). The permissible limit for $Cd^{2+}$ as described by World Health Organization is 0.01 mg $dm^{-3}$. The main anthropogenic pathway through which $Cd^{2+}$ enters the water bodies is via wastes and wastewaters from industrial processes such as electroplating, plastic manufacturing, metallurgical processes and industries of pigments and Cd/Ni batteries (Cheremisinoff 1995).

Since the 1970s, at least 200 ha of farm–land has been polluted by the cadmium (Lu et al. 2007). Consequently, the Cd pollution has led to contaminate the rice production and caused acute social panic. According to the recent investigation results performed by the Taiwan Environmental Protection Administration (TEPA), it is indicated that most of the Cd pollution incidents in Taiwan resulted from the wastewater discharge of stearate Cd factories. To prevent and control the Cd pollution incidents from spreading, the TEPA has either forced these factories to close down or assisted them in improving their production processes since the 1980s (Lu et al. 2007). Unfortunately, accidental incidents of Cd pollution have still emerged in an endless stream, despite the strict governmental controls placed on these questionable factories. Besides its high toxicity, there is some evidence that cadmium is carcinogenic (Hiatt and Huff 1975). When ingested by human beings, cadmium that is not excreted immediately has a long half–life of several hundred days, so that a low dose exposure over a long period of time can lead to a high body burden (Lin et al. 2005).

**Table 10.3** Heavy metal ions adsorption by soils and soil components

| Heavy metal ion | Type of soil(s) | Experimental conditions | Adsorption performance | Reference |
|---|---|---|---|---|
| Cd, Cr, Cu, Mn, Ni, Pb and Zn | Na–montmorillonite | Variable pH and in the presence of ligands | Adsorption pattern: Mn ≤ Pb ≤ Cd ≤ Zn < Ni < Cu < Cr | Abollino et al. (2003) |
| Chromium, nickel, and cadmium | Clayey soils – kaolin | Cr(VI), Ni(II), and Cd(II) in concentrations of 1,000, 500, and 250 mg/kg | Maximum removal of chromium, nickel, and cadmium of 68–71%, 71–73%, and 87–94%, respectively, observed for sequentially enhanced electrokinetic remediation approach | Reddy and Chinthamreddy (2003) |
| Cd(II), Cu(II), Pb(II), and Zn(II) | Kaolinite | Single– and multi–element systems | 50% adsorption: Cu < Zn < Pb < Cd in single–element systems, but Pb < Cu < Zn < Cd in the multi–element system | Srivastava et al. (2005) |
| Cadmium, copper, lead, manganese, nickel, and zinc | Vermiculite | Continuous column method was used in order to evaluate the feasibility to use the clay in wastewater purification systems | Total adsorptive capacity of vermiculite was found to decrease in the following order: Mn > Ni > Zn > Cd > Cu > Pb. The adsorption of metal ions on vermiculite decreases with decreasing pH and increasing ionic strength | Malandrino et al. (2006) |
| Nickel | Kaolinite, montmorillonite | Batch adsorption studies were carried out with various Ni(II) concentrations, amount of clay adsorbents, pH, agitation time and temperature | Langmuir monolayer capacity of 2.75–21.14 mg/g and Freundlich adsorption capacity of 0.70–3.40 $mg^{(1-1/n)}\ L^{1/n}$ /g for the clay adsorbents. $\Delta H$ in the range of −24.0–45.1 kJ/mol, $\Delta S$: −118.2–160.5 J/mol.K $\Delta G$: −34.6–49.5 kJ/mol | Gupta and Bhattacharyya (2006) |

(continued)

**Table 10.3** (continued)

| Heavy metal ion | Type of soil(s) | Experimental conditions | Adsorption performance | Reference |
|---|---|---|---|---|
| $Cr^{3+}$, $Ni^{2+}$, $Zn^{2+}$, $Cu^{2+}$and $Cd^{2+}$ | Ca–bentonite and Na–bentonite | Batch experiments | Langmuir model for sorption and maximum sorption capacities of 44.4 mg Cr/g, 6.32 mg Ni/g, 5.75 mg Zn/g, 7.72 mg Cu/g, 7.28 mg Cd/g on Ca–bentonite and 49.8 mg Cr/g, 24.2 mg Ni/g, 23.1 mg Zn/g, 30.0 mg Cu/g, 26.2 mg Cd/g on Na–bentonite | Álvarez–Ayuso and García–Sánchez (2003) |
| $Cu^{2+}$ | Saudi clay mineral (bentonite) | Different weights of the clay (0.5, 1.0, 1.5, and 2 g) | Sorption characteristics were described using two site Langmuir isotherms. Maximum adsorption obtained was 909 mg $Cu^{2+}$/g clay | Al–Qunaibit et al. (2005) |
| $Ni^{2+}$, $Cu^{2+}$, $Pb^{2+}$, and $Cd^{2+}$ | Clinoptilolite | Static conditions from single– and multicomponent aqueous solutions by raw and pretreated clinoptilolite | Maximum sorption capacity toward $Cd^{2+}$: 4.22 mg/g at an initial concentration of 80 mg/L and toward $Pb^{2+}$, $Cu^{2+}$, and $Ni^{2+}$ as 27.7, 25.76, and 13.03 mg/g at 800 mg/L. The sorption results fit well to the Langmuir and the Freundlich models | Sprynskyy et al. (2006) |
| Zinc, cadmium, copper, and lead | Three highly–weathered Brazilian soils (two Ultisols and one Oxisol with and without liming) | Equilibrium solutions were applied as a cocktail containing 700 mg/L of Zn, 20 mg/L of Cd, 200 mg/L of Cu, and 300 mg/L of Pb and its dilutions of 1:5 and 1:20 | Adsorption data for Zn and Cd did not fit the linear, Langmuir, Freundlich, and Temkin isotherm equations for most situations. Due to the competition with other metals, the equations, which offered the best fit for Zn and Cd, were quadratic polynomial models. On the other hand, for Cu and Pb, the equations, which showed the best fit were linear, Langmuir, and Temkin, for different situations | Ferreira Fontes et al. (2000) |

In view of its persistence as a cumulative poison and the low tolerance of the human body towards cadmium, it is of interest to develop schemes for its removal. For example, the immobilization and remobilization of Cd(II) by ferrihydrite, and the effect of $Cd^{2+}$ on the conversion of ferrihydrite to goethite and hematite has been reported previously (Lin et al. 2003; Sun et al. 1996).

Cadmium exists in wastewater in many forms including soluble, insoluble, inorganic, metal organic, reduced, oxidized, free metal, precipitated, adsorbed and complexed forms. Treatment processes for cadmium removal must be selected to remove the existing form, or the cadmium must be converted to a suitable form compatible with the removal process. Several studies have been conducted to remove cadmium from contaminated aqueous media by various techniques like filtration, chemical precipitation and complexation, adsorption onto activated carbon, floatation and ion exchange. Since the past couple of decades, cadmium removal by biosorption has gained momentum. Many of the studies conducted in this field have unanimously concluded on the promise of this novel technique in being efficient for the treatment of $Cd^{2+}$ bearing wastewater streams. Table 10.4 presents a digest of some selected studies on the adsorption performance of Cd/ $Cd^{2+}$ biosorbents.

**Table 10.4** Biosorbents used for cadmium removal from wastewaters

| Biosorbent(s) | Biosorption performance | Reference |
|---|---|---|
| Fungal biomass obtained from cultivation using starch wastewater as a substrate | Biosorption capacities of the pretreated fungal biomass of *R. oryzae*, *R. oligosporus*, *A. oryzae* and *R. arrhizus* were up to 0.28, 0.35, 0.40 and 0.56 mmol $Cd^{2+}$/g (dry weight), respectively | Yin et al. (1999) |
| Biomass of nonliving, dried brown marine algae *Sargassum natans, Fucus vesiculosus*, and *Ascophyllum nodosum* | Biomass of *A. nodosum* accumulated the highest amount of cadmium exceeding 100 mg $Cd^{2+}$/g (at the residual concentration of 100 mg Cd/L and pH 3.5) | Holan et al. (1993) |
| Black gram husk (*Cicer arientinum*) | With 99.99% sorption efficiency from 10 mg/L cadmium solution, the biomass required at saturation was 0.8 g/mg cadmium. Biosorption was rapid and equilibrium was achieved in 30 min | Saeed and Iqbal (2003) |
| Rice polish agricultural waste | The maximum removal of $Cd^{2+}$ was found to be 9.72 mg/g at pH 8.6, initial $Cd^{2+}$ concentration of 125 mg/L (20°C) | Singh et al. (2005) |
| Orange wastes from orange juice production processes | Adsorption process was quick and the equilibrium was attained within 3 h. The maximum adsorption capacity of orange waste was found to be 0.40, 0.41 and 0.43 mmol/g at pH 4–6, respectively | Pérez–Marín et al. (2007) |

(continued)

**Table 10.4** (continued)

| Biosorbent(s) | Biosorption performance | Reference |
|---|---|---|
| Shelled *Moringa oleifera* Lam. seed powder | Sorption studies resulted in the standardization of optimum conditions for removal of Cd (85.10%) as follows: biomass dosage (4.0 g), metal concentration (25 μg/mL), contact time (40 min) and volume of the test solution (200 mL) at pH 6.5. | Sharma et al. (2006) |
| Pretreated rice husk (RRH) | Simple and low–cost chemical modifications resulted in increasing the sorption capacity of RRH from 8.58 mg/g to 11.12, 20.24, 16.18 mg/g and reducing the equilibrium time from 10 h of RRH to 2, 4 and 1 h for epichlorohydrin treated rice husk (ERH), NaOH treated rice husk (NRH), sodium bicarbonate treated rice husk (NCRH), respectively | Kumar and Bandyopadhyay (2006) |
| Wheat bran | Maximum removal of $Cd^{2+}$ was found to be 87.15% at pH 8.6, initial Cd(II) concentration of 12.5 mg/L (20 °C) | Singh et al. (2006) |
| Agricultural residues viz. sugarcane bagasse (SCB), maize corncob (MCC) and Jatropha oil cake (JOC) | Maximum adsorption of $Cd^{2+}$ metal ions was observed at pH 6 for all the adsorbents viz; 99.5%, 99% and 85% for JOC, MCC, and SCB, | Garg et al. (2008) |
| Ethanol treated waste baker's yeast biomass | Highest metal uptake of 15.63 mg $Cd^{2+}$/g | Göksungur et al. (2005) |
| Grape stalk waste (a by–product of wine production) | Maximum sorption at initial pH 5.5: 0.248 mmol $Cd^{2+}$ /g | Martínez et al. (2006) |
| Husk of *Lathyrus sativus* (HLS) | 95% removal of $Cd^{2+}$ with >90% of the total adsorption taking place within the first 5 min | Panda et al. (2006) |
| Red algae (*Ceramium virgatum*) | Biosorption capacity for $Cd^{2+}$ ions was 39.7 mg/g | Sarı and Tuzen (2008) |

## *10.4.2 Chromium*

One of the most important chemical contaminants of concern is chromium (Cr), which exists in a series of oxidation states from −2 to +6 valence; the most important stable states are 0 (element metal), +3 (trivalent) and +6 (hexavalent). $Cr^{3+}$ and $Cr^{6+}$ are released to the environment primary from stationary point sources resulting from human activities. Contamination of groundwater by Cr at numerous localities primarily resulted from uncontrolled or accidental release of Cr–bearing solutions, used in various industrial applications (metallurgical, chemical, leather tanning, wood processing, textile and refractory), into the subsurface environment. Cr in such solutions mostly occurs as oxyacids and oxyanions of $Cr^{6+}$ (Mukhopadhyay et al. 2007; Xu and Zhao 2007). It is this oxidation state in which Cr is highly soluble, mobile and toxic.

$Cr^{3+}$ is an essential dietary mineral in low doses: it is required to potentiate insulin and for normal glucose metabolism. Biocidal properties of chromium salts to aquatic organisms are modified, sometimes by an order of magnitude or more, by a variety of biological and abiotic factors. These include the species, age, and developmental stage of the organism; the temperature, pH, salinity, and alkalinity of the medium; interaction effects of Cr with other contaminants; duration of exposure; and chemical form of Cr tested. For hexavalent chromium, $LC_{50}$ (96 h) values for sensitive freshwater and marine species were between 445 and 2,000 μg/L. The $LC_{50}$ value is the concentration of a material in air that will kill 50% of the test subjects (animals, typically mice or rats) when administered as a single exposure. Also called the median lethal concentration and lethal concentration 50, the $LC_{50}$ value gives an indication of the relative acute toxicity of an inhalable material. For trivalent chromium, $LC_{50}$ (96 h) concentrations were 2,000–3,200 μg/L for sensitive freshwater organisms and 3,300–7,500 μg/Lfor marine biota. Among warm–blooded organisms, hexavalent chromium was fatal to dogs in 3 months at 100 mg/L in their food and killed most mammalian experimental animals at injected doses of 1–5 mg Cr/kg body weight, but had no measurable effect on chickens at dietary levels of 100 mg/L over a 32–day period. Trivalent chromium compounds were generally less toxic than hexavalent chromium compounds, but significant differences may occur in uptake of anionic and cationic $Cr^{3+}$ species, and this difference may affect survival.

Acute and chronic adverse effects of chromium to warm–blooded organisms are caused mainly by $Cr^{6+}$ compounds; there is little conclusive evidence of toxic effects caused by $Cr^{2+}$ or $Cr^{3+}$ compounds (Langard and Norseth 1979). Most investigators agree that chromium in biological materials is probably always in the trivalent state, that greatest exposures of $Cr^{3+}$ in the general human population are through the diet (but no adverse effects have been reported from such exposures), and that no organic trivalent chromium complexes of toxicological importance have been described. Studies with guinea pigs fed $Cr^{3+}$ for 21 weeks at concentrations up to 50 mg/L dietary $Cr^{3+}$ showed no adverse effects. Domestic cats were apparently unaffected after exposure to aerosol levels of 80–115 mg $Cr^{3+}$ /m$^3$ for 1 h daily for 4 months, or after consuming diets with high amounts of chromic ($Cr^{3+}$) salts over a similar period (Langard and Norseth 1979).

Under laboratory conditions, chromium is mutagenic, carcinogenic, and teratogenic to a wide variety of organisms, and $Cr^{6+}$ has the greatest biological activity. However, information is lacking on the biological activities of water soluble $Cr^{3+}$ compounds, organochromium compounds, and their ionic states. Aquatic plants and marine polychaete worms appear to be the most sensitive groups tested. In exposures to $Cr^{6+}$, growth of algae was inhibited at 10.0 μg/L, and reproduction of worms at 12.5 μg/L. At higher concentrations, $Cr^{6+}$ is associated with abnormal enzyme activities, altered blood chemistry, lowered resistance to pathogenic organisms, behavioural modifications, disrupted feeding, histopathology, osmoregulatory upset, alterations in population structure and species diversity indices, and inhibition of photosynthesis. Not all sublethal effects observed were permanent, but the potential for acclimatization of organisms to Cr is not well documented. The great variability among species and tissues in the accumulation or concentration of Cr is attributed partly to the route of administration, partly to the concentration of Cr and its chemical species, and partly to numerous biotic and physicochemical modifiers. High accumulations of Cr have been recorded among organisms from the lower trophic levels, but there is little evidence of biomagnification through food chains. Marine bivalve molluscs, for example, accumulated measurable concentrations at ambient water concentrations of 5.0 μg/L of $Cr^{6+}$, but the significance of Cr residues in molluscs and other organisms is not well understood.

Chromium is causally associated with mutations and malignancy (Leonard and Lauwerys 1980; Norseth 1981). Under appropriate conditions, Cr is a human and animal carcinogenic agent; its biological effects depend on chemical form, solubility, and valence. In general, $Cr^{6+}$ compounds are hazardous to animals, whereas metallic Cr and $Cr^{3+}$ are essentially nontoxic (Gale 1978); however, exposure to water solubilized $Cr^{3+}$ has caused cancers and dermatitis in workers, and toxicity in rabbits (Hatherill 1981). In the chromate producing industry workers who developed respiratory cancer had been exposed to 30–1,100 μg/$m^3$ Cr in air for periods of 4–24 years, and workers producing chromate pigment who developed respiratory cancer had been subjected to an estimated $Cr^{6+}$ exposure of 500–1,500 μg/$m^3$ for 6–9 years. Carcinogens released in the chromate manufacturing process have not yet been identified (Post and Campbell 1980). Levels as low as 10 μg/$m^3$ of $Cr^{6+}$ in air produced strong irritation in nasal membranes, even after short exposures. In some persons whose lower respiratory tissues became Cr–sensitized, asthmatic attacks occurred at levels of $Cr^{6+}$ as low as 2.5 μg/$m^3$ (Steven et al. 1976). There is no evidence of Cr sensitization in mammals other than humans. In the only animal study demonstrating a carcinogenic effect of an inhaled chromate, adenocarcinomas were reported in the bronchial tree of mice exposed throughout life to $CaCrO_4$ dust at 13 mg/m (4,330 μg $Cr^{6+}/m^3$) for 35 h weekly (Langard and Norseth 1979). Trivalent Cr compounds did not produce respiratory cancers (Steven et al. 1976). In rabbits, both $Cr^{3+}$ and $Cr^{6+}$, given 1.7 mg/kg body weight daily for 6 weeks, adversely affected blood and serum chemistry, and both produced significant morphological changes in liver (Tandon et al. 1978); similar results were observed in rats (Laj et al. 1984). Although damage effects and residue accumulations were greater in rabbits treated with $Cr^{6+}$, water soluble $Cr^{3+}$ compounds also may have significant biological activity (Tandon et al. 1978).

Hexavalent chromium is present in the effluents produced during the electroplating, leather tanning, cement, mining, dyeing and fertilizer and photography industries and causes severe environmental and public health problems (Demirbas et al. 2004). Its concentrations in industrial wastewaters range from 0.5 to 270 mg/L and the tolerance limit for $Cr^{6+}$ for discharge into inland surface waters is 0.1 mg/L and in potable water is 0.05 mg/L (EPA 1990). In order to comply with this limit, it is essential that industries treat their effluents to reduce the $Cr^{6+}$ to acceptable levels. $Cr^{6+}$ is frequently used in plating process that can discharge wastewater with a concentration as high as 100 g Cr/L. At present the frequently used technology for treating wastewater that contains $Cr^{6+}$ is to chemically reduce $Cr^{6+}$ to $Cr^{3+}$ with $NaHSO_3$, $Fe^{2+}$ compounds, $SO_2$,and others, followed by pH adjustment with hydroxides of alkaline or alkaline earth metal to pH 10–12 to form $Cr(OH)_3$ suspended particulates. The particulates are subsequently agglomerated and precipitated with introduced agglomerating chemicals. The main shortcomings of this technology are the requirement of a lot of costly chemicals and the generation of vast sludge to be disposed of.

A potential alternative to remove heavy metals from wastewater is to use naturally–occurring humic substances as sorbent due to their richness in organic functional groups such as carboxylic acid, and others. Chromium can be removed from water with both aquatic plants and dead biomasses, or humic substances. Vegetable–based waste materials have been used as natural adsorbent for $Cr^{6+}$: coirpith (Sumathi et al. 2005), sawdust (Acar and Malkoc 2004), rice husk and rice husk carbon (Bishnoi et al. 2004), hazelnut shell carbon (Kobya 2004) and debris of aquatic plants (Hu et al. 2003). Batch tests on gravel, compost, sterilized compost and mixture of gravel + compost formerly carried out by the authors proved that the gravel contribution to the $Cr^{6+}$ removal is quite negligible. Besides, the removal observed was predominantly due to the biomass activity, which converted $Cr^{6+}$ into $Cr^{3+}$ under a stable form, such as $Cr(OH)_3$. The Freundlich isotherm has been particularly used to model the data of chromium sorption by the kitchen waste compost. It results in the following "*K*" values: 0.261 g $Cr^{3+}$/kg and 0.105 g $Cr^{6+}$/kg. The "*n*" is 1.56 in the $Cr^{3+}$ sorption and 1.35 in the $Cr^{6+}$ sorption. The "*K*" values indicate that the capacity of kitchen waste compost for sorbing $Cr^{3+}$ is greater than that for sorbing $Cr^{6+}$. Both "n" values in this study are greater than unity, therefore sorption of $Cr^{3+}$ and $Cr^{6+}$is favorable. In addition, the sorption affinities to $Cr^{3+}$ and to $Cr^{6+}$ are similar. X–ray absorption near–edge structure spectroscopic simulation has indicated that the Cr species distribution is estimated to be about: 54.1–61.0% $Cr_3(OH)_2$ $(OOCCH_3)7$ + 39.0–45.9% $Cr(NO_3)_3$ in all compost samples sorbing Cr from $Cr^{3+}$ solutions; and 54.5–69.0% $Cr_3(OH)_2(OOCCH_3)_7$ + 18.0–24.9% $Cr(OH)_3$ + 6.1–28.5% $CrO_3$ in the samples from $Cr^{6+}$solutions. No $Cr(OH)_3$ is expected in the $Cr(NO_3)_3$–sorbing compost samples due to the low solution pH. In contrast, the solution pH after the $CrO_3$–sorption experiments is always greater than 5.9, that confirms the formation of $Cr(OH)_3$ precipitate on the compost after the chemical reduction of $Cr^{6+}$ to $Cr^{3+}$ by the compost.

As a corollary, compost derived from cellulosic materials or kitchen wastes is effective as a biosorbent in removing Cr from water (Wei et al. 2005). Over the past years, various materials have been used by several researchers to remove chromium and results have been also very encouraging. The trend has now been set for adsorption of Cr using low cost biometarials. Table 10.5 summarizes the use, testing and removal performance of some biomaterials and biosorbents for the removal of chromium species from chromium–laden waters.

### 10.4.3 Copper

Copper, one of the most widely used heavy metals, is mainly employed in electrical and electroplating industries, and in larger amounts is extremely toxic to living organisms. The presence of copper (II) ions cause serious toxicological concerns, it is usually known to deposit in brain, skin, liver, pancreas and myocardium

**Table 10.5** Biosorbents used for chromium removal from Cr–laden aqueous media

| Biosorbent(s) | Biosorption performance | Reference |
|---|---|---|
| Green algae *spirogyra* species | Maximum removal of $Cr^{6+}$ was around $14.7\times10^3$ mg metal/kg of dry weight biomass at a pH of 2.0 in 120 min with 5 mg/L of initial concentration | Gupta et al. (2001) |
| *Neurospora crassa* fungal biomass | Biosorption capacity of acetic acid pretreated biomass was found to be $15.85\pm0.94$ mg/g biomass under optimum conditions. The adsorption constants were found from the Freundlich isotherm model at 25°C. The biosorbent was regenerated using 10 mM NaOH solution with up to 95% recovery and reused five times in biosorption–desorption cycles successively | Tunali et al. (2005) |
| Composite chitosan biosorbent prepared by coating chitosan, a glucosamine biopolymer, onto ceramic alumina | Experimental equilibrium data to Langmuir and Freundlich adsorption isotherms and ultimate capacity obtained from the Langmuir model was 153.85 mg $Cr^{6+}$/g chitosan | Boddu et al. (2003) |
| Mucilaginous seeds of *Ocimum basilicum* | Seeds boiled in water were found to be superior in terms of mechanical stability and exhibited fairly optimal $Cr^{6+}$ uptake kinetics. Maximum adsorption capacity from Langmuir isotherm was 205 mg Cr/g dry seeds | Melo and D'Souza (2004) |
| *Eichhornia crassipes* | Freundlich isotherm was found to represent the measured sorption data well. Fourier transform infrared spectrometry showed that the hydroxyl group was the chromium–binding site within pH range (pH 1–5) where chromium does not precipitate. Results indicated that the biomass of *E. Crassipes* is suitable for development of efficient biosorbent for the removal of chromium from wastewater of chemical and allied process industries | Mohanty et al. (2006) |
| *Sargassum* sp. algae | Capacity of removal obtained at optimum conditions was 19.06 mg of metal/g biosorbent | Vieira et al. (2008) |
| Palm flower (*Borassus aethiopum*) | For $Cr^{3+}$, maximum adsorption capacity was 6.24 mg/g by raw adsorbent and 1.41 mg/g by acid treated adsorbent. For $Cr^{6+}$, raw adsorbent exhibited a maximum adsorption capacity of 4.9 mg/g, whereas the maximum adsorption capacity for acid treated adsorbent was 7.13 mg/g. There was a significant difference in the concentrations of $Cr^{6+}$ and total chromium removed by palm flower | Elangovan et al. (2008) |

(continued)

**Table 10.5** (continued)

| Biosorbent(s) | Biosorption performance | Reference |
|---|---|---|
| Eggshells | Crushed eggshells possess relatively high sorption capacity, when comparing with other sorbents that was evaluated as 21–160 mg/g. Eggshells were able to remove the concentration of $Cr^{3+}$ ions below the acceptable level, i.e. at 40°C, at the initial concentration of metal ions 100 mg/kg, at sorbent concentration 15 g/L | Chojnacka (2005) |
| Shells of Walnut (WNS) *(Juglans regia)*, Hazelnut (HNS) *(Corylus avellana)* and Almond (AS) *(Prunus dulcis)* | Langmuir isotherm with maximum $Cr^{6+}$ion sorption capacities of 8.01, 8.28, and 3.40 mg $g^{-1}$ for WNS, HNS and AS, respectively. Percentage removal by WNS, HNS and AS was 85.32%, 88.46% and 55.00%, respectively at a concentration of 0.5 mmol/L | Pehlivan and Altun (2008) |
| *Cassia fistula* biomass | Adsorption capacity of biomass for $Cr^{3+}$and $Cr^{6+}$was found to be significantly improved by the treatments of gluteraldehyde (95.41 and 96.21 mg/g) and benzene (85.71 and 90.81 mg/g) respectively | Abbas et al. (2008) |
| Above–ground plant parts of wheat straw and grass | Biosorption was found to be a quick process. The equilibrium was reached within 10–20 min. Biosorption capacity of the studied sorbents was intermediate when compared with other sorbents of plant origin ca. 20 mg $Cr^{3+}$/g, but since these materials are commonly abundant and of minimal cost, it is possible to improve wastewater treatment efficiency by increasing the concentration of the sorbent | Chojnacka (2006) |
| *Turbinaria ornata* seaweed | Brown seaweed (*Turbinaria* spp) was pre–treated with sulfuric acid, calcium chloride and magnesium chloride and tested for its ability to remove chromium from tannery wastewater. *urbinaria* weed exhibited maximum uptake of about 31 mg of chromium for one gram of seaweed at an initial concentration of 1,000 mg/L of chromium. Freundlich and Langmuir adsorption isotherm models were used to describe the biosorption of $Cr^{3+}$ by *Turbinaria* spp. | Aravindhan et al. (2004) |
| Sugarcane bagasse, maize corn cob and Jatropha oil cake | Maximum adsorption was observed in the acidic medium at pH 2 with a contact time of 60 min at 250 rpm stirring speed. Jatropha oil cake had better adsorption capacity than sugarcane bagasse and maize corn cob under identical experimental conditions. The results showed that studied adsorbents can be an attractive low cost alternative for the treatment of wastewaters in batched or stirred mode reactors containing lower concentrations of chromium | Garg et al. (2007) |

(continued)

**Table 10.5** (continued)

| Biosorbent(s) | Biosorption performance | Reference |
|---|---|---|
| *Helianthus annuus* (sunflower) stem waste | Maximum metal removal was observed at pH 2.0. The efficiencies of boiled sunflower stem absorbent and formaldehyde–treated sunflower stem absorbent for the removal of $Cr^{6+}$ were 81.7% and 76.5%, respectively for dilute solutions at 4.0 g/L adsorbent dose | Jain et al. (2009) |
| pre–consumer processing agricultural waste: Rice husk | Maximum metal removal was observed at pH 2.0. The efficiencies of boiled and formaldehyde treated rice husk for $Cr^{6+}$ removal were 71.0% and 76.5% respectively for dilute solutions at 20 g/L adsorbent dose. The experimental data fitted with Freundlich and Dubinin–Radushkevich isotherm models | Bansal et al. (2009) |
| Low cost activated carbon (ATFAC) was prepared from coconut shell fibers (an agricultural waste) + A commercially available activated carbon fabric cloth (ACF) (control) | The maximum adsorption capacities of ATFAC and ACF at 25°C are 12.2 and 39.56 mg/g, respectively. $Cr^{3+}$ adsorption increased with an increase in temperature (10°C: ATFAC—10.97 mg/g, ACF—36.05 mg/g; 40°C: ATFAC—16.10 mg/g, ACF—40.29 mg/g). sorption capacity of activated carbon (ATFAC) and activated carbon fabric cloth is comparable to many other adsorbents/carbons/biosorbents utilized for the removal of trivalent chromium from water/wastewater | Mohan et al. (2006) |
| Japanese cedar (Cryptomeria japonica) bark | The equilibrium data at different temperatures fit well in the Langmuir isotherm model. The endothermic nature of the adsorption was confirmed by the positive value of enthalpy change (18.9 kJ/mo1). The positive value of entropy change (65.2 J/mol.K) suggested the increased randomness at the solid–solution interface during the adsorption. The studies showed that Japanese cedar bark can be used as a cost–effective adsorbent for the removal of $Cr^{6+}$ from wastewater | Aoyama et al. (2004) |
| Natural biosorbents tested: Natural sediment, chitin chitosan, *Aspergillus flavus* I–V, *Aspergillus fumigatus* I–ll, *Helmintosporium* sp, *Cladosporium* sp, *Mucor rouxii* mutant, *M. rouxii* IM–80, *Mucor* sp–I and 2, *Candida albicans* and *Cryptococcus neoformans* | *C. neoformans*, natural sediment, *Helmintosporium* sp and chitosan was more efficient to remove $Cr^{6+}$ achieving the following percentage of removals: 98%, 98% and 63%, respectively. | Ismael Acosta et al. (2004) |

(Davis et al. 2000). Copper toxicity is a much overlooked contributor to many health problems; including anorexia, fatigue, premenstrual syndrome, depression, anxiety, migraine headaches, allergies, childhood hyperactivity and learning disorders. The involvement of copper toxicity and bio–unavailability in such a wide range of health conditions may seem unusual.

Copper is an essential nutrient in humans, and has not been shown to be carcinogenic in animals or humans. However, young children, and infants in particular, appear to be especially susceptible to the effects of excess copper. Case reports have attributed adverse effects (diarrhea and weight loss) in infants to rather low levels of copper in drinking water, estimated as 0.22–1.0 mg/L or 1.0–6.5 mg Cu/L. High levels of copper in tap water in homes with copper plumbing have been linked to childhood cirrhosis in Germany. In other studies, consumption by adults of drinking water containing $>3$ mg/L ionized copper was associated with a significant increase in nausea, abdominal pain or vomiting.

Copper is a component of many naturally occurring minerals and is extensively used in industry and household products. Copper is also found in surface water, groundwater, seawater and drinking water. Surface water concentrations of copper range from 0.5 to 1,000 μg/L. Most of the copper tends to be bound to sediments. Urban runoff often contains elevated concentrations of copper due to household and industrial uses of water. Sewage is also a major source of copper input to rivers and streams, although some is removed in treatment plants because of its sediment binding properties. Copper in surface water is a well–known environmental hazard, associated with toxicity to a variety of aquatic organisms. The concentration of copper in drinking water can vary widely, depending on variations in acidity/alkalinity (pH), mineral content (hardness), and copper availability in the distribution system. Results from studies in the U.S., Europe and Canada indicate that copper levels in drinking water can range from $<0.005$ to $>30$ mg/L, with the corrosion of copper pipes serving as the most frequent cause of copper contamination.

Input of copper into aquatic ecosystems increased sharply during the past century and includes inputs from waste discharges into saline waters, industrial discharges into freshwater, and leaching of antifouling marine paints and wood preservatives. Many industries include copper in their processes and discharge it in wastewater streams. These industries include tanning, mining, metal processing and finishing, electroplating, the automobile industry, and the pharmaceutical industry to name a few. World production of copper amounts to 13 million tons a year, and is still rising. Present anthropogenic inputs of copper are two to five times higher than natural loadings; the atmosphere is a primary recipient of these inputs. In mining and industrial areas, precipitation of atmospheric fallout is a significant source of copper to the aquatic environment. In surface water copper can travel great distances and it strongly attaches to organic matter and minerals. It does not break down in the environment and thus accumulates in plants and animals. In copper–rich soil only a small number of plants can survive. This is why there is not much plant diversity near copper–disposing factories and it is a serious threat to farmlands. It negatively influences the activity of microorganisms and earthworms and thus seriously slows the decomposition of organic matter. Another important factor in decreasing the copper in industrial waste streams is the possibility of reclamation of copper. Waste stream concentrations can range up to several thousand mg/L for copper plating bath waste where the EPA's recommended limit is 1.3 mg/L. Industries already must lower their copper limits to meet their local legal guidelines and recycling the metal these companies are throwing away provides extra financial incentive for them to do so through continuous research and development in this direction.

Adsorption and biosorption technology have consequently both gained great concern in research over the last 15 years for the removal of copper ($Cu^{2+}$) ions from wastewaters. Stylianou et al. (2007) have studied the ability of natural zeolite (clinoptilolite) and exfoliated vermiculite to remove copper from aqueous solutions in fixed bed column and batch reactors. The effect of agitation speed (0, 100, 200, 400 rpm), temperature (25°C, 45°C, 60°C), and particle size [2.5–5.0 mm, dust (<0.25 mm)] and solution pH (1.00–4.00) on the removal of heavy metals was studied. Fixed bed experiments were conducted, using three different volumetric flow rates of 5–7–10BV (Bed Volumes)/h, under an initial normality of 0.01 N (317.7 mg/L), at initial pH of 4.00 and ambient temperature (25°C). Stylianou et al. (2007) found that vermiculite was more effective for the removal of copper in batch mode reactors under all the tested conditions, while the removal efficiency follows the order: vermiculite > clinoptilolite dust > clinoptilolite 2.5–5.0 mm. Kaewsarn (2002) has reported the biomass of marine algae to have high uptake capacities for a number of heavy metal ions. Kaewsarn (2002) investigated the adsorption properties of a pre–treated biomass of marine algae *Padina* sp. for $Cu^{2+}$. It was found that biosorption capacities were solution pH dependent and the maximum capacity obtained was 0.80 mmol/g at a solution pH of about 5. The biosorption kinetics was found to be fast, with 90% of adsorption within 15 min and equilibrium being reached at 30 min. The study of Kaewsarn (2002) demonstrated that the pre–treated biomass of *Padina* sp. could be used as an effective biosorbent for the treatment of $Cu^{2+}$ containing wastewater streams. Later, Basci et al. (2004) studied the adsorption capacity of wheat shell for copper (II) at various pH (2–7), agitation speeds (50–250 rpm) and initial metal ion concentrations (10–250 mg/L). The maximum biosorption of copper onto wheat shell occurred at 240 rpm agitation speed and at pH between 5 and 6 and the biosorption values of $Cu^{2+}$ were increased with increasing pH from 2 to 5 and decreased with increasing copper/wheat shell ratios from 0.83 to 10.84 mg $Cu^{2+}$/g wheat shell. Basci et al. (2004) concluded that wheat shell was a suitable biosorbent for removing $Cu^{2+}$ from aqueous solutions.

Recently, Vilar et al. (2008) have studied the biosorption of copper ions by an industrial algal waste, from agar extraction industry in a batch system. The latter biosorbent was compared with the algae *Gelidium* itself, which is the raw material for agar extraction, and the industrial waste immobilized with polyacrylonitrile (composite material). Earlier, Cho and Jian (1998) have investigated the adsorption of copper(II) in aqueous solutions by living mycelium pellets of *Phanerochaete chrysosporium*. The maximum copper adsorption capacity of the fungal mycelium estimated with the Langmuir model was 3.9 mmol Cu per gram of dry mycelium compared with 1.04 mmol Cu/g of a strong acidic ion–exchange resin. The living mycelium also showed a high affinity to copper in diluted solutions.

Table 10.6 presents a summary of the $Cu^{2+}$ removal performance of some recently studied biosorbents and low cost waste materials. It is brought to the attention of the reader that Table 10.6 (and similar tables) is not exhaustive but an attempt for reasonably representative listing of the results of research studies for $Cu^{2+}$ removal by biosorption.

### *10.4.4 Nickel*

Nickel is a naturally occurring element that exists in various mineral forms. It is used in a wide variety of applications including metallurgical processes and electrical components, such as batteries (ATSDR 1988; USAF 1990). Nickel contamination of the environment occurs locally from emissions of metal mining, smelting, and refining operations; from combustion of fossil fuels; from industrial activities, such as nickel plating and alloy manufacturing; from land disposal of sludges, solids, and slags; and from disposal as effluents.

The absorption of nickel is dependent on its physico–chemical form, with water soluble forms being more readily absorbed. The metabolism of nickel involves conversion to various chemical forms and binding to various ligands (ATSDR 1988). Nickel is excreted in the urine and faeces with relative amounts for each route being dependent on the route of exposure and chemical form. Most nickel enters the body via food and water consumption, although inhalation exposure in occupational settings is a primary route for nickel–induced toxicity. In large doses (>0.5 g), some forms of nickel may be acutely toxic to humans when taken orally (Daldrup et al. 1983; Sunderman et al. 1988). Oral $LD_{50}$ values for rats range from 67 mg nickel/kg (nickel sulphate hexahydrate) to>9,000 mg nickel/kg (nickel powder) (ATSDR 1988). Toxic effects of oral exposure to nickel usually involve the kidneys with some evidence from animal studies showing a possible developmental/reproductive toxicity effect (ATSDR 1988; Goyer 1991).

Nickel toxicity reduces photosynthesis, growth, and nitrogenase activity of algae; fermentative activity of a mixed rumen microbiota; growth rate of marine bacteria;  tion of fungi. Adverse effects of excess nickel have also been observed with yeasts, higher plants, protozoans, mollusks, crustaceans, insects, annelids, echinoderms, fishes, amphibians, birds, and mammals. In mammals, the toxicity of nickel is a function of the chemical form of nickel, dose, and route of exposure. Exposure to nickel by inhalation, injection, or cutaneous contact is more significant than oral exposure. Toxic effects of nickel to humans and laboratory mammals are documented for respiratory, cardiovascular, gastrointestinal, hematological, musculoskeletal, hepatic, renal, dermal, ocular, immunological, developmental, neurological, and reproductive systems.

Signs of nickel poisoning in fishes include surfacing, rapid mouth and opercular movements and, prior to death, convulsions and loss of equilibrium. Destruction of the gill lamellae by ionic nickel decreases the ventilation rate and may cause blood hypoxia and death. Other signs of nickel poisoning in fishes include decreased

**Table 10.6** Biosorbents used in $Cu^{2+}$ removal from wastewaters and other aqueous solutions

| Biosorbent(s) | Biosorption performance | Experimental techniques | Reference |
|---|---|---|---|
| Lichen biomass of *Cladonia rangiformis* hoffm. | Highest metal uptake Langmuir isotherm: 7.6923 mg $Cu^{2+}$/g inactivated lichen (15°C) | Batch adsorption experiments | Ekmekyapar et al. (2006) |
| Powdered *Chlorella vulgaris* algal cells | Experimental results showed that *C. vulgaris* algal cells were effective in removing copper ions from aqueous solutions. Biosorption of copper on algal cells was rapid where equilibrium was attained within 15 min | Batch adsorption experiments | Abu Al–Rub et al. (2006) |
| *Sphaerotilus natans* immobilised in polysulfone matrices | Highest value for loading was 5.4 mg/g estimated at pH 5.5 and 0.18 g of lyophilised biomass per gram of beads. | Batch adsorption experiments | Beolchini et al. (2003) |
| *Pseudomonas cepacia* | Results suggested that $Cu^{2+}$ uptake by *P. cepacia* cells involves surface binding and not intracellular accumulation by active transport. | Copper uptake measured by analysis of cell digests | Savvaidis et al. (2003) |
| Marine alga Sargassum sp. | Model in which the mass transfer is controlled by diffusion in the biosorbent best represented the fixed–bed column dynamics | Batch adsorption and fixed–bed column experiments | Da Silva et al. (2002) |
| Waste beer yeast (byproduct of brewing industry) | Maximum biosorption capacity of $Cu^{2+}$onto beer yeast was 0.0228 mmol/g (20°C) | Batch adsorption experiments | Han et al. (2006) |
| Spent–grain | Maximum adsorption capacity of spent–grain was 10.47 mg/g dry weight (pH 4.2). Spent–grain is an abundant and by–product from the whisky distilling industry and it was suggested that it could be economically and effectively applied as a biosorbent for the removal of $Cu^{2+}$ ions from distilling wastewaters | Column experiments | Lu and Gibb (2008) |
| Dried leaves from the Brazilian flora | Dried leaves are a good and commercially interesting alternative for sorption of metal ions compared to other biomasses | Electron Paramagnetic Resonance | de Carvalho et al. (2001) |
| Powdered waste sludge (PWS) | Maximum biosorption capacity was 156 mg $Cu^{2+}$/g PWS when PWS concentration was 0.25 g/L with an average particle size of 64 μm | Batch biosorption experiments | Pamukoglu and Kargi (2006) |

| | | | |
|---|---|---|---|
| Olive mill residues (OMR) | Regeneration tests permitted to demonstrate that a concentration factor of about 2 can be obtained in no–optimized conditions, highlighting the possibility of using OMR for the treatment of metal bearing effluents | Characterization of waste material performed by microanalysis and Scanning electron microscopy; batch mode experiments | Vegliò et al. (2003) |
| Marine algae *Gelidium sesquipedale*, an algal waste (from agar extraction industry) and a composite material (the algal waste immobilized in polyacrylonitrile) | Desorption kinetics was very fast for all biosorbents. The homogeneous diffusion coefficient varied between $1.0\times10^{-7}$ $cm^2/s$ for algae *Gelidium* and $3.0\times10^{-7}$ $cm^2/s$ for the composite material | Batch mode experiments | Vilar et al. (2007) |
| Grape stalks | Biosorption system was able to remove 99% of the copper from the Fl effluent (0.08 mg/L of $Cu^{2+}$ in the final effluent), using a biomass concentration of 2 g/L. For the F2 effluent, a biomass concentration of 4 g/L was required to obtain a final copper concentration 0.18 mg/L. $Cu^{2+}$ was also removed from the F3 effluent with an efficiency of 98% (final metal concentration of 0.15 mg/L) | Multistage process: biosorption was carried out with a free biomass suspension in a two–stage, countercurrent, stirred batch system. The first two effluents labeled in this work as Fl and F2 had 10 and 50 mg/L of copper, respectively. The third effluent had a complex metal mixture containing 10 mg/L Cu, 50 mg/L Zn, 5 mg/L Ni, 100 mg/L Ca, 100 mg/L Na | Machado et al. (2003) |
| *Staphylococcus* sp. bacteria | Copper could be recovered from the CMP wastewater at relatively high efficiencies | Copper Chemical Mechanical Planarization (Cu–CMP) wastewater used in Continuous–flow packed column experiments | Stanley and Ogden (2003) |

(continued)

**Table 10.6** (continued)

| Biosorbent(s) | Biosorption performance | Experimental techniques | Reference |
|---|---|---|---|
| Immobilized cells of *Pycnoporus sanguineus* (white–rot fungus) | Optimum uptake of $Cu^{2+}$ions was observed at pH 5 with a value of 2.76 mg/g. Biosorption equilibrium data were best described by Langmuir isotherm model followed by Redlich–Peterson and Freundlich models, respectively. Fourier Transform Infrared Spectroscopy analysis showed that —OH, —NH, C—H, C=O, —COOH and C—N groups were involved in the biosorption of $Cu^{2+}$ ions onto immobilized cells of *P. sanguineus* | Batch mode experiments; Fourier Transform Infrared Spectroscopy analysis | Yahaya et al. (2009) |
| Marine alga *Gracilaria Corticata* | Maximum uptake for copper with initial concentrations 50, 100 and 150 mg/L at pH 4 was obtained 74.00%, 75.10% and 79.81% respectively. The study demonstrated that the pre–treated biomass of *Gracilaria corticota* could be used as an effective biosorbent for the treatment of $Cu^{2+}$–containing wastewater streams | Batch adsorption experiments | Esmaeili et al. (2008) |
| Walnut (WNS) *(Juglans regia)*, hazelnut (HNS) *(Corylus avellana)*, and almond (AS) *(Prunus dulcis)* shell | The sorption process conformed to the Langmuir isotherm with maximum $Cu^{2+}$ ion sorption capacities of 6.74, 6.65, and 3.62 mg/g for WNS, HNS, and AS, respectively. The experimental results demonstrated that chelation and ion exchange is one of the major adsorption mechanisms for binding metal ions to the sorbents. The percentage removal $Cu^{2+}$ ion was maximum at $10^{-3}$ mol/L solution concentration and initial pH of 6.0 (80.3%, 75.6%, and 75.0% by WNS, HNS, and AS, respectively) | Batch adsorption experiments | Altun and Pehlivan (2007) |

| | | | |
|---|---|---|---|
| *Mucor rouxii* biomass (MRB) | Maximum adsorption was noted within pH range 5.0–6.0, and the process follows Langmuir adsorption isotherm ($r^2$=0.998). Amino, carboxyl and phosphate groups present on the cell surface of the biomass are involved in chemical interaction with $Cu^{2+}$ ion as revealed from Fourier Transform Infrared Spectroscopy and Scanning Electron Microscopy–Energy Dispersive X–ray. Scanning Electron Microscopy and Atomic Force Microscopy micrographs revealed the formation of metal nanostructure on the biomass surface due to copper adsorption | Scanning Electron Microscopy–Energy Dispersive X–ray, Atomin Force Microscopy and Fourier Transform Infrared Spectroscopy analyses | Majumdar et al. (2008) |
| Brown alga *Fucus vesiculosus* | Langmuir maximum metal uptake was 1.66 mmol/g. Biosorption was accomplished by ion exchange between metals in solution and algal protons, calcium and other light metals, and by complexation of the adsorbed metals with algal carboxyl groups. Fourier Transform Infrared Spectroscopy spectra showed a shift in the bands of carboxyl, hydroxyl and sulfonate groups. | Fourier Transform Infrared Spectroscopy analysis | Mata et al. (2008) |

concentrations of glycogen in muscle and liver with simultaneous increases in levels of lactic acid and glucose in blood, depressed hydrogen peroxide production in tissues and a reduction in superoxide dismutase, and contractions of vascular smooth muscle–signs similar to those associated with hypertension in mammals. Ionic nickel is lethal to sensitive species of aquatic organisms at 11–113 μg/L. Deaths occur among embryos of rainbow trout at 11–90 μg/L, daphnids at 13 μg/L, embryos of channel catfish at more than 38 μg/L, embryos of the narrowmouthed toad at 50 μg/L, and embryos of largemouth bass at 113 μg/L. Species intermediately resistant to nickel died at 150–410 μg/L, including mysid shrimp at 150 μg/L, freshwater snails at 237 μg/L, clam embryos at 310 μg/L, and embryos of salamanders at 410 μg/L. Aquatic bacteria and yeasts are comparatively tolerant to nickel. Sensitive species of freshwater eubacteria and actinomycetes show reduced growth at 5 mg Ni/L; for marine eubacteria, growth inhibition begins at 10–20 mg/L. Sensitive species of yeasts show growth inhibition at 1.0 mg Ni/L; resistant species of yeasts show a reduction in growth at 5–20 mg Ni/L.

Nickel(II) ($Ni^{2+}$) ion is one such heavy metal frequently encountered in raw wastewater streams from industries such as non–ferrous metal, mineral processing, paint formulation, electroplating, porcelain enameling, copper sulphate manufacture and steam–electric power plants (Padmavathy 2008). While the $Ni^{2+}$ ion concentration in plating rinse can approach 2–900 mg/L (Padmavathy 2008), wastewaters from paint and ink formulation, porcelain enameling and copper sulphate manufacture industries record effluent $Ni^{2+}$ ion concentrations varying over 0–40, 0.25–67 and around 20 mg/L, respectively.

A number of conventional techniques have been used to remove $Ni^{2+}$ ion from wastewater streams, such as adsorption on activated carbon, chemical precipitation, and crystallization in the form of nickel carbonate. However, these methods are either ineffective or expensive when initial metal concentrations in wastewaters are low, or when very low concentrations of heavy metals in treated water is required. In the past few decades, biosorption using microbial biomass as the adsorbent, has emerged as a potential alternative technique to the existing methods for metal removal. The biosorptive capacity for nickel has frequently been shown to be low in contrast to other transition metals, e.g., $Cu^{2+}$, $Pb^{2+}$ or $Ag^{+}$, in accordance with the Irving–Williams series, an empirically determined series of stability constants of organo–metal complexes. The biosorptive capacities of several microorganisms and chemical properties of nickel in that context have been summarised and discussed by Tsezos et al. (1995) but these data are for studies conducted before 1995. Table 10.7 summarizes the biosorption capacities of some biosorbents for $Ni^{2+}$ studied during the last 10 years. Without obvious differences between the major groups of microorganisms, the biosorptive capacities of algae, bacteria and fungi range between 5 and 50 mg Ni/g dry matter, corresponding to high equilibrium concentrations of 100 mg/L and above, and neutral to slightly acidic pH values (Pümpel et al. 2003). In the concentration range of practical interest for bioremediation studies (typically only a few mg/L) nickel biosorption only reaches 0.5–10 mg/g (Asthana et al. 1995; Ivanitsa et al. 1999; Tsezos et al.1995). Nickel can be bioprecipitated onto cell surfaces with a number of biogenic precipitant ligands, as described in the literature for other

**Table 10.7** Biosorption capacity of biosorbents used in $Ni^{2+}$ removal

| Biosorbent(s) | Biosorption capacity | Adsorption model | Reference |
|---|---|---|---|
| Fungal biosorbents, PFB1 and PFB2 | 214 mg/g (PFB1) 110 mg/g (PFB2) | Langmuir | Suhasini et al. (1999) |
| Loofa sponge–immobilized biomass of *Chlorella sorokiniana* | 60.38 mg/g | Langmuir and Freundlich | Akhtar et al. (2004) |
| Protonated rice bran | 102 mg/g | Langmuir | Zafar et al. (2007) |
| *C. Vulgaris* | 1,282.05 μg/g | Langmuir | Wong et al. (2000) |
| *Sargassum wightii* seaweed | 18.58 mg/g | Langmuir | Vijayaraghavan et al. (2005a) |
| *Pseudomonasfluorescens* 4F39 free cells or immobilized cells in beads of agar (biobeads) | 145, 37 and 7.6 mg/g dry sorbent for free cells, immobilized cells and biobeads, respectively | Langmuir | López et al. (2002) |
| Cone biomass of *Thuja orientalis* | 12.42 mg/g | Langmuir | Malkoc (2006) |
| Tea factory waste | 15.26 mg/g ( pH 4.0 at 25°C) | Freundlich and Langmuir | Malkoc and Nuhoglu (2005) |
| *Cassia fistula* (Golden Shower) biomass | 100% $Ni^{2+}$ removal for initial $Ni^{2+}$ concentration of 25 mg/L | Langmuir | Hanif et al. (2007) |
| *Chlorella vulgaris* | Increasing from 48.1 to 60.2 mg/g for 15–45°C | Freundlich, Langmuir and Redlich–Peterson | Aksu (2002) |
| Marine green alga *Ulva reticulata* | 46.5 mg/g | Thomas for column mode adsorption experiments | Vijayaraghavan et al. (2005b) |
| *Streptomyces coelicolor* A3(2) | 7.3% $Ni^{2+}$ uptake at initial concentration of 250 mg $Ni^{2+}$/L | Freundlich and Langmuir | Öztürk et al. (2004) |
| Brown algae *Sargassum* sp. | Natural algae: 181 mg/g Acid treated forms of algae: 250 mg/g | Freundlich and Langmuir | Kalyani et al. (2004) |
| Non–adapted and adapted growing cells of a non–pathogenic *Candida sp.* | 6.8 and 30.8 mg/g at 321.5 and 300.6 mg/L initial $Ni^{2+}$ concentrations, respectively | Langmuir | Dönmez and Aksu (2001) |
| *Phanerochaete chrysosporium* | 55.9 mg/g | Langmuir | Ceribasi and Yetis (2001) |
| Biomass and silica–immobilized biomass of *Medicago sativa* (alfalfa) | Batch mode: 4.1 mg/g alfalfa biomass | (Langmuir) | Gardea–Torresdey et al. (1996a) |
| Cyanophyceae *Lyngbya taylorii* (alga) | Natural alga: 38.15 mg/g dry biomass. After phosphorylation of the alga: 163.75 mg/g dry biomass | Langmuir | Klimmek et al. (2001) |

metals, for example $Ni^{2+}$ could be removed as its phosphate precipitate since metal phosphates are highly insoluble (Pümpel et al. 2003). Accordingly, heavy metals were removed as their phosphates via the activity of cell–bound phosphatase (PhoN) of *Citrobacter* sp. N14 (Pümpel et al. 2003).

Literature abounds with over 70,000 studies conducted to remove $Ni^{2+}$ by adsorption and over 3,500 research papers reporting the application for biosorption for nickel remediation. Whilst it is not feasible to discuss all these studies in this review paper, a few of the most recent advances in $Ni^{2+}$ removal by biosorption are immediately appraised. Özer et al. (2008) have recently investigated the biosorption of $Ni^{2+}$ ions on *Enteromorpha prolifera*, a green algae in a batch system. The single and combined effects of operating parameters such as initial pH, temperature, initial metal ion concentration and biosorbent concentration on the biosorption of $Ni^{2+}$ ions on *E. prolifera* were analyzed using response surface methodology (RSM). The optimum biosorption conditions were determined as initial pH 4.3, temperature 27°C, biosorbent concentration 1.2 g/L and initial $Ni^{2+}$ ion concentration 100 mg/L. At optimum biosorption conditions, the biosorption capacity of *E. prolifera* for $Ni^{2+}$ ions was found to be 36.8 mg/g after 120 min biosorption. The monolayer coverage capacity of *E. prolifera* for nickel(II) ions was found as 65.7 mg/g. In order to examine the rate limiting step of $Ni^{2+}$ biosorption, such as the mass transfer and chemical reaction kinetics, the intraparticle diffusion model, external diffusion model and the pseudo second order kinetic model were tested with the experimental data. It was found that for both contributed to the actual biosorption process and that the pseudo second order kinetic model described the $Ni^{2+}$ biosorption process with a good fitting.

Later, Akar et al. (2009) have studied that batch and dynamic flow biosorption using the waste biomass entrapped in silica–gel matrix for the removal of $Ni^{2+}$ ions from synthetic solutions and real wastewater. The batch biosorption conditions were examined with respect to initial pH, contact time, and initial $Ni^{2+}$ ion concentration. Zeta potential measurements showed that immobilized biosorbent was negatively charged in the pH range of 3.0–8.0. The immobilized biomass was also found to possess relatively high biosorption capacity (98.01 mg/g), and biosorption equilibrium was established in a short time of operation (5 min). The applicability of sorbent system was also investigated in a continuous mode, and column studies were performed under different flow rate, column size, and biosorbent dosage. Akar et al. (2009) reported the successful removal of $Ni^{2+}$ ions from industrial wastewater in dynamic flow treatment mode. Their results showed that silica–immobilized waste biomass was a low–cost promising sorbent for sequester of $Ni^{2+}$ ions from synthetic and real wastewater. Amini et al. (2009) also investigated the effects of biosorbent *Aspergillus niger* dosage, initial solution pH and initial $Ni^{2+}$ concentration on the uptake of $Ni^{2+}$ by NaOH pretreated biomass of *A. niger* from aqueous solution in batch experiments. Amini et al. (2009) deduced the optimum $Ni^{2+}$ uptake to be 4.82 mg/g biomass at pH 6.25.

Popuri Srinivasa et al. (2009) have lately developed a new biosorbent by coating chitosan, a naturally and abundantly available biopolymer, on to polyvinyl chloride (PVC) beads. The biosorbent was characterized by Fourier Transform Infrared spectra, porosity and surface area analyses. Equilibrium and column flow adsorption characteristics of $Ni^{2+}$ ions on the biosorbent were equally studied. The effect of pH, agitation time, concentration of adsorbate and amount of adsorbent on the extent of adsorption was also investigated. The maximum monolayer adsorption capacity of chitosan coated PVC sorbent as obtained from Langmuir adsorption isotherm was found to be 120.5 mg/g for the $Ni^{2+}$ ions. In addition, breakthrough curves were obtained by Popuri Srinivasa et al. (2009) from column flow experiments, and their results demonstrated that chitosan coated PVC beads could be used efficiently for the removal of $Ni^{2+}$ ions from aqueous medium through adsorption. Lastly, Singh et al. (2009) have investigated the application of an aquatic weed, *P. hysterophorous*, in ash form for removal of $Ni^{2+}$ from aqueous solutions. It was observed that the removal of Ni increased from 67.30% to 97.54%, with the $Ni^{2+}$ concentration decreasing from 477.21 to 67.83 mg/L at 25 °C, pH 11.0. The additional adsorption isotherm and thermodynamic analyses revealed that the adsorption capacity for $Ni^{2+}$ by *P. hysterophorous* was found to be much better than other common adsorbents reported for the removal of $Ni^{2+}$. Özer et al. (2008), Akar et al. (2009), Amini et al. (2009), Popuri Srinivasa et al. (2009) and Singh et al. (2009) unanimously found that the Langmuir and Freundlich isotherm models defined the equilibrium data very well.

### *10.4.5 Lead*

Lead is one of the most dangerous contaminants which is released to the environment by various anthropogenic activities (Carrasquero Durán et al. 2006). In the air, lead is in the form of particles and is removed by rain or gravitational settling. The solubility of lead compounds in water is a function of pH, hardness, salinity, and the presence of humic material. Solubility is highest in soft, acidic water. The sink for lead is the soil and sediment. Because it is strongly adsorbed to soil, it generally is retained in the upper layers of soil and does not leach appreciably into the subsoil and groundwater. Lead compounds may be transformed in the environment to other lead compounds; however, lead is an element and cannot be destroyed. Anthropogenic sources of lead include the mining and smelting of ore, manufacture of lead–containing products, combustion of coal and oil, and waste incineration. Many anthropogenic sources of lead, most notably leaded gasoline, lead–based paint, lead solder in food cans, lead–arsenate pesticides, and shot and sinkers, have been eliminated or strictly regulated due to lead's persistence and toxicity. Because lead does not degrade, these former uses leave their legacy as higher concentrations of lead in the environment.

Based on a survey of 900 public water supply systems, EPA in 1988 estimated that 99% of the 219 million people in the United States using public water supplies were exposed to drinking water with levels of lead <5 μg/L and approximately two million people are served by drinking water with levels of lead >5 μg/L. A survey of 580 cities in 47 states indicated that the national mean concentration of lead in drinking water was 29 μg/Lafter a 30 s flushing period. However, it was estimated that in 1988 the average lead content of drinking water decreased to 17 μg/L. In 1991, EPA examined the occurrences of lead in source water and distributed water.

By resampling at the entry point to the distribution system, few samples were found to contain lead at levels above 5 μg/L.

EPA now estimates that approximately 600 groundwater systems may have water, leaving the treatment plant, with lead levels above 5 μg/L. Based on several data sets, it is estimated that <1% of the public water systems in the United States have water entering the distribution system with lead levels above 5 μg/L. These systems are estimated to serve <3% of the population that receives drinking water from public systems.

Lead poisoning is a medical condition caused by increased levels of the metallic lead in the blood. Lead may cause irreversible neurological damage as well as renal disease, cardiovascular effects, and reproductive toxicity. Humans have been mining and using this heavy metal for thousands of years, poisoning themselves in the process due to accumulation, exposure and direct contact. These dangers have long been known, though the modern understanding of their full extent and the small amount of lead necessary to produce them is relatively recent; blood lead levels once considered safe are now considered hazardous, with no known threshold. The current EPA and WHO drinking water standard for $Pb^{2+}$ is 0.05 mg/L and 10 μg/L, respectively. $Pb^{2+}$ accumulates mainly in bones, brain, kidney and muscles and may cause many serious disorders like anemia, kidney diseases, nervous disorders and sickness even death (Lo et al. 1999).

Recently, the removal of $Pb^{2+}$ ions from drinking water has been focused due to its high toxicity to humans in terms of nausea, convulsions, coma, renal failure, cancer and subtle effects on metabolism and intelligence. Different approaches to remove $Pb^{2+}$ ions from wastewater, including chemicals precipitation, ion exchange, reverse osmosis (Modher et al. 2009), electrokinetic remediation, phytoremediation and adsorption have been used to reduce the concentrations to safe levels before discharge into fragile environments. Research has shown that adsorption is the simplest and best cost–effective method. Various adsorbents/biosorbents such as activated carbon, iron oxides, filamentous fungal biomass and natural condensed tannin have been explored for the removal of $Pb^{2+}$ ions and the results have been promising so far. Efforts to find low cost and biodegradable sorbents for $Pb^{2+}$ have hence constituted in making progress this field of lead remediation research.

The immediate sections revisit some selected recent advances made in $Pb^{2+}$ ion removal by biosorption and Table 10.8 additionally makes a résumé of similar studies with respect to novel biosorbents tested, the types of water and wastewater studied, the befitting adsorption model(s) and biosorption capacity for $Pb^{2+}$.

Modher et al. (2009) have used nonliving *Spirogyra neglecta* biomass, to study the effects of initial concentration and contact time, pH and temperature on the biosorption of $Pb^{2+}$. The surface characteristics of the nonliving algal biomass were examined using scanning electron microscope and Fourier Transformed Infrared spectroscopy. Modher et al. (2009) found that the non–living algal biomass exhibited a caves–like, uneven surface texture along with lot of irregular surface. The Fourier Transform Infrared Spectroscopy analysis of the alga biomass revealed the presence of carboyl, amine and carboxyl group which were responsible for adsorp-

**Table 10.8** New biosorbents for $Pb^{2+}$ removal from synthetic and natural wastewaters

| Biosorbent(s) | Media tested | Biosorption capacity | Adsorption model(s) | Reference |
|---|---|---|---|---|
| New chemical–modified lawny grass adsorbents | Synthetic aqueous solutions | 261.1–321.2 mg/g | Combined Langmuir–Freundlich | Lu et al. (2009) |
| *Candida albicans* | Synthetic aqueous solutions | 828.5 mg/g at 25°C, 831.26 mg/g at 35°C, 833.33 mg/g at 45°C | Langmuir and Freundlich | Baysal et al. (2009) |
| Pestan (fungal extracellular polysaccharide) | Wastewater | 120 mg/g | Freundlich | Moon et al. (2006) |
| *Rhodotorula glutinis* yeast | Synthetic aqueous solutions | 73.5 mg/g | Langmuir | Cho and Kim (2003) |
| Native and acidically modified *Cicer arientinum* pod biomass | Synthetic aqueous solutions | 171.28 mg/g for acid treated biomass | Dubinin–Radushkevich | Nadeem et al. (2009) |
| Powder of mature leaves of the Neem (*Azadirachta indica*) tree | Synthetic aqueous solutions | 300 mg/g | Langmuir and Freundlich | Bhattacharyya and Sharma (2004) |
| Chitosan produced from silkworm chrysalides (ChSC) | Battery manufacture wastewater | 88.3 mg/g | Langmuir and Freundlich | Paulino et al. (2008) |
| Green algae *Cladophora fascicularis* | Synthetic aqueous solutions | 198.5 mg/g at 25°C | Langmuir and Freundlich | Deng et al. (2007) |
| Industrial strain of *Saccharomyces cerevisiae* immobilized on cone biomass of *Pinus nigra* | Real industrial wastewater | 30.04 mg/g | Freundlich, Langmuir and Dubinin–Radushkevich | Çabuk et al. (2007) |
| Formaldehyde polymerized banana stem | Battery manufacturing wastewater | 125 mg/g | Freundlich | Noeline et al. (2005) |

tion of $Pb^{2+}$. The maximum adsorption capacity of $Pb^{2+}$ by the nonliving biomass of *Spirogyra neglecta* was 132 mg $g^{-1}$. This study was novel and it concluded that the maximum adsorption capacity for $Pb^{2+}$ by the nonliving biomass of *Spirogyra neglecta* was higher than reported for other biosorbents. Therefore, it has been advocated that this new biosorbent has a great potential for removing $Pb^{2+}$ from polluted water. Still searching for novel biosorbents, Lugo–Lugo et al. (2009) have evaluated natural, formaldehyde–treated and copolymer–grafted orange peels as adsorbents to remove $Pb^{2+}$ ions from aqueous solutions. Lugo–Lugo et al. (2009) reported (an optimum pH 5) that the adsorption process was fast, reaching 99% of sorbent capacity in 10 min for the natural and treated biomasses and 20 min for the grafted material. The treated biomass showed the highest sorption rate and capacity in the batch experiments, with the results fitting well to a pseudo–first order rate equation. In the continuous test with the treated biomass, the capacity at complete exhaustion was 46.61 mg/g for an initial concentration of 150 mg/L. Additional analyses by Scanning Electron Microscopy–Energy Dispersive X–ray indicated that the materials had a rough surface, and that the adsorption of the metal took place on the surface. The Fourier Transform Infrared Spectroscopy spectrums revealed that the functional groups responsible for metallic biosorption were the –OH, –COOH and $-NH_2$ groups on the surface. Rahmani et al. (2009) concluded that waste activated sludge (WAS) can be a useful product in removal of lead and perhaps other similar heavy metals from wastewater and industrial wastes. In addition, Rahmani et al. (2009) propose that it is possible to improve the biosorption treatment they studied by performing additional aeration and feeding. The study of Rahmani et al. (2009) consisted in observing during treatment of synthetic solutions of $Pb^{2+}$ by a sample WAS (mixed liquor suspended solids at 10,000 mg/L). Their results showed that the efficiency of $Pb^{2+}$ removal was reduced by increasing the initial metal concentration, but the changes of contact time had resulted in different responses depending to the test condition. Moreover, it was found that the effects of aeration and feeding in increasing the efficiency of treatment were more pronounced for the stage of treating lower concentrations of lead. The maximum removal of $Pb^{2+}$ in the first step (without aeration and feeding) was 55.2% in contact time of 75 min, in the second step (without feeding and by 12 h aeration) 63.3% and third step it was 94.2% in contact time of 48 h.

A still novel and interesting advancement was the use of Konjac glucomannan beads by Long et al. (2009) as metal biosorbent for $Pb^{2+}$ from aqueous solutions. *Konjac glucomannan* is a water–soluble dietary fiber derived from the root of the Konjac plant. The Konjac, commonly known as the corm–like tuber of Amorphophallus Konjac C. Korch, is a perennial plants native of warm tropical to tropical regions, eastern Asia, from Japan and south of China to Indonesia. Konjac is a nutrient–rich food plant food with low calorie and no fat content. In China, Konjac is called *Ju Ruo*, which has been used as a food for over 2,000 years, as described in the "Ode to the Capital of Shu Kingdom", written by *Zuo Si* of *Xijin* Dynasty. The main characteristics of this fiber are as follows: It is an extremely viscous soluble fiber in nature (Chandalia et al. 2000) and can form extremely viscous solutions (McCarty 2002; Tabatabai and Li 2000); it has a very high molecular weight (200,000–2,000,000 g/mol), it has a very high water holding capacity of up to 100 times its own water weight ands it can form either a reversible or a thermo–non–reversible gel. The results of Long et al. (2009) convincingly showed that Konjac glucomannan beads may be potentially very useful for the treatment of wastewater contaminated with lead.

### *10.4.6 Zinc*

Zn is an essential and beneficial element for human bodies and plants. Complete exclusion of Zn is not possible due to its dual role, an essential microelement on the one hand and a toxic environmental factor on the other (Brune et al. 1994). However, Zn can cause nonfatal fume fever, pneumonitis, and is a potential hazard as an environmental pollutant (Lu et al. 2004). Zinc is an essential nutrient in humans and animals that is necessary for the function of a large number of metalloenzymes. These enzymes include alcohol dehydrogenase, alkaline phosphatase, carbonic anhydrase, leucine aminopeptidase, superoxide dismutase, and deoxyribonucleic acid (DNA) and ribonucleic acid (RNA) polymerase. As such, zinc is required for normal nucleic acid, protein, and membrane metabolism, as well as cell growth and division. Zinc also plays an essential role in the maintenance of nucleic acid structure of genes (zinc finger phenomenon). Zinc deficiency has been associated with dermatitis, anorexia, growth retardation, poor wound healing, hypogonadism with impaired reproductive capacity, impaired immune function.

Even though zinc is a very essential requirement for a healthy body, excess zinc can be harmful, and cause zinc toxicity (Fosmire 1990). Excessive absorption of zinc can also suppress copper and iron absorption. The free zinc ion in solution is highly toxic to plants, invertebrates, and even vertebrate fish. The Free Ion Activity Model (FIAM) (De Schamphelaere et al. 2005; Hough et al. 2005; Markich et al. 2003; Parker and Pedler 1997; Parker et al. 2001) is well–established in the literature, and shows that just micromolar amounts of the free ion kills some organisms.

Zinc is mainly used in the production of noncorrosive alloys and brass and in galvanizing steel and iron products. Zinc undergoes oxidation on the surface, thus protecting the underlying metal from degradation. Galvanized products are widely used in construction materials, automobile parts, and household appliances. Zinc oxide is used to form white pigments in rubber processing and to coat photocopy paper. Zinc is used therapeutically in human medicine in the treatment of zinc deficiency, various skin diseases, wound healing, and to reduce pain in sickle cell anemia patients.

Zinc is discharged into the global environment at an estimated yearly rate of 8.8 million metric tons; 96% of the total is a result of human activities (Leonard and Gerber 1989). Major sources of anthropogenic zinc discharges to the environment include electroplaters, smelting and ore processors, drainage from active and inactive mining operations, domestic and industrial sewage, combustion of fossil fuels and solid wastes, road surface runoff, corrosion of zinc alloys and galvanized surfaces, and erosion of agricultural soils (Buhl and Hamilton 1990). During smelting, large amounts of zinc are emitted into the atmosphere. The removal of $Zn^{2+}$ from effluents is a major problem due to the difficulty in treating such wastewaters by conventional treatment methods (Fernando et al. 2009). Hence, substantial research has been in progress to find suitable, high efficiency and low cost natural sorbents for $Zn^{2+}$ removal by adsorption. Several of the studies reported in literature for the removal of $Zn^{2+}$ by biosorption bear testimony of the feasibility of this heavy metals removal technique. Whilst it is not possible to discuss all of these research works which exceed 5,000 in number, a few of the latest ones are summarized in Table 10.9.

## 10.5 Conclusion

The removal of heavy metals ions from our environment especially wastewater streams is now steadily shifting from the use of conventional adsorbents to the use of biosorbents for reasons of degradability and environmental sustainability. This shift is also perfectly contextualized within the concept of *Green Chemistry*. The presence of heavy metals in the environment has always been of major concern because of their toxicity, bioaccumulating tendency, and threat to human life, aquatic life and the environment, as a whole. In recent years, many low cost sorbents such as algae, fungi bacteria and lignocellulosic agricultural by–products have been pre-treated and investigated for their biosorption capacity towards heavy metals. So far, the results have been promising and the scope for the use of biosorbents for bioremediating wastewaters is widening. Composts are also being revealed as excellent biosorbents in comparison to many other adsorbents. We have reviewed the sources and toxicology of six selected heavy metals as well as the reason why they need to be removed from our environment.

Our review has also shown that the conventional methods of removal are expensive, hence the use of low cost, abundant environmentally friendly biosorbents is justified. Although biosorption is promising, its precise mechanism is not well elucidated and well established. There is to date no overall and general adsorptive mechanism which could be adapted and simplified for an overwhelming majority of biosorption reactions and subsequently used to explain the specific heavy metal(s) sorption for a particular sorbent. This is actually a gap in knowledge which researchers are earnestly up to. This knowledge is essential for understanding the process and it will serve as a basis for quantitative stoichiometric considerations, which are fundamental for mathematical modeling and scale–up of biosorption processes which find their true place in actual wastewater treatment systems.

**Acknowledgements** We wish to express our gratitude and appreciation to all the researchers whose valuable data, research findings and discussions as reported in their respective publications and reports, have been of considerable significance in adding substance to this review on survey of recent developments. We are grateful to our other colleagues and the anonymous reviewers whose criticisms have benefited the manuscript. Our sincere thanks to Brian Phelps (http://www.phelpstek.com/) for having been gracious enough to provide us with the structure of humic acid.

**Table 10.9** Digest of biosorption studies for $Zn^{2+}$ removal by novel biosorbents

| Biosorbent(s) | Media tested | Biosorption capacity and performance | Adsorption model(s) | Kinetic and/or thermodynamic behaviour | Reference |
|---|---|---|---|---|---|
| *Azadirachta indica* bark | Synthetic aqueous solutions | 33.49 mg/g | Langmuir, Freundlich and Redlich–Peterson | Pseudo–second–order model | King et al. (2008) |
| Orange peel cellulose with *Phanerochaete chrysosporium* immobilized Ca–alginate beads | Synthetic aqueous solutions | 168.61 mg/g | Langmuir and Freundlich | Box–Behnken model | Lai et al. (2008) |
| Poly(2–octadecyl–butanedioic acid), prepared from polyanhydride PA–18 | Synthetic aqueous solutions | Adsorption capacity of this water insoluble polymer for lead (II) was substantially higher than other heterogeneous adsorbants and is equivalent to those obtained with homogeneous sorbents. | Dubinin–Radushkevich isotherm | Gibb's free energy values for this process indicate that it is spontaneous | Laurino (2008) |
| Maize (*Zea mays*) leaf | Synthetic aqueous solutions | Maize leaf has potential for the removal of zinc ions from industrial wastewater | Freundlich and Langmuir | Free energy change −5.4525 kJ/mol. Kinetic studies show that the uptake of zinc ions increases with time | Adesola Babarinde et al. (2008) |
| *Gossypium hirsutum* (Cotton) waste biomass | Synthetic aqueous solutions and hazardous aqueous streams | At optimized conditions Pb(II) uptake was more rapid in case of industrial effluents in comparison to synthetic solutions | Langmuir | Pseudo–second–order model | Riaz et al. (2009) |

(continued)

**Table 10.9** (continued)

| Biosorbent(s) | Media tested | Biosorption capacity and performance | Adsorption model(s) | Kinetic and/or thermodynamic behaviour | Reference |
|---|---|---|---|---|---|
| Mature leaves and stem bark of the Neem (*Azadirachta indica*) tree | Synthetic aqueous solutions | 147.08 mg Zn/g for Neem leaves and 137.67 mg Zn/g Neem bark. Due to its outstanding zinc uptake capacity, the Neem tree is an excellent biomaterial for accumulating zinc from aqueous solution | Langmuir and Freundlich | Pseudo–second–order model | Arshad et al. (2008) |
| Clay/poly(methoxyethyl) acrylamide (PMEA) composite | Synthetic aqueous solutions | Maximum adsorption capacity 81.02 mg/g | Langmuir | Pseudo–second–order model. Thermodynamic parameters show that the adsorption was spontaneous and endothermic | öölener et al. (2008) |
| Diethylenetriamine–bacterial cellulose | Synthetic aqueous solutions | Study provides the relatively comprehensive data for the EABC application to the removal of metal ion in the wastewater | Langmuir | Pseudo–second–order model | Shen et al. (2009) |

## References

Abbas M, Nadeem R, Zafar MN, Arshad M (2008) Biosorption of chromium (III) and chromium (VI) by untreated and pretreated Cassia fistula biomass from aqueous solutions. Water Air Soil Pollut 191:139–148. doi:10.1007/s11270-007-9613-8

Abollino O, Aceto M, Malandrino M, Sarzanini C, Mentasti E (2003) Adsorption of heavy metals on Na–montmorillonite. Effect of pH and organic substances. Water Res 37:1619–1627. doi:10.1016/S0043-1354(02)00524-9

Aboulroos SA, Helal MID, Kamel MM (2006) Remediation of Pb and Cd polluted soils using in situ immobilization and phytoextraction techniques. Soil Sediment Contam 15:199–215. doi:10.1080/15320380500506362

Abu Al–Rub FA, El–Naas MH, Ashour I, Al–Marzouqi M (2006) Biosorption of copper on Chlorella vulgaris from single, binary and ternary metal aqueous solutions. Process Biochem 41:457–464. doi:10.1016/j.procbio.2005.07.018

Acar FN, Malkoc E (2004) The removal of chromium(VI) from aqueous solutions by Fagus orientalis L. Bioresour Technol 94:3–15. doi:10.1016/j.biortech.2003.10.032

Adesola Babarinde NA, Oyebamiji Babalola J, Adetunji AA (2008) Isotherm and thermodynamic studies of the biosorption of zinc(II) from solution by maize leaf. Pac J Sci Technol 9:196–202

Adriano DC (2001) Trace elements in terrestrial environments; biochemistry, bioavailability and risks of metals. Springer–Verlag, New York

Akar T, Kaynak Z, Ulusoy S, Yuvaci D, Ozsari G, Akar ST (2009) Enhanced biosorption of nickel(II) ions by silica–gel–immobilized waste biomass: biosorption characteristics in batch and dynamic flow mode. J Hazard Mater 163:1134–1141. doi:10.1016/j.jhazmat.2008.07.084

Akhtar N, Iqbal J, Iqbal M (2004) Removal and recovery of nickel(II) from aqueous solution by loofa sponge–immobilized biomass of Chlorella sorokiniana: characterization studies. J Hazard Mater 108:85–94. doi:10.1016/j.jhazmat.2004.01.002

Aksu Z (2002) Determination of the equilibrium, kinetic and thermodynamic parameters of the batch biosorption of nickel(II) ions onto Chlorella vulgaris. Process Biochem 38:89–99

Aksu Z (2005) Application of biosorption for the removal of organic pollutants: a review. Process Biochem 40:997–1026. doi:10.1016/j.procbio.2004.04.008

Al–Qunaibit MH, Mekhemer WK, Zaghloul AA (2005) The adsorption of Cu(II) ions on bentonite–a kinetic study. J Colloid Interface Sci 283:316–321. doi:10.1016/j.jcis.2004.09.022

Altun T, Pehlivan E (2007) Removal of copper(II) ions from aqueous solutions by walnut–, hazelnut– and almond–shells. CLEAN – Soil Air Water 35:601–606

Álvarez–Ayuso E, García–Sánchez A (2003) Removal of heavy metals from waste waters by natural and Na–exchanged bentonites. Clay Clay Miner 51:475–480. doi:10.1346/CCMN.2003.0510501

Alves WL, Passoni AA (1997) Compost and vermicompost of urban solid waste in Licania tomentosa (benth.) seedlings production to arborisation. Pesqui Agropecu Bras 32:1053–1058

Amini M, Younesi H, Bahramifar N (2009) Biosorption of nickel(II) from aqueous solution by Aspergillus niger: response surface methodology and isotherm study. Chemosphere 75:1483–1491. doi:10.1016/j.chemosphere.2009.02.025

Anastasi A, Varese GC, Marchisio VF (2005) Isolation and identification of fungal communities in compost and vermicompost. Mycologia 97:33–44. doi:10.3852/mycologia.97.1.33

Ansari AA (2008) Effect of vermicompost on the productivity of potato (Solanum tuberosum), spinach (Spinacia oleracea) and turnip (Brassica campestris). World J Agric Sci 4:333–336

Aoyama M, Kishino M, Jo T–S (2004) Biosorption of Cr(VI) on Japanese cedar bark. Sep Sci Technol 39:1149–1162. doi:10.1081/SS-120028576

Apak R, Tutem E, Hugul M, Hizal J (1998) Heavy metal cation retention by unconventional sorbents (red muds and fly ashes). Water Res 32:430–440. doi:10.1016/S0043-1354(97)00204-2

Arancon NQ, Edwards CA, Lee S, Byrne R (2006) Effects of humic acids from vermicomposts on plant growth. Eur J Soil Biol 42:65–69. doi:10.1016/j.ejsobi.2006.06.004

Aravindhan R, Madhan B, Raghava Rao J, Unni Nair B (2004) Recovery and reuse of chromium from tannery wastewaters using Turbinaria ornata seaweed. J Chem Technol Biotechnol 79:1251–1258. doi:10.1002/jctb.1119

Arias M, Barral MT, Mejuto JC (2002) Enhancement of copper and cadmium adsorption on kaolin by the presence of humic acids. Chemosphere 48:1081–1088. doi:10.1016/S0045-6535(02)00169-8

Arshad M, Zafar MN, Younis S, Nadeem R (2008) The use of neem biomass for the biosorption of zinc from aqueous solutions. J Hazard Mater 157:534–540

Asthana RK, Chatterjee S, Singh SP (1995) Investigations on nickel biosorption and its remobilization. Process Biochem 30:729–734. doi:10.1016/0032-9592(94)00065-4

Atiyeh R, Subler S, Edwards CA, Metzger J (1999) Growth of tomato plants in horticultural potting media amended with vermicompost. Pedobiologia 43:724–728

Atiyeh RM, Lee S, Edwards CA, Arancon NQ, Metzger JD (2002) The influence of humic acids derived from earthworm–processed organic wastes on plant growth. Bioresour Technol 84:7–14. doi:10.1016/S0960-8524(02)00017-2

ATSDR (Agency for Toxic Substances and Disease Registry) (1988) Toxicological profile for nickel, ATSDR/U.S. Public Health Service, ATSDR/TP–88/19

Azab MS, Peterson PJ (1989) The removal of Cd from wastewater by the use of biological sorbent. Water Sci Technol 21:1705–1706

Bachman GR, Metzger JD (2008) Growth of bedding plants in commercial potting substrate amended with vermicompost. Bioresour Technol 99:3155–3161. doi:10.1016/j.biortech.2007.05.069

Bansal M, Garg U, Singh D, Garg VK (2009) Removal of Cr(VI) from aqueous solutions using pre–consumer processing agricultural waste: a case study of rice husk. J Hazard Mater 162:312–320. doi:10.1016/j.jhazmat.2008.05.037

Basci N, Kocadagistan E, Kocadagistan B (2004) Biosorption of copper (II) from aqueous solutions by wheat shell. Desalination 164:135–140. doi:10.1016/S0011-9164(04)00172-9

Baszynski T (1986) Interference of $Cd^{2+}$ in functioning of the photosynthetic apparatus of higher plants. Acta Soc Bot Pol 99:291–304

Baysal Z, Çinar E, Bulut E, Alkan H, Dogru M (2009) Equilibrium and thermodynamic studies on biosorption of Pb(II) onto Candida albicans biomass. J Hazard Mater 161:62–67. doi:10.1016/j.jhazmat.2008.02.122

Benedetti MF, Van Rinniburgh P (1995) Metal ion binding to humic substances: application of the non–ideal competitive adsorption model. Environ Sci Technol 29:446–457

Benítez E, Nogales R, Masciandaro G, Ceccanti B (2000) Isolation by isoelectric focusing of humic–urease complexes from earthworm (Eisenia fetida)–processed sewage sludges. Biol Fert Soils 31:489–493. doi:10.1007/s003740000197

Beolchini F, Pagnanelli F, Toro L, Vegliò F (2003) Biosorption of copper by Sphaerotilus natans immobilised in polysulfone matrix: equilibrium and kinetic analysis. Hydrometall 70:101–112. doi:10.1016/S0304-386X(03)00049-5

Bhattacharyya KG, Sharma A (2004) Adsorption of Pb(II) from aqueous solution by Azadirachta indica (neem) leaf powder. J Hazard Mater 113:97–109. doi:10.1016/j.jhazmat.2004.05.034

Bishnoi NR, Bajaj M, Sharma N, Gupta A (2004) Adsorption of Cr(VI) on activated rice husk carbon and activated alumina. Bioresour Technol 91:305–307. doi:10.1016/S0960-8524(03)00204-9

Boddu VM, Abburi K, Talbott JL, Smith ED (2003) Removal of hexavalent chromium from wastewater using a new composite chitosan biosorbent. Environ Sci Technol 37:4449–4456

Boni MR, Sbaffoni S (2009) The potential of compost–based biobarriers for Cr(VI) removal from contaminated groundwater: column test. J Hazard Mater 166:1087–1095. doi:10.1016/j.jhazmat.2008.12.036

Brune A, Urbach W, Dietz KJ (1994) Compartmentation and transport of zinc in barley primary leaves as basis mechanisms involved in zinc tolerance. Plant Cell Environ 17:153–162

Buhl KJ, Hamilton SJ (1990) Comparative toxicity of inorganic contaminants related by placer mining to early life stages of salmonids. Ecotoxicol Environ Saf 20:325–342

Çabuk A, Akar T, Tunali Gedikli S (2007) Biosorption of Pb(II) by industrial strain of Saccharomyces cerevisiae immobilized on the biomatrix of cone biomass of Pinus nigra: equilibrium and mechanism analysis. Chem Eng J 131:293–300. doi:10.1016/j.cej.2006.12.011

Canellas LP, Olivares FL, Okorokova–Façanha AL, Façanha AR (2002) Humic acids isolated from earthworm compost enhance root elongation, lateral root emergence, and plasma membrane $H^+$–ATPase activity in maize rots. Plant Physiol 130:1951–1957

Carrasquero–Durán A, Flores I (2009) Evaluation of lead(II) immobilization by a vermicompost using adsorption isotherms and IR spectroscopy. Bioresour Technol 100:1691–1694. doi:10.1016/j.biortech.2008.09.013

Carrasquero Durán A, Flores I, Perozo C, Pernalete Z (2006) Immobilization of lead by a vermicompost and its effect on white bean (Vigna Sinenis var. Apure) uptake. Int J Environ Sci Technol 3:203–212

Ceribasi IH, Yetis U (2001) Biosorption of Ni(II) and Pb(II) by Phanerochaete chrysosporium from a binary metal system – kinetics. Water SA 27:15–20

Chandalia M, Garg A, Lutjohann D, von Bergmann K, Grundy SM, Brinkley LJ (2000) Beneficial effects of high dietary fiber intake in patients with type 2 diabetes mellitus. New Engl J Med 342:1392–1398

Chen G, Zeng G, Tu X, Huang G, Chen Y (2005) A novel biosorbent: characterization of the spent mushroom compost and its application for removal of heavy metals. J Environ Sci (China) 17:756–760

Cheremisinoff PN (1995) Handbook of water and wastewater treatment technology. Marcel Dekker Inc, New York

Cho S, Jian Y (1998) Copper adsorption and removal from water by living mycelium of white–rot fungus phanerochaete chrysosporium. Water Res 32:2746–2752. doi:10.1016/S0043-1354(98)00024-4

Cho DH, Kim EY (2003) Characterization of $Pb^{2+}$ biosorption from aqueous solution by Rhodotorula glutinis. Bioprocess Biosyst Eng 25:271–277

Chojnacka K (2005) Biosorption of Cr(III) ions by eggshells. J Hazard Mater 121:167–173. doi:10.1016/j.jhazmat.2005.02.004

Chojnacka K (2006) Biosorption of Cr(III) ions by wheat straw and grass: a systematic characterization of new biosorbents. Pol J Environ Stud 15:845–852

Cimino G, Cappello RM, Caristi C, Toscano G (2005) Characterization of carbons from olive cake by sorption of wastewater pollutants. Chemosphere 61:947–955. doi:10.1016/j.chemosphere.2005.03.042

Contreras–Ramos SM, Escamilla–Silva EM, Dendooven L (2005) Vermicomposting of biosolids with cow manure and oat straw. Biol Fert Soils 41:190–198. doi:10.1007/s00374-004-0821-8

Contreras–Ramos SM, Álvarez–Bernal D, Dendooven L (2006) Eisenia fetida increased removal of polycyclic aromatic hydrocarbons from soil. Environ Pollut 141:396–401. doi:10.1016/j.envpol.2005.08.057

Da Silva EA, Cossich ES, Tavares CRG, Filho LC, Guirardello R (2002) Modeling of copper(II) biosorption by marine alga Sargassum sp. In fixed–bed column. Process Biochem 38:791–799. doi:10.1016/S0032-9592(02)00231-5

Daldrup T, Haarhoff K, Szathmary SC (1983) Toedliche nickel sulfate–intoxikation. Berichte zur Gerichtlichen Medizin 41:141–144

Dang VBH, Doan HD, Dang–Vu T, Lohi A (2009) Equilibrium and kinetics of biosorption of cadmium(II) and copper(II) ions by wheat straw. Bioresour Technol 100:211–219. doi:10.1016/j.biortech.2008.05.031

Das K, Keener HM (1997) Numerical model for the dynamic simulation of a large composting system. Trans ASAE 40:1179–1189

Davis TA, Volesky B, Vieira RHSF (2000) Sargassum seaweed as biosorbent for heavy metals. Water Res 34:4270–4278. doi:10.1016/S0043-1354(00)00177-9

Davis TA, Volesky B, Mucci A (2003) A review of the biochemistry of heavy metal biosorption by brown algae. Water Res 37:4311–4330. doi:10.1016/S0043-1354(03)00293-8 de Carvalho RP, Guedes KJ, Pinheiro MVB, Krambrock K (2001) Biosorption of copper by dried plant leaves studied by electron paramagnetic resonance and infrared spectroscopy. Hydrometall 59:407–412. doi:10.1016/S0304-386X(00)00170-5

De Schamphelaere KAC, Stauber JL, Wilde KL, Markich SJ, Brown PL, Franklin NM, Creighton NM, Janssen CR (2005) Toward a biotic ligand model for freshwater green algae: surface–bound and internal copper are better predictors of toxicity than free $Cu^{2+}$–ion activity when pH is varied. Environ Sci Technol 39:2067–2072

Demirbas E, Kobya M, Senturk E, Ozkan T (2004) Adsorption kinetics for the removal of chromium (VI) from aqueous solutions on the activated carbons prepared from agricultural wastes. Water SA 30:533–539

Deng L, Su Y, Su H, Wang X, Zhu X (2007) Sorption and desorption of lead (II) from wastewater by green algae Cladophora fascicularis. J Hazard Mater 143:220–225. doi:10.1016/j.jhazmat.2006.09.009

Dominguez J, Edwards CA, Subler S (1997) A comparison of vermicomposting and composting. Biocycle 38:57–59

Dönmez C, Aksu Z (2001) Bioaccumulation of copper(II) and nickel(II) by the non–adapted and adapted growing Candida Sp. Water Res 35:1425–1434. doi:10.1016/S0043-1354(00)00394-8

Drasch GA (1993) Increase of cadmium body burden for this century. Sci Total Environ 67:75–89. doi:10.1016/0048-9697(83)90105-5

Dries J, Bastiaens L, Springael D, Kuypers S, Agathos SN, Diels L (2005) Effect of humic acids on heavy metal removal by zero–valent iron in batch and continuous flow column systems. Water Res 39:3531–3540. doi:10.1016/j.watres.2005.06.020

Ekmekyapar F, Aslan A, Kemal Bayhan Y, Cakici A (2006) Biosorption of copper(II) by nonliving lichen biomass of Cladonia rangiformis hoffm. J Hazard Mater 137:293–298. doi:10.1016/j.jhazmat.2006.02.003

Elangovan R, Philip L, Chandraraj K (2008) Biosorption of hexavalent and trivalent chromium by palm flower (Borassus aethiopum). Chem Eng J 141:99–111. doi:10.1016/j.cej.2007.10.026

El–Eswed B, Khalili F (2006) Adsorption of Cu(II) and Ni(II) on solid humic acid from the Azraq area, Jordan. J Colloid Interface Sci 299:497–503. doi:10.1016/j.jcis.2006.02.048

Elouear Z, Bouzid J, Boujelben N, Feki M, Montiel A (2008) The use of exhausted olive cake ash (EOCA) as a low cost adsorbent for the removal of toxic metal ions from aqueous solutions. Fuel 87:2582–2589. doi:10.1016/j.fuel.2008.01.019

EPA (Environmental Protection Agency) (1990) Environmental pollution control alternatives. EPA/625/5–90/025, EPA/625/4–89/023, Cincinnati, US

Erhan D, Kobya M, Elif S, Ozkan T (2004) Adsorption kinetics for the removal of chromium III from aqueous solutions on the activated carbonaceous prepared from agricultural wastes. Water SA 30:533–540

Esmaeili A, Beirami P, Rustaiyan A, Rafiei F, Ghasemi S, Assadian F (2008) Evaluation of the marine alga Gracilaria Corticata for the biosorption of Cu (II) from wastewater in a packed column. J Mar Environ Eng 9:65–73

Fernando A, Monteiro S, Pinto F, Mendes B (2009) Production of biosorbents from waste olive cake and its adsorption characteristics for $Zn^{2+}$ ion. Sustainability 1:277–297

Ferreira Fontes MP, Teixeira De Matos A, Da Costa LM, Lima Neves Julio C (2000) Competitive adsorption of zinc, cadmium, copper, and lead in three highly–weathered Brazilian soils. Commun Soil Sci Plant Anal 31:2939–2958

Figueira MM, Volesky B, Ciminelli VST, Roddick FA (2000) Biosorption of metals in brown seaweed biomass. Water Res 34:196–204. doi:10.1016/S0043-1354(99)00120-7

Fosmire GJ (1990) Zinc toxicity. Am J Clin Nutr 51:225–227

Gajalakshmi S, Abbasi SA (2008) Solid waste management by composting: state of the art. Crit Rev Environ Sci Technol 38:311–400. doi:10.1080/10643380701413633

Gajalakshmi S, Ramasamy EV, Abbasi SA (2005) Composting–vermicomposting of leaf litter ensuing from the trees of mango (Mangifera indica). Bioresour Technol 96:1057–1061. doi:10.1016/j.biortech.2004.09.002

Gale TF (1978) Embryotoxic effects of chromium trioxide in hamsters. Environ Res 16:101–109
Gao K, Pearce J, Jones J, Taylor C (1999) Interaction between peat, humic acid and aqueous metal ions. Environ Geochem Health 21:13–26. doi:10.1023/A:1006592627852
Gardea–Torresdey JL, Tiemann KJ, Gonzalez JH, Cano–Aguilera I, Henning JA, Townsend MS (1996a) Removal of nickel ions from aqueous solution by biomass and silica–immobilized biomass of Medicago sativa (alfalfa). J Hazard Mater 49:205–216. doi:10.1016/0304-3894(96)01757-8
Gardea–Torresdey JL, Tang L, Salvador JM (1996b) Copper adsorption by esterified and unesterified fractions of Sphagnum peat moss and its different humic substances. J Hazard Mater 48:191–206. doi:10.1016/0304-3894(95), 00156-5
Garg P, Gupta A, Satya S (2006) Vermicomposting of different types of waste using Eisenia foetida: a comparative study. Bioresour Technol 97:391–395. doi:10.1016/j.biortech.2005.03.009
Garg UK, Kaur MP, Garg VK, Sud D (2007) Removal of hexavalent chromium from aqueous solution by agricultural waste biomass. J Hazard Mater 140:60–68. doi:10.1016/j.jhazmat.2006.06.056
Garg U, Kaur MP, Jawa JK, Sud D, Garg VK (2008) Removal of cadmium (II) from aqueous solutions by adsorption on agricultural waste biomass. J Hazard Mater 154:1149–1157. doi:10.1016/j.jhazmat.2007.11.040
Gibert O, de Pablo J, Luis Cortina J, Ayora C (2005) Sorption studies of Zn(II) and Cu(II) onto vegetal compost used on reactive mixtures for in situ treatment of acid mine drainage. Water Res 39:2827–2838. doi:10.1016/j.watres.2005.04.056
Göksungur Y, Üren S, Güvenç U (2005) Biosorption of cadmium and lead ions by ethanol treated waste baker's yeast biomass. Bioresour Technol 99:103–109. doi:10.1016/j.biortech.2003.04.002
Gondar D, López R, Fiol S, Antelo JM, Arce F (2006) Cadmium, lead, and copper binding to humic acid and fulvic acid extracted from an ombrotrophic peat bog. Geoderma 135:196–203. doi:10.1016/j.geoderma.2005.12.003
Goyer R (1991) Toxic effects of metals. In: Amdur MO, Doull JD, Klaassen CD (eds) Casarett and Doull's toxicology, 4th edn. Pergamon Press, New York, pp 623–680
Green–Pedersen H, Jensen BT, Pind N (1997) Nickel adsorption on $MnO_2$, $Fe(OH)_3$, montmorillonite, humic acid and calcite: a comparative study. Environ Technol 18:807–815
Gupta SS, Bhattacharyya KG (2006) Adsorption of Ni(II) on clays. J Colloid Interface Sci 295: 21–32. doi:10.1016/j.jcis.2005.07.073
Gupta R, Garg VK (2008) Stabilization of primary sewage sludge during vermicomposting. J Hazard Mater 153:1023–1030. doi:10.1016/j.jhazmat.2007.09.055
Gupta VK, Shrivastava AK, Jain N (2001) Biosorption of chromium(VI) from aqueous solutions by green algae spirogyra species. Water Res 35:4079–4085. doi:10.1016/S0043-1354(01)00138-5
Han R, Li H, Li Y, Zhang J, Xiao H, Shi J (2006) Biosorption of copper and lead ions by waste beer yeast. J Hazard Mater 137:1569–1576. doi:10.1016/j.jhazmat.2006.04.045
Hanif MA, Nadeem R, Bhatti HN, Ahmad NR, Ansari TM (2007) Ni(II) biosorption by cassia fistula (golden shower) biomass. J Hazard Mater 139:345–355. doi:10.1016/j.jhazmat.2006.06.040
Harman G, Patrick R, Spittler T (2007) Removal of heavy metals from polluted waters using lignocellulosic agricultural waste products. Ind Biotechnol 3:366–374
Hatherill JR (1981) A review of the mutagenicity of chromium. Drug Chem Toxicol 4:185–195
Hervas L, Mazuelos C, Senesi N, Saiz–Jimenez C (1989) Chemical and physico–chemical characterization of vermicomposts and their humic acid fractions. Sci Total Environ 81–82:543–550
Hiatt V, Huff JE (1975) Environmental impact of cadmium. Overview. Int J Environ Stud 7:277–285
Holan ZR, Volesky B, Prasetyo I (1993) Biosorption of cadmium by biomass of marine algae. Biotechnol Bioeng 41:819–825. doi:10.1002/bit.260410808
Hough RL, Tye AM, Crout NMJ, McGrath SP, Zhang H, Young SD (2005) Evaluating a 'free ion activity Model' applied to metal uptake by Lolium perenne L. Grown in contaminated soils. Plant Soil 270:1–12
Hu MJ, Wei YL, Yang YW, Lee JF (2003) Immobilization of chromium(VI) with debris of aquatic plants. Bull Environ Contam Toxicol 71:840–847. doi:10.1007/s00128-003-0212-0

Ismael Acosta R, Rodríguez X, Gutiérrez C, de Guadalupe MM (2004) Biosorption of chromium (VI) from aqueous solutions onto fungal biomass. Bioinorg Chem Appl 2:1–7. doi:10.1155/S1565363304000019

Ivanitsa VO, Vasilyeva TV, Buchtiyarov AE, Lindström EB, McEldowney S (1999) Interactions between marine bacteria and heavy metals'. In: Amils R, Ballester A (eds) Biohydrometallurgy and the environment toward the mining of the 21st century – part B. Elsevier, Amsterdam, pp 317–325

Jadia CD, Fulekar MH (2008) Phytoremediation: the application of vermicompost to remove zinc, cadmium, copper, nickel and lead by sunflower plant. Environ Eng Manag J 7:547–558

Jain M, Garg VK, Kadirvelu K (2009) Chromium(VI) removal from aqueous system using Helianthus annuus (sunflower) stem waste. J Hazard Mater 162:365–372. doi:10.1016/j.jhazmat.2008.05.048

Jordão CP, Pereira MG, Einloft R, Santana MB, Bellato CR, de Mello JWV (2002) Removal of Cu, Cr, Ni, and Zn from electroplating wastes and synthetic solutions by vermicompost of cattle manure. J Environ Sci Health A37:875–892

Jordão CP, Braganç R, Fernandes A, de Lima RK, de Souza NB, Martins de Barros P (2009) Zn(II) adsorption from synthetic solution and kaolin wastewater onto vermicompost. J Hazard Mater 162:804–811. doi:10.1016/j.jhazmat.2008.05.104

Kaewsarn P (2002) Biosorption of copper(II) from aqueous solutions by pre–treated biomass of marine algae Padina sp. Chemosphere 47:1081–1085. doi:10.1016/S0045-6535(01)00324-1

Kalyani S, Srinivasa Rao P, Krishnaiah A (2004) Removal of nickel (II) from aqueous solutions using marine macroalgae as the sorbing biomass. Chemosphere 57:1225–1229. doi:10.1016/j.chemosphere.2004.08.057

Kamergam N, Alagumalai K, Daniel T (1999) Effect of vermicompost on the growth and yield of green gram (Phaseolus aurus Roxb.). Trop Agric 76:143–146

Kaushik P, Garg VK (2004) Dynamics of biological and chemical parameters during vermicomposting of solid textile mill sludge mixed with cow dung and agricultural residues. Bioresour Technol 94:203–209. doi:10.1016/j.biortech.2003.10.033

Kidwai M, Mohan R (2005) Green chemistry: an innovative technology. Found Chem 7:269–287

King P, Anuradha K, Lahari SB, Kumar YP, Prasad VSRK (2008) Biosorption of zinc from aqueous solution using Azadirachta indica bark: equilibrium and kinetic studies. J Hazard Mater 152:324–329. doi:10.1016/j.jhazmat.2007.06.101

Klimmek S, Stan H–J, Wilke A, Bunke G, Buchholz R (2001) Comparative analysis of the biosorption of cadmium, lead, nickel, and zinc by algae. Environ Sci Technol 35:283–288

Kobya M (2004) Removal of Cr(VI) from aqueous solutions by adsorption onto hazelnut shell activated carbon: kinetic and equilibrium studies. Bioresour Technol 91:317–321. doi:10.1016/j.biortech.2003.07.001

Kocasoy G, Güvener Z (2009) Efficiency of compost in the removal of heavy metals from the industrial wastewater. Environ Geol 57:291–296. doi:10.1007/s00254-008-1372-3

Krishnan KA, Anirudhan TS (2003) Removal of cadmium(II) from aqueous solutions by steamactivated sulphurised carbon prepared from sugar–cane bagasse pith: kinetics and equilibrium studies. Water SA 29:147–156

Kumar U, Bandyopadhyay M (2006) Sorption of cadmium from aqueous solution using pretreated rice husk. Bioresour Technol 97:104–109. doi:10.1016/j.biortech.2005.02.027

Kuzovkina YA, Knee M, Quigley MF (2004) Cadmium and copper uptake and translocation in five willow (Salix L.) species. Int J Phytoremediation 6:269–287. doi:10.1080/16226510490496726

Lai Y–L, Annadurai G, Huang F–C, Lee J–F (2008) Biosorption of Zn(II) on the different Ca–alginate beads from aqueous solution. Bioresour Technol 99:6480–6487. doi:10.1016/j.biortech.2007.11.041

Laj S, Jain VK, Tandon SK (1984) Comparative toxicity of trivalent and hexavalent chromium IV: biochemical changes in blood and liver of rat. J Environ Biol 5:29–35

Landgraf MD, Javaroni Rde C, Rezende MO (1998) Characterization of humic substances by capillary electrophoresis. J Capillary Electrophor 5:193–199

Laurino JP (2008) Removal of lead (II) ions by poly(2–octadecyl butanedioic acid): isothermal and kinetic studies. J Macromol Sci A: Pur Appl Chem 45:612–619. doi:10.1080/10601320802168736
Leonard A, Gerber GB (1989) Zinc toxicity: does it exist? J Am Coll Toxicol 8:1285–1290
Leonard A, Lauwerys RR (1980) Carcinogenicity and mutagenicity of chromium. Mutat Res/Rev Gene Toxicol 76:227–239. doi:10.1016/0165-1110(80)90018-4
Lin X, Burns RC, Lawrance GA (2003) Effect of Cd(II) and anion type on the ageing of ferrihydrite and its subsequent leaching under neutral and alkaline conditions. Wat Air Soil Pollut 143:155–177
Lin X, Burns RC, Lawrance GA (2005) Composition on the precipitation of cadmium(Ii) using lime and magnesia. Wat Air Soil Pollut 165:131–152
Liu A, Gonzalez RD (2000) Modeling adsorption of copper(II), cadmium(II) and lead(II) on purified humic acid. Langmuir 16:3902–3909
Lo W, Chua LWH, Lam KH, Bi SP (1999) A comparative investigation on the adsorption of lead (II) by filamentous fungal biomass. Chemosphere 39:2723–2736. doi:10.1016/S0045-6535(99)00206-4
Logsdon G (1994) Worldwide progress in vermicomposting: earthworms and composting. Biocycle 35:63–65
Loh TC, Lee YC, Liang JB, Tan D (2005) Vermicomposting of cattle and goat manures by Eisenia foetida and their growth and reproduction performance. Bioresour Technol 96:111–114. doi:10.1016/j.biortech.2003.03.001
Long X, Luo X, Wang Y, Li Z (2009) Sorption of Pb(II) from aqueous solution by konjac glucomannan beads. Sci China Ser E: Tech Sci 52:223–226
López A, Lázaro N, Morales S, Marqués AM (2002) Nickel biosorption by free and immobilized cells of Pseudomonas fluorescens 4 F39: a comparative study. Wat Air Soil Pollut 135:157–172
Loukidou MX, Zouboulis AI, Karapantsios TD, Matis KA (2004) Equilibrium and kinetic modeling of chromium(VI) biosorption by Aeromonas caviae. Colloids Surf A: Physicochem Eng Aspects 242:93–104. doi:10.1016/j.colsurfa.2004.03.030
Lu S, Gibb SW (2008) Copper removal from wastewater using spent–grain as biosorbent. Bioresour Technol 99:1509–1517. doi:10.1016/j.biortech.2007.04.024
Lu X, Kruatrachue M, Pokethitiyook P, Homyok K (2004) Removal of cadmium and zinc by water hyacinth, eichhornia crassipes. ScienceAsia 30:93–103
Lu LT, Chang IC, Hsiao TY, Yu YH, Ma HW (2007) Identification of pollution source of cadmium in soil: application of material flow analysis and a case study in Taiwan. Environ Sci Pollut Res Int 14:49–59
Lu D, Cao Q, Cao X, Luo F (2009) Removal of Pb(II) using the modified lawny grass: mechanism, kinetics, equilibrium and thermodynamic studies. J Hazard Mater 166:239–247. doi:10.1016/j.jhazmat.2008.11.018
Lugo–Lugo V, Hernández–López S, Barrera–Díaz C, Ureña–Núñez F, Bilyeu B (2009) A comparative study of natural, formaldehyde–treated and copolymer–grafted orange peel for Pb(II) adsorption under batch and continuous mode. J Hazard Mater 161:1255–1264. doi:10.1016/j.jhazmat.2008.04.087
Machado RM, Correia MJN, Carvalho JMR (2003) Integrated process for biosorption of copper from liquid effluents using grape stalks. Sep Sci Technol 38:2237–2254. doi:10.1081/SS-120021622
Majumdar SS, Das SK, Saha T, Panda GC, Bandyopadhyoy T, Guha AK (2008) Adsorption behavior of copper ions on Mucor rouxii biomass through microscopic and FTIR analysis. Colloids Surf B Biointerfaces 63:138–145. doi:10.1016/j.colsurfb.2007.11.022
Malandrino M, Abollino O, Giacomino A, Aceto M, Mentasti E (2006) Adsorption of heavy metals on vermiculite: influence of pH and organic ligands. J Colloid Interface Sci 299:537–546. doi:10.1016/j.jcis.2006.03.011
Malkoc E (2006) Ni(II) removal from aqueous solutions using cone biomass of Thuja orientalis. J Hazard Mater 137:899–908. doi:10.1016/j.jhazmat.2006.03.004
Malkoc E, Nuhoglu Y (2005) Investigations of nickel(II) removal from aqueous solutions using tea factory waste. J Hazard Mater 127:120–128. doi:10.1016/j.jhazmat.2005.06.030
Maranon E, Sastre H (1991) Heavy metal removal in packed beds using apple wastes. Bioresour Technol 38:39–43

Markich SJ, Brown PL, Jeffree RA, Lim RP (2003) The effects of pH and dissolved organic carbon on the toxicity of cadmium and copper to a freshwater bivalve: further support for the extended free ion activity model. Arch Environ Contam Toxicol 45:479–491. doi:10.1007/s00244-003-2175-x

Martínez M, Miralles N, Hidalgo S, Fiol N, Villaescusa I, Poch J (2006) Removal of lead(II) and cadmium(II) from aqueous solutions using grape stalk waste. J Hazard Mater 133:203–211. doi:10.1016/j.jhazmat.2005.10.030

Martín–Gil J, Navas–Gracia LM, Gómez–Sobrino E, Correa–Guimaraes A, Hernández–Navarro S, Sánchez–Báscones M, del Carmen Ramos–Sánchez M (2008) Composting and vermicomposting experiences in the treatment and bioconversion of asphaltens from the prestige oil spill. Bioresour Technol 99:1821–1829. doi:10.1016/j.biortech.2007.03.031

Mason IG (2006) Mathematical modeling of the composting process: a review. Waste Manag 26:3–21. doi:10.1016/j.wasman.2005.01.021

Mata YN, Blázquez ML, Ballester A, González F, Muñoz JA (2008) Characterization of the biosorption of cadmium, lead and copper with the brown alga Fucus vesiculosus. J Hazard Mater 158:316–323. doi:10.1016/j.jhazmat.2008.01.084

McCarty MF (2002) Glucomannan minimizes the postprandial insulin surge: a potential adjuvant for hepatothermic therapy. Med Hypotheses 58:487–490

Melo JS, D'Souza SF (2004) Removal of chromium by mucilaginous seeds of Ocimum basilicum. Bioresour Technol 92:151–155. doi:10.1016/j.biortech.2003.08.015

Modher A, Hussain AS, Pozi M (2009) Characterization of the adsorption of the lead (II) by the nonliving biomass Spirogyra neglecta (Hasall) Kützing. Am J Biochem Biotechnol 5:75–83

Mohan D, Singh KP, Singh VK (2006) Trivalent chromium removal from wastewater using low cost activated carbon derived from agricultural waste material and activated carbon fabric cloth. J Hazard Mater 135:280–295. doi:10.1016/j.jhazmat.2005.11.075

Mohanty K, Jha M, Meikap BC, Biswas MN (2006) Biosorption of Cr(VI) from aqueous solutions by Eichhornia crassipes. Chem Eng J 117:71–77. doi:10.1016/j.cej.2005.11.018

Mohee R, Mudhoo A (2005) Analysis of physical properties of an in–vessel composting matrix. Powder Technol 155:92–99. doi:10.1016/j.powtec.2005.05.051

Moon S–H, Park C–S, Kim Y–J, Park Y–I (2006) Biosorption isotherms of Pb (II) and Zn (II) on Pestan, an extracellular polysaccharide, of Pestalotiopsis sp. KCTC 8637P. Process Biochem 41:312–316. doi:10.1016/j.procbio.2005.07.013

Mudhoo A, Mohee R (2006) Sensitivity analysis and parameter optimization of a heat loss model for a composting system. J Environ Informatics 8:100–110. doi:10.3808/jei.200600080

Mudhoo A, Mohee R (2007) Overall heat transfer coefficients in organic substrates composting. J Environ Informatics 9:87–99. doi:10.3808/jei.200700090

Mudhoo A, Mohee R (2008) Modeling heat loss during self–heating composting based on combined fluid film theory and boundary layer concepts. J Environ Informatics 11:74–89. doi:10.3808/jei.200800113

Mukhopadhyay B, Sundquist J, Schmitz RJ (2007) Removal of Cr(VI) from Cr–contaminated groundwater through electrochemical addition of Fe(II). J Environ Manag 82:66–76. doi:10.1016/j.jenvman.2005.12.005

Mullen MD, Wolf DC, Ferris FG, Beveridge TJ, Flemming CA, Bailey GW (1989) Bacterial sorption of heavy metals. Appl Environ Microbiol 55:3143–3149

Murray K, Linder PW (1984) Fulvic acids: structure and metal binding. Predominant metal binding sites. J Soil Sci 35:219–222

Muscolo A, Bovolo F, Gionfriddo F, Nardi S (1999) Earthworm humic matter produces auxins–like effect on Daucus carota cell growth and nitrate metabolism. Soil Biol Biochem 31:1303–1311

Nadeem R, Nasir MH, Hanif MS (2009) Pb (II) sorption by acidically modified cicer arientinum biomass. Chem Eng J 150:40–48. doi:10.1016/j.cej.2008.12.001

Nagavallemma KP, Wani SP, Lacroix S, Padmaja VV, Vineela C, Babu Rao M, Sahrawat KL (2004) Vermicomposting: recycling wastes into valuable organic fertilizer, vol 8, Global theme on agrecosystems. International Crops Research Institute for the Semi–Arid Tropics, Patancheru, p 20

Naja G, Volesky B (2006) Behavior of the mass transfer zone in a biosorption column. Environ Sci Technol 40:3996–4003. doi:10.1021/es051542p

Nakasaki K, Nag K, Karita S (2005) Microbial succession associated with organic matter decomposition during thermophilic composting of organic waste. Waste Manag Res 23:48–56. doi:10.1177/0734242X05049771

Noeline BF, Manohar DM, Anirudhan TS (2005) Kinetic and equilibrium modelling of lead(II) sorption from water and wastewater by polymerized banana stem in a batch reactor. Sep Purif Technol 45:131–140. doi:10.1016/j.seppur.2005.03.004

Norseth T (1981) The carcinogenicity of chromium. Environ Health Perspect 40:121–130

Oke IA, Olarinoye NO, Adewusi SRA (2008) Adsorption kinetics for arsenic removal from aqueous solutions by untreated powdered eggshell. Adsorpt 14:73–83. doi:10.1007/s10450-007-9047-z

Özer A, Gürbüz G, Çalimli A, Körbahti BK (2008) Investigation of nickel(II) biosorption on Enteromorpha prolifera: optimization using response surface analysis. J Hazard Mater 152:778–788. doi:10.1016/j.jhazmat.2007.07.088

Öztürk A, Artan T, Ayar A (2004) Biosorption of nickel(II) and copper(II) ions from aqueous solution by Streptomyces coelicolor A3(2). Colloids Surf B Biointerfaces 34:105–111. doi:10.1016/j.colsurfb.2003.11.008

Padmavathy V (2008) Biosorption of nickel(II) ions by baker's yeast: kinetic, thermodynamic and desorption studies. Bioresour Technol 99:3100–3109. doi:10.1016/j.biortech.2007.05.070

Pamukoglu MY, Kargi F (2006) Removal of copper(II) ions from aqueous medium by biosorption onto powdered waste sludge. Process Biochem 41:1047–1054. doi:10.1016/j.procbio.2005.11.010

Panda GC, Das SK, Chatterjee S, Maity PB, Bandopadhyay TS, Guha AK (2006) Adsorption of cadmium on husk of Lathyrus sativus: physico–chemical study. Colloids Surf B Biointerfaces 50:49–54. doi:10.1016/j.colsurfb.2006.03.022

Parker DR, Pedler JF (1997) Reevaluating the free–ion activity model of trace metal availability to higher plants. Plant Soil 196:223–228. doi:10.1023/A:1004249923989

Parker DR, Pedler JF, Ahnstrom ZAS, Resketo M (2001) Reevaluating the free–ion activity model of trace metal toxicity toward higher plants: experimental evidence with copper and zinc. Environ Toxicol Chem 20:899–906. doi:10.1897/1551-5028

Patterson JW (1977) Wastewater treatment. Science Publishers, New York

Paulino AT, Santos LB, Nozaki J (2008) Removal of $Pb^{2+}$, $Cu^{2+}$, and $Fe^{3+}$ from battery manufacture wastewater by chitosan produced from silkworm chrysalides as a low–cost adsorbent. React Funct Polym 68:634–642. doi:10.1016/j.reactfunctpolym.2007.10.028

Pehlivan E, Altun T (2008) Biosorption of chromium(VI) ion from aqueous solutions using walnut, hazelnut and almond shell. J Hazard Mater 155:378–384. doi:10.1016/j.jhazmat.2007.11.071

Peng K, Li X, Luo C, Shen Z (2006) Vegetation composition and heavy metal uptake by wild plants at three contaminated sites in Xiangxi area, China. J Environ Sci Health Part A 40:65–76

Pereira MG, Arruda MAZ (2003) Vermicompost as a natural adsorbent material: characterization and potentialities for cadmium adsorption. J Braz Chem Soc 14:39–47

Pérez–Marín AB, Meseguer Zapata V, Ortuño JF, Aguilar M, Sáez J, Lloréns M (2007) Removal of cadmium from aqueous solutions by adsorption onto orange waste. J Hazard Mater 139:122–131. doi:10.1016/j.jhazmat.2006.06.008

Periasamy K, Namasivayam C (1995) Removal of nickel(II) from aqueous solution and plating industry wastewater using an agriculture waste peanut hulls. Waste Manag 15:63–68. doi:10.1016/0956-053X(94)00071-S

Pommery J, Ebenga JP, Imbenote Palavit J, Erb F (1988) Determination of the complexing ability of a standard humic acid towards cadmium ions. Water Res 22:185–189

Popuri Srinivasa R, Vijaya Y, Boddu VM, Abburi K (2009) Adsorptive removal of copper and nickel ions from water using chitosan coated PVC beads. Bioresour Technol 100:194–199. doi:10.1016/j.biortech.2008.05.041

Post MA, Campbell PG (1980) Lead chromate pigments – a literature survey on environmental and toxic effects. U.S. Dep. Comm Nat Bur Stand Rep NBSIR 80–1974, p 38

Pümpel T, Macaskie LE, Finlay JA, Diels L, Tsezos M (2003) Nickel removal from nickel plating waste water using a biologically active moving–bed sand filter. Biometals 16:567–581

Rahmani K, Mahvi AH, Vaezi F, Mesdaghinia AR, Nabizade R, Nazmara Sh (2009) Bioremoval of lead by use of waste activated sludge. Int J Environ Res 3:471–476

Ravikumar TN, Yeledhalli NA, Ravi MV, Narayana Rao K (2008) Physical, physico–chemcial and enzymes activities of vermiash compost. Karnataka J Agric Sci 21:222–226

Reddy KR, Chinthamreddy S (2003) Sequentially enhanced electrokinetic remediation of heavy metals in low buffering clayey soils. J Geotech Geoenviron Eng 129:263–277

Riaz M, Nadeem R, Hanif MA, Ansari TM, Khalil–ur–Rehman (2009) Pb(II) biosorption from hazardous aqueous streams using Gossypium hirsutum (cotton) waste biomass. J Hazard Mater 161:88–94. doi:10.1016/j.jhazmat.2008.03.096

Richard TL, Walker LP, Gossett JM (2006) The effects of oxygen on solid–state biodegradation kinetics. Biotechnol Prog 22:60–69. doi:10.1021/bp050171d

Romero E, Plaza C, Senesi N, Nogales R, Polo A (2007) Humic acid–like fractions in raw and vermicomposted winery and distillery wastes. Geoderma 139:397–406. doi:10.1016/j.geoderma.2007.03.009

Ryckeboer J, Mergaert J, Vaes K, Klammer S, De Clercq D, Coosemans J, Insam H, Swings J (2003) A survey of bacteria and fungi occurring during composting and self–heating processes. Ann Microbiol 53:349–410

Saeed A, Iqbal M (2003) Bioremoval of cadmium from aqueous solution by black gram husk (Cicer arientinum). Water Res 37:3472–3480. doi:10.1016/S0043-1354(03)00175-1

Sarı A, Tuzen M (2008) Biosorption of cadmium(II) from aqueous solution by red algae (Ceramium virgatum): equilibrium, kinetic and thermodynamic studies. J Hazard Mater 157:448–454. doi:10.1016/j.jhazmat.2008.01.008

Savvaidis I, Hughes MN, Poole RK (2003) Copper biosorption by Pseudomonas cepacia and other strains. World J Microbiol Biotechnol 19:117–121. doi:10.1023/A:1023284723636

Schroeder ED (1977) Water and wastewater treatment. McGraw Hill, Tokyo, International Student Edition

Seki H, Suzuki A (1995) Adsorption of heavy metal ions onto insolubilized humic acid. J Colloid Interface Sci 171:490–494. doi:10.1006/jcis.1995.1207

Sharma S, Pradhan K, Satya S, Vasudevan P (2005) Potentiality of earthworms for waste management and in other uses – a review. J Am Sci 1:4–16

Sharma P, Kumari P, Srivastava MM, Srivastava S (2006) Removal of cadmium from aqueous system by shelled Moringa oleifera Lam. Seed powder. Bioresour Technol 97:299–305. doi:10.1016/j.biortech.2005.02.034

Shen W, Chen S, Shi S, Li X, Zhang X, Hu W, Wang H (2009) Adsorption of Cu(II) and Pb(II) onto diethylenetriamine–bacterial cellulose. Carbohydr Polym 75:110–114. doi:10.1016/j.carbpol.2008.07.006

Singh KK, Rastogi R, Hasan SH (2005) Removal of cadmium from wastewater using agricultural waste 'rice polish'. J Hazard Mater 121:51–58. doi:10.1016/j.jhazmat.2004.11.002

Singh KK, Singh AK, Hasan SH (2006) Low cost bio–sorbent 'wheat bran' for the removal of cadmium from wastewater: kinetic and equilibrium studies. Bioresour Technol 97:994–1001. doi:10.1016/j.biortech.2005.04.043

Singh R, Sharma RR, Kumar S, Gupta RK, Patil RT (2008) Vermicompost substitution influences growth, physiological disorders, fruit yield and quality of strawberry (Fragaria x ananassa Duch.). Bioresour Technol 99:8507. doi:10.1016/j.biortech.2008.03.034

Singh RS, Singh VK, Tiwari PN, Singh JK, Sharma YC (2009) Biosorption studies of nickel on Parthenium hysterophorous ash. Environ Technol 30:355–364. doi:10.1080/09593330902753156

ŏlener M, Tunali S, Safa Özcan A, Özcan A, Gedikbey T (2008) Adsorption characteristics of lead(II) ions onto the clay/poly(methoxyethyl)acrylamide (PMEA) composite from aqueous solutions. Desalination 223:308–322. doi:10.1016/j.desal.2007.01.221

Spark KM, Wells JD, Johnson BB (1997) The interaction of a humic acid with heavy metals. Aust J Soil Res 35:89–102

Sprynskyy M, Buszewski B, Terzyk AP, Namieśnik J (2006) Study of the selection mechanism of heavy metal ($Pb^{2+}$, $Cu^{2+}$, $Ni^{2+}$, and $Cd^{2+}$) adsorption on clinoptilolite. J Colloid Interface Sci 304:21–28. doi:10.1016/j.jcis.2006.07.068

Srivastava P, Singh P, Angove M (2005) Competitive adsorption behavior of heavy metals on kaolinite. J Colloid Interface Sci 290:28–38. doi:10.1016/j.jcis.2005.04.036

Srivastava VC, Mall ID, Mishra IM (2006) Characterization of mesoporous rice husk ash (RHA) and adsorption kinetics of metal ions from aqueous solution onto RHA. J Hazard Mater 134:257–267. doi:10.1016/j.jhazmat.2005.11.052

Stanley LC, Ogden KL (2003) Biosorption of copper (II) from chemical mechanical planarization wastewaters. J Environ Manag 69:289–297. doi:10.1016/j.jenvman.2003.09.009

Steven JD, Davies LJ, Stanley EK, Abbott RA, Ihnat M, Bidstrup L, Jaworski JF (1976) Effects of chromium in the Canadian environment. Nat. Res. Coun. Canada, NRCC No. 15017. Available from Publications, NRCC/CNRC, Ottawa, K1A OR6, p 168

Stylianou MA, Inglezakis VJ, Moustakas KG, Malamis SPh, Loizidou MD (2007) Removal of Cu(II) in fixed bed and batch reactors using natural zeolite and exfoliated vermiculite as adsorbents. Desalination 215:133–142. doi:10.1016/j.desal.2006.10.031

Sud D, Mahajan G, Kaur MP (2008) Agricultural waste material as potential adsorbent for sequestering heavy metal ions from aqueous solutions – a review. Bioresour Technol 99:6017–6027. doi:10.1016/j.biortech.2007.11.064

Suhasini IP, Sriram G, Asolekar SR, Sureshkumar GK (1999) Nickel biosorption from aqueous systems: studies on single and multimetal equilibria, kinetics, and recovery. Sep Sci Technol 34:2761–2779. doi:10.1081/SS-100100803

Sumathi KMS, Mahimairaja S, Naidu R (2005) Use of low–cost biological waste and vermiculite for removal of chromium from tannery effluent. Bioresour Technol 96:309–316. doi:10.1016/j.biortech.2004.04.015

Sun Y, Paige CR, Snodgrass WJ (1996) The effect of cadmium on the transformation of ferrihydrite into crystalline products at pH 8. Water Air Soil Pollut 91:307–325

Sunderman F Jr, Dingle B, Hopfer SM, Swift T (1988) Acute nickel toxicity in electroplating workers who accidentally ingested a solution of nickel sulfate and nickel chloride. Am J Ind Med 14:257–266

Suthar S, Singh S (2008) Vermicomposting of domestic waste by using two epigeic earthworms (Perionyx excavatus and Perionyx sansibaricus). Int J Environ Sci Technol 5:99–106

Tabatabai A, Li S (2000) Dietary fiber and type 2 diabetes. Clin Excell Nurse Pract 4:272–276

Taiwo LB, Oso BA (2004) Influence of composting techniques on microbial succession, temperature and pH in a composting municipal solid waste. Afr J Biotechnol 3:239–243

Tandon SK, Saxena DK, Gaur JS, Chandra SV (1978) Comparative toxicity of trivalent and hexavalent chromium. Alterations in blood and liver. Environ Res 15:90–99

Thanabalasingam P, Pickering WF (1986) Arsenic sorption by humic acids. Environ Pollut B Chem Phys 12:233–246

Tognetti C, Laos F, Mazzarino MJ, Hernandez MT (2005) Composting vs vermicomposting: a comparison of end product quality. Compost Sci Util 13:6–13

Tsezos M, Remoudaki E, Angelatou V (1995) A systematic study on equilibrium and kinetics of biosorptive accumulation. The case of Ag and Ni. Int Biodeterior Biodegrad 129–153. doi:10.1016/0964-8305(95)00049-B

Tunali S, Kiran I, Akar T (2005) Chromium(VI) biosorption characteristics of Neurospora crassa fungal biomass. Miner Eng 18:681–689. doi:10.1016/j.mineng.2004.11.002

Tyler G, Pahlsson AM, Bengtsson G, Baath E, Tranvik L (1989) Heavy metal ecology and terrestrial plants, microorganisms and invertebrates: a review. Water Air Soil Pollut 47:189–215. doi:10.1007/BF00279327

Ulmanu M, Marañón E, Fernández Y, Castrillón L, Anger I, Dumitriu D (2003) Removal of copper and cadmium ions from diluted aqueous solutions by low cost and waste material adsorbents. Water Air Soil Pollut 142:357–373. doi:10.1023/A:1022084721990

Urdaneta C, Parra L–MC, Matute S, Garaboto MA, Barros H, Vázquez C (2008) Evaluation of vermicompost as bioadsorbent substrate of Pb, Ni, V and Cr for waste waters remediation using total reflection X–ray fluorescence. Spectrochim Acta Part B 63:1455–1460. doi:10.1016/j.sab.2008.10.004

USAF (1990) Nickel. In: Harry G (ed) Installation restoration program toxicology guide, vol 5. Armstrong Aerospace Medical Research Laboratory, Wright Patterson

Vegliò F, Beolchini F, Prisciandaro M (2003) Sorption of copper by olive mill residues. Water Res 37:895–4903. doi:10.1016/S0043-1354(03)00414-7

Vieira MGA, Oisiovici RM, Gimenes ML, Silva MGC (2008) Biosorption of chromium(VI) using a Sargassum sp. Packed–bed column. Bioresour Technol 99:3094–3099. doi:10.1016/j.biortech.2007.05.071

Vijayaraghavan K, Yun Y–S (2008) Bacterial biosorbents and biosorption. Biotechnol Adv 26:266–291. doi:10.1016/j.biotechadv.2008.02.002

Vijayaraghavan K, Jegan J, Palanivelu K, Velan M (2005a) Biosorption of cobalt(II) and nickel(II) by seaweeds: batch and column studies. Sep Purif Technol 44:53–59. doi:10.1016/j.seppur.2004.12.003

Vijayaraghavan K, Jegan J, Palanivelu K, Velan M (2005b) Biosorption of copper, cobalt and nickel by marine green alga Ulva reticulata in a packed column. Chemosphere 60:419–426. doi:10.1016/j.chemosphere.2004.12.016

Vilar VJP, Botelho CMS, Boaventura RAR (2007) Copper desorption from Gelidium algal biomass. Water Res 41:1569–1579. doi:10.1016/j.watres.2006.12.031

Vilar VJP, Botelho CMS, Boaventura RAR (2008) Copper removal by algae Gelidium, agar extraction algal waste and granulated algal waste: kinetics and equilibrium. Bioresour Technol 99:750–762. doi:10.1016/j.biortech.2007.01.042

Vinkler P, Lakatos B, Meisel J (1976) Infrared spectroscopic investigations of humic substances and their metal complexes. Geoderma 15:231–242

Volesky B (2001) Detoxification of metal–bearing effluents: biosorption for the next century. Hydrometall 59:203–216. doi:10.1016/S0304-386X(00)00160-2

Volesky B, Schiewer S (1999) Biosorption, Metals. In: Flickinger MC, Drew SW (eds) Encyclopedia of bioprocess technology: fermentation, biocatalysis, and bioseparation. Wiley, New York, pp 433–453

Wang LK, Hung YT, Hung Li K (2007) Vermicomposting process. Biosolids Treat Process 6:689–704

Wei YL, Lee YC, Hsieh HF (2005) XANES study of Cr sorbed by a kitchen waste compost from water. Chemosphere 61:1051–1060. doi:10.1016/j.chemosphere.2005.03.037

Wong JPK, Wong YS, Ta NFY (2000) Nickel biosorption by two chlorella species, C. Vulgaris (a commercial species) and C. Miniata (a local isolate). Bioresour Technol 73:133–137. doi:10.1016/S0960-8524(99)00175-3

Xi B, Wei Z, Hongliang L (2005) Dynamic simulation for domestic solid waste composting process. J Am Sci 1:34–45

Xu Y, Zhao D (2007) Reductive immobilization of chromate in water and soil using stabilized iron nanoparticles. Water Res 41:2101–2108. doi:10.1016/j.watres.2007.02.037

Yahaya YA, Don MM, Bhatia S (2009) Biosorption of copper (II) onto immobilized cells of Pycnoporus sanguineus from aqueous solution: equilibrium and kinetic studies. J Hazard Mater 161:189–195. doi:10.1016/j.jhazmat.2008.03.104

Yin P, Yu Q, Jin B, Ling Z (1999) Biosorption removal of cadmium from aqueous solution by using pretreated fungal biomass cultured from starch wastewater. Water Res 33:1960–1963. doi:10.1016/S0043-1354(98)00400-X

Zafar MN, Nadeem R, Hanif MA (2007) Biosorption of nickel from protonated rice bran. J Hazard Mater 143:478–485. doi:10.1016/j.jhazmat.2006.09.055

Zaller JG (2006) Foliar spraying of vermicompost extracts: effects on fruit quality and indications of late–blight suppression of field–grown tomatoes. Biol Agric Hortic 24:165–180

# Chapter 11
# Fe–Mn Concretions and Nodules to Sequester Heavy Metals in Soils

**Dionisios Gasparatos**

## Contents



D. Gasparatos (✉)
Laboratory of Soils and Agricultural Chemistry, Agricultural University of Athens,
Iera Odos 75, Athens, 11855, Greece
e-mail: gasparatos@aua.gr

E. Lichtfouse et al. (eds.), *Environmental Chemistry for a Sustainable World: Volume 2: Remediation of Air and Water Pollution*, DOI 10.1007/978-94-007-2439-6_11,

**Abstract** Over the last two decades, considerable attention has been paid to the management of metal-contaminated soils. Fe-Mn concretions and nodules can be used to sequester metals by adsorption. Fe-Mn concretions and nodules are discrete bodies with variable compositions formed in the soil system under alternating oxidizing and reducing conditions. This chapter highlights the high adsorption capacity of soil Fe-Mn concretions and nodules for many metal contaminants. The geochemical association of various metals with either Mn or Fe rich phase in Fe-Mn concretions and nodules are a primary environmental procedure that controls the dynamics of these contaminants in the soil system. The formation of Fe-Mn concretions and nodules is the most efficient and durable process for metal contaminants sequestration in the soils. Since the formation of soil concretions has a potentially beneficial effect on metals availability, the application of these environmental materials as geochemical reactors to improve the efficiency of *in situ* technologies for remediating metal contaminated soils is strongly recommended.

**Keywords** Soil • Contamination • Redox process • Fe – Mn oxides • Heavy metals • Remediation technologies • Fe-Mn concretions and nodules • High sorption capacity • Arsenic • Toxic metal sequestration • Environmental geochemistry • Lead • Chromium

## 11.1 Introduction

In terrestrial ecosystems the soil is of central significance because as a very important "ecological crossroad" it is the place where many kinds of interactions take place between solids, liquids, gases and the biota (Fig. 11.1). Soil, in an environmental context, is not only a sink to dispose off undesirable materials, but also a transmitter of many contaminant chemicals as metals to surface-ground water, atmosphere and living organisms (Kabata-Pendias 2001; Gasparatos et al. 2005a). Therefore the metal contaminants content of soil governs the composition of these elements in plants and animals.

As a consequence of industrialization during the last centuries, the contamination of soils with toxic metals has become a major environmental concern in many parts of the world. These metals are considered as hazardous pollutants with a long residence time in soils due to their toxicity and lack of biodegradability (Alloway 1995; Adriano 2001).

Toxic metals may be retained by soil components through a number of processes such as electrostatic adsorption, formation of inner-sphere sorption complexes or multinuclear surface complexes, and precipitation of new mineral phases. In soil, Fe and Mn oxides, oxyhydroxides and hydroxides (for the sake of brevity all are called oxides) constitute only a small fraction of the total solid phase but with their high sorption capacity often control the location, mobility and bioavailability of metal contaminants (McKenzie 1980; Contin et al. 2007; Manceau et al. 2007). Because of their high surface area and high surface-charge density, the Fe and Mn compounds

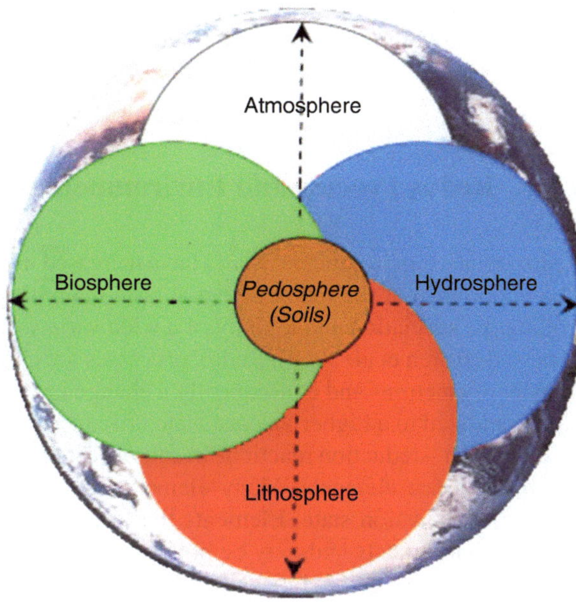


**Fig. 11.1** Schematic view of the soil as an "ecological crossroad" – it is the place where many kinds of interactions take place between solids (Lithosphere), liquids (Hydrosphere), gases (Atmosphere) and the biota (Biosphere) (Adapted from Lin et al. 2005)

are useful for retarding transport of inorganic contaminants in groundwater systems (Stipp et al. 2002) and to some extent for purifying the soil fines from heavy metals (Lombi et al. 2002, 2004). Since Fe and Mn oxides are able to bind metals, the use of Fe- Mn rich materials could be appropriate to treat soils contaminated with toxic metals (McKenzie 1980; Mench et al. 1994; Puschenreiter et al. 2005).

Natural Fe and Mn compounds in soils may display several features with a wide variety of sizes and shapes such as nodules, concretions, coatings, laths, spindles etc. (Latrille et al. 2001; Manceau et al. 2007). Among them, a special place is occupied by the group of Fe-Mn concretions and nodules due to the ability of their Fe and Mn oxides to concentrate and control the distribution and mobility of metals in soils (Childs 1975; Suarez and Langmuir 1976; Zaidelman and Nikiforova 1998).

A number of investigations have focused on soil Fe-Mn concretions and nodules (Dawson et al. 1985; Palumbo et al. 2001; Liu et al. 2002). Furthermore, several studies have shown the high adsorption capacity of soil Fe-Mn concretions and nodules for many toxic metal pollutants and in some cases they are considered the primary environmental material that controls metal dynamics in the soil system (Manceau et al. 2003; Gasparatos et al. 2005b; Gasparatos 2007).

This review was compiled in an effort to assess the general extent of scientific knowledge regarding the properties of Fe-Mn concretions and nodules and their role in the metals sequestration. A literature research was conducted in order to identify (a) the ability of Fe-Mn concretions and nodules on metals immobilization and (b) potentially viable *in situ* remediation technologies using the Fe – Mn oxides (main components of concretions and nodules) for soils where metals are the principal contaminants of concern.

Over 100 papers were reviewed and synthesized to summarize the geochemical role of Fe – Mn concretions and nodules to control metals dynamics in the soil system.

## 11.2 Redox Process and Environmental Issues

Redox-related environmental issues have increased in importance in the last decades. The importance of the redox potential, as a main biogeochemical variable, in controlling the speciation and toxicity of a wide variety of elements have been recently reviewed (Borch et al. 2010). Redox processes are chemical reactions that include a transfer of electrons and consequently a change in valence state of elements that are either oxidized to a higher valence state or reduced to a lower valence state.

Oxidation – reduction reactions in soils affect the biogeochemical cycles of many major and trace elements. Many elements can exist in nature in more than one valence or oxidation state. Elements that occur in more than one valence state in natural environments and may be present as contaminants are listed in Table 11.1. The chemical speciation, bioavailability, toxicity, and mobility of these elements in the environment are directly affected by reduction and oxidation reactions. For example, Cr is insoluble (immobile) under reducing (anaerobic) conditions. In contrast, Fe and Mn are insoluble under oxidizing (aerobic) conditions but are quite soluble (mobile) under anaerobic conditions.

Environmental mobility of other potentially hazardous metals, such as Cd, Ni, and Zn, is indirectly related to redox conditions because these metals form ionic complexes and solid precipitates with redox-sensitive elements.

Redox conditions regulate many of the biogeochemical reactions in the soil system. However, in order to achieve these conditions a number of factors must be simultaneously fulfilled in soil system (Trolard and Bourrie 2008):

- excess of water;
- restriction of the oxygen sources;

**Table 11.1** Redox-sensitive elements potentially present as contaminants in the environment (Essington 2004)

| Elements | Redox state |
|---|---|
| Arsenic (As) | III, V |
| Nitrogen (N) | –III, 0, III, V |
| Carbon (C) | –IV to IV |
| Molybdenum (Mo) | III, IV, V, VI |
| Chromium (Cr) | III, VI |
| Selenium (Se) | –II, 0, IV, VI |
| Copper (Cu) | I, II |
| Sulfur (S) | –II to VI |
| Iron (Fe) | II, III |
| Vanadium (V) | III, IV, V |
| Manganese (Mn) | II, III, IV |
| Antimony (Sb) | III, V |

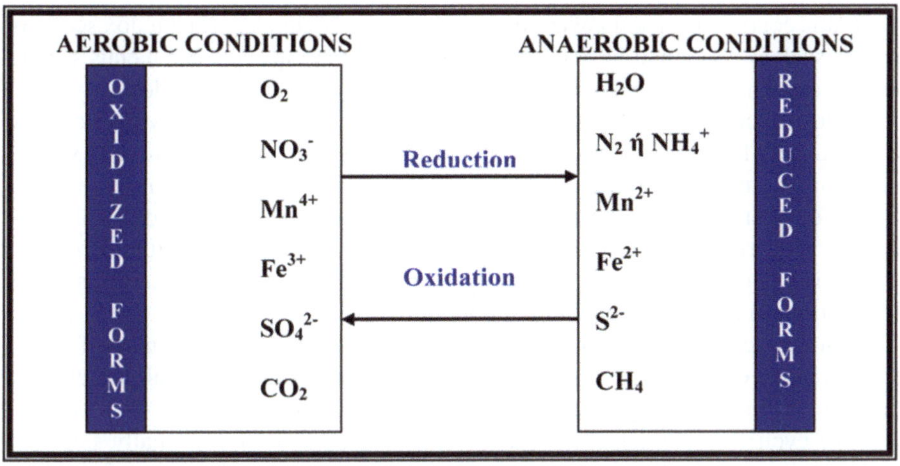


**Fig. 11.2** Transformations of the redox sensitive elements under aerobic and anaerobic soil conditions

- presence of bioavailable substrates as electron donors;
- temperature conditions favorable to microflora activity;
- presence of elements with changeable oxidation state as electron acceptors

Electron donors in soils are: (a) organic matter and organic compounds and (b) reduced forms of inorganic compounds (Fig. 11.2). Electron acceptors are oxidized forms of inorganic compounds such as $O_2$, $NO_3^-$, $Fe^{3+}$, $Mn^{4+}$ (Fig. 11.2). Anaerobic environments such as wetland soils are usually limited by electron acceptors especially $O_2$ and have an abundant supply of electron donors. Once oxygen is consumed, alternative electron acceptors are used with the follow descending sequential order: $NO_3^-$, $Mn^{4+}$, $Fe^{3+}$, $SO_4^{2-}$ and $CO_2$ (Vepraskas 2001). Redox reactions in soils are mainly controlled by microbial activity and the presence of a supply of carbon for the microbes; during respiration, these organisms use organic substances as electron donors (Fiedler et al. 2007). There is a wealth of literature on specific research topics, as well as a comprehensive review specifically directed to the field of organic matter dynamics in anaerobic soils like paddy soils (Kögel-Knabner et al. 2010).

The soil system is in a steady struggle between oxidation, driven by oxygen in the atmosphere and its diffusion in soil pores and water films, and reduction, driven by organisms and organic matter availability. When a soil environment is converted form aerobic to anaerobic drastic changes occur. Prolonged anaerobic conditions change biogeochemical processes in the soil. The solubility of some compounds increase, nutrients can be leached out and toxic compounds may be formed. Both pH and redox potential have long been considered as critical parameters controlling the fate of pollutants in the environment, especially affecting the pollutants reactions on the soil–water interface (Davranche and Bollinger 2000).

Redox conditions of soil govern metals chemistry. Many important redox sensitive metals such as As Cr and Se undergo redox transformation and some metals such as Pb, Cd and Zn are affected indirectly form dissolution of Fe-Mn oxides.

Redox processes can play a role in soil remediation by either removing inorganic contaminants from soil solution or immobilizing them in the solid phase (fixation) or enhancing mobility in the soil profile so that the contaminants can be removed from the system. Thus the redox process may directly or indirectly lead to soil remediation via: (a) changes in speciation to a lower toxicity (b) Solubility decrease to immobilize contaminants or solubility increase to leach out contaminants and (c) adsorption impacts on contaminant mobility.

Redox processes are also responsible for Fe and Mn segregation in the form of Fe–Mn oxide coatings or Fe-Mn concretions-nodules. Among natural particles, Fe and Mn oxides have long been recognized as playing an important role in controlling the cycling of nutrients and trace elements especially in seasonally saturated soil containing naturally high concentrations of trace metals (Chen et al. 2006).

Several examples from the literature will confirm that many studies has been dedicated to the development of the quantitative understanding of contaminant fate in fully aerobic soils (Alloway 1995) or permanently reducing subsurface environments (Reddy and Delaune 2008). Far less research has addressed the biogeochemical functioning of environments characterized by fluctuating redox conditions which lead to the formation of Fe-Mn concretions and nodules (Clausnitzer et al. 2003; Thompson et al. 2006; Gasparatos 2007). An understanding of soil redox processes is crucial for predicting and protecting environmental health and can provide new opportunities for engineered remediation strategies.

## 11.3 Genesis and Properties of Pedogenic Fe-Mn Concretions and Nodules

### *11.3.1 Formation Process and Environmental Conditions*

Since their discovery in marine sediments during the scientific expeditions of the H.M.S. Challenger, iron – manganese concretions and nodules have been studied in a wide range of natural geochemical systems. Fe – Mn concretions and nodules have been observed in oceans (Koschinsky and Halbach 1995; Banerjee et al. 1999), lakes (Schwertmann et al. 1987; Belzile et al. 2001), rivers (Halbach 1976) and soils (Palumbo et al. 2001; Cornu et al. 2005). They are relatively common in soils throughout the world and their presence became known in the late of 1930s with the classic work of Wheeting (1936), Winters (1938) and Drosdoff and Nikiforoff (1940).

Soil Fe-Mn concretions and nodules are discrete bodies formed in the soil system under alternating oxidizing and reducing conditions. These firm to extremely firm subrounded morphological features are formed by the processes of reduction, translocation, and oxidation of Fe and Mn (Zhang and Karathanasis 1997; Zaidelman and Nikiforova 1997; Vepraskas 1999, 2001; Gasparatos et al. 2005b; Chen et al. 2006; Zaidelman et al. 2009). During wet periods, Fe(III) and Mn(III/IV) are reduced and dispersed throughout the soil matrix, while during drying periods when the soil environment becomes more oxidized and oxygen levels gradually increase, they reprecipitate lining or filling matrix pores. The concretions and nodules always have inclusions of the soil matrix and contain soil materials such as skeletal grains, clay minerals and pores cemented together under the influence of Fe and Mn oxides. Therefore Fe-Mn concretions and nodules are characterized by a greater concentration of Fe and Mn oxides than the surrounding soil matrix (Childs 1975; Sidhu et al. 1977; Ram et al. 2001; Gasparatos et al. 2004a; Aide 2005).

Different terms are often used in the literature to describe them as glaebules, septaria, gravels, ferricrete (Brewer 1964; Singh and Gilkes 1996; Matchavariani 2005), but in recent years the terms concretions and nodules are longer accepted (King et al. 1990; White and Dixon 1996; Kanev and Kazakov 1999; Liu et al. 2002; D'Amore et al. 2004; Aide 2005).

As noted by Vepraskas (1999, 2001) these terms have been used interchangeably. However recent studies have shown that as opposed to nodules, concretions have a distinct internal structure with well -expressed concentric rings around a point or plane, and, thus these two redoximorphic features reflect different pedogenetic specifics which can describe the course of soil formation (Gasparatos 2007; Hickey et al. 2008). Soil Fe-Mn concretions and nodules have been found to be widespread in soils with some internal drainage restrictions (Vidhana Arachchi et al. 2004; Zhang and Karathanasis 1997) and have been studied in different soil types (Cornu et al. 2005; Liu et al. 2002; Pai et al. 2003a, b; Palumbo et al. 2001; Latrille et al. 2001; Sanz et al. 1996).

Soil Fe- Mn concretions and nodules formed gradually in the long-term soil pedogenetic processes and they reflect the specific features of the current soil formation. Their shape, structure and elemental composition were the products of soil – forming and environmental conditions. Therefore the elemental composition and distribution characteristics within Fe-Mn concretions and nodules may reflect their forming redox history of the pedoenvironment. However, the effect of concretion accumulations on soil chemical properties may be significant at high concentrations, depending on the chemistry and mineralogy of their oxide components (Gasparatos et al. 2005b).

### *11.3.2 Morphological Properties*

Soil Fe- Mn concretions and nodules are morphological characteristics with significant heterogeneity in relation to the surrounding soil material and can easily be isolated from it and even seen with the naked eye. There is wide variety in shape, colour and size depending on the soil properties and the conditions under which they formed.

Soil Fe-Mn concretions and nodules can occur in several forms, including spherical (Phillippe et al. 1972; Palumbo et al. 2001; Gasparatos et al. 2005b), oval (Latrille et al. 2001), subangular (Childs and Leslie 1977), tubular (Gaiffe and Kubler 1992) and irregularly shaped (Ojanuga and Lee 1973; White and Dixon 1996) with size ranging from a few millimetres (0.25 mm) up to some centimetres (4 cm). Several researchers are studying the size distribution of concretions and nodules in sand size fractions from 0.5 to 2 mm (Rhoton et al. 1991, 1993) and in gravel size fractions (>2 mm) (Zhang and Karathanasis 1997; Gasparatos et al. 2004a). Their colour also varies considerably, with black (black, dark gray) and brown (brown, rusty) colour dominating the majority of studies. Rhoton et al. (1993) found in Glossic Fragiudlafs, Fe-Mn nodules with dominant brown hue 7.5 YR. Liu et al. (2002) observed Fe-Mn nodules of reddish brown colour (2.5 YR 5/4) in an Orthic Agrudalf and Pai et al. (2003b) found Fe nodules in a colour range from reddish yellow to reddish brown in a Typic Paleudult. Manganese is the main factor affecting the colour of concretions and nodules because of the dark colour of Mn oxides (Rabenhorst and Parikh 2000). Generally the darker concretions and nodules appear to have higher Mn content than the red, yellow, orange and/or brown hues and high chromas which indicate dominance of Fe oxides (Phillippe et al. 1972; Sanz et al. 1996; Zhang and Karathanasis 1997; Ram et al. 2001).

The morphology of the Fe-Mn concretions and nodules not only varies between soils but also between the horizons of the same soil profile. Phillippe et al. (1972) found that at the surface soil horizons in a hydrosequence spherical Fe-Mn concretions prevailed whose shape became more irregular with increasing depth. Schwertmann and Fanning (1976) studying two hydrosequences in western Germany found that in soils with better drainage, Fe-Mn concretions exhibited an irregular shape, with moderate hardness, brown (rusty) coatings around a black core, while in soils with impermeable horizons, black spherical concretions with high hardness dominated in surface horizons. In Glossic Fragiudalfs, shape and size of Fe nodules is influenced by the position of the fragipan horizon in the soil profile. The size of the nodules was larger and their shape changes gradually from sub-spherical to sub-angular within and just above fragipan (Lindbo et al. 2000). Gasparatos et al. (2004a, 2006) and Gasparatos (2007) studied Fe-Mn concretions and nodules from Alfisols under the stereolight microscope. Fe-Mn concretions were somewhat spherical in shape and have a 10 YR 4/6 surface coating, encasing a very dark gray (N 3/0) core (Fig. 11.3a) while Fe-Mn nodules show darker colours (Fig. 11.3b).

Research on the role of Fe-Mn concretions and nodules in adsorption of metals has shown the need for the expansion of knowledge on the morphology and fabric characteristics of these soil constituents (Gasparatos et al. 2006). A usually optical microscope observation shows that the concretions and nodules contain grains of primary minerals cemented with an ochreous – brown or dark matrix which consists by Fe – Mn oxides and clay minerals (Fig. 11.4).

Optical microscope observations showed that the Fe-Mn concretions have a differentiated fabric and more specifically a concentric internal structure in which the constituents are arranged in zones or bands around a nucleus, often a primary mineral grain or a void (Fig. 11.5) (Latrille et al. 2001; Gasparatos et al. 2006). However, the degree of expression (number and thickness) of these concentric rings in individual concretions is quite variable. On the contrary Fe – Mn nodules lack well – developed bands and no visible zoning was observed by optical microscope (Fig. 11.6).

**Fig. 11.3** Spherical Fe–Mn concretions (**a**) and nodules (**b**) sample from Greek Alfisols (Gasparatos 2007)

Over the last 30 years the use of scanning electron microscopy (SEM) for micromorphological study of soils has become an important technique to assess the size, morphology, structure of soil minerals (primary and clay minerals) and their interaction with the soil solution (Sposito and Reginato 1992).

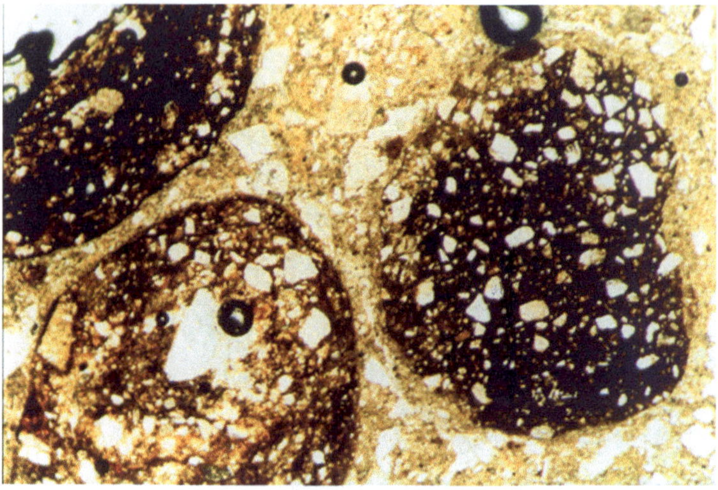

**Fig. 11.4** Thin section of Fe–Mn concretions and nodules viewed by optical microscopy. Their internal structure displays abundant primary mineral grains (Gasparatos 2007)

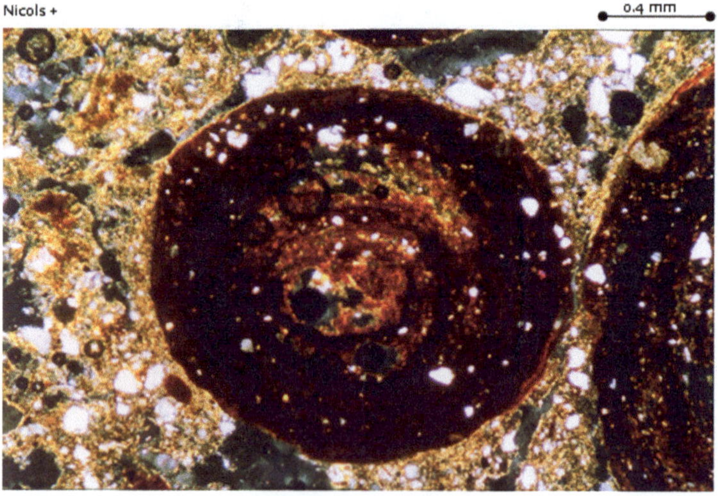


**Fig. 11.5** Photomicrograph showing Fe-Mn concretions from Greek Alfisols characterized by a differentiated fabric (Gasparatos et al. 2006)

Analysis of Fe-Mn concretions and nodules using electron microscopy was originally based on the study of their morphology, their components at high magnifications as well as the qualitative elemental analysis (Cescas et al. 1970; Pawluk and Dumanski 1973). Most recent studies includes the identification of specific

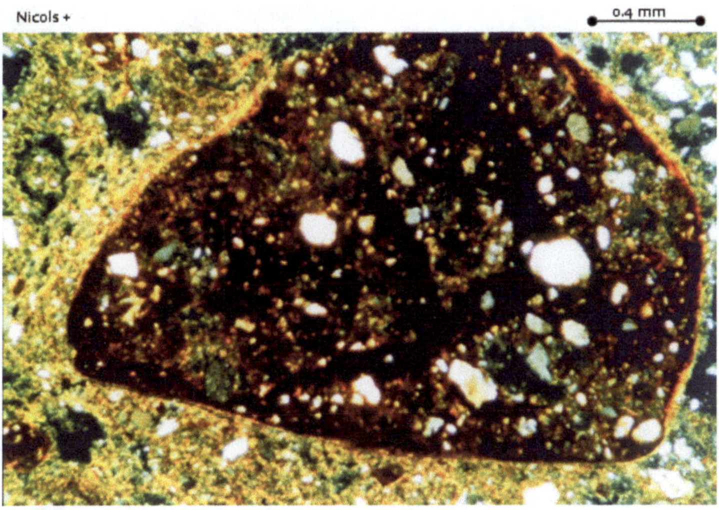


**Fig. 11.6** Photomicrograph showing Fe-Mn nodules from Greek Alfisols characterized by homogenous fabric (Gasparatos 2007)

minerals, quantitative elemental analysis and specification of the structure of Fe-Mn concretions and nodules (White and Dixon 1996; Palumbo et al. 2001; Liu et al. 2002; Cornu et al. 2005; Gasparatos et al. 2005b; Gasparatos 2007).

SEM examination shows that Fe-Mn concretions exhibited a differentiated structure by well-expressed bands whereas Fe-Mn nodules a homogeneous fabric without any specific pattern (Palumbo et al. 2001; Liu et al. 2002). Recently Gasparatos et al. (2005b) showed that Fe–Mn concretions have a distinctive concentric structure of alternating Fe- and Mn-rich zones (Fig. 11.7) while Fe-Mn nodules are characterized by homogeneous distribution of elements (Fig. 11.8). The concentric internal structure of the Fe-Mn concretions indicates a mode of formation in waves as the result of alternating wet and dry periods (Zhang and Karathanasis 1997; Gasparatos et al. 2005b). Soil redox changes significantly influence the geochemical behavior of Fe and Mn compounds which are reduced during wet periods and dispersed in pore waters, while they reprecipitate during dry periods, lining or filling soil pores. Contrasting Fe and Mn concentrations in concretions are attributed to the higher oxidation potential of Mn because of variations in the air movement into the soil pores imposing specific oxidation – reduction conditions (Huang et al. 2008; Aide 2005). The repetition of these cycles leads to the development of an onion like internal structure of Fe-Mn concretions suggestive of seasonal growth (Manceau et al. 2003). Each ring most likely represents the precipitation of soluble soil constituents due to suitable local chemical conditions and a phase in the development of the accretionary material.

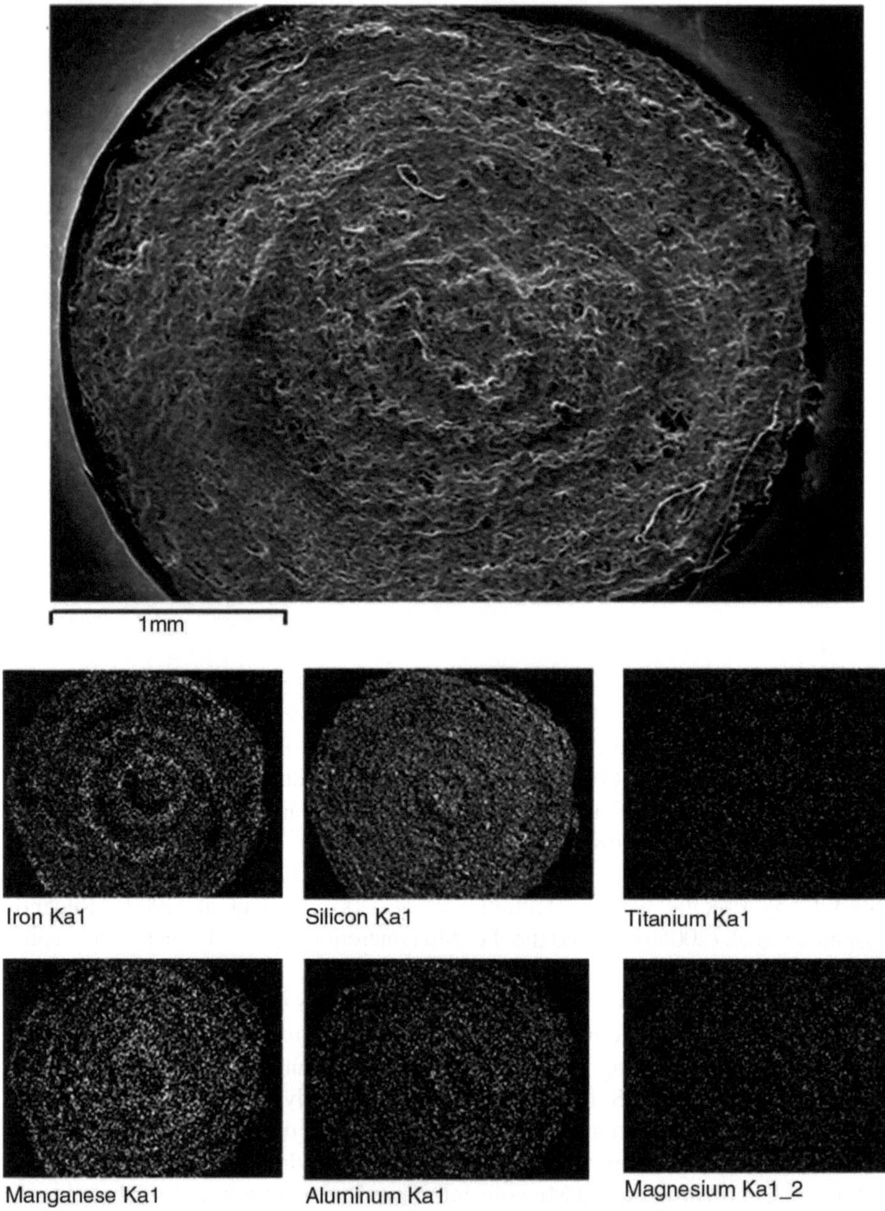


**Fig. 11.7** Representative SEM photomicrograph image showing the internal fabric of a Fe-Mn concretion sample and X-ray mapping of Fe, Si, Ti, Mn, Al and Mg (Gasparatos et al. 2005b)

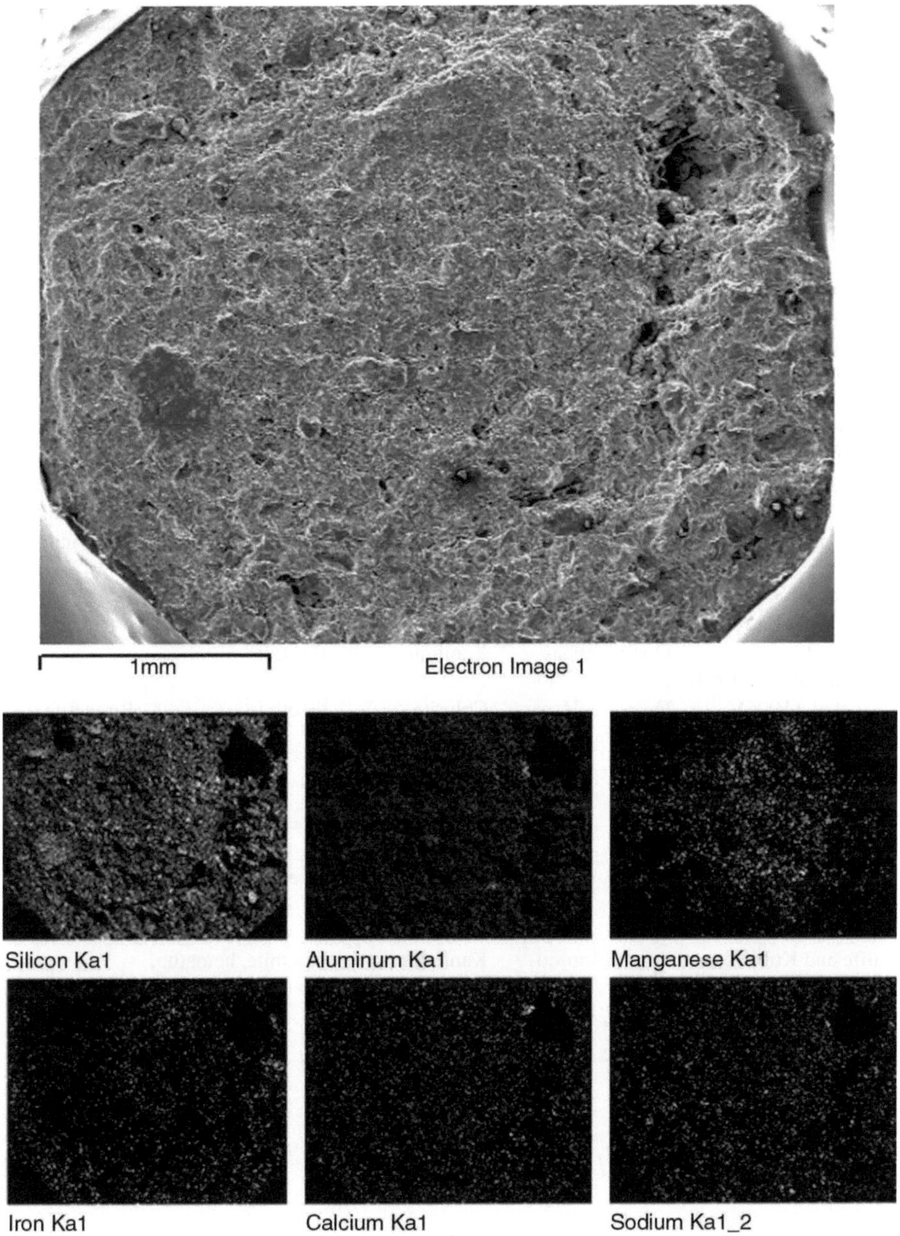


**Fig. 11.8** Representative SEM photomicrograph image showing the internal fabric of a Fe-Mn nodule sample and X-ray mapping of Si, Al, Mn, Fe, Ca and Na (Gasparatos 2007)

**Table 11.2** Literature review of the mineralogical composition of soil Fe – Mn concretions and nodules

| | | Secondary minerals | | |
|---|---|---|---|---|
| | Primary minerals | Clay minerals | Iron oxides | Manganese oxides |
| Gasparatos (2007) | Quartz, albite | Illite, kaolinite, smectite | Goethite, ferrihydrite | Vernadite |
| Aide (2005) | Quartz | – | Goethite, ferrihydrite | – |
| Cornu et al. (2005) | Quartz, feldspars, micas | Kaolinite | Goethite, ferrihydrite | – |
| Vidhana Arachchi et al. (2004) | – | Vermiculite, illite, kaolinite | Goethite | Lithiophorite, birnessite |
| Gasparatos et al. (2004a) | Quartz, feldspars | Illite | Goethite, ferrihydrite | – |
| Liu et al. (2002) | Quartz | Illite, kaolinite | Goethite, ferrihydrite | Lithiophorite, vernadite |
| Ram et al. (2001) | Quartz, feldspars, micas | Kaolinite | Goethite, ferrihydrite, lepidocrocite | Manganite |
| Palumbo et al. (2001) | Quartz | Kaolinite | – | – |
| Zhang and Karathanasis (1997) | Quartz, micas | Kaolinite | Goethite | – |
| Sanz et al. (1996) | Quartz, feldspars, micas, calcite | Chlorite | Goethite | Birnessite, vernadite |
| White and Dixon (1996) | Quartz, micas | Smectite, kaolinite | – | – |
| Singh and Gilkes (1996) | Quartz | Kaolinite | Goethite, hematite, maghemite | |
| Rhoton et al. (1993) | Quartz, feldspars | Vermiculite, illite, kaolinite | Goethite | – |
| Gaiffe and Kubler (1992) | Quartz, muscovite, feldspars | Kaolinite | Goethite, hematite, maghemite | – |
| Sidhu et al. (1977) | Quartz, feldspars | Illite | – | – |
| Childs (1975) | Quartz, feldspars | – | – | – |

## *11.3.3 Mineralogical Composition*

The mineralogical composition of Fe-Mn concretions and nodules is largely determined by the soil environment in which they are formed and the properties of their Fe – Mn oxides. Table 11.2 presents a literature review of the mineralogical composition of Fe-Mn concretions and nodules from various research papers published over the last 30 years.

The majority of studies have shown that the primary minerals of the Fe-Mn concretions and nodules were quartz, feldspars and micas. The mineralogical similarity between the Fe-Mn concretions and nodules and the surrounding soil matrix indicates their pedogenetic origin and their *in situ* formation due to cementation of soil materials by Fe and Mn oxides (Manceau et al. 2003; Gasparatos 2007). The nature of Fe and Mn oxides in soil concretions and nodules is of great interest from the viewpoints of both pedogenesis and environmental chemistry.

The iron and manganese oxides in soil concretions have been reported to appear as x-ray amorphous or very poorly crystalline (Childs 1975). In recent studies the main components of concretions where shown to be ferrihydrite and goethite or hematite with variable amounts of Al as substitutions (Liu et al. 2002; Cornu et al. 2005). Gasparatos et al. (2004a) concluded that the Fe-Mn concretions from Greek Alfisols are only enriched in thermodynamically unstable Fe oxides. The association of the poorly crystalline goethite and ferrihydrite in the Fe-rich concretions of their study reflect an environment containing $Fe^{2+}$, oxidized in the presence of factors that inhibit or retard the transformation of the poorly crystalline oxides to more stable forms. Mn oxides were found to be fine-grained with relatively diffuse XRD peaks (Rhoton et al. 1993; Zhang and Karathanasis 1997; Cornu et al. 2005). Because of their low degree of crystallinity it is difficult to distinguish Mn oxides in soil Fe-Mn concretions and nodules. However efforts combining x-ray diffraction techniques and selective dissolution procedures have been successful in identifying and characterizing Mn minerals like lithiophorite and vernadite (Liu et al. 2002; Tokashiki et al. 1986, 2003; Vidhana Arachchi et al. 2004).

### *11.3.4 Geochemistry*

The study of the geochemical composition of Fe-Mn concretions and nodules originally started with samples mainly from marine sediments (Li 1982). The researchers found that the Fe-Mn nodules from the oceans contained significant quantities of Ni, Cu and Co of economic value (Chauhan et al. 1994; Dutta et al. 2001). The geochemical study of soil Fe-Mn concretions and nodules began in the mid 1970s (Childs 1975) and was expanded during the last decade (Ram et al. 2001; Cornu et al. 2005).

Table 11.3 presents data of enrichment factor (EF) for each element which calculated as "element concentration in concretion-nodule/element concentration in surrounding soil" (Dawson et al. 1985; Gasparatos et al. 2004b). The EF data show that the degree of enrichment varies from element to element, with lower values of Si, Al concentrations in the Fe-Mn concretions and nodules than the soil matrix ($EF < 1$). On the contrary, concretions and nodules are enriched 30–60 times in Mn with respect to the host soil. They are moderately enriched in Fe, whereas they are in most cases depleted in major alkaline earth metals (Ca and Mg). According to the data in Tables 11.4 and 11.5, Fe-Mn concretions and nodules are composed mainly of Si, Al, Fe and Mn. Liu et al. (2002) showed that Si, Al, Fe, Mn, Ca, K and Ti were the main elements presents in Fe-Mn concretions and nodules isolated from a Chinese Alfisol.

**Table 11.3** A literature review of the enrichment factor (EF) for various elements in Fe-Mn concretions and nodules

| | $EF_{Si}$ | $EF_{Al}$ | $EF_{Fe}$ | $EF_{Mn}$ | $EF_{Ca}$ | $EF_{Mg}$ | $EF_{K}$ | $EF_{Na}$ |
|---|---|---|---|---|---|---|---|---|
| Tan et al. (2006) | 0.70 | 0.78 | 2.11 | 57.8 | 2.29 | 0.77 | 0.66 | 1.49 |
| Palumbo et al. (2001) | | 0.95 | 2.34 | 30.7 | 0.14 | 0.56 | 0.78 | 0.47 |
| Zaidelman and Nikiforova (1998) | 0.80 | 0.80 | 5.9 | 30.0 | 1.1 | 0.70 | – | – |
| Childs (1975) | 0.72 | 0.83 | 3.5 | 25.7 | 0.71 | – | 0.89 | – |

**Table 11.4** Major (%) and trace element (ppm) composition of Fe- Mn concretions and nodules of different sizes from Greek Alfisols (Gasparatos 2007)

| | Fe-Mn concretions | | Fe-Mn nodules | |
|---|---|---|---|---|
| | 4.76–2.00 mm | 2.00–1.00 mm | 4.76–2.00 mm | 2.00–1.00 mm |
| Total concentrations (%) | | | | |
| $SiO_2$ | 61.90 | 57.85 | 56.63 | 54.80 |
| $Al_2O_3$ | 10.50 | 10.45 | 9.88 | 11.22 |
| $Fe_2O_3$ | 12.30 | 16.55 | 10.15 | 10.30 |
| MnO | 1.82 | 0.70 | 5.17 | 5.05 |
| CaO | 0.57 | 0.58 | 0.60 | 0.70 |
| MgO | 0.95 | 1.00 | 1.06 | 1.30 |
| $K_2O$ | 1.70 | 1.57 | 1.80 | 1.70 |
| $Na_2O$ | 2.30 | 2.55 | 2.30 | 2.30 |
| Total concentrations (ppm) | | | | |
| Ba | 695–6,790 | 120–1,586 | 1,835–11,670 | 995–8,092 |
| Co | 191–622 | 115–226 | 786–1,198 | 595–1,280 |
| Pb | 302–395 | 350–640 | 375–735 | 450–943 |
| Ni | 201–592 | 186–332 | 506–1,874 | 515–3,360 |
| Cr | 155–351 | 337–529 | 144–230 | 188–302 |
| Cu | 35–48 | 35–67 | 41–96 | 54–97 |
| Zn | 80–244 | 87–109 | 80–161 | 84–526 |

Many studies have shown the general enrichment of Fe – Mn concretions and nodules in trace elements with respect to the surrounding soil matrix (Childs 1975; Sidhu et al. 1977; Dawson et al. 1985; Gasparatos et al. 2004b).

For example, high enrichment factors were observed, in nodules from Sicilian soils for Mn (296), Co (93), Ce (45), Pb (31), Ba (18), Ni (17), Cd (15) and Fe (2.6) (Palumbo et al. 2001). Figure 11.9 shows average enrichment factors of Ba, Co, Ni, Pb, Cr, Cu and Zn in the Fe-Mn concretions and nodules of different sizes from Greek soils (Gasparatos 2007). The EF values vary from element to element and have a great range for elements such as Ba and Co.

Childs and Leslie (1977) found that in each case, the Fe-Mn concretions from soils in New Zealand had high concentrations of Fe, Mn, Ti, Co, S, P, Mo, Cu, V in relation to the surrounding soil mass.

Gasparatos et al. (2004b) show that the Fe-Mn concretions absorbed significant amounts of metals and especially Pb and Cd from the soil system. According to the enrichment factors, the affinity of metals for the concretions due to the charged

**Table 11.5** Major (%) and trace element (ppm) composition of Fe- Mn nodules from a lateritic subsoil (Neaman et al. 2004)

| Element | Soil | Nodules | Enrichment factor |
|---|---|---|---|
| (%) | | | |
| $SiO_2$ | 31.94 | 14.67 | 0.46 |
| $Al_2O_3$ | 27.98 | 12.55 | 0.45 |
| $Fe_2O_3$ | 18.36 | 41.88 | 2.28 |
| MnO | 0.91 | 13.00 | 14.29 |
| MgO | 0.20 | 0.22 | 1.10 |
| CaO | <0.01 | <0.01 | – |
| $Na_2O$ | <0.01 | <0.01 | – |
| $K_2O$ | 0.17 | 0.32 | 1.88 |
| $TiO_2$ | 2.59 | 1.11 | 0.43 |
| $P_2O_5$ | 0.14 | 0.32 | 2.29 |
| (ppm) | | | |
| As | 288 | 1,252 | 4.3 |
| Ba | 176 | 2,435 | 13.8 |
| Cd | 0.3 | 2.8 | 9.3 |
| Co | 24.3 | 216 | 8.9 |
| Cr | 296 | 2,771 | 9.4 |
| Cu | 57.5 | 333 | 5.8 |
| Mo | 22.6 | 130 | 5.8 |
| Ni | 60.6 | 419 | 6.9 |
| Pb | 32.9 | 357 | 10.9 |
| Zn | 103 | 273 | 2.7 |

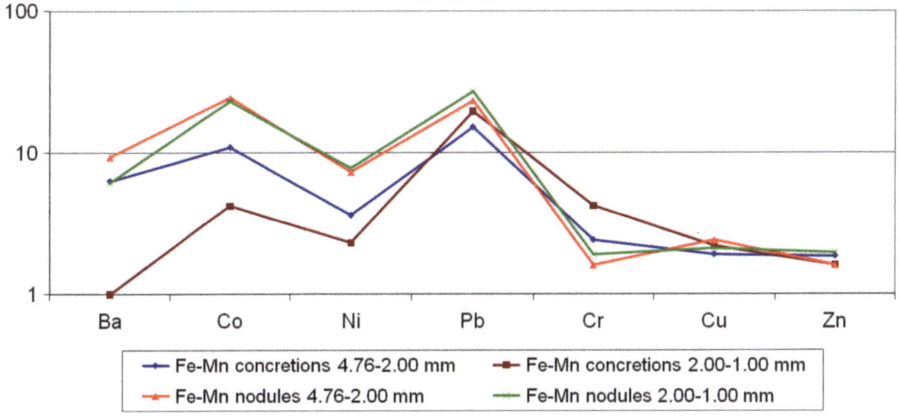


**Fig. 11.9** Average enrichment factors of elements in the Fe-Mn concretions and nodules of different sizes from Greek Alfisols (Gasparatos 2007)

surfaces of Fe and Mn oxides, follow the order Pb>Cd>Mn>Co>Ni>Fe. McKenzie (1980) also found adsorption of Co, Mn, Ni and Pb at the same order on nine synthetic Mn oxides. In Fe-Mn nodules form nine main soils in China, Pb, Cd, Ba and Co had high accumulation, Ni moderate accumulation and Cu, Zn accumulated to a minor degree (Tan et al. 2006).

## 11.4 Role of Pedogenic Fe-Mn Concretions and Nodules in the Environmental Geochemistry of the Soil

Although significant literature data exists on trace metals enrichment of Fe-Mn concretions and nodules, an efficient and robust process for their sequestration, their geochemical speciation is still rather limited (Manceau et al. 2003; Marcus et al. 2004). Researchers have often used total metal concentrations or chemical extraction techniques to provide some insight into metal speciation and bioavailability (Liu et al. 2002; Gasparatos et al. 2005b). These techniques are among the most commonly used methods because of their simplicity and adaptability for routine estimation of metals in a large number of samples. For a more vigorous discussion of these see Ure and Davidson 2002. However there are several reviews that report on the most significant criticisms and pitfalls of the extraction methods of speciation (Pickering 1986; Nirel and Morel 1990). Cutting-edge synchrotron-based techniques such as extended X-ray absorption fine structure (EXAFS) spectroscopy, a surface variant (SEXAFS), X-ray absorption near edge structure spectroscopy (XANES, also called NEXAFS), X-ray microfluorescence ($\mu$XRF) spectroscopy and X-ray microdiffraction ($\mu$XRD), have revolutionized our knowledge on metals reactions and have given detailed insights into metals phase association and speciation in complex systems such as Fe – Mn concretions and nodules without disrupting the natural soil matrix (D'Amore et al. 2005; Vodyanitskii 2006).

### *11.4.1 Iron*

Iron is the fourth element by decreasing order of abundance in the terrestrial crust after O, Si, and Al. Being the most abundant transition metal on the Earth's surface, Fe plays a particularly important role in environmental biogeochemistry (Borch et al. 2010). The biogeochemical cycles of Fe oxides and trace metals are closely linked. An important geochemical feature of Fe-Mn concretions and nodules is their abundance of Fe, particularly in the form of secondary Fe (III) oxyhydroxides such as poorly crystalline goethite (a-FeOOH) and thermodynamically unstable Fe oxides like ferrihydrite (Liu et al. 2002; Gasparatos et al. 2004a; Cornu et al. 2005). Gasparatos et al. (2004a) found that the total Fe content of Fe-Mn concretions from Greece ranged from 14.98% to 16.53% (mean 15.96%) while 78% of the total Fe content in concretions existed in the form of Fe oxides minerals. Tan et al. (2006) noted that the content of Fe in Fe-Mn nodules from China was 1.1–5.4 times greater than those in the corresponding soil matrix.

It is well known and established that Fe oxides have the capacity to sorb toxic metals (McKenzie 1980). According to Aide (2005) iron concentrations of soil nodules from two Alfisols show linear relationships with V, Cr, As, Se, Mo and Pb. Elements that form carbonates and hydroxide complexes and oxyanions were reported as preferentially bound to the slight positively charged FeOOH phases (Childs and Leslie 1977; Palumbo et al. 2001). Chromium and vanadium often occur in isomorphous substitution of Fe (III) in the crystal structure of Fe oxides (goethite, hematite). The chemical structure of many Fe oxides and particularly of amorphous Fe oxides precipitates (like ferrihydrite), which possess a large specific surface area, make them efficient sorbents and therefore sinks for many cations. The metal ions can be sorbed on the surface or incorporated into the oxide structure (Schwertmann and Taylor 1989; Cornell and Schwertmann 1996).

### *11.4.2 Manganese*

The behaviour of Mn in soil is very complex and is controlled by different environmental factors, of which pH-Eh conditions are the most important (Kabata-Pendias 2001). The solubility of Mn is highly sensitive to redox condition. Manganese occurs in two oxidation states: Mn (IV), which is the most stable in neutral to slightly alkaline conditions, and Mn (II), which is stable in reducing conditions (Post 1999).

Although the Mn (II) ion is readily soluble under reducing conditions, Mn is not very mobile, especially under oxidizing conditions, because Mn (IV) ions form insoluble hydrous oxides.

The properties of Mn oxides such as small crystal size, surface-active with large amounts of pH-dependent surface charges, very low zero point of charge (commonly <3), large surface areas (often near 200 $m^2\ g^{-1}$), and high oxidizing capacity have important geochemical implications (Post 1999).

Manganese oxide minerals commonly occur in the environment as coatings and fine-grained aggregates (concretions and nodules). Although Mn is not quantitatively a dominant element (<0.1 wt.%) in the soil environment, the usually large adsorption capacities and scavenging properties of Mn oxide minerals provide one of the primary controls of metals in soil (LaForce et al. 1999; Vidhana Arachchi et al. 2004). According to Vodyanitskii et al. (2004, 2009) when forming oxides in soils, Mn operates as an individual phase carrying metals contaminants and metalloids. Metals such as Cu, Co, Cr, Ni, Pb, and Zn associate to Mn oxides through coprecipitation and substitution so when Mn is reduced the solubility of Pb, Zn, Cu, and Ni increases. In general, the affinity of metal ions for the surface of Mn oxides is: Pb>Cu>Mn>Co>Zn>Ni (McKenzie 1989). Association of Co with Mn oxides was documented by Aide (2005), Cornu et al. (2005) and Tan et al. (2006). Childs (1975) reported that Co and Ba were associated with Mn-rich phases. Associations of Ba with Mn were reported by Ross et al. (1976) and White and Dixon (1996). These metal contaminants absorbed on the Mn oxide minerals may be entrapped within concretions and nodules during Fe-Mn concretions and nodules. Mn oxides (birnessite, todorokite and lithiophorite) are important oxidants and Cr (III) oxidation to the more toxic Cr (VI) have been studied recently (Negra et al. 2005; Dixon and White 2002).

The chemical similarity between Mn and Fe as the dominant redox-sensitive elements present in soil is reflected in their common association in the Fe-Mn

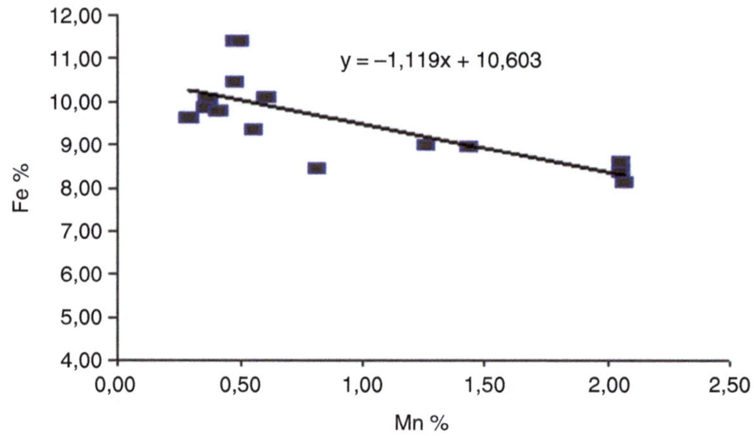


**Fig. 11.10** Correlation of Mn with Fe in soil concretions from Greece (Gasparatos et al. 2005b)

concretions and nodules. The significant negative relationship between Fe and Mn in soil concretions in many studies have shown that that while Fe may deposit on the Mn accumulations as well as on the purer Fe concentrations, Mn tends to deposit at sites where both Mn and Fe concentrations already exist (Fig. 11.10) (Sullivan and Koppi 1992; Gasparatos et al. 2005b). Mn compounds oxidize $Fe^{2+}$ and influence its site of deposition (White and Dixon 1996).

Because Mn is mobilized at higher pH's and under less reductive environments compared with Fe it seem to concentrates at a faster rate than Fe as the concretion formation progresses. Thus at the late stage of concretion development the Mn supply is largely exhausted and the smaller concretions, which have formed more recently than the large ones, showed a higher Fe and lower Mn composition. These trends indicate that the formation and the development of concretions are a dynamic process that controls the translocation and transformation of Fe and Mn in the soil system.

### *11.4.3 Copper*

Copper as the Zn, show low concentrations in Fe-Mn concretions and nodules as evidenced by the low values of enrichment factor for Cu (Palumbo et al. 2001; Cornu et al. 2005; Tan et al. 2006; Gasparatos 2007).

Copper mobility is decreased by sorption to mineral surfaces. $Cu^{2+}$ sorbs strongly to mineral surfaces over a wide range of pH values. The majority of literature shows that Cu in Fe-Mn concretions and nodules associated with the Mn phases (Palumbo et al. 2001; Latrille et al. 2001; Liu et al. 2002). Dawson et al. (1985) using microanalysis found that areas of Fe-Mn nodules rich in Mn had a higher content of Cu. However Mckenzie (1975) noted that Cu tended to be associated with Fe in the soils, but no correlations were found in the soil Fe- Mn nodules. More recently Sen et al. (2002) using batch adsorption studies showed that $Cu^{2+}$ adsorb more strongly on the naturally present colloidal fines, Fe oxide and kaolin than on the solid matrix (sand material).

### *11.4.4 Zinc*

Low Zn content of Fe-Mn concretions and nodules has been found on a number of studies (Childs 1975; Palumbo et al. 2001). Gasparatos (2007) found that the enrichment factor for Zn has the lowest values among metals, ranging from 1.60 to 1.85 for concretions and from 1.6 to 1.96 for nodules. These results shown that despite the high affinity of Zn for Fe-Mn oxides, it is weakly partitioned into Fe-Mn concretions and nodules. The low content of Zn in concretions and nodules may be due to their strong presence in different soil mineral phases such as clay minerals (Childs 1975). Apart from reversible adsorption by cation exchange mechanisms, Zn can also be sorbed in clay minerals irreversibly by lattice penetration. The latter mechanism leads to Zn fixation in excess of the cation exchange capacity and may be due to sorption of Zn in a hydrolysed form and precipitation of $Zn(OH)_2$. Recently Neaman et al. (2004) studying Fe-Mn nodules from a lateritic soil in northern Brazil failed to find any correlation between Fe – Mn oxides and Zn. Manceau et al. (2003) with the synergistic use of three noninvasive synchrotron-based techniques: X-ray microfluorescence (μXRF), X-ray microdiffraction (μXRD), and extended X-ray absorption fine structure spectroscopy (EXAFS) found that Zn in Fe-Mn nodules is bound to four different crystalline phases, in goethite, lithiophorite, phyllosilicates, and the Mn oxide birnessite. Therefore despite the strong tendency of the Fe and Mn oxides to bind and immobilize Zn, partitioning in Fe-Mn concretions and nodules is very low.

### *11.4.5 Cobalt*

High concentrations of Co (~1,000 ppm) were found in Fe-Mn concretions and nodules from many parts of the world (Palumbo et al. 2001; Liu et al. 2002; Gasparatos 2007).

Cobalt is one of the elements most strongly concentrated in marine ferromanganese nodules that can contain up to several weight percent of Co. Pedogenic Fe-Mn concretions and nodules are characterized by much lower concentrations although they are always enriched in relation to the surrounding soil matrix. Recently Gasparatos et al. (2004b) found that the Fe-Mn concretions from Central Greece are enriched in Co approximately eight times in relation to the soil. There has been a considerable debate whether Co is associated with Mn or Fe. Palumbo et al. (2001) found that Co was not very clearly associated with either Mn or Fe phases in ferromanganese nodules from soils of Italy. Other studies have shown a close relationship between Mn and Co in the nodules (McKenzie 1989; Neaman et al. 2008). In many studies the distribution of Co in Fe-Mn concretions and nodules follows that of Mn and shows that the concretions rich in Mn are also enriched in Co. These findings are consistent with the results of Sidhu et al. (1977) that have shown an increase in Co contents with increasing concretion size in soils of India. An explanation for the strong accumulation of Co in to soil concretions and nodules is based in the oxidation of absorbed Co(II) to Co(III) on Mn oxides and incorporation into the Mn oxide mineral structure (Burns 1976). Absorbed Co(II) is oxidized either by oxide-bound Mn(IV), oxide-bound Mn(III), or by $O_2$ directly to produce Co(III) and Mn(II) or Mn(III). The reduced Mn (II,III) is replaced in the Mn oxide mineral structure by Co(III), and subsequently Mn either released to solution or sorbed by the oxide surface. Recently Manceau et al. (1997) using spectroscopic methods showed that Co substitutes for up to 6% of structural Mn in synthetic birnessite.

### *11.4.6 Nickel*

Nickel can occur in a number of oxidation states, but only Ni (II) is stable in the pH and redox conditions found in the soil environment (McGrath 1995). Several studies have shown the strong geochemical affinity of Ni for Fe-Mn concretions and nodules (Palumbo et al. 2001; Liu et al. 2002). Gasparatos (2007) found that the contents of Ni in Fe-Mn concretions from Greece ranged from 186 to 592 ppm with an overall mean 312 ppm. These values compare favorably with those from Indian, Sicilian and French soils (Sidhu et al. 1977; Palumbo et al. 2001; Latrille et al. 2001). Literature data demonstrated that as for Co, high Ni content in concretions and nodules is mostly observed in the Mn rich phases while low Ni content is observed in the concretions where Mn content is low or in other words where Fe predominated over Mn (McKenzie 1975; Sanz et al. 1996). In that case the enrichment of Ni may be though to be due to its adsorption as Ni (II) on the negatively charged surface of Mn oxides (specific adsorption). Recent studies using synchrotron based spectroscopic techniques such as extended X-ray absorption fine structure (EXAFS) spectroscopy have shown the preferential uptake of Ni by birnessite and not ferrihydrite supporting the strong affinity of Ni for Mn oxides which is interpreted by the formation of a pH-dependent inner sphere sorption complex (Manceau et al. 2007). In addition, Manceau et al. (2002) show that in soil ferromanganese nodules, Ni substitutes for $Mn^{3+}$ in the manganese layer of the $MnO_2$-$Al(OH)_3$ mixed – layer oxide lithiophorite.

### *11.4.7 Lead*

Lead is a widespread pollutant in the soil environment with a long residence time compared with most other inorganic pollutants. It is not sensitive to oxidation-reduction over the normal range of Eh and pH values of soils and the geochemical behaviour is dominated by the divalent cation $Pb^{2+}$. Many laboratory experiments have demonstrated the strong affinity of Pb for both Fe and Mn compounds (Sauvé et al. 2000; Businelli et al. 2003). Fe and Mn compounds sorb $Pb^{2+}$ via formation of inner-sphere sorption complexes (Scheinost et al. 2001). In soils with low organic matter these process seems to be the major mechanism controlling the fate and transport of Pb into the aquatic environments and the biota. Fe-Mn concretions and nodules as a complex Fe (III) and Mn (III/IV) mixture accumulate large amounts of Pb.

Gasparatos (2007) found that the contents in Pb of soil Fe – Mn concretions and nodules ranged from 302 to 634 ppm with an overall mean 407 ppm. These total Pb concentrations in Greek Fe-Mn concretions are comparable with those measured by Palumbo et al. (2001) in Italy (range 60–1,181 ppm, mean 657 ppm) but lower than those by Latrille et al. (2001) in France (range 61–3,200 ppm, mean 1,234 ppm). However, there remains some uncertainty about the roles of Fe oxides vs Mn oxides in controlling the adsorption of Pb to these natural heterogeneous soil constituents. Indeed, some researches reported that the Fe-oxides were the single most important determinant of Pb adsorption in soil concretions and nodules (Latrille et al. 2001) while others indicated that the role of Mn-oxides was more significant (Liu et al. 2002; Cornu et al. 2005).

Recently Hettiarachchi et al. (2000) evaluated the effects of P and Mn oxide on bioavailable Pb in five metal-contaminated soils or mine spoils from Kansas and Missouri. Cryptomelane was used as the representative Mn oxide with nine different treatments. The addition of P or cryptomelane reduced bioavailable Pb in all five materials. Nevertheless, the addition of P and cryptomelane together was more effective in reducing bioavailable Pb than the addition of either amendment alone in all materials.

### *11.4.8 Chromium*

Chromium, a redox sensitive contaminant in soils and water, poses a serious threat to humans, plants and animals. Chromium has a broad range of oxidation states (from −2 to +6), but only the +3 and +6 states are stable under most conditions of the surface environment. The two oxidation states of Cr have very different chemical, biological, and environmental properties: Cr(VI) is both highly soluble and toxic to plants and animals, yet Cr(III) is relatively insoluble and an essential micronutrient (Fendorf 1995). Recently Cr-contaminated soils from three sites in Wisconsin were analyzed for the speciation of Cr with X-ray absorption near-edge structure (XANES) spectroscopy. In most samples, less than 10% of the total Cr was present as Cr(VI) (Szulczewski et al. 1997).

The environmental parameters that most significantly affect the speciation of Cr in soil environment are the presence of other components like organic matter, hydrous metal oxides, redox potential and pH. A number of studies focusing on the important role of Fe and Mn oxides in Cr dynamics in the soil system were carried out mostly on synthetic geochemical systems.

Recently, Gasparatos (2007) found that total Cr in Fe-Mn concretions ranged from 155 to 529 ppm with an overall mean of 353 ppm while total Cr in Fe-Mn nodules ranged from 144 to 302 ppm with an overall mean 212 ppm. These values are significantly higher than those mentioned in the literature such as Tan et al. (2006) (44.6 ppm), Aide (2005) (<100 ppm), and Latrille et al. (2001) (119 ppm).

Gasparatos (2007) also found a statistically significant positive correlation between the concentration of Fe and Cr in Fe-Mn concretions from Greece, which is described by the following equation:

$$Y_{Cr} = 0.0088X_{Fe} - 487.44 \quad (r^2 = 0.77,\ p < 0.001)$$

A strong similar linear association of Cr with Fe presented also by Aide (2005) for Fe-Mn nodules from two Alfisols with imperfect drainage (Albaqualfs and Natraqualfs), described by the equation:

$$Y_{Cr} = 4.68X_{Fe} + 6.27 \quad (r^2 = 0.81)$$

Chromium adsorption likely involves isomorphic substitution of $Cr^{3+}$ with Fe, whereas $Cr_2O_4^{2-}$ adsorption may also be expected. Tzou et al. (2003) indicate that the Fe oxide surfaces are important for adsorption of Cr and are widely used in studies remediation of soil contaminated with Cr.

### *11.4.9 Arsenic*

Arsenic is a naturally occurring toxic element and its toxicity, mobility and bioavailability in soil are highly dependent on pH and redox potential. Two inorganic forms of As, arsenite ($As[OH]_3$) and arsenate ($H_2AsO_4^-$ and $HAsO_4^{2-}$), are the main species in soils. Arsenite is commonly the dominant species in moderate to strongly anoxic soil environments and is much more toxic, soluble and mobile than the oxidized form, arsenate (Aide 2005). Several studies using pure synthetic materials have shown the strong geochemical affinity of As for Fe-Mn oxyhydroxides (Sun et al. 1999; Manning et al. 2002). However few papers have focused on the environmental fate of As in relation to pedogenic Fe-Mn concretions and nodules.

Aide (2005) found a statistically significant positive correlation between the concentration of Fe and As in Fe-Mn nodules from two Alfisols on an alluvial terrace in Missouri. Neaman et al. (2008) using electron probe microanalysis concluded that As were mainly present in Fe-oxides in local concentrations up to 13,300 ppm in nodules collected from a tropical soil in Northern Brazil.

In batch experiments, As adsorbed by a Fe-Mn nodule sample decreased with pH in the pH range 6–8 and increased with As addition up to a plateau. In addition As speciation analysis and reductive dissolution of metal oxides proved that almost all the arsenite added to untreated Fe-Mn nodule material was oxidized by manganese oxides (Chen et al. 2006).

The map of the distribution of As in the soil nodules from the Morvan Region in France was determined with the aim of X-ray synchrotron microfluorescence analysis (μSXRF) (Manceau et al. 2002). The As was highly correlated with Fe, such that the As/Fe ratio was roughly the same everywhere in the nodule.

## 11.5 Application of Pedogenic Fe-Mn Concretions and Nodules to Remediating Metal Contaminated Soils

Remediation technologies for metal contaminated soils primarily aim at reducing the risks for human health, i.e. reducing the transfer of contaminants into the food chain or to surface and ground waters. One approach to the *in situ* remediation of soil contaminated with potentially toxic elements is the application of a sorptive mineral, intended to lower the soluble fraction of the element and thus reduce uptake by plants.

Many of the amendments used in soil stabilization are by-products of industrial activities, and are therefore inexpensive and available in large amounts. Overviews on previously successfully applied amending agents and their effectiveness for different metals have been given by Knox et al. (2001) and Puschenreiter et al. (2005).

Amendments that provide sorption sites that have a strong affinity for trace metals have been shown to effectively stabilize metals in soils by limiting the soluble fraction of the metals in the soil matrix. Many reports also indicated that application of Fe and/or Mn in a variety of forms (e.g., hydrous oxides, steel shot, steel sludge, furnace slag, and zero-valent iron) mixed into contaminated soils could reduce the leachability, bioaccessibility, and phytoavailability of Pb, Cd, and Zn (McKenzie 1980; Chen et al. 2000; Hettiarachchi and Pierzynski 2002; Mench et al. 1994). Brümmer et al. (1986) described the importance of Fe hydroxides, especially at neutral to slightly alkaline pH, in lowering the metal content in the soil solution. The soluble and exchangeable fractions of Cd, Ni, and Zn in two contaminated soils were decreased by the application of hydrous Fe oxides. Contin et al. (2008) found that a treatment consisting of repeated cycles of $FeSO_4$ added in solution and followed by rapid air drying and appropriate pH corrections, is able to progressively immobilize toxic metals into precipitated Fe (hydr)oxides.

Due to their high sorption capacity for metals, Fe oxides have been widely suggested as amendments for heavy metal contaminated soils. Their sorption capacity varies with pH and depends on the Fe – Mn minerals. Mn oxides are less abundant in soils than Fe oxides but have been proven to be more reactive components in soil system. This is because their point of zero charge (PZC) is low, varying from 2 to 5 (McKenzie 1989), compared to that of Fe oxides ranging from 6.8 to 8 (Childs and Leslie 1977). Thus, Mn oxides have a negative surface charge at pH values usually encountered in soil. This favours adsorption of metal contaminants at their surface (Cornu et al. 2005; McKenzie 1989).

Although the most efficient and durable process responsible for metal contaminants sequestration in soils is the formation of Fe-Mn concretions and nodules their use as environmental materials poses a challenge. From a geochemical point of view, Fe-Mn concretions and nodules represent reactors at which the migration of metal contaminants is sharply retarded (Timofeeva and Golov 2007). Adsorption and precipitation (formation of multilayered solid species on the oxide surface) are the two most common processes for metal ion deposition in this complex natural matrix. These factors were able to retard the cycling of metal contaminants and played an important role in purifying the soil system. The accumulation of metals in the Fe-Mn concretions and nodules is accompanied by a sharp decrease of their availability. Weakly hydrolyzed cations (e.g. Co, Ni, Cu, Pb etc.) are strongly adsorbed on oxide surfaces via deprotonation of the surface hydroxyl groups.

The redox cycles are important in soils which are subjected to flooded and non flooded periods. Reduction of Fe (III) and Mn (IV) oxides surfaces increases their specific surface area thereby increasing temporarily their capability of sorption (McBride 1994). The re-precipitation of re-oxidized Fe and Mn during oxidation phenomena occludes the sorbed metals within the mineral, virtually forbidding the possibility of interaction with the aqueous phase (Contin et al. 2007). An onion shell-like mechanism of physical sequestration of this kind was demonstrated by Palumbo et al. (2001) and Gasparatos et al. (2005b), for Fe-Mn concretions. A well-defined banded internal structure of the Fe-Mn concretions is consistent with relatively slow accretion rates that provide ample opportunity for adsorption of heavy metals. In contrary the undifferentiated structure of Fe-Mn nodules reflects a rather rapid accretion rate with minor adsorption of metals.

However the changeable nature of soil environment can provoke reductive dissolution of Fe – Mn oxides thereby releasing of metals bound to them into the soil solution. Cornu et al. (2009) investigating the dynamics of Fe–Mn nodules and two associated trace elements, Co and Pb, under controlled redox conditions in a column experiment showed that Pb would not be released from soil during nodule dissolution, whereas Co, which is less toxic, would be released. According to Davranche and Bollinger (2000) the remobilization of metals was dependent on the crystalline state of the solid and on metal/solid adsorption or precipitation. Nevertheless, Fe-Mn concretions and nodules as indurated precipitates are very stable during pedogenesis and thus therefore can be used as reliable geochemical barriers for metal contaminants in the soil system (Gasparatos 2007).

## 11.6 Conclusions

Redox reactions can play a significant role in modifying metal contaminants solubility and availability. The literature review indicates that the formation of Fe-Mn concretions and nodules, as the major depositional form of Fe and Mn oxides in terrestrial ecosystems, is the most efficient and durable process for metals sequestration in soils with alternating redox issues. Since the formation of soil concretions and nodules has a potential beneficial effect on metal contaminants availability, the application of these environmental materials as geochemical reactors to improve the efficiency of *in situ* technologies for remediating metal contaminated soils is strongly recommended.

**Acknowledgments** The author would like to thank two anonymous reviewers for their helpful comments that significantly improved the manuscript as well as Ms K. Kokkinou for her valuable assistance.

## References

Adriano DC (2001) Trace elements in terrestrial environments: biochemistry, bioavailability and risks of metals, 2nd edn. Springer, New York

Aide M (2005) Elemental composition of soil nodules from two alfisols on an alluvial terrace in Missouri. Soil Sci 170:1022–1033. doi:10.1097/01.ss.0000187351.16740.55

Alloway BJ (1995) Heavy metals in soils, 2nd edn. Blackie/Academic and Professional, London/Glasgow/Weinhein/New York/Tokyo/Melbourne

Banerjee R, Roy S, Dasgupta S, Mukhopadhyay S, Miura H (1999) Petrogenesis of ferromanganese nodules from east of the Chagos Archipelago, Central Indian Basin, Indian Ocean. Mar Geol 157:145–158. doi:10.1016/S0025-3227(98)00156-X

Belzile N, Chen YW, Grenier M (2001) Freshwater metallic concretions from an acidic lake characterized by X-ray energy dispersive spectrometry. Can J Anal Sci Spectrom 46:145–151

Borch T, Kretzschmar R, Kappler A, Van Cappellen P, Ginder-Vogel M, Voegelin A, Campbell K (2010) Biogeochemical redox processes and their impact on contaminant dynamics. Environ Sci Technol 44:15–23. doi:10.1021/es9026248

Brewer R (1964) Fabric and mineral analysis of soils. Wiley, New York

Brümmer GW, Gerth J, Herms U (1986) Heavy metal species, mobility and availability in soils. Z Pflanzenernähr Bodenkd 149:382–398

Burns RG (1976) The uptake of cobalt into ferromanganese nodules, soil, and synthetic manganese (IV) oxides. Geochim Cosmochim Acta 40:95–102. doi:10.1016/0016-7037(76)90197-6

Businelli M, Casciari F, Businelli D, Gigliotti G (2003) Mechanisms of Pb (II) sorption and desorption at some clays and goethite-water interfaces. Agronomie 23:219–225. doi:10.1051/agro:2002085

Cescas MP, Tyner EH, Harmer RS (1970) Ferromanganiferous soil concretions: a scanning electron microscope study of their microscope structures. Soil Sci Soc Am Proc 34:641–644

Chauhan OS, Gujar AR, Rao ChM (1994) On the occurrence of ferromanganese micronodules from the sediments of the Bengal fan: a high terrigenous sediment input region. Earth Planet Sci Lett 128:563–573

Chen ZS, Lee GJ, Liu JC (2000) The effects of chemical remediation treatments on the extractability and speciation of cadmium and lead in contaminated soils. Chemosphere 41:235–242. doi:10.1016/S0045-6535(99)00416-6

Chen Z, Kim KW, Zhu YG, McLaren R, Liu F, He JZ (2006) Adsorption (As III, V) and oxidation (As III) of arsenic by pedogenic Fe-Mn nodules. Geoderma 136:566–572. doi:10.1016/j.geoderma.2006.04.012

Childs CW (1975) Composition of iron-manganese concretions from some New Zealand soils. Geoderma 13:141–152. doi:10.1016/0016-7061(75)90063-4

Childs CW, Leslie DM (1977) Interelement relationships in iron-manganese concretions from a catenary sequence of yellow-grey earth soils in loess. Soil Sci 123:369–376. doi:10.1097/00010694-197706000-00005

Clausnitzer D, Huddleston JH, Horn E, Keller M, Leet C (2003) Hydric soils in a Southeastern vernal pool. Soil Sci Soc Am J 67:951–960

Contin M, Mondini C, Leita L, De Nobili M (2007) Enhanced soil toxic metal fixation in iron (hydr)oxides by redox cycles. Geoderma 140:164–175. doi:10.1016/j.geoderma.2007.03.017

Contin M, Mondini C, Leita L, Zaccheo P, Crippa L, De Nobili M (2008) Immobilisation of soil toxic metals by repeated additions of Fe(II) sulphate solution. Geoderma 147:133–140. doi:10.1016/j.geoderma.2008.08.006

Cornell RM, Schwertmann U (1996) The iron oxides. VCH Publishers, Weinheim
Cornu S, Deschatrettes V, Salvador-Blanes S, Clozel B, Hardy M, Branchut S, LeForestier L (2005) Trace element accumulation in Mn-Fe-oxide nodules of a planosolic horizon. Geoderma 125:11–24. doi:10.1016/j.geoderma.2004.06.009
Cornu S, Cattle JA, Samouëlian A, Laveuf C, Guilherme LRG, Albéric P (2009) Impact of redox cycles on manganese, iron, cobalt, and lead in nodules. Soil Sci Soc Am J 73:1231–1241. doi:10.2136/sssaj2008.0024
D'Amore DV, Stewart SR, Huddleston JH (2004) Saturation, reduction and the formation of iron-manganese concretions in the Jackson-Frazier wetland, Oregon. Soil Sci Soc Am J 68:1012–1022
D'Amore JJ, Al-Abed SR, Scheckel KG, Ryan JA (2005) Methods for speciation of metals in soils. J Environ Qual 34:1707–1745. doi:10.2134/jeq2004.0014
Davranche M, Bollinger JC (2000) Release of metals from iron oxyhydroxides under reductive conditions: effect of metal/solid interactions. J Colloid Interface Sci 232:165–173. doi:10.1006/jcis.2000.7177
Dawson BSW, Ferguson JE, Campbell AS, Cutler EJB (1985) Distribution of elements in some Fe-Mn nodules and an iron-pan in some gley soils of New Zealand. Geoderma 35:127–143. doi:10.1016/0016-7061(85)90026-6
Dixon JB, White GN (2002) Manganese oxides. In: Dixon JB, Schulze DG (eds) Soil mineralogy with environmental applications, SSSA book series 7. Soil Science Society of America, Madison, pp 367–388
Drosdoff M, Nikiforoff CC (1940) Iron – manganese concretions in Dayton soils. Soil Sci 49:333–345. doi:10.1097/00010694-194005000-00001
Dutta RK, Sideras-Haddad E, Connell SH (2001) Distribution of various components in a hydrogeneous ferromanganese nodule and an Afanasity Nikitin Seamount crust from Indian Ocean – a geochemical study using micro-PIXE. Nucl Instrum Methods B 181:545–550
Essington M (2004) Soil and water chemistry – an integrative approach. CRC Press, Boca Raton
Fendorf SE (1995) Surface reactions of chromium in soils and waters. Geoderma 67:55–71
Fiedler S, Vepraskas MJ, Richardson JL (2007) Soil redox potential: importance, field measurements and observations. Adv Agron 94:1–54. doi:10.1016/S0065-2113(06)94001-2
Gaiffe M, Kubler B (1992) Relationships between mineral composition and relative ages of iron nodules in Jurassian soil sequences. Geoderma 52:343–350. doi:10.1016/0016-7061(92)90045-9
Gasparatos D (2007) Genesis of Fe – Mn concretions and nodules in alfisols of thessaly. PhD thesis, Agricultural University of Athens, Athens, Greece, 275 p
Gasparatos D, Haidouti C, Tarenidis D (2004a) Characterization of iron oxides in Fe-rich concretions from an imperfectly drained Greek soil: a study by selective-dissolution techniques and X-ray diffraction. Arch Agron Soil Sci 50:485–493. doi:10.1080/0365034042000216149
Gasparatos D, Haidouti C, Tarenidis D, Tsagalidis A (2004b) Enrichment factors of heavy metals in iron – manganese concretions from imperfectly drained soils. Bull Geol Soc Greece, vol XXXVI. In: Proceedings of the 10th international congress, Thessaloniki, April 2004, pp 158–163
Gasparatos D, Haidouti C, Adrinopoulos F, Areta O (2005a) Chemical speciation and bioavailability of Cu, Zn and Pb in soils from the National Garden of Athens, Greece. In: Proceedings of the 9th international conference on environmental science and technology, Rhodes island, 1–3 Sep 2005, vol A, pp 438–444
Gasparatos D, Tarenidis D, Haidouti C, Oikonomou G (2005b) Microscopic structure of soil Fe-Mn nodules: environmental implications. Environ Chem Lett 2:175–178. doi:10.1007/s10311-004-0092-5
Gasparatos D, Haidouti C, Haroulis A, Tsaousidou P (2006) Estimation of phosphorus status of soil Fe-enriched concretions with the acid ammonium oxalate method. Commun Soil Sci Plan 37:2375–2387. doi:10.1080/00103620600819891
Halbach P (1976) Mineralogical and geochemical investigations of Finnish lakes ore. Bull Geol Soc Finland 48:33–42
Hettiarachchi GM, Pierzynski GM (2002) *In situ* stabilization of soil lead using phosphorus and manganese oxide: influence of plant growth. J Environ Qual 31:564–572

Hettiarachchi GM, Pierzynski GM, Ransom MD (2000) In situ stabilization of soil lead using phosphorus and manganese oxide. Environ Sci Technol 34:4614–4619

Hickey PJ, McDaniel PA, Strawn DG (2008) Characterization of iron – manganese cemented redoximorphic aggregates on Wetland soils contaminated with mine wastes. J Environ Qual 37:2375–2385. doi:10.2134/jeq2007.0488

Huang L, Hong J, Tan W, Hu H, Liu F, Wang M (2008) Characteristics of micromorphology and element distribution of iron – manganese cutans in typical soils of subtropical China. Geoderma 146:40–47. doi:10.1016/j.geoderma.2008.05.007

Kabata-Pendias A (2001) Trace elements in soils and plants, 3rd edn. CRC Press, Boca Raton

Kanev VV, Kazakov VG (1999) Nodules as a reservoir of mobile elements in podzolic soils. Eurasian Soil Sci 32:308–317

King HB, Torrance JK, Bowen LH, Wang C (1990) Iron concretions in a typic dystrochrept in Taiwan. Soil Sci Soc Am J 54:462–468

Knox AS, Seaman JC, Mench MJ, Vangronsveld J (2001) Remediation of metal-and radionuclides-contaminated soils by *in situ* stabilization techniques. In: Iskandar IK (ed) Environmental restoration of metals-contaminated soils. CRC Press LLC, Boca Raton, pp 21–60

Kögel-Knabner I, Amelung W, Cao Z, Fiedler S, Frenzel P, Jahn R, Kalbitz K, Kölbl A, Schloter M (2010) Biogeochemistry of paddy soils. Geoderma 157:1–14. doi:10.1016/j.geoderma.2010.03.009

Koschinsky A, Halbach P (1995) Sequential leaching of marine ferromanganese precipitates. Geochim Cosmochim Acta 59:5113–5132. doi:10.1016/0016-7037(95)00358-4

LaForce MJ, Fendorf SE, Li GC, Rosenzweig RF (1999) Redistribution of trace elements from contaminated sediments of Lake Coeur d' Alene during oxygenation. J Environ Qual 28:1195–1200

Latrille C, Elsass F, vanOort F, Denaix L (2001) Physical speciation of trace metals in Fe-Mn concretions from a rendzic lithosol developed on Sinemurian limestones (France). Geoderma 100:127–146. doi:10.1016/S0016-7061(00)00083-5

Li YH (1982) Interelement relationship in abyssal Pacific ferromanganese nodules and associated pelagic sediments. Geochim Cosmochim Acta 46:1053–1060. doi:10.1016/0016-7037(82)90058-8

Lin H, Bouma J, Wilding LP, Richardson JL, Kutilek M, Nielsen R (2005) Advances in hydropedology. Adv Agron 85:1–89. doi:10.1016/S0065-2113(04)85001-6

Lindbo DL, Rhoton FE, Hundnall WH, Smeck NE, Bigham JM, Tyler DD (2000) Fragipan degradation and nodule formation in Glossic Fragiudalfs of the Lower Mississippi River Valley. Soil Sci Soc Am J 64:1713–1722

Liu F, Colombo C, Adamo P, He JZ, Violante A (2002) Trace elements in manganese-iron nodules from a Chinese alfisol. Soil Sci Soc Am J 66:661–670

Lombi E, Zhao FJ, Zhang G, Sun B, Fitz W, Zhang H, McGrath SP (2002) *In situ* fixation of metals in soils using bauxite residue: chemical assessment. Environ Pollut 118:435–443. doi:10.1016/S0269-7491(01)00294-9

Lombi E, Hamon RE, Wieshammer G, McLaughlin MJ, McGrath SP (2004) Assessment of the use of industrial by-products to remediate a copper and arsenic-contaminated soil. J Environ Qual 33:902–910

Manceau A, Drits VA, Silvester E, Bartoli C, Lanson B (1997) Structural mechanisms of Co2+ oxidation by the phyllomanganate buserite. Am Miner 82:1150–1175

Manceau A, Tamura N, Celestre RS, Marcus MA, MacDowell AA, Celestre RS, Sublett RE, Sposito G, Padmore HA (2002) Deciphering Ni sequestration in soil ferromanganese nodules by combining X-ray fluorescence, absorption, and diffraction at micrometer scale of resolution. Am Miner 87:1494–1499

Manceau A, Tamura N, Celestre RS, MacDowell AA, Geoffroy N, Sposito G, Padmore HA (2003) Molecular-scale speciation of Zn and Ni in soil ferromanganese nodules from loess soils of the Mississippi basin. Environ Sci Technol 37:75–80. doi:10.1021/es025748r

Manceau A, Lanson M, Geoffroy N (2007) Natural speciation of Ni, Zn Ba and As in ferromanganese coatings on quartz using X-ray fluorescence, absorption, and diffraction. Geochim Cosmochim Acta 71:95–128. doi:10.1016/j.gca.2006.08.036

Manning BA, Fendorf S, Bostick B, Suarez DL (2002) Arsenic (III) oxidation and arsenic (V) adsorption reactions on synthetic birnessite. Environ Sci Technol 36:976–981

Marcus MA, Manceau A, Kersten M (2004) Mn, Fe, Zn and As speciation in a fast-growing ferromanganese marine nodule. Geochim Cosmochim Acta 68:3125–3136. doi:10.1016/j.gca.2004.01.015

Matchavariani LG (2005) Morphogenetic typification of concretions in subtropical podzolic soils of Georgia. Eurasian Soil Sci 38:1161–1172

McBride MB (1994) Environmental chemistry of soils. Oxford University Press, New York

McGrath SP (1995) Chromium and nickel. In: Alloway BJ (ed) Heavy metals in soils, 2nd edn. Blackie/Academic and Professional, London, pp 152–174

McKenzie RM (1975) An electron microprobe study of the relationships between heavy metals and manganese and iron in soils and ocean floor nodules. Aust J Soil Res 13:177–188. doi:10.1071/SR9750177

McKenzie RM (1980) The adsorption of lead and other heavy metals on oxides of manganese and iron. Aust J Soil Res 18:61–73. doi:10.1071/SR9800061

McKenzie RM (1989) Manganese oxides and hydroxides. In: Dixon JB, Weed SB (eds) Minerals in soil environments, 2nd edn, SSSA Book Series. 1. SSSA, Madison, pp 439–465

Mench MJ, Didier VL, Loffler M, Gomez A, Masson P (1994) A mimicked in-situ remediation study of metal-contaminated soils with emphasis on cadmium and lead. J Environ Qual 23:58–63

Neaman A, Mouélé F, Trolard F, Bourrié G (2004) Improved methods for selective dissolution of Mn oxides: applications for studying trace element associations. Appl Geochem 19:973–979. doi:10.1016/j.apgeochem.2003.12.002

Neaman A, Martinez CE, Trolard F, Bourrie G (2008) Trace elements associations with Fe-Mn oxides in soil nodules: comparison of selective dissolution with electron probe microanalysis. Appl Geochem 23:778–782. doi:10.1016/j.apgeochem.2007.12.025

Negra C, Ross DS, Lanzirotti A (2005) Oxidizing behavior of soil manganese: interactions among abundance, oxidation state and pH. Soil Sci Soc Am J 69:87–95

Nirel PV, Morel FM (1990) Pitfalls of sequential extractions. Water Res 24:1055–1056

Ojanuga AG, Lee GB (1973) Characteristics, distribution, and genesis of nodules and concretions in soils of the southwestern upland of Nigeria. Soil Sci 116:282–291

Pai CW, Wang MK, Zhuang SY, King HB, Hwong J-L, Hu HT (2003a) Characterisation of iron nodules in a Ultisol of Central Taiwan. Aust J Soil Res 41:37–46. doi:10.1071/SR02034

Pai CW, Wang MK, Chiang HC, King HB, Hwong J-L, Hu HT (2003b) Formation of iron nodules in a Hapludult of central Taiwan. Can J Soil Sci 83:167–172

Palumbo B, Bellanca A, Neri R, Roe MJ (2001) Trace metal partitioning in Fe-Mn nodules from Sicilian soils. Italy Chem Geol 173:257–269. doi:10.1016/S0009-2541(00)00284-9

Pawluk S, Dumanski J (1973) Ferruginous concretions in a poorly drained soil of Alberta. Soil Sci Soc Am Proc 37:124–127

Phillippe WR, Blevins RL, Barnhisel RI, Bailey HH (1972) Distribution of concretions from selected soils of the inner bluegrass region of Kentucky. Soil Sci Soc Am Proc 36:171–173

Pickering WF (1986) Metal ion speciation—soils and sediments (a review). Ore Geol Rev 1:83–146

Post JE (1999) Manganese oxide minerals: crystal structures and economic and environmental significance. Proc Natl Acad Sci 96:3447–3454. doi:10.1073/pnas.96.7.3447

Puschenreiter M, Horak O, Friesl W, Hartl W (2005) Low-cost agricultural measures to reduce heavy metal transfer into the food chain – a review. Plant Soil Environ 51:1–11

Rabenhorst MC, Parikh S (2000) Propensity of soils to develop redoximorphic color changes. Soil Sci Soc Am J 64:1904–1910

Ram H, Singh RP, Prasad J (2001) Chemical and mineralogical composition of Fe-Mn concretions and calcretes occurring in sodic soils of Eastern Uttar Pradesh. India Aust J Soil Res 39:641–648. doi:10.1071/SR98098

Reddy KR, Delaune R (2008) Biogeochemistry of wetlands: science and applications. CRC Press/Taylor & Francis Group, Boca Raton

Rhoton FE, Meyer LD, Mcchesney DS (1991) Depth-of-erosion assessment using iron-manganese nodule concentrations in surface horizons. Soil Sci 152:389–394

Rhoton FE, Bigham JM, Schulze DG (1993) Properties of iron-manganese nodules from a sequence of eroded fragipan soils. Soil Sci Soc Am J 57:1386–1392

Ross SJ, Franzmeer DP, Roth CB (1976) Mineralogy and chemistry of manganese oxides in some Indiana soils. Soil Sci Soc Am J 40:137–143

Sanz A, Garcia-Gonzalez MT, Vizcayno C, Rodriguez R (1996) Iron-manganese nodules in a semi-arid environment. Aust J Soil Res 34:623–634. doi:10.1071/SR9960623

Sauvé S, Martinez CE, McBride M, Hendershot W (2000) Adsorption of free lead ($Pb^{2+}$) by pedogenic oxides, ferrihydrite and leaf compost. Soil Sci Soc Am J 64:595–599

Scheinost AC, Abend S, Pandya KI, Sparks DL (2001) Kinetic control of Cu and Pb sorption by ferrihydrite. Environ Sci Technol 35:1090–1096

Schwertmann U, Fanning DS (1976) Iron-manganese concretions in hydrosequences of soils in loess in Bavaria. Soil Sci Soc Am J 40:731–738

Schwertmann U, Taylor RM (1989) Iron oxides. In: Dixon JB, Weed SB (eds) Minerals in soil environments, 2nd edn. SSSA, Madison, pp 380–438

Schwertmann U, Carlson L, Murad E (1987) Properties of iron oxides in two Finnish lakes in relation to the environment of their formation. Clay Clay Miner 35:297–304. doi:10.1346/CCMN.1987.0350407

Sen TK, Mahajan SP, Khilar KC (2002) Adsorption of $Cu^{2+}$ and $Ni^{2+}$ on iron oxide and kaolin and its importance on $Ni^{2+}$ transport in porous media. Colloids Surf A Physicochem Eng Aspects 211:91–102

Sidhu PS, Sehagal JL, Sinha MK, Randhawa NS (1977) Composition and mineralogy of iron-manganese concretions from some soils of the Indo-Gangetic Plain in northhwest India. Geoderma 18:241–249. doi:10.1016/0016-7061(77)90034-9

Singh B, Gilkes RJ (1996) Nature and properties of iron rich glaebules and mottles from south-west Australian soils. Geoderma 71:95–120. doi:10.1016/0016-7061(95)00092-5

Sposito G, Reginato RJ (1992) Opportunities in basic soil science research. Soil Science Society of America, Madison, p 107

Stipp SLS, Hansen M, Kristensen R, Hochella MF Jr, Bennedsen L, Dideriksen K, Balic-Zunic T, Leonard D, Mathieu H-J (2002) Behaviour of Fe-oxides relevant to contaminant uptake in the environment. Chem Geol 190:321–337. doi:10.1016/S0009-2541(02)00123-7

Suarez DL, Langmuir D (1976) Heavy metal in a Pennsylvania soil. Geochim Cosmochim Acta 40:589–598. doi:10.1016/0016-7037(76)90105-8

Sullivan LA, Koppi AJ (1992) Manganese oxide accumulations associated with some soil structural pores. I. Morphology, composition and genesis. Aust J Soil Res 30:409–427. doi:10.1071/SR9920409

Sun X, Doner HE, Zavarin M (1999) Spectroscopy study of arsenite [AsIII] oxidation on Mn-substituted goethite. Clays Clay Miner 47:474–480

Szulczewski M, Helmke PA, Bleam WF (1997) Comparison of XANES analyses and extractions to determine chromium speciation in contaminated soils. Environ Sci Technol 31:2954–2959

Tan WF, Liu F, Li YH, Hu YO, Huang QY (2006) Elemental composition and geochemical characteristics of iron-manganese nodules in main soils of China. Pedosphere 16:72–81. doi:10.1016/S1002-0160(06)60028-3

Thompson A, Chadwick OA, Rancourt DG, Chorover J (2006) Iron-oxide crystallinity increases during soil redox oscillations. Geochim Cosmochim Acta 70:1710–1727. doi:10.1016/j.gca.2005.12.005

Timofeeva YO, Golov VI (2007) Sorption of heavy metals by iron – manganic nodules in soils of Primorskii Gregion. Eurasian Soil Sci 40:1308–1315. doi:10.1134/S1064229307120071

Tokashiki Y, Dixon JB, Golden DC (1986) Manganese oxide analysis in soils by combined X-ray diffraction and selective dissolution methods. Soil Sci Soc Am J 50:1079–1084

Tokashiki Y, Hentona T, Shimo M, Vidhana Arachchi LP (2003) Improvement of the successive selective dissolution procedure for the separation of birnessite, lithiophorite and goethite in soil manganese nodules. Soil Sci Soc Am J 67:837–843

Trolard F, Bourrie G (2008) Geochemistry of green rusts and fougerite: a reevaluation of Fe cycles in soils. Adv Agron 99:227–288. doi:10.1016/S0065-2113(08)00405-7

Tzou YM, Wang MK, Loeppert RH (2003) Sorption of phosphate and Cr (VI) by Fe (III) and Cr (III) hydroxides. Arch Environ Contam Toxicol 44:445–453. doi:10.1007/s00244-002-2090-6

Ure AM, Davidson CM (2002) Chemical speciation in soils and related materials by selective chemical extraction. In: Ure AM, Davidson CM (eds) Chemical speciation in the environment, 2nd edn. Blackwell Science Ltd, New York, pp 265–298

Vepraskas MJ (1999) Redoximorphic features for identifying aquic conditions. North Carolina Agric Res Serv Tech Bull 301. North Carolina State University, Raleigh

Vepraskas MJ (2001) Morphological features of seasonally reduced soils. In: Richardson JL, Vepraskas MJ (eds) Wetland soils: genesis, hydrology, landscapes and classification. Lewis Publishers, Boca Raton, pp 163–182

Vidhana Arachchi LP, Tokashiki Y, Baba S (2004) Mineralogical characteristics and micromorphological observations of brittle/soft Fe/Mn concretions from Okinawan soils. Clay Clay Miner 52:462–472. doi:10.1346/CCMN.2004.0520407

Vodyanitskii YN (2006) The composition of Fe-Mn nodules as determined by synchrotron X-ray analysis (review of publications). Eurasian Soil Sci 39:147–156

Vodyanitskii YN, Vasilev AA, Lesovaya SN, Sataev EF, Sivtsov AV (2004) Formation of manganese oxides in soils. Eurasian Soil Sci 37:663–675

Vodyanitskii YN, Vasilev AA, Vlasov MN, Korovushkin AV (2009) The role of iron compounds in fixing heavy metals and arsenic in alluvial and soddy-podzolic soils in the Perm area Eurasian. Soil Sci 42:738–749. doi:10.1134/S1064229309070047

Wheeting LC (1936) Shot soils of western Washington State. Soil Sci 41:35–45

White GN, Dixon JB (1996) Iron and manganese distribution in nodules from a Young Texas Vertisol. Soil Sci Soc Am J 60:1254–1262

Winters E (1938) Ferromanganiferous concretions from podzolic soils. Soil Sci 46:35–45. doi:10.1097/00010694-193807000-00005

Zaidelman FR, Nikiforova AS (1997) On some general regularities of the formation and changes in properties of Mn-Fe concretions in soils of humid landscapes. Arch Agron Soil Sci 41:367–382. doi:10.1080/03650349709366007

Zaidelman FR, Nikiforova AS (1998) Manganese-iron concretions in soils and their change under the effect if gleyification on parent materials of different genesis. Eurasian Soil Sci 31:817–825

Zaidelman FR, Nikiforova AS, Stepantsova LV, Safronov SB, Krasin VN (2009) Manganese, iron, and phosphorus in nodules of Chernozem-like soils on the Northern Tambov Plain and their importance for the diagnostics of gley intensity. Eurasian Soil Sci 42:477–487

Zhang M, Karathanasis AD (1997) Characterization of iron-manganese concretions in Kentuky alfisols with perched water tables. Clay Clay Miner 45:428–439. doi:10.1346/CCMN.1997.0450312

# Chapter 12
# Bioremediation of Arsenic in Contaminated Terrestrial and Aquatic Environments

**Beata B. Dabrowska, Meththika Vithanage, K.R. Gunaratna, Arun B. Mukherjee, and Prosun Bhattacharya**

## Contents



B.B. Dabrowska
Department of Botany, Stockholm University, SE-106 91, Stockholm, Sweden

M. Vithanage (✉)
Chemical and Environmental Systems Modeling Research Group, Institute of Fundamental Studies, Hantana Road, Kandy 20000, Sri Lanka
e-mail: meththikavithanage@gmail.com; meththika@ifs.ac.lk

K.R. Gunaratna
Department of Environmental Microbiology, School of Biotechnology, AlbaNova, Royal Institute of Technology, SE-100 44, Stockholm, Sweden

A.B. Mukherjee
Department of Biological and Environmental Sciences, Helsinki University, FI-00014, P.O. Box 27, Helsinki, Finland

P. Bhattacharya
KTH-International Groundwater Arsenic Research Group, Department of Land and Water Resources Engineering, Royal Institute of Technology, SE-100 44, Stockholm, Sweden

E. Lichtfouse et al. (eds.), *Environmental Chemistry for a Sustainable World: Volume 2: Remediation of Air and Water Pollution*, DOI 10.1007/978-94-007-2439-6_12,

**Abstract** The name Arsenic is derived from the Greek word *arsenikon,* meaning potent. This element occurs in the ecosystem in different oxidation states of which As(III) and As(V) are most common to humans, animals, plant species. As is present in drinking water and soil from natural sources as well as a pollutant from agricultural and industrial processes. Differences in arsenic uptake by different plant species is controlled by many factors such as root surface area, root exudates, and rate of evapotranspiration. Some plant species have high affinity to accumulate arsenic in tissues above ground. Hyperaccumulator plants have a threshold arsenic content above 1,000 $\mu g\ g^{-1}$ DM.

We review bioremediation studies with especial emphasis on biosorption research on different arsenic species, plants and their biomass, agricultural and industry wastes, as well as the biomass of some fungi species. Bioremediation is considered as an alternative technique for the removal of As in groundwater. One of the popular methods among bioremediation techniques, phytoremediation uses living plants to remove arsenic from the environment or to render it less toxic, in bioaccumulation processes. Phytoremediation techniques often do not take into account the biosorption processes of living plants and plant litter. In biosorption techniques, contaminants can be removed by a biological substrate as a sorbent such as bacteria, fungi, algae or vascular plants. Bioremediation assures in situ treatment of polluted soils. Biosorption characteristics, equilibrium and kinetics of different biosorbents have also been addressed here. Evaluation of the current literature suggests that arsenic bioavailability and molecular level phytoremediation processes in bioremediation are crucial for designing phytoremediation technologies with improved, predictable remedial success.

**Keywords** Arsenic toxicity • Phytoremediation • Bioconcentration • Biosorption Cell interaction • Biosorbents • Metabolic processes • Root accumulation Bioavailability • Antagonism • Synergism • Apoplast • Transfer coefficient (TC) Passive uptake • Bioconcentration factor (BCF) • CF • BF • Phytomixing Rhizofiltration • Phytovolatilization • Hydrolic barriers • Bio-oxidation Biomethylation • Phytoextraction • Phytostabilization

## 12.1 Introduction

Arsenic (atomic number 33) is an ubiquitous element in the ecosystem. Its toxicity has been known from ancient times. Man has lived with this element for thousands of years and will have to live with it for thousands more. However, during the last three decades, high concentrations of arsenic in groundwater have been reported in different regions of the world such as the Bengal Delta (West Bengal-India and Bangladesh), (>0.05–3,200 $mg{\cdot}L^{-1}$), (Mandal et al. 1996; Bhattacharya et al. 1997; Chowdhury et al. 2000), Madhya Pradesh (Pandey 2000; Pandey et al. 1999, 2009), and many countries such as China, Inner Mongolia, Iran, USA, Argentina and Canada (WHO 2001; ATSDR 2007; Bhattacharya et al. 2002; Smedley and Kinniburgh 2002; Bundschuh et al. 2009).

Various techniques have been developed for arsenic removal from aqueous systems such as precipitation, adsorption, membrane processes, coagulation-filtration, ion exchange, nano filter and low cost filters such as sand filter unit, ALCAN filter, Bishuddhya filter, SORAS method, three-pitcher filter and so on. Of these, sorption from solution has received more attention due to its high efficiency. However, due to high maintenance costs the use of adsorbents and disposal by phytoremediation has been cited as a promising approach to arsenic removal.

In this study, we present an overview of: (1) environmental behaviour of arsenic in relation to plant-soil and plant-water systems, (2) arsenic compounds in plant cell interactions, focussing on the principal biological process (bioaccumulation and translocation of arsenic), (3) phytoremediation as a bio-remediation technique for arsenic removal, and (4) research carried out on plant and fungi materials useful as biosorbents. Phytoremediation techniques usually do not consider the biosorption processes of living plants and their dead and decaying parts. Combining bioaccumulation and biosorption may improve the effectiveness of arsenic remediation techniques and may reduce the disposal problem of arsenic adsorbed material. However further studies are needed for the development of the methodology and to enhance its cost effectiveness.

## 12.2 Soil/Water – Plant Interaction

Inorganic arsenic (arsenate and arsenite) is highly toxic to plants because it detaches phosphoryzation and inhibits phosphate uptake. At higher concentrations, arsenic interferes with plant metabolic processes and reduces their growth. Under certain conditions it may lead to plant death (Dermatas et al. 2004). The sensitivity of plants to arsenic appears to be determined by the absorption ability of the plants or non-translocation of arsenic to shoots. Beans and other legumes are amongst the most sensitive plants to arsenic toxicity (NRCC 1980). The arsenic toxicity threshold limit for most plants is 40 and 200 $mg{\cdot}kg^{-1}$ in sandy and clay soils, respectively (Tu and Ma 2002). However, it is important to understand that possible toxicity of arsenic depends on its species, and it has been documented that 1 mg $L^{-1}$ soluble arsenic causes damage to plants such as cow peas and barley (Mok and Wai 1994).

Plants can be classified into three basic groups based on the growth in contaminated soil: excluders, indicators, and accumulators (Baker 1981; Prasad 2008) and they have shown four different conceptual strategies of trace element concentrations in plant crops in relation to the increasing of total metal/loids concentrations in the soil.

### *12.2.1 Excluders*

Arsenic excluders restrict uptake and translocation of arsenic to the shoots. They may prevent or retain low, constant arsenic concentration in shoots over a wide-ranging arsenic concentration in the soil solution. Indicator plants can actively accumulate trace elements in their aerial tissues and generally reflect metal levels in the soil. They tolerate the existing concentration level of metals by producing intracellular metal binding compounds (chelators), or alter arsenic compartmentalization by storing metals in non-sensitive plant parts, *Bidens pilosa* (Sun et al. 2009).

### *12.2.2 Accumulators*

Accumulator plants uptake and translocate arsenic to shoots without toxic symptoms (Ross and Kaye 1994; Prasad 2008). Mention has been made of "root accumulators", which have higher arsenic concentration in roots than in shoots, *Carex rostrata*, *Eriophorum angustifolium*, *Phragmites australis*, and *Salix* sp (Stoltz and Greger 2002a, b, 2006a, b, c) and also it has been pointed out that spontaneous vegetation species covered the large toxic mine spill contaminated areas in Spain (Del Rio et al. 2002). The most arsenic tolerant species were *Amaranthus bilitoides* (0.8–120 $mgAs{\cdot}kg^{-1}DM$), *Chamaemelum fuscatum* (7–23 mg $As{\cdot}kg^{-1}DM$), *Convolvulus arvensis* (0.1–26 mg $As{\cdot}kg^{-1}DM$), *Cynodon dactylon* (0.2–40 mg $As{\cdot}kg^{-1}DM$), and *Malva nicaensis* (1–28 mg $As{\cdot}kg^{-1}DM$).

Hyperaccumulator plants are an extreme form of accumulators (Brooks et al. 1977). They can take up and accumulate toxin arsenic up to 1,000 $mg{\cdot}kg^{-1}$ shoot DM, and concentrate more than 2.3% arsenic in above ground biomass, an example of these being Chinese brake fern (*Pteris vittata*) (Porter and Peterson 1975; Kramer 2000; Ma et al. 2001).

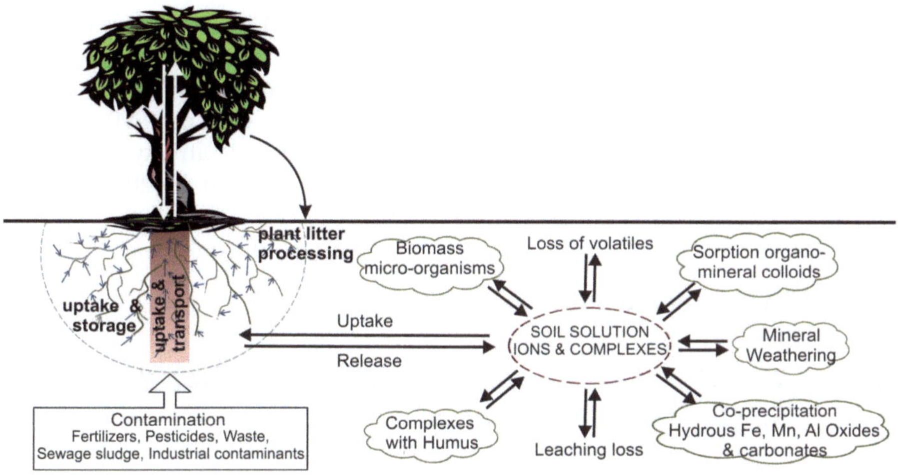


**Fig. 12.1** Arsenic and arsenic compounds in the soil-plant system

## 12.3 Factors Affecting Arsenic Uptake by Plants

The main soil factors affecting arsenic transfer from soil particles to the plant roots are: arsenic ion concentration in solution, bioavailability of arsenic, mass flow with soil solution, decrease in soil pH, soil sorption/desorption ratio, decrease in redox-potential, solubilization, co-precipitation, microbial activity, interaction with other ions (Fig. 12.1).

Ion interactions within plants are divided into three categories: (i) addition, when there is no effect on plant uptake and metabolism; (ii) antagonism, when ions compete with arsenic ions, and/or have taken the same transfer paths (phosphates, $Ca^{+}$, $Cd^{2+}$), and (iii) synergism, when other ions multiplies the stress on the plants (Siedlecka et al. 2001; Sun et al. 2009). In the presence of Cd, an inhibitory effect on arsenic absorption by the plant was also mentioned in the literature. The coexistence of Cd and arsenic decreased the ion activity on the root surfaces, while concentration of arsenic in shoots increases with increasing Cd in soil (Sun et al. 2009). Warren and Alloway (2003) described 11% reduction in arsenic uptake by lettuce (*Lactuca sativa*) after ferrous oxides soil amendment.

The geochemical behaviour of arsenic and phosphate are fairly similar, and the uptake of arsenate through the phosphate uptake system is well documented for different plants including angiosperms (Asher and Reay 1979; Ullrich-Eberius et al. 1989; Mehrag and Macnair 1990; Pickering et al. 2000), mosses (Wells and Richardson 1985), and lichens (Nieboer et al. 1984). It is known that phosphates compete with arsenates for the same uptake system via root plasma membranes (Mehrag and Macnair 1990, 1992). This competition is sensitive in sandy topsoil containing organic compounds.

Phosphate addition to the media suppressed arsenic uptake and its bioaccumulation (Barrachina et al. 1995; Creger and Peryea 1994; Meharg and Hartley-Whitaker

2002; Hoffman et al. 2004; Mkandawire et al. 2005). However Peryea (1998) reported that phosphate can stimulate arsenate desorption from soil by increasing its bioavailability, and also by increasing its water solubility. Phosphates can also reduce toxic effects of arsenate in the plant tissues (Low and Lee 1990; Boisson et al. 1999; Fayiga and Ma 2006). The phosphorous to arsenic affinity of vascular plants is calculated as the discrimination ratio of 4:1 by Naidu et al. (2006).

## 12.4 Phytoremediation Processes

### *12.4.1 Transfer of Arsenic Ions from Bulk Soil to the Root Surface – The Plant Factor*

Some plants have the capacity to alter their membrane permeability, change the metal binding capacity of cell walls, or exude various quantities of chelating substances from the roots to enhance or prevent uptake of elements (e.g. Greger 2008). The bioavailability of arsenic depends on root uptake from the bulk soil. Adsorption depends on root excretions, as well as on symbiotic microorganisms (bacteria and mycorrhizal fungi). Interactions between roots and bacteria may impact arsenic uptake (Asher and Reay 1979; Burd et al. 2000).

Two passive, non-metabolic mechanisms are responsible for arsenic transport from the bulk soil to plant roots and these are: mass flow and diffusion (Fig. 12.2). These processes take place when trace elements are present in the rhizosphere in high concentrations.

The physiological role of hyperaccumulators in the rhizosphere has been described in recent literature (Lasat 2002; Wenzel et al. 2003; Rajkumar and Freitas 2008). A number of studies were focused on metal resistant plant growth-promoting bacteria streams (Burd et al. 2000; Faisal and Hasnain 2005; Zhao et al. 2009). Certain microorganisms enhanced transformation of trace elements by means of various mechanisms including methylation, demethylation, complex formation, oxidation (Alexander 1999; Adriano et al. 2004). However, at present the mechanisms of monomethylarsonic acid and dimethylarsinic acid uptake by plant roots are unknown.

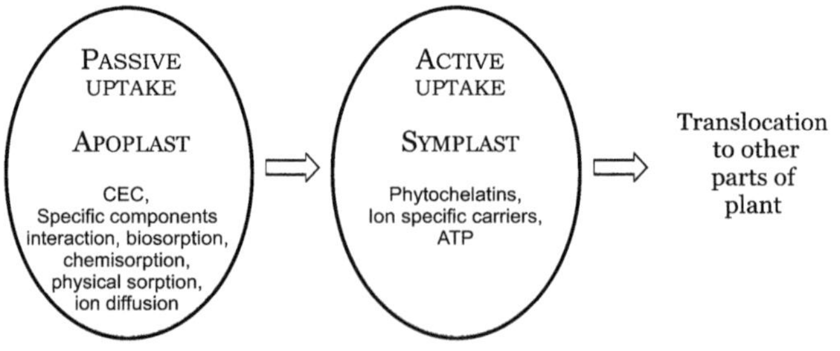


**Fig. 12.2** Active and passive cell processes of the trace elements uptake by plants

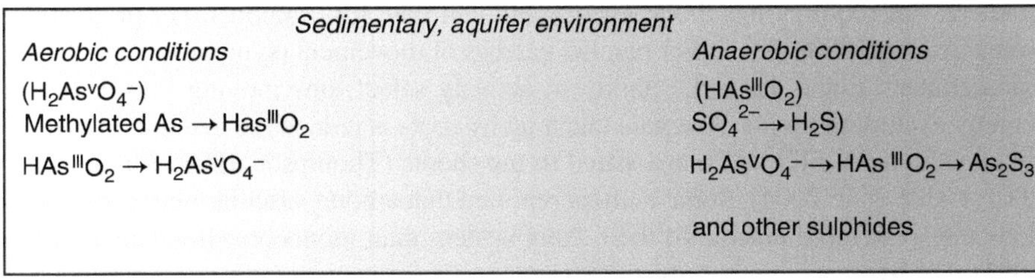


**Fig. 12.3** Arsenic species occur in different aquifer condition (Partly from Jacks and Bhattacharya 1998)

Plants use three separate systems to take up arsenic, (1) passive uptake through the apoplast, (2) direct transcellular transport from the environment to the plant vascular system, and (3) active uptake through the symplast (Ross and Kaye 1994; Marschner 1995; Siedlecka et al. 2001; Greger 2005). The active, metabolic uptake proceeds by the taking up of ions against their concentration gradient. The ATP (Adenosine- Tri-Phosphate) -driven $H^+$ pump creates a pH and electropotential gradient which enhances ion passing to the cell through the selective ion channels or carriers. There are a number of root factors controlling trace element uptake from soils. These are: soil acidification by root exudates, activity and selectivity of the translocators, root membrane activity, strategy mechanisms of avoidance, and release of redactors and/or oxidants, root excretion of chelators, phytosiderophores, acids, and hydrogen ions (Ross and Kaye 1994; Marschner 1995; Siedlecka et al. 2001; Greger 2005). Arsenic can be bound to the sulphides and is available for plant uptake under reduction conditions (Fig. 12.3) (Armstrong et al. 1992; Jacks and Bhattacharya 1998; Brix 1993; Blute et al. 2004).

The transfer of metal ions from soil particles to plant tissues has been described by Kloke et al. (1994), by means of a soil-plant transfer coefficient (TC). The arsenic TC can vary between 0.01 and 0.1 for different plant species. This will be discussed further in this review under the section on bioaccumulation.

## *12.4.2 Translocation Mechanisms from the Root to the Shoot*

Arsenic shows low mobility (unlike phosphorous) with respect to translocation from roots to shoots in plants, except for hyperaccumulators (Zhao et al. 2009). The uptake of arsenic by plant roots varies for different plants or their parts (Stoltz and Greger 2002a, b, 2005). Arsenic ions are taken up, translocated from the roots to the shoot system via the xylem and then redistributed between tissues. The translocation of arsenic ions from root to shoot depends on root pressure and leaf transpiration (Raskin et al. 1997; Caille et al. 2004; Ghosh and Singh 2005).

Passive uptake includes adsorption processes, ion exchange, and diffusion of the ions from the soil solution into the root endodermis (tissues) which can decrease the rate of transportation of some ions. Active uptake takes place against the concentration gradient, but requires metabolic energy and can therefore inhibit toxin production. Some ions have different and/or parallel patches of movement (symplastic, apoplastic, transcellular). Plants have the ability to actively select ions moving into cells and thereby excluding toxins. Arsenate taken up by roots is reduced to arsenite in cytosol by glutathione (GSH) and translocated to the shoots (Thompson 1993; Sattelmacher 2001; Caille et al. 2004). Some authors reported that among vascular plants, dicotyledons transport more arsenic to their shoot system than monocots (Bondada and Ma 2003). The hyperaccumulator fern *P. vittata* localises arsenic in epidermal cells, mesophyl cells and xylem tissues (Bondada et al. 2007). The translocation factor (TF) or shoot:root quotient (SRQ) (Baker and Whiting 2002) expresses arsenic concentration in the shoot in relation to arsenic concentration in the root as:

$$TF = \frac{[As]_{shoot}}{[As]_{root}} \tag{12.1}$$

The Translocation Factor is important in characterising plant capacity in phytoremediation techniques (Asher and Reay 1979; Tu et al. 2002; Sekhar et al. 2007).

Arsenic-tolerant plants translocate less than 2.8% arsenic to the parts above ground leading to low accumulation (Table 12.1). The TF values for wetland plants from mine tailing impoundments were between 0.01 for *Salix sp.* up to 0.2 for *C. rostrata* (Stoltz and Greger 2002a, b). Other arsenic–tolerant plants had values between 0.17 and 0.15 (Fitz and Wenzel 2002). A more comprehensive study carried out by Raab et al. (2007) including 46 plant species, reported a range from 0.01 to 0.9, with a median of 0.09, in arsenate-treated plants.

Hyperaccumulators had values about two magnitudes higher, e.g. 23.86 for *P. vittata* (Ma et al. 2001), and 90.91 for *Pityrogramma calomelanos* (Franseconi et al. 2002). TF values lower than 1 for plants indicates that arsenic transportation to the shoots is limited (Gonzaga et al. 2006). Knudson et al. (2003) reported lower arsenic concentration in the roots of basin wild rye (*Leymus cinereus*) at the evaluated phosphates level, with higher shoot concentration. Higher phosphate concentration stimulated higher root to shoot transfer and about 50% lower arsenic uptake in water fern (*Salvinia natans*) tissues has been observed in the presence of phosphates (Rahman et al. 2008) (Table 12.2). However, there is little knowledge of the extent and mechanisms of As transport in the phloem. Particularly in rice (*Oryza sativa*), As concentrations decrease distinctly from roots to grain (roots > stems and leaves > husks > grain) (Abedin et al. 2002; Liu et al. 2006; Xu et al. 2008). Zhao et al. (2009) pointed out therefore that the contributions of xylem-versus phloem-derived As to the grain need to be evaluated experimentally.

### *12.4.3 Bioconcentration and Bioaccumulation*

Definitions were adapted to differentiate 'bioconcentration' and 'bioaccumulation' in accordance with the International Union of Pure and Applied Chemistry (IUPAC 1996; Gobas et al. 1999); (1) Bioconcentration is the intake of chemical contaminants

**Table 12.1** Bioconcentration factor (BCF) and translocation factor (TF) for arsenic tolerant plant species in soil and sediments

| Species | Treatment (mg As $kg^{-1}$ soil) | BCF (shoots) | BCF (roots) | TF | Sources |
|---|---|---|---|---|---|
| *Pteris vittata* | 500 | 2.3 | – | – | Ma et al. (2001) |
| *P. vittata* | 38.9 | 0.7 | – | – | Ma et al. (2001) |
| *P. vittata* | 50, 100, 200, 500 | 63.3, 59.8, 36.8, 21.0 | – | 24.6, 15.9, 7.5, 4.5 | Tu and Ma (2002) |
| *Pityrogamma calomelanos* | 510 | 5.41–16.37 (fronds) | 0.17–0.61 | 31.36–26.94 | Franseconi et al. (2002), recalculated |
| *Carex rostrata* | Mine tailings $As_{tot}$ 152 ± 8.1 | 0.74 ± 0.15 | 3.51 ± 0,55 | 0.212 (recalculated) | Stoltz and Greger (2002a)[a] |
| *Eriophorum angustifolium* | Mine tailings $As_{tot}$ 152 ± 8.1 | 1.09 ± 0.27 | 36.0 ± 11.6 | 0.030 (recalculated) | Stoltz and Greger (2002a)[a] |
| *Salix sp.* | Tailings $As_{tot}$ 152 ± 8.1 | 0.11 ± 0.004 (stem) | 16.6 ± 3.67 | 0.013 (recalculated) | Stoltz and Greger (2002a)[a] |
| *Leymus cinereus* | 15 and 16 weeks +low P[b] | 1.44 | 72.7 | 0.020 | Knudson et al. (2003), realculated |
| *L. cinereus* | 15, 16 weeks +high P[b] | 2.23 | 24.07 | 0.083 | Knudson et al. (2003), realculated |
| *L. cinereus* | 50, 16 weeks +low P[b] | 0.86 | 25.76 | 0.033 | Knudson et al. (2003), realculated |
| *L. cinereus* | 50, 16 weeks +high P[b] | 1.18 | 18.22 | 0.065 | Knudson et al. (2003), realculated |
| *Talinum cuneifolium* | 2000; control first month | 0.62 | 0.18 | 3.9 | Sekhar et al. (2007) |
| *T. cuneifolium* | 2000 second month | 0.72 | 0.14 | 4.9 | Sekhar et al. (2007) |
| *T. cuneifolium* | 2000 third month | 0.82 | 0.13 | 6.3 | Sekhar et al. (2007) |
| *Launaea arborescens*[c] | Mine tailings | 0.0033 ± 0.004 | – | – | Oyarzun et al. (2008) |

(continued)

**Table 12.1** (continued)

| Species | Treatment (mg As $kg^{-1}$ soil) | BCF (shoots) | BCF (roots) | TF | Sources |
|---|---|---|---|---|---|
| *Lygeum spartum* | Mine tailings | 0.0087 ± 0.011 | – | – | Oyarzun et al. (2008) |
| *Salsola genistoides* | Tailings | 0.0036 ± 0.004 | – | – | Oyarzun et al. (2008) |
| *Bidens pilosa* | 25–125 | 0.27–1.07 | – | 0.03–0.06 | Sun et al. 2008 |
| *Najas marina*[c] | 10.6 (4.87–20.3) -sediment | 0.045 (stem) 0.1 (leaves) | 1.5 | 0.0301 (stem) 0.066 (leaves) | Mazej and Germ (2009), recalculated |
| *Potamogeton lucens*[c] | 10.6 (4.87–20.3) -sediment | 0.043 (stem) 0.052 (leaves) | 1.2 | 0.036 (stem) 0.044 (leaves) | Mazej and Germ (2009), recalculated |

$BF_{bio}$[$(As)_{in\ plant}$: ($NH_4$OAc-extractable As)$_{in\ environment}$] or BCF as the quotient of measured concentration of As in plant material (mg·$kg^{-1}$ dry mass) per initial As concentration in solution; and TF [$(As)_{shoot}$: $(As)_{root}$]; (w) concentration of arsenic forms in water solution

[a]Avg ± SE

[b]Low P = soil amended with 3 mg P·$kg^{-1}$ , respectively high P = 15 3 mg P·$kg^{-1}$

[c]Avg ± SD

**Table 12.2** Bioconcentration factor (BCF) and translocation factor (TF) for arsenic tolerant plant species in water solution

| Species | Treatment (mg As $l^{-1}$ solution) | BCF (shoots) | BCF (roots) | TF | Sources |
|---|---|---|---|---|---|
| *Salvinia minima* | 14.98 As(III) | 60 | – | – | Hoffman et al. (2004), recalculated |
| *S. minima* | 29.97 | 35 | – | – | Hoffman et al. (2004), recalculated |
| *Pteris vittata* | 10 MMA | 309 | 66 | 4.82 | Fayiga et al. (2005) |
| *P. vittata* | 10 As(III) | 211 | 71 | 2.97 | Faisal and Hasnain (2005) |
| *Pteris cretica* | 10 MMA | 98 | 54.3 | 1.81 | Faisal and Hasnain (2005) |
| *P. cretica* | 10 As(III) | 402 | 107 | 4.80 | Faisal and Hasnain (2005) |
| *Salvinia natans* | 0.132 As(III) | 60 | – | – | Rahman et al. (2008) recalculated |

$BF_{bio}[(As)_{in\ plant}$: $(NH_4OAc$-extractable As$)_{in\ environment}]$ or BCF as the quotient of measured concentration of As in plant material (mg·kg$^{-1}$ dry mass) per initial As concentration in solution; and TF $[(As)_{shoot}: (As)_{root}]$; (w) concentration of arsenic forms in water solution

through an organism's epithelial tissues, and the subsequent concentration of the contaminants within the organism's tissues to a level that exceeds ambient environmental concentrations; process that results in an organism having a higher concentration of a substance than is in its surrounding environmental media; (2) Bioaccumulation is the process by which chemical contamination in organisms increases with each step in the food chain.

The trace element concentrations in plant tissues can be expressed as a bioconcentration factor (BCF, BF or CF) and is the quotient of measured concentration of arsenic in plant material (mg·kg$^{-1}$ DM) per initial arsenic concentration in solution (Eisler et al. 1972; Ross 1994; Tu and Ma 2002; Sekhar et al. 2007). The shoot or root accumulation factor ($AF_{bio}$) in calculated as the ratio of the $NH_4OAc$-extractable concentration from the arsenic in plant (Stoltz and Greger 2002a):

$$AF_{bio} = \frac{As_{in\ plant}}{[NH_4OAc - extractableAs]_{in\ environment}} \quad (12.2)$$

Where As in plant is arsenic concentration in the plant root or shoot, or entire plant.

Arsenic concentrations in ranges 5–20 mg·kg$^{-1}$ DM are critical for most of the non-accumulating plants (Kabata-Pendias and Pendias 2001; Lombi et al. 2002). While Del Rio et al. (2002) described wild plants occurring on uncontaminated sites, after arsenic contaminated sludge was spilled in that area values ranged from 0.1 to 0.6 mg As·kg$^{-1}$ DM. Arsenic hyperaccumulator *P. vittata* can accumulate 2.3% of this element in fronds under conditions of arsenic spiked soil, and 0.7% in arsenic contaminated soil (Ma et al. 2001) and about 3,000 mg·kg$^{-1}$ DM As following foliar arsenic application (Bondada et al. 2007). Tu et al. (2002) reported that *P. vittata* fronds accumulated arsenic in concentrations of 7.2 mg·kg$^{-1}$ DM. Hyperaccumulation was attended by production of large biomass depicting 75–99% accumulation

(Ma et al. 2001; Tu et al. 2002). The highest bioconcentration was 63.3% (with range 21.0–63.3%) in soil arsenic contamination successively with 50, 100, 200, and 500 mgAs·kg$^{-1}$ of soil (Table 12.1). Luongo and Ma (2005) reported that *P. vittata* was able to translocate arsenic from roots to fronds, reducing arsenate to arsenite, and maintaining high concentrations of phosphate in the roots which can be contributed to arsenic tolerance and hyperaccumulation. Chandra et al. (2007) reported that the hyperaccumulator plant *Talinum cuneifolium* can uptake arsenic in increased proportion in the presence of acetate, carbonate and phosphate ions, as well as NPK combination, whereas nitrate, chloride and sulphate significantly decreased arsenic uptake.

## 12.5 Phytoremediation Methods

A large number of methods are available for the removal of toxic elements from water and soil by means of plants using this system called phytoremediation. In general, two different types of processes: bioaccumulation and biosorption are responsible for phytoremediation. Bioaccumulation is based on intra-cellular accumulation processes. Biosorption is based on extra-cellular physical, chemical, and chemi-physical processes, including diffusion processes.

Phytoremediation techniques, using living plants are useful to remove, or render harmless arsenic compounds (Fig. 12.4), and bioaccumulation has been employed in several ways to decontaminate arsenic polluted sites using phytoremediation

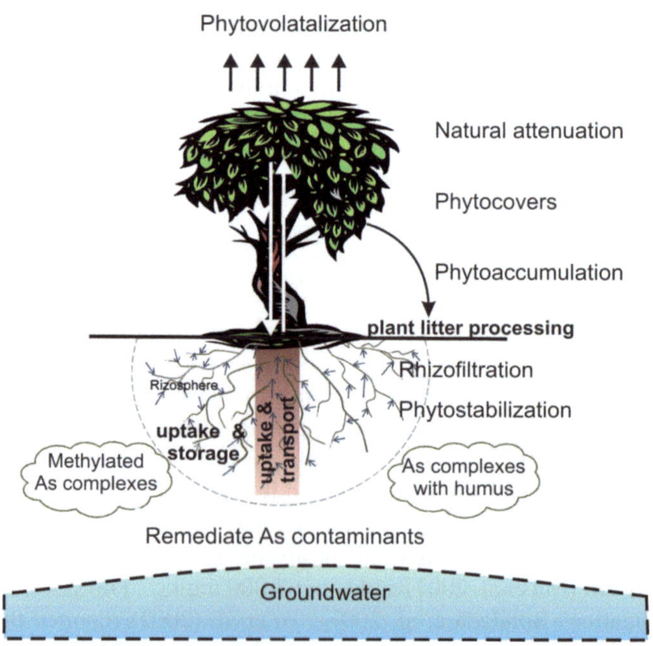


**Fig. 12.4** Phytoremediation possibly methods of arsenic in soil and water environment

techniques (Cunningham and Berti 1993; Suthersan 2002; Prasad 2004). In the roots, ions are diffused into the apoplast and then take up by the symplast which makes the pathway to translocation.

Clemente et al. (2006) showed that active Phytoremediation of *Brassica juncea* followed by spreading organic amendmends such as cow manure and compost as well as lime for fixation of arsenic in a site affected by the toxic spill of pyrites at Aznalcóllar in Spain. In aerobic soils, the retention of arsenic is mainly dependent on iron oxides and hydroxides and these could have been formed in the soil compartment (Kabata-Pendias and Pendias 2001). Sorghum was studied by Haque et al. (2009) as a potential hyperaccumulator for arsenic removal in hydrophonic media and they showed that uptake is completely dependent on arsenic speciation. Use of sedges for arsenic phytoremediation demonstrated that they can be considered as potential species for phytostabilization and rhizofiltration (Alarcon-Herrera et al. 2009).

Arsenic biomass harvested from the phytoremediation process may be a cause of concern. Appropriate storage, disposal, and management are necessary so that there is no threat to the environment (Ernst 2005; Ghosh and Singh 2005).

### *12.5.1 Phytoextraction*

Phytoextraction, phytoaccumulation or phytomining, are the processes used by hyperaccumulators for translocating metal contaminants from soil to the aboveground biomass, which is then harvested to remove contaminants. The main process used in phytoextraction techniques is bioconcentration. Stoltz and Greger (2002a) described the different types of arsenic-accumulating plants. The bioconcentration factor, mentioned as phytoextraction coefficient, is used to assess phytoextraction effectiveness and determine plant possibilities (Ghosh and Singh 2005). Accumulator and hyperaccumulator plants are used in this process. Shoot accumulators, are defined as plants with higher arsenic concentration in their shoots in comparison to the extractable fraction of water soluble arsenic in an arsenic contaminated medium (e.g. *E. angustifolium*). Root-accumulator plants, such as *P. australis, C. rostrata, Salix borealis* and *Salix phylicifolia*, have higher concentrations in the root than that in the shoot tissues (Tables 12.1 and 12.2). They have also mentioned an "intermediate" and "low-accumulator" plant species, with intermediate and lower contaminant concentration in tissues. Arsenic hyperaccumulators such as *P. vittata* and *P. calomelanos* are able to accumulate more than 2,000 mg per kg DM (Gonzaga et al. 2006).

Arbuscular mycorrhizal (AM) fungi play a strong role in uptaking phosphorus and arsenic. The AM fungi system is recognized as an important component of phytoremediation strategies for arsenic. AM plants, such as fern *P. vittata,* consist of a large surface area for adsorption with small fungal hyphae radii; however, uptake kinetics are different, being influenced by chemical alteration of the rhizosphere–hyphosphere, mineralization of organic forms or greater carbon-use efficiency (O'Keefe and Sylvia 1991; Agely et al. 2005). In the last decade, chelate-induced metal mobilization was selected as one of the promising strategies in phytoextraction technology for enhancement of metal uptake from soils. However in recent years, risks posed by EDTA have restricted the use of this method (Wenzel et al. 2003).

### *12.5.2 Rhizofiltration*

The rhizofiltration method is analogous to phytoextraction and phytofiltration methods. In this, plants are used to remove contaminants from solution surrounding the root zone. This technique is used to clean contaminated water such as groundwater, stormwater or waste water using hyperaccumulators and arsenic-tolerant macrophytes such as underwater holly-leaved Naiad (*Najas marina*), pondweed (*Potamogeton lucens*), yellow water-lily (*Nuphar lutea*) with floating leaves (Mazej and Germ 2009), floating *Pistia stratiotes*, *Salvinia herzogi* or certain macroalgaes (Anawar et al. 2008; ITRC 2009). However past studies suggest that the trace metals bioavailability at the rizosphere may improve the remediation methods. The bioavailability in the rizosphere will be critical when considering the efficiency and long term success of the remediation process (Kid et al. 2009).

### *12.5.3 Phytostabilisation*

During the phytostabilisation process, plants are used to immobilize and physically stabilize contaminants in the soil and groundwater through absorption onto roots and accumulation by roots, or precipitation within the rhizosphere. This process reduces the mobility of the contaminant and prevents significant erosion or migration of contaminants in the groundwater and air. This is carried out by arsenic-tolerant plants, soil ammendments such as lime, phosphates, Fe, Al and Mn oxyhydroxides and organic amendments which have been often used on the laboratory scale (Mench et al. 2000, 2003; Brown et al. 2005; Clemente et al. 2006). Typically this form of phytoremediation requires plants to have extensive root systems and adequate soil coverage. Stoltz and Greger (2002a, b; 2006a, b, c) described cottongrass (*Eriophorum*), common reed (*P. australis*), bottle sedge (*C. rostrata*) and Salix as suitable for stabilizing and immobilizing contaminates in mine tailings.

### *12.5.4 Phytovolatilization*

In the phytovolatilization process, plants transpire or diffuse volatile arsenic out of their roots, leaves or stems. In this mechanism, the plants absorb organic and inorganic pollutants from soil or water in the transpiration stream and volatilize them into the atmosphere in a modified or unmodified form at comparatively low concentrations. Methylated and volatile arsenic compounds are involved in this process (Frankenberger and Arshad 2002; US EPA 2002; Adriano et al. 2004). However, literature on phytovolatilized arsenic is sparce. Some fungi species which use the volatalization technique are used in soil contamination remediation. Commercial vegetable crops such as broccoli, cabbage, rice and other plants are quite effective in phytovolatilization, but it has been observed that the presence of sulphate and salinity in soils hinder the volatilization process (Terry and Zayed 1994).

### *12.5.5 Hydraulic Barriers*

Trees and wetland plants are used as hydraulic barriers to control movement of arsenic contaminated surface water and groundwater, as well as to physically stabilize the soil environment (US EPA 2000). Perennial herbs and trees having the capability of absorbing large volumes of water, are used in order to prevent or control the migration of contaminants to groundwater (Rock and Sayre 2000; ITRC 2009). The water uptake rate is given by the Eq. 12.3:

$$\mathrm{U} = (\mathrm{TSCF}) \cdot \mathrm{T} \cdot \mathrm{C} \tag{12.3}$$

where: U = uptake rate of the contaminant, in mg·day$^{-1}$; TSCF = transpiration stream concentration factor, dimensionless, where TSCF is concentration in vegetation/concentration in soil water; T = transpiration rate of vegetation in L·day$^{-1}$; C = aqueous-phase concentration in soil water or groundwater, in mg·L$^{-1}$ (Schnoor 1998).

From the above Eq. 12.3, the uptake rate of the arsenic contaminants can be estimated. First order kinetics can be assumed, as an approximation for the time needed to gain remediation. The uptake rate should be divided by the mass of contaminant remaining in the soil and this will result the first order rate constant for contaminant solution uptake.

$$K = \frac{\mathrm{U}}{\mathrm{Mo}}, \tag{12.4}$$

where: k = first-order rate constant for contaminant solution uptake, in year$^{-1}$; U = contaminant uptake rate, in kg·year$^{-1}$; $M_0$ = mass of contaminant initially in kg (ITRC 2009).

Mature deep-rooted plants e.g. poplar (*Populus*), and aspen (*Populus tremula*), willow (*Salix*), ash (*Fraxinus*), alder (*Alnus*), eucalyptus (*Eucalyptus*), mesquite (*Prosopis*), bald cypress (*Taxodium distichum*), birch (*Betula*), and cedar (*Cedrus*) can typically transport about 3,700–6,200 m$^3$ of water per year at a density of 1,500 trees per acre (4,050 m$^2$). The riparian wetlands can also play the same role and according to Suthersan (2002) and ITRC (2009) can proceed effectively.

### 12.5.6 *Phyto-Covers*

Vegetative covers constructed for surface soil remediation can be designed using plant seed mixes in order to maximize rain interception and transpiration capacities. Established canopies over remediated areas reduce infiltration from precipitation and limit percolation into the deep subsurface. Constructed wetlands can be used as a kind of phyto-cover. In this technique, phytoaccumulation, phytostabilisation and phytovolatilization are present. (ITRC 2009; Suthersan 2002). Experiments carried out in the United Kingdom on brownfield land contaminated by trace metals and arsenic indicated that woody biomass may provide an effective form of phytostabilisation (Frence et al. 2006). Riparian buffers used as hydraulic barriers in combination with phytocover can be used as a filter for agriculture runoff and aquatic protection.

### 12.5.7 *Natural Remediation*

Natural remediation, or natural attenuation (NA) or monitored natural attenuation (MNA) is an in-situ method based on non-invasive, natural processes pursuit to a decreased concentration of contaminants by original physical, chemical, and biological processes(Adriano et al. 2004; US EPA 2004). Naturally occurring rhizosphere with coexisting fungi and bacteria organisms is the most effective tool for volatilisation of methylated arsenic, (Del Rio et al. 2002). However the concept of expansion has been proposed by including acceleration of the native processes by adding inexpensive and minimally invasive amendments to the soil (Adriano et al. 2004). The main intent is not to change but to reduce the bioavailable arsenic fraction by lowering its risk.

## 12.6 Bioremediation

Bioremediation technology is another group of process based on degradation, stabilisation, or volatilization of contaminants from sites using microorganism (fungi and bacteria) (Andrews et al. 2000; Bhattacharya et al. 2002). Remediation of arsenic contaminated soil is based on the ability of micro organisms, especially soil fungal species, to transform arsenic into methylated arsenic and volatilize the arsine which could be trapped. This method can be used to volatilise arsenic aerobically in agricultural soils (Peterson et al. 1981). The main mechanisms of arsenic bioremediation are bio-oxidation and biomethylation. Bio-oxidation is used for the purification of arsenic-contaminated water under aerobic conditions with the help of isolated cultures of *Alcaligenes faecalis* and other species this genus (Osborne and Ehrlich 1976; Ahmann et al. 1997).

Arsenic can be methylated by microorganisms such as aerobic and anaerobic fungi and methanogenic bacteria to produce volatile molecules such as $CH_3As$. Also biomethylation was observed both under aerobic and anaerobic conditions with the help of fungi (Cox and Alexander 1973; Andrews et al. 2000). The rate of biomethylation depends on environmental conditions such as temperature, potential redox, and pH (Fergusson 1990; Wallschläger and London 2008). In aquatic sediments, the anaerobic condition is prevalent. Arsenic methylation typically occurs under phosphate or nitrogen deficient environments (Luongo and Ma 2005), however concentration of nitrates and sulphates can be moderately reduced by bacterial processes (Bhattacharya and Jacks 2000; Bhattacharya et al. 2002). Arsenic Biomethylation is also possible with filamentous fungi e.g. *Scrophulariopsis brevicaulis* and *Aspergillus niger*, *Penicillium* sp., *Candida humicola* and *Gliocladium roseum* (Huysmans and Frankenberger 1991; Andrews et al. 2000). Methylated arsenic compounds such as mono methylarsenic (MMA) and dimethylarsinic (DMA) may be adsorbed to oxides. Adsorption of these compounds is weaker than adsorption of arsenate, however, it is stronger than that of arsenite (Xu et al. 1991; Jing et al. 2005; Gustafsson and Bhattacharya 2008).

## 12.7 Biosorption

Biosorption is defined as a physio-chemical interaction which occur between metal(loid)s and extra-cellular compounds such as polysaccharides, proteins and lipids with amino, carboxyl, phosphate and sulphate groups (Volesky 2003; Febrianto et al. 2009). It is an alternative process for removing arsenic and trace metals from contaminated solutions and a mechanism of metal(loid) immobilization by biomass. (Veglio and Beolchini 1997; Kratochvil and Volesky 1998). Biosorption is defined also as the passive binding of trace elements on a biosorbent (biomass) containing chemically active sites or functional groups (e.g. Volesky 2003). Numerous biological materials have been tested for the removal of toxic arsenic ions from aqueous solution over the last two decades. Arsenic biosorbents were taken from inactive, dead biological biomass, such as algae (Hansen et al. 2006; Vilar et al. 2006), vascular plants (e.g. Mohanty et al. 2006; Pandey et al. 2009), fungi (Dambies et al. 2002; Say et al. 2003; Loukidou et al. 2003; Mohan et al. 2007; Čerňanský et al. 2007), and bacterial materials (e.g. Kuyucak and Volesky 1988). A plant-based biosorbent mainly consists of cellulose and lignins. Primary plant cell walls consist of micelles of cellulose, hemicellulose (including pectins) and glucoproteins. The negatively charged, carboxylic groups act as cation exchangers and anion repellers (Ross and Kaye 1994). An essential step for industrial improvement of biosorption is biomass immobilization which provides biosorbent particles with the adequate size, density and mechanical strength required by continuous systems (Volesky 2003).

### 12.7.1 Biosorption Capacity

Biosorption capacity [qmax] called also as the maximum specific uptake, is an important characteristic of an adsorbent. It is defined as "the amount of adsorbate taken up by the adsorbent per unit of adsorbent mass" (Febrianto et al. 2009). This characteristic depends on a series of properties, such as porosity, pore and particle size distribution, specific surface area, cation exchange capacity (CEC), pH, quantity and type of functional groups, and temperature. Maximum biosorbent uptake is calculated at equilibrium state. Different models have been used to describe arsenic solutions – biosorbent equilibrium, the most common being Languimir, Freundlich, BET, and Dubinin-Radushkevich isotherms (Mohan and Pittman 2007; Febrianto et al. 2009). Most of the biosorbent capacities referred in this review (Table 12.3) are low in relation to the capacity of activated carbon (Table 12.3).

### 12.7.2 Biosorption Mechanisms

Biosorption of heavy metal is a passive non-metabolically mediated process of metal binding by biosorbent (Volesky 2003). Biosorption mechanisms can be divided into two main categories; metabolism dependent (transport across cell membrane and intracellular precipitation and accumulation) and metabolism independent (precipitation, physical and chemical adsorption, ion exchange, complexation) (e.g. Veglio and Beolchini 1997). Metabolism dependent ion transport across cell membranes can take place only in living cells, and is associated with an active defence system of microorganisms and it requires time for reaction. The precipitation of arsenic may take place both in solution and on the cell surface (Ercole et al. 1994). However, it may be dependent on the metabolism of cells if, in the presence of trace metals/metal(loids), organism cells produced compounds which favour the precipitation process (Veglio and Beolchini 1997). Biosorption mechanisms which are independent of cell metabolism are relatively rapid and are reversible (Kuyucak and Volesky 1988; Hoffman et al. 2004; Mukherjee and Kumar 2005). Dead biomass has higher metal uptake capacity and the process is nutrient independent (Aksu et al. 1992).

Adsorption by plant material consists of three steps: surface adsorption (physical and chemical), diffusion into particles, and adsorption and fixation within the mineral particles. The maximum metal uptake and the affinity of the beds for a certain metal are important parameters of the biosorption process. They can be attained from the representative sorption isotherms. Arsenic transport across the cell membrane yields intracellular accumulation, which depends on the metabolism of cells. This means that this kind of biosorption may take place only with viable cells (Veglio and Beolchini 1997). It is often associated with an active microorganisms system, which react in the presence of a toxic substance.

**Table 12.3** Maximum sorption capacity for arsenic by different adsorbents

| Biosorbent | Treatment; time | pH | $C_i$ Initial As concentration (mg As·l$^{-1}$) | Maximum capacity $q_{max}$ (mg As·g$^{-1}$ DM adsorbent) | References |
|---|---|---|---|---|---|
| 1 | 2 | 3 | 4 | 5 | 6 |
| Activated carbon | – | – | – | 0.82–1.02 | Bunnaul et al. (1999) |
| Procaryota | | | | | |
| Gram negative bacteria, dead cells | 1 g biomass·l$^{-1}$, 30 min | 3.0 | 33 (AsV) | 19 (AsV) | Loukidou et al. (2001) |
| Gram negative bacteria, dead cells | 1 g biomass·l$^{-1}$, 30 min | 7.0 | 33 (AsV) | 28 (AsV) | Loukidou et al. (2001) |
| Fungi | | | | | |
| Mycan *P chrysogenum* | Dead biomass | 3 | 1–300 (AsV) | 24.52 (AsV) | Loukidou et al. (2001) |
| *Penicillium purpurogenum* | – | 5.0 | 10–750 (AsIII) | 35.6 (AsIII) | Say et al. (2003) |
| Tea fungi | 45 min; autoclaved | 7.2 | 1.3 As(III) | 1.11 As(III) | Murugesan et al. (2006) |
| Tea fungi | 60 min; $FeCl_3$ treated biomass | 7.2 | 1.3 As(III) | 5.4 As(III) | Murugesan et al. (2006) |
| Tea fungi | 75 min; autoclaved | 7.2 | 0.9 As (V) | 4.95 As (V) | Murugesan et al. (2006) |
| Tea fungi | 75 min; $FeCl_3$ treated biomass | 7.2 | 0.9 As(V) | 10.26 As (V) | Murugesan et al. (2006) |
| Tea fungi | Immersed in $FeCl_3$ | 6–8 | 4 mg·l$^{-1}$ As(V) | 29.81 (AsV) | Mamisahebei et al. (2007), recalculated |
| *Neosartorya fischeri* | Pelletized SAB medium | – | – | 0.388±0.133 | Čerňanský et al. (2007) |
| *N. fischeri* | Compact SAB medium | – | – | 0.783±0.040 | Čerňanský et al. (2007) |
| Algae | | | | | |
| *Codium taylori* | inactivated | 2–4 | – | 90 (As III) | Kuyucak and Volesky (1989) |
| *Lessonia nigrescens* | Green collected, size reduced | 2.5 | 50–600 | 45.2 (AsV) | Hansen et al. (2006) |
| *L. nigrescens* | As above | 4.5 | 50–600 | 33.3 (AsV) | Hansen et al. (2006) |
| *L. nigrescens (macroalga)* | As above | 6.5 | 50–600 | 28.2 (AsV) | Hansen et al. (2006) |
| *Inonotus hispidus* | 30 min | 6 | 4,000 l$^{-1}$ | 51.9 As(III) | Sarı and Tuzen (2009) |

(continued)

**Table 12.3** (continued)

| Biosorbent | Treatment; time | pH | $C_i$ Initial As concentration (mg As·l$^{-1}$) | Maximum capacity $q_{max}$ (mg As·g$^{-1}$ DM adsorbent) | References |
|---|---|---|---|---|---|
| 1 | 2 | 3 | 4 | 5 | 6 |
| *I. hispidus* | 30 min· | 2 | 4,000 l$^{-1}$ | 59.6 As(V) | Sarı and Tuzen (2009) |
| Vascular plants | | | | | |
| Peat (*Sphagnum* sp.) | Compacted, batch-type test | 4.0 | – | 0.06 | Koivula et al. (2008) |
| Orange wastes (*Citrus* sp.) | Temp 30°; | 7–11, 2–6 | – | 0.704 (AsIII), 0.674 (AsIII) | Ghimire et al. (2002) |
| Orange wastes (*Citrus* sp.) | Temp 30°; Phosphorylation | 10 | – | 0. 682 (AsIII) | Ghimire et al. (2003) |
| Orange wastes (*Citrus* sp.) | Phosphorylation | 3 | – | 0. 682 (AsV) | Ghimire et al. (2003) |
| *Moringa oleifera* seeds (drumstick) | 60 min | 7.5, 2.5 | 25 (AsIII) and (AsV) | 0.602 (AsIII), 0.856 (AsV) | Kumari et al. (2006) |
| Pine wood char (*Pinus* sp.) | – | 3.5 | 0.0010–0.100 | 0.0012 (AsIII) | Mohan et al. (2007) |
| Oak wood char (*Quercus* sp.) | – | 3.5 | 0.0010–0.100 | 0.006 (AsIII) | Mohan et al. (2007) |
| Pine bark char (*Pinus* sp.) | – | 3.5 | 0.0010–0.100 | 12 (AsIII) | Mohan et al. (2007) |
| Oak bark char (*Quercus* sp.) | – | 3.5 | 0.0010–0.100 | 0.0074 (AsIII) | Mohan et al. (2007) |
| Rice polish *(Oryza sativa)* | 60 min | 7.0 | 1 (AsIII); | 0.139 (AsIII) | Ranjan et al. (2009) |
| Rice polish *(O. sativa)* | 60 min | 4.0 | 1 (AsV); | 0.147 (AsV) | Ranjan et al. (2009) |
| Fresh biomass of vascular plants | | | | | |
| *Salvinia minima* | Living plants 96 h | 5.5 | 400 µM (29.97) (AsIII); | 1.05 ± 0.06 (AsIII) | Hoffman et al. (2004) |
| *S. minima* | Living plants 96 h | 5.5 | 200 µM (14.98) (AsIII); | 0.9 0 ± 04 (AsIII) | Hoffman et al. (2004) |
| *S. natans* | Living plants 120 h | 7.5 | 0.2 (AsV), 2 (AsV) | 0.748 (AsV), 0.54 (AsV) | Mukherjee and Kumar (2005) |
| *Garcinia cambogia* | Fresh biomass | 6 | 50–2,500 (AsIII) | 0.128 (AsIII) | Kamala et al. (2005) |
| *G. cambogia* | Immobilized biomass | 6 | 50–2,500(AsIII) | 0.704 (AsIII) | Kamala et al. (2005) |
| *Momordica charantia* (bitter melon) | Green collected, size reduced, 45 min | 9.0 | 0.5 (AsIII) | 0.88 (AsIII) | Pandey et al. (2009) |

#### 12.7.2.1 Adsorption

Kuyucak and Volesky (1988) hypothesized that some metal ion biosorption by dead biomass of algae, fungi and yeasts takes place through electrostatic interactions between ions in solution and cells walls.

#### 12.7.2.2 Ion Exchange

Cell walls of microorganisms and other plant material contain polysaccharides as basic building blocks. Bivalent metal ions can exchange with some ions of the polysaccharides (e.g. Veglio and Beolchini 1997). Alginates of marine algae usually occur as natural salts of $K^+$, $Na^+$, $Ca^{2+}$ and/or $Mg^{2+}$. These metallic ions can exchange with a number of other ions such as $Co^{2+}$, $Cu^{2+}$, $Cd^{2+}$ and $Zn^{2+}$, resulting in the biosorptive uptake of the metals (e.g. Veglio and Beolchini 1997; Loukidou et al. 2001, 2003). Phosphorylation of biosorbent leads to higher binding of arsenates to the cell wall (Ghimire et al. 2002, 2003).

#### 12.7.2.3 Complexation

Metal removal from solution could also take place through formation of complexes on the cell surface (e.g. Mcafee et al. 2001). Metal ions can bind on to single ligand or through chelation. Mcafee et al. (2001) described arsenic complexation to chitin and chitosan active groups. However, no other records were found on biosorption through arsenic complexation.

#### 12.7.2.4 Precipitation

As mentioned above, precipitation may be either dependent on the cellular metabolism or independent of it. Metal removal from solution is often associated with the active protective system of microorganisms. In the presence of a toxic metal, they produce compounds which favour the precipitation process (Veglio and Beolchini 1997).

### *12.7.3 Biosorption Characteristics*

This review attempts to summarise recent studies in the removal of arsenic using plants and fungi in situ and as biosorbent materials. Analysis and design of biosorption/adsorption processes involves biosorption/adsorption equilibrium. Equilibrium concentration is a function of temperature. The adsorption equilibrium at a given constant temperature is referred to as adsorption isotherm. The adsorption capacity ($q_t$), the most important characteristic of a biosorbent, is dependent on pore and particle size, specific surface area, pH, cation exchange capacity, functional groups and temperature.

#### 12.7.3.1 Adsorption Equilibrium

The amount of metal ions adsorbed on the biosorbent is determined from the difference between the initial and remaining concentrations of the metal ion in the solution at the sampling time:

$$q_t = \frac{V(C_i - C_t)}{m_B}, \tag{12.5}$$

where: $q_t$ is the metal ion adsorbed, in mg arsenic ion per gram dry matter of biosorbent, at a given time; V is the volume of solution (L); $C_i$ and $C_t$ are the initial concentration and concentration of arsenic ions remaining in the solution, in mg $L^{-1}$ at a given time; and $m_B$ is the dry weight of the biosorbent, in g.

Design of adsorption-desorption separation processes involves relevant adsorption-desorption equilibrium. The adsorption equilibrium relationship at a given adsorbent concentration is referred to as the adsorption isotherm.

#### 12.7.3.2 Equilibrium Models

Equilibrium adsorption isotherms are usually used to determine the capacity of an adsorbent and several equilibrium adsorption isotherm models are available such as Lanmuir, Brunauer-Emmer-Teller (BET), and Freundlich (e.g. Ruthven 1984; Dang et al. 2009). The most common are the mono-layer adsorption described by Langmuir (1918), and the multi-layer adsorption published by Freundlich (1906). The quantity of adsorbed metal as a function of metal concentration in solution is used to estimate the maximum capacity of the adsorbents. Biosorption involves several processes: electrostatic forces and chemical reactions between binding sites and metals; theoretically it is more correct to apply the pseudo second- order model that fits most biosorption processes (Cabral 1992).

The distribution coefficient *Kd*, or partition coefficient (*Kp*) is used as an estimation of soil solubility. Equilibrium between the solid and the aqueous phase is reached by extraction. The partition coefficient, *Kp*, for arsenic, describes the extractable fraction available from plants to the total concentration in the soil matrix, according to the formula (Sardans and Peñuelas 2006):

$$Kp = \frac{As\left(\text{in water extract from soil or sediment}\right)}{\text{As (in soil or sediment)}} \tag{12.6}$$

The main factors affecting equilibrium are: temperature, and influence of biosorption performance in the range 20–35°C (Aksu et al. 1992), pH affect on the solution chemistry of the ions, activity of functional groups in the biomass, and competition between metallic ions (e.g. Galun et al. 1987; Ramelow et al. 1992; Oyarzun et al. 2008). The biomass concentration in the solution controls the specific uptake, lower values of biomass concentration increases specific uptake (e.g. Fourest and Roux 1992).

#### 12.7.3.3 Kinetic Models

Adsorption kinetics is the removal rate which controls the residence time of the sorbate in the solid–solution interface. Kinetic studies were conducted in batch reactors at various initial sorbate concentrations, particle sizes, disturbance speeds, pH and temperature. Linear regression is used to determine the best-fitting kinetic rate equation. Additionally, linear least-squares methods can also be applied to the linearly transformed kinetic rate equations to verify experimental data and kinetic rate equations using coefficients of determination (Ho 2006).

Several adsorption kinetic models have been discussed in adsorption kinetics and rate-limiting steps. These include pseudo-first and -second-order rate models, Weber and Morris sorption kinetic model, Adam–Bohart–Thomas relation (Djeribi and Hamdaoui 2008), first-order reversible reaction model (Baral et al. 2006), external mass transfer model (Apiratikul and Pavasant 2008), first-order equation of Bhattacharya and Venkobachar (Sag and Aktay 2002), Elovich's model and Ritchie's equation (Ho 2006; Febrianto et al. 2009).

## 12.8 Plants as Biosorbents

### *12.8.1 Vascular Plants*

Plant materials such as living plants (Hoffman et al. 2004), their parts or their dried, seized, and chemically treated seeds (Kumari et al. 2006; Koivula et al. 2009; Pandey et al. 2009) and also the residue of vascular plants from industry or agriculture, such as rice polish, and orange wastes (Ranjan et al. 2009; Ghimire et al. 2003), were tested as biosorbent material (Table 12.3) for arsenic.

Maximum sorption capacity for arsenic depends on the three factors; sorbate solution characteristic, biosorbent characteristic and process duration. Sorbate solution characteristics such as initial arsenic concentration, pH of the solid solution, occurrence of other co-ions, sorption time duration, and temperature, control the quantity of adsorption. Important biosorbent characteristic are: material porosity, particle size distribution, specific surface area, cation exchange capacity and surface functional groups. Arsenic sorption is limited by phosphate concentration (Kumari et al. 2006; Ghimire et al. 2003; Ranjan et al. 2009). Cadmium, selenium, and carbonate ions in solution above 200 mg·$L^{-}1$ showed an improved arsenic biosorption. Kumari et al. (2006) reported that optimum pH for *Moringa oleifera* seeds as a biosorbent is different for arsenite and arsenate ion sorption, and was respectively 2.5 and 7.5 mg As·$L^{-1}$.

Maximum biosorption capacity for living plants was observed to be a two-stage process i.e. rapid first phase, and a slow second phase. The first, rapid phase of reaching equilibrium was found for arsenate sorption by *Salvinia minima*, to be 6–12 h (Hoffman et al. 2004), and for arsenite sorption by *Salvinia natans* 48 h,

(Mukherjee and Kumar 2005), with different arsenic ions concentrations. The second phase was up to 96 and 12 h (Table 12.3). The adsorption of metal ions onto a plant surface was conducted in a relatively short-time, while the uptake of metal ions took a long-time and was more complex (Veglio and Beolchini 1997; Hoffman et al. 2004; Mukherjee and Kumar 2005). Pandey et al. (2009) reported chelation of As(III) with the –OH groups for fresh different parts of the biomass of *Momordica charantia.*

Most of these experimental results were in good agreement with the Langmuir and Freundlich sorption models. However the Dubinin–Radushkevich (D–R) sorption isotherm, was applied to evaluate the nature of sorption and was used to explain the heterogeneity of surface energies (Ranjan et al. 2009).

## *12.8.2 Plant Litter as Biosorbent*

The biosorbent biomass constitutes mainly lignin, cellulose and other elements stored in the cell walls and vacuoles. Fibres, lignins cellulose and other substances bind to cell walls, such as phenols, cutin, suberins waxes and others (e.g. Berg and McClaugherty 2003) and can be used as biosorbent.

Certain cell wall components, specially lignin and pectin, are assumed to be connected with the sorption of metal ions (Bailey et al. 1999; Quek et al. 1998; Randall et al. 1974; Senthilkumaar et al. 2000; Volesky and Holan 1995). Plant fibres are capacious for sorption of metal ions, and can be used in water cleaning (e.g. Bailey et al. 1999; Conrad and Hansen 2007). Mesocarp fibres from *Coccos nucifera* coir consist mainly of xylem and a fibre sheath. Cell wall lignins are distributed in cell walls of the fibre sheath, but in the xylem they were smaller in the secondary walls than in the walls of the fibre sheath. Pectin was in detectable quantities only in the fibre sheath walls, middle lamella, cell corners and extracellular matrix. In the xylem, pectin was regularly distributed in the wall, with a higher concentration in the middle lamella and cell corners. All cell walls facing the lacuna had a high content of pectin. The contaminant ions were mainly loaded on the xylem and tend to be loaded on the large fibres. Contaminants were distributed on and across the whole secondary wall (Conrad 2008).

## *12.8.3 Other Plant Groups*

### 12.8.3.1 Algae

Alginate beads are most commonly used with immobilized biomass, however only a few studies, have been carried out on their metal uptake. Alginate is also well known as an immobilization supporter (Brierley 1990; Cotoras et al. 1993; Mata et al. 2008). Readily available plant residual material and their parts, algae, fruit waste material, and natural polysaccharide gelmatrixes such as alginate are used as biosorbents. Alginate is a component of the outer cell wall of brown algae, which has high biosorption capacity, and is rich in carboxyl groups, the main functional groups involved in heavy metal biosorption (Brierley 1990).

Biosorbent material is usually collected as living biomass which is then air-dried and size reduced, often pulverized. These types of materials were examined to estimate the maximum adsorption capacities and sorption kinetics of copper smelting wastewater, with 400–1,000 mg $As \cdot kg^{-1}$ (e.g. Hansen et al. 2006). Macroalga *Lessonia nigrescens*, which is abundant all along the coast of Chile, was used by Hansen et al. (2006) as a biosorbent for arsenic (V) solution. The maximum adsorption capacities were estimated to be 45.2 $mg \cdot g^{-1}$ (at pH=2.5), 33.3 $mg \cdot g^{-1}$ (at pH=4.5), and 28.2 $mg \cdot g^{-1}$ (at pH=6.5). Lower pH increased adsorption of arsenate and values found were high in comparison with other reported arsenic adsorbents (Table 12.3).

The sorption kinetics of arsenic by *L. nigrescens* was modelled by Lagergren's first-order rate equation reported by Hansen et al. 2006. The kinetics was observed to be independent of pH during the first 120 min of adsorption with the Lagergren first-order rate constant of approximately $1.07 \cdot 10^{-3}$ $min^{-1}$.

#### 12.8.3.2 Bacteria and Cyanobacteria

The *Prokaryotas* cell walls, mainly composed of polysaccharides, proteins and lipids, offer particularly abundant metal-binding functional groups, such as carboxylate, hydroxyl, sulphate, phosphate and amino groups. This physicochemical phenomenon of non-metabolism dependent metal biosorption, is relatively rapid and can be reversible (Veglio and Beolchini 1997; Kuyucak and Volesky 1988). Living cells employ both active and passive modes for heavy metal uptake. Gram-negative bacteria reached equilibrium state for arsenite after 30 min and the maximum capacity was 19 $mg \cdot g^{-1}$ at pH 3, and 28 $mg \cdot g^{-1}$ at pH 7 (Loukidou et al. 2001).

#### 12.8.3.3 Fungi

Mushrooms and other fungi contain various amounts of trace metals such as Cd, Hg, Pb, V, Se, and As (Vetter 2004). Concentrations are often much higher in lawn decomposing fungi compaired to mycorrhizal species (Lodenius and Herranen 1981), however, metals can be found in the cell walls of fungi whereas in other plants, most metals are bound in the plant tissue, as their nutrient source (Lodenius and Herranen 1981). Mushroom species collect minimal As from soil and the atmosphere, but there are also some exception such as *Agaricus sp.* and *Macrolepiota rhacodes* which absorb more arsenic (Vetter 1994).

Microscopic filamentous fungi, as a part of microbial communities, influence transformation and distribution of metal(loid)s in the environment. Some of the fungi species (*Aspergillus clavatus*, *Neosartorya fischeri*, *Talaromyces wortmannii*, *T. flavus*, *Eupenicillium cinnamopurpureum*) were tested with modified and unmodified biomass to examine arsenic adsorption and volatilization (Čerňanský et al. 2007).

The most important feature of the mycelium wall is that it is built naturally by a carbohydrate polymer – chitin and chitosan. Chitin is a long, unbranched polysaccharide derivative of cellulose, the C2 hydroxyl group has been replaced by the acetyl amino group $-NHCOCH_3$. The primary unit in the polymer chain is 2-Deoxy-2(acetyl-amino) with glucose. These units are linked by ß, (1→4) glycosidic bonds forming long linear chains with degrees of polymerization from 2,000 to 4,000. Chitosan is derived from chitin by deacetylation of chitin using concentrated alkali at high temperature (Mcafee et al. 2001; Chen and Chung 2006).

The high adsorption capacity of fungi is due to the large numbers of hydroxyl- and amino groups of chitosan having high hydrophilicity; primary amino groups providing high reactivity, and polymer chains of chitosan providing suitable configurations for efficient complexation with metal ions (Mcafee et al. 2001; Say et al. 2003). The optimal pH value for As(III) removal was between 5 and 7.2, giving adsorption capacities of 35.6 $mg{\cdot}g^{-1}$ (*Penicillium purpurogenus*) and 1.3 $mg{\cdot}g^{-1}$ (tea fungi) biomass. Tea fungi biomass treated with $FeCl_3$ had approximately a five times higher biosorption capacity than that without treatment (Say et al. 2003; Murugesan et al. 2006). Pentavalent arsenic biosorption capacity, in the optimal pH range 2–8, was 4.95 $mg{\cdot}g^{-1}$ (for tea fungi biomass). Biomass treated with $FeCl_3$ had a biosorption capacity of twice that without treatment (Murugesan et al. 2006; Mamisahebei et al. 2007). These were compared using the Freundlich and Langmuir isotherm models (Mamisahebei et al. 2007).

## 12.9 Conclusions

Based on the whole review, we find that phytoremediation techniques are good as a tool for cleaning up of soil environments and for drinking water protection. This method can be used as a relatively non-expensive tool for removing or/and stabilising arsenic, reducing environmental impact in addition to increased public acceptance. It is now a well established fact that by the growing of plants, it is possible to clean up pollutants from contaminated soils or to alter the chemical and physical nature of soil pollutants so that they no longer create risks to humans. About 400 plant species have been recorded as phytoremediation and hyperaccumulator plants for arsenic. Several species of ferns and a number of other plant species show great potential for extracting large amounts of arsenic from areas which are highly contaminated with this metalloid. The primary difficulty with using hyperaccumulators for phytoremediation of arsenic is their climate limitations and handling of the arsenic bearing wastes e.g. contaminated plants or their ashes. A few authors have attempted to asses plants species which are capable of tolerating arsenic contaminated soil, and accumulating arsenic in sufficiently large concentrations to be considered viable for phytoremediation purposes. They can be used as phytostabilisators and protecting barriers against ground water contamination. However, the gaps in knowledge regarding the molecular level behaviour of arsenic and plants, microbes etc. are limiting the use of phytoremediation for arsenic removal.

The results found in the literature of different organism groups were clarified with regard to biosorbents, their biosorption capacity and as being potential sources for future investigations of the best bio and phytoremediation techniques, and the preparation of multifunctional models of phyto and bioremediation techniques for removal of arsenic in different treatment have been summerized.

## References


Abedin MJ, Cresser MS, Meharg AA, Feldmann J, Cotter-Howells J (2002) Arsenic accumulation and metabolism in rice (*Oryza sativa* L.). Environ Sci Technol 36:962–968

Adriano DC, Wenzel WW, Vangronsveld J, Bolan NS (2004) Role of assisted natural remediation in environmental clean up. Geoderma 122:121–142

Agely AA, Sylvia DM, Ma LQ (2005) Mycorrhizae increase arsenic uptake by the hyperaccumulator Chinese brake fern (*Pteris vittata* L.). J Environ Qual 34:2181–2186

Ahmann D, Krumholz LR, Hemond HF, Lovley DR, Morel FMM (1997) Microbial mobilization of arsenic from sediments of the Aberjona Watershed. Environ Sci Technol 31(10): 2923–2930

Aksu Z, Sag Y, Kutsal T (1992) The biosorption of Cu (II) by *C. Vulgaris* and *Z. ramigera*. Environ Technol 13:579–586

Aksu Z, Kutsal T, Giia S, Haeiosmanoglu N, Gholaminejad M (1991) Investigation of biosorption of Cu (It), Ni (II) and Cr (VI) ions to activated sludge bacteria. Environ Technol 12:915–921

Alarcon-Herrera MT, Nunez-Montoya OG, Melgoza-Castillo A, Royo-Marquez MH, Almeida FAR (2009) Potential use of sedges (Cyperaceae) in arsenic phytoremediation. In: Bundschuh J, Armienta MA, Birkle P, Bhattacharya P, Matschullat J, Mukherjee AB (eds) Natural arsenic in groundwaters of Latin America. CRC Press, Boca Raton, pp 649–655

Alexander M (1999) Biodegradation and bioremediation, 2nd edn. Academic, San Diego

Anawar HM, Anawar Garcia-Sanchez A, Tari Kul Alam M, Majibur Rahman M (2008) Phytofiltration of water polluted with arsenic and heavy metals. Int J Environ Pollut 33(2–3):292–312

Andrews P, Cullen WR, Polishchuk E (2000) Arsenic and antimony biomethlyation by Scopulariopsis brevicaulis: interaction of arsenic and antimony compounds. Environ Sci Technol 34(11):2249–2253

Apiratikul R, Pavasant P (2008) Batch and column studies of biosorption of heavy metals by *Caulerpa lentillifera*. Bioresour Technol 99:2766–2777

Armstrong J, Armstrong W, Bectett PM (1992) *Phragmites australis*: venturi- and humidity-induced pressure flows enhance rhizome aeration and rhizosphere oxidation. New Phytol 120:107–207

Asher CJ, Reay PF (1979) Arsenic uptake by barley seedlings. Aust J Plant Physiol 6:459–466

ATSDR (2007) Toxicological profile for arsenic. Agency for Toxic Substances and Disease Registry (ATSDR), Atlanta, 500 pp

Bailey SE, Olin TJ, Brick RM, Adrian DD (1999) A review of potentially low-cost sorbents for heavy metals. Water Res 33(11):2469–2479

Baker AJM (1981) Accumulator and excluders – strategies in response of plant to heavy metals. J Plant Nutr 3:643–654

Baker AJM, Whiting SN (2002) In search of the holy grail—a further step in understanding metal hyperaccumulation? New Phytol 155:1–7

Baral SS, Das SN, Rath P (2006) Hexavalent chromium removal from aqueous solution by adsorption on treated sawdust. Biochem Eng J31:216–222

Barrachina AC, Carbonell FB, Beneyto JM (1995) Arsenic uptake, distribution, and accumulation in tomato plants-effect of arsenite on plant-growth and yield. J Plant Nutr 18:1237–1250

Berg B, McClaugherty C (eds) (2003) Plant litter, decomposition, humus formation, carbor sequestration. Springer, Heidelberg, p 286

Bhattacharya P, Chatterjee D, Jacks G (1997) Occurrence of as-contaminated groundwater in alluvial aquifers from delta plains, Eastern India: options for safe drinking water. Water Res Dev 13:79–92

Bhattacharya P, Frisbie SH, Smith E, Naidu R, Jacks G, Sarkar B (2002) Chapter 6. Arsenic in the environment: a global perspective. In: Sarkar B (ed) Heavy metals in the environment. Marcel Dekker, Inc., New York, pp 147–215

Bhattacharya P, Jacks G (2000) Arsenic contamination in groundwater of the sedimentary aquifers in the Bengal Delta Plains: a review. In: Bhattacharya P, Welch AH (eds) Arsenic in groundwater of sedimentary aquifers. Pre-Congress Workshop Abstract Volume, 31st International Geological Congress, Rio de Janeiro, Brazil, pp 19–21

Blute NK, Brabander DJ, Hemond HF, Sutton SR (2004) Arsenic sequestration by ferric iron plaque on cattail roots. Environ Sci Technol 38(22):6074–6077

Boisson J, Mench M, Vangronsveld J, Kopponen P, De Koe T (1999) Immobilization of trace metals and arsenic by different soil additives: evaluation by means of chemical extractions. Commun Soil Sci Plant Anal 30:365

Bondada BR, Ma LQ (2003) Tolerance of heavy metals in vascular plants: arsenic hyperaccumulation by Chinese brake fern (*Pteris vittata* L.). In: Chandra S, Srivastava M (eds) Pteridology in the new millennium. Kluwer Academic, Dordrecht, pp 397–420

Bondada BR, Underhill RS, Ma LQ, Guyodo Y, Mikhaylova A, Davidson MR, Duran RS (2007) Localization, and speciation of arsenic in the hyperaccumulating fern *Pteris vittata* L. In: Bhattacharaya P, Mukherjee AB, Bundschuh J, Zevenhoven R, Loeppert RH (eds) Arsenic in soil and groundwater environment. Trace metals and other contaminants in the environment, vol 9., pp 299–314

Brierley CL (1990) Bioremediation of metal-contaminated surface and groundwaters. Geomicrobiol J 8:201–223

Brix H (1993) Macrophyte-mediated oxygen transfer in wetlands: transport mechanisms and rates. In: Moshiri GA (ed) Constructed wetland for water quality improvement. CRC Press, Boca Raton, pp 123–152

Brooks RR, Lee J, Reeves RD, Jaffre T (1977) Detection of nickeliferous rocks by analysis of herbarium specimens of indicator plants. J Geochem Explor 7:49–57

Brown SL, Sprenger M, Maxemchuk A, Compton H (2005) Ecosystem function in alluvial tailings after biosolids and lime application. J Environ Qual 34:1–6

Bundschuh J, García ME, Birkle P, Cumbal LH, Bhattacharya P, Matschullat J (2009) Occurrence of health effects and remediation of arsenic in groundwaters of Latin America. In: Bundschuh J, Armienta MA, Birkle P, Bhattacharya P, Matschullat J, Mukherjee AB (eds) Natural arsenic in grounwaters of Latin America. CRC Press/Taylor & Francis Group, Boca Raton/London, pp 3–15

Burd GI, Dixon DG, Glick BR (2000) Plant growth-promoting bacteria that decrease heavy metal toxicity in plants. Can J Microbiol 46:237–245

Bunnaul P, Saejen R, Arrykul S, Kooptarnond K (1999) Study on the preparation of activated carbon using for arsenic removal and its adsorption characteristics. In: Proceedings of FEISEAP'99 Inter-Regional Symposium on Sustainable Development (ISSD) Felix Kanchanaburi-Swissotel-River Kwai, Thailand, 18–20 May

Cabral JPS (1992) Selective binding of metal ions to *Pseudomonas syringae* cells. Microbios 71:47–53

Caille N, Zhao FJ, McGrath SP (2004) Comparison of root absorption, translocation and tolerance of arsenic in the hyperaccumulator *Pteris vittata* and the nonhyperaccumulator *Pteris tremula*. New Phytol 165:755–761

Čerňanský S, Urík M, Ševc J, Khun M (2007) Biosorption and biovolatilization of arsenic by heat-resistant fungi. Environ Sci Pollut Res 14(Special Issue 1):31–35

Chandra KS, Kamala CT, Chary NS, Mukherjee AB (2007) Arsenic accumulation by *Thalinum cuneifolium* – application for phytoremediation of arsenic-contaminated soils of Patancheru, Hyderabad, India. In: Bhattacharaya P, Mukherjee AB, Bundschuh B, Zevenhoven R, Loeppert RH (eds) Arsenic in groundwater and environment: trace metals and other contaminants in the environment, vol 9. Elsevier, Amsterdam, pp 315–338

Chen C-C, Chung Y-C (2006) Arsenic removal using a biopolymer chitosan sorbent. J Environ Sci Health Part A Toxic/Hazard Subst Environ Eng 41(4):645–658

Chowdhury K, Biswas BK, Chowdhury TR, Samanta G, Mandal K, Basu C, Chanda CR, Lodh D, Saha KC, Mukherjee SK (2000) Groundwater arsenic contamination in Bangladesh and West Bengal, India. Environ Health Perspect 108:393–397

Clemente R, Almela C, Pilar Bernal M (2006) A remediation strategy based on active phyroremediation followed by natural attenuation in a soil contaminated by pyrite waste. Environ Pollut 143:397–406

Conrad K (2008) Correlation between the distribution of lignin and pectin and distribution of sorbed metal ions (lead and zinc) on coir (*Cocos nucifera* L.). Bioresour Technol 99:8476–8484

Conrad K, Hansen HCB (2007) Sorption of zinc and lead on coir. Bioresour Technol 98:89–97

Cotoras D, Viedma P, Pimentel J (1993) Biosorption of metal ions by attached bacterial cells in a packed-bed bioreactor. In: Torma A, Apel M, Brierley C (eds) Biohydrometalurgical technologies, vol 2, The minerals. Metals & Materials Society, Warrandale, pp 103–110

Cox DP, Alexander D (1973) Effect of phosphate and other anions on trimethylarsine formation by *Candida humicola*. Appl Microbiol 25(3):408–413

Creger TL, Peryea FJ (1994) Phosphate fertilizer enhances arsenic uptake by apricot liners grown in lead-arsenate-enriched soil. Hortic Sci 29:88–92

Cunningham SD, Berti WR (1993) Remediation of contaminated soils with green plants: an overview. Vitro Cell Dev Biol 29:207–212

Dambies L, Vincent T, Guibal E (2002) Treatment of arsenic containing solution using chitosan derivatives: uptake mechanism and sorption performances. Water Res 36:3699–3710

Dang VBH, Doan HD, Dang-Vu T, Lohi A (2009) Equilibrium and kinetics of biosorption of cadmium (II) and copper (II) ions by wheat straw. Bioresour Technol 100(1):211–219

Del Rio M, Font R, Conceptiốn A, Vélez D, Montoro R, de Bailốn AH (2002) Heavy metals and arsenic uptake by wild vegetation in the Guadiamar river area after the toxic spill of the Azalcốllar mine. J Biotechnol 98:125–137

Dermatas D, Moon DH, Menounou N, Meng X, Hires R (2004) An evaluation of arsenic release from monolithic solids using a modified semi-dynamic leaching test. J Hazard Mater 116: 25–38

Djeribi R, Hamdaoui O (2008) Sorption of copper(II) from aqueous solutions by cedar sawdust and crushed brick. Desalination 225:95–112

Eisler R, Zaroogian GE, Hennekey RJ (1972) Cadmium uptake by marine organisms. J Fish Res Board Can 29:1367–1969

Ercole C, Veglio F, Toro L, Ficara G, Lepidi A (1994) Immobilisation of microbial cells for metal adsorption and desorption. In: Mineral bioprocessing I. Snowbird, Utah

Ernst WHO (2005) Phytoextraction of mine wastes – options and impossiblities. Cheme der Erde 65(51):29–42

Faisal M, Hasnain (2005) Chromate resistant Bacillus cereus aguments sunflower growth by reducing toxicity of Cr (VI). J Plant Biol 48(2):187–194

Fayiga AO, Ma LQ (2006) Using phosphate rock to immobilize metals in soil and increase arsenic uptake by hyperaccumulator *Pteris vittata*. Sci Total Environ 359(1–3):17–25

Fayiga AO Ma LQ, Santoa J, Rathinasabapathi B, Stamps B, Littell RC, (2005) Effects of arsenic species and concentrations on arsenic accumulation by different fern species in a hydroponic system. Int J Phytoremediation 7:231–240

Febrianto J, Kosasih Sunarso J, Ju Y-H, Indraswati N, Ismadji S (2009) Equilibrium and kinetic studies in adsorption of heavy metals using biosorbent: a summary of recent studies. J Hazard Mater 162(2–3):616–645

Fergusson JE (1990) The heavy elements: chemistry, environmental impact and health effects. Pergamon Press, Oxford, 614 pp

Fitz WJ, Wenzel WW (2002) Arsenic transformations in the soil-rhizosphere-plant system: fundamentals and potential application to phytoremediation. J Biotechnol 99(3):259–278

Fourest E, Roux JC (1992) Heavy metal biosorption by fungal mycelial by-products: mechanism and influence of pH. Appl Microbiol Biotechnol 37:399–403

Frankenberger WT Jr, Arshad M (2002) Volatilisation of arsenic. In: Frankenberger WT Jr (ed) Environmental chemistry of arsenic. Marcel Dekker, New York, pp 363–380

Franseconi K, Visootiviseth P, Sridokchan W, Goessler W (2002) Arsenic species in an arsenic hyperaccumulating fern, *Pityrogramma calomonanos*: a potential phytoremediator of arsenic-contaminated soils. Sci Total Environ 284:27–35

Frence CJ, Dickinson NM, Putwain PD (2006) Woody biomass phytoremediation of contaminated brownfield land. Environ Pollut 141:387–395

Freundlich HMF (1906) Über die adsorption in Losungen. Z Phys Chem 57:385–470

Galun M, Galun E, Siegel BZ, Keller P, Lehr H, Siegel SM (1987) Removal of metal ions from aqueous solutions by *Penicillium* biomass: kinetic and uptake parameters. Water Air Soil Pollut 33:359–371

Ghimire KN, Inoue K, Makino K, Miyajima T (2002) Adsorption removal of arsenic using orange juice residue. Sep Sci Technol 37(12):2785–2799

Ghimire KN, Inoue K, Yamaguchi H, Makino K, Miyajima T (2003) Adsorptive separation of arsenate and arsenite anions from aqueous medium by using orange waste. Water Res 37(20):4945–4953

Ghosh M, Singh SP (2005) A rewiev on phytoremediation of heavy metals and utilization of its byproducts. Appl Ecol Environ Res 3(1):1–18

Gobas FAPC, Wilcockson JB, Russell RW, Haffner GD (1999) Mechanisms of biomagnification in fish under laboratory and field conditions. Environ Sci Technol 33(1):133–141

Gonzaga SMI, Santos JAG, Ma LQ (2006) Arsenic phytoextraction and hyperaccumulation by fern species. Sci Agric 63(1):90

Greger M (2005) Influence of willow (*Salix viminalis* L.) roots on soil metal chemistry: effects of clones with varying metal uptake potential. In: Huang PM, Gobran GR (eds) Biogeochemistry of trace elements in the rhizosphere. Elsevier, Amsterdam, pp 301–312

Greger M (2008) Trace elements and radionuclides in edible plants. In: Prasad MNV (ed) Trace elements as contaminants and nutrients: consequences in ecosystems and human health. Wiley, New York, pp 121–136

Gustafsson JP, Bhattacharya P (2008) Geochemical modelling of arsenic adsorption to oxide surfaces. In: Bhattacharya P, Mukherjee AB, Bundschuh J, Zevenhoven R, Loeppert RH (eds) Arsenic in soil and groundwater environment: biogeochemical interaction, health effects and remediation. Trace metals and other contaminants in environment, vol 9. Elsevier B.V., Amsterdam, pp 159–206

Hansen HK, Ribeiro A, Mateus M (2006) Biosorption of arsenic(V) with *Lessonia nigrescens*. Min Eng 19:486–490

Haque N, Mokgalaka NS, Peralta-Videa JR, Gardea-Torresdey JL (2009) Phytoremediation of arsenic by sorghum (Sorghum biocolr) under hydroponics. In: Bundschuh J, Armienta MA, Birkle P, Bhattacharya P, Matschullat J, Mukherjee AB (eds) Natural arsenic in groundwaters of Latin America. CRC Press, Boca Raton, p 643

Ho Y (2006) Second-order kinetic model for the sorption of cadmium onto tree fern: a comparison of linear and non-linear methods. Water Res 40:119–125

Hoffman T, Kutter C, Santamaria JM (2004) Capacity of *Salvinia minima* Baker to tolerate and accumulate As and Pb. Eng Life Sci 4(1):61–65

Huysmans KD, Frankenberger WT (1991) Evolution of trimethylarsine by a *Penicillium* sp. isolated from agricultural evaporation pond water. Sci Total Environ 105:13–28

ITRC (2009) Phytotechnology technical and regulatory guidance document. In: Technical/regulatory guidelines. http://www.itrcweb.org/Documents/PHYTO-3.pdf

IUPAC (1996) Glossary of terms relating to pesticides. Pure Appl Chem 68(5):1167–1193

Jacks G, Bhattacharya P (1998) Arsenic contamination in the environment due to the use of CCA-wood preservatives. Arsenic in wood preservatives, part I, Kemi report 3/98. pp 7–75

Jing C, Liu S, Patel M, Meng X (2005) Arsenic leachability in water treatment adsorbents. Environ Sci Technol 39:5481–5487

Kabata-Pendias A, Pendias H (2001) Trace elements ion soils and plants, 3rd edn. CRC Press, Boca Raton

Kamala CT, Chu KH, Chary NS, Pandey PK, Ramesh SL, Sastry ARK, Sekhar KC (2005) Removal of Arsenic (III) from Aqueous Solutions Using Fresh and Immobilized Plant Biomass. Water Res, 39:2815–2826

Kid P, Barceló J, Bernal MP, Navari-Izzo F, Poschenrieder C, Shilev S, Clemente R, Monterroso C (2009) Trace element behaviour at the root-soil interface: implications in phytoremediation. Environ Exp Bot 67:243–259

Kloke A, Sauerbeck DR, Vetter H (1994) In: Nriagu JO (ed) Changing metal cycles and human health. Springer, Berlin, p 113

Knudson JA, Meikle T, DeLuca TH (2003) Role of mycorrhizal fungi and phosphorus in the arsenic tolerance of basin wildrye. J Environ Qual 32:2001–2006

Koivula MP, Kujala KK, Rönkkömäki H, Mäkelä M (2009) Sorption of Pb(II), Cr(III), Cu(II), As(III) to peat, and utilization of the sorption properties in industrial waste landfill hydraulic barrier layers. J Hazad Mater 164(1):345–352

Kramer U (2000) Cadmium for all meals – plants with an unusual appetite. New Phytol 145:1–5

Kratochvil D, Volesky B (1998) Advances in the biosorption of heavy metals. Trends Biotechnol 16:291–300

Kumari P, Sharma P, Srivastava S, Srivastava MM (2006) Bio sorption studies on shelled *Moringa oleifera* Lamarck seed powder: removal and recovery of arsenic from aqueous system. Int J Miner Process 132(78):131–139

Kuyucak N, Volesky B (1989) Biosorbents for recovery of metals from industrial solutions. Biotechnol. Lett. 10, 137–142.

Kuyucak N, Volesky B (1988) Biosorbents for recovery of metals from industrial solutions. Biotechnol Lett 10(2):137–142

Langmuir I (1918) The adsorption of gases on plane surfaces of glass, mica and platinum. J Am Chem Soc 40:1361–1403

Lasat MM (2002) Phytoextraction of toxic metals: a review of biological mechanisms. J Environ Qual 31:109–120

Liu WJ, Zhu YG, Hu Y, Williams PN, Gault AG, Meharg AA, Charnock JM, Smith FA (2006) Arsenic sequestration in iron plaque, its accumulation and speciation in mature rice plants (*Oryza sativa* L.). Environ Sci Technol 40:5730–5736

Lodenius M, Herranen M (1981) Influence of a chlor-alkali plant on the mercury contents of fungi. Chemosphere 10:313–318

Lombi E, Zhao F, Fuhrmann M, Ma LQ, McGrath SP (2002) Arsenic distribution and speciation in the fronds of the hyperaccumulator *Pteris vittata*. New Phytol 156:195

Loukidou MX, Matis KA, Zouboulis AI (2001) Removal of arsenic from contaminated dilute aqueous solutions by biosorption. In: 7th international conference on environmental science and technology Ermoupolis, Syros Island, Sept 2001, pp 286–290

Loukidou MX, Matis KA, Zouboulis AI, Liakopoulou-Kyriakidou M (2003) Removal of As(V) from wastewaters by chemically modified fungal biomass. Water Res 37(18):4544–4552

Low KS, Lee CK (1990) The removal of cationic dyes using coconut husk as an adsorbent. Pertanica 13:83–103

Luongo T, Ma LQ (2005) Characteristics of arsenic accumulation by *Pteris* and non-*Pteris* ferns. Plant Soil 277:117

Ma LQ, Komar KMM, Tu C, Zhang W, Cai Y, Kenence ED (2001) A fern that hyperaccumulates As. Nature Lond 409:579

Mamisahebei S, Jahed-Khaniki GR, Torabani A, Nasseri S, Naddafi K (2007) Removal of arsenic from an aqueous solution by pretreated waste tea fungal biomass. Iran J Environ Health Sci Eng 4(2):85–92

Mandal BK, Roychowdhury T, Samanta G, Basu K, Chowdhury P (1996) Arsenic in groundwater in seven districts of West Bengal, India: the biggest arsenic calamity in the world. Curr Sci 70(11):976–986

Marschner H (1995) Mineral nutrition of higher plants, 2nd edn. Academic, London, p 889

Mata YN, Blázquez ML, Ballester A, González FMuñoz JA (2008) Characterization of the biosorption of cadmium, lead and copper with the brown alga *Fucus vesiculosus*. J Hazard Mat 158:316–323

Mazej Z, Germ M, (2009) Trace element accumulation and distribution in four aquatic macrophytes. Chemosphere 74:642–647

Mcafee BJ, Gould WD, Nedeau JC, da Costa ACA (2001) Biosorption of metal ions using chitosan, chitin, and biomass of *Rhizopus oryzae*. Sep Sci Technol 36(14):3207–3222

Meharg AM, Hartley-Whitaker J (2002) Arsenic uptake and metabolism in arsenic resistant and nonresistant plant species. New Physiol 154:29–43

Mehrag AA, Macnair MR (1990) An altered phosphate uptake system in arsenate-tolerant *Holcus lanatus* L. New Phytol 116:29–35

Mehrag AA, Macnair MR (1992) Supression of the high-affinity phosphate-uptake system: a mechanism of arsenate tolerance in *Holcus lanatus* L. J Exp Bot 43:519–524

Mench M, Vangronsveld J, Clijsters H, Lepp NW, Edwards R (2000) In situ metal immobilisation and phytostabilisation of contaminated soils. In: Rerry N, Bañuelos G (eds) Phytoremediation of contaminated soil and water. Lews Publishers, Bora Raton, pp 323–358

Mench M, Bussiere S, Boisson J, Castaing E, Vangronsveld J, Ruttens A, De Koe T, Bleeker P, Assuncão A, Manceau A (2003) Progress in remediation and revegetation of the barren jales gold mine spoil after in situ treatments. Plant Soil 249:187–202

Mkandawire M, Taubert B, Duel EG (2005) Resource manipulation in uranium and arsenic attenuation by *Lemna gibba* L. (duckweed) in tailing water of a former uranium mine. Water Air Soil Pollut 166:83–101

Mohan D, Pittman CU Jr, Bricka M, Smith F, Yancey B, Mohammad J, Steele PH, Alexandre-Franco MF, Serrano VG, Gong H (2007) Sorption of arsenic, cadmium, and lead by chars produced from fast pyrolysis of wood and bark during bio-oil production. J Colloid Interf Sci 310(1):57–73

Mohanty K, Jha M, Meikap BC, Biswas MN (2006) Biosorption of Cr(VI) from aqueous solutions by *Eichhornia crassipes*. J Chem Eng 117(1):71–77

Mok WM, Wai CM (1994) Mobilization of arsenic in contaminated river waters. In: Nriagu JO (ed) Arsenic in the environment, part I: cycling and characterization. Wiley, New York, pp 99–117

Mukherjee S, Kumar S (2005) Adsorptive uptake of arsenic (V) from water by aquatic fern *Salvinia natans*. J Water Supply Res Technol AQUA 54(1):47–53

Murugesan GS, Sathishkumar M, Swaminathan K (2006) Arsenic removal from groundwater by pretreated waste tea fungal biomass. Bioresour Technol 97(3):483–487

Naidu R, Smith E, Owens G, Nadebaum P, Bhattacharya P (2006) Management of arsenic-contaminated soils. In: Naidu R, Smith E, Owens G, Bhattacharya P, Nadebaum P (eds) Managing arsenic in the environment: from soil to human health. CSIRO Publishing, Melbourne, pp 419–453

Nieboer E, Padovan D, Lavoie P (1984) Anion accumulation by lichens. II. Competition and toxicity studies involving arsenate, phosphate, sulfate, and sulphite. New Phytol 96:83–94

National Research Council Committee on Animal Nutrition (NRCC) (1980) Mineral tolerance of domestic animals. National Academy of Science, Washington, DC

O'Keefe DM, Sylvia DM (1991) Mechanisms of the vesicular-arbuscular mycorrhizal plant-growth response. In: Arora DK, Rai B, Mukerji KG, Knudsen GR (eds) Handbook of applied mycology. Marcel Dekker, New York, pp 35–53

Osborne FH, Ehrlich HL (1976) Oxidation of arsenite by a soil isolated of alcaligenes. J Appl Bacteriol 41:295–305

Oyarzun R, Cubas P, Higueras P, Lillo J, Llanos W (2008) Environmental assessment of the arsenic-rich, Rodalquilar gold–(copper–lead–zinc) mining district, SE Spain: data from soils and vegetation. Environ Geol 58:761–777. doi:DOI 10.1007/s00254-008-1550-3

Pandey PK (2000) Who should be credited for the discovery and first reporting of arsenicosis in Kaudikasa in Madhya Pradesh? Curr Sci 78:1412

Pandey PK, Khare RN, Sharma S, Pandey M (1999) Arsenicosis and deteriorating ground water quality: unfolding crisis in central East Indian region. Curr Sci 77:686–693

Pandey PK, Choubey S, Verma Y, Pandey M, Chandrashekhar K (2009) Biosorptive removal of arsenic from drinking water. Bioresour Technol 100:634–637

Peryea FJ (1998) Phosphate starter fertilizer temporarily enhances soil arsenic uptake by apple trees grown under field conditions. Hortic Sci 33:826–829

Peterson PL, Benson LM, Zeive R (1981) Metalloids. In: Lepp NW (ed) Effects of heavt metal pollution on plants. Applied Science, London, pp 272–342
Pickering IJ, Prince RC, George MJ, Smith RD, George GN, Salt DE (2000) Salt reduction and coordination of arsenic in Indian mustard. Plant Physiol 122:1171–1177
Porter EK, Peterson PJ (1975) Arsenic accumulation by plants on mine waste (United Kingdom). Sci Total Environ 4:365–371
Prasad MNV (2004) Phytoremediation of metals and radionuclides in the environment: the case of natural hyperaccumulators, metal transport, soil-amending chelators and transgenic plants. In: Prasad MNV (ed) Heavy metal stress in plants: from biomolecules to ecosystems, 2nd edn. Springer, Heidelberg, pp 345–392
Prasad MNV (2008) Trace element in traditional healing plants-remedies or risk. In: Prasad MNV (ed) Trace elements as contaminants and nutrients: consequences in ecosystems and human health. John Wiley & Sons, Inc., Hoboken, NJ, USA
Quek SY, Al Duri B, Wase DAJ, Forster CF (1998) Coir as a biosorbent of copper and lead. Process Saf Environ Protect 76(B1):50–54
Raab A, Williams PN, Meharg A, Feldmann J (2007) Uptake and translocation of inorganic and methylated arsenic species by plants. Environ Chem 4:197–203
Rahman MA, Hasegawaa H, Ueda K, Makia T, Rahman MM (2008) Influence of phosphate and iron ions in selective uptake of arsenic species by water fern (*Salvinia natans* L). Chem Eng J 145:179–184
Rajkumar M, Freitas H (2008) Influence of metal resistant-plant growth promoting bacteria on the growth of Ricinus communis in soil contaminated with heavy metals. Chemosphere 71:834–842
Ramelow GJ, Fralick D, Zhao Y (1992) Factors affecting the uptake of aqueous metal ions by dried sea weed biomass. Microbios 72:81–93
Randall JM, Bermann RL, Garrett V, Waiss ACJ (1974) Use of bark to remove heavy metal ions from waste solutions. Forest Prod J 24(9):80–84
Ranjan D, Talat M, Hasan SH (2009) Biosorption of arsenic from aqueous solution using agricultural residue 'rice polish'. J Hazard Mater 166(2–3):1050–1059
Raskin I, Smith RD, Salt DE (1997) Phytoremediation of metals: using plants to remove pollutants from the environment. Curr Opin Biotechnol 8(2):221–226
Rock S, Sayre PG (2000) Regulatory considerations for phytoremediation. In: Raskin I, Ensley BD (eds) Phytoremediation of toxic metals: using plants to clean up the environment. John Wiley, New York, pp 43–49
Ross SM (1994) Toxic metals: fate and distribution in contaminated ecosystems. In: Ross SM (ed) Toxic metals in soil-plant systems. Wiley, Chichester, pp 189–243
Ross SM, Kaye KJ (1994) The meaning of metal toxicity in soil-plant systems. In: Ross SM (ed) Toxic metals in soil-plant systems. Wiley, Chichester, pp 27–61
Ruthven DM (1984) Principles of adsorption and adsorption processes. Wiley, New York/Chichester/Brisbane/Toronto/Singapore, p 432
Sag Y, Aktay Y (2002) Kinetic studies on sorption of Cr(VI) and Cu(II) ions by chitin, chitosan, and *Rhizopus arrhizus*. Biochem Eng J 12:143–153
Sardans J, Peñuelas J (2006) Introduction of the factor of partitioning in the lithogenic enrichment factors of trace element bioaccumulation in plant tissues. Environ Monit Assess 115:473–498
Sattelmacher B (2001) The apoplast and its significance for plant mineral nutrition. New Phytol 14:167–192
Say R, Yilmaz N, Denizli A (2003) Biosorption of cadmium, lead, mercury, and arsenic ions by the fungus *Penicillium purpurogenum*. Sep Sci Technol 38(9):2039–2053
Schnoor JL (1998) Phytoremediation. Technology evaluation report TE-98-01. Prepared for Ground-Water Remediation Technology Analysis Center, Pittsburg
Sekhar KC, Kamala CT, Chary NS, Mukherjee AB (2007) Arsenic accumulation by Talinum cuneifolium – application for phytoremediation of arsenic-contaminated soils of Patancheru, Hyderband, India. In: Bhattacharya P, Mukherjee AB, Bundschuh J, Zevenhoven R, Loeppert RH (eds) Arsenic in soil and groundwater environment: biogeochemical interactions, health

effects and remediation, vol 9, Trace metals and other contaminants in environment. Elsevier, Amsterdam, pp 315–338

Senthilkumaar S, Bharathi S, Nithyanandhi D, Subburam V (2000) Biosorption of toxic heavy metals from aqueous solutions. Bioresour Technol 75(2):163–165

Siedlecka A, Tukendorf A, Skórzyńska-Polit E, Maksymiec W, Wójcik M, Baszyński T, Krupa Z (2001) Angiosperms (Asteraceae, Convolvulaceae, Fabaceae and poaceae; other than Btassicaceae). In: Prasad MNV (ed) Metals in the environment. Marcel Dekker, New York/Basel, pp 171–217

Smedley PL, Kinniburgh DG (2002) A review of the source, behaviour and distribution of arsenic in natural waters. Appl Geochem 1:517–568

Stoltz E, Greger M (2002a) Accumulation properties of As, Cd, Pb, and Zn by four wetland plant species growing on submerged mine tailings. Environ Exp Bot 47:271–280

Stoltz E, Greger M (2002b) Cotton grass effects of trace elements in submersed mine tailings. J Environ Qual 31:1477–1483

Stoltz E, Greger M (2005) Effects of different wetland plant species on fresh unweathered sulphidic mine tailings. Plant Soil 276(1–2):251–261

Stoltz E, Greger M (2006a) Release of metals and arsenic from various mine tailings by *Eriophorum angustifolium*. Plant Soil 289(1–2):199–210

Stoltz E, Greger M (2006b) Root penetration through sealing layers at mine deposit sites. Waste Manage Res 24:552–559

Stoltz E, Greger M (2006c) Influences of wetland plants on weathered acidic mine tailings. Environ Pollut 144(2):689–694

Sun Y, Zhou Q-X, Liu W-T, Wang L (2009) Joint effects of arsenic, cadmium on plant growth and metal bioaccumulation: a potential Cd hyperaccumulator and As-excluder *Bidens pilosa* L. J Hazard Mater 161(2–3):808–814

Sun GX, Williams PN, Carey AM, Zhu YG, Deacon C, Raab A, Feldmann J, Islam RM, Meharg AA (2008) Inorganic As in rice bran and its products are an order of magnitude higher than in bulk grain. Environ Sci Technol 42:7542–7546

Suthersan SS (2002) Natural and enhanced remediation systems. Arcadis Lewis Publishers/CRC Press, Boca Raton

Terry T, Zayed AM (1994) Sellenium volatilization in plants. In: Frankenberger WT Jr, Benson S (eds) Sellenium in the environment. Marcel Dekker, New York, pp 343–367

Thompson DJ (1993) A chemical hypothesis for arsenic methylation in mammals. Chem Biol Interact 88(89–1):14

Tu C, Ma LQ (2002) Effects of arsenic concentrations and forms on arsenic uptake by the hyperaccumulator ladder brake. J Environ Qual 31:641–647

Tu C, Ma LQ, Bondada B (2002) Arsenic accumulation in the hyperaccumulator Chinese brake and its utilization potential for phytoremediation. J Environ Qual 31:1671–1675

Ullrich-Eberius C, Sanz IA, Novacky AJ (1989) Evaluation of arsenate- and vandate associated changes of electrical membrane potential and phosphate transport in *Lemma gibba* GI. J Exp Bot 40:119–128

US EPA (US Environmental Protection Agency) (2000) Introduction to phytoremediation. Publication nr 600/R-99/107, Cincinnati

US EPA (2002) Arsenic treatment technologies for soil, waste, and water. http://www.clu-in.org/arsenic. 6 Feb 2009

US EPA (2004) Monitored natural attenuation, chapter IX. In: How to evaluate alternative cleanup technologies for underground storage tank sites. A guide for corrective action plan reviewers. http://www.epa.gov/oust/pubs/tums.htm. Last updated 23 June 2008, visited 12 July 2009

Veglio F, Beolchini F (1997) Removal of metals by biosorption: a review. Hydrometallurgy 44:301–16

Vetter J (1994) Data on arsenic and cadmium contents of some common mushrooms. Toxicon 32:11–15

Vetter J (2004) Arsenic content of some edible mushroom species. Eur Food Res Technol 219(1):71–74

Vilar VJP, Botelho CMS, Bonaventura RAR (2006) Equilibrium and kinetics modelling of Cd(II) biosorption by algae *Gelidium* and agar extraction algal waste. Water Res 40(2):291–302

Volesky B (2003) Sorption and biosorption. BV Sorbex, Inc., Montreal

Volesky B, Holan ZR (1995) Biosorption of heavy metals. Biotechnol Prog 11(3):235–250

Wallschläger D, London J (2008) Determination of methylated arsenic-sulfur compounds in groundwater. Environ Sci Technol 42(1):228–234

Warren GP, Alloway BJ (2003) Reduction of arsenic uptake by lettuce with ferrous sulfate applied to contaminated soil. J Environ Qual 32:767–772

Wells JM, Richardson DHS (1985) Anion accumulation by the moss *Hylocomium splendens*: uptake and competition studies involving arsenate, selenate, selenite, phosphate, sulphate and sulphite. New Phytol 101:571–583

Wenzel WW, Bunkowski M, Puschenreiter M, Horak O (2003) Rhizosphere characteristics of indigenously growing nickel hyperaccumulator and excludor plants on serpentine soil. Environ Pollut 123:131–138

WHO (2001) Arsenic and arsenic compounds, environment health criteria, 2nd edn. World Health Organization (WHO), Geneva

Xu H, Allard B, Grimvall A (1991) Effects of acidification and natural organic materials on the mobility of arsenic in the environment. Water Air Soil Pollut 57:269–278

Xu XY, McGrath SP, Meharg A, Zhao FJ (2008) Growing rice aerobically markedly decreases arsenic accumulation. Environ Sci Technol 42:5574–5579

Zhao FJ, Wang JR, Barker JHA, Schat H, Bleeker PM, McGrath SP (2003) The role of phytochelatins in arsenic tolerance in the hyperaccumulator Pteris vittata. New Phytol 159:403–410

# Chapter 13
# Microbial Pathogen Inactivation Using Heterogeneous Photocatalysis

**Mohammad Abbas Mahmood, Sunandan Baruah, Anil Kumar Anal, and Joydeep Dutta**

## Contents



**Abstract** The concentration of toxic materials and infectious microorganism in the natural resources of drinking water is constantly increasing causing severe environmental pollution. Availability of safe drinking water to people in developing countries has become a serious challenge. The traditional chemical methods for the disinfection of drinking water have limitations due to their costs for the undeveloped nations. Furthermore, generation of harmful disinfection byproducts associated with the chemical disinfection processes like chlorination is a source of rising concern among the masses. Hence, there is an urgent need to work out some more reliable alternate techniques for the purpose. Heterogeneous photocatalysis, being

M.A. Mahmood • S. Baruah • A.K. Anal • J. Dutta (✉)
Center of Excellence in Nanotechnology, School of Engineering and Technology, Asian Institute of Technology, P.O. Box 4, Khlong Luang, Pathum Thani 12120, Thailand
e-mail: mohammadabbas33@gmail.com; joy@ait.asia

E. Lichtfouse et al. (eds.), *Environmental Chemistry for a Sustainable World: Volume 2: Remediation of Air and Water Pollution*, DOI 10.1007/978-94-007-2439-6_13,

considered as a promising technique to control environmental pollution has attracted the interest of researchers during the last couple of decades. The method is not only free of generation of harmful byproducts, but could also be cost effective and environmental friendly by utilizing ambient solar light.

We present an overview of photocatalytic inactivation of water borne microbial pathogens. We focus here on various factors involved in the disinfection processes and discuss how different researchers have addressed them to improve the overall efficiency of photocatalytic inactivation of microbes. Research reports on the subject show that heterogeneous photocatalysis degrades wide range of microbial pathogens including bacteria, molds, fungi, and virus. Titanium dioxide ($TiO_2$) has been found as an efficient photocatalyst to produce hydroxyl ($OH^{\bullet}$) radicals under ultraviolet irradiation. The nanoparticulate photocatalyst has been observed to adsorb at the surface of the microbe and perforate the cell envelop through redox reaction leading not only to complete disintegration/mineralization of the cell. There exists an optimum catalyst loading while the inactivation rate increases with intensity of the incident light. For enhancement of the microbial inactivation efficiency under solar light, the photocatalyst has been modified in different ways which include doping and coupling with other materials, dye sensitization, and application of voltage. Suspended particles and ions have been observed to suppress the photocatalytic inactivation rate. Furthermore, water under the photocatalytic treatment has been seen to change spontaneously from basic into acidic, while, decreasing pH of suspension from 7 to 4 during the course of photocatalytic treatment increases the inactivation rate. The photocatalytic microbial inactivation has been seen to follow Langmuir-Hinshelwood kinetic model.

**Keywords** Pathogens • Nanoparticles • Photocatalysis • Antimicrobial Decontamination • Water microbes • Trihalomethanes • Haloacetic acids • Nitrate Water quality • Humic acid • Light irradiation • $TiO_2$,

## 13.1 Introduction

Extended human activities such as agriculture, industrialization, deforestation etc., have confronted life with various challenges. The constantly rising water pollution is one of the most serious amongst them. The most common way of disinfection of water is through chemical means involving chlorination, ozonation etc. Recent studies have indicated the potential risks associated with chlorination due to the concomitant production of harmful disinfection-by-products (DBPs), such as trihalomethanes (THMs) and haloacetic acids (HAAs) (Gopal et al. 2007; Sadiq and Rodriguez 2004) which are reportedly created as a result of reaction of chlorine anions ($Cl^-$) with organic materials present in water (http://www.karinya.com/chlorine.htm). Chlorine by itself is a toxic gas and a 30 min exposure to its concentration of 500–800 ppm can be lethal (Patnaik 2007). Chlorination has also been found less effective to

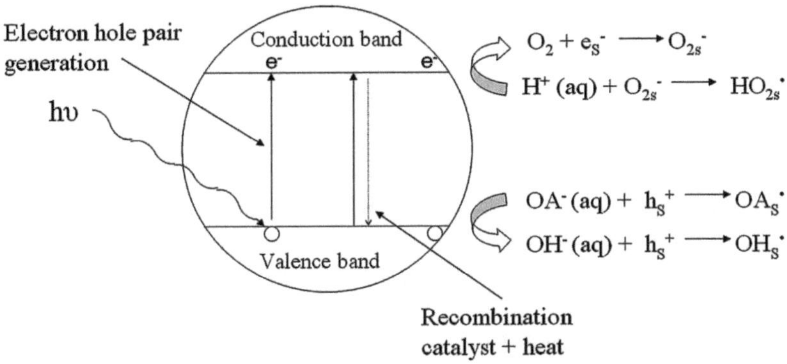


**Fig. 13.1** Schematic diagram showing photocatalysis on a wide bandgap semiconductor surface (Baruah and Dutta 2009)

degrade bacterial spores and cysts. Ozonation, on the other hand, has turned out as more powerful disinfection technique which can readily degrade a wide range of microorganisms including those which are resistant to chlorination. The process, however, has limitation in its wide spread applications due to its sophisticated system with high operational cost (http://www.water-research.net/ozone.htm). Ozone itself highly unstable and its effect vanishes shortly leaving no residuals to inhibit bacterial regrowth. This necessitates the system to be applied close the point of use otherwise water safety cannot be ensured to the end-users. Owing to the problems associated with the conventional water disinfection processes, there is a need for more effective alternate water purification techniques that could be eco-friendly, economic, and could be applicable at the point-of-use. In this context, heterogeneous photocatalysis has aroused interest among the research community for the purification of drinking water due to its robustness in degrading/mineralizing virtually all types of pollutants. Potential utilization of solar light can render the process cost effective and environmentally friendly.

The underlying mechanism of heterogeneous photocatalysis is illustrated in Fig. 13.1. It involves a photocatalyst, which is normally a wide band gap semiconductor material. Upon irradiation with light of energy not less than the band gap energy of the material, electron-hole ($e^-$–$h^+$) pairs are created due to the excitation of electrons from valance to the conduction band. Majority of these photogenerated charge carriers undergo wasteful recombination, however, a significant number of them appear at the surface initiate redox reactions in molecules adsorbed at the surface of the photocatalyst and degrade them. The photogenerated electrons and holes, due to their high redox potentials, have been found to degrade almost all types of organic, inorganic, and microbial contaminants (Gaya and Abdullah 2008). In water the photogenerated electrons and holes create a variety of other oxidizing species which dissociate the contaminant molecules and microorganisms and degrade/mineralize them.

The fundamental reaction of photocatalysis can be represented by;

$$MO + hv \rightarrow MO\left(e^- + h^+\right)$$

Where, MO stands for metal oxide photocatalyst, like titanium dioxide ($TiO_2$), tin oxide ($SnO_2$), tungsten oxide ($WO_3$), and zinc oxide (ZnO), etc. Reactions initiated by the photo-generated electrons, which lead to the formation of superoxide anions ($^{\bullet}O_2^-$), hydrogen peroxide molecules ($H_2O_2$), hydroxyl radicals ($^{\bullet}OH$), hydrogen dioxide anion ($HO_2^-$), and the hydroperoxy radicals ($^{\bullet}HO_2$), include (Banerjee et al. 2006);

$$MO\left(e^-\right) + O_2 \rightarrow MO + {}^{\bullet}O_2^-$$

$$MO\left(e^-\right) + {}^{\bullet}O_2^- + 2H+ \rightarrow MO + H_2O_2$$

$$MO\left(e^-\right) + H_2O_2 \rightarrow MO + {}^{\bullet}OH + OH^-$$

$${}^{\bullet}O_2^- + H_2O_2 \rightarrow {}^{\bullet}OH + OH^- + O_2$$

$${}^{\bullet}O_2^- + H+ \rightarrow {}^{\bullet}OH_2$$

$$MO\left(e^-\right) + {}^{\bullet}OH_2 \rightarrow MO + HO_2^-$$

$$HO_2^- + H^+ \rightarrow H_2O_2$$

$$2{}^{\bullet}OH_2 \rightarrow O_2 + H_2O_2$$

The reactions initiated by the photo-generated holes include:

$$MO\left(h^+\right) + H_2O \rightarrow MO + {}^{\bullet}OH + H^+$$

$$MO\left(h^+\right) + OH^- \rightarrow MO + {}^{\bullet}OH$$

The primary objective of this review is to provide a detailed analysis of activities in the photocatalytic inactivation of microbial organisms and the progress in the application of this technique to practical purposes. The report discusses various factors that affect the efficiency of the photocatalytic inactivation process. Owing to the wide variety of microorganisms and different experimental conditions reported in the literature, it is difficult at this stage to draw a conclusive border that can standardize all the parameters involved in the processes. The technique is attractive in regard to its potential for the abatement of environmental pollution. However, more efforts are still needed to enhance the photocatalytic efficiency to a level where it can be an obvious choice for practical applications in the inactivation of water borne microbes.

## 13.2 Factors Involved in the Photocatalytic Microbicidal Process

Photocatalytic inactivation of microorganisms is a complex process that depends on several parameters. The rate of inactivation has been reported to vary with the type, physiological state, and concentration of microorganisms (Lonnen et al. 2005; Rincon and Pulgarin 2004a). The nature, morphology, concentration, and state (slurry or immobilized) of the catalyst material, also, have a great influence on the inactivation rate (Adams et al. 2006; Huang et al. 1998; Qi et al. 2004; Sondi and Salopek-Sondi 2004). Another important factor that has a profound effect on the photocatalytic bactericidal activity is the intensity and wavelength of light irradiation (Hayakawa et al. 2007; Huang et al. 1998; Leung et al. 2008). The absorption peaks of most of the semiconductor photocatalysts lie in the ultraviolet (UV) region i.e. have wavelengths smaller than 400 nm. However, it has been reported that the photolytic effect of UV-C (200–280 nm) is more pronounced than its photocatalytic effect, While it is just the reverse in the case of UV-A (350–400 nm) irradiation (Paleologou et al. 2007). Furthermore, the inactivation rate has been reported to increase directly with the intensity of light (Benabbou et al. 2007). The pH level of the microbial suspension is also found to affect the interaction between the catalyst and microbes and hence, the inactivation rate is quite high at some specific pH value depending upon the nature of microbes (Rincon and Pulgarin 2004b). In addition to these, the photocatalytic microbicidal activity also depends on water quality. The inactivation has been reported to occur much faster in distilled water than that in natural surface water (Rincon and Pulgarin 2004a) which could be due to scattering of light by suspended particles present in natural water, thereby lowering the total optical absorption of the photocatalyst. In conclusion, the type and physiological state of microbial, type and concentration of photocatalyst, wavelength and intensity of irradiation, pH of the microbial suspension, and quality of water under treatment are the important factor upon which the rate photocatalytic microbial inactivation depends.

### *13.2.1 Photocatalytic Inactivation of Microbes*

This section is mostly devoted to studies on the efficiency of metal oxide semiconductor mediated photocatalysis in degrading all types of water. These include bacteria, viruses, molds, algae and fungi. Among the various bacterial species, *Escherichia coli* (*E. coli*) have been extensively used as a model microorganism in testing the efficacy of most of the newly synthesized disinfectant photocatalysts as well as newly

**Table 13.1** Photocatalytic inactivation of three bacterial strains with undoped and Pd-doped $TiO_2$/$SnO_2$ films

| Org. | Catalyst film | L. intensity | Irr. time (h) | %viability loss | R. constant ($min^{-1}$) |
|---|---|---|---|---|---|
| *Escherichia coli* | Nil | 10 $mWcm^{-2}$ | 2 | Nil | – |
| *Escherichia coli* | $TiO_2/SnO_2$ | Nil | 2 | Nil | – |
| *Escherichia coli* | $TiO_2$ | 10 $mWcm^{-2}$ | 2 | 85 | 0.017 |
| *Escherichia coli* | $PdO/TiO_2$ | 10 $mWcm^{-2}$ | 2 | 98 | 0.023 |
| *Escherichia coli* | $SnO_2$ | 10 $mWcm^{-2}$ | 2 | 56 | 0.007 |
| *Escherichia coli* | $PdO/SnO_2$ | 10 $mWcm^{-2}$ | 2 | 68 | 0.008 |
| *Staphylococcus aureus* | $TiO_2$ | 10 $mWcm^{-2}$ | 3 | 87 | 0.004–0.020 |
| *Staphylococcus aureus* | $PdO/TiO_2$ | 10 $mWcm^{-2}$ | 3 | 98 | 0.007–0.028 |
| *Staphylococcus aureus* | $SnO_2$ | 10 $mWcm^{-2}$ | 3 | 79 | 0.004–0.012 |
| *Staphylococcus aureus* | $PdO/SnO_2$ | 10 $mWcm^{-2}$ | 3 | 90 | 0.008–0.015 |
| *Saccharomyces cerevisiae* | $TiO_2$ | 10 $mWcm^{-2}$ | 3 | 54 | 0.004 |
| *Saccharomyces cerevisiae* | $PdO/TiO_2$ | 10 $mWcm^{-2}$ | 3 | 65 | 0.006 |
| *Saccharomyces cerevisiae* | $SnO_2$ | 10 $mWcm^{-2}$ | 3 | 31 | 0.002 |
| *Saccharomyces cerevisiae* | $PdO/SnO_2$ | 10 $mWcm^{-2}$ | 3 | 41 | 0.003 |

designed photo reactors (Krishna et al. 2008). *E. coli*, a gram negative bacterium, is abundantly found in the digestive tracts of most warm blooded animals including human beings. *E. coli* is generally referred to as opportunistic pathogen except fewer of the strains e.g. *E. coli* O157:H7. However, they are used as an indicator of the existence of fecal contamination. Apart from *E. coli* in pure water, the photocatalytic inactivation of other coliform bacteria (Total˜ $2.5 \times 10^3$/100 ml, Fecal˜ $9 \times 10^2$/100 ml) present in spring water has also been reported (Gelover et al. 2006).

$TiO_2$ Degussa P25® has been used to inactivate four genera of gram negative bacteria including *E. coli* K-12, *Pseudomonas aeruginosa*, *Salmonella typhimurium*, and *Enterobacter cloacae* (Ibanez et al. 2003). Photocatalytic inactivation of *Escherichia coli*, *Staphylococcus aureus*, *Saccharomyces cerevisiae*, and *Aspergilus niger* spores have been reported with Pd-doped $TiO_2$ and $SnO_2$ films on glass substrates (Erkan et al. 2006). A summary of the results of these experiments are tabulated in Table 13.1. Keeping all the essential parameters constant, the inactivation rate of these organisms lie in the following order: *E. coli* > *S. aureus* > S. *cerevisiae* > *A. niger* spores. The *A. niger* spores were found very resistant against the photocatalytic process and survived for up to 8 h of irradiations on both the photocatalyst surfaces.

The efficacy of $TiO_2$ (Degussa P25®) coating (250 $mg/cm^2$) to degrade bacteria (*E. coli, Pseudomonas aeruginosa*), fungi (*Candida albicans, Fusarium sloani*), and protozoa (the trophozoite stage of *Acanthamoebe polyphaga*), spores (*Bacillus subtilis*), and cysts under solar light irradiation has also been investigated (Lonnen et al. 2005). Results indicate that sensitivity of the organisms decreases from bacteria to fungi, to spores, and to cysts due, obviously, to the hardness of their respective cell envelops. Detailed results are included in Table 13.2. The negligible inactivation of cysts, in the process, excludes this organism from the list of potentially degradable microorganism by photocatalysis. However, the activity is greatly influenced by the

**Table 13.2** Solar-only disinfection (SODIS) and Solar-photocatalytic disinfection (SPC-DIS) of different water borne

| | | SODIS | | SPC-SODIS | |
|---|---|---|---|---|---|
| Pathogen | Starting population | Reduction (logunits) | Time (h) | Reduction (logunits) | Time (h) |
| *Escherichia coli DH5a* | $3\times10^5$ | −5.5 | 2.5 | −5.4 | 1.5 |
| *Candida albicans* | $2.5\times10^5$ | −5.4 | 6.0 | −5.4 | 4.0 |
| *Fusarium solani (conidia)* | $3.4\times10^5$ | −5.5 | 8.0 | −5.5 | 4.0 |
| *Pseudomonas aeruginosa* | $9.6\times10^4$ | −5.0 | 2.0 | −4.9 | 2.0 |
| *Acanthamoeba Polyphaga (Trophozoites)* | $2\times10^4$ | −4.2 | 6.0 | −4.2 | 2.0 |
| *Acanthamoeba Polyphaga (Cysts)* | $4\times10^4$ | 0.0 | 8.0 | 0.0 | 8.0 |
| *Bacillus subtilis (spores)* | $9.2\times10^6$ | −1.7 | 8.0 | −1.7 | 8.0 |

light intensity. Higher intensities achievable through suitably designed concentrators may enable the system to degrade the most resistant organisms like cysts. The microbicidal effect of $TiO_2$ encompasses the Fusarium species of soil borne pathogenic fungi that are potentially harmful for crops particularly those grown in hydroponic farming. These include *Fusarium equiseti, Fusarium oxysporum, Fusarium anthophilum, Fusarium vertcillioides*, and *Fusarium solani* (Sichel et al. 2007). The photocatalytic inactivation of the fungal genera have been observed in a suspension of $TiO_2$ (Degussa P25®) at concentrations of 0.035 g/L under solar light illumination. Results indicate that the relative sensitivity shown by these fungal species, against solar-only and solar plus $TiO_2$ lie in the order: *F. oxysporum*, > *F. solani*, > *F. verticillioides* > *F. anthophilum* > *F. equiseti* and *F. verticillioides* > *F. oxysporum* > *F. solani* > *F. anthophilum* > *F. equiseti*. The photocatalytic inactivation of *Clostridium perfringes* spores, which show resistance against chlorination, was witnessed with the use of $TiO_2$ photocatalyst. Complete inactivation was reported to occur after about 2 h of UV (360 nm) irradiation with the intensity of 192.73 $Wm^{-2}$ (Dunlop et al. 2008). Using thin film of $TiO_2$ in synergism with the UV-C irradiation a 3log diminution in the cell viability of *Legionella pneumophila* was observed to occur within 15 min (Hayakawa et al. 2007).

The bactericidal effect of $TiO_2$, under UV-illumination, on three species of foodborne pathogenic bacteria, *Salmonella choleraesuis*, *Vibrio papahaemolyticus*, and *Listeria monocytogenes*, was reported by Kim et al. (2003). Results show that UV-alone inactivated 5–20% of the cells, while $TiO_2$ plus UV caused damage to 95% cells in each strain over the same time. The report also indicates that more than 92% of *Salmonella choleraesuis* lost viability after 30 min illumination with 20 W UV at intensity of 1 $Wm^{-2}$. *Listeria monocytogenes* strain showed more resistance against the activity than *V. papahaemolyticus* and *S. choleraesuis*. Apart from its bactericidal effect, photocatalysis has been proved more efficient in deactivating viruses, like, *polio-virus* and *influenza-virus* (Watts et al. 1995). As most of the important factors of the process have been discussed in other sections, we list here most of the microorganisms that have been reported in the literature during studies on photocatalytic inactivation. The list includes; *Cryptosporidium parvum* oocysts (Ryu et al. 2008), *Legionella pneumophila* (Oguma et al. 2004), *Pseudomonas aeruginosa* (Wu et al. 2009), *Total coliform* (Rizzo 2009), *Bacillus cereus* spores (Zhao et al. 2009), *Fusarium solani* spores (Fernandez-Ibanez et al. 2009), *Bacillus anthracis* (Prasad et al. 2009), *Aspergilus niger* (Vohra et al. 2006), *Escherichia coli, Lactobacillus helveticus* (Liu and Yang 2003), *Entrobacter cloacae* SMI, *Streptococus iniae, edwardsiella tarda* (Cheng et al. 2008), *Penicillium expansum* (Maneerat and Hayata 2006), *Giardia intestinalis* cysts, *Acanthameoba castellani* cysts (Sokmen et al. 2008), marine bacteria *Alteromonas alvinellae* and *Photobacterium phosphoreum* (Leung et al. 2008) and Bacterial phage MS2 (Sato and Taya 2006b). Conclusively, bactericidal effect of photocatalysis has been tested on a wide range of microorganisms including bacteria, fungi, viruses, spores and cysts. Keeping all the parameters constant, inactivation rate depends upon the thickness of the outer covering of the cell. The inactivation rate in gram negative bacteria is higher than in the gram positive bacteria, while spores and cysts show high resistance to the process.

### *13.2.2 Photocatalyst Disinfectants*

Several nanoparticulate materials; like chitosan (Qi et al. 2004), silver (Kim et al. 2008), fullerene $nC_{60}$ (Lyon et al. 2008) etc. have been reported to exhibit microbicidal effects. The activities of metal oxides semiconductors in synergism with electromagnetic irradiation are significantly high. Nanoparticulate ZnO semiconductor shows strong cytotoxicity towards certain bacterial cells, such as *Escherichia coli* and *Lactobacillus helveticus* (Adams et al. 2006). $TiO_2$ is the most extensively studied material for its potent photocatalytic degradation of a vast variety of water and air borne microbial pathogens. In a report by Dunlop et al. (2002) the higher bactericidal activity exhibited by Degussa P25® than pure $TiO_2$ from Aldrich was attributed to the smaller particle size (~30 nm) of Degussa P25® than that of Aldrich (~1 μm) as catalyst in the form of small particles cover more surface area to initiate more redox reactions. Moreover, in $TiO_2$ from Degussa, which is a mixture of anatase and rutile in the ratio of 70:30, the electron-hole recombination is effectively inhibited due to the existence of grain boundaries in the nanoparticulate mixture. This inhibition diverts more number of photogenerated charge carriers towards oxidation/reduction reactions in the bacterial cells. $TiO_2$ nanoparticles from Aldrich, on the other hand, were virtually perfect crystalline anatase and lacking defect sites to suppress the electron-hole recombination. The effect of crystal structure on the photocatalytic inactivation of bacteriophage MS2 was studied by Sato and Taya (2006b). Catalyst composites of rutile-anatase mixture, with 0–100 wt.% anatase, were tested and maximum inactivation rate was observed for 70 wt.% of anatase phase. On the basis of the kinetic studies, it was concluded that for the maximum quantum inactivation efficiency, the two types of particles in the nano-composites must be in close contact of one another. In some cases Millennium PC500® $TiO_2$ nanoparticles displayed better photocatalytic activity than the Degussa P25® (Guillard et al. 2008).

Millennium PC500® is 100% anatase with specific surface area 320 $m^2/g$ which is several times higher than that of Degussa P25®. The faster inactivation of the PHL1273 strain with PC500® than with Degussa P25® may be due to the smaller particle size of PC500®, that is, more particles adhere on the bacterial surface and make it more permeable with their photocatalytic redox-reactions.

Gumy et al. (2006) carried out a comparative study on thirteen $TiO_2$ samples, supplied by different companies, for their relative efficiencies in the photocatalytic inactivation of *E. coli* under simulated sun light irradiation and found that Degussa P25® was the most efficient photocatalyst material among them inspite of the fact that some of the samples had higher specific area than Degussa P25®. Regarding the optimal concentration of the photocatalyst, different researchers have reported different results. This is not unexpected because of the diversified nature of all the parameters involved in the experiments. However, concentration of the photocatalyst has a predominant effect on the rate of activity. Keeping all other parameters constant, the activity, in each experiment, is seen to increase till the concentration reaches its optimal value. Beyond this, the activity decelerates. Huang et al. (1998) reported that among the tested amounts of 0.1, 0.2, 0.5, 1.0 g/L, the catalyst concentration of 1.0 g/L is the one at which the rate of bacterial inactivation was the highest, showing that the activity increases monotonically with the catalyst concentration. Kim et al. (2003), however, reported that the activity decreased as the catalyst loading was increased from 1.0 to 1.25 g/L indicating 1.0 g/L as the optimal catalyst concentration. Contrary to this, Benabbou et al. (2007) found that among the different concentrations between 0.1 and 2.5 g/L, 0.25 g/L was the optimum catalyst concentration which gave complete disinfection of *E. coli* at concentration of $10^5$–$10^6$ cfu/mL. It was hypothesized that a bacterium size is about 30 times larger than the catalyst particle. So, each bacterium cell could be surrounded by the catalyst particles. The number of particles, in the optimum quantity (0.25 g/L), was enough to cover all the available bacterial cells, completely, and degrade them. If lesser quantity of the catalyst was used, the available particles might not have covered each bacterium. On other hand, if the concentration of the catalyst, introduced in the suspension, was higher than the optimal value, then many particles might have not found any room on a bacterium surface. Such particles, obviously, could not contribute in the activity. Rather, their presence in the suspension could be counterproductive due to the scattering of incoming light. This has been illustrated in Fig. 13.2. For higher concentration of the bacteria (~$10^8$ cfu/mL), the hypothesis seems unacceptable, as, the estimated catalyst surface area, for the optimal quantity, was not enough to cover all the bacterial cells. Several reports suggested that 1 mg/ml is the optimal catalyst concentration for photo-bactericidal activity (Chawengkijwanich and Hayata 2008; Cho et al. 2004; Huang et al. 2000). Generally, the optical catalyst loading depends upon the particle as well as the intensity of light used.

Besides $TiO_2$, Ren et al. (2009) have reported the photocatalytic bactericidal effect of $Bi_2WO_6$ under visible light irradiation. Also, recent reports show that nanoparticulate ZnO exhibits high bactericidal activity under visible light irradiation (Jones et al. 2008); the reason obviously is the existence of surface defects in ZnO nanocrystals. Bactericidal effect of polyoxometalates, as a homogeneous photocatalyst, has also

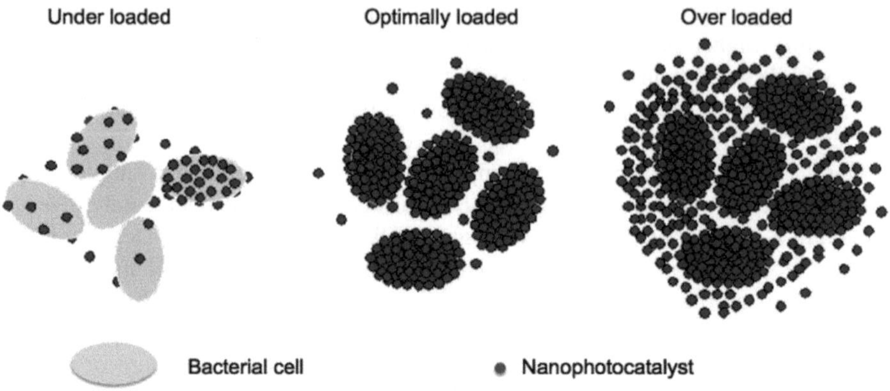


**Fig. 13.2** Schematic diagram illustrating different catalyst loadings for the photocatalytic degradation of bacteria

been reported. Bae et al. (2008) used different concentrations (below a specific value for each) of tungstosilicic acid ($H_4SiW_{12}O_{40}$), phosphotungstic acid ($H_3PW_{12}O_{40}$), and phosphomolybdic acid ($H_3PMo_{12}O_{40}$) in *E. coli* and *B. subtilis* suspensions separately. The suspensions were irradiated with UV-A which revealed that polyoxometalates mediated homogeneous photocatalysis was more effective than $TiO_2$ based heterogeneous photocatalysis. However, separation of homogeneous photocatalyst could be difficult from water after treatment. The discussion shows that the commercially available form of $TiO_2$ (Degussa P25®), has been found as the most efficient photocatalyst under UV irradiation with optimal catalyst loading depending upon the type and concentration of microbial. Nanoparticulate ZnO, on other hand, exhibits high photocatalytic bactericidal effects under visible light.

### *13.2.3 Effect of Light Irradiation*

A photocatalytic material like $TiO_2$ has been reported to show no bactericidal effect in the dark, while light alone has a considerable antimicrobial efficacy depending upon its wavelength. However, the synergistic action of $TiO_2$ and light irradiation has proven to be effective for inactivation of microorganisms (Huang et al. 1998). The mechanism of photocatalysis has been hypothesized to commence with the photo-generation of electron-hole pairs in the valance band of the $TiO_2$ semiconductor and proceeds through the migration of the electron in to the conduction band and the hole to the lower energy states in the valance band, and redox reactions due to the photo-generated electrons and holes. This is possible only if the incoming light photon is energetic enough to move the targeted electron to the conduction band. $TiO_2$ has a band gap of 3.2 eV. This corresponds to the maximum wavelength of 387.5 nm, following the Einstein's equation, which the incident radiation needs to have so that the process can be initiated. This is the reason for which $TiO_2$ exhibits high photocatalytic activity with ultraviolet light irradiation. For a wavelength below the threshold value, 387 nm for $TiO_2$, the photocatalytic microbicidal activity increases with the intensity of the irradiated light (Huang et al. 1998). This is consistent to the quantum theory of light, which states that the intensity of light measures the striking rate of photons on unit surface area. Hence, higher light intensity causes higher level photogeneration of electron-holes pairs (Cho et al. 2004). Hayakawa et al. (2007) reported that $TiO_2$ photocatalyst caused prompt lethality to the legionella pneumophila bacterial strain in the presence of UV-C ($\lambda=254$ nm) at 100 mW/cm$^2$ intensity, as compared with UV-A ($\lambda=365$ nm). This, obviously, is due to the photolytic effect of the highly energetic UV-C radiations. Benabbou et al. (2007) have also, reported a similar trend in the photocatalytic inactivation of *E. coli* under UV-A ($\lambda>340$ nm), UV-B ($\lambda>290$ nm), UV-C ($\lambda>200$ nm) irradiation as given in Fig. 13.3. Paleologou et al. (2007), while disinfecting *E. coli* in wastewater, found that UV-C irradiation was two times more efficient than UV-A plus $TiO_2$. Shang et al. (2009) have shown that the inclusion of $TiO_2$ in the coliform (fecal) suspension during UV-C irradiation effectively suppressed the photo-reactivation of the bacteria during the subsequent 3 h fluorescent light illumination. The reason might be the generation of stable oxidizing agents primarily hydrogen peroxide. Park and Hong (2006) have reported significant enhancement in the photocatalytic degradation of algae (e.g. *Chroococcus sp*) when the $TiO_2$ and UV system was subjected to additional illumination by normal fluorescent light (FL). The reason for the enhancement was the initiation of photosynthesis in algae due to the FL irradiation that increased the production of oxygen. Concomitantly more oxidizing species were generated due to which the photocatalytic inactivation got promoted.

Leung et al. (2008) reported $TiO_2$ catalyzed photo-inactivation of marine bacterial species under fluorescent light irradiation. The relative susceptibility of the strains was attributed to their intracellular profile and level of superoxide dismutase and catalase. Several species of gram-negative as well as gram-positive bacteria were photocatalytically inactivated under fluorescent light irradiation by Pal et al. (2007). The catalyst used in the study was immobilized $TiO_2$. It was observed that *E. coli* was completely inactivated in 30 min irradiation time using $TiO_2$ loading of 1,666 mg/m$^2$ and UV-A irradiance of 0.013 mW/cm$^2$. This indicates that $TiO_2$ mediated photocatalysis can be effective for indoor air disinfection as 0.013 mW/cm$^2$ of UV light is usually present in the fluorescent light spectrum. Ultrasonic irradiation (36 kHz, 300 W) exhibited bactericidal effect which was observed by Ogino et al. (2006), to increase significantly with the addition of $TiO_2$ and $Al_2O_3$. Hydroxyl radical was reported as the prime contributor in the activity. The effect of $TiO_2$ was more than that of $Al_2O_3$.

The major attraction of the application of photocatalysis, in detoxification and disinfection of water and air, is its potential utilization of sunlight. In addition to being economical, utilization of solar light for the disinfection purpose has subtle environmental impacts (Munoz et al. 2005) as can be seen in Table 13.3. However, there are concerns about the low efficiency of solar photocatalytic disinfection that arises from the small percentage of UV radiation in the solar spectrum. Lonnen et al. (2005) reported photocatalytic degradation of several bacteria protozoa and fungal

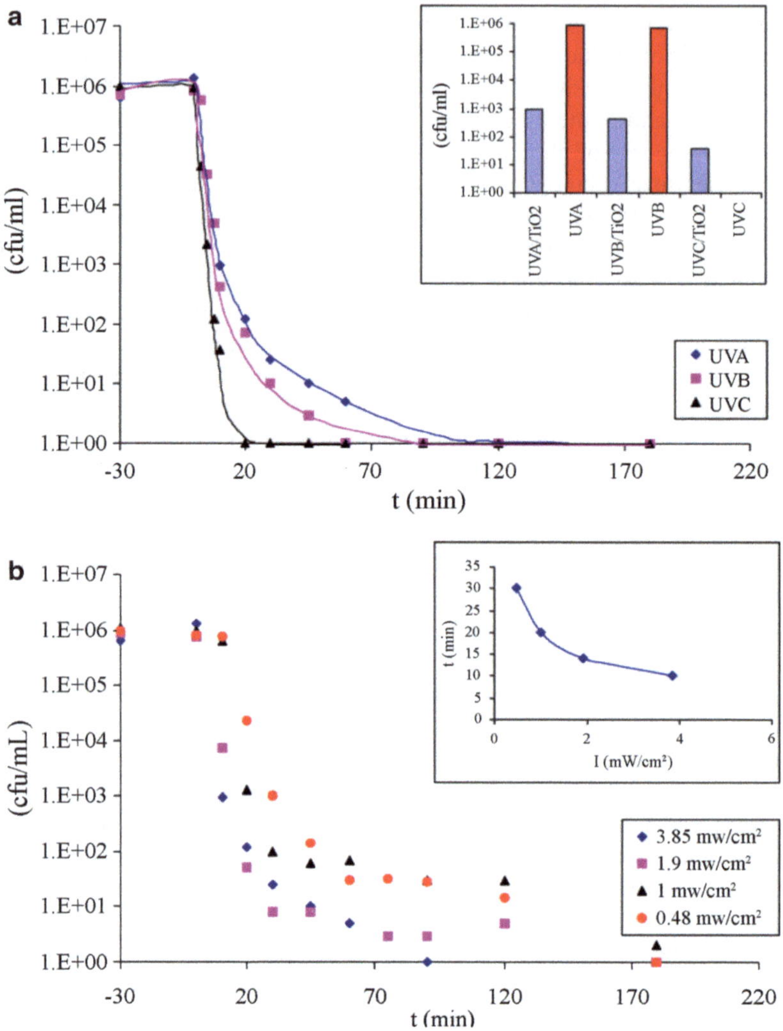


**Fig. 13.3** (**a**) Effect of the type of UV radiations on the photocatalytic inactivation of *E. coli* with $TiO_2$ (0.25 g/L); insert: survival of *E. coli* after 10 min irradiation in the process of absence of $TiO_2$ using different light intensities; (**b**) Effect of intensity of UV-A on photocatalytic inactivation of *E. coli* with $TiO_2$ (0.25 g/L). Insert: time to decrease bacterial concentration from $10^6$ to $10^2$ cfu/ml as a function of irradiation intensity (Benabbou et al. 2007)

strains in a batch-process reactor utilizing simulated solar light. However, the system left cysts of *A. polyphaga* and *B. subtilis* spores unharmed. Sichel et al. (2007) used solar light to kill agricultural fungi, Fusarium species, with $TiO_2$. The average solar UV-A irradiance during the activity was 25.78 W/m². Among the targeted five Fusarium species, *F. equiseti* spores were found highly resistant and took 6 h to reduce it to the detectable limit (i.e. from 1,200 to 2 cfu/ml). During this period, the

**Table 13.3** Life cycle impact assessment results for each advanced oxidation process (*AOP*), like, photocatalytic (*PhC*) and photofenton (*PhF*) in both energy scenario (solar and eletricity) (Munoz et al. 2005)

| | | Photocatalysis | | Photo-fenton | | Photocatalysis + $H_2O_2$ | | Photocatalysis + photo fenton | |
|---|---|---|---|---|---|---|---|---|---|
| Category | Units | Electricity | Solar | Electricity | Solar | Electricity | Solar | Electricity | Solar |
| GWP | Kg $CO_2$ equi | 90 | 0.48 | 38 | 0.76 | 54 | 1.2 | 16 | 1.2 |
| ODP | Kg CFC11 equi | $2.2\times10^{-05}$ | $7.6\times10^{-08}$ | $9.0\times10^{-06}$ | $3.8\times10^{-08}$ | $1.3\times10^{-05}$ | $8.9\times10^{-08}$ | $3.7\times10^{-06}$ | $9.0\times10^{-08}$ |
| AEP | Kg $PO_4^{3-}$ equi | 0.014 | $7.8\times10^{-05}$ | 0.006 | $3.0\times10^{-04}$ | $8.5\times10^{-03}$ | $3.7\times10^{-04}$ | $2.7\times10^{-03}$ | $3.7\times10^{-04}$ |
| AP | Kg $SO_2$ equi | 0.56 | $3.7\times10^{-03}$ | 0.24 | $7.5\times10^{-03}$ | 0.34 | $1.0\times10^{-02}$ | 0.1 | 0.01 |
| HTP | Kg $C_6H_4Cl_2$ equi | 21 | 0.029 | 8.6 | 0.031 | 12 | 0.049 | 3 | 0.049 |
| FATP | Kg $C_6H_4Cl_2$ equi | 1.5 | 0.084 | 0.60 | $1.6\times10^{-03}$ | 0.92 | 0.085 | 0.32 | 0.085 |
| POFP | Kg $C_2H_4$ equi | $2.9\times10^{-02}$ | $2.8\times10^{-04}$ | $1.3\times10^{-02}$ | $9.1\times10^{-04}$ | 0.018 | $1.1\times10^{-03}$ | $6.0\times10^{-03}$ | $1.1\times10^{-03}$ |
| ARD | Kg Sb equi | 0.67 | $1.8\times10^{-03}$ | 0.29 | $5.2\times10^{-03}$ | 0.40 | $6.4\times10^{-03}$ | 0.12 | $6.4\times10^{-03}$ |

*GWP* global warming potential, *ODP* ozone depletion potential, *AEP* aquatic eutrophication potential, *AP* acidification potential, *HTTP* human toxicity potential, *FATP* freshwater aquatic toxicity potential, *POFP* photochemical ozone formation potential, *ARD* abiotic resource depletion

UV-A irradiance was varied from 15 to 34 $W/m^2$ and the inactivation got accelerated when the UV-A irradiance exceeded the value of 20 $W/m^2$. A complete inactivation of coliform (Total & Fecal) bacteria was observed in 15 and 30 min respectively, under solar irradiance of 1,000 $Wm^{-2}$ (Gelover et al. 2006). Duffy et al. (2004) used ambient solar light for the photocatalytic disinfection of *E. coli* and observed that the inactivation rate depended upon the UV dose absorbed by the bacterial suspension. The solar photocatalytic disinfection process can be accelerated by intensifying the light with the help of concentrators (Fernandez et al. 2005). From the above discussion, it could be concluded that; $TiO_2$ shows high photocatalytic activity under UV light, the microbial inactivation rate increases with the irradiation intensity, solar is significantly effective in the photocatalytic microbial inactivation, the photocatalytic efficiency could enhanced by using suitable designed solar concentrators.

### *13.2.4 Effect of pH on Photocatalytic Microbial Inactivation*

Since, the photocatalytic disinfection proceeds through the adsorption of the nanophotocatalyst at the organism surface, a variation in the pH level of the suspension is, therefore, deemed to affect the activity by establishing electrostatic interaction between the organism and the photocatalyst. Rincon and Pulgarin (2004b) have investigated the effect of pH during the photocatalytic disinfection of *E. coli* and have found that, while complete disinfection was obtained after 120 min of illumination, the suspension transformed from neutral to acidic (pH dropped from 7 to 4). It has also been seen that the photolytic inactivation of the microorganism rises from 25% to 80% with simultaneous gradual decrease of the pH from 7 to 4 through the addition of HCl, as given in Fig. 13.4. In the absence of bacteria, $TiO_2$ modified pH of the suspension decreased from 7 to 5.5 upon light illumination, which was attributed to the formation of hydrogen ions:

$$H_2O + h^+ \left(TiO_2\right) \rightarrow {}^{\bullet}OH + H^+$$

The pH modification from 7 to 4 in the photocatalytic inactivation of bacteria is due to the concomitant production of aliphatic acids during the oxidation process. It has been hypothesized that the surface charge of $TiO_2$ is a function of the pH value of the suspension. For pH higher than point of zero charge (PZC), 6.5 for $TiO_2$, the catalyst surface is negatively charged, while for pH lower than PZC it becomes positively charged, as per the following equilibrium conditions:

$$\text{pH} < \text{PZC}: \quad TiOH + H^+ \leftrightarrow TiOH_2{}^+$$

$$\text{pH} > \text{PZC}: \quad TiOH + OH^- \leftrightarrow TiO^- + H_2O$$

Virtually 100% of the $10^5$ cfu/ml of *E. coli* lost the viability in about 40 min due to the synergistic effect of photocatalysis and pH modification. Inhibition of the activity, with the increase of pH of the suspension, was observed by Kikuchi et al.

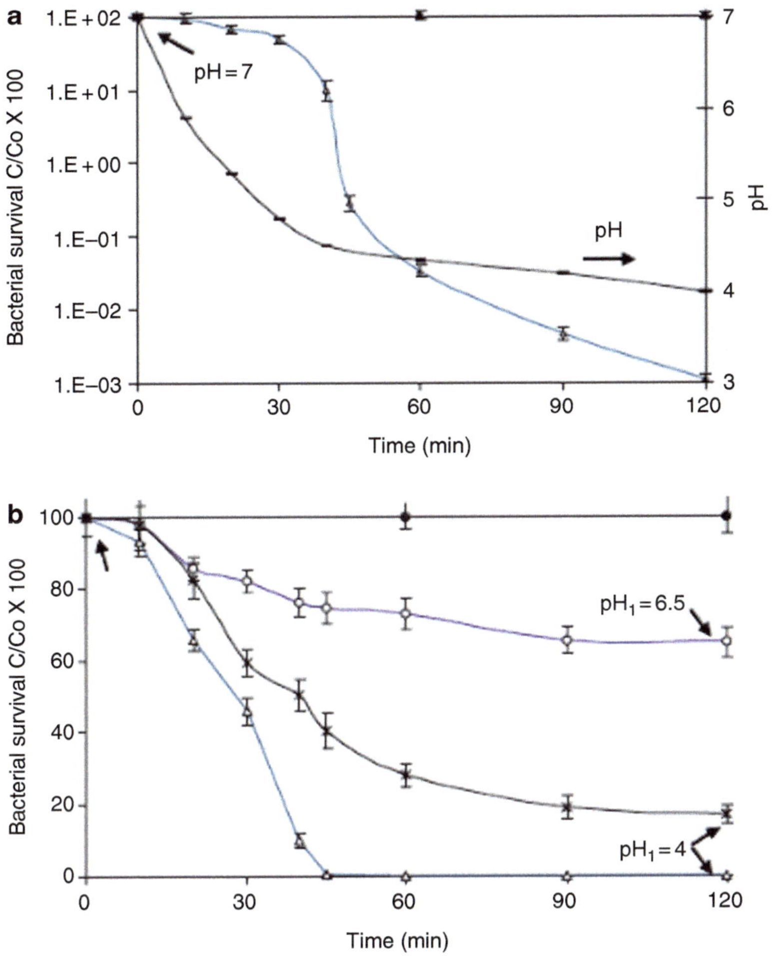


**Fig. 13.4** (**a**) Photocatalytic disinfection of water contaminated with *E. coli* K-12 (■), natural evolution of pH during treatment; initial bacterial concentration $10^5$ cfu/ml, light intensity: 400 W/m$^2$, $TiO_2$ concentration: 0.5 g/L. Dark control in the presence of $TiO_2$ (▲). (**b**) Influence of the pH modification with continuous addition of HCl (0.1 mol/L) till to reach a pH of 4.0 (*), during the photocatalytic disinfection of water contaminated with *E. coli* K-12. Photocatalytic inactivation using 0.5 g/L of $TiO_2$ (▲), photocatalytic inactivation without pH modification (○), control in *dark* (●) (Rincon and Pulgarin 2004b)

(1997), who used phosphate buffers ($Na_2HPO_4$—$NaH_2PO_4$, 8:2) to increase the pH. Koizumi and Taya (2002), while evaluating the effect of pH on the rate constant of the photocatalytic inactivation of phage MS2, observed that, within the pH range of 3–10, the rate constant attained its maximum value at pH 6. It was argued that adsorption of the phage molecules on the catalyst surface played an effective role in the enhanced activity. The interaction between the phage molecules and the catalyst depends upon the surface properties that could be varied through pH of the suspension. Alrousan et al. (2009) reported a slight enhancement in the photocatalytic inactivation of *E. coli* upon the pH reducing from basic to acidic (8.5–5.5). The discussion reveals that pH value of the bacterial suspension drops during the course of photocatalytic inactivation process; the photocatalytic bacterial inactivation rate can be enhanced with adjusting pH of the bacterial suspension at a suitable level.

### *13.2.5 Effect of Water Quality*

Nearly, all the factors affecting the efficiency of photocatalysis for water disinfection have been investigated through a number of experiments. However, most of the experiments were performed with the contaminant suspensions prepared in deionized water. Natural water, on the other hand, contains many other organic and inorganic dissolved/suspended materials. Wist et al. (2002) carried out the photocatalytic disinfection tests of *E. coli* in distilled as well as tap water for 1.5 h irradiation. Results showed that beside the high efficiency in the case of distilled water, no regrowth was observed for 48 h subsequent dark period. On the other hand, regrowth exceeded the initial concentration in 48 h post treatment period in the crude water. Rincon and Pulgarin (2004a) have examined the influence of the presence of several common inorganic and organic compounds on the photocatalytic inactivation of *E. coli* in water. The study indicated that the inclusion of anions, like, $HCO_3^-$, $HPO4^{2-}/H_2PO_4^-$, $NO_3^-$, $Cl^-$, and $SO_4^{2-}$, to the *E. coli* suspension reduces the efficacity of photocatalytic disinfection. The effect was more pronounced in the case of $HCO_3^-$ and $HPO_4^{2-}/H_2PO_4^-$than for $NO_3^-$, $Cl^-$, and $SO_4^{2-}$, which was ascribed to arise from the adsorption of anions on the catalyst surface and the subsequent inhibition of light to accede the catalyst surface, scavenging the $^{\bullet}OH$ radicals and the photogenerated holes ($h^+$) by the anions. Similar to the anions, cations like, $Na^+$ and $K^+$, were also found to suppress the activity. Photocatalytic inactivation could be accelerated with the incorporation of $H_2O_2$, as shown in Fig. 13.5. The photocatalytic disinfection of wastewater and of pure *E. coli* suspension carried out under identical conditions produced results with a striking difference between the two (Rincon and Pulgarin 2004a). The pure *E. coli* culture was completely disinfected in less than 30 min, while only 2log reduction in the bacterial survival rate was observed in the wastewater upon illumination with UV light for 2 h.

To study the effect of water quality on its photocatalytic disinfection, Alrousan et al. (2009) carried out tests on *E. coli* in distilled water and in surface water separately. The outcomes of the experiments showed that the photocatalytic disinfection took place with significantly higher rate in spiked distilled water than in the natural surface water. Further, the addition of nitrate $(NO_3)^-$, sulphate $(SO_4)^-$, and humic acids suppressed the activity in the order nitrate < sulphate < humic acids. The negative effect of anions on the activity was ascribed to their adsorption on the catalyst particles. The profound effect of humic acid, on the photocatalytic disinfection rate, showed that the main factor involved in the low efficiency of the photocatalytic disinfection of surface water was the presence of high concentration of suspended

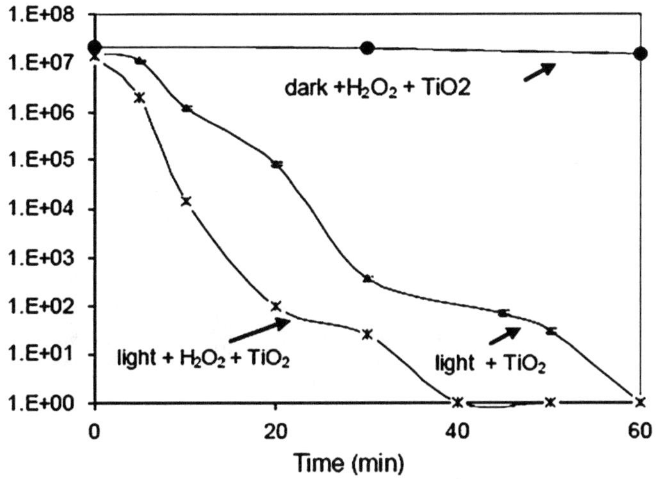


**Fig. 13.5** Influence of $H_2O_2$ (10 mg/L) on *E. coli* K-12 inactivation by sunlight in the presence of $TiO_2$ (1 g/L). The initial *E. coli* concentration was $10^7$ cfu/ml and light intensity of 400 $Wm^{-2}$ (Rincon and Pulgarin 2004b)

organic materials. Those materials were in competition with the target microorganisms for active sites on the catalyst surface, oxidants, and light photons. Paleologou et al. (2007) found that UV-C and UV-C along with ultrasound treatments were more efficient than $TiO_2$ with UV-A to disinfect spiked water. These authors proclaimed no regrowth of bacteria in the UV-C treated in the above-mentioned wastewater for 24 h of the subsequent dark period. Fernandez-Ibanez et al. (2009) showed that keeping all the conditions the same, complete photocatalytic disinfection could not be achieved in the well water. The discussion shows that the photocatalytic disinfection rate gets suppressed by the presence of organic and inorganic materials suspended in water. For the purification of surface water it is, therefore, suggested that suspended materials should be removed from water through some suitable technique prior to complete disinfection with photocatalysis.

## 13.3 Mechanism of Microbial Degradation

The microbicidal efficacy of heterogeneous photocatalysis is well documented. In order to enhance its efficiency for practical applications, detailed study of all the involved parameters is crucial. In aqueous suspension, using the catalyst in slurry form, the activity mainly proceeds through the adsorption of the catalyst particles on the organism surface. With illumination, these particles start oxidation-reduction reactions which bring about damages to the protective envelope of the organism. With prolonged illumination, the outer cell membrane is degraded to the extent that the oxidizing agents get in to the cell and attack the intracellular contents. At the

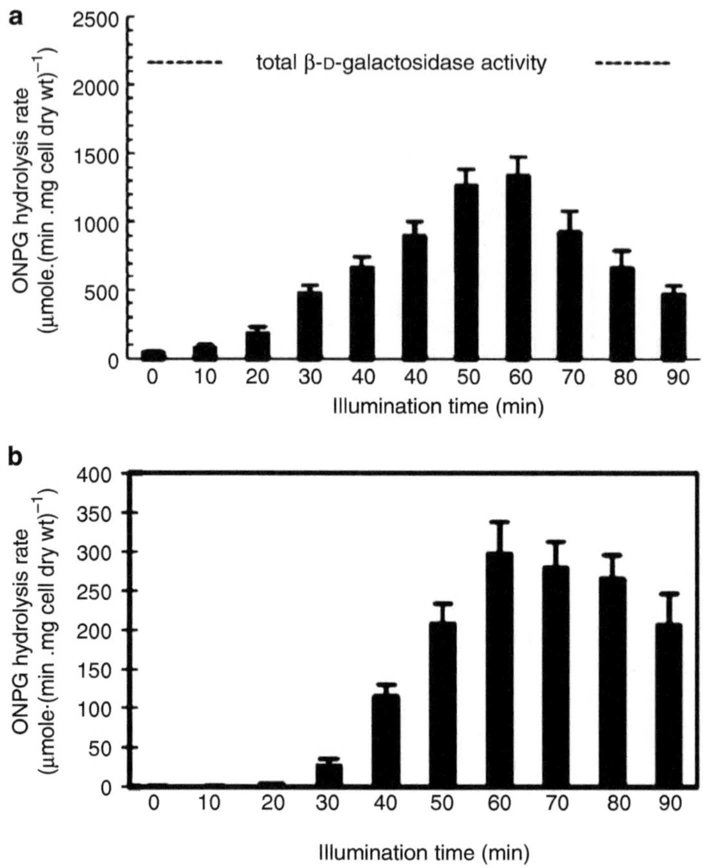


**Fig. 13.6** (**a**) Effect of $TiO_2$ photocatalytic reaction on cell permeability. $TiO_2$ and *E. coli* slurry containing $10^6$ mg/ml $TiO_2$ was illuminated with UV light (8 $W/m^2$). The o-nitrophenole β-D-galactosidase (ONPG) assay was initiated by mixing $TiO_2$-treated cells with 0.1 mM ONPG. $TiO_2$-mediated membrane damage was shown by the change in ONPG hydrolysis rate. Total β-D-galactosidase activity (*dashed lint*) represents the total accessibility of β-D-galactosidase in untreated *E. coli* cells. (**b**) The leakage of interacelluler β-D-galactosidase induced by $TiO_2$ photocatalytic reaction. Cell filtrates were obtained from cells treated with $TiO_2$ and UV light. The presence of β-D-galactosidase in the cell filtrate was determined by ONPG assay, and β-D-galactosidase activity was expressed as the ONPG hydrolysis rate (Huang et al. 2000)

same time, efflux of the intracellular contents starts and ultimately the cell dies. The gradual oxidative damage to the outer cell membrane and their impact on the viability of cells have been investigated by Huang et al. (2000). Increased cell wall permeability and the subsequent perforation of the cytoplasmic membrane were observed through the detection of the leakage of β-D-galactosidase. The results strengthened the notion that cell membrane is the primary target of photocatalytically generated oxidants in the activity, as shown in Fig. 13.6. Erkan et al. (2006) while commenting on the comparatively high resistance exhibited by *Staphylococcus aureus* (a gram-positive bacteria strain) against photocatalytic immobilization, stated that it is due to their thicker cell wall (25 nm) as compared to that of gram-negative bacteria (3 nm). The same was the cause of the prolonged viability exhibited by the bacterial spores and cysts. Sunada et al. (2003) observed that spheroplasts which lack peptidoglycan layer and some part of the outer membrane, exhibited more susceptibility to photo-killing than the intact *E. coli* cells.

Rincon and Pulgarin (2004a) have reported that the physiological state of bacteria has a marked influence on their photocatalytic inactivation. They observed that *E. coli* in the exponential phase was more susceptible to photocatalytic inactivation than in the stationary phase. This is because in the state of infancy, the organisms have very thin cell membrane and are prone to oxidative damages. Also, Dantur and Pizarro (2004) have reported that the photolytic effect of UV-A was more pronounced on *E. coli* in its exponential-growth phase than in the stationary-phase. While investigating the photocatalytic inactivation mechanism of several strains of *Legionella pneumophila*, Cheng et al. (2007) found that lipid peroxidation of the cell covering caused the hydroxyl radicals to enter the cell and destroy the intracellular components. The decrease of the C-16 branched-chain fatty acid in the treated cells confirmed the proposed mechanism. Ren et al. (2009) observed the leakage of $K^+$ ions from the photocatalytically degraded cells of *E. coli*. Sichel et al. (2007) observed the accumulation of the catalyst particles at the surface of *F. equiseti macroconidia* and *F. oxysporum* spores and showed the microscopic images of the photocatalytic degradation process of the organism. The images for *F. equiseti macroconidia* are shown in Fig. 13.7. Amezaga-Madrid et al. (2003), on the basis of Transmission Electron Microscope (TEM) micrographs of the decaying *P. aeruginosa*, have formed a different opinion that the photocatalytic treatment rendered the target bacterial cells to undergo abnormal cell divisions. Concomitantly, the daughter cells were either incapable to survive or undergo premature cell divisions, which resulted in the complete disintegration of the organism.

All the reports described above suggest that the outer cell membrane is the primary site of attack of the oxidants in the photocatalytic treatment. The oxidants inflict irreversible damages on the outer and cytoplasmic membrane due to which the microorganism ceases its life sustaining functions. The action also increases the permeability of cell envelope. This initiates influx of the oxidants into the cell, where they can degrade the intracellular contents and disrupt the whole system. At the same time the intracellular material starts leaking out and ultimately the cell disintegrates.

Cho et al. (2004) reported that the hydroxyl radicals ($^{\bullet}OH$) play the major role in the photocatalytic inactivation of *E. coli*, as there was a striking linear correlation between the $^{\bullet}OH$ concentration and the rate of *E. coli* degradation. The contribution of $^{\bullet}OH$ radicals in the activity was estimated by adding methanol ($^{\bullet}OH$ scavenger) into the suspension during the photocatalytic treatment. Kikuchi et al. (1997), on the basis of experimental evidences, suggested that it may be the $H_2O_2$ and not $^{\bullet}OH$ which play the main role in the activity. However, Rincon and Pulgarin (2004b)

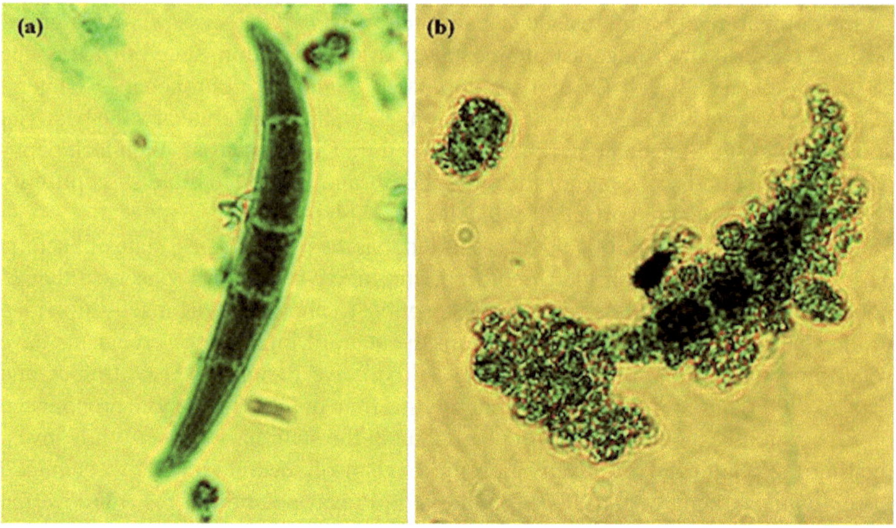


**Fig. 13.7** Macroconidia of *Fusarium equiseti* with Malachite green (**a**) before and (**b**) after 5 h photocatalytic treatment (Sichel et al. 2007)

were of the opinion that the $H_2O_2$ added to the suspension converts into $^{\bullet}OH$ radicals as per the following reactions.

$$TiO_2\left(e^-\right) + O_2 \rightarrow TiO_2 + O_2^{\bullet-}$$

$$O_2^{\bullet-} + H^+ \rightarrow HO_2^{\bullet}$$

$$O_2^{\bullet-} + HO_2^{\bullet} \rightarrow O_2 + HO_2^-$$

$$2HO_2^{\bullet} \rightarrow O_2 + H_2O_2$$

$$HO_2^{\bullet} + e^- + H^+ \rightarrow H_2O_2$$

$$H_2O_2 + HO_2^{\bullet} \rightarrow H_2O + {}^{\bullet}OH + O_2$$

$$TiO_2\left(e^-\right) + H_2O_2 \rightarrow TiO_2 + OH^- + {}^{\bullet}OH$$

Cho et al. (2004) stated that the $^{\bullet}OH$ radical is 1,000–10,000 times more effective in the inactivation of *E. coli* than the other disinfectants, like, chlorine, ozone, and chlorine dioxide. Farshbaf Dadjour et al. (2006) used ascorbic acid, histidine and glutathione and observed that $^{\bullet}OH$ was the main contributor in the *Legionella pneumophila* inactivation in the ultrasonic treatment in the presence of $TiO_2$. While investigating the role of $H_2O_2$ and $^{\bullet}OH$ radicals in the photocatalytic degradation of bacteria, using the Electron Spin Resonance (ESR) technique, Yan et al. (2009) observed a significant enhancement in the inactivation rate with the addition of $H_2O_2$. However, suppression of the activity with the addition of catalase indicated that it's the $^{\bullet}OH$ radical, which plays the prime role in the photocatalytic bacterial deactivation process. $H_2O_2$, on the other hand, acts as a source of $^{\bullet}OH$ radicals as per the above mentioned reactions. In addition to $^{\bullet}OH$, a significant number of carboxyl radical anion ($CO_2^-$) were produced during the photocatalytic process of $TiO_2$ (anatase) in the presence of UV-A, using the electron spin resonance technique (Dodd and Jha 2009). The $CO_2^-$ might form in the reaction of $^{\bullet}OH$ radicals with di- and tri-carboxylic acids and possibly amino acids. The radical itself looked unable to cross the cell membrane and degrade DNA. However, they can readily react with oxygen to form superoxide anions, as per the reaction

$$CO_2^{\bullet -} + O_2 \rightarrow CO_2 + O_2^{\bullet -}$$

The $O_2^{\bullet -}$ radicals subsequently converted into $H_2O_2$ which can diffuse through the cell membrane and form $^{\bullet}OH$ radicals, through Fenton reaction, in close proximity to the DNA and degrade it. In conclusion, the outer cell covering of a bacterium is the primary site of attack of the photocatalyst which adsorb at the cell surface and produce $^{\bullet}OH$ radicals with the irradiation of light. Through oxidation reactions $^{\bullet}OH$ radicals perforate the cell membrane and efflux of the intracellular contents starts leading to the cell death.

## 13.4 Kinetics of Photocatalytic Bacterial Inactivation

The photocatalytic microbial inactivation is a complex process and its rate is sensitive to a variety of factors, like, nature and concentration of the microorganism, type, composition and concentration of the catalyst, nature and intensity of illumination, pH level of the suspension, water quality, etc. The time profiles of the microbial viability under photocatalytic treatments are, however, alike. Three regions characterize the general curve of the photocatalytic loss of viability with time. These include; (a) the retarded inactivation at the initial stage, (b) the "log-linear" middle part which covers the major portion of the curve and at which, the process proceeds with maximum rate, (c) the decelerated end part of the curve, as given in Fig. 13.8.

In order to standardize their experimental data, different researchers used different kinetic models. Several researchers are of the opinion that their experimental data agree with the Chick-Watson kinetic model (Pal et al. 2007; Rincon and Pulgarin 2004b). The general form of the Chick-Watson kinetic model is given below:

$$log\left(\frac{N}{N_0}\right) = -k[c^n]t \tag{13.1}$$

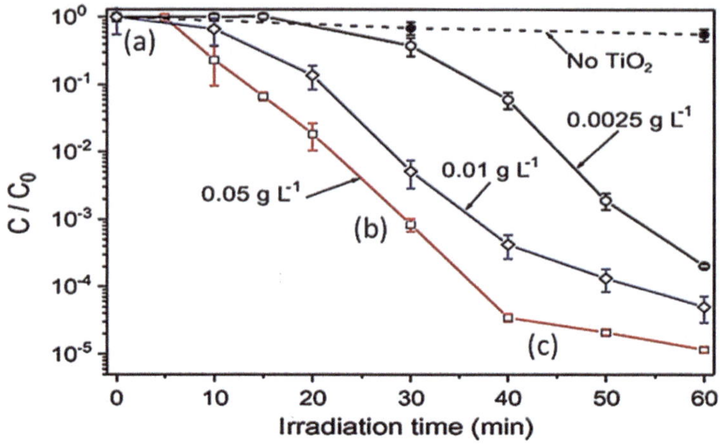


**Fig. 13.8** Photocatalytic inactivation of $10^6$ cfu/ml *E. coli* suspension with different loading of Degussa P25® (Marugan et al. 2008)

where $N_0$, N are the concentrations of microorganism at time t=0 and t=t, c is the catalyst concentration, n is the reaction order, t is the irradiation time, and k is the reaction constant. The value of n, determined from the slope of the curve of log(dN/dt) vs log(c), is nearly 1. The inactivation process is, therefore, termed as 'first-order reaction'. Equation 13.1 can also be written as

$$log\left(\frac{N}{N_0}\right) = -k[c]t = -k_1 t \tag{13.2}$$

where $k_1$ is the rate constant. The pseudo-first-order kinetic model, given by Eq. 13.2, has no provision for the shoulder and tail and hence, is applicable to the log-linear region. In order to incorporate the shoulder into it, the delayed-Chick-Watson model was introduced (Rennecker et al. 1999) which is of the form:

$$\log(N / N_0) = \left[\begin{array}{lll} 0 & t \geq t_{lag} & = \frac{1}{k_1}\log\left(\frac{N}{N_0}\right) \\ -k_1(t - t_{lag}) & t > t_{lag} & = \frac{1}{k_1}\log\left(\frac{N}{N_0}\right) \end{array}\right. \tag{13.3}$$

The delayed-Chick-Watson model can fit reasonably the microbial inactivation curve including shoulder. However, the provision for the tail is still not there. To include the consideration for all the three regions of the curve, a modified-Chick-Watson model was thus proposed. This is given as:

$$\log(N / N_0) = k_1[1 - e^{-k_2 t}] \tag{13.4}$$

Where, $k_2$ is the first order decay constant. The modified-Chick-Watson model, given in Eq. 13.4 can explain the initial delay in the activity or inhibition in the process activity at the end. However, it is unable to provide information on both the regions, simultaneously (Marugan et al. 2008). Hom's model for the microbial disinfection kinetics is frequently applied to the experimental data that departs from the Chick's inactivation law. Hom's model introduced an empirical coefficient in the Chick-Watson model given in Eq. 13.2 (Barbeau et al. 1999). The expression took the form:

$$\log(N / N_0) = -kc^n t^m \tag{13.5}$$

This expression reduces to the classic Chick-Watson model for m = 1, produces shoulder and tail regions for m > 1 and m < 1, respectively. However, it is also unable to produce both regions simultaneously. The problem was addressed by Barbeau et al. (1999), Haas and Joffe (1994) who introduced the modified Hom's kinetic model, the simplified version of which, assuming that the catalyst remains unaltered during the photocatalytic process, is given in Eq. 13.6 (Marugan et al. 2008).

$$\log(N / N_0) = -k_1(1 - e^{-k_2 t})^{k_3} \tag{13.6}$$

Where, $k_1$ is the decay constant, $k_2$ is the first order ozone decay constant and $k_3$ a model parameter. Graphical features of Eq. 13.6 indicate that the modified Hom's kinetic model could reproduce the experimental data describing all the three regions. However, the model is devoid of the physical meaning of its three parameters $k_1$, $k_2$, and $k_3$ (Marugan et al. 2008). As has been found experimentally, the photocatalytic process of bacterial inactivation mainly occurs due to the interaction of bacteria and the photocatalyst. Using the Langmuir-Hinshelwood kinetic model Marugan et al. (2008) deduced new expressions that could describe the photocatalytic inactivation of bacteria. These expressions are given by Eqs. 13.7 and 13.8.

$$\frac{dN_1}{dt} = -k_1 k_2 \frac{N_1^n}{1 + k_2 N_1^n + k_2 N_2^n} \tag{13.7}$$

$$\frac{dN_2}{dt} = k_1 k_2 \frac{N_1^n - N_2^n}{1 + k_2 N_1^n + k_2 N_2^n} \tag{13.8}$$

Where, $N_1$ is the number of undamaged colonies, $N_2$ is the number of damaged colonies and $k_1$ is a constant. Marugan et al. (2008) proclaimed that their kinetic model is capable not only to produce the bacterial inactivation curve with all its three distinct regions, but also has each of its three parameters with its physical meaning, such that; the kinetic constant $k_1$ stands for the rate of reaction between the microorganisms and the reactive oxygen species (ROS). This is a unique constant when applied to log-linear region. The pseudo-adsorption constant $k_2$ represents the adsorption of the catalyst on the bacteria and is responsible for the shoulder in the inactivation curve.

The inhibition co-efficient n is dealing with the affect of concentration of the disinfection by-products, which are in competition with the viable cells for the ROS. Thus, n is mainly concerned with the tail region of the inactivation curve. Hence, among a variety of proposed kinetic models, the one represented by Eqs. 13.7 and 13.8 is the most suitable one due its specific constant corresponding to each of the three regions in the kinetic curve of the photocatalytic inactivation of bacteria.

## 13.5 Photocatalytic Efficiency Enhancements

Microbicidal effect of $TiO_2$ mediated photocatalysis has been conceded based on numerous experimental evidences. The process, no doubt, looks very advantageous in its applications for the disinfection of drinking water, but has some limitations in its practical use. The photocatalyst $TiO_2$ used as slurry show high bactericidal efficiency, but its extraction from the treated water is a cumbersome task. The simplest possible solution is to immobilize the catalyst by coating it on a fixed support. Van Grieken et al. (2009) observed diminution of the efficiency in applying the catalyst in the immobilized form due to the loss of exposed surface area of the embedded $TiO_2$ particles. In order to improve its efficiency, modification of the catalyst material through doping has been extensively examined with a variety of dopants and doping processes. Sayilkan et al. (2009) synthesized $Sn^{+4}$-doped $TiO_2$ nanoparticles through hydrothermal process. The particles, due to their large specific area of 100 $m^2/g$, exhibited strong bactericidal effect towards *E. coli* (gram negative) and *S. aureus* (gram positive) bacterial strains, in both ways i.e. slurry and thin film on glass. The incorporation of $Sn^{+4}$ ions in to the $TiO_2$ crystals is expected to cause a red-shift in the absorption of the catalyst, which may improve its bactericidal efficacy under visible light illumination. Erkan et al. (2006) fabricated thin film of Pd-doped $TiO_2$ on glass substrates though sol-gel method and observed significant changes in its optical and electrical properties. The inactivation of *E. coli* was found to increase from 85% to 98% upon irradiation of 10 $mW/cm^2$ UV-visible light for 2 h, given in Table 13.1. Nitrogen-doped $TiO_2$ has been observed to inactivate a variety of human pathogens under visible light irradiation (Wong et al. 2006). $TiO_2$ doped with iron showed better E. coli inactivation than pure $TiO_2$ under UV-A irradiation (Wang and Gong 2007). Vohra et al. (2006) reported complete disinfection of indoor air from a variety of micro-pathogens using silver doped titania. The strong oxidative $^{\bullet}OH$ radicals can also be produced through photofenton reactions. Wu et al. (2009) prepared a composite palladium oxide (PdO) and nitrogen doped titanium oxide (TiON) via sol-gel method and disinfected gram-negative as well as gram-positive bacteria with visible light irradiation. The inclusion of $FeCl_3$, as a source of $Fe^{3+}$ ions, into the bacterial suspension generates additional $^{\bullet}OH$ radicals in the presence of visible light of wavelength up to 600 nm. Rincon and Pulgarin (2006) claimed enhanced bactericidal effect with $Fe^{3+}+TiO_2$ under UV-vis irradiation. Apart from doping, just mixing the catalyst with another suitable material could alter its bactericidal effect. The Ag-$TiO_2$ (1% w/w) composite not only showed

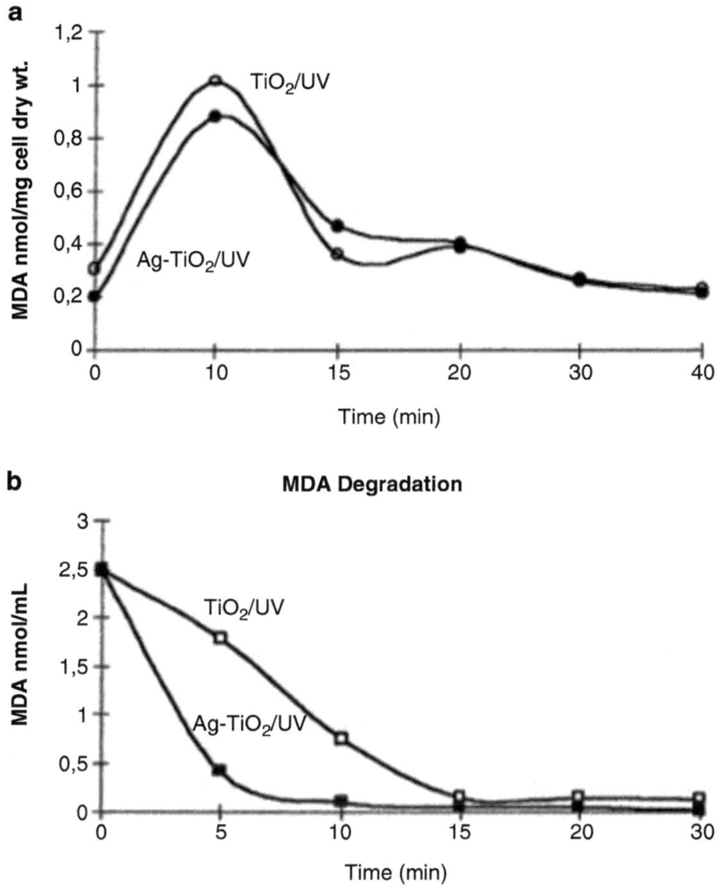


**Fig. 13.9** (**a**) Formation and (**b**) degradation of malondialdehyde (*MDA*) during the photocatalytic treatment of *E. coli* on neat $TiO_2$ and Ag-loaded $TiO_2$ (Sokmen et al. 2001)

strong bactericidal effect but also degraded malondialdehyde (MDA), which is a potent carcinogenic material produced from the lipid peroxydation during the *E. coli* cell disintegration (Sokmen et al. 2001), as shown in Fig. 13.9.

Kubacka et al. (2008) used anatase type $TiO_2$ nano-powder loaded with silver (Ag) through impregnation or with photo deposition methods. The photocatalytic efficiencies of both the samples were identical for Ag loading up to 1 wt.%, while, above this loading the efficiency of powder synthesized via photo deposition was significantly higher than the impregnated samples. The reason might be the way in which Ag got attached to the $TiO_2$ that facilitates formation and delivery of $^{\bullet}OH$ radicals. Tatsuma et al. (2003) reported that $TiO_2$-$WO_3$ composite, in conjunction with the UV-illumination, not only inactivated *E. coli* during the illumination but maintained the activity in post treatment dark period too. The report hypothesized that during the irradiation period the composite gets charged with the reductive energy, while during the subsequent dark stage it keeps discharging and sustains the activity. This effect was not observed with the film that was not charged with illumination prior to the activity. The assumption of charging was derived from the 50–40% (at 600 nm) decrease in the reflectance of the film, while no further decrease was observed with continuous irradiation. Spectroscopic analysis revealed that tungsten oxide ($WO_3$), present in the composite was electrochemically charged. Dye sensitization has gained great popularity in solar photovoltaics. Illumination with visible light the dye material gets excited and injects electrons into the conduction band of the photocatalyst substrate. Consequently, the photo-generation of electron-hole pairs occurs with visible light irradiation. Yu et al. (2003) reported enhanced photo inactivation of bacterial cells on the dye ($CuPcTs/CuPsCl_{14–15}$) coated $TiO_2$ film under visible light irradiation. The rate of inactivation increased with the time of dye adsorption (6 h was found to be optimum). Yao et al. (2007) have reported photocatalytic inactivation several species of plant pathogens with dye (TPPN)-sensitized $TiO_2$ under visible light illumination. Polyhydroxy fullerenes (PHF) have also been applied to scavenge photo-generated electrons from photocatalysts and concomitantly, enhance the generation of $^{\bullet}OH$ radicals. Krishna et al. (2008) reported that the amount of PHF used was 100 times less than that of other enhancers, while, the activity increased by 70%. The $Nd^{+3}$-doped $TiO_2$ (anatase) coated $NiFe_2O_3$ also exhibited improvement in the bactericidal activity (Rana et al. 2006). Christensen et al. (2003) found that *E. coli* inactivation by the $TiO_2$ electrodes, under UV irradiation, was accelerated with the application of a small voltage. Dunlop et al. (2008) observed a significant increase in the photocatalytic inactivation of *Clostridium perfringes* spores by applying a small bias to the $TiO_2$ coated Degussa Ti-alloy electrode. In order to avoid the cumbersome job of catalyst removal from the treated water, many of the researchers prefer to apply it in the form of coatings on solid surfaces. In this connection, Subrahmanyam et al. (2008) used pumice stone to support the $TiO_2$ photocatalyst and reported its strong bactericidal effect. Nano-composites, comprising of multiwalled carbon nanotubes (MWNT) coated with $TiO_2$ nanoparticles, synthesized by Lee et al. (2005) and the photocatalytic microbicidal efficacy of the composite was tested on endospores of *Bacillus cereus* under UV irradiation with peak at 350 nm. Results indicated that the $TiO_2$-MWNT composite was about three times more effective than the popular $TiO_2$ photocatalyst, Degussa P25®. The high efficiency was ascribed to the very large specific surface area and electron trap at the interface between MWNT and $TiO_2$ particles. Krishna et al. (2005) found two times enhancement with the $TiO_2$ (anatase) coated MWNT to inactivate Bacillus cereus endospores under UV irradiation. Further, *E. coli* contaminated water was photocatalytically disinfected with $TiO_2$ thin film irradiated with fluorescent light (28 $W/m^2$) (Sato and Taya 2006a). It was found that application of $Cu^{+2}$ (10 $mM/m^3$) conferred double photocatalytic efficiency on the film. It has been observed that $Cu^{+2}$ was reduced by the photo-generated electrons on $TiO_2$ and converted in to $Cu^+$. The reaction of $Cu^+$ with $H_2O_2$ produced OH radicals. $TiO_2$ composite with the activated carbon also exhibited enhanced photocatalytic inactivation of *E. coli* than pure $TiO_2$ (Li et al. 2008).

Owing to the recent reports about the predominantly high antibacterial activity of the illuminated ZnO nanoparticles, Rajendran et al. (2010) deposited them on fabrics which were found significantly effective against *Staphylococcus aureus*. Hence, a number of attempts have been made to improve the efficiency of photocatalysts under visible light. These include; doping and/or compositing of the photocatalysts with suitable materials, dye sensitization, and application electrical voltage etc.

## 13.6 Conclusion

Photocatalysis has been extensively studied due to its potential effects in the disinfection of drinking water. Its microbicidal effects have been tested on wide range of commonly found water borne microbial pathogens. The inactivation rate has been observed to vary with microbial species depending upon the stiffness and thickness of their cell envelops. Titanium dioxide in its commercially available form (Degussa P25®) has been found with high microbicidal efficiency under UV irradiation, while the bacterial inactivation rate increases linearly with the intensity of light. Water disinfection through photocatalysis has been observed to suppress due to presence of suspended materials and so efficiency of the process is lower for surface water than distilled water. The activity also depends upon the pH value of the water under treatment. Attempts have been made to improve solar photocatalytic disinfection efficiency through doping, coupling, and dye sensitization. Most of the researchers have claimed significant improvements in the efficiency accomplished with such modifications. An intensive research work in the field is going on and it is hoped that soon we will be having an efficient and low cost photocatalytic water purification system working under ambient solar light.

**Acknowledgements** The authors would like to acknowledge partial financial support from the National Nanotechnology Center, belonging to the National Science & Technology Development Agency (NSTDA), Ministry of Science and Technology (MOST), Thailand and the Centre of Excellence in Nanotechnology at the Asian Institute of Technology, Thailand, and NWFP university of Engineering and Technology Peshawar, Pakistan.

## References

Adams LK, Lyon DY, Alvarez PJJ (2006) Comparative eco-toxicity of nanoscale $TiO_2$, $SiO_2$, and ZnO water suspensions. Water Res 40(19):3527–3532

Alrousan DMA, Dunlop PSM, McMurray TA, Byrne JA (2009) Photocatalytic inactivation of *E. coli* in surface water using immobilised nanoparticle $TiO_2$ films. Water Res 43(1):47–54

Amezaga-Madrid P, Silveyra-Morales R, Cordoba-Fierro L, Nevarez-Moorillon GV, Miki-Yoshida M, Orrantia-Borunda E et al (2003) TEM evidence of ultrastructural alteration on *Pseudomonas aeruginosa* by photocatalytic $TiO_2$ thin films. J Photochem Photobiol B Biol 70(1):45–50

Bae E, Lee JW, Hwang BH, Yeo J, Yoon J, Cha HJ et al (2008) Photocatalytic bacterial inactivation by polyoxometalates. Chemosphere 72(2):174–181

Banerjee S, Gopal J, Muraleedharan P, Tyagi AK, Raj B (2006) Physics and chemistry of photocatalytic titanium dioxide: visualization of bactericidal activity using atomic force microscopy. Curr Sci 90(10):1378–1383

Barbeau B, Boulos L, Desjardins R, Coallier J, Prevost M (1999) Examining the use of aerobic spore-forming bacteria to assess the efficiency of chlorination. Water Res 33(13):2941–2948

Baruah S, Dutta J (2009) Nanotechnology applications in pollution sensing and degradation in agriculture. Environ Chem Lett 7(3):191–204

Benabbou AK, Derriche Z, Felix C, Lejeune P, Guillard C (2007) Photocatalytic inactivation of Escherischia coli. Effect of concentration of $TiO_2$ and microorganism, nature, and intensity of UV irradiation. Appl Catal B Environ 76(3–4):257–263

Chawengkijwanich C, Hayata Y (2008) Development of $TiO_2$ powder-coated food packaging film and its ability to inactivate *Escherichia coli* in vitro and in actual tests. Int J Food Microbiol 123(3):288–292

Cheng YW, Chan RCY, Wong PK (2007) Disinfection of *Legionella pneumophila* by photocatalytic oxidation. Water Res 41(4):842–852

Cheng TC, Chang CY, Chang CI, Hwang CJ, Hsu HC, Wang DY et al (2008) Photocatalytic bactericidal effect of $TiO_2$ film on fish pathogens. Surf Coat Technol 203(5–7):925–927

Cho M, Chung H, Choi W, Yoon J (2004) Linear correlation between inactivation of *E. coli* and OH radical concentration in $TiO_2$ photocatalytic disinfection. Water Res 38(4):1069–1077

Christensen PA, Curtis TP, Egerton TA, Kosa SAM, Tinlin JR (2003) Photoelectrocatalytic and photocatalytic disinfection of *E. coli* suspensions by titanium dioxide. Appl Catal B Environ 41(4):371–386

Dantur KI, Pizarro RA (2004) Effect of growth phase on the *Escherichia coli* response to ultraviolet-a radiation: influence of conditioned media, hydrogen peroxide and acetate. J Photochem Photobiol B Biol 75(1–2):33–39

Dodd NJF, Jha AN (2009) Titanium dioxide induced cell damage: a proposed role of the carboxyl radical. Mutat Res Fund Mol Mech Mutagen 660(1–2):79–82

Duffy EF, Al Touati F, Kehoe SC, McLoughlin OA, Gill LW, Gernjak W et al (2004) A novel $TiO_2$-assisted solar photocatalytic batch-process disinfection reactor for the treatment of biological and chemical contaminants in domestic drinking water in developing countries. Solar Energy 77(5):649–655

Dunlop PSM, Byrne JA, Manga N, Eggins BR (2002) The photocatalytic removal of bacterial pollutants from drinking water. J Photochem Photobiol A Chem 148(1–3):355–363

Dunlop PSM, McMurray TA, Hamilton JWJ, Byrne JA (2008) Photocatalytic inactivation of *Clostridium perfringens* spores on $TiO_2$ electrodes. J Photochem Photobiol A Chem 196(1):113–119

Erkan A, Bakir U, Karakas G (2006) Photocatalytic microbial inactivation over Pd doped $SnO_2$ and $TiO_2$ thin films. J Photochem Photobiol A Chem 184(3):313–321

Farshbaf Dadjour M, Ogino C, Matsumura S, Nakamura S, Shimizu N (2006) Disinfection of *Legionella pneumophila* by ultrasonic treatment with $TiO_2$. Water Res 40(6):1137–1142

Fernandez P, Blanco J, Sichel C, Malato S (2005) Water disinfection by solar photocatalysis using compound parabolic collectors. Catal Today 101(3–4 SPEC. ISS):345–352

Fernandez-Ibanez P, Sichel C, Polo-Lopez MI, de Cara-Garcia M, Tello JC (2009) Photocatalytic disinfection of natural well water contaminated by *Fusarium solani* using $TiO_2$ slurry in solar CPC photo-reactors. Catal Today 144(1–2):62–68

Gaya UI, Abdullah AH (2008) Heterogeneous photocatalytic degradation of organic contaminants over titanium dioxide: a review of fundamentals, progress and problems. J Photochem Photobiol C Photochem Rev 9(1):1–12

Gelover S, Gomez LA, Reyes K, Teresa Leal M (2006) A practical demonstration of water disinfection using $TiO_2$ films and sunlight. Water Res 40(17):3274–3280

Gopal K, Tripathy SS, Bersillon JL, Dubey SP (2007) Chlorination byproducts, their toxicodynamics and removal from drinking water. J Hazard Mater 140(1–2):1–6

Guillard C, Bui TH, Felix C, Moules V, Lina B, Lejeune P (2008) Microbiological disinfection of water and air by photocatalysis. C R Chim 11(1–2):107–113

Gumy D, Morais C, Bowen P, Pulgarin C, Giraldo S, Hajdu R et al (2006) Catalytic activity of commercial of $TiO_2$ powder for the abatement of the bacteria (*E. coli*) under solar simulated light: influence of the isoelectric point. Appl Catal B Environ 63:76–84

Haas CN, Joffe J (1994) Disinfection under dynamic conditions: modification of Hom's model for decay. Environ Sci Technol 28(7):1367–1369

Hayakawa T, Kuroiwa A, Higashi E, Nakano K (2007) Photoinduced bactericidal effect of titania thin film against *Legionella pneumophila*. Med Bull Fukuoka Univ 34(2):71–81

Huang N, Xiao Z, Huang D, Yuan C (1998) Photochemical disinfection of *Escherichia coli* with a $TiO_2$ colloid solution and a self-assembled $TiO_2$ thin film. Supramol Sci 5(5–6):559–564

Huang Z, Maness PC, Blake DM, Wolfrum EJ, Smolinski SL, Jacoby WA (2000) Bactericidal mode of titanium dioxide photocatalysis. J Photochem Photobiol A Chem 130(2–3):163–170

Ibanez JA, Litter MI, Pizarro RA (2003) Photocatalytic bactericidal effect of $TiO_2$ on *Enterobacter cloacae*. Comparative study with other Gram (–) bacteria. J Photochem Photobiol A Chem 157(1):81–85

Jones N, Ray B, Ranjit KT, Manna AC (2008) Antibacterial activity of ZnO nanoparticle suspensions on a broad spectrum of microorganisms. FEMS Microbiol Lett 279(1):71–76

Kikuchi Y, Sunada K, Iyoda T, Hashimoto K, Fujishima A (1997) Photocatalytic bactericidal effect of $TiO_2$ thin films: dynamic view of the active oxygen species responsible for the effect. J Photochem Photobiol A Chem 106(1–3):51–56

Kim B, Kim D, Cho D, Cho S (2003) Bactericidal effect of $TiO_2$ photocatalyst on selected foodborne pathogenic bacteria. Chemosphere 52(1):277–281

Kim JY, Lee C, Cho M, Yoon J (2008) Enhanced inactivation of *E. coli* and MS-2 phage by silver ions combined with UV-A and visible light irradiation. Water Res 42(1–2):356–362

Koizumi Y, Taya M (2002) Kinetic evaluation of biocidal activity of titanium dioxide against phage MS2 considering interaction between the phage and photocatalyst particles. Biochem Eng J 12(2):107–116

Krishna V, Pumprueg S, Lee SH, Zhao J, Sigmund W, Koopman B et al (2005) Photocatalytic disinfection with titanium dioxide coated multi-wall carbon nanotubes. Process Saf Environ Protect 83(4 B):393–397

Krishna V, Yanes D, Imaram W, Angerhofer A, Koopman B, Moudgil B (2008) Mechanism of enhanced photocatalysis with polyhydroxy fullerenes. Appl Catal B Environ 79(4):376–381

Kubacka A, Ferrer M, Martinez-Arias A, Fernandez-Garcia M (2008) Ag promotion of $TiO_2$-anatase disinfection capability: study of *Escherichia coli* inactivation. Appl Catal B Environ 84(1–2):87–93

Lee SH, Pumprueg S, Moudgil B, Sigmund W (2005) Inactivation of bacterial endospores by photocatalytic nanocomposites. Colloids Surf B Biointerfaces 40(2):93–98

Leung TY, Chan CY, Hu C, Yu JC, Wong PK (2008) Photocatalytic disinfection of marine bacteria using fluorescent light. Water Res 42(19):4827–4837

Li Y, Ma M, Wang X (2008) Inactivated properties of activated carbon-supported $TiO_2$ nanoparticles for bacteria and kinetic study. J Environ Sci 20(12):1527–1533

Liu HL, Yang TCK (2003) Photocatalytic inactivation of *Escherichia coli* and *Lactobacillus helveticus* by ZnO and $TiO_2$ activated with ultraviolet light. Process Biochem 39(4):475–481

Lonnen J, Kilvington S, Kehoe SC, Al-Touati F, McGuigan KG (2005) Solar and photocatalytic disinfection of protozoan, fungal and bacterial microbes in drinking water. Water Res 39(5):877–883

Lyon DY, Brown DA, Alvarez PJJ (2008) Implications and potential applications of bactericidal fullerene water suspensions: effect of $nC_{60}$ concentration, exposure conditions and shelf life. Water Sci Technol 57(10):1533–1538

Maneerat C, Hayata Y (2006) Antifungal activity of $TiO_2$ photocatalysis against Penicillium expansum in vitro and in fruit tests. Int J Food Microbiol 107(2):99–103

Marugan J, van Grieken R, Sordo C, Cruz C (2008) Kinetics of the photocatalytic disinfection of *Escherichia coli* suspensions. Appl Catal B Environ 82(1–2):27–36

Munoz I, Rieradevall J, Torrades F, Peral J, DomÃ¨nech X (2005) Environmental assessment of different solar driven advanced oxidation processes. Solar Energy 79(4):369–375

Ogino C, Farshbaf Dadjour M, Takaki K, Shimizu N (2006) Enhancement of sonocatalytic cell lysis of *Escherichia coli* in the presence of $TiO_2$. Biochem Eng J 32(2):100–105

Oguma K, Katayama H, Ohgaki S (2004) Photoreactivation of *Legionella pneumophila* after inactivation by low- or medium-pressure ultraviolet lamp. Water Res 38(11):2757–2763

Pal A, Pehkonen SO, Yu LE, Ray MB (2007) Photocatalytic inactivation of Gram-positive and Gram-negative bacteria using fluorescent light. J Photochem Photobiol A Chem 186(2–3):335–341

Paleologou A, Marakas H, Xekoukoulotakis NP, Moya A, Vergara Y, Kalogerakis N et al (2007) Disinfection of water and wastewater by $TiO_2$ photocatalysis, sonolysis and UV-C irradiation. Catal Today 129(1–2 SPEC. ISS):136–142

Park C, Hong K (2006) Optoelectrical properties of ZnO thin films grown by pulsed laser deposition. J Ceram Process Res 7(4):292–295

Patnaik P (2007) A comprehensive guide to the hazardous properties of chemical substances. Wiley, Hoboken, 1059 pp

Prasad GK, Agarwal GS, Singh B, Rai GP, Vijayaraghavan R (2009) Photocatalytic inactivation of *Bacillus anthracis* by titania nanomaterials. J Hazard Mater 165(1–3):506–510

Qi L, Xu Z, Jiang X, Hu C, Zou X (2004) Preparation and antibacterial activity of chitosan nanoparticles. Carbohydr Res 339(16):2693–2700

Rajendran R, Balakumar C, Ahammed HAM, Jayakumar S, Vaideki K, Rajesh EM (2010) Use of zinc oxide nano particles for production of antimicrobial textiles. Int J Eng Sci Technol 2(1):202–208

Rana S, Rawat J, Sorensson MM, Misra RDK (2006) Antimicrobial function of $Nd^{3+}$-doped anatase titania-coated nickel ferrite composite nanoparticles: a biomaterial system. Acta Biomater 2(4):421–432

Ren J, Wang W, Zhang L, Chang J, Hu S (2009) Photocatalytic inactivation of bacteria by photocatalyst $Bi_2WO_6$ under visible light. Catal Commun 10(14):1940–1943

Rennecker JL, Marinas BJ, Owens JH, Rice EW (1999) Inactivation of *Cryptosporidium parvum* oocysts with ozone. Water Res 33(11):2481–2488

Rincon AG, Pulgarin C (2004a) Bactericidal action of illuminated $TiO_2$ on pure *Escherichia coli* and natural bacterial consortia: post-irradiation events in the dark and assessment of the effective disinfection time. Appl Catal B Environ 49(2):99–112

Rincon AG, Pulgarin C (2004b) Effect of pH, inorganic ions, organic matter and $H_2O_2$ on *E. coli* K12 photocatalytic inactivation by $TiO_2$: implications in solar water disinfection. Appl Catal B Environ 51(4):283–302

Rincon AG, Pulgarin C (2006) Comparative evaluation of $Fe^{3+}$ and $TiO_2$ photoassisted processes in solar photocatalytic disinfection of water. Appl Catal B Environ 63(3–4):222–231

Rizzo L (2009) Inactivation and injury of total coliform bacteria after primary disinfection of drinking water by $TiO_2$ photocatalysis. J Hazard Mater 165(1–3):48–51

Ryu H, Gerrity D, Crittenden JC, Abbaszadegan M (2008) Photocatalytic inactivation of *Cryptosporidium parvum* with $TiO_2$ and low-pressure ultraviolet irradiation. Water Res 42(6–7):1523–1530

Sadiq R, Rodriguez MJ (2004) Disinfection by-products (DPBs) in drinking water and predictive models for their occurrence: a review. Sci Total Environ 321:21–46

Sato T, Taya M (2006a) Copper-aided photosterilization of microbial cells on $TiO_2$ film under irradiation from a white light fluorescent lamp. Biochem Eng J 30(2):199–204

Sato T, Taya M (2006b) Enhancement of phage inactivation using photocatalytic titanium dioxide particles with different crystalline structures. Biochem Eng J 28(3):303–308

Sayilkan F, Asilturk M, Kiraz N, Burunkaya E, Arpac E, Sayilkan H (2009) Photocatalytic antibacterial performance of $Sn^{4+}$-doped $TiO_2$ thin films on glass substrate. J Hazard Mater 162(2–3):1309–1316

Shang C, Cheung LM, Ho CM, Zeng M (2009) Repression of photoreactivation and dark repair of coliform bacteria by $TiO_2$-modified UV-C disinfection. Appl Catal B Environ 89(3–4):536–542

Sichel C, de Cara M, Tello J, Blanco J, Fernandes-Ibanes P (2007) Solar photocatalytic disinfection of agricultural pathogenic fungi: Fusarium species. Appl Catal B Environ 74(1–2):152–160

Sokmen M, Candan F, Sumer Z (2001) Disinfection of E. coli by the Ag-$TiO_2$/UV system: lipid-peroxidation. J Photochem Photobiol A Chem 143(2–3):241–244

Sokmen M, Degerli S, Aslan A (2008) Photocatalytic disinfection of *Giardia intestinalis* and *Acanthamoeba castellani* cysts in water. Exp Parasitol 119(1):44–48

Sondi I, Salopek-Sondi B (2004) Silver nanoparticles as antimicrobial agent: a case study on *E. coli* as a model for Gram-negative bacteria. J Colloid Interface Sci 275(1):177–182

Subrahmanyam M, Boule P, Durga Kumari V, Naveen Kumar D, Sancelme M, Rachel A (2008) Pumice stone supported titanium dioxide for removal of pathogen in drinking water and recalcitrant in wastewater. Solar Energy 82(12):1099–1106

Sunada K, Watanabe T, Hashimoto K (2003) Studies on photokilling of bacteria on $TiO_2$ thin film. J Photochem Photobiol A Chem 156(1–3):227–233

Tatsuma T, Takeda S, Saitoh S, Ohko Y, Fujishima A (2003) Bactericidal effect of an energy storage $TiO_2$-$WO_3$ photocatalyst in dark. Electrochem Commun 5(9):793–796

van Grieken R, Marugan J, Sordo C, Pablos C (2009) Comparison of the photocatalytic disinfection of *E. coli* suspensions in slurry, wall and fixed-bed reactors. Catal Today 144(1–2):48–54

Vohra A, Goswami DY, Deshpande DA, Block SS (2006) Enhanced photocatalytic disinfection of indoor air. Appl Catal B Environ 64(1–2):57–65

Wang X, Gong WQ (2007) Preparation and photocatalytic bactericidal action of $Fe^{3+}$ doped nanocrystalline $TiO_2$ film. Wuhan Ligong Daxue Xuebao/J Wuhan Univ Technol 29(7):50–53

Watts RJ, Kong S, Orr MP, Miller GC, Henry BE (1995) Photocatalytic inactivation of coliform bacteria and viruses in secondary wastewater effluent. Water Res 29(1):95–100

Wist J, Sanabria J, Dierolf C, Torres W, Pulgarin C (2002) Evaluation of photocatalytic disinfection of crude water for drinking-water production. J Photochem Photobiol A Chem 147(3):241–246

Wong MS, Chu WC, Sun DS, Huang HS, Chen JH, Tsai PJ et al (2006) Visible-light-induced bactericidal activity of a nitrogen-doped titanium photocatalyst against human pathogens. Appl Environ Microbiol 72(9):6111–6116

Wu P, Xie R, Imlay JA, Shang JK (2009) Visible-light-induced photocatalytic inactivation of bacteria by composite photocatalysts of palladium oxide and nitrogen-doped titanium oxide. Appl Catal B Environ 88(3–4):576–581

Yan G, Chen J, Hua Z (2009) Roles of $H_2O_2$ and OH radical in bactericidal action of immobilized $TiO_2$ thin-film reactor: an ESR study. J Photochem Photobiol A Chem 207(2–3):153–159

Yao KS, Wang DY, Chang CY, Weng KW, Yang LY, Lee SJ et al (2007) Photocatalytic disinfection of phytopathogenic bacteria by dye-sensitized $TiO_2$ thin film activated by visible light. Surf Coat Technol 202(4–7):1329–1332

Yu JC, Xie Y, Tang HY, Zhang L, Chan HC, Zhao J (2003) Visible light-assisted bactericidal effect of metalphthalocyanine-sensitized titanium dioxide films. J Photochem Photobiol A Chem 156(1–3):235–241

Zhao J, Krishna V, Hua B, Moudgil B, Koopman B (2009) Effect of UVA irradiance on photocatalytic and UVA inactivation of *Bacillus cereus* spores. J Photochem Photobiol B Biol 94(2):96–100

# Index



E. Lichtfouse et al. (eds.), *Environmental Chemistry for a Sustainable World: Volume 2: Remediation of Air and Water Pollution*, DOI 10.1007/978-94-007-2439-6,

MIX
Papier aus verantwortungsvollen Quellen
Paper from responsible sources
FSC® C105338

If you have any concerns about our products,
you can contact us on
**ProductSafety@springernature.com**

In case Publisher is established outside the EU,
the EU authorized representative is:
**Springer Nature Customer Service Center GmbH**
**Europaplatz 3, 69115 Heidelberg, Germany**

Printed by Libri Plureos GmbH
in Hamburg, Germany